# AGRIBUSINESS
## PRINCIPLES *of* MANAGEMENT

# AGRIBUSINESS

## PRINCIPLES *of* MANAGEMENT

DAVID D. VAN FLEET

ELLA W. VAN FLEET

GEORGE J. SEPERICH

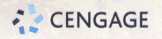
CENGAGE

Australia • Brazil • Canada • Mexico • Singapore • United Kingdom • United States

**Agribusiness: Principles of Management**
David D. Van Fleet, Ella W. Van Fleet, and George J. Seperich

Vice President, Careers & Computing: Dave Garza

Senior Acquisitions Editor: Sherry Dickinson

Director, Development, Careers & Computing: Marah Bellegarde

Editorial Assistant: Scott Royael

Developmental Editor: Julie M. Vitale (iD8-TripleSSS)

Director, Market Development: Deborah Yarnell

Senior Market Development Manager: Erin Brennan

Director, Brand Management: Jason Sakos

Senior Brand Manager: Kristin McNary

Senior Production Director: Wendy A. Troeger

Production Manager: Mark Bernard

Senior Content Project Manager: Elizabeth C. Hough

Senior Art Director: David Arsenault

Cover Image credits: Globe background: Caribbean, Pacific and Atlantic Oceans © Anton Balazh

Credits for stacked images: (top image) Farmer and researcher analysing corn plant: © Goodluz/www.shutterstock.com; (middle image) Bread factory surveillance: © Zurijeta.shutterstock.com; (bottom image) Supermarket clerk: © iStockphoto/Steve Debenport

For product information and technology assistance, contact us at **Cengage Customer & Sales Support, 1-800-354-9706**

For permission to use material from this text or product, submit all requests online at **www.cengage.com/permissions.**

Library of Congress Control Number: 2012951193

ISBN-13: 978-1-111-54486-7

ISBN-10: 1-111-54486-7

**Cengage**
200 Pier 4 Boulevard
Boston MA 02210
USA

Cengage is a leading provider of customized learning solutions with employees residing in nearly 40 different countries and sales in more than 125 countries around the world. Find your local representative at: **www.cengage.com.**

To learn more about Cengage platforms and services, register or access your online learning solution, or purchase materials for your course, visit **www.cengage.com.**

**Notice to the Reader**

Publisher does not warrant or guarantee any of the products described herein or perform any independent analysis in connection with any of the product information contained herein. Publisher does not assume, and expressly disclaims, any obligation to obtain and include information other than that provided to it by the manufacturer. The reader is expressly warned to consider and adopt all safety precautions that might be indicated by the activities described herein and to avoid all potential hazards. By following the instructions contained herein, the reader willingly assumes all risks in connection with such instructions. The publisher makes no representations or warranties of any kind, including but not limited to, the warranties of fitness for particular purpose or merchantability, nor are any such representations implied with respect to the material set forth herein, and the publisher takes no responsibility with respect to such material. The publisher shall not be liable for any special, consequential, or exemplary damages resulting, in whole or part, from the readers' use of, or reliance upon, this material.

Printed Number: 7    Print Year: 2023
Printed in Mexico

## DEDICATION

Our spouses and children have always been fundamental to our professional and personal lives. They give meaning to our lives, and so we dedicate this book to them: Dirk and Marijke Van Fleet and Barbara and Ilya Seperich

This book is also dedicated to our students from over the years, as they have helped shape this endeavor.

In addition, we would be remiss if we did not dedicate this book to future students joined through the passage of time by this text. You are the future of this discipline – students today, leaders tomorrow.

# BRIEF CONTENTS

# CONTENTS

## PART ONE   MANAGEMENT IN AGRIBUSINESS: AN INTRODUCTION

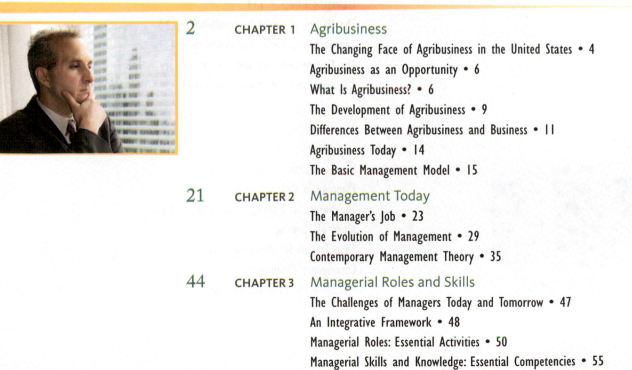

©iStockphoto/Jacob Wackerhausen

## PART FIVE LEADING IN AGRIBUSINESS

## PART FOUR ORGANIZING IN AGRIBUSINESS

© iStockphoto/Studiovision

# PREFACE

Agribusiness: *Principles of Management*, 1st edition, is a general "business management" book written especially for students in agribusiness courses. The book includes coverage of all segments of the agribusiness food chain, including agricultural producers, processors, distributors, and farmers and ranchers. It consists of management material generally found in conventional introductory management textbooks, but most of the examples focus explicitly on the agribusiness industry. This coverage also enables the book to be used in traditional AACSB-accredited business schools as they offer specialized sections for agribusiness students.

The authors appreciate the increasing need for managers who can harness resources to move a perishable product around the globe to sustain another equally perishable entity on our planet—people. Understanding how to do that is what agribusiness management is all about. Individuals who comprehend both the principles of business and the nuances of agribusiness will be in even greater demand in the future.

## Why Offer a Management Book for Agribusiness?

Effective agribusinesses must employ the same concepts of management that other segments of the economy use. While traditional management books tend to focus on large corporations of no specific type and with little regard for product perishability, most of our examples focus explicitly on the agribusiness industry. Rather than trying to cover all business topics such as marketing, finance, and accounting, our content focuses on management. It is based on current management literature and accepted management practice. With a writing style that is quite readable—friendly in tone without heavy research discussion—it is applicable for entry- or higher-level courses in agribusiness and traditional business programs.

*Agribusiness: Principles of Management*, 1st edition, is built around three major concepts that respond to market needs and emerging trends. First, it views agribusiness as a technology-oriented industry that includes production, processing, distribution, sales, and all manner of related businesses. Second, it recognizes that agribusiness organizations range in size from small, family-owned farms or businesses to some of the largest corporations in the world. Third, it reflects the changing face of agribusiness—the fact that most people who work in agribusiness do not work on farms or ranches but instead are in, or will work in, the many other agribusiness organizations.

The continued growth of agribusiness programs and the growing collaboration with traditional business schools suggests the need for a book tailored specifically to that market. Increasingly, faculty members trained in business rather than agriculture or agricultural economics are teaching agribusiness courses. These faculty members may be more accustomed to, and therefore more receptive to, texts that are like traditional business texts rather than agricultural texts with a business emphasis. As a result, agribusiness books are emerging in business areas such as accounting, finance, risk analysis, marketing, and communication.

## Logical Chapter Organization

*Agribusiness: Principles of Management*, 1st edition, consists of 24 chapters divided into six major parts, plus two appendices. The first three chapters

(Part One) are introductions to agribusiness and to the field of management in general. Part Two (Chapters 4 to 7) introduces readers to the environment in which managers must operate, including the organizational, competitive, global, and ethical and social environments. Part Three (Chapters 8 to 11) discusses planning and decision making, one of the four primary functions of a manager. Part Four (Chapters 12 to 15) covers organizing, which is another primary function of a manager. Included are designing jobs, organizing the company, recognizing changes that call for reorganizing, and staffing and compensating employees. Part Five (Chapters 16 to 20) includes five chapters on leadership, motivation, groups and teams, and managerial communication. Part Six (Chapters 21 to 24) focuses on control, including establishing a control system, maintaining quality management, operations control techniques, and types of information systems.

Appendix 1 is a list of large, mostly well-known agribusiness organizations that learners may not have heretofore considered as agribusinesses. Appendix 2, Control Techniques and Methods, provides additional material for instructors to supplement the basic management material with budgeting, financial statements, ratios, financial analyses, and other control techniques.

## Features of the Book

### An Engaging Style & Approach

As authors, our job is to read and interpret research, then write so that the material flows smoothly and logically from one point to another throughout the book. We use straightforward language and a logical sequencing of material, along with vignettes, cases, and examples from both small and large businesses to make the material clear, understandable, and interesting. More specifically, in our writing we try to adhere to the following criteria.

- **Readable**—To assure readability, we present the material in a friendly, straightforward, easy-to-read style without heavy research discussion, unnecessary jargon, or detailed summaries of research findings.

- **Interesting**—Undisguised vignettes, cases, and the use of numerous examples from both small and large businesses, plus the inclusion of little-known "Food For Thought" facts make the book more interesting and understandable to the reader.

- **Realistic**—We try to make the learning job easier and more enjoyable by incorporating realistic material. Content is tied to organizations from all over the globe and all aspects of life, including small agribusiness examples that are frequently carried forward to several chapters. Examples for one smaller agribusiness, Summer Farms, are shaded to call special attention as we follow that company throughout the book.

- **Accurate**—The book is firmly grounded in both recent and historical research as well as accepted management practices.

- **Current**—Being on the cutting edge means having the most up-to-date material. While we use classic references to material, the research-based material is timely to enhance its utility to readers.

- **Appealing**—Color, photographs, graphics, and tables enhance the learning experience and make the book more appealing to learners.

## Pedagogical Features That Keep Students Involved

*Agribusiness: Principles of Management*, 1st edition, is a comprehensive introduction to management because it does not attempt to cover other business topics, such as marketing, finance, and accounting in depth. It employs several pedagogical features to facilitate learning and the application of knowledge and skills, to improve students' critical thinking skills, and to develop a managerial vocabulary:

- **Learning Objectives**—Each chapter begins with a set of learning objectives to serve as a guide to studying and reviewing the chapter.

- **Manager's Vocabulary**—Key terms in each chapter contribute to a complete manager's vocabulary that is important in the real world of management, beyond this management course. These vocabulary words are combined and alphabetized in the *Glossary*.

- **Opening Vignettes**—Each chapter begins with a short, real-world agribusiness story that illustrates some of the concepts featured in the chapter.

- **Examples**—References to both small and large businesses, especially the agribusiness sector, make the book come alive.

- **A Focus on Agribusiness Boxes**—Short, real-world examples of key points or important lists help keep the material in each chapter interesting, relevant, and easier to remember.

- **Food for Thought**—Several "little-known fun facts" related to the chapter material are interspersed throughout the book to stimulate thinking and discussion.

- **Closing Case Studies**—Each chapter closes with a case designed to focus on main points of the chapter and to initiate class discussion.

- **End-of-Chapter Questions**—For review and reinforcement purposes, each chapter contains three sets of questions: straightforward review questions, analysis questions that require more careful thought about the content of the chapter, and completion questions built around the *Manager's Vocabulary*.

- **Research References**—Both recent and historical research as well as general business literature document the material in each chapter.

- **Additional Features**—In addition to the above features, color identifies key elements of each chapter, and tables, charts, and photographs bring a strong visual aspect to the learning process.

## Additional Resources for the Instructor

The **Instructor Resource CD-ROM to Accompany Agribusiness: Principles of Management, 1st edition**
ISBN-13: 978-1-1115-4487-4

For instructors, this CD-ROM offers invaluable assistance allowing you access to all of your resources—anywhere and at any time!

- The *Instructor's Manual* suggests answers to the review, analysis, and fill-in-the-blank questions as well as the Closing Case questions that appear at the end of each chapter. The *Instructor's Manual* also suggests additional activities or questions.

- The *Computerized Testbank in ExamView®* makes generating tests and quizzes a snap. With hundreds of questions you can create customized assessments for your students with a click of a button. Also, you can add your own unique test questions!

- Customizable instructor support slide presentations in *PowerPoint®* format focus on key points in each chapter. Use for in-class lectures, as handouts, and for student reviews.

## The Management Model in the Text

As pointed out in the first chapter, the basic management model in this book indicates the following about managers:

1. Managers perform a variety of roles requiring several different skills as they strive to effectively and efficiently accomplish the goals of their organizations.

2. Managers must develop an understanding of the complex environments in which they function.

3. Managers develop strategies and plans to achieve organizational objectives.

4. Managers use those strategies and plans to help shape the organization.

5. Managers must understand people, so they need to develop skills in leadership, motivation, communication, and teamwork.

6. Managers have to measure performance and adjust operations to adapt to changing conditions.

Readers should use this model as a sort of "road map" as they go through the text.

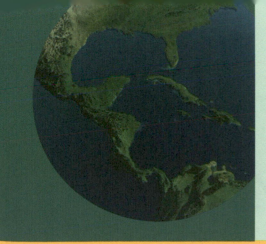

# ABOUT THE AUTHORS

Currently available texts in agribusiness typically are not authored by management scholars. The principal author of *Agribusiness: Principles of Management*, 1st edition, is an established management scholar and previous management textbook author. His coauthors contribute their own experiences in teaching, business, and agribusiness.

**Dr. David D. Van Fleet**, a Fellow of the Academy of Management, is an experienced book author who has also published numerous journal articles. A Professor of Management in the Morrison School of Agribusiness and Resource Management at Arizona State University, he teaches and conducts research on agribusiness. He is author or coauthor of the following titles: *Contemporary Management, Organizational Behavior, Behavior in Organizations, Military Leadership, The Violence Volcano: Reducing the Threat of Workplace Violence,* and *Workplace Survival: Dealing with Bad Bosses, Bad Workers, Bad Jobs.*

**Dr. Ella W. Van Fleet**, Founder and President of Professional Business Associates, is an experienced author with an impressive background including more than 35 years of experience in teaching, training, managing, and consulting, plus three interdisciplinary degrees in Business and Higher Education. She has firsthand knowledge of agribusiness, having grown up on a dairy farm. In addition to several professional journal articles, she is coauthor of two books, including *The Violence Volcano: Reducing the Threat of Workplace Violence and Workplace Survival: Dealing with Bad Bosses, Bad Workers, Bad Jobs.*

**Dr. George J. Seperich** is a Professor in the Morrison School of Agribusiness and Resource Management at Arizona State University. His academic interests include identifying necessary conditions for agribusiness economic development at the state level, corporate strategy and the development of management, and marketing agribusiness case studies as teaching tools and developing agribusiness as an academic discipline in Mexico. He is author or coauthor of six books and several editions, including, *Introduction to Agribusiness Marketing, Cases in Agribusiness Management, Food Science and Technology,* and *Managing Power and People.*

# ACKNOWLEDGMENTS

The authors express their appreciation to the reviewers of the manuscript for the valuable ideas and feedback:

**Thorsten Egelkraut**
Assistant Professor
Agricultural and Resource Economics
Oregon State University
Corvallis, OR

**Sierra Howry, PhD**
Assistant Professor
Department of Agriculture
Angelo State University
San Angelo, TX

**Jennifer Keeling Bond, PhD**
Assistant Professor
Department of Agricultural and Resource
    Economics
Colorado State University
Fort Collins, CO

**Rick Mathias**
Lecturer, Food Marketing
California State Polytechnic University, Pomona
Pomona, CA

**James Sterns**
Associate Professor, Food and Resource
    Economics Department
University of Florida
Gainesville, FL

The authors would also like to thank those who had input into this book, either as a resource for course content or consultant:

**Ricky W. Griffin**
Texas A&M University

**Tim O. Peterson**
North Dakota State University

For their help and hard work in bringing this book to fruition we also thank:

**Marah Bellegarde**
Delmar Cengage Learning

**Sherry Dickinson**
Delmar Cengage Learning

**Christina Gifford**
Delmar Cengage Learning

**Scott Royael**
Delmar Cengage Learning

**Julie Vitale**
ID8TripleSSS Media Development, LLC

**Marion Waldman**
ID8TripleSSS Media Development, LLC

We would be remiss if we didn't also acknowledge our families. Our spouses and children are fundamental in our professional and personal lives. They give meaning to our lives and so we dedicate this book to them.

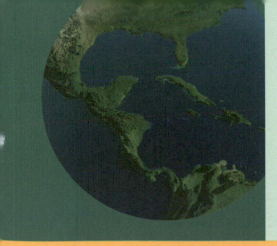

# HOW TO USE THIS BOOK

How to Use This Book, *Agribusiness: Principles of Management, 1st edition, provides a variety of features to aid your learning, including focusing your attention, building your vocabulary, introducing and reinforcing concepts, making your reading easy and interesting, and reviewing the material.*

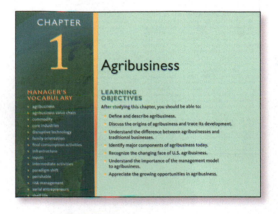

## LEARNING OBJECTIVES

Consider these objectives before you begin reading a chapter to help you focus your study. When you have completed the chapter, review these objectives to ensure that you understand the key points of the chapter.

## MANAGER'S VOCABULARY

These key terms are the critical vocabulary words you will need to learn. Use this listing as part of your study and review.

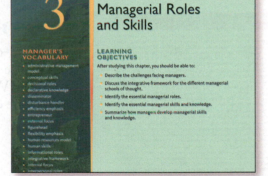

## OPENING VIGNETTE

Read the vignette to set the stage for the concepts that follow in the chapter.

# FOOD FOR THOUGHT

These informative boxes contain agribusiness-related facts that are intended to break the more serious text with information that is simply interesting, intriguing, or fun. Read and use them as a basis for conversation with your colleagues and fellow learners.

## A FOCUS ON AGRIBUSINESS

As you continue through the chapter, read these interesting mini-cases to reinforce the chapter content.

## CHAPTER CASE STUDIES

These end-of-chapter case studies are designed to enhance your comprehension of the chapter concepts. Work through each case study and answer the questions that follow.

## END-OF-CHAPTER QUESTIONS

Use these questions for review and reinforcement of chapter content.

# MANAGEMENT IN AGRIBUSINESS: AN INTRODUCTION

**1**

# Agribusiness

## LEARNING OBJECTIVES

After studying this chapter, you should be able to:

- Define and describe agribusiness.
- Discuss the origins of agribusiness and trace its development.
- Understand the difference between agribusinesses and traditional businesses.
- Identify major components of agribusiness today.
- Recognize the changing face of U.S. agribusiness.
- Understand the importance of the management model to agribusiness.
- Appreciate the growing opportunities in agribusiness.

## McDonald's Story Is No Crock

**Y**es, Virginia, there was a McDonald—two of them, in fact. In 1937, two brothers, Richard (Dick) and Maurice (Mac) McDonald, began their carhop drive-in business in San Bernardino, CA. They did not even serve hamburgers, only hot dogs. They were not even on the cutting edge of drive-in development. (This honor went to the Pig Stand, a carhop drive-in restaurant, located in Hollywood at the corner of Sunset and Vermont Streets.) Dick and Mac were simply looking for their market "niche" and trying to capitalize on America's growing love affair with the automobile.

**FIGURE 1.1** McDonald's restaurants are easy to spot with their Golden Arches.
© iStockphoto/Ivana Star

They eventually built a larger restaurant in San Bernardino. They expanded their menu, increased the number of carhops to 25, and serviced 125 cars in their parking lot. By 1948 they had achieved a level of success neither brother dreamed possible. They were rich and "bored." Dick McDonald said, "The money was pouring in, and there wasn't much for us to do."

Around 1948, they opened a new restaurant, the Dimer, where every item on the menu cost 10 cents ... drinks, fries, and hamburgers. For maximum efficiency, they organized their kitchen like an assembly line and changed their "concept" from carhop to self-service windows where customers placed their orders and had them filled. Rather than wait for a carhop to take the order and bring it to the car, the customer could leave the car and pick up the order. They called it the "Speedy Service System."

They also raised the price to 15 cents per hamburger. Initially, they lost customers but after only six months they regained all the business they lost and more. Their customers adapted to the new concept. Dick and Mac went from following trends to inventing one of their own.

Business was great—so great that they needed new equipment to keep up with the demand. They ordered new grilling equipment and new mixers to make their famous milk shakes. The shakes were made with four multi-mixers. They became the best customer of the multi-mixer manufacturer located in Chicago.

Dick and Mac's success attracted attention. Their dairy product supplier, the Carnation Corporation, even offered financial support to help expand the franchise. However, Dick and Mac were not very good at franchising and, in fact, were happy with their level of success.

In 1954 their multi-mixer salesman, Ray A. Kroc, of Oak Park, IL, called Dick McDonald and flew to California to see the business first-hand. He sat in his rented car and watched as lines formed in front of each service window before the restaurant opened at noon. He asked Dick McDonald when this rush of customers would end and was told sometime late at night. Ray Kroc said, "These guys have got something. How about if I open some of these places?"

The McDonald's were not interested in taking on a national business; they did not like the hassle of franchising. So, Ray Kroc became the franchising agent with the license to take the company national. He opened his first restaurant, which he named "McDonald's," on April 15, 1955, in Des Plaines, IL. He also established the McDonald's Corporation. But he felt he was constrained by the relationship with Dick and Maurice. In 1961, he asked the McDonald brothers to name their price and to sell him the company outright. Their price was $2.7 million for the company and the name.

The McDonald's Corporation has become a business legend, and its restaurants have become a symbol of American presence in other countries (Figure 1.1). In fact, McDonald's ubiquity led to the development of the "Big Mac Index." This index by *The Economist* magazine is used as a substitute for the classical economics measure, purchasing power parity. In place of a "basket" of staples, the Big Mac hamburger is used to identify economic anomalies. For instance, a Big Mac in China costs (2009) $1.83 (12.5 yuan) considerably less than the cost in the United States $3.57 (2009). To economists this means the Chinese yuan is under-valued compared with the U.S. dollar.

In agribusiness, as in other forms of business, success is measured differently by different people. All businesses must define their comfort levels with success and risk. A journey from business's simple beginnings and principles to an economic measurement from a fast-food stand to an international icon—this is the story of a unique and often overlooked agribusiness. Thanks to Ray Kroc, the McDonald's story is no "crock."[1]

## INTRODUCTION

The McDonald brothers, Dick and Mac, started a simple business and employed common business principles, but with a product that has agricultural roots, mostly wheat and swine or cattle. As such, their story is certainly a business story, but more importantly, it is an agribusiness story. "So what's the difference?" you ask. The purpose of this chapter is to explain the difference and demonstrate the breadth and scope of a unique global industry, agribusiness. The vignettes, examples, and case studies employed in this book will demonstrate both the uniqueness of agribusiness and its simultaneous common attributes with traditional businesses. To fully understand the growth and the challenges of agribusiness, throughout this book we will also refer frequently to much smaller businesses, especially Summer Farms, which is introduced in this chapter.

## The Changing Face of Agribusiness in the United States

The growth from raising crops for personal consumption to becoming part of the agribusiness industry has been a common story in the United States, as is the shrinking of farm employment and the growth of non-farm agribusinesses. As shown in Figure 1.2, farm and closely related employment has fallen and leveled off, while peripherally farm-related employment (e.g., food processing and distribution) has risen dramatically. So in today's economic structure agribusiness jobs are predominantly non-farm jobs.

Family farming and ranching operations still exist. These enterprises are agribusinesses, but the face

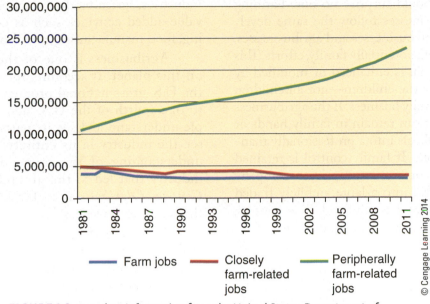

FIGURE 1.2 Based on information from the United States Department of Agriculture Economic Research Service Farm and farm-related employment.

of agribusiness is changing. According to the U.S. Department of Agriculture, 98 percent of U.S. farms are family operations; however, it is a mistake to confuse the term "family" with a given size of operation. When we compare contributions to agricultural production at the national level, "family" designates the type of ownership, not size of operation. Even large farms can be family operations. Of course, some large farms also may be non-family farms, where the operator or persons related to the operator do not own a majority of the business. As Table 1.1 indicates, the total number of U.S. farms varies substantially with and inversely to both total sales and percent contribution to agricultural production.

In farming as in other forms of business, consolidation of operations is also occurring. Although most of the 2.1 million U.S. farms are "small," approximately

47,600 farms have annual sales of $1 million or more. About 11 percent or 5,200 farms have sales equal to or exceeding $5 million, and 64 percent of these farms are family owned. Seventy-one percent of these family farms specialize in beef, high-value crops, or dairy, where they capture economies of scale. For example, dairy production costs decrease with increases in herd size. Total costs per hundredweight for operations with 1,000 cows or more are less than half of the costs of farms with fewer than 50 cows.[2]

Even the criterion of ownership can be deceptive since 277,500 farm operators rented 62 million acres of farmland to others to farm. And there is also an age factor, since 28 percent of farm operators are at least 65 years old. Family operations are also "home" operations, in that the operators live on the premises, regardless of the farm's size.[3]

TABLE 1.1 Farm Contribution to Agricultural Production*

| SIZE OF FARM OPERATION | SALES IN THOUSANDS OF DOLLARS | PERCENT OF TOTAL U.S. FARMS | PERCENT OF TOTAL U.S. AGRICULTURAL PRODUCTION |
|---|---|---|---|
| Small Farms | <$100 | 86.6 | 9.8 |
| Medium Farms | $100–$250 | 5.1 | 6.6 |
| Large Farms | $250–$500 | 4.3 | 12.2 |
| Very Large Farms and Non-Family Farms | >$500 | 7.4 | 71.4 |

* Based on Robert A. Hoppe and David E. Banker, "Small Farms = Family, Retirement, Residential/Lifestyle, and Low Sale," *Structure and Finances of U.S. Farms: Family Farm Report*, U.S. Department of Agriculture, Economic Research Service, Economic Information Bulletin, No. 66, July 2010.

These figures should surprise no one. Farming and ranching agribusinesses follow the same developmental pattern as all businesses and agribusinesses. They begin as personal, potentially family, efforts. This family-control potential is enhanced or modified by the interest shown by the children—not all children want to go into the family business. If the agribusiness survives and grows, it can remain in family hands or undergo a metamorphosis into a professionally managed company, or both. If family control is retained the business can become a large private company (e.g., Cargill) or evolve into a large, publicly traded firm, such as Archer Daniels Midland. Size matters, but it is not an indication of ownership status.

## Agribusiness as an Opportunity

Agribusiness is an industry with tremendous potential for growth and development as well as monumental and consequential issues it must address.[4] The Chinese symbol that represents both crisis and opportunity could well serve as the logo for agribusiness.

For centuries, agricultural productivity came from the application of two resources: land and labor. More recently a substantial proportion of increased productivity in agriculture output in the United States has come from the use of another resource: capital, primarily in the form of technological change including biotechnology, improved animal husbandry, and improvements in machinery and chemicals.[5] Future improvements will come from the application of a fourth resource: management.

Agribusiness is an industry that is becoming chronically short of smart, clever, and concerned individuals. Thus, individuals who understand the principles of business and the nuances of agribusiness are in greater need than ever before. The industry needs individuals who can harness resources to move a perishable product around the globe to sustain another equally perishable entity on our planet: people. Understanding how to do that is what management is all about.

## What Is Agribusiness?

**Agribusiness** involves the production, distribution, and consumption of food, clothing, and even shelter. It includes all economic activity in the food and fiber system, which encompasses the input supply industries, agricultural production, and post-harvest, value-added activities such as commodity processing, food manufacturing, and food distribution.[6]

Agribusiness is one of the largest industries on this planet. It accounts for nearly one-fifth of the U.S. gross national product and employs close to one-fourth of the U.S. labor force.[7] Most people understand and see its parts, but relatively few see the industry in its entirety. It was this lack of understanding of the tremendous scope of this industry that led to the ground-breaking book, *A Concept of Agribusiness*, in 1957.[8] The book debuted the term "agribusiness."

Today, a half-century later, agribusiness firms ranging from large multinational corporations to emerging food manufacturing and input supply firms present a strong demand for more and better employees trained in both management and agricultural sciences. Students who have mastered economic and business concepts along with agricultural sciences do extremely well in finding rewarding careers following graduation. It is this labor demand and the resulting career opportunities that led to the writing of the book that you are now reading.

Through one of its subsets—agriculture— agribusiness is older than business. For centuries the two entities, agribusiness and agriculture, were considered separate; only in the mid-twentieth century were their commonalities uncovered. Despite its uniqueness, the agribusiness industry uses standard business terms for many of its transactions. This use is a reflection of a very different heritage.

### FOOD FOR THOUGHT 1.1

Agriculture is the nation's largest employer, with more than 21 million people involved in some phase.

## The Core Industries

Knowing the origin of the term *agribusiness* is helpful but does not explain what an agribusiness is. We assume that everyone knows what "business" is, yet many students are surprised by the length of the term's definition in any good standard collegiate dictionary.

Agribusiness involves (1) **inputs**—seeds, fertilizer, financing, equipment—that are used in

## A FOCUS ON AGRIBUSINESS
## Summer Farms

Joshua and Katherine Summer left the cold New England climate in 1956 and moved westward to a small community where they purchased mostly forested acreage. They grew vegetables and raised chickens to feed their family, sold their meager excess production to buy the things they did not produce, and sold timber from their acreage to acquire and clear more land. Over the years the Summer family also grew, providing more family members to pitch in during the growing and harvesting seasons, enabling them to produce even more than the family needed for their own personal consumption. Throughout these early years they thought of themselves as a small family farm that provided most of their food supply, but already they were also functioning as a small agribusiness by selling their farm produce and natural resources to purchase other family needs and to expand their production.

The growing demand plus the increasing interest of family members motivated the Summer family to plant more crops. A few years later Joshua and Katherine began to turn over most of the farm work and a lot of the business decisions to the children who maintained an interest in staying on the farm. They adopted the name "Summer Farms" and began to more formally organize the expansion of the business to include additional crops and the subsequent division of responsibilities for their different crops or fields, which they referred to as their "farms" (plural). Various ideas for expansion have since been considered, including organic produce and meat, hay and cotton, and raising poultry or livestock. Some have been successful and some have not.

The founders have officially retired, and their aging children are now facing some sensitive decisions about the long-term future of this family business. They had hoped to turn the management over to their children (the founders' grandchildren), but today more and more grandchildren are getting college degrees and leaving the farm, often to work in large agribusiness corporations. Some of the grandchildren who chose to stay out of the family business are enjoying success as attorneys or as executives in large agribusinesses, paving their way to enjoy the old homestead as "gentleman farmers" or "hobby farmers." So, to keep the agribusiness property in the family rather than see it gobbled up by a large "agricorporation," the parents may need to consider this new type of ownership succession instead of "writing off" the grandchildren permanently.

Today, Summer Farms is too small to provide the large quantities of farm products needed by a large corporation like McDonald's. However, this family-owned business may someday become part of a giant corporation that does count McDonald's as one of its customers.

It is important to note that Summer Farms was never a "hobby farm," which is a small farm not intended as a primary source of income. Hobby farms provide some recreational land for the owner and/or the owner's children, or in some cases to provide a hobby or, at most, a sideline income. Some hobby farms are even run at an ongoing loss for financially secure owners who can afford this country-home lifestyle choice instead of a business.

NOTE: Throughout this book, look for the color shading shown here when Summer Farms is used as an example to represent smaller agribusinesses.

**TABLE 1.2** Agribusiness Industries

| CORE INDUSTRIES | | | SUPPORT INDUSTRIES |
|---|---|---|---|
| INPUT INDUSTRIES | AGRICULTURAL PRODUCTION INDUSTRIES | VALUE-ADDED INDUSTRIES | |
| Seed | Farms | Commodity processing | Banking |
| Chemical | Ranches | Food manufacturing | Transportation |
| Water | Forest products | Food distribution | Government |
| Machinery | Aquaculture | | Insurance |
| | | | Equipment supply/service |
| | | | Industry associations |
| | | | Education |

SOURCE: Adapted from G. Seperich, *2000–2001 Arizona Agribusiness Directory* (Mesa, AZ: Arizona State University East and The Agri-Business Council of Arizona).

production; (2) **intermediate activities**—grading, storage, processing, packaging, distribution, pricing, marketing; and (3) **final consumption activities**—restaurants, groceries. The input supply industries, production agriculture, and value-added activities are considered the agribusiness **core industries**. Similarly, these core industries use the material and services of certain ancillary industries which are considered agribusiness **support industries**. Table 1.2 contains examples of both core and support industries.[9]

Farming and ranching are part of the core of agribusiness, representing its production process stage. So, while they are part of agribusiness, they are not the end; in fact, they are not even the beginning.[10]

## The Chain of Agribusiness— The Value Chain

The linkages between the agribusiness industry activities are frequently referred to as the **chain of agribusiness** (Figure 1.3). This chain can be further refined as the **agribusiness value chain**.[11] It begins with raw materials and commodities of relatively low value; then each step in the chain modifies the material or commodity and adds value to the product derived from its step. For example, growers

harvest raw fruits or vegetables that are transported in bulk to processors or wholesalers, where they are modified or repackaged and transported to grocery chains in processed form or in smaller batches, and finally to local grocers who display and sell the products in quantities more suitable for consumers to purchase.

It should be noted also that both food and non-food products are involved, and the distinction between them is not always clear. Cotton may seem like a non-food product, it is also considered a food product as it can be used in the manufacture of some human foods and in cottonseed oil used for cooking. Similarly, corn for human consumption is a food, but corn grown for livestock consumption is regarded as a non-food product.

Another way of looking at the term *agribusiness* is much simpler. Agribusiness involves all the value chain activities usually associated with a business or industry with the additional provision that most of the materials, commodities, and end products are **perishable**. The perishable nature of a product, along with variability in such factors as color, texture, size, and shape, is the distinguishing characteristic of agribusiness products and industries from other businesses. This aspect of the agribusiness industry

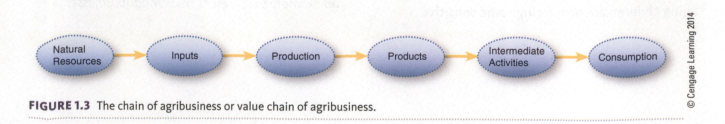

**FIGURE 1.3** The chain of agribusiness or value chain of agribusiness.

is also a key factor in all business decisions. For this reason the story of McDonald's is an agribusiness story. Its entire business system involved a highly perishable commodity: food—hamburgers, French fries, shakes, etc.

Appendix A provides a partial list of agribusiness firms based on the U.S. Department of Agriculture's identification.

## The Development of Agribusiness

Agribusiness is thoroughly entwined in all of the cultures and countries of our planet regardless of economic orientation. Only the human enterprise, hunter-gatherer, is older than agriculture/agribusiness. Few cultures are based on the "hunter-gatherer" model today; its last bastion is the "wild" fishery industry. The fact that an adjective, "wild," had to be added to that industry's description indicates that it, too, is now subject to "farming," as in salmon, shrimp, or catfish farming.

### The Early Influence of Agriculture

It is difficult to discuss the development of agribusiness without involving the development and maturation of agriculture. The Egyptians and Mayans built great cities and pyramids because agriculture afforded them the time and food to initiate these huge "human resource" rich projects. Hunter-gatherer societies left no such monuments; they did not possess the resources, though they may have had the time. The switch from a hunter-gatherer culture to an agricultural-based culture was one of the great "leaps" in technology made by the human species. Agriculture was among the first of the "disruptive" technologies that fostered a paradigm shift.

### The Early Shift Toward Controlling Nature

A **disruptive technology** radically transforms markets, creates wholly new markets, or destroys existing markets. The **paradigm shift** involved a very different way of existence—a radical change in thinking from the accepted point of view to a new one. It shifted from following nature to controlling nature, or at least attempting to control it. Agricultural development was more than "just" growing a crop. It necessitated developing a rule of law, land ownership, acquiring possessions beyond transportable implements and weapons, needing large families (cheap labor), using astronomy to understand the seasons rather than serving as a primitive "Global Positioning System (GPS)" which the hunter-gatherers used, recognizing the need for leadership hierarchies or management, establishing towns, and using subsidiary or support tradespeople to assist the primary crop production activity.[12]

## The Mutually Supportive Relationship Between Agriculture and Local Communities

This book does not dwell on the development of agriculture, but it is important to understand the contribution of agriculture to the development of "civilization." The clue to the emergence of agribusiness is contained in the list of developments originating from agriculture—crop production and tradespeople. The support industries were critical to agricultural development. The early agriculturalists were self-reliant. If a tool was broken they repaired it or made a new one. However, as more individuals became involved in crop production and centers of crop production developed, it was natural that some individuals became recognized as support people rather than agriculturalists.

This handy complementarity among the "town" people and the people of the soil served to presage the development of agribusiness. It is a benchmark of our species that towns and farms grew in conjunction with each other. Each needed the other to survive and thrive. Each defined the potential of the other. Each limited the potential of the other. What was needed was the appearance of another disruptive technology to expand this potential.

In the 1700s, virtually all households farmed. If they produced more food than needed to feed the family, it was sold or bartered to obtain items not readily produced on the farm, for example, kerosene for home lanterns or plows for the fields. Poor transportation required that communities be relatively self-sufficient. During the 1700s and 1800s, all farms and businesses were small with limited production technology, no professional managers, little access to capital, and no large-scale distribution networks. Limited transportation and communication infrastructures restricted them to local markets. *The whole value chain was essentially carried out on each farm.*

This *status quo* was sufficient because *all* markets were local with little need to go beyond the local environs. Certainly, some early farms located near rivers or canals had access to other larger markets through the use of boat and barge traffic. However, this approach was prohibitively expensive since the transportation people controlled the market. The cost of transporting an individual farm crop negated any advantage to accessing a more affluent market and putting that crop in the hands of "middle men."

## Early Technological Innovations

The development of the railroad and the telegraph in the nineteenth century made up the next node of disruptive technology. These two technologies led to enormous improvements in the **infrastructure** for conducting business—the basic physical and organizational structures needed to facilitate the production of goods and services for the operation of a society or enterprise.[13] This improved infrastructure resulted in an enlarged definition of "local market." The marketing radius for a town or farm was no longer limited by horse or oxen endurance; the train, also known as the "iron horse," now defined it. Market prices were now set by markets located at the nexus of railroad lines and transmitted to "middlemen" by telegraph. New towns began to support these technologies with some growing into cities and metropolises because of this extended support.

This was the beginning of agribusiness. As transportation improved, specialists began offering their wares and services over greater areas, and farmers could offer less perishable items. Advances in distance communication began creating market-exchange possibilities fostering the development of more distant markets. These advances led to developments in financing allowing farmers and others to borrow money for downtimes and invest in ways to improve productivity. It was also the advent of greater competition within and among production regions. The seasonal nature of locally grown agricultural products could be augmented through transportation. Bananas grown in the Caribbean or in Central America could be made available in places that could not grow bananas except maybe in greenhouses.

Continued improvements in transportation, communication, and financing coupled with improved machinery, better fertilizers and pesticides, and crop varieties led to tremendous increases in agricultural productivity.[14] Those increases were felt throughout the chain of agribusiness.

Farming was no longer limited to traditional lands. Farmers were encouraged to seek the fortunes in other, hitherto undeveloped regions of the country. People without land migrated to other regions that were made accessible by the new railroads to seek their fortunes and pursue the dreams of "homesteads." Immigration was encouraged to attract the land-limited farmers of Europe to seek their destinies in land beyond the Mississippi. Once again, agriculture established the "beach heads" of civilization in previously undeveloped regions. First the farms were established and then the towns.

Unfortunately, these incursions led to resentment on the part of indigenous peoples located in those regions. They had developed their own form of agriculture or were still living in hunter-gatherer societies. The resolution of these issues was not always pleasant and was often one-sided. Everyone does not always view "progress" or change as a positive activity. Nevertheless, change did occur.

## Increased Specialization from Innovations Related to Chemistry and Biology

Relationships changed—agriculture was now part of the agricultural sciences as chemistry and biology came to the aid of crop production. This was the source of the new chemistries of fertilizers, pesticides, etc. and the new varieties of crops allowing for early maturation, larger yields, and greater capacity to withstand transportation. In the United States Abraham Lincoln signed the Morrell Act, granting each state the land and funds to establish "land grant universities" dedicated to the advancement of agricultural production and rural development.[15]

To take advantage of these productivity changes, more and more specialization occurred. Farms focused on fewer crops, suppliers focused on particular products, and workers narrowed the range of their skills. This led to larger organizations. So fewer and fewer aspects of the agribusiness chain were completed on the farm, and more and more organizations came to provide each link in the chain.

### FOOD FOR THOUGHT 1.2

The number of farmers' markets in the United States increased from 1,755 to 5,274 between 1995 and 2009. The number jumped another 16 percent to 6,132 the following year (2010).

## Agribusiness Goes Global

As agribusiness and agribusinesses became larger and more globally focused, it was no longer necessary to be content with the limited land resources of the country of origin. Agribusiness firms began to seek their competitive niches elsewhere. This first wave of globalization allowed the development of one of the largest food companies in the world, Nestlé SA. This giant company is based in Switzerland, a land-locked country with no access to the sea, an historic and prodigious national work ethic, and a significant dairy origin. However, its success is not based on "milk chocolate" as it once was but on other agribusinesses not possible in its home country.[16]

## Differences Between Agribusiness and Business

In the earlier section on the development of the agribusiness industry, we began a discussion of the differences that exist between agribusinesses and traditional businesses. Now we will discuss other significant differences between these two entities aside from the perishable nature of the agribusiness product.

### A FOCUS ON AGRIBUSINESS
### NAPI and Navajo Pride

Navajo Agricultural Products Industry (NAPI) located seven miles south of Farmington, New Mexico, was formed in 1970 to oversee and operate the Navajo Indian Irrigation Project (NIIP) and assist the Navajo nation in developing its agricultural resources. The Navajo nation covers more than 27,000 square miles and is the homeland of the Navajo people, with its own history, culture, and language. It is located where four states (Arizona, Colorado, New Mexico, and Utah) meet. This strategic location enables NAPI to easily reach local, regional, and international markets. Several major interstates, including I-40 and I-10, are close by and the nearby Gallup Railroad transports NAPI's Navajo Pride brand products to a number of markets throughout the nation.

NAPI raises alfalfa to feed herds of cattle, providing ranchers with much-needed hay during the harsh winter. Other products are sold under the "Navajo Pride" brand, including potatoes, popcorn, and pinto beans. The top crop is pinto beans, which surprisingly are exported mostly to Mexico. NAPI's international customers are not just south of the border, however. Most of the popcorn, for example, is exported primarily to Asia, the Middle East, and South America. Its experience in international business has led NAPI to identify its next big venture: raising black beans for Cuba. Export restrictions to Cuba do not include "staples" such as beans and rice, so Cuba is a potentially strong market.

**FIGURE 1.4** Bales of hay await pick up by farmers.
© iStockphoto/Chapin 31

One key employee, Roselyn Yazzie, says "I live here. I literally live here. I do have a life on the outside. I own cattle, horses. I own my own home and so forth, but this is my home. And I guess it's the Navajo reservation. It's my home. I take a lot of pride when the bags leave here with the name saying 'Navajo Pride' and with the saying that it was produced here on the Navajo nation. That makes me very proud." (See Figure 1.4.)

NAPI is also proud to be a valued member of the New Mexico community, contributing more than $30 million annually into the Four Corners economic base.[17]

## Family Orientation or Intent

Almost all agribusiness ventures begin as family endeavors—an orientation that not all of today's traditional businesses have. (See Figure 1.5.) While many traditional businesses in the early and mid-twentieth century had a **family orientation**, this is no longer the case. Indeed, many website-based companies or application-oriented companies were developed by lone individuals to demonstrate the feasibility of the business concept, seek out investors, sell the company, and move on to other entrepreneurial ventures. But there is little intent to keep the business in the family. Individuals with this approach to business development have even acquired a label, **serial entrepreneurs**. There is nothing wrong with this approach: it is one of the great engines of economic growth. But it does define a difference between agribusiness and traditional business.

Serial entrepreneuring is not the orientation of most agribusinesses which begin as family enterprises, like Summer Farms, with the intent of the entrepreneurial founder to keep the agribusiness in the family. This is easy to envision with a farm or ranch, but even large agribusinesses have this plan. Anhueser-Busch Company, the "King of Beers," for example, was a small brewing company initiated by a German immigrant, Augustus Busch, in St. Louis. It grew over time into a large domestic agribusiness with a domestic share of the U.S. beer market approaching 50 percent. Even though it was sold in 2009 to another large international brewer, InBev NV, it is still under the helm of the family represented by Augustus Busch III, the great-grandson. Now he sits on the board of a larger agribusiness

firm. The name of the new firm is Anhueser-Busch InBev, which is a salute to multiple family/country origins. It is now a Belgian-Brazilian-American company, an example of globalization at work. One of its major competitors is SAB Miller SA, a South African and American firm.[18]

Eventually the family-orientation may be lost, as is evident from the examples given, but the intent generally is to keep the business in the family as the Busch family did for three generations. Today, some agribusiness companies, for example, Kraft and Kellogg, retain only the family name and are managed by professional business people.[19] A few agribusinesses remain that continue to involve the family, such as Brown-Forman (diversified distiller) and Cargill Company (commodity-oriented company).[20] According to the Form 10-K that Tyson Foods, Inc., filed with the S.E.C. in 1009, family members still own most of that multi-national company's stock and exert substantial influence over management affairs.

---

### FOOD FOR THOUGHT 1.3

The Burpee Seed Company was founded in Philadelphia in 1876 by an 18-year-old man. Within 25 years, he had developed the largest, most progressive seed company in America. Still a family-owned company, it refuses to offer GMO seeds, only open-pollinated and hybrid varieties.

---

## Initially Tied to Location or Commodity

Another distinction of agribusinesses involves a unique tie-in: *its initial competitive advantage was tied to a location and/or a commodity*. The nature of the agribusiness was related to its location. E&J Gallo Winery, for example, is located in the California counties (Napa, Sonoma, and Mendicino) that are famous for their vineyards. That location determined the nature of the business. A California-based wine initially provided a competitive advantage for the company because of its natural conditions. This advantage has largely disappeared as a result of globalization, where other countries discovered similar natural attributes for wine production. The location of the original agribusiness, however, has remained.[21]

Similarly, an agribusiness is closely *related to an agricultural commodity*. A **commodity** is a fungible

**FIGURE 1.5** Agribusiness frequently involves family business, such as this one.

© iStockphoto/Morgan L

good with little or no qualitative differences (i.e., it is essentially the same no matter who produces it). The Gallo Wine Company's commodity was grapes. Everyone produced grapes in these California counties, but not everyone had the ability to move the commodity up the value chain to produce wine. Even now, the agribusinesses in these California counties continue to seek their niches in the value chain. There is a difference in quality, quantity, and price between table grapes, grape juice, cooking wine, table/everyday wine, premium wine, and super premium wines.

Agribusiness firms are generally tied to location and expand their expertise in a commodity or are tied to the commodity and seek their locations internationally. Some agribusinesses (Summer Farms, for example) are tied to their locations and commodities permanently, with little need to move. This is the case with Scotch whiskies which are tied to Scotland—their competitive advantage. They must therefore export their product. Other countries or agribusinesses may copy their "product," but it will never be a Scotch whisky; it can be only "Scotch Whisky-like." Our American forefathers tacitly surrendered this competitive advantage to Scotland by labeling all American whiskies as "Whiskey"—with an "e"—to delineate their non-Scottish or Irish origins.

This example applies also to champagne which can be produced only in certain regions of France known for this fermenting process. Other countries may produce sparkling wines, but not champagne. This form of delineation is a significant and sustainable competitive advantage.

In the semi-conductor, computer chip, athletic shoe, software, or hardware businesses, location may be only a secondary concern. This is why it is easy for these businesses to seek a less sustainable competitive advantage and outsource their production facilities to places where labor is relatively inexpensive. This strategy may engender continuous movement of production facilities as various emerging nations attract direct foreign investment through factories for its people who are willing to work for less money to have jobs and futures.

Of course, some countries try to emulate this location/commodity status and are not successful. New Zealand attempted this approach with "kiwifruit." Unfortunately, while we know it as "kiwifruit," which implies an origin in New Zealand, in reality it was adopted from the Chinese gooseberry. Other countries now produce "kiwifruit." In fact,

many South American countries and Italy outperform New Zealand.

## Needs Risk Mitigation

The location/commodity orientation of an agribusiness creates another distinguishing characteristic: risk management. **Risk management** is an approach to control uncertainties and potential dangers by assessing what those are, then developing strategies and tactics to reduce them.[22] All companies regardless of orientation consider risk management or perhaps more correctly, risk-mitigation, an essential operational concern. The entire "hedge" industry had its origin in the late nineteenth century as a service to agribusiness operations. A business tied to a particular commodity or location was highly susceptible to the vicissitudes of weather and other natural calamities.

An effort to mitigate this risk witnessed the formation of the commodity trading industry, the development of forward sales, the introduction of "puts and calls" and the introduction of derivatives. The need of the agribusiness for risk mitigation also attracted investors willing to purchase or "cover" the risk. If the agribusiness does not experience loss, its "risk insurance" is not needed and this expense becomes a gain for the investor. However, if the agribusiness does experience a loss, then its "risk insurance" covered that loss and this time the investor was on the "wrong end of the risk." This beginning has fostered the development of an entire industry of modern finance.

An agribusiness such as Coca-Cola, Inc. may "hedge" the price of corn because of its need for high-fructose corn syrup as an essential ingredient in syrup production. Southwest Airlines will "hedge" the price of jet fuel to "control" fuel costs because of oil supply volatility. This hedge, if performed correctly, can provide a competitive advantage. The entire hedging industry has become so complex (there are now weather derivatives) that it is easy to forget the need of agribusiness that led to the industry's development.

## Produces Ingestible or Wearable Products

Agribusiness firms generally produce products that are *personal*. You may like the car you drive, maybe even love it; you may enjoy the laptop computer you use (though it has become a commodity, also) and you may treasure your entertainment system, but these products rarely get as personal as agribusiness products. Agribusiness products are personal because you literally consume them or wear them. You may enjoy the taste of wine or beer or distilled spirit, but in this enjoyment you are consuming it and it becomes part of you. So it is with Birdseye frozen foods, Kellogg's ready-to-eat cereals, Hershey's chocolates, and others. Nothing is more personal than ingesting a product. Agribusiness consumption is very personal.

As a result of this very human activity, we develop loyalties to products. Sometimes these loyalties assume "cult" status as in In-N-Out Burgers, White Castle hamburgers, or Levi's jeans. This is a difficult status for a non-agribusiness to achieve, although some have morphed from company names to verbs, as in Google, or from company trademarked products to generic names like Scotch tape. We do not search the Internet—we "Google," regardless of whether or not we use Google Search.

## Produces Highly Perishable Products

As mentioned earlier, the perishability of materials, commodities, and products is a characteristic of many agribusiness products and industries that distinguish agribusiness from other businesses. To make a profit, Summer Farms, for instance, must use or sell its products before they become unusable or unfit to sell. This aspect of the agribusiness industry is a key factor in all business decisions. Of course, all products have life cycles, but agribusiness products by their nature are far more perishable and variable in size and quality. The apple progressing through this value chain to yield applesauce gains both value and an extended shelf life, but it remains perishable. Ford cars and tractors may have useful product lives, but the manufacturer does not consider its products "perishable." The term **shelf life** as developed by the agribusiness industry defines the usable or safe life of a product. It was in use long before the consumer became aware of "use by" and "sell by" dates commonly seen on food packages today.

These five characteristics, then, should be present in agribusiness firms: family-orientation/intent, location/commodity-based, risk-management/mitigation, an ingestible/wearable product, and relatively high perishability. Not all five differentiating characteristics may be evident, but generally most of them are present. Before the current wave of globalization

began, almost all agribusinesses bore these characteristics regardless of the country of origin.

## Agribusiness Today

Though this story of agribusiness development was a little parochial and primarily limited to the United States, this dynamic change has also occurred elsewhere. And in the process, agribusiness has been changed greatly. It has evolved into an industry that encompasses the globe. It is responsible for feeding and clothing the population of this planet.[23] Today, agribusiness touches on health, nutrition, safety, science, and politics.

### Connection to Agriculture

The agricultural connection within agribusiness remains evident. Oddly, there are fewer individuals involved in production. The United States had its start in the labors of its yeoman farmers. As stated earlier, nearly everyone was involved in farming. Today, in the United States about 1 percent of the population produces all of the food and fiber needs for the country, plus an excess that is globally traded.[24] The remaining millions of agribusiness people are all employed in pre-production support and post-production sectors. It is an industry that began on an unpaved farm road, continues through Wall Street (United States) and Fleet Street (United Kingdom), and eventually ends up on Main Street. It is an industry that still retains tremendous potential. It is an industry that must be able to feed the population of this planet as it grows and ages (See Figure 1.6).

**FIGURE 1.6**  Ravioli production in a modern food-processing plant.

© iStockphoto/Galanter

## Connection to the Sciences

At the same time, agribusiness has broadened its connection to the sciences. Originally it was chemistry and biology; today it is biotechnology and nanotechnology. Large agribusinesses involved in the pre-production sector (e.g., Monsanto) have become leaders in biotechnological advancements.

## Connection to Financial Structures

Agribusiness also has become an industry involved with the development and evolution of business financial structures. The original derivatives that are now developed by the "rocket scientists" on Wall Street were originally financial vehicles developed in Chicago to distribute production risks more broadly beyond the farm, thus allowing the farm to concentrate on production.

## Subject to Political Considerations

Agribusiness today is a global industry with a national anchor that is often affected by political forces. Though our planet has witnessed remarkable progress, national food security remains a significant issue. Perhaps more than money or access to money, the issue of food security most clearly demarcates the "haves" from the "have-nots." It has long been the goal of this industry to diminish the gap between those who have access to adequate nutritional resources and those who do not. However, the problem that plagues the industry is that this imbalanced access is rarely a scientific, engineering, or financial issue as much as it is a political issue.

## Characterized as Economic Paradox

Similarly, agribusiness is an industry involved in a paradox. While many economically advanced countries (e.g., the United States and the United Kingdom) have a large and productive agribusiness industry; they possess also a very efficient production sector operated by a declining percentage of individuals. The reverse is true also, as there remain countries with a small agribusiness component and an extremely large but very inefficient production sector. The latter are countries that use a significant portion of their economic resources simply for securing an adequate food supply (e.g., Rwanda and Nepal). If the purpose of writing and studying history is to provide lessons for the benefit of others, it is a purpose that still needs explanation in this sector.

## Close Ties to Natural Resources

Agribusiness is an industry that remains tied to natural resources. The most important of these resources is water—fresh water. Certainly, Earth is a beautiful blue marble when viewed from space, but most of that "blue" is salt water. The amount of available fresh water is finite. Again, like many issues water lends itself to separating populations into "haves" and "have-nots." Access to clean and plentiful freshwater is the hallmark of agribusiness and a personal necessity, yet it still remains an unfulfilled need for a fourth of the world's population. Technical solutions abound, but political considerations preclude application.

## The Basic Management Model

Figure 1.7 presents the basic management model used in this book. That model indicates that:

1. In their efforts to accomplish effective and efficient goals, managers must perform a variety of roles requiring several different skills.

2. Using those skills, managers must develop an understanding of the complex environments in which they must function and the numerous forces that affect them.

3. Using that understanding, managers must develop strategies and plans to achieve organizational objectives.

4. Those strategies and plans help to shape the organization—its members, structure, and culture.

5. To provide the basis for successful organizational functioning, managers must understand people and therefore must develop skills in leadership, motivation, communication, and teamwork.

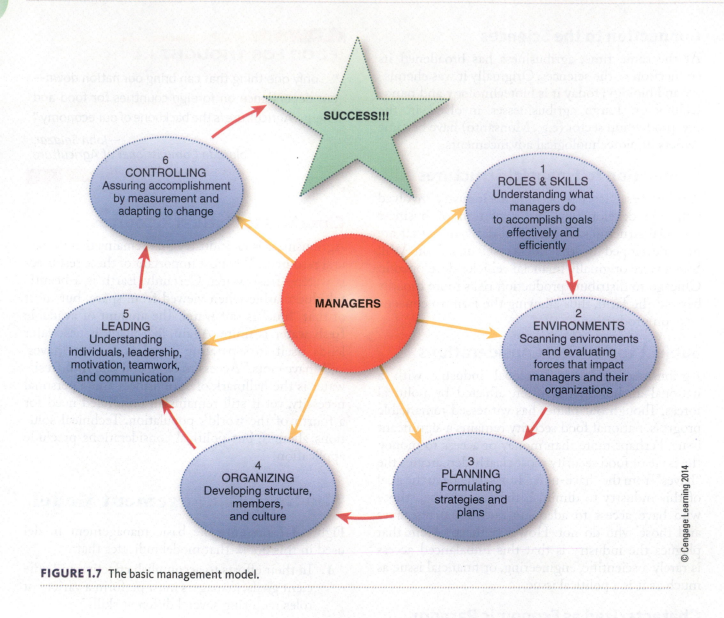

**FIGURE 1.7** The basic management model.

© Cengage Learning 2014

6. To assure that their organizations are following plans and maximizing the potential of their members, managers must measure performance and adjust operations to adapt to changing conditions.

The material in the book follows the management functions illustrated in Figure 1.7 and described in the next chapter.

## CHAPTER SUMMARY

Agribusiness is a highly diverse segment and a major industry in the United States. It includes all economic activity in the food and fiber system, which encompasses the input supply industries, agricultural production and post-harvest value-added activities, such as commodity processing, food manufacturing, and food distribution.

The agribusiness value chain starts with natural resources as inputs to a production process, including farming and ranching, that in turn leads to products that travel through various intermediate activities before reaching consumers. The value chain includes both food and non-food items.

In this industry shelf life is important because agribusiness products are perishable; that is, they have a relatively short usable or safe life. Perishability is, then, a major distinguishing characteristic of agribusiness.

Agribusiness began to emerge as agriculture and went through a disruptive technology that radically transformed markets, creating a paradigm shift. That shift involved a movement from following nature to controlling nature. As agriculture replaced hunting and fishing, towns grew and civilization developed. As infrastructure developed, agricultural products could be moved to more distant markets, thus expanding agribusiness. With improvements in agriculture through science and improvements in infrastructure through technology, agribusiness has grown tremendously.

Differences between agribusiness and other forms of business include perishability and the tendency for many agribusiness firms to remain in the hands of the owners' families. The commodity and location orientation of agribusiness are other distinguishing characteristics and lead to risk management.

Agribusiness is an industry with tremendous potential for growth and development. In the future, individuals who understand the principles of business and the nuances of agribusiness will be in greater need than ever before.

## CHAPTER ACTIVITIES

### REVIEW QUESTIONS

1. Define agribusiness and differentiate it from other forms of business.
2. What is the agribusiness value chain?
3. Identify several agribusiness organizations. What makes them agribusinesses?
4. What are the four characteristics generally found in an agribusiness product?
5. What are serial entrepreneurs and how do they differ from other entrepreneurs?

### ANALYSIS QUESTIONS

1. Do you think that agribusiness is a necessary component of any economic system? Why or why not?
2. Think about the development of agribusiness. Can you identify a company that has developed in such a way that it encompasses all of the stages of the value chain in the one company?
3. Why is the distinction of a commodity important to understanding agribusiness?
4. Think of someone you know who is employed in agribusiness. Describe that person's job position and daily activities.
5. Select one agribusiness product (e.g., corn, soybeans, cotton, poultry) and, using library and Internet sources, discuss the market for that product over the next ten years.

## FILL IN THE BLANKS

1. An approach to control uncertainties and potential dangers by assessing what those are, then developing strategies and tactics to reduce them, is known as _____ _____.

2. The input supply industries, production agriculture, and value-added activities are considered the _____ industries of agribusiness.

3. Banking, transportation, and insurance are considered as the _____ industries of agribusiness.

4. Food manufacturing and distribution are considered as _____ _____ industries of agribusiness.

5. Seed and chemicals are two of the _____ industries of agribusiness.

6. Beginning with raw materials and commodities of relatively low value then modifying the material or commodity to add value to the product is known as the _____ _____ _____ or the _____ _____ _____.

7. The initial competitive advantage of agribusinesses was tied to a _____ or a _____.

8. _____ refers to the various businesses involved in the production of ingestible or wearable products, including farming and contract farming, seed supply, agrichemicals, farm machinery, wholesale and distribution, processing, marketing, and retail sales.

9. A fungible good (the same no matter who produces it) for which there is demand, but which is supplied without qualitative differentiation across a market is known as a _____.

10. The basic physical and organizational structures needed for the operation of a society or enterprise (e.g., roads and communication) are known as the _____.

## ▼ CHAPTER I CASE STUDY
### WILL THE FUTURE SUCCESS OF AGRIBUSINESS BE DETERMINED BY GM?

"**F**ew companies excite such extreme emotions as Monsanto. To its critics, the agribusiness giant is a corporate hybrid of Victor Frankenstein and Ebenezer Scrooge, using science to create foods that threaten the health of both people and the planet and intellectual property laws to squeeze every last penny out of the world's poor. To its admirers, the innovations in seeds pioneered by Monsanto are the world's best hope of tackling a looming global food crisis." This lengthy quote from *The Economist* magazine nicely depicts the plight of the Monsanto Company of St. Louis.

The Monsanto Company was founded in 1901 by John F. Queeny, who had only a sixth-grade education, and his wife, Olga Monsanto. The company's first product was saccharine, the artificial sweetener. However, it was not until the end of World War II that the company got its "second wind" and recognized that the future of the chemical industry was in boosting crop production. The company grew and developed its experience in chemicals to develop pesticides. These chemicals, now called *crop protection chemicals*, changed the face and capacity of agriculture. These adjuncts provided a means to control the ravages of nature and increase global crop yields. As the company grew it began to acquire large seed genetic firms. Early on Monsanto recognized that if the business of agriculture was to increase yields and minimize costs, it all begins with seed stock.

Before the 1970s two kinds of companies were evident: agricultural chemical companies and agricultural genetics companies. The chemical companies were Dupont Co. Inc., BASF SE, and others. Among the genetics companies were Dekalb Seed Company and Asgrow Seed Company. However, significant discoveries in biotechnology changed this comfortable duality.

The demarcation between chemistry and biology became blurred. It became possible to produce staple crops, like corn, that not only grew well and yielded a high amount of corn but had their own genetic resistance to a crop protection chemical. Because of Monsanto's merger with Pharmacia and UpJohn, it had a competitive advantage in expertise. As a result it developed a strain of corn, Bt corn, which carried a resistance to the herbicide Roundup®, produced by Monsanto. This allowed a farmer to plant corn seed and as the crop and weeds that accompanied it grew the farmer could spray Roundup. The Roundup killed the weeds but did not harm the corn. By minimizing chemical applications, there was less strain on the natural ecology of the soil but greater yields of corn, which met the farmer's agenda to maximize returns on investment.

This was a significant advancement for farming. It allowed the farmer to control weeds, gain a large corn harvest, and minimize herbicide costs. Since farmers do not set the price of their crops (this is done in open market commodity exchanges around the world), anything that reduces the cost of production is very attractive.

Monsanto Co. expanded its competitive advantage by purchasing seed companies for their genetic stock and developing what is now known as "genetically modified" organisms (GMOs) or crops. They have been successful at this activity. Monsanto has been producing GM crops for around 13 years (2010). The entire agricultural seed/chemistry industry changed in 1980 when Monsanto won a United States Supreme Court decision that allowed patents on living organisms. Monsanto's success has attracted significant competitors: BASF SE (Germany), Bayer CropScience AG (Germany), Dupont (United States), and Syngenta AG (Switzerland).

The first GM crops entered the market in 1996. It took ten years before more than a billion acres of these crops were planted around the world. It took only three more years before the second billion acres were planted.

It is this success that produced the controversy described in the opening paragraph of this chapter. Not since the "Green Revolution" fostered by Nobel Prize winner, Norman Borlaug, has the promise of increased crop production been more evident, especially in the developing world, Africa and Asia. This technology has been adopted in Asia, South America, and North America. Farmers grumble about the price of the seed; however, there is no doubt about its efficacy.

It is interesting that Europe finds itself on both sides of this controversy. It is the world leader in fostering suspicions about GM crops and has banned them from supermarkets. Yet, as is apparent, its companies are competitors to Monsanto in developing these products.

Companies, especially agribusiness firms, must develop a strategic mission that fosters adaptability and opportunism in order to sustain their economic viability. As Monsanto Company demonstrates, a sustainable competitive advantage attracts competition and controversy.

Furthermore, a competitive advantage established by a court decision may not be sustainable. Court rulings are pushing back against the ability to patent genes, at least human genes.[25]

### ► Case Study Questions

1. Boosting crop production to feed the hungry seems like a noble company mission. At what point does this cause become more important or less important than the argument that genetic modifications may lead to health or environmental problems?

2. Think about your own value system. Would you be eager or reluctant to work for Monsanto or its competitors?

3. Monsanto and its competitors legally own the seeds that they genetically modify as well as all of the future generations of seeds that are yielded by the original GM crops. Should this be the case or should the growers own those succeeding generations of seeds?

# REFERENCES

1. John F. Love, *McDonald's: Behind the Scenes* (New York: Bantam Books, 1986), 13; Stacy Perman, *In-N-Out Burger: A Behind the Counter Look at the Fast-Food Chain That Breaks All of the Rules* (New York: Collins Business, 2009), 64; "Cheesed Off: Burgernomics Points to Uncompetitive Currencies in Continental Europe," *The Economist* (July 18, 2009).

2. Robert A. Hoppe and David E. Banker, Structure and Finances of U.S. Farms: Family Farm Report, U.S. Department of Agriculture, Economic Research Service, Economic Information Bulletin, No. 66, July, 2010.

3. Ibid.

4. N. Scott, ed., *Agribusiness and Commodity Risk: Strategies and Management* (London: Risk Books, 2003); and R. E. Just and R. D. Pope, eds., *A Comprehensive Assessment of the Role of Risk in US Agriculture* (Norwell, MA: Kluwer Academic Publishers, 2001).

5. J. Wilkinson, "The Globalization of Agribusiness and Developing World Food Systems," *Monthly Review*, 2009, at www.monthlyreview.org (accessed July 6, 2010).

6. Cliff Ricketts and Kristina Ricketts, *Agribusiness: Fundamentals & Applications, 2nd ed.* (Mason, OH: Cengage Learning, 2008); George Seperich, *2000–2001 Arizona Agribusiness Directory* (Mesa, AZ: Arizona State University East and Agri-Business Council of Arizona, Inc., 1997).

7. "Introduction: About the U.S. Department of Agriculture (USDA)," United States Department of Agriculture at www.aphis.usda.gov (accessed July 6, 2010) and various Bureau of Labor Statistics reports at www.bls.gov (accessed July 6, 2010).

8. John H. Davis and Ray A. Goldberg, *A Concept of Agribusiness* (Boston: Harvard Business School, 1957); see also Ray A. Goldberg, *Agribusiness Coordination: A Systems Approach to the Wheat, Soybean, and Florida Orange Economics* (Boston: Harvard Business School, 1968).

9. Michael Porter, *The Competitive Advantage of Nations* (New York: Free Press, 1990).

10. M. W. Woolverton, G. L. Cramer, and T. M. Hammonds, "Agribusiness: What Is It All About?" *Agribusiness*, 1985, 1(1): 1–3.

11. Porter, op. cit.

12. J. S. Lee, *A Reference Unit on the Meaning and Importance of Agribusiness. Research and Curriculum Unit for Vocational-Technical Education* (Mississippi State, MS: Mississippi State University, 1974).

13. D. D. Besanko, M. Shanley Dranove, and S. Schaefer, *Economics of Strategy, 4th ed.* (New York: John Wiley & Sons, 2007).

14. Lee, op. cit.

15. M. R. Gisolfi, "From Crop Lien to Contract Farming: The Roots of Agribusiness in the American South, 1929–1939," *Agricultural History*, 2006, 80(2): 167–189.

16. Bruce L. Gardner, *American Agriculture in the Twentieth Century: How It Flourished and What It Cost* (Cambridge, MA: Harvard University Press, 2002); see also M. Mazoyer and L. Roudart, *A History of World Agriculture: From the Neolithic Age to the Current Crisis* (New York: Monthly Review Press, 2006).

17. "Navajo Pride," America's Heartland Stories, Episode 306, at www.americasheartland.org (accessed July 6, 2010); "What's New at NAPI," Navajo Price at www.navajopride.com (accessed July 6, 2010); "NAPI: Navajo Oasis in the Desert," *New Mexico Business Journal*, June 1995, at findarticles.com (accessed July 6, 2010); Erny Zah, "NAPI Hay a Hit with Reservation Ranchers," *Navajo Times*, February 11, 2010, at www.navajotimes.com (accessed July 6, 2010).

18. "A Brief History of Anheuser-Busch," Anheuser-Busch website at www.anheuser-busch.com/briefHistory.html (accessed July 7, 2010).

19. "History," Kraft Foods at www.kraftfoodscompany.com (accessed July 6, 2010) and "Our History," Kellogg Company at www.kelloggcompany.com (accessed July 6, 2010).

20. "Heritage," Brown-Forman at www.brown-forman.com (accessed July 6, 2010) and "Our History," Cargill at www.cargill.com (accessed July 6, 2010).

21. See http://gallo.com.

22. J. McMahan, *Professional Property Development* (New York: McGraw-Hill, 2007), 158.

23. "2010–19 Long-Term Agricultural Projections," United States Department of Agriculture at www.ers.usda.gov (accessed June 30, 2010).

24. K. O. Fuglie, James M. MacDonald, and Eldon Ball. "Productivity Growth in U.S. Agriculture," United States Department of Agriculture, Economic Research Service, EB-9, September 2007. (Available at: www.ers.usda.gov/publications/eb9/.)

25. "Briefing: Monsanto, the Parable of the Sower," *The Economist* (November 21, 2009): 71–73; "Annual Report," Monsanto at www.monsanto.com (accessed June 30, 2010); K. Drlica, *Understanding DNA and Gene Cloning, 4th ed.* (New York: John Wiley & Sons, 2004); and "Genetic Shock: A Surprising Court Ruling in America May Loosen the Drug Industry's Grip on Important Genes," *The Economist* (November 30, 2010).

# CHAPTER

# 2

# Management Today

## MANAGER'S VOCABULARY

- administrative management
- administrative managers
- behavioral school
- bureaucracy
- classical school
- contingency approach
- controlling
- effectiveness
- efficiency
- entropy
- equifinality
- finance managers
- first-line managers
- Hawthorne studies
- high-involvement management
- human relations
- human resource managers
- human resources
- leading
- management
- management information systems (MIS)
- management science
- marketing managers
- middle managers
- operations management

*(Continues)*

## LEARNING OBJECTIVES

After studying this chapter, you should be able to:

- Define management and describe its complexity and pervasiveness.

- Define successful management in terms of efficiency and effectiveness.

- Discuss the importance of history and theory to the field of management.

- Discuss the origins of management and trace its development.

- Describe and assess the classical, behavioral, and quantitative schools of management theory.

- Identify major components of contemporary management theory.

## Where Have All the Cowboys Gone?

There are few icons as readily identifiable as the American cowboy. His history has been celebrated in print and celluloid, his story has generated a whole genre of literature, and his image has become a marketable tourist commodity. Behind that image was the work performed by the cowboy: nurturing, protecting, and transporting cattle from their feeding grounds to the "railhead" and eventually to the packing plants located in the Midwest. His job was to herd cattle (see Figure 2.1). His job was to produce beef. He and the ranch owner were at the beginning of the agribusiness "value chain."

**FIGURE 2.1** Getting the herd to market.
© PD Loyd/www.Shutterstock.com

"Driving" the cattle to the railhead provided the cowboy with his job, established his image, and contributed to his legendary status. But it was slow, the losses were large, and in the end the cattle had to ride railcars to cities like Chicago. These disadvantages eventually led to the demise of the American cowboy as he had once existed and to the development of a new meat-producing agribusiness industry.

It was practically an economic axiom of faith that led to this new industry: growing personal affluence or wealth accompanied by the desire to upgrade caloric consumption. The American steak was the poster icon for this affluence. But in the cowboy era, cattle had to be produced locally, so meat was supplied only locally. These nascent agribusinesses were small farms and therefore limited in their abilities to supply the large packing plants. More and large ranches were needed to produce cattle to meet the growing consumption pattern. The ranches in the Southwest and West were large and produced large numbers of cattle, but they were distant from the market. Simply driving and shipping cattle was inefficient. Better and more efficient means were necessary to satisfy the demand for beef.

When consumer demand exceeds supply, someone (often more than one) is sure to take advantage of the opportunity presented, often leading to a new industry. And someone did.

As a result, the meat-producing industry was changed by the introduction of a new way to get meat to the consumer: ship the meat, not the cattle, to the packing houses. In 1867, J. B. Sutherland of Detroit, Michigan, gained a patent for a refrigerated rail car. But it was Gustavus Swift, one of the original five meat-packer families in Chicago (Armour, Swift, Morris, Wilson, and Cudahy), who saw the potential for this development and commissioned his own design in 1867. This changed the entire meat-producing agribusiness industry. It allowed the shipment of meat rather than cattle to any market in the country and also encouraged the development of refrigerated ships, which opened an international market for American beef. This development allowed other agribusiness industries with highly perishable products to ship their products nationally. The refrigerated railroad car joined the railroad as a significant disruptive technology of the nineteenth century.

Fast forward into the early twenty-first century to see how this American industry has changed. The ranches in the Southwest and West remained, but the cowboy had morphed into a ranch hand. The American/Western agribusiness industry was in different hands as meat rather than cattle traveled to markets. The bold move by "Gus" Swift gave his firm a competitive advantage in the nineteenth century, but it was not sustainable through the twentieth century.

In 2007, a small company—founded by Jose Batista Sobrinho in 1953 to process five head of cattle in Brazil—purchased Swift & Company for several billion dollars in cash. In 2010, the newly named JBS Swift & Company celebrated its 150th anniversary. Today, the company is the second largest pork and beef processor in the world behind JBS Friboi SA.

Securing the number one and number two spots has required management skill and left the cowboy behind.[1]

# INTRODUCTION

The growth of organizations such as Swift & Company illustrates the world of management and managers quite well and also provides useful insights for small companies, such as Summer Farms (Chapter 1). This chapter begins with an overview of management. First, it describes the manager's job by defining management—noting the importance of both efficiency and effectiveness—and identifying managerial functions (Figure 2.2). Following that is an outline of the history of management, noting the importance of history, origins, major schools of management, and contemporary developments in the field. Together we will explore the dynamic nature of the management process in all its forms—how it started and how it has changed.

## The Manager's Job

The chief executive officer (CEO) of Archer Daniels Midland calls upon his experience and training every day as he makes decisions that affect thousands of people and involve millions of dollars. Rick Holley, who once worked for General Electric as well as Burlington Northern, now is president of Plum Creek, the largest private landowner and timber management company in the United States. Leslie E. Whitfill and his wife Floris' moved from Lubbock, TX to Phoenix, AZ in 1928, where they began a citrus tree business in 1943 that became Whitfill Nursery. The family still manages its three Arizona locations. Joshua and Katherine Summer (Chapter 1) gradually reduced their managerial tasks and decision making as they began turning over their growing farm operations to their heirs.

**FIGURE 2.2** A manager explains an idea to co-workers.

A manager in each of these organizations leads and directs the their activities, and each manager is accountable for the organization's performance.

Moreover, each of these organizations and the people who manage them illustrate the complexity of managing.

## Definition of Management

**Management** can be defined as a set of activities directed at the efficient and effective use of resources in the pursuit of one or more goals. Three distinct elements of this definition, illustrated in Figure 2.3, warrant a special discussion.

First, the manager's job involves activity.[2] Managers do not sit around all day thinking. Instead, their days are filled with action, including talking, listening, reading, writing, meeting, observing, and participating. As we will discuss later, most managerial activities can be classified into one of four categories: planning, organizing, leading, and controlling.

Second, management involves the efficient and effective use of resources. For example:

- Human resources (tree planters and sales agents for Whitfill Nursery; assembly-line workers, managers, and dealers for Hormel).
- Physical resources (fertilizer and trucks for Summer Farms; buildings, office furniture, and raw materials for Breyers Ice Cream).

- Financial resources (retained earnings, product sales, and bank loans for West Side Feed Lots and John Deere).
- Informational resources (sales projections and market research for Procter & Gamble; residential and commercial building plans by developers for Green Things Landscaping).

Third, our definition of management notes the importance of organizational goals. Goals are targets for which organizations aim. They can be developed for many different areas and levels of an organization. The setting of appropriate goals is an important part of the manager's job.

**The Complexity of Management.** Management is an extremely complex process, part of which stems from the different activities in which managers engage.[3] Managers must also change activities frequently. In the United States, about half of a manager's activities require less than nine minutes to perform. That is, the typical manager stops doing one thing and starts something else every nine minutes.[4] This pattern contrasts sharply with the duration of activities of managers in other countries, however.

The complexity of the managerial job continually serves as a source of new excitement, and its dissimilarity is attractive to those seeking new challenges. The manager's job requires enormous energy and has many rewards. One of those rewards, financial compensation (particularly for top executives), has come under fire in recent years.[5]

**The Pervasiveness of Management.** Another characteristic of managerial work is its

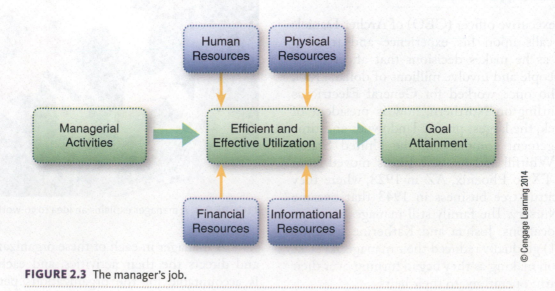

**FIGURE 2.3** The manager's job.

© Cengage Learning 2014

pervasiveness—its applicability in many different situations and its influence in contemporary society.[6] Imagine, for example, how many organizations influence your daily life. You wake in the morning to a G.E. alarm clock, shower with Procter & Gamble toiletries, eat General Foods breakfast products with milk from a local dairy, get into your Ford automobile and fill it with Texaco gas, then drive to your place of business. All of this is before noon, and all of those businesses are run by managers. What we often do not think about is that most of the large corporations, agribusinesses or not, buy raw materials or products from smaller businesses, including small agribusinesses, which are managed by very different managers.

Management is not limited to businesses, however. It is also practiced in universities, government agencies, health-care organizations, social organizations, religious institutions, and families. Indeed, management can be found in virtually every collection of people who find it necessary to coordinate their activities. Thus, it is well worth developing a better understanding of how these organizations function and of the individuals who lead them.

**Levels of Management.**    There are many kinds of managers. If we take a hypothetical organization and draw two horizontal lines through it, as in Figure 2.4, we have a common way of classifying managers: top, middle, and first-line. While it is easy to see these levels of management in a large organization, such as Hormel Foods, in a small family farm only one or a few individuals perform the work of all these levels. The dividing lines are arbitrary, but **top management** always refers to those at the upper levels of the organization and usually includes the chief executive officer and the vice presidents. Irene Rosenfeld, Chairman and CEO of Kraft Foods, is a top manager. So are Ralph Webb, president and owner of Webb Dairy and its 100-acre farm, and his son Jerry Webb, who tend to make joint decisions and do not have middle or first-line managers. Top managers set overall organizational goals, determine strategy and operating policies, and represent the organization to the external environment. The job of CEO is demanding, and turnover is common.[7]

**Middle managers** make up the largest group in most companies. Their ranks extend from top management all the way down to those immediately above first-line management. They include such positions as plant manager, division manager,

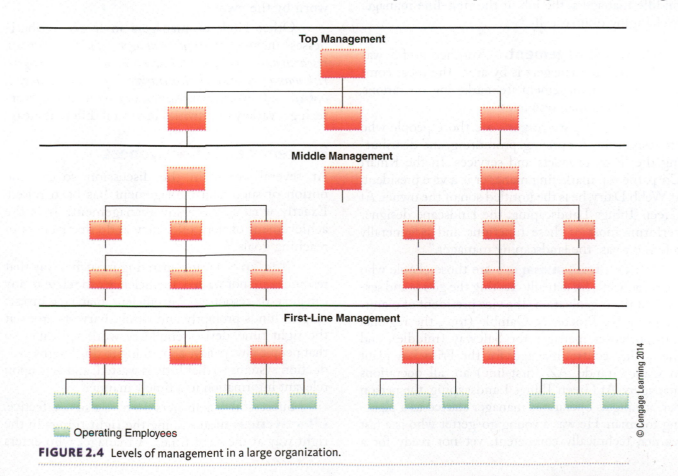

Top Management

Middle Management

First-Line Management

Operating Employees

**FIGURE 2.4**  Levels of management in a large organization.

© Cengage Learning 2014

and operations manager regardless of whether they actually hold such a title in their small agribusiness. At Summer Farms, the founder's daughter and her husband oversee the vegetable farms, and two sons manage the soybean, corn, hay, and timber production. All members act as middle managers. The middle manager's job is changing and some experts warn that the future of such positions is questionable, but these are the managers who generally implement the strategies and policies set by top managers and coordinate the work of lower-level managers.[8]

**First-line managers** are those people who supervise operating employees. In a very small agribusiness, this manager may be the owner of the company. A medium-size business like Green Things Landscaping may give its first-line managers titles such as foreman or team leader, or refer to this person as a "right-hand man." In a larger organization first-line managers are called supervisors, department managers, office managers, and foremen. A first-line manager is usually that individual's first management position. In contrast to middle and top managers, first-line managers spend much of their time directly overseeing the work of operating employees. As is the case with middle managers, the job of the first-line manager is changing dramatically.[9]

## Areas of Management.

Another useful way to differentiate managers is by area. The most common areas of management are marketing, operations, finance, and human resources.

**Marketing managers** are those people who are responsible for pricing, promoting, and distributing the firm's products and services. In the Kmart Corp. the top marketing manager is a vice president; at Webb Dairy, he is the untitled son of the owner. At Green Things Landscaping, the landscape designer performs most of these functions and is generally referred to as "the landscaping manager."

**Operations managers** are those people who are responsible for actually creating the goods and services of the organization. The vice president of manufacturing for Procter & Gamble (top), the regional transportation manager for Safeway (middle), and the quality-control manager for the Frito-Lay plant in Casa Grande, AZ (first-line) are all operations managers. At Green Things Landscaping, the person performing as operations manager started as a planting foreman. He was a young go-getter who is a fast learner, technically competent, yet not ready for a

managerial slot in a larger corporation. At Summer Farms, the eldest son of the founders, Josh, performed this role before replacing his father as CEO, but is usually referred to as Chief Trouble Shooter.

**Finance managers** are responsible for managing the financial assets of the organization. They oversee the firm's accounting systems, manage investments, control disbursements, and are responsible for maintaining and providing relevant information to the CEO about the firm's financial health. In a family business, this slot is often filled by the founder's wife or daughter, or someone else who has earned the utmost trust of the owner. Otherwise, an outside bookkeeper may be hired on a part-time or as-needed basis.

Another important area of management is human resources management (HRM). **Human resource (HR) managers** are responsible for determining future human resource needs, recruiting and hiring the right kind of people to fill those needs, designing effective compensation and performance appraisal systems, and ensuring that the firm follows various legal guidelines and regulations. In small businesses, these are mostly unofficial roles that represent just another hat worn by the owner.

Other kinds of managers in larger agribusinesses include *strategic managers*, *public relations*, *research and development*, *international*, *sales managers*, *risk managers*, and *administrative* or *general managers*. **Administrative managers** are generalists, overseeing a variety of activities in several different areas.

## Efficiency and Effectiveness

At several points in the discussion so far, the notion of successful management has been raised. Exactly what is successful management? It is the achievement of both efficiency and effectiveness in reaching goals.[10]

**Efficiency** means operating in such a way that resources are not wasted. Inefficiency can relate to any category of resources. An efficient manager invests surplus funds promptly and wisely, harvests crops at the right time, devises employee work schedules so that people always have something to do, designs production systems so that little is wasted, and acts upon relevant information in a timely manner.

Successful managers are also effective. **Effectiveness** means doing the right things in the right way at the right times. A manager who enters a new market just before it starts expanding, gets out of a market before it collapses, and maintains an appropriate competitive posture is more effective than one who enters a market as it starts to decline, gets out of a market just before it starts growing, and does not maintain an appropriate competitive posture. For example:

- Green Things Landscaping was a small, one-man retail nursery that expanded into a large landscaping business after accurately predicting a building boom in its region.
- Summer Farms is still uncertain about how much of its scarce resources to commit to organic farming.
- With no shortage of capital to test new inventory, Walmart managers started carrying under-the-cabinet kitchen appliances well before most other merchants, dropped home computers just before the market slumped, and always stayed in touch with competitors' pricing policies.[11]

Effectiveness combined with efficiency is the hallmark of successful management. It is important to recognize, however, that what might appear to be effective in the short term could prove to be ineffective in the long term.

## FOOD FOR THOUGHT 2.2

Using what he had learned as an Army cook serving 20,000 soldiers a day in the Korean War, Dave Thomas founded Wendy's, the first chain to serve fast foods that were not pre-cooked and to sell through a drive-thru window.

## Managerial Functions

Management functions are the sets of activities inherent in most managerial jobs. Many of these activities can be grouped into one of four general functions: planning and decision making, organizing, leading, and controlling. The relationships among these functions are shown in Figure 2.5. All managers enact each of these, although the exact mixture varies over time and is based on the manager's position in the organization.

**Planning and Decision Making.** As noted in Figure 2.5, **planning and decision making** involve determining the organization's goals and deciding how best to achieve them. Just as a carpenter looks at a blueprint to determine which rooms to put in a new house and how they must be configured, a manager looks at plans to determine what course has been charted for the organization.

For Ralph Webb, this may mean determining how much money he can and should borrow to replace milking equipment or to buy additional cows and a new truck, then determining which brands or models to purchase. Top managers at Monsanto might decide to increase its market share of Roundup® Agricultural Herbicide by 10 percent by the year 2020 and to develop ten new herbicides within that same time frame. These targets serve as the blueprints for other managers to follow. For example, marketing managers must launch new advertising campaigns, production managers must figure out how to cut costs (so that prices may be reduced), and finance managers must determine the effects of increased promotional expenditures and product revenues on cash flows. Similarly, market research managers must begin to determine what

| **Plan and Make Decisions:** Determine organization's goals and how best to achieve them | **Organize:** Group activities and resources to facilitate goal attainment | **Lead:** Guide and direct employees toward goal attainment | **Control:** Monitor and adjust organizational activities toward goal attainment |

**FIGURE 2.5** Basic management functions.

new products will be successful and R&D managers must start searching for ways to create them.

In general, the planning process consists of three steps. First, goals and objectives are established usually by top management. Then, top managers also develop strategic plans that serve as broad, general guidelines that chart the organization's future. Finally, tactical plans are developed, often by middle managers. Decision making pervades each of these activities.[12] It is important that plans are consistent and integrated across both organizational levels and management areas. It is also important for managers to juggle and improve on decision making and planning, but not to exclude other primary responsibilities.

**Organizing.**     The second basic managerial function is **organizing**—the process of grouping activities and resources in a logical and appropriate fashion. In its broadest sense, organizing is creating the organizational shape of a firm. The determination of how much power to give each division head, the number of subordinates under each division head, and the kinds of committees are necessary to the organizing process.[13] In like manner, when a supervisor groups the activities to be performed by team members, she is organizing.

Green Things went through the organizing procedure when it expanded from a one-man nursery to a landscaping business. Ralph Webb has relatively little organizing to do until he starts planning for his son to take over the major part of the dairy business and possibly expand it. Summer Farms needs to reorganize as it continues to expand, especially going into new ventures such as organic farming. Quaker Oats has a domestic foods division, an international foods division, and a division for Fisher-Price toys. Decisions that brought about the creation of this specific set of divisions, as opposed to a different set, were a part of Quaker's organizing process.

**Leading.**     The third basic function inherent in the manager's job is **leading**—the set of processes associated with guiding and directing employees toward goal attainment. Key parts of leading are motivating employees, managing group processes, and dealing with conflict and change. Note that each of these activities relates to behavioral concepts and processes.[14] In most family businesses, leading becomes

extremely important and challenging as the businesses are passed down to younger family members whose views for managing and growing the business may clash with the founder's. When Michael Eisner became president of Disney, he sold the real estate unit to clearly show his intended direction for the company, he convinced old and new employees to work together rather than to battle one another, he improved communication and motivation, and he linked rewards to performance. Each of these activities was part of the leading function. Eisner served as an effective role model and as a prime motivator for people at Disney.

**Controlling.**     The final basic managerial function is controlling. **Controlling** is the process of monitoring and adjusting organizational activities toward goal attainment. Consider Monsanto's attempt to increase its market share by 10 percent by the year 2020, assuming that this goal was set in 1990, and that managers at Monsanto believed that the market share growth would increase about 1 percent each year. In 1991 an increase of 1.2 percent would indicate that things are on target. No increase in 1992, however, would signify a problem and increased advertising may be required. An increase of 3 percent during 1996 might suggest that advertising can be reduced somewhat.

This method of monitoring progress and making appropriate adjustments is controlling.[15] When a supervisor monitors the performance of group members and assists them in achieving their objectives, that supervisor is also engaged in controlling. Monitoring may not always involve personnel, however. The manager at Webb's Dairy would be monitoring the production of grass and hay to feed the cows, the volume of milk from his cows to determine when to sell and buy cows, and the demand for that milk. Summer Farms should monitor planting and harvesting conditions including both immediate and long-term weather forecasts, increases or decreases in area population and grocery stores, and more.

The sets of activities inherent in most managerial jobs, then, can be grouped into the four general functions of planning and decision making, organizing, leading, and controlling. All managers enact each of these, although the exact mixture will vary over time and with the manager's position in the organization.

## A FOCUS ON AGRIBUSINESS
## Claytie

Manager, entrepreneur, farmer, rancher, failed gubernatorial candidate, and part-time educator—that's Claytie. Clayton Wheat "Claytie" Williams, Jr. was born and raised in west Texas. He graduated from Texas A&M University with a degree in animal husbandry and then served in the Army. While looking for a better job, he waited tables until an insurance executive told him that he would make a successful insurance salesman. So Claytie sold life insurance until he heard about a farm for sale. He talked the farm owner into forming an oil and gas partnership and then, as they say, "the rest is history."

Working as a lease broker, Claytie developed a series of oil and gas companies. As of 2010, he was Chairman of the Board, President and Chief Executive Officer of Clayton Williams Energy, Inc. He has stayed in that business despite (or perhaps because of) its boom and bust nature. He says that the excitement and achievement are the allure. When asked about the most difficult business decision he has ever made, he responded, "Laying-off people. Nothing else is even close. When you have to lay off a man or a woman when it is not their fault, then you've hurt people. And that's the hardest thing I have ever had to do."

But Claytie is not just an oil man; his operations have been highly diversified. He is or has been involved in farming, ranching, real estate, and banking as well as cow-calf operations, a safari company, and for several years he was co-teaching an award-winning course on entrepreneurship at Texas A&M with Dr. Ella Van Fleet, a co-author of this book.

Claytie's entrepreneurial spirit shows up in his ranching and farming. He introduced a strain of South African Klein Grass developed at Texas A&M as well as more heat-resistant Brangus cattle. His farm grows alfalfa, oats, and the special strain of grass. He was involved with grass conservation and invented a more efficient hay baler. He staggers dams across washes to deflect water to grass instead of having it erode the hillside. He cross-fences and rotates his grassland to keep it and his cattle healthy. As a result, his agribusiness ventures remain profitable.

While both entrepreneurs and managers plan, organize, and operate organizations, entrepreneurs, such as Claytie, are keenly aware from the beginning of the personal risks they assume and therefore find innovative ways of reducing their risks. Some entrepreneurs never become managers and some managers were never entrepreneurs, but Claytie is both!

With an estimated net worth of $100 million, Claytie's name was added to the Forbes 400 list of the richest people in America. He was ranked No. 62 among the top 100 landowners in America in 2009 with 146,655 acres, primarily in Texas and Wyoming.[16]

## The Evolution of Management

To understand how any organization arrived at its current approach to management, it is useful to know the organization's origins. For instance, a Polaroid plant in upstate New York began to suffer from problems of low morale and declining productivity a few years ago. Company officials tried to solve those problems but were unsuccessful because they could not account for the causes. Then the corporate historian went to work and soon explained what was going on: Over a 15-year period, management had gradually increased its control over the work, which in turn resulted in negative reactions from workers. No single manager had seen the problem because each had taken a narrow view of the situation.[17]

What can we learn from this example? Among other things, we can learn to appreciate better the value of history. That is the topic to which we next turn.

## The Importance of History and Theory

An understanding of history serves two primary purposes. First, it helps managers understand current developments better and why "we do it this way." (This is what happened at Polaroid.) History also helps managers avoid repeating mistakes. If a particular course of action did not work well several years ago, it may not be any more successful today. Of course, if circumstances have changed, it may work well now.

In similar fashion, it is helpful to understand management theory and to have an appreciation of the value of theory in general. Theory helps the manager organize information. Systems theory, for example, allows the manager to categorize a large, complex network of variables into a single framework. This framework in turn helps the manager develop a better understanding of how the variables are related to one another (systems theory is covered later in this chapter). Theory also helps the manager approach problems in a systematic fashion. Using a framework like the one just mentioned, the manager is able to classify variables as causes or as effects. The manager may then be able to predict certain effects based on certain causes (e.g., how the unemployment rate affects the demand for all kinds of products). For example, if advertising is a cause and sales increases are effects, the manager can develop a theory (or model) of how various advertising increases or decreases will affect sales.

Knowing both history and theory helps the manager understand and appreciate current conditions and developments. It also facilitates the ability to predict various future conditions. For example, a plant manager at Welch Foods who remembers the bad decisions made by the company in the past and who understands why those decisions were made can avoid repeating them today. This is true also of Joshua Summer's long-ago decision to cut costs by reducing irrigation in his fields during a lengthy period of drought. Similarly, managers who understand the interrelationships among critical conditions affecting the business (its culture) can better predict the future effects of changes in those conditions (organizational culture is discussed in Chapter 13). Thus, both history and theory are valuable parts of the manager's tool kit.[18]

### FOOD FOR THOUGHT 2.3

A number of agricultural crops are 90%–100% dependent on honeybee pollination, including almonds, apples, avocados, blueberries, cranberries, cherries, kiwifruit, macadamia nuts, asparagus, broccoli, carrots, cauliflower, celery, cucumbers, onions, legume seeds, pumpkins, and squash.

## Origins of Management Theory

As a scientific discipline, management is only a few decades old. Examples and illustrations of management in use, however, go back thousands of years. In this section we first consider several of these ancient examples and then trace the conditions that led to the emergence of contemporary management.[19]

**Ancient Management.** If we look at the accomplishments of many ancient civilizations, it is clear that they must have used management concepts and techniques. One example includes the complexities inherent in building the Egyptian pyramids or managing the vast Roman Empire. It is doubtful that such things could have been accomplished without using effective management. One of the earliest recorded uses of management is the Egyptians' construction of the pyramids; but the Babylonians, the Greeks, the Chinese, the Romans, and the Venetians also practiced management. Management concepts were also discussed by Socrates, Plato, and Alfarabi, who developed one of the earliest codes of government operations.[20]

In spite of this widespread practice of management, however, there was little interest in management as a scientific field of study until about a hundred years ago. There are several reasons for this. For one thing, there were few large businesses until the late nineteenth century. Governments and military organizations were not interested in increasing profits and therefore paid little attention to efficiency or effectiveness. For another thing, the first field of commerce to be studied was economics, and economists were initially more concerned with macroeconomic issues than they were with micromanagement concerns. This pattern began to change during the nineteenth century, however.

## Precursors of Modern Management.

During the Industrial Revolution the factory system began to emerge. Factories brought together for the first time large numbers of workers performing a wide variety of different jobs. Managers in charge of these factories had to cope with new kinds of problems that related to coordinating and supervising this kind of arrangement. They also had to contend with emerging societal concerns about child labor, working hours and conditions, and minimum wage levels.

One of the first people to confront these issues was British industrialist Robert Owen, who improved working conditions in his plants, set a higher minimum working age for children, provided meals for his employees, and shortened working hours. Charles Babbage, developer of a mechanical digital calculator—forerunner of modern computers—recognized the importance of efficiency and the human element in the workplace. In America, Daniel McCallum saw the need for systematic management in the railroad industry and implemented many innovative practices.[21]

Thus, throughout the nineteenth century there was a growing awareness of the need for more systematic approaches to management. As a result of these early efforts, three schools of management thought have been developed since the changes brought about by the Industrial Revolution. These three theories are known as the classical school, the behavioral school, and the quantitative school.

## The Classical School

The **classical school** of management is actually composed of two distinct subareas: scientific management and administrative management. Historically, scientific management focused on the work of individuals whereas administrative management was concerned with how organizations should be put together.

## Scientific Management.

The goal of **scientific management** in the early days was to determine how jobs should be designed to maximize the output of an employee. The pioneers of scientific management were Frederick W. Taylor and Frank and Lillian Gilbreth.

Taylor was an industrial engineer interested in labor efficiency. At his first job, for Midvale Steel Company in Philadelphia, Taylor observed a phenomenon he subsequently labeled "soldiering": laborers working at a reduced pace. Other managers were unaware of its existence because they actually knew very little about the jobs their employees were performing. Taylor studied each job and determined the most efficient way to perform it. He then installed a piece-rate pay system in which a worker is paid according to what he or she has actually produced at the end of the day. Encouraged by his results at Midvale, Taylor left the company and became an independent consultant. At Simonds Rolling Machine Company, he studied and redesigned jobs, introduced rest breaks, and adapted the piece-rate pay system. The results were improved output and morale. At Bethlehem Steel Corporation he applied his ideas to the tasks of loading and unloading railcars, with equally impressive results.

Over the years Taylor gradually solidified his thinking about work and developed what came to be called scientific management.[22] The practice of scientific management rests on the four steps summarized in Figure 2.6. First, the manager should develop a science for each element of the job; that is, study the job and determine how it should be done. This replaced the "rule-of-thumb" methods that managers had been using. Second, the manager should scientifically select and then train, teach, and develop workers. This replaced the previous system in which each worker trained himself or herself. Third, the manager should ensure that workers are using the

**Scientific Management**

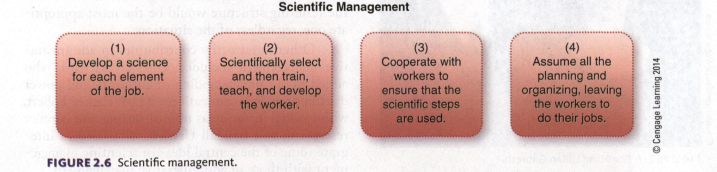

| (1) | (2) | (3) | (4) |
|---|---|---|---|
| Develop a science for each element of the job. | Scientifically select and then train, teach, and develop the worker. | Cooperate with workers to ensure that the scientific steps are used. | Assume all the planning and organizing, leaving the workers to do their jobs. |

© Cengage Learning 2014

**FIGURE 2.6** Scientific management.

scientific steps already developed, including monitoring their work to ensure they adhere to the one best way. Fourth, the manager assumes all planning and organizing responsibilities while the workers perform their tasks. Taylor's ideas and methods have had a profound influence on contemporary business in areas ranging from assembly-line technology to compensation systems.[23]

Frank and Lillian Gilbreth were also notable pioneers in the scientific management movement (Figure 2.7). Their work was popularized first in a book and later in a movie entitled *Cheaper by the Dozen*, a reference to their application of scientific management practices to their family of 12 children.

Frank's work contributed to the craft of bricklaying and to medicine. He observed that even though bricklaying was one of the oldest construction technologies known, there were no generally accepted work guidelines on how to lay bricks efficiently. He applied the principles of scientific management by first studying and then standardizing the steps involved. His methods reduced the total number of steps undertaken by the bricklayer from 18 to 5, and more than doubled output. He also made major contributions to the medical field by helping the handicapped and by streamlining operating room procedures, thus greatly reducing the time the average patient spent on the operating table.[24]

Lillian also made a variety of important contributions. She was primarily interested in ensuring that the welfare of the worker was not forgotten. She assisted Frank in the areas of time and motion studies and industrial efficiency and was an early contributor to personnel management.

Henry Gantt, an associate of Taylor's, developed the Gantt Chart (Chapter 10), a device for scheduling work over a span of time. It is still used today. He also worked in the area of pay systems.[25]

**Administrative Management.**  The second important subarea of classical management theory is called **administrative management**. Whereas scientific management focused on the work of individual employees, administrative management was concerned with how organizations should be structured. The primary contributors to this area include Henri Fayol and Max Weber.

Henri Fayol was the Taylor of administrative management; that is, he was perhaps its greatest contributor and most visible proponent. Drawing on more than 50 years of industrial experience, Fayol developed fourteen general guidelines or principles of management, including "authority" (managers have the right to give orders so that they can get things done), "remuneration" (compensation for work should be fair to both the employee and the employer), and "esprit de corps" (promoting team spirit will give the organization a sense of unity).[26] Fayol believed that these principles were universally valid and if applied and followed, they would always enhance managerial effectiveness.

Max Weber, a German sociologist considered to be a founder of the field of sociology, was the first to describe the concept of **bureaucracy**.[27] The bureaucratic form of organization is based on a comprehensive set of rational rules and guidelines with an emphasis on technical competence as the basis for determining who would get what jobs. Weber's guidelines were similar in concept to Fayol's 14 principles and were designed for managers to use in structuring their organizations. Weber assumed that the resulting structure would be the most appropriate one, regardless of the situation.

Other noteworthy contributors to administrative management included Chester Barnard, who added to our understanding of authority and power distributions in organizations; Mary Parker Follett, who worked in the areas of goal setting and conflict resolution; and Lyndall Urwick, who tried to integrate some of the central ideas of scientific management with those of administrative management.

© mustafamemisoglu.blogspot.com

**FIGURE 2.7** Frank and Lillian Gilbreth.

### Assessment of the Classical School.

The classical school of management has a number of strengths and weaknesses. On the plus side, managers today are still using many of the insights and developments of these pioneers who also helped bring the study of management to the forefront as a valid scientific concern. On the negative side, many of their ideas now seem quite simplistic and relevant only in isolated settings. For example, many people are motivated by a variety of factors beyond economic incentive. Moreover, the classical school tended to underestimate the role of the individual. This flaw was primarily responsible for the growth of the second school of thought: the behavioral school.[28]

## The Behavioral School

Although many early theorists ignored, or at least neglected, the human element in the workplace, there were a few scattered voices in the wilderness. Mary Parker Follett, for example, recognized the potential importance of the individual. So did Hugo Munsterberg, a German psychologist who published a pioneering book in 1913 that subsequently became the cornerstone of industrial psychology.[29] The real catalyst for the emergence of the **behavioral school**, however, was a series of research studies conducted at the Hawthorne plant of Western Electric between 1927 and 1932.[30] This research has come to be known as the **Hawthorne studies**.[31]

The Hawthorne studies actually consisted of several experiments that had little impact on performance but nevertheless are particularly noteworthy. In the lighting study, researchers manipulated the lighting for a group of workers and compared their performance with that of a group whose lighting had not been changed. Quite surprisingly, performance changed in both groups. In the second experiment, the researchers established a piecework pay system for a group of nine men to test the assumption that people are motivated solely by money and therefore will produce as much as possible to get as much pay as possible. Instead, the group established a standard level of acceptable output for its members. Workers who fell below this standard were called chiselers and were pressured to do more, and those above the standards were labeled rate-busters and were pressured to bring their output into line with that of the rest of the group. In the third study, a group of female workers was paid to participate in a series of experiments while assembling relays (See Figure 2.8). The group's output increased gradually over time apparently due to knowledge of results

and their paid participation, although some influence of social factors also may have been involved.[32]

Based on these studies, researchers concluded that a variety of social factors previously unknown to managers were important. For example, the researchers attributed the results in the lighting study to the special attention the workers were receiving for the first time. From the piecework experiment they concluded that social pressure was a powerful force to be reckoned with. The relay experiments were widely misinterpreted for years but clearly showed not only the impact of social forces but also, and more importantly, the effects of feedback and rewards.[33]

### Human Relations.

The Hawthorne studies gave birth to a new way of thinking about workers by focusing on the importance of the individual in the workplace. Whereas previous views ignored the role of the individual, the **human relations** model recognized that people bring their own unique needs and motives into the workplace. While at work, an individual is exposed to the task, a supervisor, and so forth; but she or he also experiences a social context. This context includes the possible satisfaction of social needs, membership in the work group, and the possible satisfaction of special needs, such as the need to be with others and to be liked and accepted by them. These factors then combine to influence such responses as satisfaction and performance.

Two writers are particularly identified with the human relations movement. Abraham Maslow proposed an approach to the understanding of human needs.[34] The other primary proponent of human

**FIGURE 2.8** Women in the Relay Assembly Test Room.
Source: Women in the Relay Assembly Test Room, ca. 1930. Western Electric Hawthorne Studies Collection. Baker Library Historical Collections, Harvard

**TABLE 2.1** Theory X and Theory Y Assumptions About Workers

| THEORY X ASSUMPTIONS | THEORY Y ASSUMPTIONS |
|---|---|
| 1. People do not like work and try to avoid it.<br>2. People do not like work, so managers have to control, direct, coerce, and threaten employees to get them to work toward organizational goals.<br>3. People prefer to be directed, to avoid responsibility, to want security; they have little ambition. | 1. People do not naturally dislike work; work is a natural part of their lives.<br>2. People are internally motivated to reach objectives to which they are committed.<br>3. People are committed to goals to the degree that they perceive personal rewards when they reach their objectives.<br>4. People will both seek and accept responsibility under favorable conditions.<br>5. People have the capacity to be innovative in solving organizational problems.<br>6. People are bright, but under most organizational conditions their potentials are underutilized. |

Source: D. McGregor and W. Bennis, The Human Side of Enterprise: 25th Anniversary Printing, 1960, Copyright © 1960 The McGraw-Hill Companies, Inc. Reprinted with permission.

relations is Douglas McGregor, who described two quite different opinions of workers that managers might hold. These opinions, called **Theory X** and **Theory Y**, are summarized in Table 2.1. According to McGregor, Theory X typifies pessimistic managerial thinking. The more optimistic Theory Y was the view he felt was more appropriate.[35]

**Contemporary Behavioral Science.** Although the views espoused by early human relations theorists had some validity, they were also somewhat naive and simplistic. For example, these theorists believed that if managers made workers happier, the employees would work harder. Contemporary behavioral science, known as the **human resources** approach, takes a more complex view. It acknowledges and attempts to explain a variety of individual and social processes as both determinants and consequences of human behavior. For example, performance is caused by many things, including motivation and ability. As a consequence of performing at a high level, people may achieve a variety of rewards. These rewards may consequently affect future motivation that in turn influences performance again. Thus, instead of presenting simple, universal principles, contemporary behavioral science presents complex, contingency views, which are discussed in later chapters. Today, an understanding of human behavior is seen as an important tool that managers can use to do a better job of drawing on the important human resources that all organizations have.

**Assessment of the Behavioral School.** The behavioral school yielded some significant but simplistic insights into the role of the individual in the workplace. Behavioral scientists today continue to work toward a better understanding of human behavior in organizational settings. Like the classical school, the behavioral view is an important but incomplete theory of management. Another important piece of the puzzle is the quantitative school of management thought.[36]

## The Quantitative School

The third school of management thought is the **quantitative school**. As the term implies, this approach focuses on quantitative or measurement techniques and concepts of interest to managers. It has its roots in World War II, when the military sought new and better ways to deal with troop movement, arms production, and similar problems. There are three branches of the quantitative school: management science, operations management, and management information systems.

**Management Science.** **Management science** is concerned with the development of sophisticated mathematical and statistical tools and techniques that the manager can use, primarily to enhance efficiency.[37] For example, a manager might use a management science model to help decide how large a new plant should be. The model would contain a number of equations related to such things as projected production volume at the plant, construction cost per square foot, utility costs for plants of different dimensions, and so forth; and the solutions would give the manager useful guidelines as to how big to make the plant. Large agribusinesses, such as Frito-Lay, use management science models to plan schedules, set prices, schedule maintenance, perform other activities. Quantitative management tools are also helping in the war on terrorism.[38]

Advancements in the management science area have been greatly helped by breakthroughs in computers and other forms of electronic information processing. Such innovations as the personal computer enhance a manager's access to the tools and techniques of management science. Managers of small and medium-size agribusinesses today have access to many free or low-cost computer programs to assist them.

### Operations Management.
**Operations management** is somewhat like management science but focuses more on application.[39] It concerns the various processes and systems that an organization uses to transform resources into finished goods and services. Decisions about where a plant should be located, how it should be arranged, the amount of inventory it should carry, and how its finished goods should be distributed are all elements of operations management. This approach is used by manufacturers of parts for irrigation equipment and large agribusinesses like Heinz, Hershey, and Hormel in managing their assembly lines. Since many aspects of operations management are related to control, all of Chapter 23 in this book is devoted to it.

### Management Information Systems.
**Management information systems (MISs)** make up the third branch of the quantitative school. An MIS is a system created specifically to store and provide information to managers. For instance, an MIS for a large processor, such as Tyson Fresh Meats, may contain information about everything from the finished goods inventory of a plant in Pasco, Washington, to the number of operating employees at its headquarters

in Dakota Dunes, South Dakota. The data are kept as current as possible, making it possible for a marketing manager to tap into the system and check the inventory levels in Pasco while a human resource manager in World Headquarters at Springdale, AR, is verifying the number of employees in Dallas. Of course, most systems of this type make extensive use of computer technology. We will learn more about MIS in Chapter 24, "Information Systems."

### Assessment of the Quantitative School.
The primary value of the quantitative school lies in the portfolio of tools it provides for management. These tools can greatly enhance a manager's decision making, planning, and control. At the same time, we should remember that tools cannot replace human intuition and insight. A manager should choose the right tools for the job, apply them properly, and then understand the results.

## Contemporary Management Theory

In recent years several new perspectives on management have emerged. These have not yet attained the stature of schools of thought, but they still provide useful techniques and approaches that managers should understand. In this section we explore systems theory, the contingency approach, and total quality management (TQM) and high-involvement management.

### Systems Theory

**Systems theory**, illustrated in Figure 2.9, is an approach to understanding how organizations function and operate.[40] A system is an interrelated set of elements that function as a whole. It has four basic

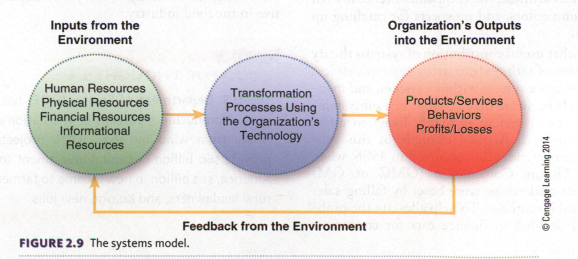

**Inputs from the Environment**

Human Resources
Physical Resources
Financial Resources
Informational Resources

Transformation Processes Using the Organization's Technology

Products/Services
Behaviors
Profits/Losses

**Organization's Outputs into the Environment**

**Feedback from the Environment**

© Cengage Learning 2014

**FIGURE 2.9** The systems model.

parts. First, the system receives from the environment the four kinds of inputs or resources that were included in the definition of management. For an organization like Hormel, human resources include managers and field personnel, physical resources include processing plants and refrigeration systems, financial resources are derived from product sales and stockholder investment, and informational resources include federal, state, and local food processing controls and demand forecasts.

Second, these various resources are transformed into outputs through a variety of processes, which represent the organization's technology. The third part of the model, outputs, includes products or services, behaviors, and profits or losses. For Hormel, products are grocery products and refrigerated foods, behaviors are the effect of its employees on the environment, and profits or losses are reflected in the level of funds put back into or taken from the environment. Finally, feedback from the environment provides the system with additional information about how well its actions are being accepted.

Interaction with the environment is one of the major contributions that systems theory makes to the manager. This notion comes from the concept of open systems. An open system is one that actively interacts with its environment, whereas a closed system does not. Because all organizations are open systems, managers should always monitor and be sensitive to their environments. Organizations with managers who forget this are almost invariably left behind. For example, for years American automobile companies assumed they could just go on producing cars as usual and not worry about what car manufacturers were doing in other countries. As a result of this attitude, the companies eventually fell behind competitors, and prospects for catching up were dim.

Another useful contribution of systems theory is the notion of **subsystem interdependencies**. A subsystem is a system within a system, and managers should be aware that a change in a subsystems within a parent system is highly likely to affect other subsystems. A good example of this subsystem interdependencies occurred in 1986, when General Motors Corporation (GMC or GM) automobile dealerships were beset by falling sales and bloated inventories. To help alleviate the problem, GM decided to finance cars for consumers at a previously unheard of 2.9 percent interest rate. However, GM's financing division, General Motors Acceptance Corp., or GMAC, could not handle this level of financing. Such interest rates would have hurt its profit margins and consequently lowered its bond ratings. GM itself had to underwrite the low interest rates, even though this would ordinarily have been the responsibility of GMAC. Thus, the needs of one subsystem, GM's dealerships, necessitated a certain course of action, but the environmental circumstances surrounding another subsystem, GMAC, kept it from responding appropriately.[41]

Three other concepts of systems theory are also useful. **Synergy** suggests that two people or units can achieve more working together than working individually.[42] When a retail chain like Kmart purchases Walden Book Co., it is partially because managers recognize that the two businesses complement one another. When ConAgra's frozen food works with a small wine producer to add its wine (1 percent) to its gourmet frozen meals, ConAgra enhances its product and the small producer has an outlet for its production. **Entropy** is what happens when organizations take a closed-system perspective; they falter and die. Quaker Oats and Snapple made this mistake. The key is to stay in tune with the environment and work hard at keeping the organization stimulated and vital. Hershey's chocolate, because of its unique trust arrangement, is severely constrained in its ability to merge with and acquire other companies; therefore, it must be constantly aware of the global chocolate market for both threats and opportunities. Finally, **equifinality** is the idea that two or more paths may lead to the same place. Del Monte and Dole can pursue different strategies, for example, and yet be equally effective in the food industry.

## FOOD FOR THOUGHT 2.4

The U.S. Department of Energy (DOE) has set a goal of producing 5 percent of the nation's electricity from wind by 2020. DOE projects will provide $60 billion in capital investment to rural America, $1.2 billion in new income to farmers and rural landowners, and 80,000 new jobs.

## Contingency Approach

A second important contemporary perspective on management is the **contingency approach**, which argues that appropriate managerial actions in a situation depend or are contingent on certain major elements of that situation.[43] Early approaches to management problems sought universal answers. As illustrated in the top portion of Figure 2.10, the premise is that a given problem or situation can be solved or acted upon in "one best way," which is rarely true. In the middle portion is the extreme contrast, the case approach, which argues that each situation is completely unique so that common solutions do not exist, which is also rarely true. A synthesis of these two extremes, the contingency approach, dominates management theory today.

The contingency approach recognizes that there are few, if any, "one best ways" in management because the complexities of human behavior and social systems like organizations make situations different. Yet not every situation is unique, either. As shown in the bottom of Figure 2.10, the contingency approach acknowledges these differences and suggests that, when a manager is confronted with

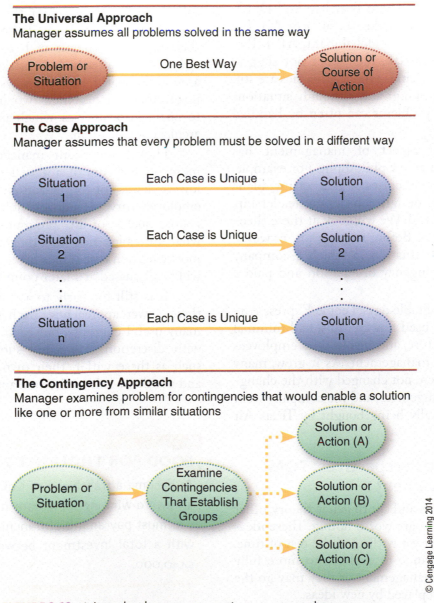

**FIGURE 2.10** Universal and case versus contingency approaches.

a problem or situation, he must examine important contingencies to determine which of several potential solutions or actions may be appropriate. What will work best in one situation may not work best in another, but groups of similar situations can be identified by contingencies such that within those groups similar solutions hold while across groups they do not. Those contingencies may be organizational size, technology, the competitive environment, the experience and knowledge of workers, or other factors depending upon the particular situation. Anheuser-Busch (AB) was a successful national and global beer producer; however, a changing economic environment favoring consolidation and global exposure favored InBev, a company whose products AB distributed. InBev (a Brazilian and Dutch conglomerate) tendered a takeover offer of AB. The result was the creation of a global giant, AB InBev. The contingency approach, then, does not suggest that every single problem is unique. Theories are still useful guides to thinking; and although situations may vary, different problems can often be handled quite similarly.

Almost every aspect of management has embraced the contingency philosophy. For example, there are contingency theories of goal setting, planning, organizational design, job design, leadership, motivation, and control. We will look at these theories in later chapters. Before moving on, however, let's examine a real situation in which a company did not take a contingency orientation and paid a high price.

In its early, profitable days, People Express Airlines was small and used few controls, had minimal organizational structure, and involved its employees in many aspects of management. As it grew, many of these practices were not changed with the changing situational contingencies, and the company went into decline, eventually being bought by Texas Air in 1986.[44]

## Other Emerging Perspectives: TQM and High Involvement

In addition to systems and contingency theory, other emerging perspectives are worth noting. These ideas are new and have not yet withstood the test of time. Thus, they may continue to evolve into more fully framed theories of management or they may go the way of fads and be replaced by new ideas.

**TQM** is used here to refer to the entire quality movement that has become ubiquitous in business organizations worldwide.[45] It began with the emergence of statistical quality control in the United States during the 1920s and 1930s. However, the real impetus was its adoption by Japan following World War II.[46] This topic has become so important in contemporary management that Chapter 22 is devoted exclusively to it.

**High-involvement management** includes not just all forms of participative management but also a fundamentally different approach to management than has been traditional.[47] The traditional, control-oriented approach to management—which is based on the assumption that hierarchical organizations represent the best way to assure performance—will not easily be displaced.[48] Nevertheless, many elements of high-involvement management are already being tried in organizations around the world. In many ways it seems that high-involvement management would particularly apply to small agribusinesses, which are typically nonhierarchical organizations.

High-involvement management relies much more on self-control and self-management at the lowest levels of organizations.[49] Quality circles, employee survey feedback, job enrichment, work teams, quality-of-life programs, gainsharing, and what are called new design plants are among the most common techniques being tried. Each of these will be discussed at a later point in this book.

It is still too early to assess the full impact of these newer approaches. Each offers the manager many useful concepts, but each should be followed with discretion. Managers should consider ideas such as these within their own situational context and use them only as they are appropriate.[50]

### FOOD FOR THOUGHT 2.5

To become a franchisee of Cold Stone Creamery, an award-winning ice cream maker and retailer, you must pay an initial franchise fee of $42,000, with a total investment between $294,000 and $440,000.

# CHAPTER SUMMARY

Management can be defined as a set of activities directed at the efficient and effective use of resources in the pursuit of one or more goals. Management is complex because of the large number of different activities managers must engage in and because they must change activities frequently. Management is also pervasive, occurring in all organizations.

Managers exist at all organizational levels. Although the actual number of levels varies among organizations, most companies have three general levels: top, middle, and first-line. Additionally, there are different kinds of managers, such as marketing managers, operations managers, finance managers, human resource managers, and administrative managers.

All managers perform four general functions, to varying degrees: planning and decision making, organizing, leading, and controlling. Successful management involves achieving both efficiency and effectiveness. Efficiency means operating in such a way that resources are not wasted. Effectiveness means doing the right things in the right way at the right times. Most managers are not able to achieve equal amounts of efficiency and effectiveness. Some organizations might require more of one than the other, but a combination is always the hallmark of success.

Knowledge of history will help you better understand current developments and avoid repeating mistakes. An understanding of theory helps you organize and further comprehend information that is available to you as a manager. It also enables you to be more systematic in your approach to problems so that you can reach better solutions.

The classical school of management emerged around the turn of this century and consists of two major subareas: scientific management and administrative management. Scientific management focused on the work of individuals and sought to determine how jobs could be designed to maximize employee output. Administrative management focused on how organizations should be structured for best performance.

The behavioral school actually began at about the same time as the classical school, but it did not have a significant impact on the field until later. This movement did much to advance the view that more humanistic approaches to the treatment of workers were needed. Contemporary behavioral science takes a more complex view of the worker, acknowledging and attempting to explain a variety of individual and social processes as both determinants and consequences of human behavior.

The quantitative school of management theory emerged during World War II, when the military sought new and better ways to deal with the managerial problems of modern warfare. Management science is concerned with the development of mathematical and statistical tools and techniques for managers. Operations management is similar but tends to focus on the application of techniques to operational aspects of management. MISs are created to store and provide information for managers.

The major components of contemporary management theory include systems theory, the contingency approach, and the emerging perspectives of TQM and high-involvement management. Among the useful contributions of systems theory are the study of an organization's interaction with its environment, subsystem interdependencies, and the concepts of synergy, entropy, and equifinality. The contingency approach suggests that there is no "one best way" to manage; however, it does not hold that every case is completely unique.

# CHAPTER ACTIVITIES

## REVIEW QUESTIONS

1. What is management? Identify different kinds of managers, both by levels and by areas.
2. What are efficiency and effectiveness?
3. What are the major functions of management? Briefly describe each.
4. Briefly explain why history and theory are important to the field of management.
5. What are the important contributions and weaknesses of the three schools of management?

## ANALYSIS QUESTIONS

1. Think of someone you know who is a manager. Describe that person's management position in terms of the type of organization in which he or she works, his or her level in the organization, and the area of management in which he or she practices.
2. Given the importance of history and theory, how can many managers operate successfully without a good understanding of either? Would they do better with such an understanding? Why?
3. Which school of management theory do you feel most closely matches your present ideas about how management should function? Why?
4. Why were management theories of the past relatively simple? Was it because the theorists did not know much, or because the managerial world that the theorists were describing was relatively simple? Defend your view.
5. Illustrate the concepts of synergy, entropy, and equifinality in a system other than a business one. (You may need to go to the library and find some articles on systems to answer this question fully.)

## FILL IN THE BLANKS

1. Executives between top management and first-line management, including plant managers, division managers, and operation managers, are called _____ managers.

2. Those who supervise operating employees, including supervisors, department managers, office managers, and foremen are called _____ _____ managers.

3. The subarea of the classical school of management that attempted to design jobs to increase individual output is known as _____ management.

4. The _____ school of management is focused on the potential importance of the individual in the workplace.

5. The _____ _____ model recognizes that people have their own unique needs and motives that they bring to the workplace.

6. The branch of the quantitative school of management that develops advanced mathematical and statistical tools and techniques for managers is called _____ _____.

7. _____ management focuses on the application of mathematical and statistical tools to managing an organization's processes and systems.

8. A system created specifically to store and provide information to managers is known as a/an _____ _____ _____.

9. An approach to understanding how the different elements of an organization function and operate is called _____ _____.

10. The _____ approach argues that appropriate managerial actions depend on certain major elements of situations.

## CHAPTER 2 CASE STUDY
### SAUDI ARABIA AND THE OUTSOURCING OF FARMLAND

Agribusinesses have always been secure in the knowledge that the agricultural sector is land-dependent. This dependency was a form of security. You cannot move a farm or ranch to another country. Already, two waves of globalization have involved the outsourcing of jobs and finances. Needless to say, this form of outsourcing raised serious political issues in many developed countries.

Now a new wave of outsourcing has been initiated, and it strikes at the heart of traditional agribusiness security. It involves outsourcing farmland. Countries that are net food importers and financially well-off have begun to approach developing countries that are land-rich but less well-off financially about the acquisition of farmland. Traditionally, the land-rich country would produce the crop and sell it to the food-importing country. However, the poor, land-rich country does not have the financial resources to exploit this market potential, so it decides to make a deal that calls for the wealthy but land-poor country to supply the finances and necessary materials. Thus, the poor but land-rich country supplies the land and the cheap labor to produce a food crop that is then taken wholly or in part to the well-off, financing country.

At first glance, this could be a "win-win" situation for both parties. It has not turned-out that way, though. The investing country views this arrangement as another form of direct foreign investment, i.e. providing jobs and finances to a country that needs them. However, often the population of the poor country perceives this arrangement as a "sell out" of resources or another form of colonialism. The poorer country has the perception that it has yielded something valuable (land) to another country and has lost some amount of food security. A major issue of all countries is food security—a country's ability to control resources to provide food for its citizenry. It is a concern of all countries, as only a very few countries are self-sufficient agriculturally. Self-sufficiency means that the country is able to meet food consumption demand from its own production without importation. Few countries are able to do this.

In 2009, King Abdullah of Saudi Arabia held a ceremony for a batch of rice that was grown for Saudi Arabia by Ethiopia under the arrangement described above. It was part of the King Abdullah initiative for Saudi agricultural investment abroad. This ceremony brought the issue to the attention of the global agribusiness and business community. It seems Saudi Arabia was not alone. Other investing countries were Kuwait, India, Djibouti, Egypt, Qatar, Bahrain, China, and South Korea. They have invested in Cambodia, Ethiopia, Malawi, Mozambique, Sudan, Turkey, Philippines, Indonesia, and Madagascar.

However, these arrangements were between sovereign entities, i.e. country to country—not between publicly held corporations. Although described as investments, these arrangements were in reality a way of insuring a supply of food for the investing country that it could not produce itself. In time, the country supplying the land and people may use the investments to bolster their ability to export this crop in a more traditional trade arrangement. All of these arrangements are for a stated period of time and are not perpetual.

Interestingly, not all of these arrangements have involved sovereign entities, that is, country-to-country arrangements. The United States and the United Kingdom were involved in this form of foreign direct investment through publicly held companies. Goldman Sachs (U.S. investment company/bank) is involved with pig and poultry ventures in China, and Sun Biofuels (U.K. biotechnology firm) has invested in crop production in Ethiopia and Mozambique.

Like many globalization initiatives the intent, form and perception of the investment is open to interpretation and viewed differently by all involved. Agribusiness ventures attract greater scrutiny because of the concern about food security.[51]

### ▶ Case Study Questions

1. When a poor but land-rich country is paid to grow food for a wealthy but land-poor country, how is this different from having a U.S. company grow a crop on U.S. soil and export it to another country?

2. Do you agree or disagree that a poor but land-rich country in this type of outsourcing arrangement has given up some of its food security? Has what the poor country given up been offset by what it has gained?

3. How may the concern for food security be affected by such an outsourcing arrangement?

## REFERENCES

1. "Swift Celebrates 150 Years of Meat Industry Excellence," Press Release, Swift & Company, June 24, 2005, at www.jbsswift.com (accessed July 9, 2010); "Made in Chicago: The Refrigerated Rail Car," PBS, n.d., at www.pbs.org (accessed June 30, 2010); "Annual Report," JBS Swift and Company, 2010; George Borgstrom, *Principles of Food Science, Vol. 2* (Westport, CT: Food and Nutrition Press, 1968), 348.

2. Henry Mintzberg, "The Manager's Job: Folklore and Fact," *Harvard Business Review* (July–August 1975): 49–61; see also Brian Dumaine, "What the Leaders of Tomorrow See," *Fortune*, July 3, 1989, 48–62.

3. Ibid. See also Peter Drucker, "What Makes an Effective Executive," *Harvard Business Review* (June 2004): 58–68; and Henry Mintzberg, *The Nature of Managerial Work* (New York: Harper & Row, 1973).

4. Don Hellriegel, John Slocum, and Richard Woodman, *Organizational Behavior, 6th ed.* (St. Paul, MN: West, 1992); see also Ford S. Worthy, "How CEOs Manage Their Time," *Fortune*, January 18, 1988, 88–97, and Thomas A. Stewart, "How to Manage in the New Era," *Fortune*, January 15, 1990, 58–72.

5. "Executive Pay," *Business Week*, April 15, 2002, 80–100; Thomas M Carroll, "Rolling Back Executive Pay," *Time*, March 1, 1993, 49–50; "Can We Put the Brakes on CEO Pay?" *Management Review* (May 1, 1992): 10–15; "Pay for Skills: Its Time Has Come," *Industry Week*, June 15, 1992, 22–31.

6. Page Smith, *The Rise of Industrial America* (New York: McGraw-Hill, 1984).

7. Carrie Gottlieb, "And You Thought You Had It Tough," *Fortune*, April 25, 1988, 83–84.

8. Rosabeth Moss Kanter, "The Middle Manager as Innovator," *Harvard Business Review* (July–August 2004): 150–161; "Caught in the Middle," *Business Week*, September 12, 1988, 80–88; see also Rosemary Stewart, "Middle Managers: Their Jobs and Behaviors," in *Handbook of Organizational Behavior*, ed. Jay W. Lorsch (Englewood Cliffs, NJ: Prentice-Hall, 1987), 385–391.

9. See Steven Kerr, Kenneth D. Hill, and Laurie Broedling, "The First-Line Supervisor: Phasing Out or Here to Stay?" *Academy of Management Review* (January 1986): 103–117.

10. Fred Luthans, "Successful vs. Effective Real Managers," *Academy of Management Executive* (May 1988): 127–132; see also "The Best Performers," *Business Week*, Spring 2006 Special Issue, 61–140; and Jim Collins, "The Ten Greatest CEO's of All Times," *Fortune*, July 21, 2003, 54–68.

11. "Two Wal-Mart Officials Vie for Top Post," *The Wall Street Journal*, July 23, 1986, 6; see also Sarah Smith, "America's Most Admired Corporations," *Fortune*, January 29, 1990, 58–63.

12. For a classic discussion of planning, see George Steiner, *Top Management Planning* (New York: Macmillan, 1969).

13. For a review of the organizing function, see Robert C. Ford, Barry R. Armandi, and Cherrill P. Heaton, *Organization Theory* (New York: Harper & Row, 1988).

14. For a recent review of the leadership function, see Gary Yukl, *Leadership in Organizations, 7th ed.* (Upper Saddle River, NJ: Prentice-Hall, 2010).

15. For a general overview of control, see William H. Newman, *Constructive Control* (Englewood Cliffs, NJ: Prentice-Hall, 1975).

16. Mike Shea, "Clayton Williams," *Texas Monthly*, October 2007 at www.texasmonthly.com (accessed July 7, 2010); Land Report Editors, "Land Report 100: No. 62 Clayton Williams Jr.," *The Land Report: The Magazine of the American Landowner*, January 21, 2010, at www.landreport.com (accessed July 7, 2010); "Clayton W. Williams," Forbes.com, n.d., at people.forbes.com (accessed July 7, 2010); Mike Cochran, *Claytie: The Roller-Coaster Life of a Texas Wildcatter* (College Station, TX: Texas A&M University Press, 2007); Carrie Moline Steenson, "The Next Billion Dollar Aggie," *Texas Business*, (August 1981): 4–7 and 21.

17. "Profiting from the Past," *Newsweek*, May 10, 1982, 73–74.

18. Alan L. Wilkins and Nigel J. Bristow, "For Successful Organization Culture, Honor Your Past," *Academy of Management Executive* (August 1987): 221–227; and Alan M. Kantrow, "Why History Matters to Managers," *Harvard Business Review* (January–February 1986): 81–88.

19. David D. Van Fleet, "Doing Management History: One Editor's Views," *Journal of Management History* (2008): 237–247.

20. For general discussions of ancient management practices, see Daniel A. Wren and Arthur G. Bedeian, *The Evolution of Management Thought, 6th ed.* (New York: Wiley, 2008), and Claude S. George, Jr., *The History of Management Thought* (Englewood Cliffs, NJ: Prentice-Hall, 1968).

21. See Wren and Bedeian, *The Evolution of Management Thought*, for details about these and other management pioneers.

22. Frederick W. Taylor, *Principles of Scientific Management* (New York: Harper and Brothers, 1911).

23. See Charles D. Wrege and Ann Marie Stotka, "Cooke Creates a Classic: The Story Behind F. W. Taylor's Principles of Scientific Management," *Academy of Management Review* (October 1978): 736–749.

24. J. Michael Gotcher, "Assisting the Handicapped: The Pioneering Efforts of Frank and Lillian Gilbreth," *Journal of Management* (March 1992): 5–13.

25. See Wren and Bedeian, *The Evolution of Management Thought*.

26. Henri Fayol, *Industrial and General Management*, trans. J. A. Conbrough (Geneva: International Management Institute, 1930).

27. Max Weber, *Theory of Social and Economic Organization*, trans. T. Parsons (New York: Free Press, 1947).

28. Stephen J. Carroll and Dennis J. Gillen, "Are the Classical Management Functions Useful in Describing Managerial Work?" *Academy of Management Review* (January 1987): 38–51.

29. Hugo Munsterberg, *Psychology and Industrial Efficiency* (Boston: Houghton Mifflin, 1913).

30. Paul R. Lawrence, "Historical Developments of Organizational Behavior," in *Handbook of Organizational Behavior*, ed. Jay W. Lorsch (Englewood Cliffs, NJ: Prentice-Hall, 1987), 1–9.

31. R. G. Greenwood, A. A. Bolton, and R. A. Greenwood, "Hawthorne a Half Century Later: Relay Assembly Participants Remember," *Journal of Management*, vol. 9, no. 2 (1983): 217–231. For details of the work, see also Elton Mayo, *The Human Problems of an Industrial Civilization* (New York: Macmillan, 1933); and Fritz Roethlishberger and William Dickson, *Management and the Worker* (Cambridge, MA: Harvard University Press, 1939). For a summary, see Wren and Bedeian, *The Evolution of Management Thought*.

32. Greenwood, Bolton, and Greenwood, "Hawthorne a Half Century Later: Relay Assembly Participants Remember," *Journal of Management*.

33. Ibid.

34. Abraham Maslow, "A Theory of Human Motivation," *Psychological Review* (July 1943): 370–396.

35. Douglas McGregor, *The Human Side of Enterprise* (New York: McGraw-Hill, 1960).

36. For reviews of recent happenings in the field of organizational behavior, see David D. Van Fleet, *Behavior in Organizations* (Boston: Houghton Mifflin, 1991); and Ricky W. Griffin and Gregory Moorhead, *Organizational Behavior, 9th ed.* (Mason, OH: South-Western, Cengage Learning, 2010).

37. See Robert Markland, *Topics in Management Science, 2nd ed.* (New York: Wiley, 1983).

38. "Quantitative Analysis Offers Tools to Predict Likely Terrorist Moves," *The Wall Street Journal*, February 17, 2006, B1.

39. For a review, see Everett E. Adam, Jr., and Ronald J. Ebert, *Production and Operations Management, 4th ed.* (Englewood Cliffs, NJ: Prentice-Hall, 1989); see also Richard B. Chase and Eric L. Prentis, "Operations Management: A Field Rediscovered," *Journal of Management* (Summer 1987): 339–350.

40. See Donde P. Ashmos and George P. Huber, "The Systems Paradigm in Organization Theory: Correcting the Record and Suggesting the Future," *Academy of Management Review* (October 1987): 607–621; and Fremont Kast and James Rosenzweig, "General Systems Theory: Applications for Organization and Management," *Academy of Management Journal* (December 1972): 447–465.

41. "Buyers Respond to New GM Incentives: Questions Remain on Company Strategy," *The Wall Street Journal*, August 29, 1986, 2.

42. Kathleen M. Eisenhardt and D. Charles Galunic, "Coevolving—At Last, a Way to Make Synergies Work," *Harvard Business Review* (January–February 2000): 91–103.

43. For an early summary of contingency theory, see Fremont E. Kast and James E. Rosenzweig, *Contingency Views of Organization and Management* (Chicago: Science Research Associates, 1973).

44. "Airline's Ills Point Out Weaknesses of Unorthodox Management Style," *The Wall Street Journal*, August 11, 1986, 15.

45. Lloyd Dobyns and Clare Crawford-Mason, *Quality or Else* (Boston: Houghton Mifflin, 1991).

46. David D. Van Fleet and Ricky W. Griffin, "Quality Circles: A Review and Suggested Future Directions," in *International Review of Industrial and Organizational Psychology 1989*, eds. C. L. Cooper and I. Robertson (New York: John Wiley & Sons, 1989), 213–233.

47. E. E. Lawler III, *High Involvement Management* (San Francisco: Jossey-Bass, 1990) and E. E. Lawler III, *The Ultimate Advantage* (San Francisco: Jossey-Bass, 1992). For a review of participative management, see M. R. Weisbord, *Productive Workplaces: Organizing and Managing for Dignity, Meaning, and Community* (San Francisco: Jossey-Bass, 1987).

48. Lawler, *The Ultimate Advantage*, 25.

49. Lawler, *High Involvement Management*.

50. Michael A. Hitt, "Transformation of Management for the New Millennium," *Organizational Dynamics* (Winter 2000): 7–17.

51. M. Fitzgerald, "The new breadbasket of the world?" *Irish Times*, January 30, 2010, at www.irishtimes.com (accessed July 9, 2010); "Outsourcing's Third Wave: Rich Food Importers are Acquiring Vast Tracts of Poor Countries' Farmland. Is This Beneficial Foreign Investment or Neocolonialism?" *The Economist* (May 23, 2009): 61–63; S. McCrummen, "The Ultimate Crop Rotation," *Washington Post Foreign Service* (November 23, 2009), at www.washingtonpost.com (accessed July 8, 2010).

# CHAPTER

# 3

# Managerial Roles and Skills

## LEARNING OBJECTIVES

After studying this chapter, you should be able to:

- Describe the challenges facing managers.
- Discuss the integrative framework for the different managerial schools of thought.
- Identify the essential managerial roles.
- Identify the essential managerial skills and knowledge.
- Summarize how managers develop managerial skills and knowledge.

- procedural knowledge
- process knowledge
- resource allocator
- scientific management model
- spokesperson
- technical skills

## Making Stars at Starbucks

"**H**ey Buddy, can you spare a cup of coffee and some health benefits? I will work hard for you. I promise this."

"Eight years earlier when Crystal had been on the street, she could never have imagined that in the future she would have a Waspy guy, the proverbial "Man" himself, all dressed up in a two-thousand dollar suit, begging her for a job."

**FIGURE 3.1** Professional Coffee Barista at Work.
© iStockphoto/Warren Goldswain

This event related by Michael Gill in his book, *How Starbucks Saved My Life: A Son of Privilege Learns to Live Like Everyone Else*, underscores the importance of managerial roles and skills. Crystal was the manager of a Starbucks store in New York City. Michael was a recently released advertising executive. Michael's former salary level exceeded Crystal's easily; however, the word *former* is a significant adjective. They were engaged in this conversation that would lead to Michael's employment as a barista because the CEO of Starbucks, Howard Schultz, had unique insight into the "employment experience" and acted on it. Successful business leaders understand what it takes to develop a business model and to provide the resources necessary to achieve success.

The Starbucks Corporation has 17,651 stores, 160,000 employees, and in 2012 generated net revenues of $13.3 billion. To the average person Starbucks' success depends on ubiquity—they seem to be everywhere—and great coffee (Figure 3.1). This is only part of the story.

Starbucks' success is dependent upon the "coffee experience" its baristas provide. These individuals interact with store customers, establishing rapport as part of "the experience" and providing the exact, unique product requested by the patron. Howard Schultz's insight was that it takes more than a unique job title to encourage an over-the-top service attitude and work ethic in its employees (Starbucks calls them "partners") beyond any other coffee establishment.

If an organization wants its employees and partners to believe in the business and its mission, it must first earn their belief and respect. How does Starbucks achieve this loyalty? The CEO's letter in its annual report spells it out: "For our partners, despite economic pressure, we continued to offer health care benefits to eligible full- and part-time partners, something we have done for over 20 years. We were also proud to be able to contribute a match under our 401(k) program this year—another benefit from which many companies retreated in 2009."

This course of action for its partners has given Starbucks a distinct advantage over competitors, but it is an expensive advantage. It is an advantage Howard Schultz, CEO, insists on. "I always wanted to build the kind of company that my father never got a chance to work for."

More specifically, Schultz said, "It's an ironic fact that, while retail and restaurant businesses live or die on customer service, their employees have among the lowest pay and worst benefits of any industry. These people are not only the heart and soul but also the public face of the company."

How simple is that statement? If you want people to believe in your business and product, you have to demonstrate that you believe and trust them. It is essential for managers to work for the success of their enterprises, to provide appropriate resources, and to motivate employees. Starbucks store managers understand these principles and apply them.

Michael Gill put it best when he said, "After finding out about the health benefits that Starbucks offered, I really wanted this job."

Others will utter these words in many languages as Starbucks expands beyond its U.S. borders. Howard Schultz is betting that the Starbucks business model and its partners' strategic advantages have "legs." All because of two very simple axioms that continue to serve Starbucks well: If your business model revolves around human capital, then it is imperative that you invest in it. If your customers are looking for experience, provide it.[1]

## INTRODUCTION

The managers at Starbucks have demonstrated that they have the knowledge and skills to allocate resources and motivate employees in a way that allows the organization to compete successfully. Judging from the length of time they have been in business, the owner and managers of Summer Farms have demonstrated similar abilities. Developing managerial knowledge and skills for people entering agribusiness is what this book is all about. This chapter begins with a brief discussion of the challenges facing managers today and tomorrow. Next, we present an integrative framework that synthesizes the different managerial schools of thought presented in Chapter 2. Using the integrative framework, we will explore the different roles that managers play. Then we present the essential managerial skills and knowledge every manager needs to activate the managerial roles. Finally, we discuss how managers develop managerial skills and acquire managerial knowledge. After completing this chapter, you will have started to develop important conceptual skills for the field of management.

# The Challenges of Managers Today and Tomorrow

Today's manager faces many challenges and will continue to do so in the future. When Joshua Summer started his little family farm, about the only competition he faced was other small family farms within, say, 50 miles. Today, competition can come from halfway around the globe (Figure 3.2). One of the more pressing challenges is how to compete successfully not just against local or national competition but also against *global competition*. The challenge of global competition has managers rethinking how organizations should be structured and how work should be done.[2] You will learn more about these challenges in Chapter 6. Self-managing work teams, total quality management, and just-in-time inventory are all examples of managers trying to find better ways to compete.

## International and Global Competition

As international competition heats up, organizations will need to turn to their most valuable resource for new ways to compete. That resource is people. But this means all people, not just a select class or group of people. One of our greatest resources and challenges is the diversity of our culture, but in the past many businesses have often wasted this resource. As Jim Preston, president and CEO of Avon Products, Inc., said, "Talent is color blind. Talent is gender blind. Talent has nothing to do with dialects, whether they're Hispanic or Irish or Polish or Chinese."[3] He went on to say that if the United States is to regain its competitive advantage, managers will have to harness the human power of all the diverse groups that make up the work force. Agribusinesses are no exception.

**FIGURE 3.2** Managers today compete globally by shipping and by importing products from around the world.

Taking advantages of the opportunities that diversity can bring to the business will be one of the challenges of management, including harnessing the abilities of the disabled as required by the Americans with Disabilities Act.

## Higher Quality

Another challenge that managers face is producing goods and services of high quality. The concept of total quality management has become a hallmark of progressive companies. Total quality management (TQM) has to do with making an organization-wide commitment to continuous improvement and to completely meeting the customers' needs. In Chapter 22, the topic of TQM is discussed more fully. For example, DuPont's Pioneer unit was able to increase the quality of hybrid seed corn and reduce the cost of the mechanical and labor-intensive processes for creating it by more than 20 percent.[4] Summer Farms improved the quality of its vegetables when it installed an automatic irrigation system in response to the challenge of a continuing drought. But a commitment to total quality does not just happen. Every manager in the organization must accept the challenge for this to occur.

## Increased Productivity

At the same time that managers are trying to satisfy customer needs, they are also trying to *increase productivity* within the organization. Productivity is a measure of output (product or service) as it relates to inputs (people, materials, money, and information). The United States is the productivity leader in the world. However, workers in the United States work more hours per year than many other countries, so if one adjusts for that, Norway is first in productivity, the United States is second, and France is third.[5]

## Risk Management

Risk management presents still another challenge to managers, especially in agribusiness.[6] In addition to the risks associated with changes in weather, consumer buying habits, regulatory policies, pests, and bacteria, they also must plan for risks such as safety, security, workplace violence, and even terrorism, including bioterrorism. Violence in the workplace is an increasingly common problem for managers.[7] The emergence of the global economy is leading to a collision of cultures, ideologies, doctrines, and national-security measures, which has increased the likelihood of terrorism for all organizations.[8] While that risk may be small, its impact would be great.

These are only a few of the many challenges facing managers today and tomorrow. Some companies, such as Body Shop International (cosmetic industry), take pride in balancing the challenges of making a profit and being socially responsible. To be able to effectively deal with these challenges, a manager must know which managerial roles to play and which managerial skills to apply.[9]

## FOOD FOR THOUGHT 3.1

Corn experts estimate that about 90,000 kernels make up one bushel and there are about 600 kernels/plant. So the loss of 150 plants in a field of 30,000—one half of 1 percent—can mean one bushel lost.

## An Integrative Framework

Chapter 2 presented a historical perspective of management. Three different schools of thought were discussed: the classical school, the behavioral school, and the quantitative school. In addition, contemporary theories such as systems theory, the contingency approach, TQM, and high involvement were introduced. While these schools of thought may appear unrelated, each represents one part of the whole story of management.

Many of these may be thought of as specific **managerial models**. Models are representations of a more complex reality. A toy train is a physical model that represents a real train. Social models like the managerial schools of thought represent a set of assumptions and a way of viewing the world. The contingency approach to management argues that there is no one best way to manage across all situations. Therefore, each management model explains only one part of the whole management process. But each model does contribute its own unique insight to the process. Together the different models provide a manager with a useful framework.

For example, the challenge of coordinating an organization's activities increases as the organization increases in size. A large agribusiness such as ConAgra may therefore have more policy manuals and scheduled meetings than Webb Dairy or even Arcadia Farms of California. Some organizations face complex and turbulent environments while others face definable and stable environments. An organization that faces a turbulent environment must use open-systems thinking to remain adaptive to these external forces. On the other hand, an organization with a more stable environment will most likely use

the administrative management school—thinking to routinize the work. The important point here is that the application of different managerial schools of thought (managerial models) is contingent on key factors experienced by the organization.

An **integrative framework** has been developed to help managers better understand this relationship.[10] The framework is built around two continua (Figure 3.3). The horizontal continuum runs from an internal focus to an external focus. The manager moves along this continuum trying to satisfy both the internal needs or forces of the organization and the external needs or forces of the environment. The **internal focus** helps to maintain the organization. It might include activities such as employee participation in decision making and the management of information. At the other end of the spectrum, an **external focus** helps to make the organization competitive. This focus might include such things as organization growth and productivity.

The vertical continuum extends from an efficiency emphasis to a flexibility emphasis. The manager moves along this continuum trying to make the organization as efficient as possible yet simultaneously flexible enough to adapt to environmental forces. The **efficiency emphasis** helps the manager use resources in the best way possible. The **flexibility emphasis** helps the manager to adapt to change. While the efficiency dimension seeks stability and direction, the flexibility dimension seeks employee

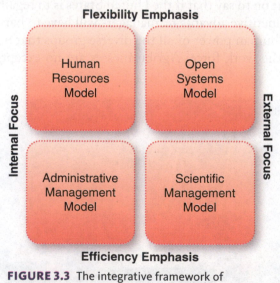

**FIGURE 3.3** The integrative framework of management.
Source: Adapted from Robert E. Quinn, Sue R. Faerman, Michael P. Thompson, and Michael R. McGrath, *Becoming a Master Manager: A Competency Framework*. New York: John Wiley & Sons, 1990, p. 13. Reproduced with permission of John J. Wiley & Sons, Inc.

participation and innovation. The two axes demonstrate the dynamic tension placed on a manager. As you can see in Figure 3.3, each management model fits into one of the quadrants.

## Scientific Management Model

The **scientific management model** focuses on making the best product in the most efficient way. In some ways, it was a forerunner to the current quality movement. The central idea was to identify the best way to do a task and then to make sure that the task was done that way.[11] One reason for making a product as efficiently as possible is to keep its cost low. By doing so, the product is more attractive to customers. Therefore, the scientific management model focuses on satisfying the customer (an external focus) by producing the product the best way possible at the lowest cost.

## Administrative Management Model

The **administrative management model** focuses on the internal operation of the organization. It attempts to make administrative procedures more efficient by finding better ways to coordinate and monitor the work of the organization. By doing this, the organization becomes more stable and predictable. Just as the scientific management model pleases customers, the administrative management model satisfies managers. One reason for making the work of the organization as predictable as possible is so managers can more easily evaluate its productivity. In addition, when the work is predictable, it is clear when management must take action to correct deficiencies.

## Human Resources Model

The **human resources model** emphasizes flexibility in adapting to internal organizational changes by focusing on the people inside the organization. It attempts to build both team orientation and employee involvement among members of the organization. By doing so, the organization hopes that employees will become committed to the organization and willing to adapt to organizational changes. In that way the organization becomes more cohesive and adaptable. Just as the administrative management model pleases managers, the human resources model satisfies organizational members. One reason for this emphasis on people is that, while people are an organization's most valuable resource, they are also its most expensive resource. In addition, organizations that have shown that they value their employ-

ees' involvement are also widely respected and are considered good places to work.[12]

## Open Systems Model

The **open systems model** emphasizes flexibility in adapting to external changes. It attempts to focus on adapting to the feedback from external forces in a creative and innovative manner. In addition, it requires the application of the entrepreneurial role. Chapter 5 explores the topic of entrepreneurship in more detail. By seeking new ideas and developing new ways of producing old products, organizations try to remain competitive in the global economy. If, on the other hand, an organization does not adapt, it runs the risk of failing. Just as the human resources model pleases organizational members, the open systems model satisfies organizational owners such as stockholders. One reason for emphasis on adapting to environmental forces such as economic, competitive, and sociocultural elements is the survival of the organization. In addition, while worldwide boundaries of business are disappearing rapidly,[13] organizations that do adapt to environmental changes will continue to be able to offer employment and improved standards of living.

## An Analysis of the Integrative Framework

The **integrative framework** provides the manager with a way of making sense out of the different managerial models. If you examine Figure 3.3 again, you can see that it includes both complementary and opposing schools of thought.

For example, both the administrative management and the scientific management models emphasize efficiency and stability. The human resources and open systems models share a common emphasis on flexibility and adaptation. The human resources and administrative management models focus on internal processes in the organization. The open systems and scientific management models focus on external forces and their effect on the organization.

In addition, each model has its opposing point of view. The human resources model with its emphasis on team orientation and employee involvement is in direct contrast to the scientific management model, which emphasizes worker efficiency and task simplification. Likewise, the models of administrative management and open systems are diametrically opposed. The administrative management model emphasizes developing stable and efficient organizations, while the open systems model argues for an adaptive and fluid organization.

These differences among the managerial models reflect the challenges managers face every day. Remember, no one model is always right or always wrong. Effective managers use all of the different managerial models daily to perform their jobs.

Is it enough just to know which managerial model applies to be an effective manager? The answer to this question is an emphatic no. But knowing the appropriate managerial school is an excellent place to start because it provides a conceptual framework for focusing the manager's attention. Once a manager has determined the appropriate managerial model for the specific challenge at hand, the manager must next be able to act. Managers activate the different managerial models through **managerial roles**. A manager's role in an organization is similar to an actor's role in a play. It consists of certain actions that the manager is expected to perform and ways in which he or she is expected to behave.

### FOOD FOR THOUGHT 3.2

The black-and-red checkerboard trademark of Nestlé Purina PetCare, formerly Ralston Purina, was inspired by founder William Danforth's memory of the cloth his mother used to sew shirts and dresses for her children.

## Managerial Roles: Essential Activities

Managers work long hours at an intense pace.[14] They are frequently interrupted. Most of their work is disjointed and fragmented. Most of their encounters are brief and seldom with the same person or group. While managers use many different communication methods, they usually prefer face-to-face communication with subordinates, peers, superiors, and others inside and outside the organization. In addition, managers play definite managerial roles to get their work done. A manager's role in an organization is similar to an actor's role in a play. It consists of certain actions that the manager is expected to perform and ways in which he or she is expected to behave. Managers activate the different managerial models through the managerial roles that they play.

Managerial roles fall into three general categories. Roles begin with the formal authority bestowed on managers by their organization. This authority is accompanied by a certain amount of status. This status causes all managers to be involved in interpersonal relationships with subordinates, peers, and superiors, who in turn provide managers with information they need to make decisions.[15] Table 3.1 presents the ten roles grouped according to those primarily concerned with either interpersonal relationships, information exchange, or decision making.

### Interpersonal Roles

There are three **interpersonal roles** in the manager's job. The first is that of **figurehead**. When serving as a figurehead, the manager simply puts in an appearance as a representative of the organization. Although there is little serious communication and no important decision making in the figurehead role, it should not be overlooked. At the interpersonal level, the figurehead role symbolizes what the organization is all about and its attitude toward others. When a manager takes a visitor to dinner, attends a ribbon-cutting ceremony, serves as a company representative at a wedding or funeral, or participates in local fund-raising events, he or she is playing the figurehead role.

The second interpersonal role is that of **leader.** As a leader, the manager hires employees, motivates them to work hard, and deals with behavioral processes. The manager is responsible for coordinating the efforts of subordinates. When a manager inspires employees by creating a vision of the future, emphasizes the importance of high quality performance, and clarifies the roles of each subordinate, the manager is performing the leadership role.

The third interpersonal role is that of **liaison,** which involves dealing with people outside the organization on a regular basis. Managers must be able to work well with outside agencies that can help them achieve their organizational goals. In a sense, the liaison role is about developing a network of mutual obligation with others. For example, a vice president of human resources might serve as a trustee on your university's board of trustees. By doing this, the manager develops a network of contacts outside of her organization. Some of these contacts may prove to be useful to her organization. In addition, when the university asks for information on what industry is seeking in new undergraduates, the manager successfully fills the

**TABLE 3.1** Managerial Roles

| CATEGORY | ROLES | DESCRIPTION | ACTIVITY EXAMPLE |
|---|---|---|---|
| **Interpersonal** | Figurehead | Performs routine duties as the symbolic head | Cutting ribbon at groundbreaking for new building |
| | Leader | Responsible for staffing, training, motivating and activating subordinates | Inspiring employees with praise or bonus for quarterly performance |
| | Liaison | Serves as a link with outsiders who provide favors and information | Serving as an advisor or judge for a local group such as the FFA |
| **Informational** | Monitor | Obtains wide variety of internal and external information for understanding the organization and environment | Scanning industry reports and both external and internal publications for information relevant to this company |
| | Disseminator | Transmits information received from outsiders or subordinates to members of the organization | Forwarding appropriate information to personnel within the company |
| | Spokesperson | Transmits information to outsiders on organization's plans, policies, actions, results, etc. | Board meetings and speeches, etc., involving transmission of information to outsiders |
| **Decisional** | Entrepreneur | Searches for opportunities and initiates "improvement projects"; also supervises design of certain projects | Developing or approving new ideas and strategies |
| | Disturbance Handler | Responsible for corrective action when organization faces unexpected disturbances | Intervening to resolve internal conflicts |
| | Resource Allocator | Responsible for allocating various kinds of organizational resources | Activities involving budgeting or budget requests |
| | Negotiator | Responsible for major negotiations with other groups or organizations | Attempting to reach agreement with union, supplier, or customer |

Sources: M. J. Martinko and W. L. Gardner (1990) Structured observation of managerial work: A replication and synthesis. *Journal of Management Studies*, 27: 329–357; Stephen J. Carroll and Dennis J. Gillen (1987) Are the classical management functions useful in describing managerial work? *The Academy of Management Review*, 12: 38–51; and Henry Mintzberg, *The Nature of Managerial Work*. New York: HarperCollins, 1973. Table "Mintzberg's Managerial Roles" from The Nature of Managerial Work by Henry Mintzberg. © 1973.

liaison role when she provides insightful information about necessary skills of undergraduates.

## Informational Roles

The manager's job also consists of three basic **informational roles** (Figure 3.4). The first is the role of **monitor**, in which the manager actively watches the environment for information that might be relevant to the organization. A manager is acting as a monitor when he or she reads *The Wall Street Journal*, asks employees about work-related problems they are having, closely scrutinizes television commercials, or scouts area stores for competitive information.

The opposite of the monitor role is that of **disseminator**. In this role the manager relays information gleaned through monitoring to the appropriate

**FIGURE 3.4** Managers serve as spokespersons for their units and organizations and may be called upon to give speeches as part of their jobs.

people in the organization. The owner of Green Things Landscaping finds magazine photos of unique landscaping ideas that he brings back to the artist and other managers. For Frito-Lay an article in *Fortune* may be of little direct use to one manager but of considerable value to a colleague. To fill the role of disseminator successfully, the first manager clips the article and passes it on to the person who may find it useful.

The third informational role is that of **spokesperson**. The spokesperson role is similar to the figurehead role, but the manager in the spokesperson role presents information of meaningful content and/or answers questions on the firm's behalf. For example, during the Union Carbide disaster in India in 1985 when a plant's poisonous gas leak killed thousands of people, managers regularly appeared at news conferences to make statements about new events and to answer reporters' questions. More recently, the BP oil spill in 2010 showed how crucial the spokesperson role can be. In this high-profile case, comments by BP spokespersons, especially their CEO, damaged the public's trust in the company to such an extent that the CEO was forced to resign. As a result, the organization itself may experience the fallout from a tainted public image for many years. In contrast, Johnson & Johnson's response in 1982, 1986, and again in 2010 to the Tylenol incidents demonstrated how the spokesperson role can be a company's greatest asset in a company crisis. If Summer Farms should experience a misfortune similar to the *Listeria*-contaminated Rocky Ford cantaloupes in Colorado in 2011, one of its managers will be forced to serve as company spokesperson, ready or not.

## FOOD FOR THOUGHT 3.3

Potato chips were invented accidentally in 1852 by George Crum in an angry response to a criticism of his cooking at a posh resort, Moon's Lake House, in Saratoga Springs, New York.

## Decisional Roles

The final category of managerial roles are **decisional roles** in nature; that is, these are the roles that managers play when they make decisions. First, there is the role of **entrepreneur**, in which the manager looks for opportunities that the organization can pursue and takes the lead in doing so. If a manager at Cargill notices that a particular waste by-product can easily be transformed into a new, potentially marketable product and then submits a proposal offering to take charge of the initiative, the manager is being an entrepreneur. Whole Foods was started by John Mackey because he was frustrated trying to find sources for natural foods. It is now one of the largest natural-foods retailers in the United States.[16] Sometimes smaller companies have the most successful entrepreneurs because they are not flush with capital and therefore are more motivated to "invent." For example, the wooden parquet floor tile that is so popular today was the brainchild of the owner of a lumber mill in a small Tennessee town who wanted to find a productive use for pieces of wood that accumulated on the mill property and would have otherwise been discarded. Hartco Hardwood Tiles boosted the company's cash flow for several years before it was sold to Armstrong for a large sum.

The role of **disturbance handler** is also important. This role involves the manager in mediating or resolving conflicts between, for example, two groups of employees, a sales representative and an important customer, or another manager and a union representative. If a manager at the Ford glass plant or Purina resolves a grievance, he or she is being a disturbance handler.

The **resource allocator** role focuses on determining how resources will be divided among different areas within the organization. If there is $275,000 to divide among three departments, each of which has requested $100,000, the manager has to decide how to distribute the funds. Although this task may involve using some quantitative procedure to analyze the best use of the funds, it also requires interpersonal skill in explaining the decision to each department. In this situation, the manager is combining the resource allocator role with the leader role. The example here talks about financial resources, but the manager also allocates material, people, and information as part of the resource allocator role.

Finally, there is the role of **negotiator**. In this role, the manager attempts to work out agreements and contracts that operate in the best interests of the organization. Such agreements may be labor contracts, purchasing contracts, or sales contracts. The negotiator's role occurs both inside and outside of the organization. For example, a company president may negotiate a contract with a consulting firm, or a procurement officer may negotiate for janitorial services for the organization. Both managers are performing the negotiator role. Inside the organization, a manager may facilitate the settlement of an equal

## A FOCUS ON AGRIBUSINESS
### Management Roles at Oppenheimer

The Oppenheimer Groups traces its origins to 1858 when four Oppenheimer brothers followed the gold rush to British Columbia and founded Oppenheimer Bros. & Co. in Victoria. Oppenheimer is now one of North America's top fresh-produce companies which brings in more than 100 varieties of produce from more than 25 different countries. It is a major distributor of Ocean Spray fresh cranberries and has a cherry partnership with Orchard View Farms of The Dalles, Oregon.

An organization this size has numerous management roles. You can get a feel for some of them through the following brief biographies of some of its managers.

Nolan Quinn is the Director of the Berry Category. Quinn has a degree in agribusiness and joined Oppenheimer in 1995 as a marketing assistant. He moved to quality control the following year, then worked in brand management, and eventually ended up in a sales role. In 2001, his responsibilities grew to include oversight of Oppenheimer's business with Walmart. More recently, Nolan oversaw the market introduction of Ocean Spray fresh blueberries.

James Milne is Director of Citrus & Avocado Categories. He is from Wellington, New Zealand, and has a bachelor's degree in commerce and administration. He began his food-marketing career with New Zealand-based ENZA in 1987. He joined Oppenheimer as pipfruit category director in 1996. Having spent 15 years bringing internationally grown apples and pears to market, James was chosen for Oppenheimer's top marketing spot in 2001.

David Nelley is the Director of the Pipfruit and Pineapple Categories. He has a bachelor's degree in international management and marketing. His career in produce began in April 1993. During his tenure, David was heavily involved in building demand for the New Zealand Braeburn apple in North America. More recently, he has been instrumental in introducing ENZA Jazz™ apples and the Linda brand pineapple to North America.

Steve Woodyear-Smith is Director of the Kiwifruit and Mango Categories. A New Zealander with a bachelor's of commerce degree in marketing and international business, he joined Oppenheimer in 2000, following ten years in various management roles with kiwifruit marketer ZESPRI International in North America. In 2002, Steve added the mango category to his role at The Oppenheimer Group.[17]

opportunity grievance or negotiate a labor contract. Each of these activities also represents the negotiator role. The head of Summer Farms is a negotiator when he referees a heated discussion between two managers, who are also members of his family.

## An Analysis of Managerial Roles

Since the original managerial roles were identified, there have been many follow-up studies attempting to validate the roles.[18] The results generally support the earlier findings. However, managers at different levels within the organization place different emphases on some of the roles.[19] For example, top managers say they place great importance on the roles of liaison, spokesperson, and resource allocator. On the other hand, middle managers and first-line managers list the leader role as one of their most important roles.

How do these ten managerial roles fit into the integrative framework? The figurehead, liaison, spokesperson, and entrepreneur roles all fit nicely into the open systems model. Each of these roles has something to do with an effort to improve the effectiveness and competitiveness of the organization. The roles of leader, disseminator, and disturbance handler fit into the human resources model. These roles deal primarily with people inside the organization. Negotiator is a role that fits into both the open systems and human resources models because it deals with negotiating both inside and outside of the organization. Finally, monitor and resource allocator are roles necessary in the administrative management model. These two roles focus on the internal functioning of the organization.

As you can see, these roles do not satisfy all four of the managerial models in the integrative framework. One reason is that the integrative framework point of view was not used in developing these roles. Other management scholars have identified additional managerial roles believed to be critical for a successful and effective manager.[20]

More recently, two essential managerial roles were identified for each of the four models in the integrative framework.[21] Table 3.2 presents the eight roles grouped by managerial model. Studies using this newer set of managerial roles have confirmed that the eight roles do indeed appear in the four indicated

quadrants of the integrative framework.[22] If you compare the two lists of roles, you can see that some of the roles, such as the monitor role (Figure 3.5), are the same. In other cases some of the earlier roles, such as spokesperson and entrepreneur, are combined into a single role such as innovator. Thus, the two lists of managerial roles do complement each other even though they were created from different viewpoints.

However, because the newer set of managerial roles was created specifically with the integrative framework in mind, we will focus on these roles in this book. This does not mean that the older set of managerial roles is unimportant or wrong. You can find each of these roles in the integrative framework, but they do not provide complete coverage of all the quadrants in the integrative framework. To be a mentor to an employee means to help develop that employee by providing career information and performance feedback. Good mentors

**TABLE 3.2**  Managerial Roles of the Integrative Framework

| CATEGORIES/ROLES | DESCRIPTION | IDENTIFIABLE ACTIVITIES |
|---|---|---|
| **Scientific Management** | | |
| Producer | Task oriented; usually involves motivating others to increase production and to accomplish stated goals | High interest, energy, and personal drive |
| Director | Goal setter; a take-charge and decisive individual; usually involves planning and initiating goals, defining roles, and giving instructions | Develops goals, defines tasks to be accomplished, and provides direction to accomplish the tasks |
| **Administrative Management** | | |
| Coordinator | Reliable; expected to maintain the structure and flow of the organization | Schedules, organizes, and coordinates staff efforts |
| Monitor | Desire for details; knows what is going on in the unit and checks on progress | Reviews, analyzes, and responds to work processes |
| **Human Resources** | | |
| Facilitator | Process oriented; fosters teamwork and manages interpersonal conflict | Develops a cohesive and group problem-solving atmosphere |
| Mentor | Caring oriented; helpful, considerate, approachable, open, and fair | Develops people through personal involvement |
| **Open Systems** | | |
| Innovator | Creative; envisions new ways, packages them invitingly, and convinces others of the need | Facilitates adaptation and change |
| Broker | Powerful; represents, negotiates, and acquires resources for the organization | Politically astute, persuasive, and influential |

Source: Robert E. Quinn, Sue R. Faerman, Michael P. Thompson, and Michael R. McGrath, *Becoming a Master Manager: A Competency Framework.* New York: John Wiley & Sons, 1990.

© iStockphoto/Pali Rao

**FIGURE 3.5** The monitor role involves going over the work of others, as this manager is doing with one of her subordinates.

understand how they achieved the position they currently hold. They understand their strengths and weaknesses. In other words, they are self aware.

In summary, managers activate the different managerial models through enacting different managerial roles. Each role is important and must be properly executed. Whether a manager performs a role successfully is determined by the extent to which he or she possesses the necessary management skills and knowledge.

---

### FOOD FOR THOUGHT 3.4

Globally, Heinz uses more than 5 billion pounds of tomatoes every year in its products.

---

## Managerial Skills and Knowledge: Essential Competencies

**Managerial skills** reflect the ability to perform the various behaviors managers need to execute their roles effectively. Effective managers tend to possess a certain mix of skills that sets them apart from others. Skills, then, are behavioral in nature, such as the ability to present a strategic plan to the board of directors or conduct a performance feedback session with an employee. **Managerial knowledge**, on the other hand, is the special information and mental activity a manager uses to decide how to behave. Again, effective managers have a mix of different types of knowledge that they combine in unique ways to be effective. Knowledge, then, is cognitive in nature,

such as formulating a new way to solve the company's parking problem or calculating the estimated return on investment for a new venture.

From these examples, it should be clear that skills and knowledge work together. The process comprises three components: (1) the existence of knowledge, (2) a method for accessing this knowledge, and (3) the ability to enact a set of behaviors using the retrieved knowledge to perform the given task. The third component is what people observe and label as a skill. Actually, the first two components are indispensable prerequisites to the actual execution of the observable actions, and these components are knowledge.

## Managerial Skills

The core managerial skills have been identified as technical, human, and conceptual and can be developed in managers by providing them with managerial knowledge.[23] It is believed that if managers are to perform effectively in their managerial roles, they need specific managerial skills and knowledge. Figure 3.6 illustrates the relative importance of these skills by level of management.

**Technical skills** are the skills a manager needs to perform specialized tasks within the organization. On a small cotton farm, technical skills could include operating the picking machine, packing the bales, and operating the carding machines. At Webb Dairy the manager must know how to fill and set the feeding equipment, hook up the milking machines, and perform other tasks between the milking and shipping operations. The dairy manager also may need to know how to attach and read estrogen monitors on the cows. In the Mayo Clinic, technical skills are those possessed by physicians, nurses, and lab technicians. At Dean Witter Reynolds, they are the skills associated with understanding investment opportunities, tax regulations, and so forth. The electrical and mechanical engineering skills the professional engineers at Hewlett-Packard possess are the relevant technical skills for that organization. At Walmart, a manager who decides on an advertising campaign, the promotional items, and pricing for a new product is using technical skills. And in major agribusiness firms such as Archer Daniels Midland, Bunge, or Cargill, there are positions that require each of these sets of technical skills.

As indicated in Figure 3.6, technical skills are important for first-line managers. Because they spend much of their time working with operating

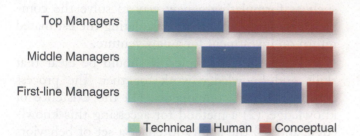

| Top Managers | | | |
| Middle Managers | | | |
| First-line Managers | | | |

■ Technical  ■ Human  ■ Conceptual

**FIGURE 3.6** The importance of managerial skills by level.
Sources: Based on Tim O. Peterson and David D. Van Fleet, "The Ongoing Legacy of R. L. Katz: An Updated Typology of Management Skills," *Management Decision* incorporating the *Journal of Management History*, (2004), Vol. 42, No. 10, pp. 1297–1308; David D. Van Fleet, D. D. (1991). *Behavior in Organizations*. Boston: Houghton Mifflin; and Robert L. Katz, "Skills of an Effective Administrator," *Harvard Business Review*, January–February, 1955, pp. 33–42.

employees, they must have a good understanding of the work those employees are doing. Technical skills are slightly less important for middle managers, because a greater proportion of their time is devoted to other managerial activity; they are even less important (although not unimportant) for top managers.

**Human skills** are the skills a manager needs to work well with other people. They include the ability to understand someone else's position, to present one's own position in a reasonable way, to communicate effectively, and to deal effectively with conflict. Managers with well-developed human skills can create a climate of trust and security in which organizational members feel free to express themselves without fear of punishment or humiliation.[24]

Because all managers must interact with others, human skills are important at all managerial levels. In fact, some say they are the most critical but the least developed skills in managers.[25] In general, the better any manager's human skills are, the more effective the manager is likely to be. While some managers have been successful in spite of limited human skills, it is rare. If it happens, it is only because another managerial skill is critically needed by the organization at the time. More often managers such as Ellen Kullman, CEO of DuPont, and Patrick McGinnis, CEO of Nestlé Purina, demonstrate that human skills are very important in running successful organizations.

**Conceptual skills** relate to a manager's ability to think in the abstract. Managers need to be able to see relationships between forces that others may not see, to understand how a variety of factors are interrelated, and to take a global perspective of the

organization and its environment. A manager who recognizes an opportunity that no one else has seen and then successfully exploits that opportunity has drawn on conceptual skills. For example, Howard C. and Leatha Morrison tested their early visions during challenging times, first in Gilbert, Arizona, in the early 1920s, then back in Oklahoma where they were born, and next in California during the depression era. They concluded that Gilbert offered the greatest opportunity and they returned to stay in 1934. They recognized the opportunities in Arizona, and over the years the Morrison family built one of the area's largest farming operations, including sizeable dairy, cotton, and cattle businesses—Morrison Farms, Morrison Brothers Ranch, Arizona Dairy Company, and Windmill Ranch. All are family businesses. And while those continue, the family has moved into real estate development as well.[26]

Conceptual skills also include the ability to define and understand situations (Figure 3.7). If a plant manager at Alcoa notices that turnover at the plant is increasing, the manager needs to address the situation in some way. The first step is to define the problem: unacceptable turnover. Next, the manager must determine what is causing the problem and identify one or more ways to reduce it. For example, closer inspection may show that only one department is affected, which suggests a

© Galina Barskaya/www.Shutterstock.com

**FIGURE 3.7** One of the most important managerial skills is the ability to think conceptually, in the abstract.

problem specific to that department. Appropriate action might include discussions with the department manager, the entire work group, or both.

As shown earlier in Figure 3.6, conceptual skills are most important for top managers because their job is to identify and exploit new opportunities. Recall the lumber mill owner's idea to use scrap wood to create a new product, parquet floors, in the Hartco Hardwood Tile example given earlier in this chapter. Conceptual skills are moderately important for middle managers and less important for first-line managers.

Technical, human, and conceptual skills are all necessary for effective management. As we have seen, top managers need a large measure of conceptual skills and a lesser amount of technical skills, whereas first-line managers need the reverse combination. Middle managers need equal measures of technical and conceptual skills. Finally, all managers need to have human skills.

## Managerial Knowledge

To enact a skill, a manager must possess managerial knowledge. Cognitive psychologists have identified three types of knowledge that a person has to have to perform any task: declarative, procedural, and process.[27] For example, for a manager to make an investment requires knowledge of investment terms (declarative), of tax procedures (procedural), and of the stock market process (process).

**Declarative knowledge** deals with facts and definitions. It is sometimes called the "what" of a topic. For example, the boldface words in this text are definitions and therefore a kind of declarative knowledge. Without this type of knowledge, we could not think or communicate. This form of knowledge assists a manager in describing a situation. It also allows managers to interact intelligently about a specific problem within the organization.

**Procedural knowledge** outlines a set of steps to be taken to accomplish a task. It has often been called the "how to" of a task. A manager who sits down to prepare a budget is following a set of steps to accomplish this task. Procedural knowledge activates technical skills. For example, when a manager at British Petroleum prepares a scheduling chart, procedural knowledge is being used to carry out this technical skill.

**Process knowledge** provides a manager with a mental map for a specific topic. It is often referred to as knowledge about "how something works." Just like a road map helps you get around town by providing you with a model, a mental map helps the manager understand the relationships between different factors by providing a mental picture of the situation. The integrative framework (see Figure 3.3) is a good example of process knowledge. It provides a mental map that helps you understand the relationship among the different managerial models. Throughout this book, the figures provide excellent mental maps to assist you in developing process knowledge. Process knowledge activates the conceptual skills. A product manager at Armour-Eckrich who conceptualizes a new use or a new market for a product line is using process knowledge. The owner manager of Hartco Hardwood Tiles acquired a different type of process knowledge when he added the hardwood tiles to his long-time lumber milling business.

## An Analysis of Managerial Skills and Knowledge

Since the original work on managerial skills, many studies have been conducted that confirm the three broad skill areas identified in this early research.[28] The more recent research done on knowledge types also seems to be consistent with the original conceptualization of skills.

Declarative knowledge is a foundation. A great deal of what is covered in this book is declarative knowledge. The vocabulary list at the beginning of each chapter is an example. By learning the definitions and meanings of these words, you will be able to better understand the lectures in class. You also will be able to ask better questions and formulate more intelligent answers during class discussions.

Process knowledge and procedural knowledge build on declarative knowledge. Process knowledge helps us explain how something works. For example, being able to explain how the brake system in a car works would indicate that you have process knowledge about a car's brake system, just as being able to explain how an oil company converts oil into gasoline would indicate process knowledge of the refining method. But to be able to explain how anything works, we also rely on our declarative knowledge about the topic.

Procedural knowledge specifies how to do a task. When you do multiplication or long division, you are using procedural knowledge. A manager redesigning a job is using procedural knowledge. In fact, anytime a person executes a set of steps to complete a task, he is using procedural knowledge. Procedural knowledge builds on both declarative knowledge and process knowledge.

For example, imagine that you were asked to prepare a PERT chart for an upcoming project in your organization. Not only would you have to know the steps in the PERT chart procedure (procedural knowledge), but you would also have to be able to formulate a work flow diagram of the project (process knowledge). In addition, you would need to know the definition (declarative knowledge) of a node, an activity, and the critical path. You would also have to know facts (declarative knowledge) such as the constants in calculating the expected time of completion for each activity in the chart. As you can see, procedural knowledge is dependent on both process knowledge and declarative knowledge. When actually preparing the PERT chart, you would be demonstrating both a conceptual skill (diagramming the work flow) and a technical skill (preparing the PERT chart).

Skills are what we actually see people do. In this chapter, we have identified three broad skills. Each relies on different types of knowledge. Knowledge is critical to performing any skill, but generally it goes unseen. Knowledge is stored either internally in our minds or externally in books or computer systems. When the knowledge is stored externally, the manager must know where and how to access the knowledge. By reading this book, you are not only storing some of the knowledge internally, you are also learning where to access knowledge that is stored externally in this book.

Conceptual skills rely heavily upon process knowledge. A manager formulating a new strategic plan for Perdue Farms may use an analysis to determine the strengths and weaknesses of Perdue, but this skill is based on process knowledge of the relationship between the organization's structure and current environmental forces which impact on the organization. We will talk more about the impact of environmental forces in Part Two of the book.

While conceptual skills depend heavily on process knowledge, technical skills rely most heavily on procedural knowledge. Technical skills generally have some procedure to follow to complete the task. For example, the planning and decision-making tools in Chapter 11 have defined procedures.

Finally, human skills depend equally on all of the knowledge types. This could be because the human element in an organization is the most complex and difficult to manage. For example, when a manager prepares to do a performance appraisal of an employee, the manager must conceptualize how the performance appraisal system relates to organizational entry, development, performance, and exit. In this case, the manager is using both process and declarative knowledge. When the manager prepares the performance appraisal form, he is using both procedural and declarative knowledge. Finally, when the manager sits down with the employee to discuss past performance and to discuss future performance, the manager is using all three types of knowledge.

In summary, whenever a manager exhibits a managerial skill, it is done to satisfy some managerial role. Skills build on knowledge and roles build on skills to form the managerial storehouse.

## FOOD FOR THOUGHT 3.5

Gatorade was formulated in 1965 to help counteract the effects of dehydration better than water for the Florida Gators football team in hot, humid weather. The Kansas City Chiefs were the first team besides the Gators to adopt the drink.

# Developing Managerial Skills and Acquiring Managerial Knowledge

Is it just luck when managers combine effectiveness with efficiency? It is true that managers are sometimes just plain lucky, but more often than not they become successful because they are prepared.[29] This preparation usually consists of a combination of education and experience that gives managers the technical, human, and conceptual skills and the declarative, procedural, and process knowledge necessary to contribute to an organization's efficiency and effectiveness.

## Education

An education leading to a college degree is the first step in most people's managerial careers. In fact, a college degree is almost a requirement for promotion to upper levels of management in most organizations today. In 1990, more than 90 percent of the CEOs of *Business Week*'s top one thousand corporations held a college degree.[30] Education does not stop with an undergraduate degree, however. Many managers return to school and get a graduate degree, usually an advanced degree in agribusiness or a Master of Business Administration (MBA). Many of these managers return to school because their undergraduate degree was in some other area such as chemistry or engineering. But many managers who majored in agribusiness or business return to update their managerial knowledge and skills (Figure 3.8).

In addition, managers attend advanced training and professional development programs sponsored by universities, private consulting firms, or their own companies.[31] These activities are apt to continue throughout a manager's career. This phenomenon is referred to as *lifelong learning*. It is a philosophy that learning, gaining new knowledge, and developing new skills is a lifelong process, not something that ends with a degree. The major accrediting organization for business schools (American Assembly of Collegiate Schools of Business, AACSB) has recognized the importance of this lifelong learning philosophy and has incorporated it into the new standards governing business schools' curricula.[32]

The primary advantage of education and professional development as sources of managerial knowledge and skills is that learning can be well organized and planned. The process focuses the learner's attention on a specific body of knowledge. This reduces the time it takes to learn the material.

One drawback to this method of learning is that the knowledge must be made more general in nature to satisfy a wide variety of learners.

## Experience

An education alone, of course, does not promise a person an executive position. Making an "A" in your management class will not guarantee you a management position. Most people must still work their way up an organization, sometimes making mistakes and suffering setbacks along the way.

Experience adds to your managerial knowledge. It provides you with specific declarative knowledge about your industry. It provides process knowledge in the form of specific work flows for your company. It provides specific procedural knowledge for tasks such as budgeting, hiring, and goal setting.

Experience comes in a variety of forms. Some people work while in college. Others may even hold full-time jobs before they enter college. Quite often, these people return to school as non-traditional students seeking a degree in agribusiness or business. In these cases, the individuals have experience, but not the more structured and well-organized learning that an education provides. Others enter college directly from high school. After finishing college they might accept an initial job assignment with a company that offers them an entry-level position plus a management

**FIGURE 3.8** Managers engage in lifelong learning on the job and by attending professional development programs and courses.

training program. Later, they may accept transfers to different departments to broaden their experience within the company.

Company changes are also common; and many people leave the first firm they work for and take a new job with another. Indeed, companies such as General Foods and General Mills have such good training programs that other organizations actively recruit people who have worked there for a few years.[33] Over the course of a lifelong career, a person is likely to work in a number of different jobs, usually for more than one company. Throughout this process, the individual develops managerial knowledge and skills through a combination of education and experience. Both are necessary to producing a successful and effective manager.

## CHAPTER SUMMARY

Managers face many new challenges in the business world today. One of those challenges is global competition. Another challenge is how to manage the diverse work force. Besides these two challenges, managers will also have to increase productivity while at the same time increasing quality. To meet these challenges, managers must develop knowledge and skills to perform their roles efficiently and effectively.

The integrative framework organizes the different managerial schools of thought into a coherent whole. The framework revolves around two continua. The horizontal continuum is anchored at one end by an external focus and at the other end by an internal focus. The vertical continuum emphasizes efficiency and flexibility. Each managerial model fits into the contingency framework. Managers should be aware of the different situations in which each model will be most effective.

Managers activate the managerial models through enacting different managerial roles. A role is a set of behaviors expected in a given situation. Earlier work identified ten managerial roles in three categories: interpersonal roles, informational roles, and decisional roles. Interpersonal roles are those of figurehead, leader, and liaison. Monitor, disseminator, and spokesperson are the informational roles. The four decisional roles are entrepreneur, disturbance handler, resource allocator, and negotiator.

More recent research identified eight managerial roles—two for each managerial model. The managerial models include scientific management, administrative management, human resources, and open systems. The two essential roles for the scientific management model are those of producer and director; for the administrative management model, coordinator and monitor; for the human resources model, facilitator and mentor; and for the open systems model, innovator and broker. Other management scholars have identified their own lists of managerial roles. The key point is that managers perform roles to accomplish their managerial duties.

Whether a manager performs a role successfully is determined by the extent to which he or she possesses the necessary management skills and knowledge. Managerial skills consist of technical, human, and conceptual skills. These three skills provide the manager with the essential skills to enact the managerial roles. Managerial knowledge consists of declarative, procedural, and process knowledge. Each of these knowledge types assists the manager in performing some specific skill. It takes both knowledge and skill to enact a managerial role.

Managers obtain skill and knowledge by combining education and experience. Education may consist of schooling such as college or professional development conducted by the company. Experience may be gained by job rotation, lateral movement within the company, or by moving from one company to another. Both education and experience allow managers to acquire skills and knowledge they need to be successful and effective.

## CHAPTER ACTIVITIES

### REVIEW QUESTIONS

1. What are some of the challenges facing managers?
2. Sketch the integrative framework of management and describe the managerial models of thought that are opposing forces.
3. Compare and contrast the original roles with those in the integrative framework.
4. List and describe the three essential managerial skills and the three essential managerial knowledge types. Give examples of each.
5. What are the roles of education and experience in managerial skills and knowledge?

### ANALYSIS QUESTIONS

1. Identify ten advantages to workforce diversity.
2. It has been argued that there is no one best way to manage. Do you agree or disagree with this assertion? Why?
3. Which roles do you think managers spend most of their time playing? Is an effective manager more likely to use one or two roles exclusively, or does an effective manager go back and forth between all of the roles? Why?
4. For each of the integrative framework roles, provide an example of a managerial skill or managerial knowledge that an effective manager might need.
5. You hear two classmates arguing over which is more important for a manager: education or experience. How would you help them see that it is not an either/or question? Which managerial role are you enacting during this discussion?

## FILL IN THE BLANKS

1. Different schools of thought about management, each of which represents a set of assumptions and explains one part of the management process, are known as _____ _____.

2. Operating in a way that uses resources wisely and seeks stability and direction is known as _____.

3. The roles that managers are expected to play include _____, _____, and _____.

4. The symbolic role where the manager acts as a representative of the organization by performing routine duties of a legal or social nature is referred to as _____.

5. Dealing with people outside the organization on a regular basis is the manager's _____ role.

6. When the manager is communicating with people inside or outside the organization, she is performing the _____ role.

7. When presenting information and answering questions on the firm's behalf to the public, a manager is functioning as _____.

8. When determining how resources will be divided among different areas within the organization, the manager is acting as _____ _____.

9. The skills a manager needs to perform specialized tasks within the organization are called _____ skills.

10. The skills a manager needs to work with other people are known as _____ skills.

## ▼ CHAPTER 3 CASE STUDY
### WHEN COMPANIES CAN CAN AND CAN'T CAN

So what's the price of tuna, Charlie? Packed in water or not?

Wild Bluefin Tuna will sell for almost $25 per pound in Tokyo's Tsukiji market; "ranched" Spanish Bluefin Tuna will sell for $11 per pound in the same market; and White Albacore Tuna (canned in water) is available for $4 per pound at your local supermarket. This is quite a price range for tuna. Either the economic forces of the tuna market are unique or there is a serious pricing strategy disconnect.

This broad spectrum of prices reflects the many aspects of the tuna business. It also demonstrates the effects of market and economic influences of a business. Thus, this simple example also illustrates an important management lesson: Know your business precisely, and you will understand your market to determine "significant" price points and see the influence of a changing economic and regulatory landscape. This may seem like an obvious lesson, but sometimes the consequences are far reaching.

The first two prices cited above are not for the "canned" tuna business but for sushi restaurants. The most expensive wild bluefin tuna will be served at a very exclusive high-end Tokyo restaurant. Tokyo, Japan, is a city and country where the nuances of different bluefin tuna species are appreciated and attract individuals willing to pay for the experience. The two different prices reflect the fact that a market also exists for individuals with discriminating tastes but not an unlimited expense account. Thus, the "ranched" bluefin has a market—not the very upscale market that the producers imagined but a very good market nevertheless. And, of course, the price of this product precludes its use in a "canned" product, as the albacore tuna is marketed.

With its island nation heritage, Japan is a country that appreciates the different varieties of tuna. Thus, two avenues have developed to meet this demand: traditional tuna fishing to gather the wild species, and the "ranched" species provided by other countries attempting to provide a product and develop a new market. However, both are "fresh markets" because Japan is a country that knows the "tuna" experience is best enjoyed fresh and not in any "canned" state. A willingness by the Japanese people to pay and pay well for this gustatory experience provides a very attractive market for tuna providers.

This is not the case for the U.S. market where, even though it is a country that has two coastlines, most individuals live in the hinterland. As a result, tuna is viewed as just another species of fish. Given this perception, price is a discriminating factor: too high and a purchase will not be made. In addition, the state of freshness is relative and therefore a canned product is very suitable. All of this is reflected in the price and the product. Americans are not very discriminating tuna consumers.

It may not seem fair to compare a restaurant business with a food-processing business, but both are in the "tuna" business. Both must deal with the business and marketing nuances of their business sector, and both are agribusinesses because they are very dependent upon a highly perishable product. Indeed, the restaurant business may be endangered because the bluefin tuna, or at least its Atlantic cousin, is perceived as an endangered species by most of the European community and the United States, but not by Japan.

Why is this important? It was entrepreneurial Japanese restaurateurs who developed the Atlantic bluefin market. In an effort, to find the best tuna for their customers, they opened this market through their willingness to pay a premium for the best fish and pay for air transport to Japan to guarantee freshness. Now this market may be threatened by new regulations, over which Japan has little sway.

Because U.S. consumers know or care little about the kinds of tuna they consume, what does this regulatory dispute have to do with the United States? They could become concerned about the dispute over bluefin tuna even though this species is not used in their (canned) tuna. The result of this concern has led managers to make interesting business decisions.

For example, U.S. firms no longer make StarKist (home of the famous cartoon mascot tuna "Sorry, Charlie") and Chicken of the Sea tuna products. Chicken of the Sea is now produced by Thai Union Frozen Products PCL (Thailand) after its acquisition in 2009, and StarKist was acquired from Del Monte in 2010 by Dongwon Industries Co. Ltd. (South Korea). This trend includes Bumblebee Tuna, which is owned by Connors Bros. Income Fund, an Ontario, Canada, company headquartered in San Diego, California. Thai Union Frozen Products PCL is now the largest seafood firm in the world. This title was bestowed on the company after it acquired John West tuna from MW Brands, formerly a HJ Heinz Co. MW Brands is a French company.

The tuna business can teach managers important lessons. In the case of tuna, American businesses did not understand the finer tastes of different kinds of tuna and the way tuna is presented to the buyer so they had to get out of the tuna business. Know your business...know your potential...realize your limitations. Otherwise, the consequences may be far reaching—tuna may be more expensive than you think it is.[34]

### ▶ Case Study Questions

1. Are there industries or products other than tuna for which the U.S. firms have decided not to compete with Japanese firms because Japanese consumers constitute such a major part of the consumer market?

2. What other reasons may the U.S. firms have had for opting not to compete with Japanese firms for the tuna market? How might U.S. firms competitive decisions now be influenced by the radioactive contamination from Japan's crippled nuclear plant in 2011?

3. Name at least two other products where the relatively non-discriminating taste of American consumers causes manufacturers/producers to sell "lower-status" products to the U.S. market. Is the reverse also true; that is, that a more discriminating taste of American consumers results in higher-quality, higher-priced goods being sold to the U.S. market?

## REFERENCES

1. Michael Gill, *How Starbucks Saved My Life: A Son of Privilege Learns to Live Like Everyone Else* (New York: Gotham Books, 2007), 30–33; "Starbucks Corporation Fiscal 2009 Annual Report," www.starbucks.com (accessed August 2, 2010); Nancy Koehn, "Howard Schultz and Starbucks Coffee Company," *Harvard Business School Case Studies*, 9–801–361 (Boston: Harvard Business School Publishing, 2001), 16; Mariko Sanchanta, "Starbucks Plans Big Expansion in China," *The Wall Street Journal Asia*, April 14, 2010, 1.

2. For an overview, see Ricky W. Griffin and Michael Pustay, *International Business: A Managerial Perspective, 6th ed.* (Upper Saddle River, NJ: Prentice-Hall, 2010).

3. As quoted in Sheryl Hillard Tucker and Kevin D. Thompson, "Will Diversity = Opportunity + Advancement for Blacks?" *Black Enterprise*, November 1990, 60.

4. Doyle Karr, "New Technology to Reduce DuPont Agricultural Seed Production Costs, Increase Product Quality," Pioneer Media Room, at www.pioneer.com (accessed July 17, 2010).

5. Laetitia Dard, "New ILO report says US leads the world in labour productivity, some regions are catching up, most lag behind," International Labour Organization, www.ilo.com (accessed July 19, 2010).

6. Lowell B. Catlett and James D. Libbin, *Risk Management for Agriculture* (Clifton Park, NY: Thomson Delmar Learning, 2007).

7. David D. Van Fleet and Ella W. Van Fleet, *The Violence Volcano: Reducing the Threat of Workplace Violence* (Charlotte, NC: Information Age Publishing, 2010).

8. D. D. Van Fleet and E. W. Van Fleet "Internal Terrorists: The Terrorists Inside Organizations," *Journal of Managerial Psychology*, 21(8) (2006), 763–774; Jeremy M. Weinstein, "A New Threat of Terror in the Western Hemisphere," *SAIS Review* 23 (2003): 1-17; D. D. Van Fleet and E. W. Van Fleet, "Terrorism and the Workplace: Concepts and Recommendations," in *Dysfunctional Behavior in Organizations: Violent and Deviant Behavior, Vol. 23, Part a.*, eds. R. W. Griffin, A. O'Leary-Kelly, and J. Collins (Greenwich, CN: JAI Press, 1998), 165–201.

9. See Magali A. Delmas and Michael W. Toffel, "Organizational Responses to Environmental Demands: Opening the Black Box," *Strategic Management Journal* (2008), 1027–1055.

10. Robert E. Quinn and John Rohrbaugh, "A Spatial Model of Effectiveness Criteria: Towards a Competing Values Approach to Organizational Analysis," *Management Science* (1983), 363–377; Robert E. Quinn, *Beyond Rational Management: Mastering the Paradoxes and Competing Demands of High Performance* (San Francisco: Jossey-Bass, 1988); Robert E. Quinn, et al., *Becoming a Master Manager: A Competency Framework* (New York: John Wiley & Sons, 1990), 2–12.

11. Frederick W. Taylor, *The Principles of Scientific Management* (New York: Harper and Brothers, 1911), 44.

12. James O'Toole, *Vanguard Management* (New York: Doubleday, 1985).

13. John R. Schermerhorn, Jr., *Management for Productivity, 4th ed.* (New York: John Wiley & Sons, 1993), 17.

14. Henry Mintzberg, *The Nature of Managerial Work* (New York: Harper & Row, 1975).

15. James A. F. Stoner and R. Edward Freeman, *Management, 5th ed.* (Englewood Cliffs, NJ: Prentice Hall, 1992), 13.

16. Wendy Zellner, "Peace, Love, and the Bottom Line," *Business Week*, December 7, 1998, 79–82.

17. Various pages of the Oppenheimer Group website at www.oppyproduce.com (accessed July 9, 2010); "The Oppenheimer Group," Bloomberg Businessweek at investing.businessweek.com (accessed July 9, 2010); "The Oppenheimer Group," Hoover's, Inc., at www.hoovers.com (accessed July 9, 2010).

18. See, for example, Larry D. Alexander, "The Effect Level in the Hierarchy and Functional Area Have on the Extent Mintzberg's Roles Are Required by Managerial Jobs," *Academy of Management Proceedings* (San Francisco, 1979), 186–189; Alan W. Lau and Cynthia M. Pavett, "The Nature of Managerial Work: A Comparison of Public and Private Sector Managers," *Group and Organization Studies* (December 1980), 453–466; Cynthia M. Pavett and Alan W. Lau, "Managerial Work: The Influence of Hierarchical Level and Functional Specialty," *Academy of Management Journal* (March 1983), 170–177; and Allen I. Kraut, Patricia R. Pedigo, D. Douglas McKenna, and Marvin D. Dunnette, "The Role of the Manager: What's Really Important in Different Management Jobs," *Academy of Management Executive* (November 1989), 286–293.

19. Pavett and Lau, "Managerial Work: The Influence of Hierarchical Level and Functional Specialty," 170–177.

20. Fred Luthans, Stuart A. Rosenkrantz, and Harry W. Hennessey, "What Do Successful Managers Really Do? An Observation Study of Managerial Activities," *The Journal of Applied Behavioral Science* (1985), 255–270; Fred Luthans, "Successful vs. Effective Real Managers," *Academy of Management Executive* (May 1988),

127–132; Fred Luthans, Richard M. Hodgetts, and Stuart A. Rosenkranz, *Real Managers* (Cambridge, MA: Ballinger Publishing, 1988); Fred Luthans, Diane H. B. Welsh, and Lewis A. Taylor III, "A Descriptive Model of Managerial Effectiveness," *Group and Organization Studies* (June 1988), 148–162; and Gary Yukl, *Skills for Managers and Leaders* (Englewood Cliffs, NJ: Prentice Hall, 1990).

21. Quinn, et al., *Becoming a Master Manager: A Competency Framework*.

22. Ibid., 19–20.

23. Robert L. Katz, "Skills of an Effective Administrator," *Harvard Business Review* (January–February 1955), 33–42.

24. John A. Wagner III and John R. Hollenbeck, *Management of Organizational Behavior* (Englewood Cliffs, NJ: Prentice Hall, 1992), 44.

25. L. Shullman, *Skills of Supervision and Staff Management* (Itasca, IL: F. E. Peacock Publishers, Inc., 1982).

26. Richard N. Morrison, personal correspondence, June 25, 2011; Kelly Mixer, "A Family's Story: Farmers to Developers," *Independent Newspapers*, n.d., at www.morrisonranch.com (accessed June 29, 2010).

27. Donald A. Norman, *Learning and Memory* (San Francisco: W. H. Freeman and Company, 1982); John R. Anderson, *Cognitive Psychology and Its Implications* (San Francisco: W. H. Freeman and Company, 1980); Morton Hunt, *The Universe Within: A New Science Exploring the Human Mind* (New York: Simon and Schuster, 1982); Lyle E. Bourne, et al. *Cognitive Processes, 2nd ed.* (Englewood Cliffs, NJ: Prentice Hall, 1986).

28. P. J. Guglielmino, "Perceptions of the Skills Needed by Mid-Level Managers in the Future and the Implications for Continued Education: A Comparison of the Perceptions of Mid-Level Managers, Professors of Management, and Directors of Training," *Dissertation Abstracts International*, 39(3); 1260A; P. J. Guglielmino and A. B. Carroll, "The Hierarchy of Management Skills: Future Professional Development for Mid-Level Managers," *Management Decision*, 17(4), 341–345; F. C. Mann, "Toward an Understanding of the Leadership Role in Formal Organizations," in *Leadership and Productivity*, eds. R. Dubin, et al. (San Francisco: Chandler Publishing Company, 1965), 68–103.

29. Fred Luthans, Stuart A. Rosenkrantz, and Harry W. Hennessey, "What Do Successful Managers Really Do? An Observation Study of Managerial Activities," *The Journal of Applied Behavioral Science* (1985), 255–270; Fred Luthans, "Successful Vs. Effective Real Managers," *Academy of Management Executive* (May 1988), 127–132; Fred Luthans, Diane H. B. Welsh, and Lewis A. Taylor III, "A Descriptive Model of Managerial Effectiveness," *Group and Organization Studies* (June 1988), 148–162.

30. "The Corporate Elite," special issue of *Business Week*, 1990.

31. Robert W. Mann and Julie M. Staudenmier, "Strategic Shifts in Executive Development," *Training and Development*, July 1991, 37–40.

32. L. W. Porter and L. E. McKibbon, *Future of Management Education and Development: Drift or Thrust Into the 21st Century?* (New York: McGraw-Hill, 1988).

33. "Desperate to Know Where Grads of Procter & Gamble Are Hiding?" *The Wall Street Journal*, August 20, 1986, 15.

34. Sasha Issenberg, *The Sushi Economy: Globalization and the Making of a Modern Delicacy*. (NY: Gotham Books, 2007); "France joins fight to save bluefin tuna," (February 8, 2010) at Foodnavigator.com (accessed August 1, 2010); "EU agrees plan to ban bluefin tuna trade," (May 11, 2010) at Foodnavigator.com (accessed August 1, 2010); Leigh Murray and Piyarat Setthasiriphaiboon, "Thai Union Frozen to Buy MW Brands for $883 million," *The Wall Street Journal* (July 28, 2010) at online.wsj.com (accessed August 2, 2010); "Dongwan Annual Report" at www.dwml.co.kr (accessed August 3, 2010); "Press Room," at www.bumblebee.com (accessed August 2, 2010).

# THE ENVIRONMENT OF AGRIBUSINESS MANAGEMENT

2

# CHAPTER

# 4

# Organizational Environments

## LEARNING OBJECTIVES

After studying this chapter, you should be able to:

- Discuss the nature of the organizational environment.
- Identify and describe the components of the general environment.
- Identify and describe the components of the task environment.
- Discuss the internal environments of organizations.
- Identify and describe how the environment affects organizations and how organizations respond to their environments.

## The Real Thing Is Really for Everybody

"**I**f people knew the good qualities of Coca-Cola as I know them, it would be necessary for us to lock the doors of our factories and have a guard with a shotgun to make the people line up to buy it." As the owner of Coca-Cola Company, Asa Chandler did not lack for exuberance or belief in his product.

**FIGURE 4.1** Coca Cola may need to change its containers, but consumers will not necessarily accept changes in the original formula of its iconic drink.
© Monkey Business Images/www.Shutterstock.com

Chandler was not the inventor, however. Dr. John Stith Pemberton was granted the patent to the formula for Coca-Cola on June 28, 1887. He had hoped to provide an "elixir" that would cure the world's ills and also provide a livelihood for his son, Charley. He also hoped that Charley would take the invention to new heights—but not the one the doctor had achieved. Dr. Pemberton was addicted to morphine.

On March 24, 1888, Asa Chandler, Charley Pemberton, Woolfolk Walker and his sister filed incorporation papers for the Coca-Cola Company in Fulton County, Georgia. This is the "real" beginning of the company and its rise, though not as meteoric as some had expected, to a global "iconic" brand (Figure 4.1). It was the diligent application of agribusiness principles, responding to changing environments, and the desire to be successful that led to the development of this company.

The Coca-Cola story demonstrates that it is not enough to "invent" a good product; it is more important to understand its efficacy and utility. Once a product's efficacy and utility is determined, it can be marketed. Marketing and distribution—that is, taking advantage of economies of scope—then carries a product and company to its destiny. These lessons must be relearned every so often.

It is a long way from the United States to India, but understanding "where you come from" and "where you want to go" does not change. Success over time and over geography requires constant attention and reinvention. Iconic brands are made; they do not just happen. This is why the $31 billion multinational company has struggled to establish and reestablish itself in India, one of the great consumer markets on the planet.

As Coca-Cola learned to compete successfully for its national market in the United States, it must grapple with the complexities of new markets in China and India. Experts believe that Coca-Cola's future success in these countries will be determined in the countryside—where the new consumers are. "Tried and true" marketing axioms do not work; sometimes the discoveries are surprising. "People think Indian consumers want low-priced products. There cannot be a bigger myth. They want good-quality products at a reasonable price," explains Jagdeep Kapoor, chairman and managing director of Samsika Marketing Consultants.

The product attribute that Asa Chandler liked least, bottling and pricing, has emerged as the most significant marketing factor for his product on the other side of the planet. In India, and soon in rural China, the Coca-Cola Company is marketing a 150ml bottle (about six ounces), sold chilled for five Indian rupees (around US$0.11). Meeting these criteria means developing a special cooler for areas with no or very undependable electricity. Five rupees is important, as it is a "psychological price point" for rural consumers who often do not want to "break" a ten rupee bill. This is almost the same problem Asa Chandler had with the nickel drink in the United States: no one wanted to break a dime in the early days of Coke.

It is this willingness to adapt to changing environments that keeps a product global and iconic. In the words of *The Wall Street Journal*, "Thanks to its larger emerging-markets exposure and higher-margin products, Coca-Cola historically has enjoyed a healthy premium to stodgier packaged-food makers…"[1]

## INTRODUCTION

Every organization in the world today—large, small, local, or international—is profoundly affected by its organizational environment. As Coca-Cola has successfully demonstrated, suppliers, customers, and competitors must all be closely monitored and care must be taken to learn what the market really wants and to stay abreast of changing tastes and trends in every phase of the firm's operation. Indeed, managers who fail to keep pace with changes in their environments are doomed to fall behind their competitors and may suffer irreparable damage to their organizations' revenues and profit margins.

This is the first of four chapters devoted to the environmental context of management. We begin by introducing the nature of organizational environments. Next we discuss the external environment in two parts—the general environment of organizations and the task environment. Then we provide an analysis of the internal organization and conclude with an examination of organization-environment relationships.

## The Nature of Organizational Environments

To underscore the importance of organizational environments, consider the situation of a swimmer crossing a wide river, where the organization is like a swimmer and the environment is like a river. The swimmer must first assess the current, obstacles, and distances before setting out. If these elements are properly evaluated, the swimmer will arrive at the expected point on the far bank of the river. If they are not properly understood, the swimmer might end up too far upstream or downstream. The organization's managers must likewise understand the basic elements of its environment to properly maneuver among them.[2]

The **external environment** consists of everything outside an organization that might affect it. Of course, the boundary that separates the organization from its external environment is not always clear and precise. In one sense, for example, stockholders or shareholders are part of the organization, but in another sense they are part of its environment. As shown in Figure 4.2, external environment is composed of two layers: the general environment and task environment. An organization's **internal environment**, also shown in Figure 4.2, is best reflected by its culture.

Of course, not all aspects of the environment are equally important for all organizations. A small, non-union agribusiness may not need to concern itself very

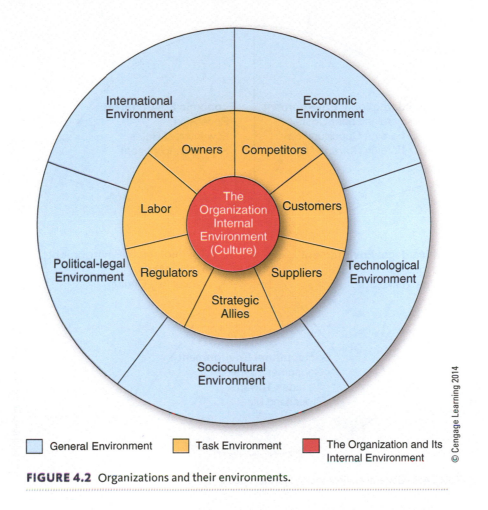

**FIGURE 4.2** Organizations and their environments.

much with unions, for example. A private university with a large endowment (like Harvard) may be less concerned about general economic conditions than might a state university (like the University of Iowa) that depends on state funding from tax revenues. Still, organizations should fully understand which environmental forces are important and how the importance of others may increase. Furthermore, as Coca-Cola has demonstrated and smaller organizations such as Summer Farms and Webb Dairy have learned, acquiring information about the environment takes place all over again as an organization expands its product line or the market to which it sells.[3]

## The General Environment of Organizations

One layer of an organization's external environment is the **general environment**. It consists of the broad dimensions and forces in an organization's surroundings that provide opportunities and impose constraints on the organization. These elements are not necessarily associated with other specific organizations. Weather, for example, affects many different types of organizations. The cost or the availability of fertilizer and gasoline for farm equipment would affect some but not all agribusinesses. The general environment of most organizations is composed of economic, technological, sociocultural, political-legal, and international dimensions. Each dimension of the general environment has the potential to influence the organization in significant ways. The general environment of Hormel Foods is shown in Figure 4.3.

## The Economic Environment

An organization's **economic environment** is the overall health of the economic system in which the organization operates.[4] Particularly important economic factors are inflation, interest rates, unemployment, and demand. During times of inflation, for example, a company pays more for resources and must raise its prices to cover the higher costs. When interest rates are high, consumers are less willing to borrow

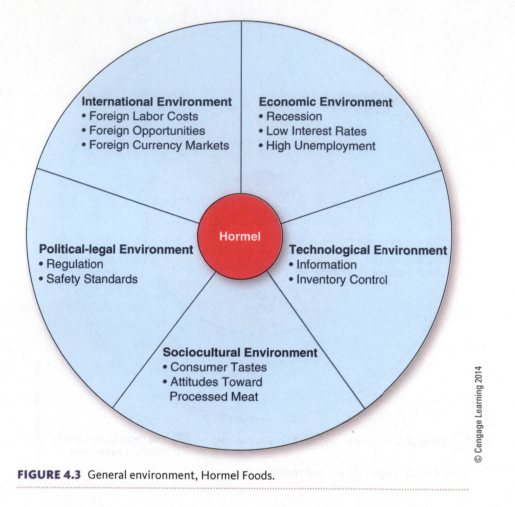

**FIGURE 4.3**  General environment, Hormel Foods.

© Cengage Learning 2014

money and the company itself must pay more when it borrows money. When unemployment is high, the company can be selective about whom it hires, but consumer buying may decline. The economic recession in the United States during the early 1990s and again in the 2000 no doubt contributed to the short-term leveling off of sales at Hormel as well as smaller agribusinesses, such as Green Things Landscaping and Hartco hardwood floors (Figure 4.4). Fortunately for Summer Farms during the same time period and even continuing today, a growing concern nationally about health has had a positive effect on the demand for locally grown foods.

The economic environment is also important to nonbusiness organizations. For example, poor economic conditions affect funding for state universities. Charitable organizations like the Salvation Army are asked to provide greater assistance during bad times, while their incoming contributions simultaneously dwindle. Hospitals are affected by the availability of government grants and the number of charitable cases they must treat without charge.

### FOOD FOR THOUGHT 4.1

Despite the economic downturn, retail sales of pet food were up 4.5 percent in 2009 at about $18 billion and projected to top $21 billion by 2013. However, organic food spending (human food) is forecasted to fall by 7 percent between 2010 and 2015 as budgets tighten further.

## The Technological Environment

The **technological environment** of an organization refers to the methods available for converting resources into products or services. Although technology is applied within the organization, the forms and availability of that technology come from the general environment. Computer-assisted manufacturing and design techniques, for example, allow McDonnell Douglas to simulate the three miles of hydraulic tubing that run through a DC-10. The

© huyangshu/www.Shutterstock.com

**FIGURE 4.4** The availability and price of oil affects the entire chain of agribusiness.

results include decreased warehouse needs, higher quality tube fittings, fewer employees, and significant time savings.[5] New innovations in robotics (e.g., robotic milkers for dairies), computers (e.g., higher powered units with more complex designing software), and other manufacturing techniques (e.g., radio frequency identification or RFID) also have implications for managers.

## The Sociocultural Environment

The **sociocultural environment** of an organization includes the customs, mores, values, and demographic characteristics of the society in which the organization functions. Sociocultural processes are important because they determine the products, services, and standards of conduct that the society is likely to value. In some countries, for example, consumers are willing to pay premium prices for designer clothes. But the same clothes may have little or no market in other countries. Consumer tastes also change over time. Drinking hard liquor and smoking cigarettes are far less acceptable than they were just a few years ago whereas organic foods have been increasing in popularity. And sociocultural factors influence how workers in a society feel about their jobs and organizations.

Appropriate standards of business conduct also vary across cultures. (Global cultures are discussed more fully in Chapter 6.) In the United States accepting bribes and bestowing political favors in return are considered unethical and are illegal as well. In other countries, however, payments to local politicians may be expected in return for a favorable response to common business transactions, such as applications for zoning and operating permits. The ethics of political influence and attitudes in the workforce are only two of the many ways in which culture can affect an organization. The sociocultural environment partially shapes consumer tastes and standards, which in turn affects most agribusinesses at some point.

### FOOD FOR THOUGHT 4.2

Avocado consumption in the United States has doubled as the population of Mexican immigrants has increased. Exports of avocados from Michoacán, Mexico, have risen fivefold since 2004, allowing many Mexican farmers to earn a living without having to migrate north.

## The Political-Legal Environment

An organization's **political-legal environment** includes government regulation of business and the relationship between business and government. It is important for three basic reasons. First, the legal system partially defines what an organization can and cannot do. Although the United States is basically a free market economy, there is still significant regulation of business activity, particularly in agribusiness.[6] Hormel and Tyson Farms, for example, are subject to strict rules about cleanliness and safety during their manufacturing processes; landscapers like Green Things must comply with various codes and obtain permits in most cities.

Second, pro- or anti-business sentiment in government influences business activity. For example, during periods of pro-business sentiment, firms find it easier to compete and have fewer concerns about regulatory issues. On the other hand, during less favorable periods firms may find their competitive strategies more restricted and have fewer opportunities for mergers and acquisitions because of anti-trust concerns. Finally, political stability has ramifications for planning. No company wants to set up shop in another country unless trade relationships with that country are relatively well defined and stable. Hence, U.S. firms are more likely to do business with England and Canada than with Iran and El Salvador. Similar issues are also relevant to assessments of local and state governments. A change in a mayor's or a governor's position can affect many organizations, especially small firms that do business in only one location and are susceptible to deed and zoning restrictions, property and school taxes, and the like.

## The International Environment

A final component of the general environment for many organizations is the **international environment**—forces that extend beyond national boundaries. As we discuss in Chapter 6, multinational firms such as Cargill, Nestlé, and Monsanto clearly affect and are affected by international conditions and markets. Philip Morris, for example, earns more than half of its revenues from outside the United States. Hormel Foods International has investments in the Philippines, Vietnam, Mexico, and Japan.[7] Even smaller firms that do business in only one country may face foreign competition at home, and they may use materials or production equipment imported from abroad. Relatively small feedlot operations may experience decreased demand when a foreign country bans the import of beef from America. On the other hand, an increase in foreign demand for certain types of lumber may result in increased revenue for U.S. lumber exporters but decreased revenue for our furniture makers.

The international dimension also has implications for not-for-profit organizations. For example, the Peace Corps sends representatives to underdeveloped countries. Medical breakthroughs achieved in one country spread rapidly to others, and cultural exchanges of all kinds take place between countries. As a result of advances in transportation and communication technology in the past century, almost no part of the world is cut off from the rest. The international dimension affects virtually every organization.[8]

## The Task Environment of Organizations

Because the impact of the general environment is often ambiguous and gradual in its effects, most organizations focus more on their task environment. Although it is also quite complex, the task environment provides useful information more readily than does the general environment. The **task environment** consists of other specific organizations or groups that are likely to influence an organization. The task environment may include competitors, customers, suppliers, regulators, labor, owners, and strategic allies. A manager can identify environmental factors of specific interest to the organization rather than having to deal with the more abstract dimensions of the general environment.

It may be easier for you to understand this concept if you think about an organization with which you are familiar. Figure 4.5 depicts the task environment of Hormel. As noted earlier, this environment consists of seven dimensions. In the case of Hormel, competitors include Armour Eckrich, Campbell Soup, ConAgra, Heinz, Kraft, Pilgrim's Pride, and Tyson Foods among others. Suppliers include Fieldale Farms, Neyers, and Hydrite Chemical as well as packaging and trucking companies. Hormel pursues strategic alliances with other food companies. Key regulators of Hormel are the United States Department of Agriculture, the Food and Drug Administration, and the Equal Employment Opportunity Commission. Important labor unions include the United Food and Commercial Workers International. Owners include The Vanguard Group, State Street Corporation, officers, and other individual investors. Sometimes the interests of these various groups are not aligned. For instance, several years ago when the management of Ben & Jerry's Homemade Holdings turned down a profitable deal with a Japanese distributor because the company did not have a strong social agenda, shareholders were quite upset.[9]

## Competitors

An organization's **competitors** are other organizations that compete with it for resources. The most obvious resources that competitors vie for are customer dollars. Hormel's competitors were noted above. Reebok, Adidas, and Nike are competitors in the athletic shoe industry. Grocery competitors include A&P, Safeway, and Kroger as well as Walmart and Target. Competition also occurs between substitute products. Thus, Chrysler competes with Yamaha (motorcycles) and Schwinn (bicycles) for your transportation dollars; and Walt Disney, Club Med, and Carnival Cruise Lines compete for your vacation dollars.

Competition is not limited to business firms. Universities compete with trade schools, the military, other universities, and the job market to attract good students. Art galleries compete with each other to attract the best exhibits. Organizations may also compete for different kinds of resources besides consumer dollars. Two totally unrelated organizations may compete to acquire a loan from a bank that has only limited funds to lend. In a large city, the police and fire departments may compete for the same tax dollars. Firms compete for quality labor, technological breakthroughs and patents, and sometimes for scarce raw materials like pesticides or fertilizer, particular types of lumber, and even prescription drugs.

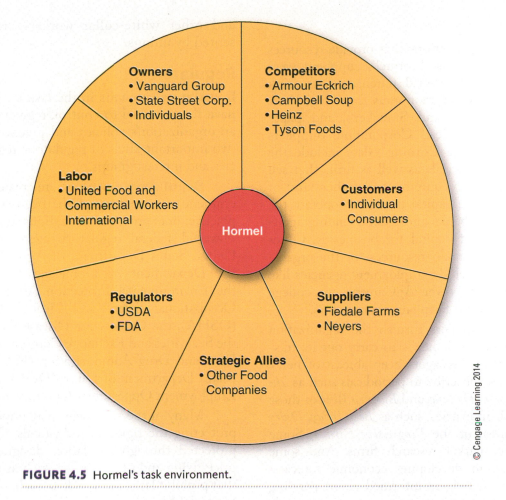

**FIGURE 4.5** Hormel's task environment.

© Cengage Learning 2014

Information about competitors is often quite easily obtained. Grocer Albertson's can monitor Food Lion's prices by reading its newspaper advertisements or by sending someone to a store to inspect price tags. Other kinds of information may be more difficult to obtain. Research activities, new product developments, and future advertising campaigns, for example, are often closely guarded secrets.[10]

## Customers

A second dimension of the task environment consists of **customers**. The customer is whoever pays money to acquire an organization's product or service. In many cases, however, the chain of customer transactions is deceivingly complex. As consumers, for example, we do not buy a bottle of Coke from Coca-Cola. We buy it from Safeway, which bought it from an independent bottler, which bought the syrup and the right to use the name from Coca-Cola. Customers need not be individuals. Schools, hospitals, government agencies, wholesalers, retailers, and manufacturers are just a few of the many kinds of

organizations that may be customers of other organizations. Common sources of information about customers include market research, surveys, consumer panels, and reports from sales representatives.

Dealing with customers has become increasingly complex in recent years. Many firms have found it necessary to focus their advertising on specific consumer groups or regions. General Foods, for example, promotes its Maxwell House coffee differently in different regions of the country, even though doing so costs two or three times what a single national advertising campaign would cost.[11] Pressures from consumer groups about packaging and related issues also complicate the lives of managers.

## FOOD FOR THOUGHT 4.3

More than half of total spending on all organic products in 2010 was made by just 8 percent of households.

## Suppliers

**Suppliers** are organizations that provide resources for other organizations. Hormel's suppliers were noted earlier. Disney World purchases soft-drink syrup from Coca-Cola, monorails from Dae-Woo, food from Sara Lee and Smucker's, and paper products from The Mead Corporation. Suppliers for manufacturers like Corning Glass include the suppliers of raw materials as well as firms that sell machinery and other equipment. Another kind of supplier provides the capital needed to operate the organization. Both banks and federal lending agencies are suppliers of capital for businesses. Other suppliers provide human resources for the organization. Examples include employment agencies like Kelly Services, Manpower, AgCareers.com, college placement offices, and farm labor contractors.

Still other suppliers furnish the organization with the information it needs to carry out its mission. To help their managers keep abreast of news, many companies subscribe to periodicals such as *The Wall Street Journal*, *Fortune*, and *Business Week* or those specific to their industries, such as *Beef Today*, *Dairy Today*, *The Packer*, or the *Progressive Farmer*. Some companies use market research firms. And some firms specialize in developing economic forecasts and in keeping managers informed about pending legislation. Most organizations try to avoid depending exclusively on one particular supplier. A firm that buys all of a certain resource from one supplier may be crippled if the supplier goes out of business or is faced with a strike. Most organizations try to develop and maintain relationships with a variety of suppliers.[12]

## Labor

Organizations must also concern themselves with labor. **Labor** refers to people who work for the organization, especially when they are organized into unions. The National Labor Relations Act of 1935 requires an organization to recognize and bargain with a union if that union has been legally established by the organization's employees. Presently, around 23 percent of the American labor force is represented by unions. Some large firms, such as Archer Daniels Midland, Cargill, and Safeway, must deal with a great many unions. Even when an organization's labor force is not unionized, its managers do not ignore unions even if they seek to avoid them. And even though people think primarily of blue-collar workers as union members, many government employees, teachers, and other white-collar workers are also represented by unions.

## Regulators

**Regulators** are units in the task environment that have the potential to control, regulate, or influence an organization's policies and practices.[13] There are two important kinds of regulators: regulatory agencies and interest groups.

**Regulatory agencies** are created by the government to protect the public from certain business practices or to protect organizations from one another. Powerful federal regulatory agencies include the United States Department of Agriculture (USDA), the Department of Homeland Security (DHS), the Environmental Protection Agency (EPA), the Occupational Safety and Health Administration (OSHA), the Centers for Disease Control (CDC), the Securities and Exchange Commission (SEC), the Food and Drug Administration (FDA), the United States Department of Labor (DOL), and the Equal Employment Opportunity Commission (EEOC).

Many of these agencies play important roles in protecting the rights of individuals. Consumers are protected through regulation designed to prevent price fixing and other abuses, such as antitrust rules that prohibit large meatpackers such as Tyson Foods, Smithfield Foods, and Hormel from buying animals from each other and sets limits on exclusive contracts that they negotiate with suppliers.[14] The FDA helps ensure that the food we eat is free from contaminants.

The costs that a firm incurs in complying with government regulations may be substantial, but these costs are usually passed on to customers. Even so, many organizations complain that there is too much regulation at the present time. One study found that, because of stringent government regulations, 48 major companies spent $2.6 billion in one year over and above normal environmental protection, employee safety, and similar costs. On the basis of these findings, the extra costs of government regulations for all businesses have been estimated at more than $100 billion per year.[15] Although the results of that regulation may benefit society through reducing pollution or protecting the rights of workers, for instance, the impact of regulatory agencies on organizations is considerable.

Federal regulators get a large amount of publicity, but the effects of state and local agencies are also significant. California has more stringent automobile emission requirements than those established by the EPA. And not-for-profit organizations must also deal with regulatory agencies. Most states, for example, have coordinating boards that regulate the operation of colleges and universities. Some of the major regulators in agribusiness are the United States Department of Agriculture, the Food and Drug Administration, and the United States Department of Labor (Figure 4.6).

The other basic form of regulator is the **interest group**. An interest group is organized by its members to attempt to influence organizations. Prominent interest groups include the National Organization for Women (NOW), Mothers Against Drunk Drivers (MADD), the National Rifle Association (NRA), the League of Women Voters, the Sierra Club, Ralph Nader's Center for the Study of Responsive Law, Consumers Union, and industry groups like the Council of Better Business Bureaus, Grocery Manufacturers of America, National Council of Farmer Cooperatives, National Cattlemen's Beef Association, or the National Cotton Council. Interest groups lack the official power of government agencies, but they can exert considerable influence by using the media to call attention to their positions. MADD, for example, puts considerable pressure on alcoholic-beverage producers (to put warning labels on their products), automobile companies (to make it more difficult for intoxicated people to start their cars), local governments (to stiffen drinking ordinances), and bars and restaurants (to limit sales of alcohol to people drinking too much).

**FIGURE 4.6** Governmental regulations require nutrition information be placed on food products.

influencing the management of companies in which they hold stock. This is especially true of owners who hold large blocks of stock.[17]

Another group of owners exerting more influence includes the managers of large corporate pension funds. These enormous funds control large blocks of shares in numerous companies, and, as a result, give their managers substantial power.[18] And given the increased power wielded by owners (and their willingness to use that power), some people fear that managers are sacrificing long-term corporate effectiveness for the sake of short-term results. For example, managers at Carnation were afraid to increase advertising costs too much for fear of attracting the attention of institutional investors. As a result, sales declined. Once Nestlé took over Carnation and loosened the purse strings, sales took off again.[19] Thus, while organizations should never ignore their owners, they have to be considerably more concerned about them now than in the past.

Cooperatives (discussed more fully in Chapter 13) are common in agribusiness. These organizations are nonprofit corporations in which the owners provide benefits to one another. The cooperative may be formed around supply (a purchasing cooperative), demand (a marketing cooperative), or service; but in each of these, the owners control the organization. That control frequently takes the form of democratic voting, thus making the owners a dominant force in cooperatives (Figure 4.7).

The owners of many agribusinesses are family members. This presents advantages but it certainly isn't problem free. Which family member is actually in control? Who gets the last say in day-to-day

### FOOD FOR THOUGHT 4.4

*The U.S. government regulates the size of the holes (3/8 to 13/16 inches for Grade A) in Swiss cheese sold in the United States, thus resulting in inferior flavored Swiss cheese, according to the Swiss who produce non-factory-made cheese with larger holes.*

## Owners

**Owners**—the people, organizations, and institutions that legally control an organization—are also becoming a major concern of managers in many businesses.[16] Stockholders of major corporations are generally happy to sit on the sidelines and let top management run their organizations, but not always. More and more of them are taking active roles in

**FIGURE 4.7** These stockholders expect to hear positive news at their annual meeting.

decisions? Who will assume control if something happens to that decision maker? What happens to ownership and control when a family member marries, divorces, dies, or just leaves the business? These and other issues must be understood and managed in any family-owned business.

## Strategic Allies

A final dimension of the task environment involves **strategic allies**—two or more companies that work together in joint ventures. In 2008 AgroFresh Inc., a subsidiary of Rohm and Haas, and Syngenta AG entered into an exclusive global strategic alliance to develop and commercialize Invinsa™ technology, a unique product for crop stress protection in field crops.[20] Alliances such as these have been around for a long time, but they have become increasingly popular.[21] As a result, the organic industry structure has become quite complex, as indicated in Figure 4.8.[22]

Strategic alliances help companies get from other companies the expertise they may lack. They are formed primarily for the purpose of improving operational efficiency or learning and technology transfer.[23] They also help spread risk. Managers must be careful, however, not to give away sensitive competitive information.

Strategic alliances need not always involve businesses. Arizona State University and the University of Arizona, for example, often work together to secure government grants. And some churches sponsor joint missionary projects.

## The Internal Environment: Corporate Culture

The **culture** of an organization is the set of values that helps its members understand what the organization stands for, how it does things, and what it considers important. Culture is an amorphous concept that defies objective measurement or observation. Nevertheless, because it is the foundation of the organization's internal environment, it plays a major role in shaping managerial behavior.[24]

## The Importance of Culture

Several years ago, executives at Levi Strauss & Co., an agribusiness noted worldwide for its clothing fabric, felt that the company had outgrown its 68-year-old building. Even though everyone enjoyed its casual and relaxed atmosphere, they needed more space. So Levi Strauss moved into a modern office building in downtown San Francisco, where its new headquarters spread over 12 floors in a skyscraper. It quickly became apparent that the change was affecting the corporate culture, and that people did not like it. Executives felt isolated, and other managers missed the informal chance meetings in the halls. Within just a few years, Strauss moved out of the skyscraper and back into a building that fosters informality. For example, there is an adjacent park area where employees converge for lunchtime conversation. Levi Strauss has a culture that is important to everyone who works there.[25]

Culture determines the "feel" of an organization. The stereotypic image of the IBM executive is a physically fit male wearing a white shirt and dark suit. In contrast, Texas Instruments and several major computer-related businesses like to talk about their "shirt-sleeve" culture, in which ties are avoided and few managers wear jackets. Of course, the same culture is not necessarily found throughout an entire organization. For example, the sales and marketing department may have a culture quite different from that of the operations and manufacturing department. Regardless of its nature, however, culture is a powerful force in organizations—one that can shape the firm's overall effectiveness and long-term success. Companies that can develop and maintain a strong culture, such as Archer Daniels Midland, Hewlett-Packard, Southwest Airlines, and Apple Computers, tend to be more effective than companies that have trouble developing and maintaining such a culture.[26]

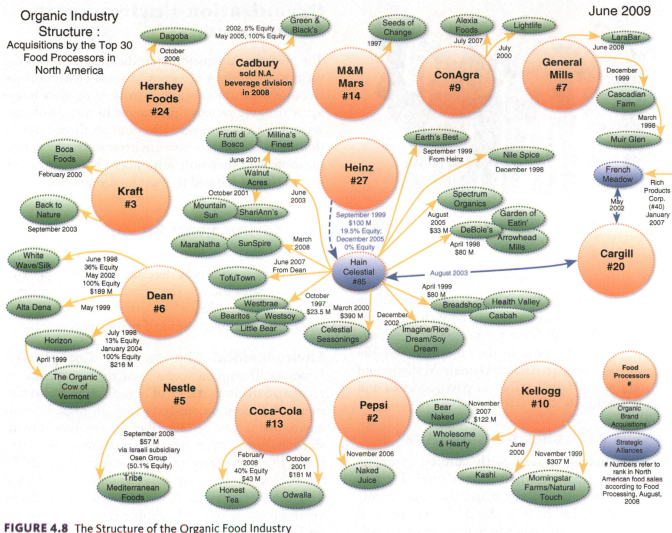

**FIGURE 4.8** The Structure of the Organic Food Industry

Source: Philip H. Howard, Organic Industry Structure, Philip H. Howard, Department of Community, Agriculture, Recreation and Resource Studies, Michigan State University. Reprinted with permission.

## Origins of Culture

Where does a culture come from? Typically, it develops and blossoms over a long period of time. Its starting point is often upper management, or even the organization's founder. For example, James Cash Penney believed in treating employees and customers with respect and dignity. Employees at J. C. Penney are still called associates rather than employees (to reflect partnership), and customer satisfaction is of paramount importance. Small agribusinesses may be unaware that they even have established cultures, but chances are that customers and suppliers have perceived some type of culture (eager to please, tight-fisted, first or last to carry a new product, etc.) based on their dealings with the businesses.

As an organization grows, its culture is modified, shaped, and refined by symbols, stories, heroes, slogans, and ceremonies. For example, a key value at Hewlett-Packard (HP) is the avoidance of bank debt. A popular story still told at the company involves a new project being considered for several years. All objective criteria indicated that HP should incur bank debt to finance it, yet Bill Hewlett and David Packard rejected it out of hand simply because "HP avoids bank debt." This story, involving two corporate heroes and based on a slogan, dictates corporate culture today.[27]

Corporate success and shared experiences also shape culture (Figure 4.9). For example, Archer Daniels Midland has a strong culture derived from its years of success in the seed and food industry.

**FIGURE 4.9** Corporate cultures build from rituals, such as this award ceremony.

At Atari, in contrast, the culture is quite weak. The management team and ownership have changed frequently, making it difficult for people to sense any real direction or purpose in the company. The differences in culture at Archer Daniels Midland and Atari are in part attributable to past successes and shared experiences.

## Managing Organizational Culture

How can managers deal with culture, given its clear importance but intangible nature? The key is for the manager to understand the current culture and then decide if it should be maintained or changed. By understanding the organization's current culture, managers can take appropriate actions. At Hewlett-Packard, the values represented by "the HP way" still exist. Moreover, they guide and direct most significant activities undertaken by the firm. Culture can also be maintained by rewarding and promoting people whose behaviors are consistent with the existing culture and by articulating the culture through slogans, ceremonies, and so forth.

To change culture, managers must have a clear idea of what it is they want to create. Many organizations today are attempting to create a strong culture. One way is to bring outsiders into important managerial positions. The choice of a new CEO from outside the organization is often a clear signal that things will be changing. Adopting new slogans, telling new stories, staging new ceremonies, and breaking with tradition can also alter culture. Methods for changing culture are discussed in Chapter 14.

## Organization-Environment Relationships

The preceding discussion identifies and describes the various dimensions of organizational environments. Because organizations are open systems, they interact with these various dimensions in many different ways. We now turn our attention to these interactions. We first discuss how environments affect organizations and then note a number of ways in which organizations respond to their environments.

## How Environments Affect Organizations

Three basic frameworks describe how environments affect organizations. The first is environmental change and complexity. The other two are competitive forces and environmental turbulence.

### Environmental Change and Complexity.

James D. Thompson was one of the first people to recognize the importance of organizational environments.[28] Thompson suggests that an organization's environment can be described along two dimensions: its degree of change and its degree of homogeneity. The *degree of change* is the extent to which the environment is relatively stable or relatively dynamic. The *degree of homogeneity* is the extent to which the environment is relatively simple (few elements, little segmentation) or relatively complex (many elements, much segmentation). These two dimensions interact to determine the level of uncertainty faced by the organization. Uncertainty, in turn, is a driving force that influences many organizational decisions. Figure 4.10 illustrates a simple view of the four levels of uncertainty defined by different levels of homogeneity and change.

No environment is totally without uncertainty, but the least environmental uncertainty is faced by organizations with simple and stable environments. A beer distributor might be in this environment, unless he or she is carrying only one of two strongly competing brands. Many franchised food operations, such as McDonald's, Baskin-Robbins, Subway and Taco Bell, experience relatively low levels of uncertainty. Taco Bell, for example, focuses on a certain segment of the consumer market that prefers Mexican food, produces a somewhat limited product line, has a constant source of suppliers, and faces comparatively consistent competition.

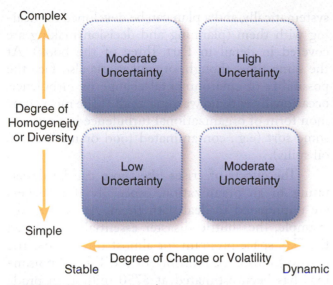

**FIGURE 4.10** Environmental uncertainty.
Source: Based on P. Kreiser and L. Marino, "Analyzing the historical development of the environmental uncertainty construct," *Management Decision* (Vol. 40, No. 9, 2002), 895–905; J. D. Thompson, *Organizations in Action: Social Science Bases of Administrative Theory* (New Brunswick, NJ: Transaction Publishers, 2003 (originally published in 1967)); H. Leblebici and G. R. Salancik, "Effects of Environmental Uncertainty on Information and Decision Processes in Banks," *Administrative Science Quarterly* (Vol. 26, No. 4, 1981), 578–596; and R. B. Duncan, "Characteristics of Organizational Environments and Perceived Environmental Uncertainty," *Administrative Science Quarterly* (Vol. 17, No. 3, 1972), 313–327.

Organizations with simple but dynamic environments normally face a moderate degree of uncertainty. The fashion industry is an example of this environment, as are detergent manufacturing (targeting a certain kind of buyer but aware of fabric content changes) and music production (catering to certain kinds of music buyers but alert to changing tastes in music). However, these relatively simple environments also change quite rapidly as competitors adjust prices and styles, consumer tastes change, and new chemicals become available.

The third combination of factors is one of stability and complexity. Again, a moderate amount of uncertainty results. Hospitals and universities face these basic conditions. Overall, the organization must deal with a myriad of suppliers, regulators, consumer groups, and competitors. Change, however, occurs quite slowly. Despite many changes, for instance, universities still have subject matter departments, libraries, and relatively conventional classrooms.

Finally, very dynamic and complex environmental conditions yield a high degree of uncertainty.

The environment has a large number of elements, and the nature of those elements is constantly changing. Intel, IBM, and other firms in the electronics field, as well as Amgen and Genzyme in the biotechnology field, face these conditions because of the rapid rate of technological innovation and change in consumer markets that characterize their industry, their suppliers, and their competitors. At the production level of agribusiness, weather is the major factor leading to uncertainty. Many agribusiness companies do not integrate backward into farm production because of its highly uncertain nature.

**Five Competitive Forces.** Although Thompson's general classifications are useful and provide some basic insights into organization-environment interactions, in many ways they lack the precision and specificity needed by managers who must deal with their environments on a day-to-day basis. Michael E. Porter, a Harvard professor and expert in strategic management, proposes a more refined way to assess environments. In particular, he suggests that organizations view their environments in terms of the following five **competitive forces** (these are related to an organization's strategy in Chapter 9).[29]

The *threat of new entrants* is the extent to which new competitors can easily enter a market or market segment. It takes a relatively small amount of capital to open a dry-cleaning service or a pizza parlor, but it takes a tremendous investment in plant, equipment, and distribution systems to enter the chemical business. Thus, the threat of new entrants is fairly high for a local pizza parlor but fairly low for Monsanto and DuPont.

*Competitive rivalry* or jockeying among contestants is the nature of the competitive relationship between dominant firms in the industry. In the soft-drink industry, Coke and Pepsi often engage in intense price wars, comparative advertising, and new-product introductions. And grocery companies continually try to outmaneuver each other with coupons and sales. Local car-washing establishments, in contrast, seldom engage in such practices.

The *threat of substitute products* is the extent to which alternative products or services may supplant or diminish the need for existing products or services. The electronic calculator eliminated the need for slide rules. The advent of microcomputers, in turn, has reduced the demand for calculators as well as for typewriters and large mainframe computers. And wireless computers using cloud technology are replacing desktop computers. Another example is Sweet'N Low, Equal, NutraSweet, and Splenda as viable substitute products threatening the sugar industry.

The *power of buyers* is the extent to which buyers of the products or services in an industry have the ability to influence the suppliers. For example, there are relatively few potential buyers for a Boeing 747. Only companies such as American Airlines, United Airlines, and KLM can purchase them; hence, they have considerable influence over the price they are willing to pay, the delivery date for the order, and so forth. On the other hand, Japanese carmakers charged premium prices for their cars in the United States during the late 1970s energy crisis. If the first customer wouldn't pay the price, there were two more consumers waiting in line who would.

The *power of suppliers* is the extent to which suppliers have the ability to influence potential buyers. The local electric company is the only source of electricity in your community; hence, subject to local or state regulation (or both), it can charge what it wants for its product, provide service at its convenience, and so forth. Likewise, even though Boeing has few potential customers, those same customers have few suppliers that can sell them a 300-passenger jet. So Boeing, too, has power. On the other hand, a small vegetable wholesaler has little power in selling to restaurants because if they don't like his or her produce, they can easily find an alternative supplier.

These five forces deal primarily with the task environment. As such they do not explicitly consider the role of government. In agribusiness, governmental regulation is critically important and so might be considered a sixth force. This approach also doesn't recognize that rivals can sometimes be allies rather than competitors. Such strategic alliances (see Chapter 6) can benefit all organizations involved. A further shortcoming of the five forces approach is the lack of specific recognition of environmental turbulence.

**Environmental Turbulence.**   Although always subject to unexpected changes and upheavals, the five competitive forces can be studied and assessed

systematically, and a plan can be developed for dealing with them (planning and decision making are covered in detail in Part Three of the book). At the same time, though, organizations also face the possibility of environmental change or turbulence, occasionally with no warning at all.[30] The most common form of organizational turbulence is a crisis of some sort (e.g., contaminated food or medication, oil spills).

The effects of crises like those can be devastating to an organization, especially if managers are unprepared to deal with them. At NASA, for example, the shuttle disaster essentially paralyzed the U.S. space program for almost three years. The cost to Johnson & Johnson for the Tylenol poisonings has been estimated at $750 million in product recalls and changes in packaging and product design.[31] BP's legal problems arising from the Gulf of Mexico oil spill will not be settled for years. At the time this book goes to press, the future of a Granada, Colorado farm that grew *Listeria*-contaminated cantaloupes leading to several deaths is undetermined.

Such crises affect organizations in different ways, and many organizations are developing crisis plans and teams. When a Delta Air Lines plane crashed in 1988 at the Dallas–Fort Worth airport, for example, fire-fighting equipment was at the scene in minutes. Only a few flights were delayed, and none had to be canceled. In 1987, a grocery store in Boston received a threat that someone had poisoned cans of its Campbell's tomato juice. Within six hours, a crisis team from Campbell's removed two truckloads of juice from all 84 stores in the grocery chain. In 2008 supermarkets reacted extremely fast to word of a suspected outbreak of *Salmonella* in fresh tomatoes.[32] Still, fewer than half of the major companies in the United States have a plan for dealing with major crises.

## How Organizations React to Their Environment

Given the myriad issues, problems, and opportunities in an organization's environments, how should the organization respond? Obviously, each organization must assess its own unique situation and then react according to the wisdom of its senior management.[33] There are, however, six basic ways in which organizations react to their environments: social responsibility (discussed in Chapter 7); information management; strategic response; mergers, takeovers, acquisitions, and alliances; organization design; and direct influence.

### Information Management.

One way organizations respond to the environment is through information management. This is especially important in forming an initial understanding of the environment and in monitoring the environment for signs of change. Organizations use several techniques for managing information. One is defining boundary spanners. A **boundary spanner** is someone like a sales representative or a purchasing agent who spends much of his or her time in contact with others outside the organization. Such people are in a good position to learn what other organizations are doing. All effective managers engage in **environmental scanning**, the process of actively monitoring the environment through observation, reading, and so forth. Within the organization, Bunge, Federal Express, Ford Motor Company, Monsanto, and many other firms have also established elaborate information systems to gather, organize, and summarize relevant information for managers and to assist in summarizing that information in the form most pertinent to each manager's needs (information systems are covered more fully in Chapter 24).

### Strategic Response.

Another way organizations respond to their environment is through a strategic response. The response may involve doing nothing (e.g., if the company feels it is doing well with its current approach), altering its strategy a bit or adopting an entirely new strategy. If the market that a company currently serves is growing rapidly, the firm might decide to invest more heavily in products and services for that market. Likewise, if a market is shrinking or does not provide reasonable possibilities for growth, the company may decide to cut back. For example, when Tenneco's managers decided that oil and gas prices were likely to remain depressed for some time to come, they decided to sell the company's oil and gas business and invest the proceeds in its healthier businesses, like Tenneco Automotive.[34] Many farm owners are waiting for more information before deciding whether to lease their farmland to the builders of windmills to generate clean electricity.

### Mergers, Takeovers, Acquisitions, and Alliances.

A merger occurs when two or more firms combine to form a new firm. For example, Time and Warner merged to create Time-Warner, and two agricultural equipment firms, Case and New Holland, merged to form CNH. A takeover occurs when one firm buys another, sometimes against its will (a hostile takeover). Usually, the firm that is taken over ceases to exist and becomes part of the other company, as when Kraft took over Cadbury. An acquired firm may continue to operate as a subsidiary of the acquiring company, as when Hartco was bought by Bruce Flooring. However, the acquired company may instead be folded into the buyer's existing operations, as when Hartco was later bought by Armstrong from Bruce Flooring and the Hartco name disappeared. In an alliance, as already discussed, the firm undertakes a new venture with another firm. Companies engage in these kinds of strategies for a variety of reasons, including gaining easy entry into new markets or expanding a firm's presence in a current market.

### Organization Design.

Another organizational response to environmental conditions is through structural design. For example, a firm that operates in an environment with relatively low levels of uncertainty might choose to use a design with many basic rules, regulations, and standard operating procedures. Alternatively, a firm that faces a great deal of uncertainty might choose a design with relatively few standard operating procedures, instead allowing managers considerable discretion over how they do things. The former type is characterized by formal and rigid rules and relationships. The latter type is considerably more flexible and permits the organization to respond more quickly to environmental change.[35] We learn much more about these and related issues in Chapter 13.

## Direct Influence of the Environment.

Organizations are not necessarily helpless in the face of their environments.[36] Indeed, many organizations are able to directly influence their environments in a number of different ways. For example, firms may influence their suppliers by signing long-term contracts with fixed prices as a hedge against inflation, which is common in agribusinesses. Or they may become their own suppliers. Sears, for example, owns some of the firms that produce the goods it sells. Du Pont bought Conoco a few years ago partially to ensure a reliable source of petroleum for its chemical operations.

Almost any major activity a firm engages in affects its competitors. When Cargill lowers the prices of its salt products, Morton may be forced to follow suit. When Dole adds a new product, Del Monte is likely to do the same. Organizations may also influence their customers by creating new uses for a product, forcing customers to buy only their patented seeds, finding entirely new customers, and

taking customers away from competitors. Developing new kinds of software, for example, expands the customer base of computer firms. Organizations also influence their customers by convincing them that they need something new (Figure 4.11). Automobile manufacturers use this strategy in their advertising to convince people that they need a new car every two or three years.

Organizations influence their regulators through lobbying and bargaining. Lobbying involves sending a company or industry representative to Washington or to a state capitol in an effort to influence relevant agencies, groups, and committees. For example, the U.S. Chamber of Commerce lobby, the nation's largest business lobby, has an annual budget of more than $100 million. The automobile companies have been successful on several occasions in bargaining with the EPA to extend deadlines for compliance with pollution control and mileage standards.[37] Budweiser tries to influence public opinion and government action through an ongoing

## A FOCUS ON AGRIBUSINESS
### Influencing Government

While government regulates business, business reciprocates and influences government. That influence may be local and low key by organizations, such as the Agri-Business Council of Arizona (ABC). These organizations exist as non-profits and serve to work with and inform governmental bodies of the impact of existing and pending legislation. They help to ensure the vitality and sustainable profitability of the industry and to further ensure that it remains a strong contributor to the overall diversified economic health of the country. Other organizations are actually registered lobbyists.

The diversity of agribusiness means that numerous organizations participate in efforts to influence government. Organizations representing crop producers, livestock and meat producers,

poultry and egg companies, dairy farmers, timber producers, tobacco companies, food manufacturers, and retailers are involved. Those organizations spend more than $100 million annually for lobbying.

In 2009 the Food Marketing Institute spent more than $4 million; the Grocery Manufacturers Association spent more than $3 million; the American Farm Bureau Federation, $2 million; the U.S. Beet Sugar Association and CropLife America, nearly $2 million each; the American Sugar Alliance, the National Pork Producers Council, the Fertilizer Institute, and United Fresh Produce Association, more than $1 million each. In addition to such organizational activities, individuals and political action committees (PACs) related to the agribusiness industry contributed $65 million at the federal level during 2008.[39]

**FIGURE 4.11** The communications industry is known for influencing its competitive environment by continually advertising new products, such as a Smartphone that reads QR codes.

© iStockphoto/franckreporter

series of ads about responsible drinking and driving and Dow's support for the Habitat for Humanity is also an attempt to influence government.[38]

Most bargaining sessions between management and unions are also attempts at mutual influence. Management tries to get the union to accept its contract proposals, and the union tries to get management to sweeten its offer. When a union is not represented in an organization, management usually attempts to keep it out. When Honda opened its first plant in the United States, it helped establish a plant union to head off efforts by the United Auto Workers to set up a branch of its own union in the plant. Corporations influence their owners with information contained in annual reports by meeting with large investors and by pure persuasion. And strategic alliance agreements are almost always negotiated through contracts. Each party tries to get the best deal it can from the other as the final agreement is hammered out.

## CHAPTER SUMMARY

Environmental factors play a major role in determining an organization's success or failure. All organizations have both external and internal environments.

The external environment consists of general and task environment layers. The general environment of an organization is composed of the broad elements of its surroundings that might affect the activities of the organization. The general environment includes the economic, technological, sociocultural, political-legal, and international environments. The effects of these dimensions on the organization are broad and gradual.

The task environment consists of specific dimensions of an organization's surroundings that are very likely to influence the organization. It consists of seven elements: competitors, customers, suppliers, regulators, labor, owners, and strategic allies. Since these dimensions are associated with specific organizations in the environment, their effects are likely to be direct.

The internal environment of an organization is its culture. Managers must understand not only its importance but also how it is determined and how it can be managed.

Organizations and their environments affect each other in several ways. Environmental influence on the organization can occur through uncertainty, competitive forces, or turbulence. Organizations, in turn, use information management, organization design, strategic response, mergers, takeovers, acquisitions, alliances, direct influence, and social responsibility to influence their task environment, and they occasionally try to influence broader elements of their general environment as well.

## CHAPTER ACTIVITIES

### REVIEW QUESTIONS

1. What is an organization's general environment? Identify and discuss each of the major dimensions of the general environment.
2. What is an organization's task environment? What are the major dimensions of that environment?
3. What is an organization's internal environment? How is it formed?
4. Describe the basic ways in which the environment can affect an organization.
5. Describe the basic ways in which an organization can respond to its environment.

### ANALYSIS QUESTIONS

1. Identify five types of companies that are likely concerned about all five dimensions of the general environment.
2. Identify examples of organizations in each dimension of the task environment of your college or university.
3. How are the general and task environments interrelated?
4. Think of an organization with which you have some familiarity. Describe its culture and suggest how that culture was created.
5. What are some recent examples of how the environment has affected specific organizations?

### FILL IN THE BLANKS

1. People who work for an organization, especially when they are organized into unions, are known as _____.
2. Organizations that compete with other organizations for resources are called _____.
3. Two or more companies that work together in joint ventures are known as _____ _____.
4. The set of values that helps an organization's members understand what the organization stands for, how it does things, and what it considers important is referred to as the organization's _____.
5. The process of actively monitoring the environment through observation, reading, and so forth is known as _____ _____.
6. Everything outside an organization that might affect an organization is known as its _____ environment.
7. The overall health of the economic system in which an organization operates is known as the _____ environment.
8. The broad dimensions and forces that provide opportunities and impose constraints on an organization are referred to as the _____ environment.
9. The customs, mores, values, and demographic characteristics of the society in which an organization functions is known as the _____ environment.
10. The methods available for converting resources into products or services is called the _____ environment.

## CHAPTER 4 CASE STUDY
### DO "ROCKET SCIENTISTS" SHOVEL MANURE?

"They are financial weapons of mass destruction," Warren Buffett wrote in his popular annual "Berkshire Hathaway Letter from the Chairman," in 2003. Mr. Buffett was referring to *financial derivatives*.

Michael Phillips has noted that, "Derivatives are financial instruments whose value 'derives' from something else, such as interest rates or heating oil prices. The first derivatives were crop futures, which appeared in the United States at the end of the Civil War and became a standard facet of business for companies across America."

So derivatives did not start out in the domain of the financial "rocket scientists," Wall Street parlance for the mathematicians and physicists hired by financial firms to develop newer forms of derivatives. They began as simple hedges—a means for the farmer to gain some protection for the price of his crop—sold on the Chicago Board of Trade. They were called *futures contracts*.

The farmer or rancher has always had to live with the possibility of the ravages of nature. Hailstorms could destroy a crop, as could a seasonal drought or a torrential downpour if it occurred at harvest time. A heavy snowstorm could prevent a rancher from reaching his cattle with feed and the cattle could perish, or a tornado could wipe out the family homestead, livestock, and equipment in seconds. These "acts of nature" are risks to be expected, feared, and tolerated in this sector of agribusiness. Little can be done to prevent them.

However, this is not the case for the price of the farmer's crop. He can exert some control over this aspect of his agribusiness. The successful and savvy farmer or rancher can use the Chicago Board of Trade futures market. Using the mechanism provided by this "securities market" a farmer can sell a futures contract on his crop. This "locks in" a price at a future date—harvest time. The futures contracts were sold in units that were already familiar to the farmer (bushels) and also at a relevant time (he knew when his crop would be harvested).

If the futures price goes down before harvest, the farmer can sell the crop for that price, buy a futures contract (to close out his position) at that price, and after settling with the broker realize a "profit" from the futures transaction. The "profit" gained and the price of the crop approximate the price he wanted in the first place. Similarly, if the price of the crop increases, he would have to pay a "margin" call to the broker, sell his crop for the higher price, and again realize the price he wanted initially, although he would miss out on the gains. The result of this venture is a price for his crop that the farmer feels is equitable.

The farmer/rancher engaged in this business does so to "lock in" a price and take control or deliver a crop. Another participant in this activity at the Chicago Board of Trade is the *speculative investor*. This individual attempts to "guess" the direction of prices, then buys and sells contracts accordingly. In times of heavy financial strain these individuals (traders) are often vilified for the speculative ventures; however, they serve a purpose. These traders provide "liquidity" because, if there are more individuals and companies willing to buy and sell futures contracts, it insures that any trader can find a "counter party" to his or her trade.

This is the reason Jim Kreutz, Giltner, NE, uses derivatives to soften the effect of a decline in price of feed corn (5,000 bushels/contract, Chicago Board of Trade) used to feed cattle, not people; and his brother-in-law, Jon Reeson, hedges the price of his feed lot steers (40,000 pounds/contract, Chicago Mercantile Exchange). Similarly, their local "farmers co-op" (a company owned "cooperatively" by farmers who invest and share the profits of the cooperative together) hedges the price of diesel fuel.

Futures contracts as sold on the Chicago Board of Trade have a distinct and valid purpose in agribusiness: They help mitigate environmental risks. These financial instruments are used by small independent farmers/ranchers and huge multi-national agribusinesses, such as Archer-Daniels Midland Company and

Cargill, as an everyday part of their agribusiness ventures. Hedging activities through derivatives has created a $300 trillion ($300,000,000,000,000) U.S. derivatives market.

Risk management is an essential feature in any business; it is vital to the viability of an agribusiness. Thus, it is unfortunate that the newer, distant cousins of these financial instruments—the ones developed by Wall Street's "rocket scientists"—are tainting a risk-management tool vital to agribusiness to function successfully in its risky environment.[40]

### ▶ Case Study Questions

1. Why does Warren Buffett say that Wall Street's "rocket scientists" give financial derivatives a bad name? (Hint: Consult the Internet and publications such as *The Wall Street Journal, Business Week*, and *Fortune*.)

2. Explain how futures contracts are risk-management tools?

3. What career choices and company choices are available to someone who wants to work with futures contracts? What types of courses would be appropriate to develop the knowledge and skills needed in these jobs? (Hint: Consult career information on the Internet or through your university's Career Placement Services.)

## REFERENCES

1. Mark Pendergrast, *For God, Country and Coca-Cola: The Unauthorized History of the Great American Soft Drink and the Company That Makes I* (New York: MacMillan Publishing, 1993), 50; "Coca-Cola India: Winning Hearts and Taste Buds in the Hinterland," *The Wall Street Journal*, May 7, 2010, at online.wsj.com (accessed July 30, 2010); John Jannarone, "Coke Still the Real Thing," *The Wall Street Journal*, July 22, 2010, C6.

2. Daniel A. Levinthal, "Organizational Adaptation and Environmental Selection—Interrelated Processes of Change," *Organization Science* (February 1991), 140–151.

3. Danny Miller, "Environmental Fit Versus Internal Fit," *Organization Science* (May 1992), 159–173.

4. For an overview of current thinking about linkages between economics and organizations, see *Organizational Economics*, eds. Jay B. Barney and William G. Ouchi (San Francisco: Jossey-Bass, 1986).

5. Robert H. Hayes and Ramchandran Jaikumar, "Manufacturing's Crisis: New Technologies, Obsolete Organizations," *Harvard Business Review* (September–October 1988): 77–85.

6. Andrew Schmitz, Charles B. Moss, Troy G. Schmitz, Hartley W. Furtan, and H. Carole Schmitz (eds.), *Agricultural Policy, Agribusiness, and Rent-Seeking Behaviour, 2nd ed.* (North York, Ontario: University of Toronto Press, 2010); "Regulation Rises Again," *Business Week*, June 26, 1989, 58–59.

7. "About Hormel Foods," Hormel Foods website at www.hormelfoods.com (accessed July 20, 2010).

8. Philip M. Rosenzweig and Jitendra V. Singh, "Organizational Environments and the Multinational Enterprise," *Academy of Management Journal* (June 1991), 340–361.

9. "Yo, Ben! Yo, Jerry! It's Just Ice Cream," *Fortune*, April 28, 1997, 374.

10. P. Rajan Varadarajan, Terry Clark, and William M. Pride, "Controlling the Uncontrollable: Managing Your Market Economy," *Sloan Management Review* (Winter 1992), 39–50; Ming-Jer Chen and Ian C. MacMillan, "Nonresponse and Delayed Response to Competitive Moves: The Roles of Competitor Dependence and Action Irreversibility," *Academy of Management Journal* (September 1992), 539–570.

11. "National Firms Find That Selling to Local Tastes Is Costly, Complex," *The Wall Street Journal*, July 9, 1987, 17. See also Regis McKenna, "Marketing in an Age of Diversity," *Harvard Business Review* (September–October 1988), 88–95.

12. Susan Helper, "How Much Has Really Changed Between U.S. Automakers and Their Suppliers?" *Sloan Management Review* (Summer 1991), 15–28.

13. "Political pendulum Swings Toward Stricter Regulation," *The Wall Street Journal*, March 24, 2008, A1, A11.

14. Alan Bjerga and Whitney McFerron, June 18, 2010, "Tyson, Meat-Company Pricing Power Target of New Rulesk," *Business Week* at www.businessweek.com (accessed July 7, 2010).

15. "Many Businesses Blame Governmental Policies for Productivity Lag," *The Wall Street Journal*, October 28, 1980, 1, 22.

16. Grant T. Savage, Timothy W. Nix, Carlton J. Whitehead, and John D. Blair, "Strategies for Assessing and Managing Organizational Culture," *The Academy of Management Executive* (May 1991), 61–75.

17. William E. Gillis and James G. Combs "How Much Is too Much? Board of Director Responses to Shareholder Concerns About CEO Stock Options," *Academy of Management Perspectives* (May 2006), 70–72. See also "The Best and Worst Boards," *Business Week*, (October 7, 2002), 104–114.

18. Joann S. Lublin and Kris Maher, "Massey Directors Opposed By Funds," *The Wall Street Journal*, May 12, 2010, at online.wsj.com (accessed July 20, 2010); Heidi N. Moore"Wall Street's Great Enablers: Pension Funds and Endowments," *Fortune* July 8, 2010, at wallstreet.blogs.fortune.cnn.com (accessed July 18, 2010).

19. John J. Curran, "Companies That Rob the Future," *Fortune*, July 4, 1988, 84–89.

20. "Stress Relief," *Farm Industry News*, March 1, 2008, at farmindustrynews.com (accessed July 20, 2010).

21. "More Competitors Turn to Cooperation," *The Wall Street Journal*, June 23, 1989, B1.

22. Phil Howard, "Consolidation in Food and Agriculture: Implications for Farmers & Consumers," *The Natural Farmer* (Spring 2006), 17–20.

23. Thomas L. Sporleder, "Strategic Alliances and Networks in Supply Chains," *Quantifying the Agri-Food Supply Chain*, eds. C.J.M. Ondersteijn, et al. (Dordrecht, The Netherlands: Springer, 2006), 159–169.

24. Terrence E. Deal and Allan A. Kennedy, *Corporate Cultures: The Rights and Rituals of Corporate Life* (Reading, MA: Addison-Wesley, 1982).

25. Gurney Breckenfield, "The Odyssey of Levi Strauss," *Fortune*, March 22, 1982, 110–124. See also "Levi Strauss …at $3 Billion Plus," *Daily News Record*, October 10, 1988, 44; John P. Kotler and James L. Heskett, *Corporate Culture and Performance* (New York: Macmillan, 1992).

26. G. Sadri and B. Lees "Developing Corporate Culture as a Competitive Advantage," *Journal of Management Development* (2001), 853; Jay B. Barney, "Organizational Culture: Can It Be a Source of Sustained Competitive Advantage?" *Academy of Management Review* (July 1986): 656-665.

27. "Hewlitt-Packard's Whip-Crackers," *Fortune*, February 13, 1989, 58–59.

28. James D. Thompson, *Organizations in Action* (New York: McGraw-Hill, 1967).

29. Michael E. Porter, *Competitive Strategy: Techniques for Analyzing Industries and Competitors* (New York: Free Press, 1980).

30. "Are You Ready for Disaster?" *MAPI Economic Report*, Washington DC, 1990.

31. Ian I. Mitroff, Paul Shrivastava, and Firdaus E. Udwadia, "Effective Crisis Management," *The Academy of Management Executive* (August 1987), 283–292.

32. Anne Holcomb, "Tomato Scare Gets Fast Response from Supermarkets, Restaurant Chains," June 11, 2008, Kalamazoo News Archive at blog.mlive.com (accessed July 19, 2010); "Getting Business to Think About the Unthinkable," *Business Week*, June 24, 1991, 104–107.

33. For recent discussions of how these processes work, see Barbara W. Keats and Michael A. Hitt, "A Causal Model of Linkages Among Environmental Dimensions, Macro Organizational Characteristics, and Performance," *Academy of Management Journal* (September 1988), 570-598; and Danny Miller, "The Structural and Environmental Correlates of Business Strategy," *Strategic Management Journal* (Vol. 8, 1987), 55–76.

34. "Why the Street Isn't Moved by Tenneco's Big Move," *Business Week*, September 26, 1988, 130–133.

35. Tom Burns and G. M. Stalker, *The Management of Innovation* (London: Tavistock, 1961).

36. Keats and Hitt, "A Causal Model of Linkages Among Environmental Dimensions, Macro Organizational Characteristics, and Performance."

37. David B. Yoffie, "How an Industry Builds Political Advantage," *Harvard Business Review* (May–June 1988), 82–89.

38. Anita Grace Day, "Public Affairs Advertising: Corporate Influence, Public Opinion and Vote Intentions Under the Third-Person Effect." Louisiana State University Dissertation in Mass Communication (2006), 2.

39. The Agri-Business Council of Arizona, Inc. Website at www.agribusinessarizona.org (accessed on July 5, 2010); "Farm, Agribusiness Lobbying Costs Top $100 Million in 2009," *Agri-Pulse*, February 17, 2010, Vol.6, No.7, at www.northamericandevon.com (accessed on July 5, 2010); and Steve Spires, "Background," OpenSecrets.org, June 13, 2010, at www.opensecrets.org (accessed on July 5, 2010).

40. David Segal, "In Letter, Warren Buffett Concedes Tough Year," *The New York Times*, February 28, 2009, A16; Warren Buffett, "Berkshire Hathaway Letter from the Chairman" Berkshire Hathaway website (2002) at www.berkshirehathaway.com (accessed July 29, 2010); Michael Phillips, "Finance Overhaul Casts Long Shadow on the Plains," *The Wall Street Journal* (July 13, 2010) at online.wsj.com (accessed July 29, 2010); Michael Crittenden and Victoria McGrane, "How the CFTC Got Power," *The Wall Street Journal* (July 15, 2010) at online.wsj.com (accessed July 28, 2010).

# CHAPTER
# 5

## The Competitive Environment

### LEARNING OBJECTIVES

After studying this chapter, you should be able to:

- Characterize the changing environment of management.
- Identify and describe the basic economic challenges of managers today.
- Identify and describe the basic competitive challenges managers face today.
- Identify and describe the basic workforce diversity challenges managers face today.
- Identify and describe the basic workplace challenges managers face today.
- Summarize the legal and social challenges managers face today.

## The Invisible Agribusiness

**"W**ho is this guy Penney—part-owner of some little store in a hick town that nobody's ever heard of—to tell us, the heads of Sears and Montgomery Ward, how to run the largest retail companies in the world?"

**FIGURE 5.1** Department stores, such as Walmart and Target, entered the world of agribusiness by adding groceries to their inventories.
© iStockphoto/Kyoungil Jeon

This conversation was imagined by Robert Sowell, a Stanford economist, to illustrate the impact of James Cash Penney on retail competition when Penney moved from mail-order retailing to build a nationwide chain of stores that delivered merchandise to consumers at a lower price. Although usually thought of as only a retail store, in a broad sense J.C. Penney's is an agribusiness because it involves the use of fiber products.

Sowell's point was simple: Business models are necessary; however, a successful model does not eliminate the need to change with economic forces, particularly competitive forces. What J.C. Penney saw that Sears and Montgomery Ward did not see was that the consumer in the 1920s–1930s was changing, the market was changing, and demographics favored a younger population. The big retailers were large enough to recover from their lack of vision and maintain their market share against J.C. Penney.

But that is not the lesson of this vignette. The real lesson was learned by a J.C. Penney clerk working in Arkansas. The clerk was more aware of the meaning of these changes than the CEOs of the large retailers. That clerk was Sam Walton.

In 1962 Sam Walton and his brother, Bud, opened a small retail store in Rogers, Arkansas. They applied the principles of the retail business trade model that Sam had learned at the J.C. Penney store; but rather than compete against the giant retailers in urban areas, they provided their service in a rural area. Their service was big store retailing at small store service and price. The concept was successful; and on October 31, 1969, Walmart Stores, Inc. was founded. The rest is history at its fullest.

Slowly Walmart Stores, Inc. grew into a major retailing force. And while the company continued to employ the "JCP" business model in rural areas, the learning and growth did not stop with this successful model. As it began to compete in foreign markets, Walmart learned to employ a model developed by Carrefour, a French retailing chain that invented the hypermarket. Carrefour began using the distribution and supply chain system not only to deliver retail goods at low prices (like others), but also to expand into food sales (Figure 5.1).

This simple and obvious addition of food retailing to its fiber connection made Walmart's ties to agribusiness classification more obvious and it became a key to Walmart's success. Following the model ever more closely, the company also added drugstore sales. If the concept had been invented in 1969, we would say they employed a "blue ocean strategy" (BOS). An organization pursuing a BOS attempts to simultaneously differentiate itself from competitors and to operate at low costs. So the purpose of a BOS is not so much to beat the competition in the existing industry, but rather to create a whole new market or a blue ocean, thereby making the competition irrelevant.

The "clerk" learned his lessons well. Walmart serves around 200 million customers per week. In 2010, it generated sales of $408 billion and employed 2.1 million people. Yet in some ways, Walmart remains invisible. For example, when the U.S. government issues its cost of living statistics, Walmart's food sales are not included because the company is recognized as a "general retailer," not a "food retailer." Yet Walmart is the No. 1 food retailer, with the Kroger Co. of Cincinnati only a distant second. Furthermore, as far back as 2003, Walmart controlled around 15 percent of domestic sales of general merchandise and food, and closer to 30 percent in some categories such as household staples and basic apparel (agribusinesses). Major agribusiness firms such as Del Monte, Sara Lee, and Nabisco do 15–30 percent of their annual business with Walmart.

And Walmart has no plans to decrease its presence in agribusiness sales. The company is still determined to keep groceries a priority because they are such a powerful draw for other store items and because food goes up as a percent of sales in a recession.

Why does Walmart remain invisible as an agribusiness? How long will this oversight remain, and how significant is it?[1]

## INTRODUCTION

Walmart managers, like managers in hundreds of various types of agribusinesses, have been very successful in confronting the basic competitive environment faced by all managers. Among other hurdles, they have dealt with changing economic conditions, numerous competitive challenges, and a number of fundamental workplace challenges. The ability to deal with these challenges has been, and will continue to be, a primary ingredient in the effectiveness of this organization as well as smaller agribusinesses, such as Summer Farms or Green Things Landscaping.

These issues and challenges are the focus of this chapter. Many of them are detailed extensions of some of the basic ideas introduced in Chapter 4. First, we will characterize the changing competitive environment facing managers. Then we will introduce and explore particular economic, competitive, workplace, and legal and social challenges (and opportunities) that organizations and their managers must address.

Of ConAgra's 13 million packages of food products totaling $12+ billion in annual sales, about 16 percent go to Walmart.

## The Changing Environment of Managers

Although small businesses have existed for centuries, most big companies have been around for less than two centuries. In the grand scope of human history, then, professional managers and large business organizations are relative newcomers, especially in agribusiness. But consider the vast changes that they have had to confront in their brief lifetimes. Electric motors, automobiles, airplanes, and telephones have each been around for less than 150 years.[2] Early industrialists like Vanderbilt and Carnegie could not take a transcontinental flight from New York to San Francisco. Messages were sent by telegraph. Workers did not have sophisticated equipment to perform their jobs and they worked long hours, often with little pay.

Now consider the tools and equipment that modern managers have come to rely on. Global positioning systems (GPSs), personal computers (including email and the Internet), microwave appliances, cellular phones, copiers, overnight delivery services, facsimile machines, industrial robots, automatic machines to do numerous chores, and many other things that have become commonplace in organizations have been around for only a brief period of time (Figure 5.2).

To bring the impact of these changes into sharper focus, consider the changes managers face in a single organization such as Ford Motor Company. When Henry Ford first started his company in 1903, he did not have to contend with governmental regulation or organized labor because there were few regulations and no autoworkers unions. He had a handful of domestic competitors, but demand for his automobiles was so great that quality was not a major issue. Indeed, most early Ford automobiles were not of particularly high quality. The firm made a single make of car—the Model T—from 1908 until 1927. Ford himself also owned the company so he did not have to concern himself with hostile stockholders. And because his firm was such a large

**FIGURE 5.2** Like all businesses, agribusinesses have had to invest in and adapt to major innovations. This high-tech farmer is in touch with the world while working the fields.

purchaser of raw materials, he could virtually dictate when the prices he would pay, when his suppliers would deliver materials, and so forth.

Over the years, Ford's situation has changed dramatically and in every way imaginable. The company now has operations around the world and must contend with a myriad of domestic and international competitors. The United Auto Workers is a major union that bargains aggressively for wages, benefits, and worker rights. Various governmental agencies regulate and control many of Ford's activities in areas ranging from pollution to hiring practices and consumer safety. Quality is of paramount importance. Suppliers often have several large customers, thus limiting Ford's importance to them. Consumer tastes and demands continue to change, necessitating new models and new features on a regular basis. Today, Ford makes more than 20 different models, each of which may be redesigned every year or so. The price of gasoline dramatically affects demand for different models—when gas prices go up, so too does demand for small, fuel-efficient cars like the Escort and hybrids like the Fusion; but when the price of gas goes down or stabilizes, demand jumps for larger cars like the LTD, the Lincoln series, or SUVs. In short, then, the environment Ford managers must cope with is vastly more complex and more prone to change today than it was just a few decades ago.

In a sense, today's managers can view the various forces and changes that affect them as challenges. If competing for resources were easy, all organizations would be successful; none would ever fail. Thus, confronting major economic, competitive, workplace, and legal and social changes are indeed challenging. Managers who do not deal effectively with them suffer and perhaps fall by the wayside.[3] At the same time, we should also remember that a natural complement to challenge is opportunity. Just as failing to meet a challenge can have negative consequences, overcoming it or meeting it successfully can lead to increased profitability, effectiveness, and reputation.

---

**FOOD FOR THOUGHT 5.2**

"With over 50 foreign cars already on sale here, the Japanese auto industry isn't likely to carve out a big slice of the U.S. market." [From *Business Week,* August 2, 1968]

---

# Economic Challenges of Managers

**Economic challenges** of managers consist of various forces and dynamics associated with the economic systems within which their organizations compete. Thus, they reflect the economic environment of organizations, as described in Chapter 4. The real economic world of managers is fraught with perils and opportunities, challenges, and payoffs—far more complex than the simple supply and demand curves that students learn to interpret. General economic factors such as inflation rates, levels of unemployment, interest rates, budget deficits or surpluses, and the international balance of trade are all major dimensions of an economic system that affects managers. Four somewhat more specific areas that have become particularly significant are entrepreneurship, downsizing and cutbacks, the emerging service sector, and corporate ownership.

## The Trend Toward Downsizing

Over the past several years many firms have been forced to go through a period of downsizing and/or cutbacks. **Downsizing** is a planned reduction in organizational size (e.g., number of employees, number of businesses, number of markets served, size of product line). **Cutbacks** are reductions in the scope of an organization's operations (i.e., operating budgets, travel expenses, research and development, expansion plans). For years managers of U.S. firms in particular had so much demand

for their products and services that they had few concerns about costs. Likewise, quality was of only token importance (a Ford had to be only as good as a Chevrolet, not foreign brands), and prices could be raised as necessary with consideration for only a handful of competitors. This atmosphere changed dramatically, however, when global competition became more widespread and productivity and quality took on greater importance as a competitive advantage. At that point many organizations began to experience loss of market share and declining income.

Many firms have found themselves with excessive payroll costs (too many employees), excess capacity (too many offices and production facilities), and technology that was far too inefficient and outdated to rise to the competitive challenge. In response, they've closed plants, refurbished plants, and slashed payrolls. In the process they've also eliminated hundreds of thousands of jobs. For example, Circuit City alone accounted for 34,000 layoffs when it closed in 2009. Cargill laid off 7 percent of its workforce at five U.S. beef plants.[4] Agribusiness giant J.R. Simplot in Grant County, Washington, laid off 600 employees.[5] The situation has not improved in 2012 as this book is being written. Both London-based bank HSBC and Bank of America, for example, have begun eliminating up to 30,000 workers or 10 percent of their workforce by 2013. Although downsizing has helped many businesses regain competitive positions, it has also undermined worker confidence in the traditional job security they have long expected from large organizations (Figure 5.3).

**FIGURE 5.3** Both large and small businesses, retailing and manufacturing, find it difficult to survive in tough economic times.

© iStockphoto/Wendell and Carolyn

## Entrepreneurship and New Careers

People decide to engage in **entrepreneurship** and choose to go into business for themselves for a variety of reasons. Many want the freedom of setting their own goals and objectives. Others want the challenge of creating something new—their own product or firm, for example. Still others simply have attractive opportunities in family-owned businesses, such as Summer Farms and Webb Dairy, that may seem safer or more lucrative than working for big corporations. In addition, many former employees of large businesses that have cut back on their payrolls (as discussed above) start their own businesses or go to work in other small businesses.

Small businesses play a vital role in the U.S. economy and in the economies of most industrialized nations around the world. For example, companies with fewer than 500 employees were responsible for the creation of 64 percent of net new U.S. jobs from 1992–2010.[6] And in many Eastern European countries just opening their markets to a free enterprise-based system, small businesses are also expected to play a major role in economic growth and development.

Unfortunately, many new businesses fail within a relatively short period of time. While there are sometimes significant consequences of failing (financial setback, social censure, damage to reputation, and so forth), it is also true that many successful entrepreneurs fail one or more times before they finally succeed (Figure 5.4). Henry J. Heinz went bankrupt before he finally succeeded, as did Phineas Taylor Barnum. Milton Snavely Hershey failed four times before he was successful.[7]

**FIGURE 5.4** This entrepreneur chose to start his own computer service business rather than work in a larger company.

© Monika Wisniewska/www.Shutterstock.com

Many other small businesses thrive from the very beginning, however. They establish a comfortable niche for themselves—serving a well-defined market—and allow the owner-entrepreneur to remain in business and set her or his own course. These businesses may remain small or grow slowly over a period of years. Finally, a few businesses take off and become large businesses within just a few years. For example, Noodles & Company grew from its start in 1995 to more than 4,000 employees by 2009. Hy-Vee slowly grew from its founding in 1930 to a $3 billion company in 1998 and then spurted to more than $6 billion by 2009.[8]

On the other hand, a landscape firm that we are familiar with (not Green Things) grew rapidly before deciding to remain small. The owner was blissfully happy playing the role of designer without having much "managing" to do until his wife convinced him to allow her to add a plant nursery, which she would manage. The nursery grew rapidly because of the designer's reputation, but his manager wife did not have an ideal personality for either retailing or overseeing crews. Soon he was so busy that designing took a back seat to managing the nursery and overseeing planting crews, leaving him with both money problems and stomach ulcers. Aware of his own desire to offer only the best, the designer turned down the idea of running only the design business and hiring horticulture managers for everything else. He sold the nursery, returned to landscape designing, and got rid of both his wife and his ulcers.

## The Service Sector

For many decades, U.S. industry was based primarily on manufacturing. The manufacturing sector is composed of firms such as automobile companies, steel mills, oil refineries, computer makers, and agribusinesses such as fiber and fabric manufacturers, vegetable oil processors, and lumber dealers—all of which make tangible products that are sold for profit. In recent decades, however, the service sector has taken on increased importance. **Services** provide utility for consumers, as opposed to providing a tangible product. Examples of service firms range from small architectural or accounting partnerships to neighborhood beauty shops, aerobics studios, and pizza parlors to giant service firms like AT&T (2010 revenues of $123 billion), Walmart Stores (2010 revenues of $408 billion), and UPS (2010 revenues of $45 billion).[9]

In 1947 the service sector accounted for less than half of America's gross national product (GNP). By 1975, however, its proportion had grown to 65 percent; and by 1985 it had climbed to more than 70 percent. By 2006 the service sector accounted for more than 80 percent of the GNP.[10] As shown in Figure 5.5, in 2010 more than 100 million workers were employed in service-producing industries while fewer than 20 million workers were employed in goods-producing industries.

## Ownership Challenges

A final economic challenge we will discuss relates to **corporate ownership**. Corporations sell stock to investors, who then own a share of the business. Until the last few decades, each corporation had so many owners (individual investors) that no single one of them could exert much influence over the firm. Two significant changes, however, have led to significant shifts in how managers respond to owners.

The first change in ownership patterns involves the emergence of the mutual fund market and the growth of institutional investing. Mutual funds are collections of stocks handled by professional fund managers. Individual investors who put their money into mutual funds gain the expertise of a financial professional in deciding what stocks to buy and sell. That is, the fund itself buys and sells stocks, passing along dividends and resale gains to the investors. Since mutual funds may have the resources to

buy and sell hundreds of millions of dollars in stock, they have much more power than do the individual investors themselves. At the same time, managers who handle investments for institutions (universities, for example) and retirement funds can also control large blocks of stock. However, since these managers themselves are responsible for showing a return to their investors, there has been some concern that managers may feel pressured to focus too much on short-term returns and not enough on long-term strength.[11]

Another ownership challenge has been the growth in hostile takeovers. A **takeover** occurs when one corporation or group of investors buys or trades for enough stock in a company to gain control over it. Such a takeover is considered a **hostile takeover** when the target company does not wish to be taken over. The acquiring firm may want the target firm as a way of entering a new market, assuring dependability of a potentially low supply or a tight market, buttressing existing positions, or for other reasons. After it is acquired, the target firm may be left alone to continue doing what it was already doing, integrated into the acquiring firm, or have its assets sold individually for profit.

Another ownership challenge arises in larger corporations where important decisions are made by managers who are not owners. The managers (agents) act for the owners (principals) but may have different goals or motivations. One way to address

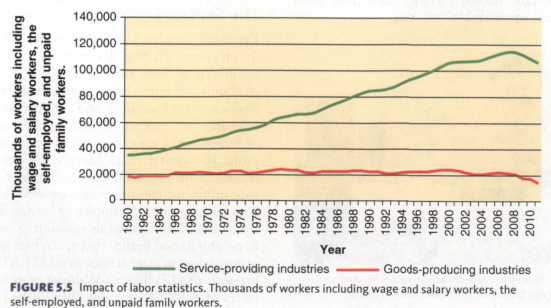

**FIGURE 5.5** Impact of labor statistics. Thousands of workers including wage and salary workers, the self-employed, and unpaid family workers.

Source: United States Bureau of Labor Statistics.

this challenge is to use a stock option plan of some sort whereby the managers have ownership interests. However, unethical managers may manipulate financial information to increase the value of their options artificially and then cash them in, leaving the organization in potential financial ruin, as happened at Enron and a few other major firms.

But even without a corporate structure, there are ownership challenges. For example, as mentioned in Chapter 4, numerous challenges arise when the owners are family members. From a purely ownership standpoint, the key issue involves who benefits—which family member receives the profits or must absorb the losses from the business. A closely related ownership issue in family businesses is succession—who takes over when the owner dies or leaves the business. In partnership businesses it is imperative that legal agreements be made so the surviving partner is not responsible for dealing with the former partner's heirs on such decisions as buyout rights, valuation of the business, division of profits, etc. after changes are made to reflect the new ownership.

## Competitive Challenges of Managers

**Competitive challenges** of managers involve efforts to gain an advantage in acquiring scarce resources. Competition can cause the fortunes of enterprises to have ups and downs. A highly recommended Bar-B-Q establishment in Phoenix, Arizona, had a good reputation but began losing it when new managers took charge. The original owners took it back over to "protect their name." Although it did regain much of its former quality and distinction, it no longer could bring back enough of its previous customers to survive. In the soft drink and bottled water markets, Coca-Cola and Pepsi-Cola are close competitors. Competing in the bottled water market involves differentiation, so Coca-Cola pushes Dasani for its fresh taste while Pepsi pushes Aquafina for its purity.[12] IBM and Dell compete in the computer industry. Likewise, Stanford and Harvard compete for the best students, and neighboring states may compete for tourist dollars or industrial development opportunities. Starbucks has increased rivalry in the coffeehouse business.[13] Virtually all organizations compete with other organizations in some way or another, be it for consumer dollars, good students, or budget appropriations.

There are several different aspects to effective competition. Many of the broader and more general ones are covered later in Chapter 9. Our concern here is with three specific sets of challenges that affect managers: productivity and quality, technology and automation, and innovation and intrapreneurship.

### FOOD FOR THOUGHT 5.3

Dr. Pepper cola is a blend of 23 spice and fruit flavors. Created in Texas in 1885, it is the oldest major soft drink brand in America.

## Productivity and Quality

During the last few decades managers have increasingly come to recognize the importance of productivity and quality as ingredients in successful competition. **Productivity** is a measure of efficiency—how much is created relative to the resources used to create it; for example, how much milk is produced relative to the food eaten by a cow. **Quality**, on the other hand, is a measure of value. For example, consider two inexpensive watches with the same selling price. One runs flawlessly for five years while the other must be discarded after only three years. The longevity of the first watch is evidence of its higher quality. Or consider the experience of returning a gallon of milk to a store. If you are treated cordially and receive a prompt refund with no questions asked, you may assess the quality of the service you received as high. On the other hand, if you are treated rudely and asked a number of probing questions implying that you did not refrigerate the milk promptly or properly, you would likely feel the quality of service you received was poor. Thus, productivity and quality are relevant for both manufacturing and services.

Managers once believed that quality and productivity were inversely related—that spending more to achieve higher quality resulted in higher total costs and therefore lower productivity. Now, however, managers recognize that just the opposite is true. Higher quality means fewer defects, more efficient use of resources, and fewer quality inspections, thereby actually boosting productivity.[14] Of course, quality is a relative concept. Among automobiles, for example, most knowledgeable consumers would expect a Lexus LS10 to be of higher quality than a KIA Rio. But there are major price differences between the two cars. Each may actually be of high quality relative to its price. When assessing quality, then, we must understand that it can have both absolute (as compared with some objective standard) and comparative (as compared with substitute products or services) dimensions.

Managers are seeing an important competitive role for both productivity and quality. An organization that falls behind its competitors in either area will be hard pressed to catch back up. Indeed, more and more organizations are attempting to compete on the basis of productivity and quality. We will return to quality and productivity in Chapter 22.

## Technology and Automation

**Technology** consists of processes and steps used to transform various inputs such as raw materials and component parts into something else—a stereo, a book, a pound of pepper jack cheese, a table, or a shirt, for example. **Automation** is the use of machinery,

especially computers and robots, to replace human labor as part of the technological transformation process. Technology has created a major competitive battlefield around the world. Indeed, many managers are finding that focusing on technology allows them both to introduce new products and to improve existing products more effectively than in the past. Much of this advantage flows from automation. Managers of both manufacturing and service firms are recognizing that robotics and other approaches to automation can help them boost productivity and quality while also lowering costs. At the same time, however, it sometimes eliminates jobs and makes other jobs less challenging than before.

For example, when Steve Jobs launched his new computer firm Next, he decided early on that manufacturing and technology would be the core of his business. He created an almost fully automated plant that operated with only six hourly-workers. Whenever company engineers designed a new circuit board on their office computers, they could transmit the

---

### A FOCUS ON AGRIBUSINESS
### Organic?

Did you take a chemistry class in high school or college? Did you learn that there are two basic types of matter, organic and inorganic? Traditionally, *organic* refers to matter produced by living beings (containing carbon) and *inorganic* refers to matter not related to living things (no carbon). When it comes to food, forget all of that (or at least be prepared to modify it) because by that definition all food is organic. That is not the "official" definition used in describing food.

First, the apparently clear distinction in chemistry is really not so clear, particularly for manmade substances. Second, for food labeling, only relatively recently have standards been adopted and those standards are not uniform throughout the world. However, the basic idea is that organic foods are grown and/or processed in such a way that inorganic and/or synthetic materials are limited or excluded altogether.

Organic products have strict production and labeling standards. The USDA recognizes "100 percent organic," which (excluding water and salt) means just what it says. "Organic" means that 95 percent (again excluding water and salt) is organic and that the other 5 percent are from an approved list of ingredients. These products cannot have been produced using sewage sludge or ionizing radiation or other excluded methods and can display the USDA seal. "Made with organic ingredients" means that at least 70 percent of the product is organic and, while they cannot display the seal, they can list up to three ingredients explicitly on the label. Finally, foods that contain less than 70 percent organic ingredients cannot use the term "organic" on the main label although specific ingredients can be identified on the information panel portion of the label. Terms, such as "all-natural," "free-range," or "hormone-free" are marketing labels; and while they may be accurate, they are not certified by the USDA.[15]

specifications to the plant by modem and pick up the completed board in 20 minutes. Moreover, the boards were made with such precision that there were less than 20 defects per one million units produced.[16]

**FOOD FOR THOUGHT 5.4**
Organic cotton fiber sales in the United States grew in 2009 by 10.4 percent over 2008, to reach $521 million. U.S. organic cotton growers increased plantings of organic cotton acreage by 26 percent in 2009 over what was planted the previous year.

## Innovation and Intrapreneurship

**Innovation** is the process of creating and developing new products or services and/or identifying new uses for existing products and/or services. Without innovation, firms become stagnant as competitors continue to introduce new products into the marketplace. 3M, based in Minnesota, has a vision of "Innovative technology for a changing world." That vision has led to numerous successful product innovations including Post-It Notes, removable Scotch magic tape, and Scotch Transparent Duct Tape.

U.S. firms have typically been among the world leaders in innovation. Frequently, after U.S. firms achieve a product breakthrough, businesses in other countries find other ways of making a similar product. This has been the case for a number of products ranging from televisions to cameras to hard disks. In recent years, however, many U.S. businesses have renewed their commitments not only to developing new products but also to applying their innovative methods to manufacturing and then marketing them.[17] Monsanto, Kodak, IBM, Syngenta, and Merck & Company are all examples of companies that have significant commitments to spending on R&D, automating plants, and striving for greater innovation.

Many managers are also finding that **intrapreneurship** is an effective approach to stimulating innovation. An intrapreneur is like an entrepreneur but is employed by and works within the framework of a larger organization. Instead of going out and starting a new business, an intrapreneur starts new ventures within a larger organization. That is, he or she develops new ideas and then champions the ideas through the various organizational channels that lead to their introduction into the marketplace. At 3M it was an intrapreneur who developed the Post-It Note and then fought for it over the objections of his boss, who did not think it was a good idea. Now, many managers actively encourage and reward intrapreneurship. For example, some firms give engineers and scientists a fixed amount of time each week to pursue pet projects. It is not unusual, though, for an intrapreneur to seek his employer's permission to resign and become an entrepreneur by building his own new business if the employer turns down the idea.

## Global Competition

**Global challenges** of managers are challenges that come from international competition. Many managers today really have no choice but to adopt a global perspective on doing business. Even if a company wants to compete only within a single economic system, it is very likely that competitors drawing on global financial resources, design and technology breakthroughs, and production efficiencies will be at a marked advantage. The small candy manufacturer that thinks it is not competing globally may be caught off guard when it finds that its long-time grocery chain is now buying a competing candy product instead at a lower price from another country (Figure 5.6). Thus, the globalization movement is such that many organizations must choose either to participate in that movement or to gradually lose their ability to compete effectively.

Organizations that are competing in a global environment face a myriad of challenges and opportunities. At the simplest level, managers must determine which market to enter and how to enter it. One

**FIGURE 5.6** As American products and dollars compete across the world, organizations must consider their ability to compete.

way is to hire a foreign broker to sell the firm's goods in the chosen market. Another is to license a company already in the market both to make and to sell the firm's products—we produce the milk, you bottle it and make the ice cream. Some managers might decide to build a new plant in that foreign market and make their own products there. Others might enter into a joint venture with another firm to help get a foothold. We discuss competition in the global environment more fully in Chapter 6.

## The Workforce Diversity Challenge of Managers

Workforce diversity has become a very important issue in many organizations. **Workforce diversity** exists in a group or organization when its members differ from one another along one or more important dimensions.[18] As more women and minorities enter the labor force, for example, the available pool of talent from which organizations hire employees is changing in both size and composition.[19] A related factor that has contributed to diversity has been the increased awareness by managers that they can improve the overall quality of their workforce by hiring and promoting the most talented people available.[20] Another reason for diversity is legislation and legal actions that have forced organizations to hire more broadly.

### Dimensions of Workforce Diversity

Many different dimensions of diversity can be used to characterize an organization, including age distribution, gender, and ethnicity.

**Age Distribution.**   A key dimension of diversity is the age distribution of workers. The average age of the U.S. workforce is gradually increasing and will continue to do so for the next several years. How does this trend affect managers? For one thing, older workers tend to have more experience, may be more stable, and can make greater contributions to productivity. On the other hand, despite improvements in health and medical care, older workers are nevertheless likely to require higher levels of insurance coverage and medical benefits. And the declining labor pool of younger workers will continue to pose problems for managers as they find fewer potential new entrants into the labor force.[21]

**Gender.**   Most organizations have also experienced changes in the worker gender dimension—the relative proportions of male and female employees. In the United States, for example, the percentage of male employees shrank from 55 percent in 1988 to fewer than 53 percent in 2009. Simultaneously, the percentage of female employees rose from 45 percent in 1988 to just over 47 percent in 2009.[22] The number of female top managers in male-dominated businesses and males in female-dominated businesses is gradually changing as well.[23]

**Race/Ethnicity.**   Another major dimension of cultural diversity in organizations involves race or ethnicity. Race relates to appearance, most commonly skin color but also other biological or genetic characteristics such as eye or hair color, bone structure, or physical aspects. Ethnicity, on the other hand, relates to cultural factors such as nationality, culture, ancestry, language, and beliefs. **Ethnicity**, then, refers to the ethnic composition of a group or organization.

Within the United States, most organizations reflect varying degrees of ethnicity comprised of whites, blacks, Hispanics, and Asians. In 2009, the percentage of whites in the U.S. workforce was just over 70 percent. At the same time, the percentage of Hispanics had climbed to 14 percent. The percentage of blacks, Asians, and others was more than 15 percent.[24]

In addition to age, gender, and ethnicity, managers are confronting other dimensions of diversity as well. Some of the more important ones include country of national origin, physically challenged employees, single parents, dual-career couples, marital status, sexual preferences, sexual orientation, religion, level of education, special dietary preferences (vegetarians, for example), and different political ideologies.[25]

### The Positive Impact of Diversity

There is no question that organizations are becoming ever more diverse, but what is the impact of this diversity on organizations? As we will see, diversity provides both opportunities and challenges for managers. For example, many managers are finding that diversity can be a source of competitive advantage in the marketplace.[26] There are six arguments that have been proposed for how diversity contributes to competitiveness.[27] These arguments are summarized in Table 5.1.

**Cost Argument.**   The cost argument suggests that managers who learn to cope with diversity will generally have higher levels of productivity and lower

**TABLE 5.1** How Diversity Contributes to Competitiveness

| ARGUMENT | RATIONALE |
|---|---|
| Cost | Effective management of diversity leads to higher levels of productivity and lower levels of turnover and absenteeism. |
| Resource acquisition | Effective management of diversity leads to women and minorities wanting to work for the organization. |
| Marketing | Effective management of diversity helps managers understand different market segments. |
| Creativity | Effective management of diversity fosters creativity in organizations. |
| Problem solving | Effective management of diversity leads to more effective problem solving because of more information being brought to bear on problems. |
| Systems flexibility | Effective management of diversity helps organizations become more flexible. |

Source: Based on data in Taylor H. Cox and Stacy Blake, "Managing Cultural Diversity: Implications for Organizational Competitiveness," *The Academy of Management Executive* (August 1991), 45–56.

levels of turnover and absenteeism. Ortho Pharmaceuticals estimates that it has saved $500,000 by lowering turnover among women and ethnic minorities.[28]

**Resource Acquisition Argument.** The resource acquisition argument for diversity suggests that organizations that manage diversity effectively will become known among women and minorities as good places to work (Figure 5.7). Such organizations will thus be better able to attract qualified employees from among these groups.

**Marketing Argument.** The marketing argument suggests that managers with diverse workforces will be better able to understand different market segments, especially since women and men as well as different cultures have different shopping, selecting, and spending habits. Traditionally, for example, men's wear has occupied the first floor of department stores

because men want to find their product the fastest way possible, whereas women's clothing tends to make the shopper wander through different departments. (We should note, however, that men's wear has been moving upstairs to make the egress more difficult for shoplifters.) A cosmetics firm like Avon that sells products to women of different races can better understand how to create such products and to effectively market them if white, black, and Latino managers are all available to provide inputs into product development, design, packaging, advertising, and so forth.[29]

**Creativity and Problem-Solving Argument.** The creativity argument for diversity suggests that organizations with diverse workforces will generally be more creative and innovative than will less diverse firms. Related to the creativity argument is the problem-solving argument. In a more diverse organization, there is more information that can be brought to bear on a problem and therefore a higher probability that better solutions will be identified.

**Flexibility Argument.** Finally, the systems flexibility argument for diversity suggests that organizations must become more flexible as a way of managing a diverse workforce. As a direct consequence, the overall organization will also become more flexible.

### The Not-So-Positive Impact of Diversity

Despite the overwhelmingly positive impact of diversity, the merging of workers of different ages, gender, and ethnicity can also present unique challenges for managers.

© Losevsky Pavel/www.Shutterstock.com

**FIGURE 5.7** Daycare centers on workplace campuses help organizations compete for employees.

**Improper or Unfair Use.**   Diversity in an organization can become a major source of conflict when it is perceived as being used improperly or unfairly; that is, when people think that someone has been hired, promoted, or fired only to promote diversity. For example, suppose a male executive loses a promotion to a female executive. If he believes that she was promoted because the firm simply wanted to have more female managers rather than because she was the better candidate for the job, he will likely feel resentful toward both her and the organization.

**Misunderstood or Inappropriate Interactions.**   Another source of conflict stemming from diversity is through misunderstood or inappropriate interactions between people. For example, suppose a male manager tells a sexually explicit joke to a new female manager. He may be trying intentionally to embarrass her, he may be clumsily trying to show her that he treats everyone the same, or he may think he is making her feel like part of the team. Regardless of his intent, however, if she finds the joke offensive she will feel anger and hostility. Intent does not matter. If someone acts in such a way as to create a hostile work environment or if a manager permits such an environment to exist, the law is being violated. A hostile work environment is one in which one or more workers find it impossible or nearly so to perform their duties satisfactorily. In this instance, the female employee's feelings may be directed only at the offending individual or more generally toward the entire firm if she believes that its culture facilitates such behaviors. And of course, sexual harassment itself is both unethical and illegal, and it can be committed by females as well as males.

**Cultural Misunderstandings.**   Conflict can also arise as a result of misunderstandings regarding cultural phenomena. For example, suppose a U.S. manager publicly praises the work of a Japanese employee for his outstanding work. The manager's action stems from the cultural belief in the United States that recognition is important and rewarding. But because the Japanese culture places a higher premium on group loyalty and identity, the employee may feel ashamed and embarrassed. In some cultures, unlike the United States, gifts or favors are given or received under a variety of conditions that we might consider unnecessary, questionable, improperly causing a conflict of interest, or bribery. Thus, a well-intentioned action may backfire and result in unhappiness.

**Other Sources of Conflict.**   Conflict may also arise as a result of fear, distrust, or individual prejudice. Members of a dominant group in an organization may worry that newcomers from other groups pose a personal threat to their own position in the organization. For example, when U.S. firms have been taken over by Japanese firms, U.S. managers have sometimes been resentful or hostile to Japanese managers assigned to work with them. In some cases, the U.S. worker fears that he is training his replacement. In other cases, an individual may be unwilling to accept people who are different from themselves. Personal bias and prejudices are still very real among some people today and can lead to potentially harmful conflict.[30]

## Managing Diversity Through Individual Strategies

Because of the tremendous potential that diversity holds for competitive advantage, as well as the possible consequences of diversity-related conflict, in recent years much attention has been focused on how individuals and organizations can better manage diversity.[31] Some of these are summarized in Table 5.2. The four basic individuals strategies include understanding, empathy, tolerance, and communication.[32]

**Understanding.**   The first individual strategy is *understanding* the nature and meaning of diversity. Some managers have taken the concepts of equal employment opportunity to an unnecessary extreme. They know that, by law, they cannot discriminate against people on the basis of sex, race, and so forth. In following this mandate, they attempt to treat everyone the same. But this belief can cause problems when it is translated into workplace behaviors among people after they have been hired. The fact is that people are not the same. While individuals must be treated fairly and equitably, managers must

**TABLE 5.2**  Ways to Manage Diversity

| INDIVIDUAL STRATEGIES | ORGANIZATIONAL STRATEGIES |
| --- | --- |
| Understand the nature of diversity | Policies and procedures |
| Empathy | Diversity training |
| Tolerance | Language training |
| Communication | Corporate culture |

© Cengage Learning 2014

understand that differences do exist among people. Thus, any effort to treat everyone the same, without regard to their fundamental human differences, is another form of discrimination and will only lead to problems.

**Empathy.**   Related to understanding is *empathy*. People in an organization should try to understand the perspectives of others. For example, suppose a group that has traditionally been comprised of white males is joined by a female. Hopefully, the males will be interested in making her feel comfortable and welcome. They may be able to do this most effectively by empathizing with how she may feel. For example, she may feel disappointed or elated about her new assignment, she may be confident or nervous about her position in the group, and she may be experienced or inexperienced in working with male colleagues. By learning more about her situation, the existing group members can further facilitate their ability to work together.

**Tolerance.**   A third related individual approach to dealing with diversity is *tolerance*. Even though managers may learn to understand diversity, and even though they may try to empathize with others, they may still not accept or enjoy some aspect of their interactions with others. For example, one firm recently reported considerable conflict among its U.S. and Israeli employees. The Israeli employees always seemed to want to argue about every issue that arose. The U.S. managers preferred a more harmonious way of conducting business and became uncomfortable with the conflict. Finally, after considerable discussion it was learned that many Israeli employees simply enjoyed arguing and just saw it as part of getting work done. The firm's U.S. employees still do not enjoy the arguing but are more willing to tolerate it as a fundamental cultural difference between themselves and their colleagues from Israel, and the Israelis have learned to be more tolerant of those from the United States as well.[33]

**Communication.**   A final individual approach to dealing with diversity is communication. Problems often get magnified over diversity issues because people are afraid or otherwise unwilling to openly discuss issues that relate to diversity. For example, suppose a younger employee has a habit of making jokes about the age of an older colleague. Perhaps the younger colleague means no harm and is just engaging in what she sees as good-natured

kidding. But the older employee may find the jokes offensive. If there is no communication between the two, the jokes will continue and the resentment will grow. Eventually, what started as a minor problem may erupt into a much bigger problem. For communication to work, it must be a two-way street. If a person wonders if a certain behavior on her or his part is offensive to someone else, the curious individual should probably just ask. Similarly, if someone is offended by the behavior of another person, he or she should explain to the offending individual how the behavior is perceived and request that it be stopped. As long as such exchanges are handled in a friendly, low-key, non-threatening fashion, they will generally have a positive outcome.

## Managing Diversity Through Organizational Strategies

Although individuals can play an important role in managing diversity, the organization itself must also play a fundamental role through its policies and practices, language, diversity training, and culture.

**Policies and Practices.**   The starting point in managing diversity is the organization's policies and practices that directly or indirectly affect how people are treated. Through an organization's policies and practices, its people come to understand which behaviors are and are not appropriate. For instance, the extent to which an organization embraces the premise of equal employment opportunity will to a large extent determine the potential diversity within an organization. But there are differences in an organization that follows the law to the letter and only practices passive discrimination and an organization that actively seeks a diverse and varied workforce. Diversity training (discussed later in this section) is an even more direct method for managing diversity.

Organizations can also help manage diversity through a variety of on-going personnel practices and procedures. Benefit packages, for example, can be structured to better accommodate individual situations. An employee who is one of a dual-career couple and who has no children may require relatively little insurance (perhaps because the spouse's employer provides more complete coverage) and would like to be able to schedule vacations to coincide with those of the spouse. On the other hand, an employee who happens to be a single parent may need a wide variety of insurance coverage and prefer to schedule vacation time to coincide with school holidays. Flexible working hours are another useful organizational practice to accommodate diversity. Differences in family arrangements, religious holidays, cultural events, and so forth may dictate that employees have some degree of flexibility regarding when they work. For example, a single parent may need to leave the office every day at 4:30 to pick up the children from their daycare centers. An organization that truly values diversity will make every reasonable attempt to accommodate such a need.

**Diversity Training.**  Many organizations are finding that diversity training is an effective means for managing diversity and minimizing its associated conflict.[34] **Diversity training** is specifically designed to better enable members of an organization to function in a diverse workplace (Figure 5.8). This training can take a variety of forms. For example, many organizations find it useful to help people learn more about their similarities to and differences from others. Men and women can be taught to work together more effectively and can gain insights into how their own behaviors affect and are interpreted by others. In one organization, a diversity training program helped male managers gain insights into how various remarks they made to one another could be interpreted by others as being sexist. In the same organization, female managers learned how to point out their discomfort with those remarks without appearing overly hostile.[35]

Similarly, white and black managers may need training to better understand each other. Managers at Mobil noticed that four black colleagues never seemed to eat lunch together. After a diversity training

**FIGURE 5.8** Diversity training may emphasize both similarities and differences among people of different backgrounds.

program, they came to realize that the black managers felt that if they ate together, their white colleagues would be overly curious about what they might be talking about. Thus, they avoided close associations with one another because they feared calling attention to themselves.[36]

**Language Training.**  Some organizations even go so far as to provide language training for their employees as a vehicle for managing diversity. Motorola, for example, provides English language training for its foreign employees on assignment in the United States. At Pace Foods in San Antonio, staff meetings and employee handbooks are translated into Spanish for the benefit of the company's large Hispanic employee population.[37] At Johns Hopkins Medical Institutions in Baltimore, medical professionals and support staff hail from all over the world. Diversity training is done in all orientations and all management training classes to help people understand themselves and their behaviors.[38]

**Corporate Culture.**  An organization's culture is the ultimate context from which diversity must be addressed and the ultimate test of an organization's commitment to managing diversity. Regardless of what managers say or put in writing, unless there is a basic and fundamental belief that diversity is valued, it cannot ever become truly an integral part of an organization. An organization that really wants to promote diversity must shape its culture so that it clearly underscores top management's commitment to and support of diversity in all of its forms throughout every part of the organization. With top management support, however, and reinforced with a clear and consistent set of organizational policies and practices, diversity can become a basic and fundamental part of an organization.

Part of that culture is created and revealed by the manner in which the organization addresses and responds to problems that arise from diversity. For example, consider the case of a manager (male or female) charged with sexual harassment. If the organization's policies put an excessive burden of proof on the individual being harassed and invoke only minor sanctions against the guilty party, it is sending a clear signal as to the importance of such matters. But the organization that has a balanced set of policies for addressing questions like sexual harassment sends its employees a different message as to the importance of diversity and individual rights and privileges. The culture at some firms is more-or-less do what you are told and keep your nose clean-no suggestions wanted. Several firms with which we are familiar have a culture that tells the employees not so subtly, "Don't bring me a problem; bring me a solution (but I don't really want that either)."

Culture is also determined and revealed by the examples that an organization sets, such as making sure that there is diversity in its key committees and executive teams. Even if diversity exists within the broader organizational context, an organization that does not reflect diversity in groups like committees and teams implies that diversity is not a fully ingrained element of its culture. In contrast, if all major groups and related work assignments reflect diversity, the message is a different one.

**FIGURE 5.9** Health care is a company benefit that employees have come to expect. Many workers will choose an employer that provides excellent health insurance rather than a higher salary.

## FOOD FOR THOUGHT 5.5

If you look at the two modified "T"s and the 'i' in the center of the Tostitos logo, you can see two people enjoying a Tostitos chip with a bowl of salsa. Supposedly, this conveys an idea of people connecting with each other.

# Other Workplace Challenges of Managers

**Workplace challenges** involve the relationships among organizations, their managers, and their operating employees. In addition to the challenges presented by increased workforce diversity, two other workplace issues today are employee expectations and rights and workplace democracy.

## Employee Expectations and Rights

For many years managers generally believed that workers were motivated only by opportunities for economic gain. Later, with more emphasis on human relations, it was thought that personal satisfaction was the driving force in motivation. Eventually, managers came to see that employee motivation is actually a very complex process. Each individual has his or her own unique set of needs and perceptions of how best to fulfill them. These individual needs continue to change (Figure 5.9). For example, more fathers want to participate in the raising of their children and more women are seeking professional careers. Thus, managers today are finding it necessary to be more flexible in how they

treat their employees.[39] They are allowing workers to have more say in how they do their jobs, providing more information about what the organization has planned, and allowing workers more freedom in selecting job assignments.

Managers seeking to motivate a diverse workforce and enhance employee performances must work harder than ever to understand their employees. At the same time, however, new concerns are being raised about worker rights and privacy.[40] For example, some managers argue that organizations should seek to help employees with drug or alcohol problems. They believe that helping with such problems is the socially responsible thing to do and that organizations have an obligation to fulfill this function. On the other hand, some managers argue that organizations should essentially "mind their own business" as long as their employees are meeting performance expectations. They reason that what a person does outside the workplace is his or her own business and that the organization should not attempt to intervene.

## Workplace Democracy

Many managers also contend with issues of workplace democracy—the practice of giving workers a greater voice in how the organization is managed. Some managers have come to believe that letting workers have a say in what the organization does will enhance worker commitment to the organization while also improving its effectiveness. The workers, in turn, get to have a voice in determining what happens to their employing organizations.

As we will see in Chapter 18, many managers have started increasing the participation they allow workers to have in deciding how they do their jobs. Along with this participation, however, is an expectation on the part of employees that they will have a greater say in a wide variety of organizational issues, including, but not limited to, working hours, organizational practices, hiring decisions, and compensation decisions. Sometimes this voice gets legitimized in very specific ways. The United Auto Workers union has had a seat on Chrysler's board of directors since wage concessions made during the automaker's financial crisis in the early 1980s.

In some parts of the world, workplace democracy has been part of organizational life for a long time. In Germany, for example, organizations are required by law to have a specified number of operating employees and managers on their governing boards. In the United States, however, workplace democracy is a relatively new concept that managers are just beginning to address. Of course, as noted in Chapter 4, cooperatives are an organizational form in which democratic processes govern the leaders and all major decisions.

## Legal and Social Challenges of Managers

**Legal challenges** of managers reflect the judicial context in which an organization operates. **Social challenges** relate more to prevailing social customs and mores. Recent concerns about legal and social challenges were in part stimulated by several widely publicized ethical scandals during the 1980s and early 1990s. For example, Ivan Boesky, David Levine, and Michael Milken were all involved in Wall Street scandals.[41] Major ethical scandals also plagued Japan, Great Britain, and West Germany. And they don't seem to go away. Two former executives of Enron were sentenced to prison for their roles in the company's collapse. Merck agreed to pay $671 million to settle claims that it overcharged Medicaid programs for four drugs. Four former executives of AIG/Berkshire Hathaway were charged with participating in a scheme to manipulate AIG's financial statements. Hopefully these executives did not learn that in business school—MBA students at a leading business school, Duke's Fuqua School of Business, were found guilty of cheating.[42]

Virtually all managers face ethical dilemmas as an inherent part of their jobs. Thus, both

organizations and the managers who work for them must strive to better understand the ethical context from which decisions are made. For example, consider the case of a manager deciding what to do about a minor pollution problem. His plant operates within governmental guidelines for his industry. Even though the manager knows his firm is generating a small amount of "acceptable" pollution, one choice he has is to do nothing. In years past, such companies simply denied that they emitted pollutants or were the cause of pollution. Another alternative is to invest in modest equipment that might further lower but not eliminate the pollution. Finally, he could spend heavily on state-of-the-art equipment that will eliminate all of his firm's pollution. None of these three choices is technically the best or the worst one. Different people can make very compelling arguments for each of these three choices.

Managers must make a variety of decisions every day about how they treat their employees, how they interact with suppliers, customers, lenders, regulators, and competitors, and so forth. Virtually all of these decisions have an ethical component. We return to issues of ethics and social responsibility in Chapter 7.

## Governmental Regulation

In theory, free market economies are characterized by relatively little governmental regulation. That is, businesses and their managers are free to compete however they see fit. Even though the United States has a free market economy, there is still an abundance of governmental regulation that both proscribes and prescribes business activity. There are a variety of reasons why government regulation has been considered necessary during the history of U.S. business. For one, large and powerful firms have sometimes been known to try to drive their weaker competitors out of business by using unfair business practices. For another, governmental regulation is sometimes necessary to support various laws such as the Occupational Safety and Health Act, the Environmental Protection Act, and so forth. There is also a feeling among some people that the government should strive to maintain a reasonable level of competition for the public good. Finally, many people believe that unscrupulous business people would resort to unethical and illegal behavior on a regular basis if they were not regulated.

On the other hand, critics of regulation argue that, if we are indeed a free market, businesses should be free to do whatever they want. The logic behind this is that if a business does something people object to, they will "punish" that business by not buying its products or services. In the United States, there has been a trend toward reducing or lowering the regulation of business. For example, in recent years many regulations regarding the airline, financial, and trucking industries have been softened. How has it worked? In the airline industry, many weaker carriers have been absorbed by larger ones, but fares have been considerably lower than they were during the days of extreme regulation. But during the recession in the mid-2000s, some airlines began finding ways to increase their revenues without increasing ticket prices: charging fees for bags, for example. In response to airline fees for bags, UPS introduced luggage boxes so that travelers can ship their luggage and avoid security hassles and waits at the baggage carousel.[43] In the financial industry, managers in many savings and loan firms took advantage of decreased regulation and promoted one of the biggest financial crises in history. We also discuss regulation more fully in Chapter 7.

## The Natural Environment

Although concerns about pollution and the environment have been raised for decades, they are increasingly coming to center stage.[44] Greater consumer awareness, growing alarm about problems ranging from global warming to the scarcity of landfills for trash, and media attention have all served to sensitize everyone about these issues. Now organizations of all types must address an increasing variety of controversial environmental issues, from air pollution to toxic waste disposal to water pollution. Nowhere is pollution of the natural environment more important than in agribusinesses, where products are produced for direct consumption by humans and other animals. For example, some environmentalists protest the plastic containers that many firms use to package their products.

Another major issue is oil pollution. Contrary to popular belief and despite the tremendous publicity accorded major oil spills, such spills account for only about 3 percent of the total global marine oil pollution. Natural oil seepage accounts for more than 40 percent of yearly oil pollution, with the remainder coming from "down the drain"—petroleum transportation and industrial oil consumption.[45] On the horizon are additional issues such as pollution through natural gas extraction (fracking) and using government funding to solve our energy needs by funding private, profit-making companies to build wind farms and solar systems.

Still another issue that has emerged is the increasingly widespread organizational practice of promoting products as environmentally sound. Sometimes these claims are legitimate, while on other occasions they are overstated. For instance, Procter & Gamble had to drop claims that its disposable diapers were biodegradable. A growing global concern is the extent to which businesses around the world are defying conventional wisdom or generally accepted practices concerning the environment. Many environmentalists are concerned about the unregulated fishing practices in Japan that result as Japanese fishing fleets hunt whales or destroy dolphins while catching tuna. The Canadian government has for years tried to get the United States to curb the pollution that results in forest-destroying acid rains. Others worry about how Brazil is destroying its rain forests for the sake of industry and expansion. And many countries in Eastern Europe are among the most polluted on Earth. For example, 70 percent of the rivers in Czechoslovakia are heavily polluted, one-third of Bulgaria's forests are damaged by unrestricted air pollution, and East Germany has major toxic waste problems.[46] Again, we return to these issues in Chapter 7.

## CHAPTER SUMMARY

Because managers today operate in an ever-changing world, they must confront a variety of issues. This confrontation carries with it both challenges and opportunities. The basic sets of challenges include economic, competitive, workplace, and legal and social issues.

Economic challenges of managers include various forces and dynamics associated with the economic system within which their organizations function. Some of the more important economic considerations are downsizing and cutbacks, the role of entrepreneurship and new careers, the emerging role of service organizations, and corporate ownership.

Competitive challenges are associated with the efforts of managers to gain an advantage for their organizations in acquiring scarce resources. Four specific sets of competitive challenges that organizations have to address today are productivity and quality, technology and automation, innovation and intrapreneurship, and global competition.

Workplace challenges are those associated with the relationships among organizations, their managers, and their operating employees. Workforce diversity is perhaps the most significant of these challenges today. Employee expectations and rights and workplace democracy are also important.

Legal challenges of organizations are those reflecting the judicial context of the organization. Social challenges involve prevailing social customs and mores. Key legal and social issues today include the ethical standards of managers, governmental regulation, and the relationship between business and the natural environment.

## CHAPTER ACTIVITIES

### REVIEW QUESTIONS

1. Identify and briefly describe the four basic sets of managerial challenges discussed in this chapter.
2. What can managers do to avoid another period of downsizing?
3. How are productivity and quality different, and how are they related? Can you change one without changing the other?
4. Why is innovation important to managers?
5. What are some arguments for and against governmental regulation of business? Can there really be a "free market"? Why or why not?

### ANALYSIS QUESTIONS

1. How is entrepreneurship affected by downsizing and cutbacks? Is there anything resembling a logical cycle between the two?
2. Suppose you are a plant manager considering the purchase of an automated assembly line. What issues would you consider before proceeding?
3. It is fairly easy to identify U.S. firms with stiff international competition (i.e., Ford). Identify some U.S. firms that have no direct foreign competitors, at least in North America.
4. What are some recent controversies regarding the natural environment and business? (Give specific examples and cite the basic issues involved in each.)
5. Suppose you have a subordinate with a drug problem. The subordinate is capable of doing her or his work just fine, but is having personal problems resulting from the drug use. Should you try to help, or should you stay out of it? What if your boss has the drug problem?

## FILL IN THE BLANKS

1. A _____ occurs when one corporation or group of investors buys or trades for enough stock in a company to gain control over it.
2. How much is created relative to the resources used to create it is a measure of efficiency known as _____.
3. A planned reduction in organizational size is called _____.
4. Reductions in the scope of an organization's operations are known as _____.
5. The process of starting a new business is known as _____.
6. Processes and steps for transforming various inputs such as raw materials and component parts into something else is called _____.
7. The use of machinery, especially computers and robots, to replace human labor as part of the technological transformation process is known as _____.
8. The process of creating and developing new products or services and/or identifying new uses for existing products and/or services is called _____.
9. The process of starting new ventures within a larger organization is referred to as _____.
10. When members of a group or organization differ from one another along one or more important dimensions it is called _____ _____ _____.

## CHAPTER 5 CASE STUDY
### PEPSI AND THE QUAKER

The waitress at a local breakfast place asked in her best professional, monotone voice, "So, do you want a 'Pepsi' with your oatmeal?" She was responding to an article on the page her customer was reading in *The Wall Street Journal*. The article was about PepsiCo's purchase of Quaker Oats a decade ago for $13.8 billion and the challenges it faced to boost sales in a very competitive ready-to-eat hot cereal market.

Pepsi-Cola and Quaker Oats may not be a new-wave breakfast, but the idea does illustrate circumstances that arise as a result of corporate acquisitions to boost overall revenue, especially in the agribusiness sector. In fact this demonstrates two important lessons for managers.

The first lesson: *It is imperative for a company to clearly differentiate itself from its competitors.* A more than casual look at a soft-drink display in the supermarket yields an array of products. Not only are the usual products present (Dr. Pepper, Coca-Cola, Pepsi-Cola, to name the large marketers) but also private label brands under the supermarket chain's logo and other private labels. This marketing array is exactly what the supermarket chain wants to provide: variety and selection.

The soft-drink industry is very competitive. The top four market-share holders in carbonated beverages are Coca-Cola, Pepsi-Cola, Dr. Pepper Snapple Group, and Cott, in descending order. Combined, these four competitors held 92 percent of the 2011 soft-drink market.

"The big four" beverage makers compete against each other primarily through marketing and packaging. The primary competition is between Coca-Cola and Pepsi-Cola. Each would prefer that customers purchased their respective products outright, but they find it advantageous to assist smaller competitors with market access to prevent the other brand from gaining more market share. For example, to keep its market share above Pepsi-Cola's, Coca-Cola bought the rights to distribute Dr. Pepper for the Dr. Pepper Snapple Group for $715 million. Dr. Pepper, in turn, was happy to have access to Coca-Cola's new touch-screen fountain system. PepsiCo markets Mountain Dew and its diet counterpart for similar reasons.

This highly competitive market environment and the need to differentiate introduce the second point that managers must understand: *The necessity to differentiate may lead to a need to master a very different market category.* This category may be out of the differentiator's area of expertise. Thus, to market Quaker Oats products, PepsiCo not only must be competitive in non-alcoholic beverage category (carbonated beverages, bottled water, sports drinks, fruit drinks, energy and other drinks) but also hot, ready-to-eat cereal products. Certainly it will bring to PepsiCo's bottom line some revenues that are not available to Coca-Cola Inc., but yet another competitive product category in which to compete. While Quaker Foods has the top spot in its market category over two private-label brands (Cream of Wheat and Bob's Red Mill), it has been a weak spot for PepsiCo.

Mastery of a competitive environment is an easy mantra to express but difficult to achieve. Mastery is even more difficult when it involves several very different product categories.

Sometimes competition has unintended consequences for the competitors. Consider the case for Cott, one of the top four companies mentioned above. This Canadian company gained entry into the carbonated beverage category at a time when many supermarkets were looking for a carbonated beverage that was a "level above" the standard supermarket private label or generic brand to attract customers away from Coke and Pepsi. Cott was the supermarket's answer. Unfortunately for the supermarkets, Coca-Cola and Pepsi-Cola were so effective at their trade that another stand-alone competitor was introduced to the carbonated beverage category.

And from all of this, another obvious lesson for managers: *Competition begets competitors.*[47]

## ▶ Case Study Questions

1. Assuming that supermarkets are in business to sell whatever consumers want to buy, why do you think supermarkets wanted another beverage competitor to attract customers away from Coke and Pepsi?

2. Assuming that Pepsi-Cola was in business to provide carbonated beverages to a market that consumes its lowest amount of that product at breakfast time, what could possibly have persuaded Pepsi-Cola to get into the ready-to-eat hot cereal breakfast market?

3. Why does competition beget competitors? Under what circumstances would this not be true?

## REFERENCES

1. Thomas Sowell, *Basic Economics: A Citizen's Guide to the Economy* (New York: Basic Books, 2000), 65–66; "History Timeline," (2009) at walmartstores.com (accessed July 28, 2010); W. Chan Kim and Renee Mauborqne, *Blue Ocean Strategy: How to Create Uncontested Market Space and Make Competition Irrelevant* (Boston: Harvard Business School Press, 2005); Nelson Lichtenstein, *Wal-Mart: The Face of Twenty-First Century Capitalism* (New York: The New Press, 2006), 128–131; Steve Painter, "In Grocery Sales, Walmart Sacks Competition," in NWAnews.com (June 21, 2009), reported in wakeupwalmart.com/news/article.html?article=2212 (accessed September 6, 2010); Ann Zimmerman, "A Fashion Identity Crisis at Wal-Mart," *The Wall Street Journal* (July 28, 2010) at online .wsj.com (accessed July 27, 2010); Sean Gregory, "Walmart vs. Target: No Contest in the Recession," *Time* (March 14, 2009) at www.time.com (accessed September 6, 2010); Karen Talley "Wal-Mart's Grocery Sales Expand," *The Wall Street Journal* (March 31, 2010) at online.wsj.com (accessed September 6, 2010).

2. Alvin Toffler, *Future Shock* (New York: Random House, 1970).

3. Mark Gottfredson, Steve Schaubert, and Herman Saenz, "The New Leader's Guide to Diagnosing the Business," *Harvard Business Review* (February 2008), 63–72.

4. "USA: Cargill's Excel Announces Layoffs After US BSE Case Cuts Beef Demand," just-food.com, January 12, 2004, at www .just-food.com (accessed July 20, 2010).

5. Rami Grunbaum, "Job Losses Adding Up Badly in the Seattle Area," *The Seattle Times,* October 26, 2008, at seattletimes. nwsource.com (accessed July 19, 2010).

6. Carl Bialik, "Sizing Up the Small-Business Jobs Machine," *The Wall Street Journal*, October 15, 2011 at www.wsj.com (accessed December 1, 2012.

7. Laura J. Margulies, "Bankruptcy Can Happen to Anyone—Bankruptcies Involving Famous People," Laura Margulies & Associates LLC at www.law-margulies.com (accessed July 21, 2010).

8. "2009 Inc.500/5000 Top Lists," *Inc.* at www.inc.com (accessed July 19, 2010) and "History," Hy-Vee corporate website at www .hy-vee.com (accessed July 20, 2010).

9. "Fortune 500," *Fortune* at money.cnn.com (accessed July 20, 2010) and "The Service 500," *Fortune*, June 1, 1992, 201.

10. "Service Sector Grows," *USA Today* (March 3, 2006), 1B.

11. John J. Curran, "Companies That Rob the Future," *Fortune*, July 4, 1988, 84–89.

12. "In a Water Fight, Coke and Pepsi Try Opposite Tacks," *The Wall Street Journal*, April 18, 2002, A1, A8.

13. Kevin Helliker and Shirley Leung, "Despite the Jitters, Most Coffeehouses Survive Starbucks," *The Wall Street Journal*, September 24, 2002, A1, A11.

14. Joel Dreyfuss, "Victories in the Quality Crusade," *Fortune*, October 10, 1988, 80–88.

15. "Food Labeling," United States Department of Agriculture, April 2008 at www.ams.usda.gov on July 7, 2010; Mayo Clinic Staff, "Organic Foods: Are They Safer? More Nutritious?" December 20, 2008, at www.mayoclinic.com at July 6, 2010; Jessica DeCostole, "The Truth About Organic Foods," *Redbook*, 2007 at www .redbookmag.com at July 7, 2010.

16. Joel Dreyfuss, "Getting High Tech Back on Track," *Fortune*, January 1, 1990, 74–77.

17. Christopher Knowlton, "What America Makes Best," *Fortune*, March 28, 1988, 40–53.

18. Marlene G. Fine, Fern L. Johnson, and M. Sallyanne Ryan, "Cultural Diversity in the Workplace," *Public Personnel Management* (Fall 1990), 305–319.

19. Badi G. Foster, et al. "Workforce Diversity and Business," *Training and Development Journal* (April 1988), 38–42.

20. Sam Cole, "Cultural Diversity and Sustainable Futures," *Futures* (December 1990), 1044–1058.

21. "How to Manage an Aging Workforce," *The Economist* (February 18, 2006), 11; "The Coming Job Bottleneck," *Business Week,* March 24, 1997, 184–185; Louis S. Richman, "The Coming World Labor Shortage," *Fortune*, April 9, 1990, 70–77.

22. Bureau of Labor Statistics at ftp.bls.gov (accessed July 20, 2010) and *Occupational Outlook Handbook* (Washington DC: U.S. Bureau of Labor Statistics, 1990–1991).

23. Patricia Sellers, "The 50 Most Powerful Women in Business," *Fortune* (November 14, 2005), 125–170; "What Glass Ceiling?" *USA Today* (July 20, 1999), 1B, 2B; L. Atwater and D. D. Van Fleet, "Another Ceiling? Can Males Compete for Traditionally Female Jobs?" *Journal of Management* (1997), 603–626.

24. Bureau of Labor Statistics at ftp.bls.gov (accessed July 20,2010); "Hispanic Nation," *Business Week*, March 15, 2004, 58–70; *Occupational Outlook Handbook* (Washington DC: U.S. Bureau of Labor Statistics, 1990–1991).

25. Cliff Edwards, "Coming Out in Corporate America," *Business Week*, December 15, 2003, 64–72.; "The Power of Diversity: Who's Got the Clout?" *Fortune*, August 22, 2005, special issue; Michael Chisholm, "Cultural Diversity Breaks the Mold," *Geographical Magazine*, November 1990, 12–16.

26. Orlando C. Richard, "Racial Diversity, Business Strategy, and Firm Performance: A Resource-Based View," *Academy of Management Journal* (2000), 164–177; Jacqueline A. Gilbert and John M. Ivancevich, "Valuing Diversity: A Tale of Two Organizations," *Academy of Management Executive* (2000), 93–103.

27. Based on Taylor H. Cox and Stacy Blake, "Managing Cultural Diversity: Implications for Organizational Competitiveness," *The Academy of Management Executive* (August 1991), 45–56.

28. Ibid.

29. For an example, see "Get to Know the Ethnic Market," *Marketing*, June 17, 1991, 32.

30. Patti Watts, "Bias Busting: Diversity Training in the Workforce," *Management Review* (December 1987), 51–54.

31. Stephenie Overman, "Managing the Diverse Work Force," *HR Magazine*, April 1991, 32–36.

32. Lennie Copeland, "Making the Most of Cultural Differences at the Workplace," *Personnel*, June 1988, 52–60.

33. "Firms Address Workers' Cultural Variety," *The Wall Street Journal*, February 10, 1989, B1.

34. For an excellent look at how to do diversity training (and what not to do), see Lawrence Otis Graham, *Proversity: Getting Past Face Value and Finding the Soul of People—A Manager's Journey* (New York: John Wiley & Sons, 1997).

35. "Learning to Accept Cultural Diversity," *The Wall Street Journal*, September 12, 1990, B1, B9.

36. "Firms Address Workers' Cultural Variety."

37. "Firms Grapple With Language," *The Wall Street Journal*, November 7, 1989, B1.

38. Donna M. Owens, "Multilingual Workforces: How Can Employers Help Employees Who Speak Different Languages Work in Harmony?" *HR Magazine,* September 2005, at www.shrm.org (accessed July 20, 2010).

39. "Flexible Formulas," *The Wall Street Journal*, June 4, 1990, R34–35.

40. "Is Your Boss Spying on You?" *Business Week*, January 15, 1990, 74–75.

41. "Guilty, Your Honor," *Business Week*, May 7, 1990, 32–37.

42. "Corporate scandals and their consequences," Josephson Institute at josephsoninstitute.org (accessed July 21, 2010).

43. Eva Vasquez, "Luggage Box an Alternative to Toting Bags," CNN Travel, June 30, 2010, www.cnn.com (accessed June 30, 2010).

44. Jeremy Main, "Here Comes the Big New Cleanup," *Fortune,* November 21, 1988, 102–118; David Kirkpatrick, "Environmentalism: The New Crusade," *Fortune*, February 21, 1990, 44–52.

45. Laurence O'Sullivan, "Oil Rig Spills Not the Major Sources of Marine Oil Pollution," suite101.com, June 18, 2010, at pollution-control.suite101.com (accessed July 20, 2010).

46. "Eastern Europe's Big Cleanup," *Business Week*, March 19, 1990, 114–115.

47. Anjali Cordeiro, "PepsiCo Pushes Breakfast in Bid to Heat Up Oatmeal," *The Wall Street Journal* (July 28, 2010) at online.wsj.com (accessed July 27, 2010); Valerie Bauerlein, "Soda-Pop Sales Fall at a Faster Rate," *The Wall Street Journal* (March 31, 2009) at online.wsj.com (accessed July 27, 2010); Paul Ziorro and Valerie Bauerlein, "Coke Buys Dr. Pepper Rights," *The Wall Street Journal* (June 8, 2010) online.wsj.com (accessed July 28, 2010); Anjali Cordeiro, "PepsiCo Profit Slips on Bottler-Deal Costs," *The Wall Street Journal* (July 21, 2010) online.wsj.com (accessed July 28, 2010).

# The Global Environment

## LEARNING OBJECTIVES

After studying this chapter, you should be able to:

- Describe the nature of the global environment, including its meaning, recent trends, managing internationalization, and managing in an international market.

- Discuss the structure of the international economy and how it affects international management.

- Identify and discuss the basic challenges inherent in international management.

- Describe the basic issues involved in competing in an international economy, including organization size and the management challenges in a global economy.

## Danone and the Fortune at the Bottom of the Pyramid

The French food company, Danone SA, wanted to increase its global customer base. Like many companies it needed to expand beyond its home country. Global expansion is easy to conceive but difficult to achieve. Generally, it happens in one of two fashions: (1) export or produce a standard product in the country of choice and look for an affluent market segment to purchase this product, or (2) adapt a standard product to the expectations of the consumer in the country of interest. All global companies respond to these two choices (Figure 6.1).

**FIGURE 6.1** Yogurt is considered a healthy food that fits into the lifestyles and budgets of many cultures.
© severesid/www.Shutterstock.com

However, C.K. Prahalad pointed to another way. Companies could be successful by providing their products in highly modified ways to the less affluent classes; that is, seeking the "fortune at the bottom of the pyramid." Danone SA is pursuing this strategy.

Senegalese webmaster Demba Gueye treats himself twice a week after work to a snack: a 10-cent tube of "Dolima" drinkable yogurt. Such spending is a splurge, considering his two-dollar-a-day food budget and the teeny size of the 50-gram sachets (2 ounces). This is the market sought by Danone SA.

Danone SA is one of the fastest growing food companies in the world, thanks to an array of high-end healthy food products with its yogurt, Evian water, and Bledina baby food. In 2009, 42 percent of its sales were from emerging markets. In Indonesia where the average daily income (ADI) is $6.13, it markets "Aqua cup" of water (8.5 oz) for 5 cents; in Mexico (ADI = $27.99), it markets "Dany Xprime" pouch of jelly (1.4 oz) for 6 cents, and in Bangladesh (ADI = $1.43), Danone sells "Shotki Doi" pot of yogurt (2.1 oz) for 8 cents.

To be successful it was imperative for Danone to understand the culture of a country and discern its wants and needs. Danone yogurt was an easy fit with this lifestyle . For example, in Senegal people like to have three or four snacks a day and only one big meal. The company needed only to provide an affordable product that was also profitable.

This strategy is not Danone's invention. The company is actually using the strategy initiated by the largest food company in the world, Nestlé SA. That strategy simply considers that "all global markets are local," an approach developed many years ago. An analyst specializing in the food industry, Warren Ackerman of Evolution Securities, has provided a label for this approach: "global" marketing.

Markets are not where you find them; they are where you seek them. Or as C.K. Prahalad said, "If we stop thinking of the poor as a burden and start recognizing them as entrepreneurs and value-conscious consumers, a whole new world of opportunity can open up."[1]

## INTRODUCTION

Firms, such as Danone, from around the world are entering new markets, taking on new challenges, and forming alliances with other firms to compete more effectively in the international business environment. To be successful today, managers should understand the international context within which they function. This holds true regardless of whether the manager runs a Fortune 500 firm or a small independent agribusiness concern like Summer Farms.

This chapter explores the global environment of management. We start by describing the influence of the global environment, and then discuss the international economy in terms of different economies and economic systems. Basic challenges of international management are introduced and discussed next. Then we focus on managing in the international economy.

## The Influence of the Global Environment

You probably woke up this morning to the sounds of an alarm clock that was made in Japan. The clothes you put on may have been made in Korea, Mexico, or Taiwan. The coffee you drank was probably made from beans grown in South America. For breakfast, you may have had oatmeal from Ireland. At lunch your sandwich may have come with a Mrs. Klein's Pickle that was produced in Arizona from cucumbers grown in Mexico. The seeds that eventually produced the sandwich bread may have come from an international company such as Syngenta (Switzerland), Groupe Limagrain (France), KWS AG or Bayer Crop Science (Germany). To get to school or work, you may have driven a Japanese car. Even if you drove a Ford or Chevrolet, some of its parts were manufactured abroad. Perhaps you didn't drive a car to school but rather rode a bus manufactured by Volvo (a Swedish company) or a motorcycle manufactured by Honda or Yamaha, both Japanese firms.

Our daily lives are strongly influenced by businesses, particularly agribusinesses, from around the world. But we are not unique in this respect. People living in other countries have much the same experience. They drive Fords in Germany, use IBM computers in Japan, eat McDonald's hamburgers in France, dine on our rice in Mexico, eat our soybeans in Canada, and snack on Mars candy bars in England. They drink Pepsi, eat soybeans, and wear Levi Strauss jeans in China. The Japanese import enormous amounts of American pharmaceuticals and use American Express credit cards. People around the world fly on United or American Airlines in planes made by Boeing, work in buildings constructed with Caterpillar equipment, eat Kellogg's cereals, and buy Mobil oil. As Starbucks has expanded globally, its coffee is becoming known as the "American" drink.[2]

In truth, we have all become part of a global village and have a global environment in which no organization is insulated from the effects of foreign markets and competition. More and more firms are viewing themselves as international or multinational businesses.[3] What do these terms mean, and why has this pattern developed? These and related questions are addressed first.

### International Business Defined

There are many different forms and levels of international business. Though the lines that distinguish one from another are perhaps arbitrary, we will identify four forms of international business.[4] These are illustrated in Figure 6.2. A **domestic business** acquires essentially all of its resources and sells all of its products or services within a single country. Many small but few large businesses are predominantly domestic. The few that do exist are large domestic businesses in the world today.

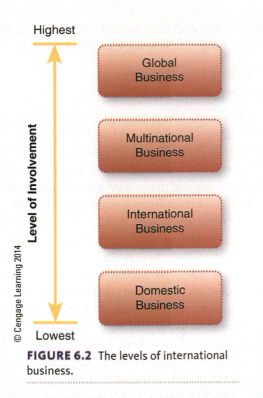

Highest

Level of Involvement

Lowest

Global Business

Multinational Business

International Business

Domestic Business

© Cengage Learning 2014

**FIGURE 6.2** The levels of international business.

Most large firms today are either international or multinational operations. An **international business** is primarily based in a single country but acquires some meaningful share of its resources or revenues from other countries. Walmart might fit this description. Most of its stores are in the United States, but many of the products it sells are manufactured abroad. Mrs. Klein's Pickles also fits here. Its processing facilities are all in the United States but it acquires many of its cucumbers from Mexico. A **multinational business** has a worldwide marketplace where it buys raw materials, borrows money, manufactures its products, and subsequently sells its products. Kraft is an excellent example of a multinational company. It has sales and production facilities around the world. Although it is a U.S. company, Kraft makes and sells products in other countries that are never seen in the United States. For example, its Milka brand of chocolate is a leading seller in Europe, *LU* biscuits are a major brand in France, and Trakinas is Brazil's favorite sandwich cookie. Kraft brands are designed and produced for and sold in individual markets, wherever they are and without regard for national boundaries.[5] Multinational businesses are often called **MNEs** or **multinational enterprises**.

The final form of international business is the global business. A **global business** transcends national boundaries and isn't committed to a single home country. While no business has truly achieved this level of international involvement, Nestlé comes close. Nestlé is based in Vevey, Switzerland, has a Belgian CEO, and holds most of its assets and gets almost all of its revenues outside Switzerland. The firm has 13 managers on its executive board, only three of whom are Swiss. About the only things that make Nestlé a Swiss firm are that its headquarters are in Switzerland and Swiss investors still own over half of the firm's stock.[6]

## The Growth of International Business

To understand why these different levels of international business have emerged, we must briefly look back to the past. After World War II, the United States was by far the dominant economic force in the world. Many countries in Europe and Asia had been devastated. There were few passable roads, few standing bridges, and even fewer factories dedicated to the manufacture of peacetime products. Places less affected by the war—Canada, South and Central America, and Africa—did not have the economic muscle to threaten the economic pre-eminence of the United States. The same was true for food and other agricultural products. Thus, when anyone in the world wanted to buy automobiles, electronic equipment, foodstuffs, or machine tools, there was only one place to shop: the United States.

In 1954, the Agricultural Trade Development Assistance Act, Public Law 480 (PL-480), was passed. Renamed Food for Peace in 1961 and expanded since, this established the framework for an expansion of U.S. agricultural exports. As a result, the United States became the "breadbasket of the world" through its exports of grains and other foodstuffs. Small farms as well as large ones benefited from this development. This clearly indicates the importance of government policy to agribusiness.

Companies in war-torn countries had no choice but to rebuild, sometimes from scratch. They were in the unfortunate but eventually advantageous position of having to rethink every facet of their operations, from technology and production to finance and marketing. Although it took many years for these countries to recover, when they eventually did, they were poised for growth. During the same era, U.S. companies grew complacent. Increased population spurred by the post-war baby boom provided growth, and increased affluence resulting from the postwar economic boom greatly raised the average individual's standard of living and expectations. The

U.S. public continually wanted new and better products and services. Companies in the United States profited greatly from this pattern; some may have been guilty of taking it for granted.

But U.S. firms are no longer isolated from global competition or the global market.[7] A few simple numbers help tell the full story. First, the volume of international trade increased more than 2,000 percent from 1960 to 1990. Foreign investment in the United States was about $300 billion in 2000, but then dropped off sharply. In 1960, 70 of the world's 100 largest firms were American. This figure dropped to 64 in 1970, to 45 in 1985, and to 32 in 2010.[8] The U.S. dominance of the global economy is a thing of the past.

U.S. firms are also finding that international operations are an increasingly important element of their sales and profits. For example, in 2002 Exxon realized 69.4 percent of its revenues abroad. For Chevron, this percentage was 55.9 percent. For Philip Morris it was 50.7 percent; and for Walmart it was 16.8 percent.[9] From any perspective it is clear that we live in a truly global economy. The days when U.S. firms could safely ignore the rest of the world and concentrate only on the U.S. market are gone forever. Now these firms must concern themselves with the competitive situations they face in lands far from home and with how companies from distant lands are competing in the United States. In recent years, for instance, U.S.

exports of soy and rice have been declining due to increased competition from Central America.

## Managing the Process of Internationalization

Managers should also recognize that the international environment dictates three related but distinct sets of challenges. Each challenge carries its own unique set of skill demands. First is the question of whether an organization should have any international involvement. That is, does it have a strategic "fit" with international operations at any level, and are international operations consistent with the organization's long-term strategy? Once this question is answered in the affirmative, a set of challenges must be confronted in terms of the level of international involvement. For example, a firm that wants to move from being domestic to importing and exporting has to manage the transition as does a firm moving from an international to a multinational business. The other set of challenges occurs when the organization has achieved its desired level of international involvement and must then function effectively within that environment. This section highlights the first set of challenges, while the next section introduces the second set.

When an organization makes the decision to increase its level of international activity, it can adopt several alternative strategies. The most basic ones are shown in Figure 6.3.[10]

**FIGURE 6.3** Alternative strategies for international business.

© Cengage Learning 2014

**Importing and Exporting.** Importing or exporting is usually the first type of international business in which a firm gets involved. **Exporting** means making the product in the firm's domestic marketplace and selling it in another country. Both merchandise and services can be exported. **Importing** occurs when a good, service, or capital is brought into the home country from abroad. For example, Inca Quality Foods founder, Luis Espinoza, teamed up with Kroger to import and distribute food from Mexico.[11] Imports into the United States include lamb from New Zealand; asparagus from Peru; apples and apple juice from Argentina, Canada, Chili, and New Zealand; fresh flowers from Columbia and The Netherlands; and wine (Riunite, Dom Perignon, Zeller Schwartz Katz) from Italy, France, and Germany, respectively. U.S. firms routinely export grain to the Russia processed foods to Japan, Canada, Mexico, and Korea; and farm machinery all over the globe (Canada, Australia, Mexico, Germany, France, United Kingdom, China, Brazil, South Africa, Russia, Japan, and Kazakhstan, to name some of the larger ones).

An import/export operation has many advantages. For one thing, it is the easiest way to enter a market with only a small outlay of capital. Because the products are usually sold "as is," there is no need to adapt the product to the local conditions, and very little risk is involved. There are also disadvantages. For example, imports and exports are subject to taxes, tariffs, and higher transportation expenses. Furthermore, because the products are not adapted to local conditions, they may miss the needs of a large segment of the market. Finally, some products may be restricted and thus can be neither imported nor exported. Agricultural products, in particular, may be subjected to strong regulation for safety reasons and to protect home country markets.

---

**FOOD FOR THOUGHT 6.1**

During the first quarter of 2010, overall imports of dog and cat food from all countries were up by an average 6.7 percent to $150.4 million. The top five exporters were China, Canada, Thailand, Australia, and Ireland. Mexican exports of dog and cat food to the United States fell by more than a third.

---

**Licensing.** There are times when a company may prefer to permit a foreign company to manufacture and/or market its products under a **licensing agreement.** In return, the licensee pays a royalty, usually based on sales. Factors that may lead the product owners to this decision include excessive transportation costs, government regulations, and home production costs. Kirin Brewery permits Charles Wells Brewery in England and Molson in Canada to produce the Japanese beer, Kirin, under Kirin's guidelines.[12] Unlike exporting, actual goods do not pass from one country to another under licensing agreements. Thus, a dotted line is used to show this relationship in Figure 6.3.

Two advantages of licensing are increased profitability and extended profitability. This strategy is frequently used for entry into less-developed countries where second-generation technology is still acceptable and maybe even considered as state of the art. A primary disadvantage of licensing is inflexibility. A firm can tie up its product or expertise for a long period of time. And if the licensee does not develop the market effectively, the licensing firm can lose profits. A second disadvantage is that licensees can take the knowledge and skills to which they have been given access for a foreign market and exploit them in the licensing firm's home market. When this happens, what once was a business partner becomes a business competitor.

**Joint Ventures/Strategic Alliances.** An increasingly common form of international business is the joint venture or strategic alliance. When two firms enter into a **joint venture**, they share in the control and ownership of a new enterprise.[13] Recently, Jeffrey Ettinger, CEO of Hormel, the maker of Spam lunchmeat, indicated that it wants to expand into China with its acquisition of Skippy peanut butter from Unilever.[14] Deere & Company has already formed a joint venture in India for the manufacture of backhoes and four-wheel-drive loaders. A **strategic alliance**, on the other hand, is a cooperative agreement that does not necessarily involve ownership. Some years ago Kodak, Fuji, Canon, and Minolta agreed to work together to develop new types of film. After the new types of film were developed, Kodak and Fuji manufactured them, while Canon and Minolta produced cameras to use them. Suiza Foods and Hershey also formed a strategic alliance. Hershey is responsible for contributing enhanced flavor technologies while Suiza's Morningstar Division is responsible for contributing enhanced packaging technologies.[15]

Joint ventures and strategic alliances have both advantages and disadvantages.[16] They allow quick entry into a market by taking advantage of the existing strengths of the participants. Japanese automobile manufacturers used this strategy to their advantage to enter the U.S. market by taking advantage of the already established distribution systems of U.S. automobile manufacturers. They are also an effective way of gaining access to technology or raw materials, and they allow firms to share the risk and cost of the new venture. The major disadvantage of these approaches lies with the collaborative nature of the operation. Although it reduces the risk for each participant, it also limits the control and the financial return that each firm can enjoy.

**Direct Investment.** Another level of commitment to internationalization is direct investment. **Direct investment** occurs when a firm headquartered in one country builds or purchases operating facilities or subsidiaries in a foreign country. The foreign operations then become either a natural part of the organization or a wholly owned subsidiary of the firm. Bunge (headquartered in St. Louis, MO) recently made a direct investment when it bought Brazilian sugar mills.[17] Deere & Company invested in a manufacturing and parts distribution facility south of Moscow in Domodedovo, Russia, in 2010.[18]

Like the other approaches for increasing a firm's level of internationalization, direct investment carries with it a number of benefits and liabilities. Managerial control is more complete, and profits need not be shared as they do in joint ventures and strategic alliances. Purchasing an existing organization provides additional benefits in that the human resources, plant, and organizational infrastructure are already in place. Acquisition is also a way to purchase the brand-name identification of a product. This could be particularly important if the cost of introducing a new brand is high. When Nestlé bought the U.S. firm Carnation a few years ago, it retained the firm's brand names for all of its products sold in the United States. Notwithstanding these advantages, the company is now operating a part of itself entirely within the borders of a foreign country. The additional complexity in the decision making, the economic and political risks, and so forth may outweigh the advantages that can be obtained by international expansion.

---

## A FOCUS ON AGRIBUSINESS
### Global Issues

Typical of major global agribusiness companies, Mosaic was formed in 2004 through the combination of two existing businesses: IMC Global and the crop nutrition business of Cargill. Mosaic is a leading producer of phosphate and potash, primary nutrients that are essential to growing food. Mosaic's customers include wholesalers, retail dealers, and individual growers in more than 30 countries. To deal with that customer base, the company has operations in ten countries.

Named as one of the 100 Best Corporate Citizens by *Corporate Responsibility Magazine*, Mosaic is noted for its conservation and environment efforts. It recycles 95 percent of the water used in its phosphate business and generates 100 percent of its Brazilian power from hydroelectric sources.

Mosaic recently expanded its global operations with a major stake in a fertilizer project in Peru. The project is the Bayóvar mine, an open-pit phosphate mine in the Sechura desert in Piura, Peru, on the northern coast. In addition, Mosaic Crop Nutrition, a wholly owned subsidiary of The Mosaic Company, is a member company of Phosphate Chemicals Export Association, Inc. (PhosChem), the largest exporter of concentrated phosphate from North America. In 2010, PhosChem entered into an historic three-year agreement with Indian Farmers Fertiliser Cooperative Limited and Indian Potash Limited (IPL), two large Indian customers. This agreement will help India's expanding crop nutrient industry.[19]

One special form of direct investment is called outsourcing. **Outsourcing**, sometimes referred to as global sourcing, involves transferring production to locations where labor is cheap. Japanese businesses have moved much of their production to Thailand because labor costs are much lower there. Many U.S. firms use maquiladoras for the same purpose. **Maquiladoras** are light-assembly plants built in northern Mexico close to the U.S. border. The plants are given special tax breaks by the Mexican government, and the area is populated with workers willing to work for very low wages. There are now more than one thousand plants in the region employing 300,000 workers, and more are planned. The plants are owned by major corporations, primarily from the United States, Japan, South Korea, and major European industrial countries. This concentrated form of direct investment benefits Mexico, the companies themselves, and workers who might otherwise be without jobs (Figure 6.4). Some critics argue, however, that the low wages paid by the maquiladoras amount to little more than slave labor.[20]

These approaches to internationalization are not mutually exclusive. Indeed, most large firms use all of them simultaneously. MNEs and global businesses have a global orientation and worldwide approach to foreign markets and production. They search for opportunities all over the world and select the best strategy to serve each market. In some settings they may use direct investment; in others, licensing; in others, joint ventures and strategic alliances; and in still others, they may limit their involvement to exporting and importing.

**FIGURE 6.4** A Maquiladora factory producing automotive parts.

Corbis

## Managing in International Markets

Even when a firm is not actively seeking to increase its desired level of internationalization, its managers are still responsible for seeing that it functions effectively within whatever level of international involvement the organization has achieved. In one sense, the job of a manager in an international business may not be much different from the job of a manager in a domestic business. Each may be responsible for acquiring resources and materials, making products, providing services, developing human resources, advertising, or monitoring cash flow.

In another sense, however, the complexity of these activities is much greater for many managers in international firms. Thus, a wider array and sophistication of management skills are necessary for effectively managing in the international economy. Rather than buying raw materials from sources in California, Texas, and Missouri, an international purchasing manager may buy materials from sources in Peru, India, and Spain. Rather than train managers for new plants in Michigan, Florida, and Oregon, the international human resources executive may be training new plant managers for facilities in China, Mexico, and Scotland. And instead of developing a single marketing campaign for the United States, an advertising director may be working on promotional efforts in France, Brazil, and Japan. Figure 6.5 illustrates the complexities associated with a single firm operating in three international markets.

The key question any manager trying to be effective in an international market must address is whether to focus on globalization or regionalism.[21] A global thrust requires that activities be managed from an overall global perspective and as part of an integrated system. Regionalism, on the other hand, involves managing within each region with less regard to the overall organization. In reality, most larger MNEs manage some activities globally (for

© Cengage Learning 2014

**FIGURE 6.5** A single firm competing in international markets.

example, finance and manufacturing) and others locally (such as human resources management and advertising). We will explore these approaches more fully later in this chapter.

## The International Economy

One thing that can be helpful to managers seeking to operate in a global environment is to better understand the structure of the international economy. While each country—and indeed many regions within any given country—is unique, some basic similarities and differences can be noted. We will describe three different elements of the global economy: mature market economies and systems, developing economies, and other economies.

### Mature Market Economies and Systems

A **market economy** is based on the private ownership of business and allows market factors such as supply and demand to determine business strategy. Mature market economies include the United States, Japan, the United Kingdom, France, Germany, and Sweden. These countries have several things in common. For example, they tend to employ market forces in the allocation of resources. They also tend to be characterized by private ownership of property, although there is some variance along this dimension. France, for example, has a relatively high level of government ownership.

U.S. managers have relatively few problems operating in market economies. Many of the business "rules of the game" that apply in the United States, for example, also apply in Germany or England; and consumers there often tend to buy the same kinds of products that U.S. consumers do. For these reasons it is not unusual for U.S. firms seeking to expand geographically to begin operations in yet another market economy. Although the task of managing an international business in an industrial market country is somewhat less complicated than operating in some other type of economy, it still poses some challenges. Perhaps the foremost challenge is that the markets in these economies are typically quite mature. Many industries, for example, are already dominated by large and successful companies. Thus, competing in these economies poses a major challenge.[22]

The map in Figure 6.6 highlights three relatively mature market systems. **Market systems** are clusters of countries that engage in high levels of trade with each other (Figure 6.6). One mature market system is Europe. Until recently, Europe was two distinct economic areas. The Eastern region consisted of communist countries such as Poland, Czechoslovakia, and Rumania. These countries relied on government ownership of business and greatly restricted trade. In contrast, Western European countries with traditional market economies have been working together for decades to promote international trade.

In particular, the **European Union** (or **EU** as it is often called) has long been a formidable market system. In 1958 Belgium, France, Italy, Luxembourg, the Netherlands, and West Germany formed the European Economic Community (EEC). As other countries joined in, the EEC became the European Union in 1993. By 2007 there were 27 member countries but the financial crisis that began in 2008 led to discussion of some members dropping out. For years these countries have been following a basic plan that called for the elimination of most trade barriers and the adoption of a single currency. The European situation has recently grown more complex, however. Communism has collapsed in most eastern countries, and they are trying to develop market economies. They also want greater participation in trade with the Western European countries. In some ways the emergence of the east has slowed and complicated business activities in the west. In the long run, however, the new markets in the east are likely to make Europe an even more important part of the world economy.

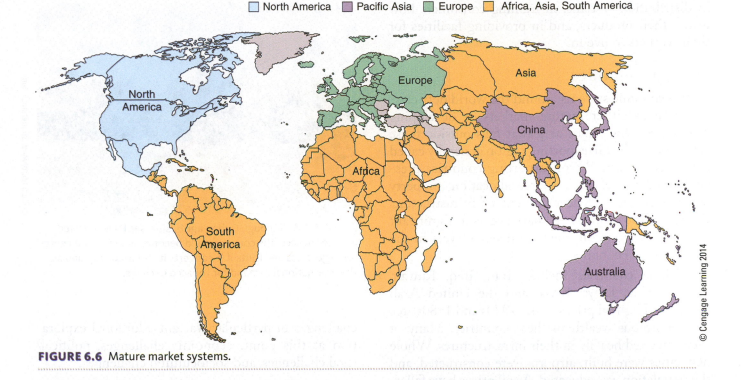

**FIGURE 6.6** Mature market systems.

*© Cengage Learning 2014*

A second mature market system is APEC, the Asia–Pacific Economic Cooperation, which includes countries that lie along the Pacific Ocean (the **Pacific Rim**). APEC was formed in 1989 by 12 Asia Pacific economies: Australia, Brunei Darussalam, Canada, Indonesia, Japan, Korea, Malaysia, New Zealand, the Philippines, Singapore, Thailand and the United States. Joining later were China, Hong Kong, Chinese Taipei, Mexico, Papua New Guinea, Chile, Peru, Russia and Viet Nam.[23]

A third mature market system is North America. The United States, Canada, and Mexico are major trading partners with one another. For example, over 80 percent of Mexico's exports go to the United States, and over 40 percent of what Mexico imports comes from the United States. Over the last several years these countries have negotiated a variety of agreements to make trade even easier, including the recent **North American Free Trade Agreement (NAFTA)**.[24]

---

### FOOD FOR THOUGHT 6.3

From 1999 to 2004, the market share for orange juice in the United Kingdom fell each year as apple juice and other blended juices increased.

---

## Developing Economies

In contrast to the highly developed and mature market economies described above, other countries have **developing economies**. These economies are relatively underdeveloped and immature. They are generally characterized by weak industry, weak currency, and relatively poor consumers. However, the government in each of these countries is actively working to strengthen its economy by opening its doors to foreign investment and by promoting international trade. Some of these countries have only recently adopted market economies, while others still use a command (i.e., government controlled) economy. Even though it is technically part of the Pacific Rim, The People's Republic of China is largely underdeveloped. Many of the countries in South America and Africa are only now developing in an economic sense. And the various states and republics that previously comprised the U.S.S.R. are also viewed as developing economies.

The primary challenges presented by the developing economies to those interested in conducting international business there are the lack of wealth on the part of potential consumers and the underdeveloped infrastructure. Developing economies have enormous economic potential, but much of it remains untapped. Thus, international firms entering these markets often must invest heavily in distribution systems, in training consumers how to use their products, and in providing facilities for their workers to live in.

## Other Economies

Some economic systems around the world defy classification as either mature markets or developing economies. One major area that falls outside these categories is the oil-exporting region generally called the Middle East. The oil-exporting countries present mixed models of resource allocation, property ownership, and the development of infrastructure. Because these countries all have access to significant amounts of crude oil, they are major players in the international economy.

These countries include Iran, Iraq, Kuwait, Saudi Arabia, Libya, Syria, and the United Arab Emirates. High oil prices in the 1970s and 1980s created enormous wealth in these countries. Many of them invested heavily in their infrastructures. Whole new cities were built, airports were constructed, and the population was educated. As oil prices have fallen, many of the oil-producing countries have been forced to cut back on these activities. Nevertheless, they are still quite wealthy. The per capita incomes of the United Arab Emirates and Qatar, for example, are among the highest in the world. Although the oil-producing nations are areas of great wealth, they provide great challenges to managers. Political instability (as evidenced by the Persian Gulf War in 1991 and the uprisings in 2011–2013) and tremendous cultural differences, for example, combine to make doing business in the Middle East both risky and difficult.

Other countries pose risks to business of a different sort. Politically and ethnically motivated violence, for example, still characterizes some countries. Foremost among these are Peru, El Salvador, India, Turkey, Columbia, and Northern Ireland.[25] Cuba presents special challenges because it is so insulated from the outside world. Because of the fall of Communism, some experts believe that Cuba will eventually join the ranks of the market economies (Figure 6.7). If so, its strategic location will quickly make it an important business center.

## Challenges of International Management

We noted earlier that managing in an international environment poses additional challenges and creates additional opportunities for the manager. Three

**FIGURE 6.7** Although Cuban agribusinesses have revolved around tobacco, the country could become an important center for vegetables and fruits if it converts its tobacco cropland as the international demand for tobacco declines.

challenges in particular warrant additional exploration at this point: economic challenges, political/legal challenges, and sociocultural challenges.[26]

## Economic Challenges of International Management

Every country is unique and creates a unique set of challenges for managers trying to do business there. However, three characteristics in particular can help managers anticipate the kinds of economic challenges they are likely to face in working abroad.[27]

**Economic System.** The first characteristic is the economic system used in a country. As we described earlier, most countries today are moving toward market economies. In a mature market economy, the key element for managers is freedom of choice. Consumers are free to decide which products to purchase and firms are free to decide which products and services to provide. As long as both the consumer and the firm can decide to be in the market, then supply and demand determine which firms and which products will be available.

A related characteristic of market economies that is relevant to managers is the nature of property ownership. There are two pure types: complete private ownership and complete public ownership. In systems with private ownership, individuals and companies—not the government—own and operate the companies that conduct business. In systems with public ownership, the government directly owns the companies that manufacture and sell products. Few countries have pure systems of private ownership or pure systems of public ownership. Most tend toward one extreme or the other, but usually a mix of public and private ownership exists.

**Natural Resources.** Another important dimension for understanding the nature of the economic environment in different countries is the availability of natural resources. There is a very broad range of resource availability in different countries. Some countries, like Japan, have relatively few natural resources of their own. Japan is thus forced to import virtually all of the oil, iron ore, lumber, and other natural resources it needs to manufacture products for its domestic and overseas markets. Its scarcity of land available for agriculture also means that Japan imports food.[28] The United States, in contrast, has enormous natural resources and is a major producer of oil, natural gas, coal, iron ore, copper, uranium, and other metals vital to the development of a modern economy. One natural resource that is particularly important in the modern global economy is oil. As noted earlier, a small set of countries in the Middle East, including Saudi Arabia, Iraq, Iran, and Kuwait, controls a very large percentage of the world's total known reserves of crude oil. Access to this single natural resource has given these oil-producing countries enormous clout in the international economy.

---

**FOOD FOR THOUGHT 6.4**

More than a dozen nations receive most of their water from rivers that cross borders of hostile upstream neighbors.

---

**Infrastructure.** A third important attribute of the economic environment of relevance to international management is infrastructure. A country's **infrastructure** includes its schools, hospitals, power plants, railroads, highways, ports, communication systems, air fields, commercial distribution systems, and so forth. The United States has a highly developed infrastructure. For example, we have modern communication, electrical-generating, and educational systems; roads and bridges are well developed; and most people have access to medical care. Overall, we have a relatively complete infrastructure sufficient to support most forms of economic development and activity.

Many countries, on the other hand, lack a well-developed infrastructure. In some countries there is not enough electrical-generating capacity to meet demand. Such countries often schedule periods of time during which power is turned off. These planned power failures reduce power demands but can be an enormous inconvenience to business. In the extreme, when a country's infrastructure is greatly underdeveloped, firms interested in beginning business may have to build an entire township, including housing, schools, hospitals, and perhaps even recreation facilities, to attract a sufficient overseas workforce.

**Purchasing Power.** A final attribute involves consumer purchasing power and income distribution in different countries. Consumer purchasing power refers to the ability of consumers in a country to purchase goods and services. If consumer purchasing power is low in a country, opportunities to develop markets in that country are not as promising as they would be had the opposite been the case. Related to this is the distribution of income in a country. Income distributions are frequently measured by the GINI Index, which measures the degree of inequality in the distribution of family income in a country (0 = total equality, 1 = total inequality). Extremely high inequality can lead to unrest and social disorder in a country. On the other hand, some degree of inequality is inevitable and perhaps even desirable as it may serve to motivate individuals to acquire skills, start business ventures, and work hard to acquire a greater share of the country's income.

Knowledge of a country's consumer purchasing power and its income distribution can help an organization in international operations involving that country. Approaching countries with large but poor populations (like Danone did with Senegal in the opening case of this chapter) would be quite different from approaching one with a very unequal distribution of income with a small but highly affluent market.

## Political/Legal Challenges of International Management

A second set of challenges facing international managers is the political-legal environment in which they will do business. Four important aspects of the political-legal environment of international management are government stability, incentives for multinational trade, controls on international trade, and the influence of economic communities on international trade.

**FIGURE 6.8** Political and economic instability threaten the international operations of organizations.

## Government Stability.

The stability of a given government can be viewed in two ways: its ability to stay in power in spite of opposing factions in the country and the permanence of its policies toward business. A country that is stable in both respects is preferable because managers then have a higher probability of successfully predicting how government will affect their businesses. Civil war in countries such as Lebanon has made it impossible for international managers to predict what government policies are likely to be and whether the government will be able to guarantee the safety of international workers. Consequently, international firms have been reluctant to invest in Lebanon (Figure 6.8).

In many countries—the United States, Great Britain, and Japan, for example—changes in government occur with very little disruption. In other countries—India, Argentina, and Greece—changes are likely to be more disruptive. Even if a country's government remains stable, there are risks that the policies adopted by that government might change. In some countries, foreign businesses may be nationalized (taken over by the government) with little or no warning. For example, the government of Peru nationalized Perulac, a domestic milk producer owned by Nestlé, because of a local milk shortage.

## International Trade Incentives.

Another facet of the political environment includes incentives to attract foreign business. For example, municipal governments in Texas have offered foreign companies like Fujitsu huge tax breaks and other incentives to build facilities there.[29] In like fashion, the French government sold land to Disney far below its market value and agreed to build a connecting freeway in exchange for the company agreeing to build a theme park outside Paris. Incentives can take a variety of forms, including reduced interest rates on loans, construction subsidies, and tax incentives. Less-developed countries tend to offer different packages of incentives. In addition to lucrative tax breaks, for example, they can also attract investors with duty-free entry of raw materials and equipment, market protection through limitations on other importers, and the right to take profits out of the country.

## International Trade Controls.

A third element of the political environment that managers should consider is the extent to which there are controls on international trade. In some instances, the government of a country may decide that foreign competition is hurting domestic trade. To protect domestic business, such governments may enact barriers to international trade. These barriers include tariffs, quotas, export restraint agreements, and "buy national" laws.[30] The Japanese market is one of the world's most difficult to enter.

A **tariff** is a tax collected on goods shipped across national boundaries. The exporting country, countries through which goods pass, and the importing country can collect tariffs. Import tariffs, which are the most common, can be levied to protect domestic companies by increasing the cost of foreign goods. For example, Japan charges U.S. tobacco producers a tariff on cigarettes imported into Japan as a way to keep those prices higher than domestic cigarette prices. Less-developed countries sometimes levy tariffs to raise money for their governments.

Quotas are the most common form of trade restriction. A **quota** is a limit on the number or value of goods that can be traded. The quota amount is typically designed to ensure that domestic competitors can maintain a certain market share. The United States, for example, has had sugar quotas for over 70 years although they have recently been eased. **Export restraint agreements** are designed to convince other governments to voluntarily limit the volume or value of goods exported to a particular country. They are export quotas, in effect. Japanese steel producers voluntarily limit the amount of steel they send to the United States each year.

**"Buy national" legislation** gives preference to domestic producers through content or price restrictions. Several countries have this type of legislation. Brazil requires that Brazilian companies purchase only Brazilian-made computers. The United States requires that many Department of Defense purchases be only from manufacturers in the United States, even though the price would be half as much if manufactured outside the country. However, the Secretary of Defense can waive the restrictions on the basis of reciprocity—if the partner country reciprocally waives its similar buy national legislation for procurements from U.S. sources.

### Trade Agreements.

Just as government policies can either increase or decrease the political risk facing international managers, trade relations between countries can either help or hinder international business. If these relations are dictated by quotas, tariffs, and so forth, they can hurt international trade. However, periodically movements to reduce many of these barriers arise. These movements generally take the form of international trade agreements.

A **trade agreement** (or trade pact) is a legal arrangement between two or more nations indicating that they will work together cooperatively in terms of trade. There are three levels of such agreements: bilateral (two countries), multilateral (more than two countries), and regional (trading blocs). There are numerous types of trading blocs depending upon the level of cooperation ranging from Free Trade Areas to Customs Unions to Economic Unions. Free Trade Areas involve two or more countries agreeing to eliminate trade barriers among themselves (e.g., the North American Free Trade Agreement). A Customs Union goes one step further in that the members also agree on a single, external tariff on goods imported from countries outside the Union. An Economic Union is a Common Market, which permits the free movement of labor and capital among member nations, that also includes an integration of economic policies and other financial regulations (e.g., the European Union).

An **international economic community** is, then, a set of countries that agrees to significantly reduce or eliminate trade barriers among its member nations. The first, and in many ways still the most important, of these economic communities is the European Union (EU), discussed earlier. Other important economic communities include the Latin American Integration Association (Bolivia, Brazil, Colombia, Chili, Argentina, and other South American countries) and the Caribbean Common Market (the Bahamas, Belize, Jamaica, Antigua, Barbados, and 12 other countries).

In 1995, the World Trade Organization (WTO) was formed in an international effort to facilitate international trade. The WTO attempts to deal with negotiating and formalizing trade agreements as well as providing a dispute resolution process among the parties. Five principles underlie the functioning of the WTO—non-discrimination between imported and domestically produced goods, reciprocal concessions among participants, binding agreements, transparency through published regulations and periodic reviews, and "safety valves" or escape clauses for unusual circumstances.

## Sociocultural Challenges of International Management

The final set of challenges for the international manager is the cultural environment and how it affects business. A country's culture includes all the values, symbols, beliefs, and language that guide behavior.

### Values, Symbols, and Beliefs.

Cultural values and beliefs are often unspoken, even taken for granted, by those who live in a particular country. Cultural factors do not necessarily cause problems for managers when the cultures of two countries are similar. Difficulties can arise, however, when there is little overlap between the home culture of a manager and the culture of the country in which business is to be conducted. For example, most U.S. managers will find the culture and traditions of England familiar. People in both countries speak the same language, they share strong historical roots, and there is a history of strong commerce between the two countries. However, when U.S. managers begin operations in Japan or the People's Republic of China, most of those commonalities disappear.[31]

Cultural differences between countries can have a very direct impact on business practice (Figure 6.9). For example, the religion of Islam teaches that people should not make a living by exploiting the misfortune of others and that making interest payments is immoral. In practice, this means that in Saudi Arabia there are no businesses that provide auto-wrecking services to tow one's car to the garage should it break down (because that would be capitalizing on misfortune). It also means that in the Sudan banks cannot pay or charge interest. Given these cultural and religious constraints, those two businesses—auto towing and banking—do not seem to hold great promise for international managers in those particular countries.

Some cultural differences between countries can be even subtler and yet have a major impact on business activities. For example, in the United States there is a very clear agreement among most managers

FIGURE 6.9 Cultural differences in food, personal values, fashion, and the like directly affect exports and imports.

about the value of time. Most U.S. managers schedule their activities very tightly and then adhere to their schedules. Other cultures don't put such a premium on time. In the Middle East, managers do not like to set appointments; and they rarely keep appointments set too far into the future. U.S. managers interacting with managers from the Middle East might misinterpret the late arrival of a potential business partner as a negotiation ploy or an insult, when it is rather a simple reflection of different views of time and its value.

**FOOD FOR THOUGHT 6.5**

Sushi originated in Southeast Asia in the fourth century B.C. as a method for preserving fish. The fermentation of the rice prevented the fish from spoiling.

**Language.**   Language itself can be a significant factor in the sociocultural environment. Beyond the obvious and clear barriers posed by people who speak different languages, subtle differences in meaning can also play a major role. For example, Gerber realized it was in trouble when it learned that its name meant "vomit" in French. Ford began to understand why its profits were lower than expected in Spain when it realized that some Spaniards read its name as "Fabrico Ordinaria Reparaciones Diaviamente," meaning "ordinarily, make repairs daily" (in the United States

this became "Fix or Repair Daily"). In many countries colors have a language of their own, which has implications for packaging and advertising as well as product color. Green, for example, is used extensively in Moslem countries, but it signifies death in some other countries. Red, white, and black also signify different moods or send different messages in different countries.

Even when the cultures of two countries are similar, there is still substantial room for misunderstanding and embarrassment. For example, when someone from the United Kingdom tells you that he is going to knock you up, take a lift, and put the telly in the boot, he has told you that he will (1) wake you in the morning, (2) take an elevator, and (3) put a television set in the trunk of your car.

Things become even more complicated when the cultures and language are truly different. In Japanese, for example, the word *hai* (pronounced "hi") means "yes." In conversation, however, this word is used much like people in the United States use "uh-huh." That is, it moves a conversation along or shows the person with whom you are talking that you are paying attention. So when does *hai* mean "yes," and when does it mean "uh-huh"? This turns out to be a relatively difficult question to answer. If a U.S. manager asks a Japanese manager if he agrees to some trade arrangement, the Japanese manager is likely to say *hai*, which may mean "yes, I agree," or "yes, I understand," or "yes, I am listening." Many U.S. managers become very frustrated in negotiations with the Japanese because they feel that the Japanese managers have said "yes" but then continue to raise issues that have been agreed upon. What many of these managers fail to recognize is that "yes" does not always mean "yes" in Japan.

## Managing in the International Economy

As already noted, managing in the international economy is both a significant challenge and an opportunity for businesses today. The nature of these challenges depends upon a variety of factors, including the size of the organization. In addition, international management also has implications for the basic functions of planning, organizing, leading, and controlling.

### The Influence of Organization Size

Though organizations of any size may compete in international markets, there are some basic differences in the challenges and opportunities faced by MNEs, medium-size organizations, and smaller organizations. But each also requires the application of appropriate management skills to succeed.

### Multinational Organizations.

The large MNEs have long since made the choice to compete in global marketplaces. In general, these firms take a global perspective on everything they do. They transfer capital, technology, human resources, inventory, and information from one market to another. They actively seek new expansion opportunities wherever feasible. MNEs tend to allow local managers a great deal of discretion in addressing local and regional issues. At the same time, each operation is ultimately accountable to a central authority. Managers at this central authority (called headquarters, central office, or some other term) are responsible for setting the overall strategic direction for the firm, making major policy decisions, and so forth. MNEs need senior managers who understand the global economy and who are comfortable dealing with executives and government officials from a variety of cultures.

### Medium-Size Organizations.

Many medium-size businesses are still primarily domestic organizations. But they still may buy and sell products made abroad and compete with businesses from other countries in their own domestic markets. Increasingly, however, medium-size organizations are expanding into foreign markets as well. One example is the Navajo organization, NAPI, discussed in Chapter 1. Another example is Molex Inc., a medium-size firm based in Chicago that manufactures electronic connectors and operates several plants in Japan and derives over half its sales from the Pacific Rim.[32] In contrast to MNEs, medium-size organizations doing business abroad are much more selective about the markets they enter. They also depend more on a few international specialists to help them manage their foreign operations.

### Smaller Organizations.

More and more smaller organizations are also finding they can benefit from the global economy. Some, for example, serve as local suppliers for MNEs. A dairy farmer who sells milk to Carnation, for example, is actually transacting business with Nestlé. Local parts suppliers also have been successfully selling products to the Toyota and Honda plants in the United States. Beyond serving as local suppliers, some small businesses also buy and sell products and services abroad.

For example, the Collin Street Bakery, based in Corsicana, Texas, ships fruitcakes around the world. In 1990, the firm shipped 145,000 pounds of fruitcake to Japan and purchased a working pineapple plantation in Costa Rica.[33] Most small businesses rely on simple importing and/or exporting operations for their international sales, thus requiring only a few specialized management positions. Collin Street Bakery, for example, has one local manager who handles international activities. Mail-order activities within each country are subcontracted to local firms in each market.

## The Management Functions in a Global Economy

The management functions that constitute the organizing framework for this book—planning, organizing, leading, and controlling—are just as relevant to international managers as to domestic managers. International managers need to have a clear view of where they want their firms to be in the future; organize to implement their plans; motivate those who work for them; and develop appropriate control mechanisms.

### Planning in a Global Economy.

To plan effectively in a global economy, managers must have a broad-based understanding of both environmental issues and competitive issues. They need to understand local market conditions and technological factors that will affect their operations. At the corporate level, executives need a great deal of information to function effectively. Which markets are growing? Which markets are shrinking? What are our domestic and foreign competitors doing in each market? Managers must also make a variety of strategic decisions about their organizations. For example, if a firm wishes to enter the market in France, should it buy a local firm there, build a plant, or seek a strategic alliance? Critical issues include understanding environmental circumstances, the role of goals and planning in a global organization, and how decision making affects the global organization. We will note special implications for global managers as we discuss planning in Chapters 8–11.

### Organizing in a Global Economy.

Managers in international businesses must also attend to a variety of organizing issues. Monsanto, for example, has operations scattered around the globe. The firm has made the decision to give local managers a great deal of responsibility for how they run their businesses.

In contrast, many Japanese firms give managers of their foreign operations relatively little responsibility. As a result, those managers must frequently travel back to Japan to present problems or get decisions approved. Managers in an international business must address the basic issues of organization structure and design, managing change, and dealing with human resources. We will address the special issues of organizing in Chapters 12–15.

**Leading in a Global Economy.**   We noted earlier some of the cultural factors that affect international organizations. Individual managers must be prepared to deal with these and other factors as they interact with people from different cultural backgrounds. Supervising a group of five managers, each of whom is from a different state in the United States, is likely to be much simpler than supervising a group of five managers, each of whom is from a different culture. Managers must understand how cultural factors affect individuals, how motivational processes vary across cultures, the role of leadership in different cultures, how communication varies across cultures, and the nature of interpersonal and group processes in different cultures. In Chapters 16–20, we will note special implications for international managers that relate to leading and interacting with others.

**Controlling in a Global Economy.**   Finally, managers in international organizations must also be concerned with control. Distances, time zone differences, and cultural factors also play a role in control. For example, in some cultures close supervision is seen as being appropriate, and in other cultures it is not. Likewise, executives in the United States and Japan may find it difficult to communicate vital information to one another because of the time zone differences. Basic control issues for the international manager revolve around operations management, productivity, quality, technology, and information systems. These issues are integrated throughout our discussion of control in Chapters 21–24.

## CHAPTER SUMMARY

International business has grown to be one of the most important features of the world's economy. Learning the skills necessary to operate in an international economy is a significant challenge facing many managers today. Businesses can be primarily domestic, international, multinational, or global in scope. Managers must understand both the process of internationalization as well as how to manage within a given level of international activity.

To compete in the international economy, managers must understand its structure. Mature market economies and systems dominate the global economy today. North America, Europe, and the Pacific Rim are especially important. Developing economies in Eastern Europe, South America, and Africa may play bigger roles in the future. The oil-exporting economies in the Middle East are also important.

Many of the challenges of international management are unique issues associated with the international environmental context. Economic, political/legal, and sociocultural challenges of international management are especially critical.

Basic issues of competing in the international economy vary according to whether the organization is a MNE, a medium-size organization, or a smaller organization. In addition, the basic managerial functions of planning, organizing, leading, and controlling must all be addressed in international organizations.

# CHAPTER ACTIVITIES

## REVIEW QUESTIONS

1. What are the four basic levels of international business activity? Give examples for each.

2. Why has international business grown so much in recent years?

3. Summarize the basic structure of the international economy. What are some of the major changes today occurring within the international economy?

4. Identify and briefly describe some of the basic challenges of international management.

5. What are some of the competitive differences for MNEs, medium-size organizations, and smaller organizations?

## ANALYSIS QUESTIONS

1. Identify industries that are most global and least global.

2. In what ways are the processes of managing an increase in internationalization versus managing in a steady state of international involvement likely to be the same and likely to be different?

3. In which situations might an organization want to decrease its level of international activity? Comment on some of the basic issues that might be involved here.

4. Some experts argue that free trade eventually helps everyone. Others, however, believe that domestic governments have an obligation to protect local businesses from foreign competition. What do you believe and why?

5. Identify several businesses and/or products that affect you regularly and are owned by foreign companies.

## FILL IN THE BLANKS

1. A business that acquires essentially all of its resources and sells all of its products or services within a single country is known as a/an _____ business.

2. An organization that is primarily based in a single country but acquires some meaningful share of its resources or revenues from other countries is known as a/an _____ business.

3. An organization that has a worldwide marketplace from which it buys raw materials, borrows money, manufactures its products, and to which it subsequently sells its products, is known as a/an _____ business.

4. A business that transcends national boundaries and is not committed to a single home country is called a _____ business.

5. A company can allows another company to use its brand name, trademark, technology, patent copyrights, or other assets under a _____ agreement.

6. An agreement to share in the control and ownership of a new enterprise is known as a/an _____ _____ .

7. A cooperative agreement that allows companies to work together but does not necessarily involve ownership is called a/an _____ _____ .

8. An economic agreement between the United States, Canada, and Mexico that aims to allow easier movement of goods and services among the three countries is the _____ _____ _____ _____ _____ .

9. The _____ _____ is a mature market system that consists of Denmark, the United Kingdom, Portugal, the Netherlands, Belgium, Spain, Ireland, Luxembourg, France, Germany, Italy, and Greece.

10. Agreements to voluntarily limit the volume or value of goods exported to a particular country are known as _____ _____ agreements.

# CHAPTER 6 CASE STUDY
## STRATEGIC POSITIONING IN THE GLOBAL LIQUOR INDUSTRY

In 2009, Prince Charles in kilts presided over the opening of a newly expanded Glenlivet whisky distillery. This facility would allow Pernod Ricard SA to expand production by 75 percent. A more apt "poster child" for globalization could not be found than in this relationship. Glenlivet, a single-malt scotch whisky first distilled by George Smith in 1824, is a Scottish icon, but its current fortunes are directed by a decidedly French firm, Pernod Ricard SA.

Diageo PLC and Pernod Ricard SA jointly bid $8.15 billion for the wines and spirits portfolio of Seagram Company Ltd of Canada on December 19, 2000. Pernod Ricard invested $3.15 billion and Diageo PLC provided the remaining $5 billion. The two companies agreed before the bid to split the liquor portfolio of the purchase. Diageo PLC was not interested in Scotch whisky holdings, and Pernod Ricard SA was very interested. It would allow them to expand their liquor portfolios in areas of future interest. It was a unique opportunity.

Scotch whisky can be produced only in Scotland. No other whiskey (spelled with an "e") can claim to be a Scotch whisky (spelled without an "e"); they are simply other forms of "whiskey." This single investment provided a unique, sustainable, competitive advantage to the holder of these assets.

This purchase and division of spoils was highly fortuitous and a major strategic gambit by Pernod Ricard SA. Ten years after this strategic move, its strategic importance was evident in growing worldwide sales. A global interest in Scotch whisky was developing. The major Scotch whisky-importing countries were the United States, France, Spain, Singapore, Venezuela, South Korea, South Africa, and Germany. This market amounted to $2.45 billion. In the United States spirits consumption (9-liter cases) was highly diversified, but Scotch whiskies were second in consumption. In fact, Vodka (Russian for "little water") was number one in market share with 30.9%, Scotch whisky 13.6%, rum 13.3%, American whiskey 10.7%, gin 6.0%, tequila 6.0% and others, 10.5%. The American competitor to Pernod Ricard SA and Diageo PLC is Constellation Brands Inc.

This market share scenario explains why Diageo PLC did not opt for the Scotch whiskies in the purchase of Seagram's spirits assets. The British company concentrates on the vodka, American whiskey, and rum portion of the United States market. Even though Pernod Ricard SA appeared to pay the smaller portion of the purchase price and gained the iconic labels, it was not as if Diageo PLC did not have its own strategic plan and, in fact, exercised it.

Strategic positioning is very important in any industry. It should be a major consideration in any acquisition and merger. Sometimes it leads to strange results, as in this case where the major exporter of Scotch whisky is a now French firm.[34]

### Case Study Questions

1. How has the global market for alcoholic beverages changed over time? Hint: search the Internet or library for information.

2. Are "whisky" and "whiskey" legally defined or merely traditionally defined? Hint: search the Internet or library for information.

3. Why is it important to position and protect brands?

# REFERENCES

1. Christina Passariello, "Danone Expands its Pantry to Woo the World's Poor," *The Wall Street Journal*, June 29, 2010, A1; Beth Kowit, "Nestlé: Tailoring Products to Local Niches," *Fortune*, July 5, 2010, at money.cnn.com (accessed July 8, 2010); M. Elliot, "The New Global Opportunity," *Fortune*, June 23, 2010, at money.cnn.com (accessed July 6, 2010); C. K. Prahalad, *The Fortune at the Bottom of the Pyramid: Eradicating Poverty Through Profits; Enabling Dignity and Choice Through Markets* (Upper Saddle River, NJ: Wharton School Publishing, 2005); H. Maucher, *Leadership in Action: Tough-Minded Strategies from the Global Giant* (New York: McGraw Hill, 1992).

2. John Pastier, "Starbucks: Selling the American Bean," *Business Week*, December 1, 2005, www.businessweek.com (accessed June 30, 2010).

3. Thomas Begley and David Boyd, "The Need for a Global Mind-Set," *MIT Sloan Management Review* (Winter 2003), 25–36; Richard M. Steers and Edwin L. Miller, "Management in the 1990s: The International Challenge," *The Academy of Management Executive* (February 1988), 21–22; David A. Ricks, Brian Toyne, and Zaida Martinez, "Recent Developments in International Management Research," *Journal of Management* (June 1990), 219–254.

4. For a more complete discussion of forms of international business, see Ricky W. Griffin and Michael Pustay, *International Business*, 6th ed. (Upper Saddle River, NJ: Prentice-Hall, 2010).

5. "Our Brands," Kraft website at www.kraftfoodscompany.com (accessed July 18, 2010).

6. "Corporate Governance Report June 2010," Nestlé's website at www.nestle.com (accessed July 18, 2010); "The Stateless Corporation," *Business Week*, May 14, 1990, 98–104.

7. Philip M. Rosenzweig and Jitendra V. Singh, "Organizational Environments and the Multinational Enterprise," *Academy of Management Review* (April 1991), 340–361.

8. "Global 500," *Fortune*, July 26, 2010, at www.fortune (accessed July 27, 2010); John Labate, "Gearing Up For Steady Growth," *Fortune*, July 29, 1991, 83–102; "The Fortune Global 500," *Fortune*, July 27, 1992, 175–232.

9. Diana Kendall, *Sociology in Our Times: The Essentials*, 7th ed. (Belmont, CA: Wadsworth, Cengage Learning, 2010), 447; and "U.S. Corporations With the Biggest Foreign Revenues," *Forbes*, July 22, 1991, 286–288.

10. John D. Daniels and Lee H. Radebaugh, *International Business*, 6th ed. (Reading, MA: Addison-Wesley, 1992).

11. Nancy J. Lyons, "Moonlight Over Indiana," *Inc.*, January 2000, 71–74.

12. "Creating a Worldwide Yen for Japanese Beer," *Financial Times*, October 7, 1994, 20.

13. Kenichi Ohmae, "The Global Logic of Strategic Alliances," *Harvard Business Review* (March-April 1989), 143–154.

14. Mark Clothier and Matthew Boyle, "Hormel Seeks Chinese Acquisition or Joint Venture," *Business Week*, March 5, 2010, at www.businessweek.com (accessed June 30, 2010); and Ian Berry and Drew Fitzgerald, "Hormel Buys Skippy, With an Eye on China," *The Wall Street Journal*, January 4, 2013, B3.

15. *The Wall Street Journal* (2000). Suiza food group: Unit will develop products in an alliance with Hershey, August 30, p. 1.

16. Dovev Lavie, "Capturing Value from Alliance Portfolios," *Organizational Dynamics* (January-March 2009), 26–36; Paul Beamish and Nathaniel Lupton, "Managing Joint Ventures," *Academy of Management Perspectives*, 2009 (23), 75–84.

17. Jack Kaskey, "Bunge to Buy Brazilian Sugar Mills for $416 Million." *Business Week*, December 24, 2009, at www.businessweek.com (accessed June 30, 2010).

18. "John Deere Officially Opens New Manufacturing Facility in Russia," John Deere website, April 27, 2010, at www.johndeere.com (accessed June 18, 2010).

19. "About Mosaic," Mosaic website at www.mosaicco.com (accessed July 5, 2010); "100 Best Corporate Citizens," *Corporate Responsibility Magazine*, at www.thecro.com (accessed July 5, 2010); Helder Marinho, "Vale Sells Stakes in Peru Fertilizer Project to Mosaic, Mitsui," *Business Week*, March 31, 2010, at www.businessweek.com (accessed July 5, 2010).

20. "The Magnet of Growth in Mexico's North," *Business Week*, June 6, 1988, 48–50; "Will the New Maquiladoras Build a Better Manana?" *Business Week*, November 14, 1988, 102–106.

21. Allen J. Morrison, David A. Ricks, and Kendall Roth, "Globalization Versus Regionalism: Which Way For the Multinational?" *Organizational Dynamics* (Winter 1991), 17–29.

22. Ben L. Kedia and Rabi S. Bhagat, "Cultural Constraints on Transfer of Technology Across Nations: Implications for Research in International and Comparative Management," *Academy of Management Review* (October 1988), 559–571; Carla Rapoport, "Japan's Growing Global Reach," *Fortune*, May 22, 1989, 48–56.

23. Louis Kraar, "The Growing Power of Asia," *Fortune*, October 7, 1991, 118–131.

24. "In the Wake of NAFTA, a Family Firm Sees Business Go South," *The Wall Street Journal*, February 23, 1999, A1, A10.

25. "Where Killers and Kidnappers Roam," *Fortune*, September 23, 1991, 8.

26. Daniels and Radebaugh, *International Business*.

27. "The Face of the Global Economy," *Business Week/Reinventing America* (A special issue of *Business Week*), 1992, 150–159.

28. Kazuaki Nagata, "Japan Needs Imports to Keep Itself Fed," *The Japan Times*, February 26, 2008, at search.japantimes.co.jp (accessed July 18, 2010).

29. John Paul Newport, Jr., "Texas Faces Up to a Tougher Future," *Fortune*, March 13, 1989, 102–112.

30. "Nations Rush to Establish new Barriers to Trade," *The Wall Street Journal*, February 6, 2009, A1, A6.

31. "Firms Address Worker's Cultural Variety," *The Wall Street Journal*, February 10, 1989, B1.

32. "You Don't Have to Be a Giant to Score Big Overseas," *Business Week*, April 13, 1987, 62–63.

33. "Famous Bakery Keeps Business Thriving," *Corsicana Daily Sun*, June 9, 1991, 1C.

34. Paul Sonne, "As World Develops a Taste, Scotch Whiskies Pour it On," *The Wall Street Journal*, June 8, 2010, B1; David Kesmodel, "Diageo Pulls Out Rokk in U.S. Vodka Wars," *The Wall Street Journal*, March 24, 2010; F. Paul Pacult, *A Double Scotch: How Chivas Regal and the Glenlivet Became Global Icons* (New York: John Wiley and Sons), 261–164.

# The Ethical and Social Environment

## MANAGER'S VOCABULARY

- codes of conduct
- conflict of interest
- corporate social audit
- ethical compliance
- ethics
- legal compliance
- lobbyist
- managerial ethics
- organizational constituents
- philanthropic giving
- political action committees (PACs)
- social involvement
- social obligation
- social obstruction
- social reaction
- social responsibility
- whistle blowing

## LEARNING OBJECTIVES

After studying this chapter, you should be able to:

- Define ethics and discuss how they are formed.

- Describe the relationship between ethics and management.

- Understand the management of ethics, including the necessity for top-management support and codes of conduct.

- Define social responsibility, discuss arguments both for and against it, and identify the general areas of social responsibility.

- Discuss approaches to social responsibility.

- Discuss the role of government in business's social responsibility.

- Describe how organizations manage social responsibility.

## "Smucker" Not an Ugly Name; "Reverse Morris Trust" Not a Football Play

The J.M. Smucker Company of Orrville, Ohio, is not a Fortune 500 company, but it's close (No. 540 in 2010). It is well known for its iconic brands (Smucker's jam, jellies, and preserves) and less known for its other brands (Jif Peanut Butter, Hungry Jack Pancakes, and Folgers Coffee). If one were pressed to name what business Smucker's is in, a glib answer, "the breakfast table," would not be incorrect. J.M. Smucker has two other distinctions besides its unique product mix, and they are lessons in corporate survival.

**FIGURE 7.1** Smucker's has been making a variety of top-selling jams and jellies for more than a century.
© Isantilli/www.Shutterstock.com

In 1897 an Ohio farmer, Jerome Monroe Smucker, made his living selling apples, but his neighbors also were selling apples. He recognized that he could make a greater return on his orchard investment if he could take a product from the orchard directly to the consumer. His problem was simple, it seems: how to distinguish or differentiate his apples from his neighbors' (competitors') apples. As an astute agribusiness person, he also knew the meaning of "value-added"—doing something to a product to make it sell at a higher profit rate. The answer was not so obvious to anyone but him: Don't sell apples; sell apple butter.

Smucker was also a clever entrepreneur. He understood that you could put a label on the apple butter to brand the product and develop loyalty to the brand. His horse-drawn wagon soon became a familiar and welcome sight in Ohio because he produced a good quality product. The family name became etched in consumers' minds with the advertising slogan, "With a name like Smucker's, it has to be good!" (Figure 7.1).

Today, the horse-drawn wagon is gone, but the Smuckers are not. Smucker's is a $5.5 billion international company with over 5,000 employees. Since its inception, the company has had five chief executives—all Smuckers. In 2012 two Smuckers share the chief executive office—Timothy and Richard, brothers and great-grandsons of the company's founder. Waiting their turns are two cousins, who are division presidents. Lesson One: it takes a huge amount of cooperation, understanding, patience, and tenacity to keep a firm growing, focused, and "in the family."

Because Smucker's is a publicly traded company, the family had to work together so it could demonstrate to shareholders that the family understood how to operate and grow a business. It did, and the shareholders are happy. Over the past ten years Smucker's has delivered a return on its investment of 309 percent vs. 15 percent for Standard and Poors. The Smucker heirs also have had to demonstrate that they were just as clever as J.M. Smucker. They did so in a rather unique maneuver called a "Reverse Morris Trust." This is Lesson Two: control.

A "Reverse Morris Trust" is not a football play; it is a complicated financial arrangement. When Smucker's purchased Jif Peanut Butter and Crisco from Procter & Gamble (P&G) for close to $800 million and Folgers for $3.7 billion, it was through this mechanism. P&G wanted to sell three firms—Jif, Crisco, and Folgers—so they were spun-off as separate companies. The separate companies then merged with Smucker's as a merger of equals with a stock swap. The merged companies' owners then owned stock control of the new company in excess of 50 percent.

Realistically, Smucker's did not buy these companies from P&G. If it had, then P&G would have had to pay capital gains tax. So Smucker's purchased newly independent companies created by P&G in the spin-offs. The companies were owned by former P&G shareholders. This is the tax-free part of the transaction. Smucker's used newly printed shares of stock to purchase the new companies, again not from P&G but in deals with the companies themselves. Since it was stock-for-stock merger of equals, this arrangement is tax free, also. Smucker's then owned Folger's, Jif Peanut Butter, and Crisco. The shareholders of these companies, formerly P&G shareholders, then owned 53.5 percent of Smucker's. They were required to own a majority of the company or else the tax-free status would have been lost.

But how do you own something if you are owned? Look at the end result of these transactions. For P&G the sale of the three companies was tax free. The Smucker's company gained the number one brands in jams and jellies, peanut butter, coffee, and cooking oil—also tax free. And the Smucker family retained control of a much larger multinational company although it experienced a dilution of ownership from 30 percent to 6 percent.

How do you control a company with only 6 percent of the stock? The J.M. Smucker and Company corporate charter is the answer. It gives long-term stockholders (family) ten votes per share on key corporate issues; short-term stockholders (not family), one vote per share. If a family member sells family shares, those shares revert to one-vote shares. "This is the first time that two companies have done a second Reverse Morris Trust with each other," said Robert Willens. "They must have really gotten along well the first time." This is why J.M. Smucker was a "clever" entrepreneur—which is fortunate when your name is more memorable than beautiful.

Not everyone thought this was clever. While perfectly legal, it seemed to run counter to much social sentiment. Because of the tax-free nature of this transfer of assets, new legislation initiated in 2010 known as the Dodd-Frank Bill, or "Finreg" (Financial regulation and reform), the Reverse Morris Trust may have breathed its last breath. The next generation of Smuckers will have to try much harder to stay in control and recognize their social responsibilities. J.M would have understood.[1]

## INTRODUCTION

The debate about the role of business in protecting the natural and social environment clearly demonstrates the increasingly important relationship between business and its environment. If a company, large or small, is not socially responsible, it may receive well-deserved criticism. But even companies that try their best to be socially responsible can nevertheless quickly become embroiled in an event that captures the public's attention (like the projected demise of the spotted owl). Some organizations know how to function within their social environments; others think they know how to function, only to occasionally stumble, as did Smucker's when they used the Reverse Morris Trust. Still others have little clear understanding of what it means to function within the social environment.

We introduced the nature of the social environment of business in Chapter 4. This chapter deals more specifically with that environment. We first examine individual ethics and then discuss managerial ethics and how organizations can manage the ethical behavior of their members. We then look at the nature of corporate social responsibility and approaches to it. Next, we analyze relationships between the government and business in a social context. Finally, we explore how organizations go about managing social responsibility.

# The Nature of Ethics

One critical component necessary to understanding the social environment of business is understanding the nature of ethics. Ethics are an individual-level phenomenon. By this we mean that the concept of ethics does not apply to an entire organization. It makes no sense, for example, to talk about an organization's ethics. But each individual manager within an organization has his or her own personal set of ethics, and the ethical standards and behaviors of those managers can affect the organization in profound ways.

As indicated in Table 7.1, there are numerous approaches to ethics. There does not appear to be any one approach that is consistently applied in management, but elements of several generally appear.

## What are Ethics?

What are ethics? Ask any four experts and you'll probably get four different answers—and maybe even more. For our purposes, let us define **ethics** as those standards or morals a person sets for himself or herself regarding what is good and bad or right and wrong.[2] Thus, while some behavior is clearly ethical or unethical, much of the behavior in an organization tends to be relatively ethical or relatively unethical.

It is important to note the distinction between something that is ethical and something that is legal. The law defines various kinds of acts as acceptable or unacceptable. In contrast, ethics often go beyond the law and are based more on prevailing societal norms and expectations. Thus an action can be (1) both legal and ethical, (2) legal but unethical, or (3) both illegal and unethical. Experts differ in their opinions as to whether an action can be ethical but illegal.

For example, suppose you are a manager for a large food processing company. Today you found a $100 bill on the floor and turned it in to the company lost-and-found department. This action is both legal and ethical. You also suspect that another manager may be stealing company property for his own use, but you decide to ignore it. Although your action here is not illegal, it is probably unethical. The actions of the manager who is stealing, though, are both unethical and illegal. Suppose you take $20 from petty cash to help a janitor who has no money to feed his family. Some would argue that this action, though illegal, is ethical. Clearly, then, the determination of ethical behavior is complex and clouded by individual values, opinions, and logic.

**TABLE 7.1** Foundations of Ethics

| | |
|---|---|
| **Deontological** | The action or decision itself rather than its results determine whether it is ethical. |
| **Egoism** | Personal self-interest determines whether an action or decision is ethical. |
| **Justice** | The type of fairness (distributive, procedural, or interactional) determines whether an action or decision is ethical. |
| **Relativism** | Whether or not an action or decision is ethical depends upon the situation. |
| **Teleological** | The results of an action or decision determine whether or not it is ethical. |
| **Utilitarianism** | The determination is based on "The greatest good for the greatest number." |

Source: Based on Ferrell, O.C., John Fraedrich, and Linda Ferrell. *Business Ethics.* Boston, MA: Houghton Mifflin Company, 2002, p. 57. Reprinted by permission.

## The Formation of Ethics

Where do ethics come from? How are they formed? Figure 7.2 illustrates the most common factors that determine individual ethics.

**Family Influences.** Family influences play a key role in determining an individual's beliefs as to what is and is not right. For example, a person who grows up in a family with high ethical standards and whose members adhere consistently to those standards is likely to develop higher ethical standards herself than is someone who grows up in a family environment characterized by low or inconsistently practiced ethical standards. Clearly the influence of the family is particularly significant when the family is also the business, as with many agribusinesses.

**Peer Influences.** Peer influences are also quite important in determining a person's ethics. Childhood friends, classmates, and others in a person's social network can shape his or her ethics. Peer pressure, for instance, can help determine how much a person will engage in such questionable activities as shoplifting, experimenting with drugs, and so forth.

**Past Experiences.** As individuals grow, their past experiences can also play a role in determining the evolution of their ethical standards. If they behave unethically in given situations and suffer negative consequences (feelings of guilt, getting caught), their behaviors will probably be more ethical next time. Conversely, if their unethical behaviors do not lead to feelings of guilt but instead lead to

rewards, they may choose to behave the same way when confronted with similar situations.

**Values and Morals.** At a more general level, basic values and morals influence ethics. A person who is profoundly religious, for example, will almost certainly have strong feelings about what is right and wrong—although these may differ from other people's definitions of right and wrong. Either way, such beliefs will probably carry over to help shape his or her personal ethics.

**Situational Factors.** Finally, situational factors are important. These are events that occur in a perhaps random way and that have the potential to determine behavior that may or may not be consistent with a person's ethics. For example, consider an employee who is honest and hard working. His wife loses her job, and the family begins to have trouble making ends meet. One day, when things look especially bad, an opportunity arises for him to make some extra money by selling a company secret to a competitor. The employee's financial situation and his despair at seeing his family suffer might cause him to accept the unethical offer. This factor is situational because if his wife had not lost her job or if he had not been offered the opportunity to sell company secrets, he might well have remained a dedicated, honest, and loyal employee throughout his career.

© Cengage Learning 2014

**FIGURE 7.2** The formation of an individual's ethics.

**FOOD FOR THOUGHT 7.1**

Ancient Egyptian governments controlled the populace by strictly controlling the production and distribution of bread, the primary food source for most people. During the eighteenth century one of the triggers for the French Revolution was a shortage of bread.

## Managerial Ethics

Managers of organizations are not robots. They are not programmed to always do the same thing regardless of the circumstances. Indeed, one of the most important factors in the behavior of managers is their ethics, because of the multitude of situations that confront managers and the ethical context of their jobs.[3] **Managerial ethics**, then,

**TABLE 7.2**  The Ethical Dilemmas of Management

| SITUATIONS INVOLVING THE RELATIONSHIP OF ⇩ | MAY LEAD TO ETHICAL DILEMMAS REGARDING THESE ISSUES OR GROUPS ⇩ |
|---|---|
| The firm to the employee | Human Resources<br>　Hiring and firing<br>　Wages and working conditions<br>　Privacy |
| The employee to the firm | Issues<br>　Conflicts of interest<br>　Secrecy and espionage<br>　Honesty in all matters<br>　　(e.g. absences, expense<br>　　accounts, supplies, etc.) |
| The firm to the environment | Groups/Organizations<br>　Community<br>　Customers<br>　Competitors<br>　Government<br>　Stockholders<br>　Suppliers and dealers<br>　Unions |

Source: GARRETT, THOMAS M.; KLONOSKI, RICHARD J., BUSINESS ETHICS, 2nd Edition, © 1986, pp.vii-x. Adapted by permission of Pearson education, Inc., Upper Saddle River, NJ

refers to the ethics of a person performing in a managerial role.

## Ethics and Management

Managers face ethical dilemmas almost daily (see Table 7.2). They occur when a manager faces two or more conflicting ethical issues. Those dilemmas occur in relationships of the firm to its employees, of the employees to the firm, and of the firm to its environment.

### Relationship of the Firm to the Employee.

The relationship of the firm to the employee involves the ways in which the organization chooses to treat its employees in different situations where ethics can come into play. In the area of hiring and firing, for instance, managers must make ethical decisions regarding who is the most qualified, how to treat minorities, and so forth. Highly qualified workers are sometimes reluctant to accept jobs in a family business, such as a family farm, because they realize that primary loyalty will most likely be extended first to family members.

Wages and working conditions must also be considered. Managers must establish a pay level that will satisfy employees but that is not excessive.

A major ethical issue today involves the compensation of top managers. Citing only one of many examples, the CEO of H.J. Heinz received more than $14 million in annual compensation in 2010. Critics argue that this is excessive.[4] The employer must also provide a work environment that is relatively safe and free from hazards, including incidents of workplace violence. A reasonable degree of job security is also something most people consider to be an ethical concern, but unfortunately that seems to be changing.

There are also ethical issues regarding the private lives of employees.[5] Employees who have drinking or drug problems may be of concern to the organization, even if they seem to be keeping their problems separate from the workplace. An area of related interest is garnishment of wages, which happens when a creditor forces the organization to pay a portion of an employee's wages toward that employee's debts.

### Relationship of the Employee to the Firm.

Other issues relate to the relationship of the employee to the firm. The focus here is how the individual behaves vis-à-vis the organization. Conflicts of interest are one major consideration. A **conflict of interest** exists when an employee is put into a situation in which his decisions may be compromised because of competing loyalties. Suppose a purchasing manager accepts a free vacation from a major equipment supplier. The next time the company needs to buy a new piece of equipment, the manager may feel obligated to give the contract to that supplier. Walmart feels so strongly about avoiding conflicts of interest that it will not allow a merchandise buyer to accept meals or gifts from sales representatives.[6] Moonlighting is also an issue. When an employee has another job in addition to her primary one, fatigue may hinder her performance in both jobs.

Secrecy and espionage are also valid considerations. For example, an employee of a computer firm may have plenty of opportunities to sell information about new products to other companies. Finally, basic issues, such as stealing and dishonest handling of expense accounts, are relevant. When an employee takes home a pad of paper from the office or makes a long-distance call on the company telephone to an old friend or family member, he is technically stealing. Likewise, a manager is stealing when she has a non-business lunch with a friend and writes the bill off on her company expense account.

## Relationship of the Firm to the Environment.

There are broad ethical considerations in how the organization interacts with various elements of its environment.

A critical component of this relationship is the *customer*. Managers must contend with a number of issues involving customers, answering such questions as when to advertise, what the warranty should be, how to price its products, and how concerned it should be with product safety.[7] In agribusiness, an issue that can be particularly important is the accuracy of the company's scales or other measurements.

Relations with *competitors* are also important. Price cutting and unfair competition can drive smaller firms out of business. A group of small, independent druggists in Arkansas sued Walmart, claiming that the discounter sold many products below cost in order to drive them out of business.[8] Price fixing—developing an agreement with competitors to sell competing products or services for the same price—is also both unethical and illegal.

Relations with *stockholders* or other owners are obviously crucial since, as owners, they hold the power. Managers have the responsibility of working in owners' best interests and of reporting appropriate information to them on a timely basis. Similarly, questions about the appropriate levels of executive compensation and benefits (as noted earlier) affect owners. Large companies with considerable power over their suppliers and dealers are often in situations in which ethical dilemmas arise. For example, if Chrysler does not keep its dealers properly informed about upcoming model changes, price adjustments, and so forth, the dealers can become resentful. On the other hand, too much information might leak into the hands of competitors.

Relations with *unions* also involve ethical issues. If a firm divulges too much information about its profitability, the union might increase its wage demands. Too little disclosure, however, is not conducive to ethical bargaining.

Another factor is the *community and surrounding environment*. In recent years, there has been a trend for communities that are hungry for industry to offer concessions on taxes and utilities, free land, and other incentives. When a company plays competing communities against one another, it may be violating ethical norms. Where building business networks is critical, managers may be tempted to grant favors or provide inside information to establish those connections.[9]

Additional issues of particular interest might be pollution and other environmental impacts, participation in United Way activities, and so forth. In 2008, an outbreak of *salmonella* linked to tomatoes led McDonald's, Burger King, and several other U.S. restaurants to dispose of tomatoes. Walmart stopped selling certain varieties, too. This sort of quick response serves to indicate the firms' ethical standards regarding customers and the environment.[10]

Monsanto claims to be doing its part to help the environment with new products and genetically engineered crops.[11] New products aren't always met with favorable responses. In 2010, West Virginia's attorney general warned agribusiness giant Monsanto that its claims about farmers reaping higher yields from a new generation of the company's genetically modified soybean seed might violate a state law on deceptive advertising.[12]

Relations with *governments* involve ethical issues, too. Taxes must be accurately determined and paid on time. Regulatory agency rules and standards must be adhered to. Lobbyists must carefully follow laws. Organizations that deal with regulators should avoid even the suspicion of conflict of interest. Yet not all managers avoid bribery, "cooking the books," keeping two sets of books, and other forms of unethical behavior.

### FOOD FOR THOUGHT 7.2

Ethics is more than one's feelings about right or wrong, one's religious beliefs, doing what the law requires, or the standards of behavior that society accepts.

## The Ethical Context of Management

It is obvious from this discussion so far that many management activities occur within an ethical context. The key dimensions of this context are illustrated in Figure 7.3.[13] First, the manager's *personal ethics*, as explored earlier in this chapter, are a major determinant of his or her ethical context. Values, predilections about right and wrong, and sense of justice and fairness all come into play. Susceptibility to situational factors is also relevant.

Second, the specific *organizational context* is important to the ethical managerial context. Of special interest are organizational practices and the behavior of leaders and peers. Organizational practices are ways in which the organization deals with

**FIGURE 7.3** The ethical context of management.

© Cengage Learning 2014

the ethical situations it encounters. Some organizations reward people who report improprieties and punish those guilty of committing them. Others tend to punish those who make misdeeds public ("whistle blowers") and do little or nothing to the guilty parties.[14] For instance, a manager at Citibank reported to his superiors that one division of the bank was engaging in illegal activities to increase its profits. He was fired.[15] The higher good here was the company policy—the company, not the individual, was given precedence. The behavior of leaders and peers can be a big influence on the individual manager's ethical context. If the people surrounding a manager routinely engage in unethical behavior, the manager is likely either to start such practices himself or herself or to leave the organization.

The *environmental context* is the third vital factor. Competition, as we have seen, is a major force to consider. When competition is keen, there is strong pressure to resort to whatever means are available to get an advantage. Regulation by the government also determines the ethical context. Too much regulation can handcuff the manager so much that she feels forced to bend the rules or the law to compete. By the same token, too little regulation can provide too many opportunities to engage in questionable ethical

## A FOCUS ON AGRIBUSINESS
### Ethics

Archer Daniels Midland has taken strong steps to correct past mistakes, thanks to Matt Damon's portrayal of Mark Whitacre in the Hollywood movie, *The Informant*. Yet, lots of people are aware of ADM's role in price fixing during the 1990's. Others may have noted the more recent tomato price-fixing scandal. Still others may recall that Gage's Fertilizer & Grain and Minn-Chem filed suit against Potash Corp., Agrium, Cargill's Mosaic Co. and Mosaic Crop Nutrition units, as well as Russian and Belarusian companies accusing them of price fixing.

In 2010 Land O'Lakes, its egg-producing subsidiary Moark, and Moark's Norco Ranch Inc., agreed to a $25 million settlement in an egg price-fixing case. Claims were still active at that time against Cal-Maine Foods Inc., Michael Foods Inc., and Rose Acre Farms Inc. Those bringing suit include T.K. Ribbing's Family Restaurant in Falconer, NY, Lisciandro's Restaurant in Jamestown, NY, Solovy Foods Inc. in Vernon, CA, and Karetas Foods Inc. in Reading, PA.

But price fixing certainly is not the only ethical issue in agribusiness. Archer Daniels Midland, Cargill, Bunge, Kraft Foods, Kellogg's, and other U.S. food producers and agricultural firms came under attack in 2008 for their usage of palm oil. The issue is that native forests are eliminated in favor of groves of palm trees, drastically altering the ecology of certain regions. Consider these also: pesticide and herbicide use, food additives, genetically modified seeds and foods, changing rain forests into agricultural land, selective breeding, animal welfare, and the list goes on. And this list does not even mention "cooking" accounting/financial books, discrimination, bullying/abusive bosses, and other issues. Clearly, ethics enter into everything that happens in agriculture and agribusiness.[16]

practices. The norms of the sociocultural environment make up the last major part of the ethical context of management. In some countries bribes, price gouging, and industrial espionage are normal business practices. In some countries managers are expected to follow accepted ethical behavior.

## Managing Ethics

One ethical dilemma that managers face is whether to enforce a company policy that is wrong, or that they feel is wrong. By enforcing it, they are condoning the wrong. By not enforcing it, they are violating company policy and are therefore subject to dismissal. By remaining silent, they become in effect a participant and therefore a supporter of the policy. The same holds true for individuals who are asked to perform unethical acts—they cannot justify their participation by hiding behind the excuse of just doing what they were told to do.

Because managers have become increasingly aware of the importance of ethics, they have also taken a greater interest in how they and their organizations should attempt to manage this area. In fact, a retired executive gave $30 million to the Harvard Business School to fund the teaching of ethics. And the Graduate School of Business at Bentley College has made ethics an integral part of its curriculum for years.[17] Today the two most common approaches to the management of ethics are through top-management support and formal codes of conduct.

### Top-Management Support

For organizations to develop and maintain cultures in which ethical managerial behavior can thrive, top management must support such behavior. In the case of Summer Farms, that means that Joshua Summer and his successors must support ethical behavior on the part of everyone in the organization. Executives can do several things to demonstrate such support.[18]

First, they can adhere to ethical standards themselves. This is probably the most important thing that can be done to promote ethical behavior throughout an organization. If middle and lower-level managers see top managers behaving unethically, they are likely to follow suit.

Another important action is to provide and encourage training in ethics.[19] In 2005, Citigroup announced it was undergoing an "ethics overhaul" with mandatory ethics training for everyone.[20]

Boeing has a program to "sensitize" employees to ethical conflicts; the Sun Company, Inc. supplements a compliance questionnaire with talks by its corporate counsel; and at Hercules, Inc. managers must sign forms saying that they have abided by company policy.[21] In addition, Office Depot, General Dynamics, McDonnell Douglas, Chemical Bank, and American Can Company have initiated ethics training for their employees.[22]

### Codes of Conduct

Another important step in the management of ethics is to establish **codes of conduct**, which usually state the importance of following ethical business practices in all areas of the organization's activities. These codes are symbolic but meaningful statements of the company's concern.

Professional groups such as the International Food and Agribusiness Management Association (IFMA), the National Grain and Feed Association (NGFA), and the American Registry of Certified Professionals in Agronomy Crops and Soils (ARCPACS) have long had codes of conduct for their members. More and more often, however, individual companies are developing their own codes of ethical conduct. One survey found that 75 percent of the 1,200 largest companies in the United States had formal ethics codes, such as Johnson & Johnson's shown in Figure 7.4.[23]

## The Nature of Social Responsibility

Whereas ethical behavior is a phenomenon primarily at the individual level, social responsibility applies more to the organizational level. **Social responsibility** refers to the obligations of the organization to protect and/or enhance the society in which it functions.[24] As we will see, however, people hold different opinions as to its real nature.[25]

### Historical Evolution

Over the years both society and organizations have taken many different views of social responsibility.[26] In general, the current view of corporate social responsibility has evolved through an historical development over three distinct periods in the United States.[27]

The first period occurred between 1860 and 1890. During this era the so-called captains of industry—Andrew Carnegie, John D. Rockefeller,

# Our Credo

We believe our first responsibility is to the doctors, nurses and patients,
to mothers and fathers and all others who use our products and services.
In meeting their needs everything we do must be of high quality.
We must constantly strive to reduce our costs
in order to maintain reasonable prices.
Customers' orders must be serviced promptly and accurately.
Our suppliers and distributors must have an opportunity
to make a fair profit.

We are responsible to our employees,
the men and women who work with us throughout the world.
Everyone must be considered as an individual.
We must respect their dignity and recognize their merit.
They must have a sense of security in their jobs.
Compensation must be fair and adequate,
and working conditions clean, orderly and safe.
We must be mindful of ways to help our employees fulfill
their family responsibilities.
Employees must feel free to make suggestions and complaints.
There must be equal opportunity for employment, development
and advancement for those qualified.
We must provide competent management,
and their actions must be just and ethical.

We are responsible to the communities in which we live and work
and to the world community as well.
We must be good citizens—support good works and charities
and bear our fair share of taxes.
We must encourage civic improvements and better health and education.
We must maintain in good order
the property we are privileged to use,
protecting the environment and natural resources.

Our final responsibility is to our stockholders.
Business must make a sound profit.
We must experiment with new ideas.
Research must be carried on, Innovative programs developed
and mistakes paid for.
New equipment must be purchased, new facilities provided
and new products launched.
Reserves must be created to provide for adverse times.
When we operate according to these principles,
the stockholders should realize a fair return.

*Johnson & Johnson*

**FIGURE 7.4** Johnson & Johnson's code of ethical conduct.
Source: Used with permission of Johnson & Johnson.

J. P. Morgan, Cornelius Vanderbilt, and others—were creating the giant steel, oil, banking, and railroad corporations that came to dominate U.S. industry. In contrast to earlier organizations, these mammoth entities held enormous power in our emerging industrial society. Abuses of this power—labor lockouts, kickbacks, discriminatory pricing, and predatory business practices—caused both a public outcry and governmental action, and several laws were passed to regulate the way in which business was carried out. These early laws indicated for the first time the interdependence of business, government, and the general public.[28]

The second period spanned the few years following the stock market crash of 1929. Mergers and general business growth had continued, and by the 1920s big business had truly come to dominate the U.S. economy. Because of this, most Americans blamed large corporations for the Great Depression, and President Roosevelt and other supporters of the New Deal succeeded in passing more legislation targeted at those corporations. In particular, laws from this era specifically delineated the social responsibilities of businesses and reinforced the importance of fairness and ethical practices at all levels.

The third period began during the 1960s and early 1970s, an era characterized by a great deal of social unrest and public awareness. Young people lashed out at the government, big business, and other dimensions of what they called *The Establishment*. There were campus sit-ins, business boycotts and bombings, and protest marches about dozens of social causes. Government began to take a greater role in business, getting involved in everything from regulations for packaging over-the-counter drugs to consumer warnings on cigarettes; and business tried to respond by espousing a greater commitment to benefiting society (Figure 7.5). The effects of this period of crisis are still being felt today.

More recently Depression-era banking legislation was repealed in 1999. The intent was to level the domestic playing field so U.S. financial firms could compete more effectively in the growing international market. Banks, securities firms, and insurance companies were allowed to merge and sell each other's products. The resulting mergers lead to a collapse of financial markets with banks becoming "too big to fail," thus requiring substantial support to continue functioning.[29] As this book goes to press, discontent is being expressed by the people toward Wall Street and the economic problems associated with continued movement of jobs to other countries. One of the results will likely be further regulation and tax code changes regarding business practices, executive pay, and profits earned in foreign manufacturing plants.

© An Nguyen/www.Shutterstock.com

**FIGURE 7.5** Following the U.S. Government's actions to reduce smoking, more and more states and cities are banning smoking in buildings, and some companies extend the ban to their entire properties.

## FOOD FOR THOUGHT 7.3

About 95 percent of the world's cities still dump raw sewage into their waters.

## Arguments About Social Responsibility

As we might infer from this brief history, the debate about the proper role of business organizations in society is far from over. In fact, today there are several factors that argue against social responsibility and several others that argue for a high level of social responsibility. The most prominent of these are summarized in Table 7.3.[30]

**Arguments For Social Responsibility.**   Several compelling arguments can be made in favor of social responsibility. These views tend to be broader and have a longer time perspective. One argument here is that corporations are citizens, in the same way that individuals are. As such, they have the same responsibilities as individual citizens to improve society as a whole (Figure 7.6).

Proponents of social responsibility also argue that the great power enjoyed by business carries with it great responsibility. In particular, business has the power to produce products, set prices, influence consumer preferences, pay employees, and so forth. Further, since business creates some problems (water, soil, and air pollution, for example), it can be argued that business should help solve them. Social responsibility can be seen as an important way to limit or constrain the power of business.

**TABLE 7.3**  Arguments For and Against Corporate Social Responsibility

| FOR | AGAINST |
|---|---|
| Like individuals, corporations are citizens. | Corporate social responsibility decreases profits, thus contradicts the real reason for corporations' existence. |
| Since business creates some problems, it should help solve them. | Corporate social responsibility gives corporations too much power. |
| Organizations have ample resources to help society. | Corporations are not accountable for the results of their actions. |
| Business, government, and the general public are partners in our society. | Corporations may lack the expertise to be socially responsible. |
| Arguments against corporate social responsibility can be logically refuted. | Corporations may have conflicts of interest in how they spend their money. |

© Cengage Learning 2014

It can also be argued that the vast resources available to companies can be used most effectively by being returned to society, at least in part. By and large, the business sector is in at least as good a position as the government to return wealth to its users in some way. The idea of partnership is also important. If business, government, and the public are indeed partners in our society, each group must attempt to protect, maintain, and nourish that society.

Last, it is possible to defend social responsibility with simple arguments of logic. In particular, logic can be used to refute each of the arguments against social responsibility. For example, it can be argued that existing laws sufficiently constrain corporate power so that socially responsible corporate behavior provides business with no additional power. With all such arguments out of the way, social responsibility seems quite desirable.

**FIGURE 7.6**  Should companies who profit from the natural resources of this country take some responsibility to provide a better quality of life for people in shanty towns like this?

### Arguments Against Social Responsibility.

One major argument against social responsibility is that, by definition, it decreases corporate profits. When Exxon gives thousands of dollars to support the arts, for instance, the money is actually being taken out of the pockets of stockholders. Many economists argue that such practices run counter to the basic premises underlying U.S. capitalism.[31] This argument is an economic view that tends to be narrow, focuses on the short term, and assumes competitive markets with little impact on one another.

Another argument against social responsibility is that it may give big business even more power, destroying the checks and balances among the government, business, and the general public. Increased activities of a socially responsible nature may tip the scales in favor of business. Accountability is also an issue: Since a business can use its money in any way it wants, the company is not accountable for the results of its activities.

Some people also argue that business organizations have no expertise in the area of social responsibility. Therefore, it would be better to leave such activities in the hands of people more skilled in social programs, such as teachers, social workers, and art administrators, for instance. Finally, some critics of social responsibility argue that such activities lead to conflicts of interest. Suppose that a large chemical manufacturing firm were contemplating a donation of $100,000 to charity. If one of its managers learns that a member of Congress who favored a certain charity is working on legislation affecting the chemical industry, this knowledge, theoretically, could influence the company's decision about where to spend the money.

## Areas of Social Responsibility

These various arguments aside, in the United States at least a relatively strong norm dictating social responsibility has evolved. That is, U.S. organizations are generally expected to behave in socially responsible ways. Those organizations may exercise social responsibility toward their constituents and the natural environment and in promoting general social welfare. Social entities closer to the organization will have a clearer and more immediate stake in what the organization does, while those further removed will have a more ambiguous and longer-term stake in the organization and its practices.[32]

**Organizational Constituents.**    In Chapter 4, we characterized the task environment as those specific elements of the environment that directly affect a particular organization. Another view of that same network is in terms of **organizational constituents**, those people and organizations that are directly affected by the practices of an organization and that have a stake in the organization's performance. Major constituents are depicted in Figure 7.7.

Virtually anything the firm does affects the interests of people who own and invest in the organization. If the firm's managers are caught committing criminal acts or violating acceptable ethical standards, the resulting bad press and public outcry will likely hurt the organization's profits and stock prices. Organizations also have responsibilities to their creditors. If poor social performance hurts an organization's abilities to repay its debts, the organization's creditors and its employees will also suffer.

A firm that engages in socially irresponsible practices toward some of its constituents is asking for trouble. For example, managers at Allegheny International once spent a half million dollars to buy a lavish Pittsburgh home in which to entertain clients. It maintained a fleet of five corporate jets so its managers could travel anywhere, anytime. While entertaining and travel are normal parts of doing business, Allegheny went too far. The company also made large loans to employees at a 2 percent interest rate. Nepotism in hiring was rampant. A close analysis of Allegheny's performance during this period suggested that it spent too much on executive perquisites,

**FIGURE 7.7** Organizational constituents.

that conflicts of interest clouded executive judgment, that improper accounting methods were employed, that managers withheld information from shareholders, and that the board of directors inadequately monitored top management. Consequently, other constituents, such as investors (who received lower dividends), the government (which received fewer tax dollars), the court system (which was eventually forced to deal with Allegheny's improprieties), and employees (who might have been paid higher wages under other circumstances) all were affected.[33]

On the other hand, consider the case of Ben and Jerry's Homemade, Inc. The company gives a portion of its pretax earnings to social causes, treats its employees and suppliers with dignity and respect, plays an active role in important trade associations, has had no major ethical scandals, is respected by its competitors, maintains good relations with government regulatory agencies, and contributes to college and university scholarship programs. This record suggests that managers at Ben and Jerry's are doing an excellent job of maintaining good relations with the firm's constituents.[34]

Not all organizations can do as well as Ben and Jerry's in attending to constituents, but most make an effort to take a socially responsible stance toward three main groups: customers, employees, and investors. Land's End, a mail order firm, is a good example of a company that has profited from good customer relations. The company trains its telephone operators to be completely informed about its policies and products, to avoid pushing customers into buying unwanted merchandise, to listen to complaints, and to treat customers with respect. As a result, the company's sales have been increasing 20 percent each year.[35] Organizations that are socially responsible in their dealings with employees treat workers fairly, make them a part of the team, and respect their dignity and basic human needs. Companies such as McDonald's, Denny's, PepsiCo, Yum! Brands, Coca-Cola, Darden Restaurants, and Safeway go to great lengths to find, hire, train, and promote qualified minorities.[36]

To maintain a socially responsible stance toward investors, managers should follow proper accounting procedures, provide appropriate information to shareholders about the financial performance of the firm, and manage the organization so as to protect shareholder rights and investments. Insider trading, illegal stock manipulation, and the withholding of financial data are examples of recent wrongdoings attributed to many different businesses.

**The Natural Environment.** A second critical area of social responsibility relates to the natural environment. Not long ago, many organizations indiscriminately dumped sewage, waste products from production, and trash into streams and rivers, into the air, and on vacant land. Now, however, many different laws regulate the disposal of waste materials. In many instances, companies themselves have seen the error of their ways and have become more socially responsible in their release of pollutants. Consequently, most forms of air and water pollution have decreased, although there is still widespread ocean dumping of sewage sludge; and much remains to be done.

Of course, it would be unethical for a company to move operations to a country with fewer pollution restrictions to avoid actually cleaning up its operations (Figure 7.8). Companies need economically feasible ways to avoid contributing to acid rain, depletion of the ozone layer, and global warming. Alternative methods are required for handling sewage, hazardous wastes, and ordinary garbage.[37] Procter & Gamble, for example, is an industry leader in using recycled materials for containers, and Hyatt has a company to help recycle waste products from its hotels.

Companies need safety policies that cut down on accidents with potentially disastrous environmental results. When one of Ashland Oil's storage tanks ruptured, spilling over 500,000 gallons of diesel fuel into Pennsylvania's Monongahela River, the company moved quickly to clean up the spill but was still indicted for violating U.S. environmental laws.[38] After the Exxon oil tanker *Valdez* spilled millions of

**FIGURE 7.8** The smokestacks at a company's power plant create dirty air that others breathe.

© Dudarev Mikhail/www.Shutterstock.com

gallons of oil off the coast of Alaska, it adopted new and more stringent procedures to keep another disaster from happening. Yet 2010 saw the worst oil spill in U.S. history when the Deepwater Horizon, a BP drilling platform in the Gulf of Mexico, exploded on April 20, and the oil began pouring into the Gulf.[39]

---

### FOOD FOR THOUGHT 7.4

Recycling one aluminum can saves enough energy to run a TV for three hours.

---

**General Social Welfare.** Some people feel that in addition to treating constituents and the environment responsibly, business organizations should promote the general welfare of society. Examples include making contributions to charities, philanthropic organizations, and not-for-profit foundations and associations; supporting museums, symphonies, and public radio and television; and taking a role in improving public health and education.[40] Some people also believe that organizations should act so as to correct, or at least not contribute to, the political inequities that exist in the world. A well-publicized expression of this viewpoint in the late 1980s was the argument that U.S. businesses should end their operations in South Africa to protest that nation's policies of apartheid.[41] Companies, such as Kodak and IBM, responded to these concerns by selling their operations in South Africa. As we shall see, this area of social responsibility is a source of much controversy for the managers of modern organizations.

## Approaches to Social Responsibility

It comes as no surprise to most people that businesses have dramatically different views of how they should behave, given the persuasive arguments both for and against social responsibility plus the varied interpretations of socially responsible behavior toward organizational constituents and the natural environment and in promoting general social welfare. In general, there are four basic approaches that characterize business postures (See Figure 7.9).[42]

### Social Obstruction

The few organizations that take what might be called a **social obstruction** approach to social

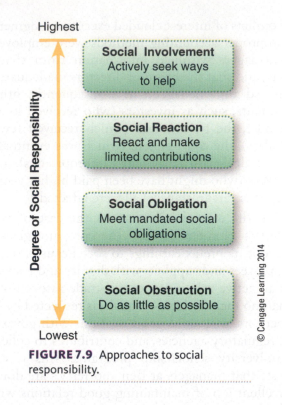

**FIGURE 7.9** Approaches to social responsibility.

responsibility usually do as little as possible to solve social or environmental problems. When they cross the ethical or legal line that separates acceptable from unacceptable practices, their typical responses are to deny or cover up their actions. A few years ago managers at Beech-Nut learned that the firm was mixing chemical additives to apple juice while advertising it as being pure juice with no additives. Rather than losing money on the existing supply of mix, the managers chose to continue selling it under false pretenses until the existing inventory was gone.

Ashland Oil also has an unfortunate history of alleged social wrongdoing followed by less-than-model responses. For example, Ashland was found guilty of rigging bids with other contractors in order to charge higher prices for highway work in Tennessee and North Carolina. It was also charged with wrongfully firing two employees because they refused to cover up illegal payments the company made.[43] BP has had similar problems—the 2010 oil spill, a 2005 refinery explosion at Texas City, and 1993 North Sea oil problems.[44]

### Social Obligation

The **social obligation** view is most consistent with the argument that any business activity that is not directly aimed at profits is inadvisable. The company that takes this approach is willing to meet its social obligations as mandated by societal norms and government regulation, but nothing more. Thus, it meets its economic and legal responsibilities but does not go beyond them. Tobacco companies such as Philip Morris have reduced their advertising in the United States and have put consumer warnings on every package of cigarettes they sell. However, they did not choose to take these measures; they were forced by government regulation. In other parts of the world where such regulation is not in force, tobacco companies are still heavily promoting their products. Hence, although they are doing nothing illegal in those countries, the companies are following the letter of the law and not its spirit.[45]

## Social Reaction

The firm using a **social reaction** approach is one that meets its social obligations but is also willing to react to appropriate societal requests and demands. That is, the company will make limited and specific positive contributions to social welfare. For example, many large corporations such as Exxon and IBM routinely match employee contributions to worthwhile charitable causes with contributions of their own. These actions fall under the heading of social reaction. So do the actions of the company that agrees to help local charities and civic organizations by providing free meeting space or donating funds to support Little League baseball teams. The key point to note is reaction; the normal pattern is for the civic group to knock on the company's door and ask for help, which the firm then agrees to give.

## Social Involvement

The firm using the **social involvement** approach to social responsibility not only fulfills its obligations and responds to requests (just as the social reaction and social obligation types of companies do), but also truly gets involved, actively seeking other ways in which to help. For example, McDonald's has established Ronald McDonald houses to help the families of sick children. This is clearly above and beyond the call of corporate duty. Another example of social involvement is the recent trend toward corporate support of the arts. Some corporations, such as Sears and General Electric, have taken active roles in supporting artists and cultural performers. Such actions may well lead to higher profits but this is by no means guaranteed, and the corporate motive appears to be primarily altruistic. Another example is Land's End, a mail-order house that some business analysts think goes beyond conventional levels of customer responsiveness.[46]

## The Government and Social Responsibility

Another area of significance is the link between the government and social responsibility. Generally, the government is actively working to regulate and control business while business is attempting to influence the government, as summarized in Table 7.4.

### Government Regulation of Business

Since the late 1800s, as already discussed, the U.S. government has taken an active role in the regulation of business. Much of government regulation has been concerned with enhancing the social responsiveness and awareness of business and with protecting the best interests of society from abuse by big business.

Government regulation has generally focused on four basic areas. First, the government has attempted to ensure fair labor practices by passing legislation regarding hiring wages, union relations, and so forth. Second, it has worked for environmental protection from pollution by business and other organizations. Third, consumer protection has theoretically been achieved by numerous laws dealing with truth in advertising, pricing, and warranties.

**TABLE 7.4**  Business and Government Interaction

| GOVERNMENT REGULATION OF BUSINESS | BUSINESS INFLUENCE ON GOVERNMENT |
| --- | --- |
| Ensures fair labor practices. | Gains favorable legislation and lessened scrutiny through personal contacts. |
| Protects the environment. | Enhances the company image to government officials and the general public. |
| Protects consumers. | Influences legislation by using lobbyists. |
| Guarantees safety and health. | Aids political candidates through donations. |

© Cengage Learning 2014

Fourth, a set of regulations has been developed to guarantee the safety and health of both employees and consumers. Currently, the federal government is especially involved in determining what it can do to further ensure the protection of individuals, businesses, and the environment from bioterrorism and other potential homeland security threats. Whatever transpires will no doubt affect small agribusiness firms, such as Summer Farms and Webb Dairy, as well as the giant corporations.

In general, the enforcement of these regulations has been assigned to several different and sometimes conflicting governmental agencies. Businesses today must contend with the Occupational Safety and Health Administration (OSHA), the Environmental Protection Agency (EPA), the Fair Labor Standards Board, the Equal Employment Opportunity Commission (EEOC), the Federal Trade Commission (FTC), and the Food and Drug Administration (FDA).[47]

Critics argue that this level of regulation is excessive and does more harm than good. For example, Goodyear Tire & Rubber once had to generate 345,000 pages of computer reports to comply with one new OSHA regulation. Moreover, the company reported that it spent more than $35 million each year to comply with federal regulations and that it required 34 employee-years just to fill out the necessary forms.[48] OSHA seems to be responding to criticisms, however, and advocates of government regulation argue that we would all be subjected to business abuses without it.[49]

## Business Influence on Government

Just as the government regulates business, so does business attempt to influence the government. Such efforts, of course, must be relatively subtle and are often quite indirect. In general, there are four common approaches businesses use to influence government.

First, company managers try to develop personal contacts with influential government leaders because such contacts can lead to favorable legislation and less scrutiny from the government.

Second, many companies, especially large ones, employ public relations firms to enhance their image with government officials and the general public.

Third, many businesses employ lobbyists. A **lobbyist** is someone based in a seat of government (either Washington or a state capital) for the express purpose of influencing the legislative body. A lobbyist for the oil industry, for example, may work to persuade key congressional leaders to vote for an upcoming bill that would result in increases in oil prices.

Fourth, some organizations make direct contributions to political candidates. Such contributions are heavily restricted and can be made only under certain circumstances. In recent years, organizations have started creating **political action committees (PACs)**, that solicit money from a variety of organizations and then make contributions to several candidates for office in order to gain their favor. For example, between 1987 and 1988 Federal Express's political action committee, Fepac, contributed more than $200,000 to Democratic campaigns. In 1988, as the congressional session came to a close, Democratic sponsors pushed a controversial bill through both the House and the Senate that allowed certain tax benefits for Federal Express employees.[50]

Opposition to business influence on government appears to be growing because at times such influence has gone too far. Some financial contributions have been so large that they almost constituted a direct bribe, and some have been made with the expectation that, or on the condition that, the candidate would take a certain position on a specific issue. Clearly, such practices are unethical and violate all premises of social responsibility. In all likelihood business and government will continue to work hard to influence one another legally and publicly, but occasional abuses will probably still occur.

## Managing Social Responsibility

The demands for social responsibility placed on contemporary organizations by an increasingly sophisticated and educated public are probably stronger than ever. As we have seen, there are pitfalls for managers who fail to adhere to high ethical standards and for companies that try to circumvent their legal obligations. Organizations therefore should fashion an approach to social responsibility the way they would develop any other business strategy. That is, they should view social responsibility as a major challenge that requires careful planning, decision making, consideration, and evaluation. They may accomplish the effective management of social responsibility through both official and unofficial approaches.[51]

## FOOD FOR THOUGHT 7.5

PetSmart Charities, the largest funder of animal welfare efforts in North America, has provided more than $109 million in grants and programs benefiting animal welfare organizations. Its in-store adoption program is believed to have helped save the lives of more than four million pets.

## Official Approaches to Social Responsibility

Formal organizational dimensions that can help manage social responsibility include legal compliance, ethical compliance, and philanthropic giving.

**Legal compliance** is the extent to which an organization complies with local, state, federal, and international laws. The task of managing legal compliance is generally assigned to the appropriate functional managers. For example, the organization's top human resource executive is generally responsible for ensuring compliance with regulations concerning recruiting, selection, pay, and so forth. Likewise, the top finance executive generally oversees compliance with securities and banking regulations. The organization's legal department is also likely to contribute to this effort by providing general supervision and answering queries from managers about the appropriate interpretation of laws and regulations. However, it is possible to follow the law and still engage in deceptive, unethical practices.[52]

**Ethical compliance** is the extent to which the members of an organization follow basic ethical (and legal) standards of behavior. We have already noted that organizations have started doing more in this area by providing training in ethics and developing guidelines and codes of conduct, for example. These activities serve as vehicles for enhancing ethical compliance. Many organizations also establish formal ethics committees, which may be asked to review proposals for new projects, help evaluate new hiring strategies, or assess new environmental protection plans, for example. Committees may also serve as peer review panels to evaluate alleged ethical misconduct by employees.

Finally, **philanthropic giving** is the awarding of funds or other gifts to charities or other social programs. Indeed, most organizations give a specified amount of their pre-tax income to social causes. Common causes include charities, animal welfare programs, colleges and universities, and arts programs like museums and symphonies. Summer Farms might choose to contribute some of its acreage to build soccer and softball fields for the local community. Unfortunately, during the recent period of cutbacks and retrenchment, many corporations have had to decrease their charitable gifts. Firms that do engage in philanthropic giving usually have a committee of top executives who review requests for grants and decide how much and to whom money will be allocated.[53]

## Unofficial Approaches to Social Responsibility

In addition to these official dimensions for managing social responsibility, there are also unofficial dimensions. Two of the more effective ways to clarify the organization's approach are to provide appropriate leadership and culture and to allow for whistle blowing.

**Organization leadership practices and culture** can go a long way toward defining the social responsibility stance an organization and its members will adopt. For example, for years Johnson & Johnson executives provided a consistent message to employees that customers, employees, communities where the company did business, and shareholders were all important, but only in that order. Thus, when packages of poisoned Tylenol showed up on store shelves in 1982, Johnson & Johnson employees did not need to wait for orders from headquarters to know what to do: They immediately pulled all the packages from the shelves before any other customers could buy them.[54] By contrast, the irresponsible behavior by top managers of Beech-Nut, as mentioned earlier, sent a far different message to that firm's employees.

**Whistle blowing** is the disclosure by an employee of illegal or unethical conduct on the part of others within the organization.[55] How an organization responds to this practice often indicates its stance toward social responsibility. Whistle blowers may have to proceed through a number of channels to be heard and may even get fired for their efforts. Many organizations, however, welcome their contributions. Individuals who observe questionable behavior typically report the incident to their bosses at first. If nothing is done, the whistle blower may then inform higher-level managers or an ethics committee if one exists. Eventually, the person may go to a regulatory agency or even the media in order to be heard.[56]

The apple juice scandal at Beech-Nut, for example, started with a whistle blower. A manager in the firm's R&D department began to suspect that its apple juice was not "100% pure." His boss, however, was unsympathetic; and when the manager went to the president of the company, the president also turned a deaf ear. Eventually, the manager took his message to the media.[57]

In another case, a lab director at SmithKline noticed that billings for fees from Medicare were abnormally high. When management ignored his information he went to the government. It cost SmithKline more than $300 million.[58]

## Evaluating Social Responsibility

Any organization that is serious about social responsibility must ensure that its efforts are producing the desired benefits, which means applying the concept of control to social responsibility.

Many organizations now require current and new employees to read their guidelines or codes of ethics and then sign statements agreeing to abide by them. An organization should also evaluate how it responds to instances of questionable legal or ethical conduct. Does it follow up immediately? Does it punish those involved? Or does it use delay and cover-up tactics? Answers to these questions can help an organization form a picture of its approach to social responsibility.

Additionally, some organizations occasionally conduct a **corporate social audit**—a formal and thorough analysis of the effectiveness of the firm's social performance. The audit requires that the organization clearly define all its social goals, analyze the resources devoted to each goal, determine how well the various goals are being achieved, and make recommendations about which areas need additional attention. Unfortunately, such audits are not conducted very often because they are expensive and time consuming. Indeed, most organizations probably could do much more to evaluate the extent of their social responsibility than they currently do.[59]

## CHAPTER SUMMARY

Ethics are the standards or morals a person sets for himself or herself about what behavior is good and bad or right and wrong. Ethics are formed by a variety of factors: family influences, peer influences, past experiences, values and morals, and situational factors.

Managers face ethical dilemmas or conflicts every day. Ethical conflicts occur when a manager is faced with two or more conflicting ethical issues. These dilemmas occur in the relationship of the firm to the employee, the relationship of the employee to the firm, and the relationship of the firm to the environment. The key dimensions of the ethical context of management are personal ethics, the organizational context (organizational practices, leader and peer behavior), and the environmental context (competition, regulation, sociocultural norms).

Managers must understand ethics and know how their organizations should attempt to manage them. For organizations to develop and maintain cultures that lead to ethical managerial behavior, top management must support such behavior. In addition, the organization can develop a code of conduct to serve as a guide to managers and a reminder of the importance of ethical behavior to the organization.

Social responsibility is the obligation of the organization to protect and/or enhance the society in which it functions. There are several arguments both for and against social responsibility. The basic areas of social responsibility include organizational constituents, the natural environment, and general social welfare.

The four basic approaches to social responsibility are social obstruction, social obligation, social reaction, and social involvement.

Government regulation of business has generally focused on four areas: ensuring fair labor practices, protecting our environment, protecting consumers, and guaranteeing the health and safety of employees and customers. Critics feel that the burden of regulation is too great; proponents of regulation feel it is the lesser of two evils. Business also tries to influence government in various ways through company and executive contacts and political contributions, public relations departments or firms, and lobbyists. These attempts to influence government are likely to continue in the future, even though abuses occasionally are made public.

Because of its obvious importance, organizations proactively attempt to manage social responsibility through both official and unofficial approaches. In addition, most businesses make an effort to evaluate the effectiveness of their social responsibility programs and activities.

## CHAPTER ACTIVITIES

### REVIEW QUESTIONS

1. What are the three general types of situations in which ethics are important to managers? What are the key dimensions of the manager's ethical context?
2. How can ethics be managed by organizations?
3. What is meant by social responsibility? What are the arguments for and against it?
4. What four general approaches to social responsibility do organizations take?
5. What is the government's role with regard to business's social responsibility? Why has it assumed that role? How does business influence government?

### ANALYSIS QUESTIONS

1. List five ways in which you feel it is ethical for students to behave. Then list five ways in which you feel it is unethical for students to behave. What are the strongest influences on your feelings about what is ethical and unethical student behavior?
2. Identify four actions that are (1) both legal and ethical, (2) legal but unethical, (3) illegal but ethical, and (4) both illegal and unethical.
3. How useful are codes of conduct or codes of ethics to business firms? Explain your response.
4. The proponents of social responsibility claim that the arguments against responsibility are flawed. Study those arguments and explain how each might be flawed. Give details to support your claim.
5. Which area of social responsibility do you feel is most in need of action? Which is least in need of action? Why?

## FILL IN THE BLANKS

1. Those standards or morals a person sets for himself or herself regarding what is good and bad or right and wrong are called _____ .

2. An employee who is put into a situation in which his decisions may be compromised because of competing loyalties is said to have a/an _____ _____ _____ .

3. Meaningful symbolic statements that emphasize the importance of ethical behavior in business are called _____ _____ _____ .

4. The obligations of an organization to protect and/or enhance the society in which it functions are known as _____ _____ .

5. _____ _____ includes not only fulfilling the organization's social obligations and responding to requests but also actively seeking ways to benefit society.

6. A formal and thorough analysis of the effectiveness of a firm's social performance is called a _____ _____ _____ .

7. The disclosure by an employee of illegal or unethical conduct on the part of others within the organization is called _____ _____ .

8. Groups that solicit funds from organizations and then makes contributions to political candidates in order to gain their favor are called _____ _____ _____ .

9. People and organizations that are directly affected by the practices of an organization and that have a stake in its performance are known as _____ _____ .

10. _____ _____ refers to meeting economic and legal responsibilities but not going beyond them.

## ▼ CHAPTER 7 CASE STUDY
### BABY STEPS TOWARD A FOOTHOLD ON ETHICS

Horace Greeley, newspaper editor, observed in 1838, "If any young man is about to commence the world, we say to him, publicly and privately, Go to the West." Apparently this advice applies today; however, the West is so far west as to be the east or Far East. Many companies are seeking their fortunes and future interests through investments in China. The Carlyle Group, a U.S. investment firm, is no different as it seeks a foothold in the Chinese food system.

In 2009, the company chose to buy a 17.3 percent stake in Yashili Group Company, a Chinese infant-formula maker. The numbers looked good. Yashili Group was the third-largest baby-formula firm in China in an industry with $5-6 billion (35–40 billion-yuan) in 2010 sales. With a new national Chinese policy away from one baby per family to more traditional family expectations, this industry is set for phenomenal growth.

This might have been your standard B-school success story, except for the events of 2008. In that year several thousand Chinese babies became ill, developed kidney stones and further complications. By the end of this crisis more than 300,000 babies would be sickened and six would die. The cause of this malady was not a viral epidemic but the milk powder used in the production of baby formulas. When authorities investigated this contamination, the story took a nasty turn. The contamination was not inadvertent or accidental but deliberate. Melamine was added to the milk powder to deceive food processors and garner a premium for their products with little regard for the food-safety consequences. There was no other way to explain the presence of melamine in the milk powder.

Melamine is a chemical; it is not natural. It was invented in 1830s by a German chemist and used for many things from plastics to fertilizers to a stabilizer for concrete—not food. It is not intended for consumption since it can cause kidney stones and kidney damage. It found its way into the milk powder because it was used to enhance the apparent "protein" content of milk powder. In the food industry, protein content is not measured directly but through the presence of a chemical bond that is unique to protein, the amide bond. The problem is that, when a compound is analyzed for protein, any amide bond may be counted as protein whether it originates in protein or a non-natural chemical. The original and standard test assumes no adulterants are present.

So why add melamine to baby-formula powder? The cost of melamine is far less than the cost of milk protein, and its addition to milk powder raises the apparent protein content of the powder and hence its value. It would not be recognized as melamine unless a very different test was used specifically looking for it. Since it is not natural and not a potential accidental contaminant in food processing, its presence indicates a deliberate addition or adulteration. The only time melamine and food were connected legitimately was during World War II when melamine was used to manufacture durable dishware for the U.S. Navy. It was durable, washable, unbreakable, but not edible. In fact, it found its way into American homes in the 1950s and 1960s as Melmac, and no doubt is making a retro return today.

The food safety authorities in China reacted to this crisis very slowly—to avoid "sullying China's international image during the Beijing Summer Olympics." It took prompt action by the food safety authorities in Hong Kong to spur their mainland colleagues into action. China's central government took dramatic steps to mollify public anger. The government "sacked the mayor of Shijazhuang after allegations that the city government covered up the reports of contamination." This is the home city of Sanlu, a Chinese dairy giant, partially owned (43 percent) by a New Zealand dairy cooperative, Fonterra. Sanlu was found to have tons of contaminated milk powder. Fonterra, its junior partner, knew about the situation one month before the Chinese government acted. The company, Sanlu, reported that it did not come forward with the information weeks earlier because it was waiting for the recall process to move through the Chinese system. Andrew Ferrier, CEO of Fonterra, said, "I can look at myself in the mirror and say that Fonterra acted absolutely responsibly."

In the end it was determined that more than three million pounds of milk products were contaminated; and because milk powder was an ingredient in many products other than baby formula, the contaminated products may have spread to Europe and Africa, and possibly the United States. Twenty-two dairies were involved or contaminated with this product. Of the 19 executives who were arrested, 15 received sentences between 2–15 years, one received a suspended death sentence, and three received life sentences. The two biggest culprits, Zhang Yujun and Geng Jinping, were executed. It may have taken time for the Chinese authorities to act but when they did act, they were decisive.

"The scandal in 2008 wiped out all profitability in the entire dairy industry of China," said Patrick Siewert, Senior Director of the Carlyle Group. Yashili did not contaminate its products, but the contaminant crept into its supply chain. To prevent reoccurrence and to rebuild confidence, the Carlyle Group has appointed a six-person committee headed by the former director of the U.S. Food and Drug Administration's Food Safety and Applied Nutrition Center to oversee Yashili's food product-quality efforts. However, according to Wang Xixin, professor of constitutional law at Peking University, not all of the culprits were in the dairy industry. Professor Xixin says that "The government (quality supervision bureaus, state and national) hid the truth from the public and behaved irresponsibly to public safety."

Food safety situations are no longer a single country's problem. Because of globalization one country's problem may be inadvertently the problem of other countries. A good example is the United States. China is the fourth-largest source of imported agricultural products for the United States. In 2009 these imports amounted to $2.9 billion. While the FDA has three offices in China, it depends on China to do the policing. "The Chinese government has enormously and effectively responded with new laws and new regulations … but the sheer size of the Chinese economy and the number of people makes it virtually impossible to check everything," said Rio Praaning Prawira Adiningrat, secretary general of the Public Advice International Foundation, a foundation that works on food-safety issues.

Times have changed; past is prologue? Not necessarily. Horace Greeley's words of advice were also followed by caveats concerning danger. This history will force the Carlyle Group to consider everything before going forward with an IPO (Initial Public Offering) for Yashili Group, the question of profitability notwithstanding.[60]

> ### ▶ Case Study Questions
>
> 1. Detecting the presence of melamine would have required a test that was specifically looking for it, but no one would have thought about a manufacturer adding such a non-food ingredient to baby formula. So how much responsibility can we expect of our government for inspecting foods for all harmful ingredients? How much is enough?
>
> 2. Can you think of a better way for our government to oversee the inspection of imported foods rather than depending on another country to police its own manufacturers?
>
> 3. China acted slowly but eventually decisively. What might they have done to send an even stronger message that food safety should be a producer's number one priority?

## REFERENCES

1. Mark Gunther, "The Making of a Future 500 Company," *Fortune,* August 16, 2010, 94–98; Robert Willens, "Reviving the Reverse Morris Trust for Mergers," (July 3, 2002) at www.cfo.com (accessed August 4, 2010); Allen Sloan, "Smucker Adds Coffee to its Breakfast Lineup—and Does it Tax Free," *The Washington Post* (June 10, 2008) at www.washingtonpost.com (accessed August 4, 2010).

2. See F. Neil Brady, "Aesthetic Components of Managerial Ethics," *Academy of Management Review* (April 1986): 337–344.

3. Thomas M. Barrett and Richard J. Kilonski, *Business Ethics, 3rd ed.* (Englewood Cliffs, NJ: Prentice-Hall, 1990).

4. "Heinz CEO's compensation up more than 6 percent in 2010," *Star Tribune* (July 9, 2010) at www.startribune.com (accessed July 20, 2010); Eleanor Bloxham "Are Compensation Committees Covering for High CEO Pay?" *Fortune* (July 15, 2010) at money.cnn.com (accessed July 20,2010); "What, Me Overpaid? CEOs Fight Back," *Business Week*, May 4, 1992, 142–148.

5. Gene Bylinsky, "How Companies Spy on Employees," *Fortune*, November 4, 1991, 131–140.

6. John Huey, "Walmart—Will It Take Over the World?" *Fortune*, January 30, 1989, 52–61.

7. Patricia Sellers, "Winning Over the New Consumer," *Fortune*, July 29, 1992, 113–125

8. "Not Everyone Loves Walmart's Low Prices," *Business Week*, October 12, 1992, 36–38.

9. "The Dark Side of Japan Inc.," *Newsweek*, January 9, 1989, 41.

10. John Carey, "Where's the Food Safety Net?" *Business Week*, June 12, 2008, at www.businessweek.com (accessed July 1, 2010).

11. Linda Grant, "There's Gold in Going Green," *Fortune*, April 14, 1997, 116–118.

12. Scott Kilman, June 26, 2010. "Monsanto Is Warned About Seed Claims," *The Wall Street Journal* at online.wsj.com (accessed July 2, 2010).

13. See Kenneth Labich, "The New Crisis in Business Ethics," *Fortune*, April 20, 1992, 167–176.

14. Erik Jansen and Mary Ann Von Glinow, "Ethical Ambivalence and Organizational Reward Systems," *Academy of Management Review* (October 1985): 814–822.

15. Roy Rowan, "The Maverick Who Yelled Foul at Citibank," *Fortune*, January 10, 1983, 46–56.

16. P. J. Huffstutter, "Tomato King Frederick Scott Salyer's Journey From Boardroom to Jail Cell," *Los Angeles Times* (June 17, 2010) at dailyme.com (accessed July 5, 2010); Stephen B. Young and Jeanette Leehr, "It's Time to Talk about the Ethics of Food," *Star Tribune* (October 18, 2009) at www.StarTribune.com (accessed July 5, 2010); Wayne Pacelle, "Dairy Industry: Got Ethics?" The Humane Society of the United States (January 27, 2010) at hsus.typepad.com (accessed July 5, 2010); Christopher Donville and Andrew Harris, "Potash Corp., Fertilizer Makers Sued for Price-Fixing Scheme," *Business Week,* September 12, 2008, at www.businessweek.com (accessed July 5, 2010); Marc Gunther, "Eco-police find new target: Oreos," *Fortune*, August 21, 2008; Mae Anderson, "Land O'Lakes to pay $25M in egg price-fixing case," *Business Week* (June 8, 2010) at www.businessweek.com (accessed July 1, 2010).

17. "Can Ethics be Taught? Harvard Gives it the Old College Try," *Business Week*, April 6, 1992, 34.

18. "CEOs Report Stricter Rules,' *USA Today*, March 20, 2006, 1B; Catherine Daily, Dan Dalton, and Albert Cannella, "Corporate Governance: Decades of Dialogue and Data," *Academy of Management Review* (2003): 371–382.

19. "Training Managers to Behave," *Time*, May 25, 2009, 41.

20. Annalisa Burgos, "Citigroup to Boost Ethics Training, Controls," *Forbes* (February 16, 2005) at www.forbes.com (accessed July 20, 2010).

21. "Ethics Training at Work," *The Wall Street Journal*, September 9, 1986, 1.

22. Suzanne Kapner, "Tough Times at Office Depot," *Fortune* (February 16, 2010) at money.cnn.com (accesses July 20,2010); "Ethics on the Job: Companies Alert Employees to Potential Dilemmas," *The Wall Street Journal*, July 14, 1986, 17.

23. "Companies Get Serious About Ethics," *USA Today*, December 9, 1986, 1B–2B.

24. Frederick D. Sturdivant, *Business and Society: A Managerial Approach, 4th ed.* (Homewood, IL.: Richard D. Irwin, 1989).

25. Donald S. Siegel and Donald F. Vitaliano, "An Empirical Analysis of the Strategic Use of Corporate Social Responsibility," *Journal of Economics & Management Strategy* (2007): 773–792; Abagail McWilliams and Donald Siegel, "Corporate Social Responsibility: A Theory of the Firm Perspective," *Academy of Management Review* (2001): 117–127; Abagail McWilliams and Donald Siegel, "How Fund Managers can Contribute to Academic Research on Corporate Social Responsibility," in *The Investment Research Guide to Socially Responsible Investing,* ed. Brian Bruce (New York: Investment Research Forums, 1998), 83–103.

26. For a review of the evolution of social responsibility, see Archie Carroll, *Business and Society: Ethics and Stakeholder Management* (Cincinnati, OH: Southwestern, 1989).

27. Archie B. Carroll, "Corporate Social Responsibility: Evolution of a Definitional Construct," *Business Society* (September 1999): 268–295; Stahrl W. Edmunds, "Unifying Concepts in Social Responsibility," *Academy of Management Review* (January 1977): 38–45.

28. Page Smith, *The Rise of Industrial America* (New York: McGraw-Hill, 1984).

29. Frederic S. Mishkin, (2006). "How Big a Problem is Too Big to Fail? A Review of Gary Stern and Ron Feldman's *Too Big to Fail: The Hazards of Bank Bailouts.*" *Journal of Economic Literature* 44 (4): 988–1004.

30. Keith Davis, "The Case For and Against Business Assumption of Social Responsibility," *Academy of Management Journal* (June 1973): 312–322.

31. "Rethinking the Social Responsibility of Business: A Reason Debate Featuring Milton Friedman, Whole Foods' John Mackey, and Cypress Semiconductor's T.J. Rodgers," *Reason* (October 2005) at reason.com (accessed July 19, 2010); Milton Friedman, *Capitalism and Freedom* (Chicago: University of Chicago Press, 1962).

32. Dirk Matten and Jeremy Moon, "'Implicit' and 'Explicit' CSR: A Conceptual Framework for a Comparative Understanding of Corporate Social Responsibility," *Academy of Management Review* (2008): 404–424.

33. "Big Trouble at Allegheny," *Business Week*, August 11, 1986, 56–61.

34. "Social and Environmental Assessment Reports," Ben and Jerry's website at www.benjerry.com (accessed July 20, 2010); Edwin M. Epstein, "The Corporate Social Policy Process: Beyond Business Ethics, Corporate Social Responsibility, and Corporate Social Responsiveness," *California Management Review* (Spring 1987): 99–114.

35. "A Mail-Order Romance: Land's End Courts Unseen Customers," *Fortune*, March 13, 1989, 44–45.

36. Cora Daniels, "50 Best Companies for Minorities," *Fortune* (June 28, 2004) at money.cnn.com (accessed July 19, 2010).

37. Jeremy Main, "Here Comes the Big New Cleanup," *Fortune*, November 21, 1988, 102–118.

38. "Ashland Just Can't Seem to Leave its Checkered Past Behind." *Business Week*, October 31, 1988, 122–126.

39. Alton Parrish, "Timeline of Events in BP Oil Spill: Day by Day, April 20 to July 13–Day 84 BP Costs Exceed $3.5 Billion," Before It's News (May 27, 2010) at beforeitsnews.com (accessed July 24, 2010).

40. Nancy J. Perry, "The Education Crisis: What Business Can Do," *Fortune*, July 4, 1988, 71–81.

41. Anthony H. Bloom, "Managing Against Apartheid," *Harvard Business Review* (November-December 1987): 49–56.

42. See S. Prakash Sethi, "A Conceptual Framework for Environmental Analysis of Social Issues and Evaluation of Business Response Patterns," *Academy of Management Review* (January 1979): 63–74. See also Steven L. Wartick and Phillip L. Cochran, "The Evolution of the Corporate Social Performance Model," *Academy of Management Review* (October 1985): 758–769.

43. "Ashland Just Can't Seem to Leave its Checkered Past Behind."

44. Loren Steffy, "How Could it Happen to BP Again?" *Houston Chronicle* (May 2, 2010) at www.chron.com (accessed July 23, 2010); Peter Rodgers, "Brokers convicted in BP corruption trial," *The Independent* (March 27, 1993) at www.independent.co.uk (accessed July 23, 2010).

45. "In 'Tobacco Smoker's Paradise' of Japan, U.S. Cigarettes are Epitome of High Style," *The Wall Street Journal*, September 23, 1991, B1, B6.

46. "A Mail-Order Romance: Land's End Courts Unseen Customers," *Fortune*, March 13, 1989, 44–45.

47. "Make the Punishment Fit the Corporate Crime," *Business Week*, March 13, 1989, 22.

48. "Many Businesses Blame Governmental Policies for Productivity Lag," *The Wall Street Journal*, October 28, 1980, 1.

49. Greg Densmore, "Scannell Brings New Look to OSHA," *Occupational Health & Safety*, January 1, 1990, 18–21; William J. Rothwell, "Complying With OSHA," *Training and Development Journal*, May 1, 1989, 52–54.

50. "How to Win Friends and Influence Lawmakers," *Business Week*, November 7, 1988, 36.

51. Wartick and Cochran, "The Evolution of the Corporate Social Performance Model"; Jerry W. Anderson, Jr., "Social Responsibility," *Business Horizons*, July-August 1986, 22–27; and Epstein, "The Corporate Social Policy Process: Beyond Business Ethics, Corporate Social Responsibility, and Corporate Social Responsiveness."

52. "Legal—But Lousy," *Fortune*, September 2, 2002, 192.

53. "To Give or Not to Give," *Time*, May 11, 2009, Global 10; Michael Porter and Mark Kramwe, "The Competitive Advantage of Corporate Philanthropy," *Harvard Business Review* (December 2002): 57–66.

54. "Unfuzzing Ethics for Managers," *Fortune*, November 23, 1987, 229–234.

55. "The Complex Goals and Unseen Costs of Whistle-Blowing," *The Wall Street Journal*, November 25, 2002, A1, A10.

56. Michael Grundlach, Scott Douglas, and Mark Martinko, "The Decision to Blow the Whistle: A Social Information Processing Framework," *Academy of Management Review* (2003): 107–123.

57. Vern Modeland, "Juiceless Baby Juice Leads to Full-Length Justice," *FDA Consumer* (June 1988) at findarticles.com (accessed July 15, 2010).

58. "A Whistle-Blower Rocks an Industry," *Business Week*, June 24, 2002, 126–130.

59. Donna J. Wood, "Corporate Social Performance Revisited," *Academy of Management Review* (October 1991): 691–718.

60. Fred Shapiro, "Who Said, 'Go West, Young Man'—Quote Detective Debunks Myth," (December 24, 2007) at LLRX.com (accessed August 15, 2010); Alison Tudor, "Chinese Formula Maker Prepares for Stock Offering," *The Wall Street Journal.* (August 7, 2010) at online.wsj.com (accessed August 15, 2010); Sharon LaFraniere, "2 Executed in China for Selling Tainted Milk." *The New York Times*, November 25, 2009, A10; Austin Ramzy and Lin Yang, "Tainted-Baby-Milk Scandal in China" (September 16, 2008) at Time.com (accessed August 16, 2010); Kate Pickert, "Melamine," (September 17, 2008) at Time.com (accessed August 16, 2010); Michael Wines, "Tainted Dairy Products Seized in Western China," *The New York Times*, July 10, 2010, A6; Jessie Jiang, "China's Rage Over Toxic Baby Milk," (September 19, 2009) at Time.com (accessed August 16, 2010); David Barboza, "China Admits New Tainted-Milk Case is Older," *The New York Times*, January 7, 2010, A6; James Areddy, "Amid Chinese Food Scares FDA has Limited Scope," *The Wall Street Journal* (August 15, 2010) at blogs.wsj.com (accessed August 15, 2010).

# PART 3

# PLANNING AND DECISION MAKING IN AGRIBUSINESS

# Basic Managerial Planning

## LEARNING OBJECTIVES

After studying this chapter, you should be able to:

- Discuss the nature of planning, including its purpose and where the responsibilities for planning lie within the organization.

- Define goals, note their purpose, and identify the steps in the goal-setting process.

- Identify and define three major kinds of plans.

- Describe three major time frames for planning and how these time frames are integrated within organizations.

- Define contingency planning and describe contingency events.

- Discuss how to manage the planning process by avoiding the road-blocks to effective planning.

# Nobody Doesn't Like Sara Lee

In 1951, Charles Lubin, a bakery entrepreneur, started the Kitchens of Sara Lee specializing in great cheesecake. Unlike many entrepreneurs, he named the cheesecake line and the company after his eight-year-old daughter, whose name also was crafted into the company's long-running jingle, "Everybody doesn't like something, but nobody doesn't like Sara Lee!" According to Sara Lee, her father told her that the product "had to be perfect because he was naming it after me." In 1956, Consolidated Foods purchased the $9 million company as a "straight stock purchase"—and part of then-CEO Nathan Cummings' plan (Figure 8.1).

**FIGURE 8.1** Sara Lee, known primarily for bakery items such as these, has expanded greatly since it began more than 60 years ago.
© Gordon Swanson/www.Shutterstock.com

Cummings' plan was to build a giant conglomerate through a long line of acquisitions. Consolidated Foods, under his leadership, acquired a number of different firms, such as 34 Piggly Wiggly stores, Jonker Fis (Dutch producer of canned goods), Oxford Chemical Corporation, E. Kahn and Sons Company (meat packer), Bryan Foods, Electrolux (vacuum cleaners), Gant and Country Set apparel, Candelle (women's apparel), Hillshire Farm and Rudy's Farm, Aris Gloves (Isotoner), and Erdal (another Dutch company). The pattern of these acquisitions was that there was no pattern; that was the point.

A popular business idea at that time was to assemble a "conglomerate"—a "portfolio" of companies gathered together. The brands were not changed to reflect a "single" brand or label. They were separate but different companies under one roof. This business entity would have the strength of its diverse offerings as a "hedge" against an economic downturn. It seemed intuitive that an economic downturn or business "slump" would not affect all businesses equally.

This plan appeared to work for Consolidated Foods. Sales reached $l billion in 1967, and $5 billion in 1983 under a new CEO, John Bryan. The organization continued its acquisitive pace with Chef Pierre (frozen prepared desserts), Douwe Egberts (Dutch coffee and grocery company), Hanes (underwear), L'eggs, Bali, and Lerin (hosiery), Gallo Salame, Productos Cruz Verde (Spanish household products company), Jimmy Dean (meats), and Nicholas Kiwi Ltd. (Australian manufacturer of home products and medicines).

The umbrella name, Consolidated Foods, no longer seemed to fit, so in 1985 it was changed to the Sara Lee Corporation to reflect its consumer-marketing orientation. (Why not? Nobody doesn't like Sara Lee!) In 1988, Sara Lee achieved $10 billion in sales and acquired more firms; in 1994, $15 billion after acquiring Playtex Apparel Inc. and other companies; and in 1998, $20 billion before acquiring Hills Bros. Coffee two years later.

For business and agribusiness professors the Sara Lee Corporation became a favorite teaser question to ask their students: "What is the business of Sara Lee?" Most students would reply bakery goods and cheesecake, unaware of the other entities in the Sara Lee conglomerate fold. The question provided an instant revelation that companies are not always as they appear.

When Brenda Barnes succeeded CEO Bryan in 2004, she quickly noted that the "old" plan was not working. In an era of focused specialist companies striving for number one or two global market positions, operating a company with the diversity of Sara Lee was difficult. The diversity that had served the company well during its growth phase was becoming a major impediment in allocating resources appropriately, encouraging managerial development, and comparing the performances of various businesses against each other. It was not an apple-orange-grape comparison; it was an Izod-Isotoner-cheesecake comparison. Barnes realized also that not all companies were contributing to the Sara Lee Corporation bottom line equally. Something had to be done, and it had to be performed rationally.

Thus, in 2005 the company executed "a bold, ambitious, multiyear plan to transform Sara Lee into a company focused on its food, beverage, and household body-care businesses. "In a major deconsolidation move, the company divested itself of 40 percent of its revenue in the same year that Barnes became president and CEO. In 2009, Barnes wrote in her shareholder's letter: "At Sara Lee, we are market leaders in large important categories—coffee, meat, and bakery—and are even more relevant in today's challenging economic times." Sales for 2009 were more than $12 billion.

Sara Lee Corporation is a more focused entity today than in the past. Barnes' efforts have helped Sara Lee to substantially boost its operating margins, its profits, and through a stock repurchase plan, its share price. It takes a strong person to announce to a corporation and its board of directors that the business plan that was once successful is now "broken" and must change. Barnes was that person. Barnes was correct and the company is better, stronger, slimmer, and almost as profitable. Barnes was the right person with the right insight and the ability to convince an entire corporation that it needed to change.

However, Barnes, 56 years old, is no longer with the Sara Lee Corporation. She was not "felled" by company infighting or politics but by a blood clot. In May 2010, she suffered a stroke. She tried to recover but not as quickly as she would have liked and resigned after six years at the helm of Sara Lee. As poet Robert Burns wrote (translated) in 1875: "The best laid plans of mice and men go often askew … And forward, though I cannot see, I guess and fear!"[1]

## INTRODUCTION

Among the larger, fast-growing organizations in the United States are two well-recognized agribusiness corporations, Sara Lee and ConAgra. As just indicated, Sara Lee planned to become diverse. Then, when it became too diverse, even if not too big, it planned to narrow its focus. As you read in Chapter 5, ConAgra is one of the largest independent food processor in the United States.[2] Its success lies in innovation and effective planning. Managers at ConAgra looked carefully at their environment and purchased Beatrice to gain access to its grocery-store distribution network and its brands like Hunt's tomato products, Peter Pan peanut butter, Armour meats, and Banquet frozen foods.[3] In doing so, ConAgra developed a path to success.[4] On a smaller scale, so did Summer Farms, especially due to the highly seasonal nature of its product.

The focus of this chapter is on the basic elements of planning, one of the management functions identified in Chapter 2, which is absolutely essential for organizations of any size. First, we look at the nature of goals, which are the basis of planning. Then the major kinds of plans and time frames for planning are discussed. After investigating contingency planning, we discuss ways to effectively manage the planning process, and finally identify some important planning tools and techniques. In the next chapter (Chapter 9) we focus on one very important type of planning: strategic planning.

# Planning in Organizations

What is planning? To better understand planning, we must first define and clarify its purpose, then identify who in the organization is responsible for it.

## Why Managers Plan

A **plan** is a blueprint or framework used to describe how an organization expects to achieve its goals. **Planning** is the process of determining which path, among several possibilities, to follow in attempting to reach a particular goal.[5] Planning activities also serve to project an image of managerial competence to the organization's constituencies.[6] The steps and procedures of the formal planning process become a symbol of the effectiveness of management.[7]

For example, when Harry Hoffman became CEO of Waldenbooks in 1979, he developed several plans for achieving his goal of making the bookseller the leader in its industry. His plans called for rapidly increasing the number of stores in the chain, increasing promotional activities to boost the sales of each store, and expanding the product line. Each of these activities was part of a systematic effort to dramatically increase the company's sales. In terms of planning, Hoffman could have chosen to maintain the status quo, branch out into other markets, maintain a pattern of slow growth, or any of several other alternatives. His choice of rapid growth, then, was the creation of his particular blueprint for action.[8] However, the Borders Group, which owns Waldenbooks, began shutting most of its stores during 2010 due to a recession.[9]

Although far smaller than the companies mentioned previously, Summer Farms had to plan extensively. Again, the seasonal nature of its products, both planting and harvesting, makes planning crucial for Summer Farms to secure the necessary equipment and labor at certain times of the year. Also, when expanding the company must first consider the highly seasonal time frame of any crops they might add. Can it handle more of the same crop in the same time frame? Does it need a crop to help maximize labor and equipment during a slower period of the year? The timber crop gave more flexibility in Summer's expansion plans since it could be harvested in the colder seasons. Planning also involves determining how much timber to cut and when to cut it, to assure sufficient capital to purchase additional crop acreage. Timing becomes especially important if Summer also expects its crop employees to participate in timber-cutting operations.

Any goal may be approached in several ways. Planning is the process of determining which is the best approach to a particular goal (Figure 8.2). Waldenbooks chose the path that seemed best for it. The same holds true for other corporations, such as ConAgra and Intel. Of course, it is also possible to choose a wrong path. This fact makes planning all the more important.

**FIGURE 8.2** Planning is a large part of a manager's job.

© iStockphoto/Troels Graugaard

Summer Farms discovered this when it sought to decrease waste and increase revenue (the company's goals) by adding a processing and packaging facility for the red beets it could grow and harvest so easily. Fortunately, by planning ahead Summer realized that, although it could reduce waste somewhat, it had neither the expertise nor a genuine interest to expand by adding processing and packaging at that time.

Green Things Landscaping had a similar awakening when its planning figures revealed that they were not likely to increase its customer base enough to offset the costs of adding an additional artist and another planting crew.

---

**FOOD FOR THOUGHT 8.1**

American wheat growers were caught off guard when wheat markets tumbled in August 2010, after Russia suspended grain shipments due to its worst drought in a century.

---

## Responsibilities for Planning

Given the obvious importance of planning, it is essential to identify planning responsibilities; that is, who does an organization's planning? The answer is quite simple: All managers are involved in the planning process.

In a large organization, planning starts with top management. The top managers work with the board of directors to establish the broad goals and strategies for the firm. The usual approach is for the top management team to develop these goals and strategies and then submit them to the board for approval. In much smaller organizations, of course, the top managers do the planning alone or with the participation of others in the organization.

Many large organizations, such as Monsanto, Phillip Morris, Coca Cola, Tenneco, General Motors, General Electric, Boeing, and Ford, make use of a planning staff.[10] As noted in the figure, a planning staff is a group of professional planners at the top level of an organization. They assist line managers, providing expertise and various resources necessary to develop appropriate kinds of plans. They also coordinate and integrate the planning activities of other levels of the organization. Even some smaller organizations use more formal approaches to planning.[11]

Middle managers play several roles in the planning process.[12] They work together to assist with strategic planning, and they undertake tactical planning[13] (discussed later in this chapter). They also work individually to develop and implement planning activities within their respective divisions or units.[14]

First-line managers also must be actively involved in planning. Like middle managers, they work together to make plans that affect more than one department or unit and work individually to plan for their own units. Their efforts mostly involve assisting with tactical planning and developing operational plans.[15] A well-integrated planning system that links all levels of the organization and takes into account the other managerial functions (i.e., organizing, leading, and controlling) can be a major ingredient in organizational success. In smaller agribusinesses, for example, first-line managers may have no formal say-so, but it can be important for them to feel that they are involved at least indirectly.

## The Nature of Organizational Goals

What are goals? Why are they important in planning? How are they established? Before discussing planning, you must first answer these basic questions.

### Definition and Purpose of Goals

A **goal** is a desired state or condition that the organization wants to achieve—a target the organization wants to hit. When General Electric says that it wants to be number one in all of its markets, when Kellogg declares that it wants to control 50 percent of the cereal market, or when Summer Farms decides to go organic, they are all specifying a marketing goal they want to achieve and then maintain.

Like targets, goals provide a clear purpose or direction for an organization (Figure 8.3). When Stanley Gault became CEO of Rubbermaid, he decided to make the company much more innovative. He set a goal of increasing sales by 15 percent, annually. He also pledged to increase employee participation and promised greater rewards for innovation. Rubbermaid's employees enthusiastically accepted these ideas and met the sales goal every year. As a result, the company became recognized for its innovation.[16] At Rubbermaid, then, goals served their intended purpose—they provided guidance and direction. This success at Rubbermaid

**FIGURE 8.3** Goal setting provides direction for all employees and should involve all employees.

(Newell Rubbermaid Inc. since 1999) led to Gault's later being brought out of retirement to head Goodyear.[17] In any organization consisting of more than one person, it is essential for both management and employees to know the company goals that determine where the organization hopes to go and where they personally fit in.

## Types of Goals

Goals come in a variety of types and are called by many different names, including missions, purposes, objectives, ends, and aims. While there is no universal set of definitions, goals can be readily differentiated by organizational level, area, and time frame and specificity.[18]

**Goals by Management Level.** One useful perspective for describing goals is their level in the organization. For example, we noted in Chapter 2 the various levels of management that characterize most organizations. It follows logically, then, that each such level is likely to have its own goals.

At the top are the purpose and mission of the organization as determined by its board of directors. An organization's **purpose** is its reason for existence. For instance, the purposes of Archer Daniels Midland, Toigo Orchards, and Purina Mills are to make a profit, whereas the purpose of the Mayo Clinic is to provide health care. An organization's **mission** is the way it attempts to fulfill its purpose.[19] Honda Motor Company attempts to fulfill its purpose (making a profit) by manufacturing and selling automobiles, motorcycles, and lawn mowers. Walmart's is to sell large quantities of merchandise at a small markup. Unfortunately, however, organizations don't always use these terms in this way and most really have more than one purpose.

Top-management goals define the strategy of the organization, or the broad plans that set its overall direction (Chapter 9). Several years ago the former CEO of Citicorp, Walter Wriston, set three broad goals for the bank that determined the strategy the bank would follow: earnings growth of 15 percent per year, a 20 percent return on stockholders' equity, and becoming the world's first truly international banking system. Wriston's successor, John Reed, continued to work toward the same set of goals through a merger with the Travelers Group that formed Citigroup.[20] For the landscape artist/nursery owner that we described in an earlier chapter (not Green Things), his goals as a manager included earning enough money for a comfortable living, but keeping the business small enough to allow him to manage it himself and to maintain his excellent reputation in the community.

Middle managers also set goals (Figure 8.4). These goals follow logically from the strategic goals set by top managers. A plant manager for Dow Chemical, for example, might have goals for reducing costs and increasing output by a certain amount for the next year. Likewise, at Citicorp the head of one of the corporation's large banks will have goals designed to contribute to the three general goals noted above. The Development (Sales) Manager at Green Things may have a goal of adding a specific number of clients over the summer months, or to encourage clients to favor more complex designs that would increase the company's total revenue. At Summer Farms the manager's goal may be to decrease planting time by two days. At Green Things

**FIGURE 8.4** A middle manager works with his staff to set departmental goals.

Landscaping, the owner-designer may set a goal of increasing the number of clients he serves, the number of orders he designs over the next 12 months, or the amount of time he personally needs to spend interacting with clients.

Of course, first-line managers also have goals. These may relate to specific projects or activities pertinent to the manager's job. For instance, a first-line supervisor in a Dow Chemical plant might have the goal of reducing costs in her unit by 5 percent. And a Citicorp banker sent to open a small branch in a new overseas location will have goals consistent with the corporation's goal of internationalization. At Summer Farms, a first-line manager may have a goal of 20 percent reduction in damage to the vegetable product during the harvesting process.

## Goals by Organizational Area.

It is possible to differentiate goals by areas of management, also discussed in Chapter 2. That is, goals can be established for each organizational area. Managers in the marketing area might develop goals for sales, sales growth, market share, and so forth. Operations managers can establish goals for costs, quality, and inventory levels. Financial goals can relate to return on investment and liquidity. Human resource goals relate to turnover, absenteeism, and employee development. Research and development goals may include innovations, new breakthroughs, and so forth. Similarly goals could be established

in product or geographic areas. Finally, in a slightly different vein, managers might set social goals, such as contributions to the community through the United Way.

As shown in Figure 8.5, goals can also be established for each area across different levels of an organization. Within the marketing area of General Mills there are goals for top managers (e.g., increase total sales by 10 percent), for middle managers (e.g., increase sales of three different products by 8, 10, and 13 percent, respectively), and for first-line managers (e.g., increase sales of one product within a certain territory by 6 percent).

## Goals by Time Frame and Specificity.

The last dimension along which we can classify goals is time frame and specificity. Most organizations establish long-range, intermediate, and short-range goals. At Chiquita, long-range goals might extend ten years ahead; intermediate goals, for the next five years; and short-term goals, for the next year. At Greenwood Resources these timeframes could be much longer as its primary area of forest products includes growing trees. Summer Farms was engaged in the initial stage of long-range planning when it sought to decrease waste and increase revenue (two of its goals) by adding a processing and packaging facility for the red beets it could grow and harvest so easily. As a result of planning ahead, the company confirmed that it could reduce waste somewhat but

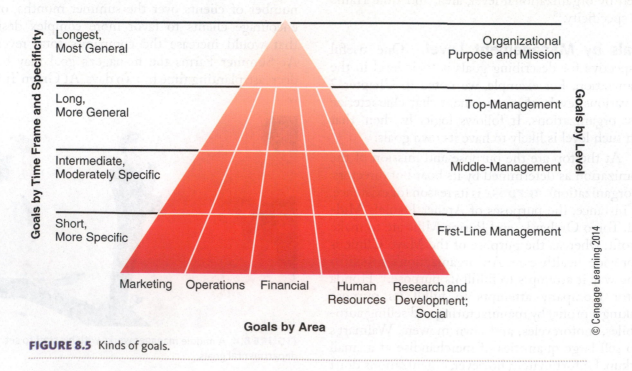

**FIGURE 8.5** Kinds of goals.

© Cengage Learning 2014

discovered it had neither the expertise nor even a genuine interest from the family members to make this a profitable way to expand at this time. Green Things Landscaping had a similar awakening when its planning figures revealed that it was not likely to increase its customer base enough to offset the costs of adding an additional designer and another planting crew.

**Specificity** refers to the extent to which the goal is precise or general. A precise goal for McDonalds might be to increase sales in a certain store by 14 percent next year. Note that the goal specifies the unit, the target amount, and a time frame. Nestlé has a goal of providing products worldwide. This is a very general goal in that it specifies no time frame and establishes only a broadly stated target.

As shown in Figure 8.5, goals tend to take longer and are more general at the top level of the organization, while they usually take less time and are more specific at the lower levels.

## Steps in Setting Goals

**Goal setting** is a six-part process, as shown in Figure 8.6. The end result of this process should be a set of consistent and logical goals that permeate the entire organization.

**Scan the Environment.**   The managers in an organization first scan the environment for opportunities and threats (step 1) and then assess organizational strengths and weaknesses (step 2). For a company such as NBC Universal, opportunities may include new foreign markets for its theme parks and new ideas for movie projects, and threats could include other entertainment companies, movie producers, and television and Internet movies. NBC Universal's organizational strengths may involve marketing savvy in the theme park industry, a solid reputation, and surplus capital. Its weaknesses may be the firm's dependence on its theme parks for operating funds and an image that does not appeal to teenagers and young adults.

**Establish General, Unit, and Subunit Goals.**
The next step in goal setting is to establish general organizational goals that match strengths and weaknesses with opportunities and threats (step 3). NBC Universal managers might decide to open three new theme parks in other countries by the year 2020, and make 15 new movies per year for the next ten years. This leads to step 4 and step 5: setting unit

and subunit goals. The theme park unit might decide that one new park every ten years is the best target, and its subunits would then set specific goals for new rides and attractions, attendance levels at existing parks, and so forth.

> **FOOD FOR THOUGHT 8.2**
> Annual U.S. ethanol production is expected to grow from 10.5 billion gallons in 2009 to 15 billion gallons by 2015.

**Monitor Progress.**   The last step (step 6) is for managers to monitor progress toward goal attainment at all levels of the organization. This progress subsequently affects all of the other steps as the cycle repeats itself. For example, a string of unprofitable movies might cause it to halt film production until managers can straighten out the problems.[21]

## Guidelines for Setting Goals

Managers can follow several guidelines to ensure effective organizational goal setting. First, to ensure they set appropriate and realistic goals, managers should understand the purpose of goal setting. Second, managers should state goals as specifically and briefly as possible for easy understanding by others. A well-stated goal should include a time frame as well as desired quantitative and qualitative results. Third, goals across areas and levels of the organization should be consistent to ensure managers are working together to achieve optimization (discussed below). Fourth, managers should communicate goals to others in the organization. Again, this helps to ensure that everyone in the organization is working together. And fifth, managers at all levels should take care to reward effective goal setting. By rewarding appropriate behavior, an organization reinforces what is important and helps to ensure that effective behavior will continue.

One approach to developing goals is to make them SMART. S.M.A.R.T. is an acronym that stands for specific, measurable, attainable, realistic, and timely, although it has a number of slightly different variations for different people and organizations. For example:

**S** for specific, significant, or stretching.

**M** for measurable, meaningful, or motivational.

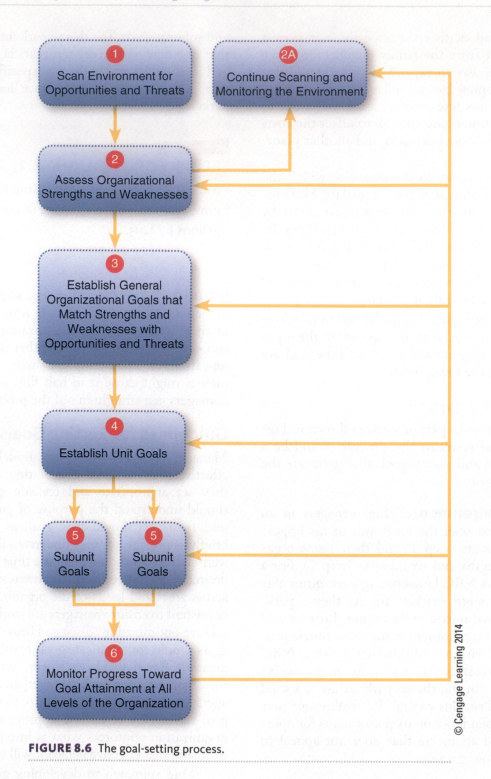

**FIGURE 8.6**  The goal-setting process.

**A** for agreed upon, attainable, achievable, acceptable, or even action-oriented.

**R** for realistic, relevant, reasonable, rewarding, or results-oriented.

**T** for timely, time-based, time-targeted, tangible, trackable, or time-bound.

## Managing Multiple Goals

Regardless of its size or diversity, any organization must pursue a variety of goals to survive. Suppose a plant manager in an organization sets a goal of reducing costs by 10 percent. One way to do this may be to buy less expensive materials and put more pressure on workers. Suppose also that a marketing manager in that organization decides to increase sales by 5 percent by promoting product quality while a human resource manager decides to cut turnover by 20 percent. In all likelihood, the actions of the first manager will hinder the other two managers from achieving their goals.

This is where goal optimization becomes important. **Goal optimization** is the process of balancing and trading off between different goals for the sake of organizational effectiveness. Achieving the balance called for in goal optimization is difficult and is never fully realized. The manager usually starts with a set of conflicting, disparate, and diverse goals. Using talent, insight, and experience, and with the appropriate level of flexibility and autonomy, she then arranges them into a unified, consistent, and congruent set of organizational goals. Doing this well takes many skills, particularly cognitive and problem-solving skills, which are the skills involved in strategic management (see Chapter 9).

The optimization process allows an organization to pursue a unified vision and helps managers maintain consistency in their actions. In the example mentioned earlier, a middle manager might help the marketing, plant, and human resource managers arrive at a new set of goals so that each can make progress in his or her area without getting in the way of the others. For the kinds of optimizing necessary at Intel, for instance, the chief executive officer and his vice presidents were no doubt involved in making the decisions.

# Kinds of Planning

Given that organizations naturally have large numbers of goals, they also have several plans at any one time. Given the variety of areas for which these plans can be developed, it is not surprising that planning falls into different categories. This section identifies and discusses the three major kinds of planning activities that go on in organizations. As shown in Figure 8.7, these activities can be described in terms of different levels of scope and different time frames.

## Strategic Planning

**Strategic planning** formulates the broad goals and plans developed by top managers to guide the general directions of the organization.[22] As illustrated in Figure 8.4, strategic plans are broad in scope and have extended time frames. Strategic planning follows from the major goals of the organization and indicates the businesses the firm is in or intends to be in and the kind of company its top managers want it to be. The key components of a strategy, then, outline how the organization will deploy resources and how it will position itself within its environment. Strategic planning at Summer Farms would include, for example, deciding when and how much additional acreage to buy, whether to add new areas such as food processing and packaging, or how to increase the demand for any one of its products if it has the capacity to do so.

PepsiCo provides an excellent example of a firm with a well-conceived and well-executed strategic plan.

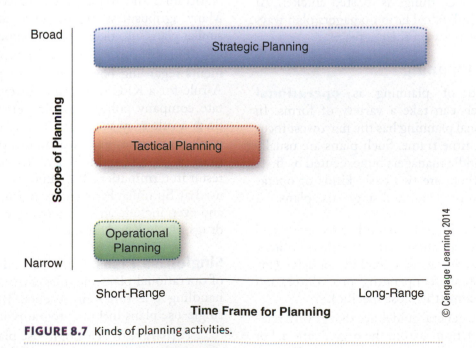

**FIGURE 8.7** Kinds of planning activities.

Specifically, managers at PepsiCo decided to organize the company into four general areas: PepsiCo Americas Beverages, PepsiCo Americas Foods, PepsiCo Europe, and PepsiCo Middle East and Africa.[23] The beverages include highly successful brands such as Pepsi, Slice, Mountain Dew, Ocean Spray, Lipton, and Tropicana. Equally successful are PepsiCo's food brands, including Frito-Lay, Quaker, Sabritas, and Gamesa. The European unit has had success with those, as well as Walkers Crisps, Paw Ridge, Copella, Snack-a-Jacks, Twistos, and Solinki. The Middle East and Africa units also have the beverages and foods, plus Kurkure, Chipsy, Smith's, Red Rock Deli and Mirinda.[24]

## Tactical Planning

Unlike strategic planning which has a broad scope and extended time frame, tactical planning has a moderate scope and an intermediate time frame. **Tactical planning** tends to focus on people and action and how to implement the strategic plans that have already been developed. It also deals with specific resources and time constraints. Tactical planning is more closely associated with middle management than with top management.

For example, KFC managers developed several tactical plans to implement an aggressive growth strategy. First, managers noted that most of KFC's business was done at night whereas McDonald's does most of its business at lunch. So tactical plans were developed to increase the restaurant's lunchtime business while retaining current levels of evening business. Second, managers planned to expand KFC's menu to include such things as roasted chicken. At Summer Farms, individual farm managers plan ways to increase productivity on their acreage.

## Operational Planning

The third kind of planning is **operational planning**, which can take a variety of forms. In general, operational planning has the narrowest focus and the shortest time frame. Such plans are usually supervised by middle managers but executed by first-line managers. There are two basic kinds of operational plans: standing plans and single-use plans.

**Standing Plans.** Plans for handling recurring and relatively routine situations are **standing plans**. Basic kinds of standing plans include policies (the most general), standard operating procedures, and rules and regulations (the most specific).

**Policies** are general guidelines that govern relatively important actions within the organization. For example, Yum! Brands—which operates or licenses Taco Bell, KFC, Pizza Hut, Long John Silver's, and A&W Restaurants in the United States—might establish a policy that does not allow any individual owning a stake in a McDonald's franchise to acquire a Pizza Hut franchise. Similarly, the company could (and generally does) also establish policies concerning the control of advertising campaigns, restaurant appearance, and sources of cooking supplies.

**Standard operating procedures (SOPs)** are more specific guidelines for handling a series of recurring activities. The manager of a Taco Bell restaurant, for example, may have a set of SOPs for inventory management. Following the SOPs, the manager could set desired levels of ingredients for each menu item, determine appropriate reorder schedules and amounts, and line up local suppliers. Following a set of SOPs, then, is fairly mechanical.

One particular form of SOP that is increasingly relevant to farms in particular and the food portion of agribusiness in general, involves Good Agricultural Practice (GAP) or variants known as Better Agricultural Practice (BAP) or even Best Agricultural Practice. As stakeholders became increasingly concerned about safety, security, and quality, they wanted assurances that producers and processors were following good, safe practices.[25] Government bodies began to develop GAPs in response to those concerns. The United States Department of Agriculture developed both GAP and Good Handling Practices (GHPs) to set voluntary standards. The state of Michigan went one step further and also established Generally Accepted Agriculture and Management Practices (GAAMPs) Many agribusiness organizations are incorporating similar plans into their own operations.[26]

Finally, **rules and regulations** are statements regarding how to perform specific activities. A rule for a KFC restaurant, for example, may dictate company policy regarding employee tardiness. Such a rule may state that if an employee is late to work three times in a two-month period, she or he must be warned and told that the next incident may result in termination. The same type of rule could be used at Summer Farms, Green Things, and Hartco, and even include an additional rule about the use of drugs, alcohol, and tobacco.

**Single-Use Plans.** The second major category of operational plans includes **single-use plans** for handling one-time-only events. The two types of single-use plans include programs and projects.

A **program** is a single-use plan for a large set of activities. The integration of KFC into the Yum!

Brand system is a good example of a program. This integration was a major operation involving thousands of people and hundreds of operating systems. In all likelihood, several task forces were created to handle the transition, and millions of dollars were spent to make KFC an integral part of the Yum! Brand organization.

A **project** is similar to a program but usually has a narrower focus. A menu addition at KFC, for example, can be considered a project. Market research determines that a particular new product will sell well, a recipe is developed and tested, relevant information is relayed to restaurant managers, the product is advertised and becomes a part of the regular menu, and managers move on to new things.

## Time Frames for Planning

Regardless of which kind of plan a manager is developing, it is important for him or her to recognize the role of the time factor. Similar to the time frames set for goals, plans focus on long-range, intermediate, and short-range time frames.

### Long-Range Planning

**Long-range planning** covers a period that can be as short as several years to as long as several decades. Virtually all large companies have long-range plans. Suppose, for example, that Boeing is planning to introduce a new generation of airplanes in ten years. This is a long-range plan. Given this plan, Boeing's managers can begin to develop other pertinent long-range plans. They may need to find a site for a new factory in six years so that the plant can be operational in eight years. They must arrange financing for the purchase of plant materials, construction costs, and raw materials for the new planes. Because they must be ready to take over operations at the plant in eight years, after four years the managers may want to identify who will be in charge and subsequently train those people.

The long-range plans for Summer Farms should include, for one thing, the amount of total acreage they hope to add for each crop, not counting timberland, within a specified period of time, Also, the family must plan its timber harvesting so that it has the finances to purchase new land.

Long-range plans are primarily associated with activities such as major expansions of products or facilities, development of top managers, large issues of new stocks and bonds, and the installation of new manufacturing systems. Top managers are responsible

for long-range planning in most organizations. As explained in 1988 by Michael D. Eisner, former CEO of Disney, "I think in terms of decades. The Nineties are EuroDisneyland. I've already figured out what we can do for 1997, and I've got a thing set for 2005."[27]

---

### FOOD FOR THOUGHT 8.3

Over 1 billion people lack access to water, and over 2.4 billion lack access to basic sanitation.

---

### Intermediate Planning

**Intermediate planning** generally involves a time perspective of between one and five years. Because of the uncertainties associated with long-range plans, intermediate plans are the primary concern of most organizations. Accordingly, top managers working in conjunction with middle managers usually develop them.

Intermediate plans are often seen as building blocks in the pursuit of long-range plans. If Syngenta has a long-range plan to have ten major new seeds available for sale in ten years, its managers may begin by developing an intermediate plan to get four under way within the next three years. At the end of three years, they would assess the situation and devise a new intermediate plan for the next time period to assure achievement of the long-range plan. Summer Farms would be planning how to plant its new acreage for organic foods and provide the necessary irrigation, drainage, and equipment.

### Short-Range Planning

Finally, **short-range planning** covers time periods of one year or less. These plans focus on day-to-day activities and provide a concrete base for evaluating progress toward the achievement of intermediate and long-range plans. To use the Syngenta example, a short-range plan might be to get two new seeds under way within the next year. The managers can thus focus on a specific set of activities (getting two new products under development) that need to be accomplished within the time frame (1 year). When considering organic farming, one of the first goals of Summer Farms was to learn the various legal regulations, starting with proper soil drainage, fertilizer and other chemical usage, product storage, and so on. In other words, it had much to learn before getting too deeply involved with the actual planning of its new enterprise. One danger to short-range planning is that it

## A FOCUS ON AGRIBUSINESS
### Water, the Next Oil

For years it has been said that water is the next oil. If so, companies had better start planning now (Figure 8.8). What that expression means is that water is increasingly scarce and, as a result, it is increasingly expensive to use and more valuable to control. So, just as for any scarce resource, companies should have plans on how to manage water.

Many firms already have such plans, in some cases as part of sustainability and/or environmental efforts. With proper planning, companies can save money and have positive environmental impacts while contributing to the concept of sustainable enterprises. Clearly a win-win for all concerned.

What have companies done? Consider Dow Chemical. The Freeport, Texas, site of Dow had difficulty getting enough water from the river to run its processes, particularly during droughts. Working with Nalco, Dow audited the site and installed automation technology to monitor the whole site's water system 24/7. The result was the company's ability to reduce water consumption

**FIGURE 8.8** Planning is an ongoing process.
© iStockphoto/Troels Graugaard

by approximately one billion gallons, annually. The whole site now uses less energy, requires less maintenance, and operates with far less water, so that droughts are no longer a concern.

Nalco also worked with a Marriott hotel property in Mumbai, India, to save water. The company's technology enabled Marriott to recycle cooling-system water instead of using fresh water, resulting in a savings of 300 million of glasses of water a year. In a country where water quality and quantity are poor, this sort of saving is especially important.

With these successes, Nalco continues a long history of treating water and water issues. A merger of the Chicago Chemical Company and Aluminate Sales Corporation in 1928 led to the formation of Nalco. Both companies sold sodium aluminate to treat water—the former to cities and industry and the latter to railroads for treating water used in steam locomotives. By 2010, Nalco had grown to over 10,000 employees in over 130 countries with over $4 billion of sales.[28]

---

becomes separated from longer-term planning and leads to a dysfunctional over-emphasis on short-term results.[29]

### FOOD FOR THOUGHT 8.4

Most Americans don't need to plan what to eat on Thanksgiving Day, as 93 percent traditionally eat turkey. In Japan, eating Kentucky Fried Chicken is a young tradition ["Kurisumasu ni wa kentakkii!" or "Kentucky for Christmas!"] that requires more conscious planning.

### Integrating Time Frames

We have seen that intermediate plans should build toward the pursuit of long-range plans, but all three time frames ideally should be integrated. A company may develop a long-range plan spanning ten years, an intermediate plan for five years, and a short-range plan for one year. Conceptually, at least, the short-range plan should be identical to the first year of the intermediate plan, which in turn should correspond to the first five years of the long-range plan.

At the end of one year, the short-range plan may or may not have been fulfilled, of course, so managers must develop a new short-range plan and modify the intermediate and long-range plans as appropriate. Suppose, for example, that Dr. Pepper has developed a short-range plan to increase total sales by 5 percent and a long-range plan to increase sales by 25 percent. At the end of the first year, however, sales have increased by only 2 percent. Dr. Pepper's managers might then drop their long-range plan down to 22 percent. This process of monitoring and adjusting plans relates to another important part of the process, contingency planning.

## Contingency Planning

**Contingency planning** is the part of the planning process in which managers identify alternative courses of action that the organization may follow if various conditions arise.[30] Again, this is something that most companies do although preparing for the unexpected is particularly important in agribusiness and will be covered again in Chapter 14. Russia is one of the top two major importers of U.S. chicken (China is the other). When in 2009 Russia banned U.S. chicken imports, companies like Tyson had to quickly develop contingency plans to deal with the uncertainty in that market. Thankfully, the ban was lifted in 2010.[31]

### The Nature of Contingency Planning

The general nature of contingency planning follows the process shown in Figure 8.9. As illustrated, managers develop an initial plan (A) that specifies possible contingency events that may dictate modification of the original plan. The organization then monitors ongoing activities so it will know if and when the events occur. Depending on the nature of the contingencies, the organization may continue its original plan (A), change to contingency plan (B), or change to another contingency plan (C).[32]

## Contingency Events

Obviously, the identification of contingency events is a critical part of contingency planning. If the events are not properly identified, or if their relevant indicators are poorly understood, the entire process of contingency planning can fall apart.[33] Scenarios are developed to suggest the more probable events upon which plans need to be developed. Does the university have contingency plans for increases or decreases in student enrollments? Does the organic grocer have plans for increases in sales? Does Summer Farms have adequate plans for preventing crop failure in a period of drought or a freeze? How would an insect devastation of its timber affect Summer Farms? In the case of Webb Dairy, does it have plans for replacing cows that show a premature reduction in milk? Is it prepared to save itself from total loss if one of its cows should someday test positive for Brucellosis (Bang's Disease) or bovine spongiform encephalopathy (Mad Cow Disease)? Contingency plans deal with just those sorts of questions.

In general, critical contingency events relate either to the extent to which the ongoing plan is being accomplished or to environmental events that might change things in the future. For example, assume that Burger King is in the midst of a major expansion program in the United States and Europe. Its managers may break down their intermediate plan into a series of five short-range plans, each of which calls for 100 new U.S. restaurants and 50 new European restaurants at the end of each calendar year. One contingency event would then be the extent to which the short-range plans are being realized. If there are 96 new U.S. restaurants and 51 new European restaurants at the end of the first year, the managers may conclude that things are going well. In contrast, 60 new U.S. restaurants and 15 new European restaurants would indicate a problem. The managers' contingency

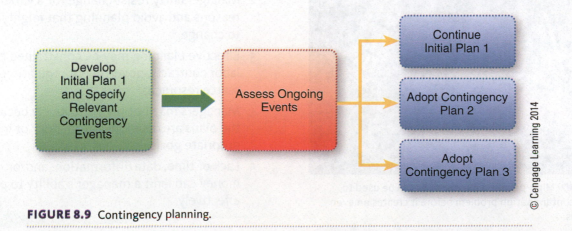

**FIGURE 8.9** Contingency planning.

© Cengage Learning 2014

plans might call for a revision of the original plan or increased efforts to catch up in the second year as well as efforts to determine why the plan was not met.

The other kind of contingency event relates to the company's environment. Burger King's managers may have based their European expansion plans on the assumption that McDonald's and Wendy's would continue to expand at current rates. If McDonald's unexpectedly begins to double its rate of European expansion, Burger King may be forced to change its own plans. In the case of Webb Dairy, if economic conditions reduce its total revenue to an unacceptable level, does it have plans such as purchasing additional cows, cutting its feed costs, or diversifying the business in some way?

Clearly, contingency planning is an important part of any organization's overall planning process. It should never be neglected. In addition, there are other things managers should do to facilitate effective planning. The next section addresses some of these.

## Crises

One particular type of contingency event is the unexpected, although not necessarily unanticipated, event. As noted in Chapter 3, Johnson & Johnson may not have had a contingency plan in place in 1982 when the first Tylenol scare occurred, but it did when the others happened in 1986 and 2010. Because, as noted in Chapter 4, the Campbell Soup Company had contingency plans along with crisis teams to handle just such events (Figure 8.10). So in 1987 when a grocer was told that someone had poisoned cans of

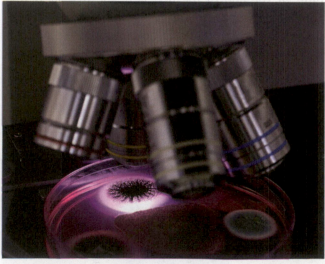

© ggw1962/www.Shutterstock.com

**FIGURE 8.10** High-powered microscopes can be used to find a food contamination problem before it creates an even greater crisis.

Campbell's tomato juice, Campbell had a crisis team there in just a few hours. In agribusiness, especially the food industry, crisis planning is a must.

## Managing the Planning Process

As we have seen several times already, planning is a vital part of all managerial jobs and requires considerable skill. Two additional dimensions of planning—understanding the roadblocks to effective planning and knowing how these roadblocks can be avoided—are critical if planning is to be carried out properly.

### Roadblocks to Effective Planning

What can go wrong when managers set out to develop plans? In truth, any number of factors or roadblocks can disrupt effective planning.[34] Let's look at Box 8.1 and consider five of the most common ones.

**The Environment.** In Chapters 4–7, we looked at the importance of the environment in goal setting and planning. Most organizations operate in environments that are both complex and dynamic. That is, managers must contend with a great many environmental forces and dimensions (complexity) and with the fact that these forces and dimensions change rapidly (dynamism). In combination, these factors make it harder to develop effective plans. For

---

**BOX 8.1**
**ROADBLOCKS TO EFFECTIVE PLANNING**

**Roadblocks**

1. The environment may be so complex and/or changing that managers fail to plan effectively.

2. Managers may resist change for a variety of reasons and avoid planning that might lead to change.

3. Effective planning may be constrained by labor contracts, government regulations, or scarce resources.

4. Managers may not plan effectively because the plans are developed from poor or inappropriate goals.

5. Lack of time, data/information, and/or money can limit a manager's ability to plan effectively.

example, when General Electric plunged into factory automation, demand dropped, other automation systems turned out to be more advanced than GE's, and soaring costs drove prices up. Each of these factors resulted from environmental complexity and dynamism. Consequently, GE's original plans for sales and income from automation were not realized.[35]

**Resistance to Change.** Another factor that can impede effective planning is resistance to change. By its very nature, planning involves change. Fear of the unknown, preferences for the status quo, and economic insecurity can combine to cause managers to resist change and, as a result, to avoid planning that might begin that change. One or more of the Summer family members could have bristled at the idea of getting into something as "new and different" as organic farming. They did reject the idea of expanding into vegetable processing and packaging, but for good reason. Webb Dairy may be hesitant to change to artificial insemination of its cows instead of the natural way. Assume Heinz has the goal that all of the company's businesses should be number one or two in their respective industries. A Heinz manager who had little hope that her business would be able to meet this standard may recognize that planning could undermine the managerial role in the company and might therefore avoid planning activities. Compensation system incentives may also not be changed to be consistent with revised plans.

**Situational Constraints.** Similarly, various situational constraints can hinder effective planning. Suppose that Hormel would like to undertake a full-scale expansion program, but its human resource managers learn that the company does not have enough managerial talent to support expansion at the desired rate. As a consequence, the rate of expansion must be scaled back to fit the availability of managers. Other constraints that can affect planning include labor unions and contracts, government regulations, a scarcity of raw materials, an Act of God (e.g., hailstorm, drought, flood) and a shortage of operating funds—all of which are well known to most small- and medium-sized agribusinesses.

**Poor Goal Setting.** Poor goal setting is a significant roadblock to effective planning. Since goal setting is the first step in the planning process, any of the things that might impede proper goal setting by definition also hinders effective planning. The ability to develop specific, measurable, achievable, realistic, and temporal goals is an important technical skill that all managers must possess.

**Time and Expense.** Finally, time and expense can limit effective planning. Like most things worth doing, planning takes a considerable amount of time. Because of the pressures that affect them, most managers occasionally find it difficult to undertake planning activities. Similarly, acquiring the necessary information to develop effective plans can cost both time and money. Managers at Polaroid who want to know about future trends in amateur photography, for instance, may need to conduct extensive market research or buy the information from private market research firms. Faced with such expenses, they may be tempted to take shortcuts or evade the issue altogether. Gathering data can be expensive, but managers may not even know the right data to get or it may not even exist, particularly for radically new products or services.

## Avoiding the Roadblocks

Fortunately for managers, there are some useful guidelines that can help them avoid roadblocks to effective planning. Six of the most basic ones are summarized in Box 8.2.

**Start at the Top.** For planning to be effective, it must start at the top. Top managers must set the goals and strategies that lower-level managers will follow. While CEO of Intel, Andrew Grove, led each of his company's planning groups. Such leadership conveys a strong and clear message to everyone that planning is important. Strategy and strategic

**BOX 8.2**

**GUIDELINES FOR AVOIDING ROADBLOCKS TO EFFECTIVE PLANNING**

**Guidelines**

1. Planning should start and be led by the top level of the organization.
2. Managers should recognize the limits and uncertainties of planning.
3. Managers should communicate what they are doing to other levels in the organization.
4. Managers should actively participate in planning.
5. Managers should make sure that long-range, intermediate, and short-range plans are well integrated.
6. Managers should develop contingency plans.

planning are critically important so the whole next chapter is devoted to that topic.

**Recognize the Limits.**    Managers also must recognize that no planning system is perfect. Because of its very nature, planning has limits and cannot be carried out with absolute precision. Coca-Cola spent years planning to introduce a new formula (New Coke) for its flagship product (Classic Coke). Even though the company did everything "by the book," the change was disastrous and Classic Coke was reintroduced within a short time. In another example, Hartco, like managers in millions of companies, never thought a national economic crisis would be serious enough to wreck the demand for its popular parquet floor product.

**Communicate.**    Communication, especially vertical communication within the organizational hierarchy, can facilitate effective planning. Top managers who know what middle- and first-line managers are doing, are better able to continue their own planning activities. At Intel, great efforts are made to ensure that every manager who might potentially be affected by a decision or plan is made aware of the plan. This should be relatively easy for managers to do in smaller companies.

**Participate.**    In similar fashion, participation aids effective planning because managers who are fully involved in planning are more likely to know what is going on, to understand their own place in the organization, and consequently to be motivated to contribute. One of the keys to the success recently enjoyed by Ford, for example, has been increased participation by managers in all aspects of planning.

**Integrate.**    We have already noted that long-range, intermediate, and short-range plans must be properly integrated. The better these plans are integrated, the more effective the organization's overall planning system will be. Until problems emerged in 2010, for instance, Toyota was known for doing a good job of integrating the various time frames across which it plans. Short-term model changes, for example, almost always mesh nicely with longer-term new model introductions.

**Develop Contingency Plans.**    The final technique to enhance planning is to develop contingency plans. As we have discussed, contingency plans are alternative actions that a company might follow if conditions change. For example, Monsanto will no doubt have several different contingency plans from which to choose, if competitors such as DuPont and Syngenta change their approaches to marketing seeds.

Recent years have seen a revolution in information technology based largely upon developments in personal computing. That revolution has added a powerful tool to every manager's kit for use in planning. It has also increased the need for technical skills in the computer area. Tomorrow's managers will be even better equipped to handle the complex and dynamic nature of the planning process although the needed skills will be greater for those managers.

## CHAPTER SUMMARY

A plan is a blueprint or framework used to describe how an organization expects to achieve its goals. Planning is the process of developing plans or of determining how best to approach a particular goal. Because planning is so important, every manager must be involved in the process.

A goal is a desired state or condition that an organization wants to achieve. Goals may be differentiated by level of management, area, and time frame and specificity. The steps in the goal-setting process are (1) scan the environment for opportunities and threats; (2) assess organizational strengths and weaknesses; (3) establish general organizational goals; (4) establish unit goals; (5) establish subunit goals; and (6) monitor progress toward goal attainment and provide feedback as the cycle repeats. Goal optimization is the process of balancing and trading off between different goals for the sake of organizational effectiveness.

The three major kinds of plans are strategic, tactical, and operational. Strategic plans are the broad, long-term plans developed by top managers to guide the general directions of the organization. Tactical plans are designed to implement strategic plans and hence have moderate scopes and intermediate time frames. Operational plans have the narrowest focus and the shortest time frames and are executed by first-line managers.

The three major time frames for planning are long-range, intermediate, and short-range. Long-range planning covers a period of time is as short as several years or as long as several decades. Intermediate planning generally takes about one to five years. Short-range planning covers time periods of one year or less and focuses on day-to-day activities, thus providing a basis for evaluating progress toward the achievement of intermediate and long-range plans. Each of these time frames should be integrated with the others to ensure smooth functioning of the organization.

Contingency planning is the part of the planning process that identifies alternative courses of action that an organization may follow if various different conditions arise. Critical contingency events relate either to the extent to which the ongoing plan is being accomplished or to environmental events that might change things in the future.

Roadblocks to effective planning include the environment, resistance to change, situational constraints, poor goal setting, and the time and expense of planning. Guidelines for avoiding these roadblocks include starting at the top, recognizing the limits to planning, communicating, participating, integrating, and developing contingency plans.

## CHAPTER ACTIVITIES

### REVIEW QUESTIONS

1. Identify and describe the three basic kinds of planning done by most organizations.
2. What are the six steps in the organizational goal-setting process?
3. How should organizations integrate plans that span different periods of time?
4. What are contingency plans? How are they developed?
5. Identify several roadblocks to effective planning. Suggest ways around these roadblocks.

### ANALYSIS QUESTIONS

1. Do you think an organization could function effectively without planning? Why or why not?
2. What kinds of goals would a not-for-profit organization such as your college or university have? What might be the similarities and differences between those goals and the goals of a business operating for a profit? Why?
3. If all managers are supposed to plan, why should planning start at the top? Could it start at the bottom? How?
4. Describe a contingency plan you once developed. How closely did the steps you followed coincide with those outlined in this chapter?
5. Why might different kinds of organizations have different conceptions of what long-range planning means?

## FILL IN THE BLANKS

1. A blueprint or framework an organization uses to describe how it expects to achieve its goals is known as a/an _____.

2. A desired state or condition that the organization wants to achieve is called a/an _____.

3. The way an organization attempts to fulfill its purpose is known as its _____.

4. The process of achieving an effective balance among the different goals of an organization is known as _____ _____.

5. Formulating the broad goals and plans of the organization is called _____ _____.

6. Plans that are designed to handle recurring and relatively routine situations are called _____ plans.

7. Identifying alternative courses of action that might be followed if various conditions arise is known as _____ _____.

8. General guidelines that govern relatively important actions within an organization are known as _____.

9. Specific guidelines for handling a series of recurring activities are referred to as _____ _____ _____.

10. Plans that are developed to handle events that only happen once are known as _____ _____ plans.

## ▼ CHAPTER 8 CASE STUDY
### ARE THE GODS WITH US?

The priestly procession entered the "holy enclosure" followed at a respectful distance by everyone else. Excitement and apprehension were palpable; everyone felt that today would be the day. A quiet fell over everyone as the sky began to brighten and the sun began to "peek" over the horizon.

This scene or some variation of it was enacted at Stonehenge (England), Chaco Canyon (United States), Abu Simbel (Egypt), and Angor Wat (Cambodia) to name a few places. The ceremony had a simple objective: determine the time of solstices and equinoxes. Defined by the Earth's orbit around the Sun, the solstices and equinoxes divide the year into four nearly equal segments. Today these segments bear the names spring, summer, fall, and winter, in order of their importance.

Why is "this order of seasons" so important? In hunter-gatherer societies, mastery of heavens was important for travel to and from their camps. In agrarian societies, mastery of the heavens also was vital to farmers' existence. The spring equinox, around March 21, marked the end of winter and a time to consider planting. Planting too early or too soon could lead to crop failure, a meager harvest, and a very long winter. Plans depended on the information.

In 2010, an old agricultural nemesis made its presence felt in Russia: drought. Rabobank, a Netherlands-based bank, lowered its estimate of the Russian wheat harvest accordingly; and Russia's President Vladimir Putin announced a ban on Russian wheat exports for the remainder of 2010. Kazakhstan and Ukraine announced possible export restrictions due to drought.

In spite of our world agronomic sophistication, drought can still manifest itself on the world stage. To the average consumer, asking the price of wheat in Russia sounds like a silly question. However, when it affects the world supply of wheat, Russia's price has consequences. The principles of supply and demand begin to manifest themselves, and suddenly potential price increases are possible for bread in the United Kingdom or pizza dough at Domino's Pizza in the United States.

What about the gods? Well, the gods have been replaced by the science of economics. Stonehenge has been replaced by the Internet and the Chicago Board of Trade wheat trading pit. But the decisions are still made by the individual farmers.

For Gary Millershaski, a Kansan with a 6,000-acre farm, this set of circumstances presents an interesting situation. "I'm an opportunist," he said as he thought about his plan for the year. He contemplated whether to increase his wheat production by planting on soil he intended to leave fallow (unused). The extra 1,000 bushels could generate thousands of dollars in extra income. He knows about the situation in Russia and Central Europe; he is aware of that Canada's harvest is also off; he knows that his state and the rest of the United States had a great harvest; and he knows that the wheat importing countries (Egypt, Brazil, European Union, Indonesia, Japan, Algeria, Nigeria, Iraq, South Korea, and Morocco) will have to purchase 52.3 million metric tons someplace. It may as well be Kansas.

But suppose the world has enough wheat in storage and there will be no price increase. Or suppose his fellow Kansans all do the same thing, creating a surplus, causing prices to fall and thus nullify his investment in time and land. He knows, also, that the top three wheat exporters account for less than half of the world's total production, according to Deutsche Bank.

All of this information is available to everyone, everywhere. Gary Millershaski has become sophisticated and has powerful decision tools at his disposal to aid in planning, but so does everyone else. This is the result of the democratization of technology and information that defines the current world. Today's agribusiness managers rarely make decisions in an information void; it is often the opposite case. Only experience and risk management can assist in an information and technology rich environment.

In our high-tech 21st Century we no longer have to rely on these ancient astronomical devices or calendars, of course. However, this does not mean that all agronomic decisions are automatic. In fact, we may now have a very different problem: too much information. The uncertainty today is not astronomical but economical. It was also an economical problem in the past, but the astronomical prediction predominated. Where are the gods when you need them?[36]

### ▶ Case Study Questions

1. How might an organization plan for an unexpected event, such as a drought, and its effect?

2. What would you do if you were the Kansas farmer? Why?

3. How does what the farmers do affect other agribusinesses?

## REFERENCES

1. "About Sara Lee" at www.saralee.com (accessed August 1, 2010); "Sara Lee CEO steps down following stroke," *News in Brief* (August 10, 2010) at www.foodnavigator-usa.com (accessed August 1, 2010); "Shareholders Letter." Sara Lee Corporation Annual Report (2009) at www.saralee.com (accessed August 2, 2010); Anjali Cordeiro and Ilan Brat, "Sara Lee's CEO Resigns Amid Medical Problems," *The Wall Street Journal,* August 10, 2010, B3.

2. "ConAgra Foods ranks No. 434 on FORTUNE's list of the World's Largest Companies," *Fortune* (October 6, 2006) at money.cnn.com (accessed July 23, 2010); "Conagra Turns Up the Heat," *Business Week,* September 2, 1991, 58–63.

3. Russell Mitchell, Lois Therrien, and Gregory L. Miles, "ConAgra: Out of the Freezer," *Business Week,* June 25, 1990, 24–25.

4. Ned Douthat, "ConAgra is good Eatin'," *Forbes,*(June 25, 2010, at blogs.forbes.com (accessed July 24, 2010); Seth Lubove, "ConAgra," *Forbes,* July 20, 1992, 114–123; and Ronald Henkoff, "Conagra: A Giant That Keeps Innovating," *Fortune,* December 16, 1991, 101.

5. Peter J. Brews and Michelle R. Hunt, "Learning to Plan and Planning to Learn: Resolving the Planning School/Learning School Debate," *Strategic Management Journal* (1999), 889–913; see also Arie P. De Geus, "Planning as Learning," *Harvard Business Review* (March–April 1988), 70–74.

6. Harrison M. Trice and Janice M. Beyer, *The Cultures of Work Organizations* (Englewood Cliffs, NJ: Prentice-Hall, 1993), 117.

7. Henri Broms and Henrick Gahmberg, "Communications to Self in Organizations and Cultures," *Administrative Science Quarterly,* 28:482–495.

8. "Waldenbooks Peddles Books a Bit like Soap, Transforming Market," *The Wall Street Journal,* October 10, 1988, A1, A4.

9. Tom Van Riper, "Where You Might Not Shop in 2010," *Forbes* (January 20, 2010) at www.forbes.com (accessed July 24, 2010).

10. George A. Steiner, *Top Management Planning* (New York: Macmillan, 1969).

11. Charles R. Schwenk and Charles B. Shrader, "Effects of Formal Strategic Planning on Financial Performance in Small Firms: A Meta-Analysis," *Entrepreneurship: Theory and Practice* (1993): 53–64; Jeff Bracker and John Pearson, "Planning and Financial

Performance of Small Mature Firms," *Strategic Management Journal* (1986), 503–522.

12. Rosabeth Moss Kanter, "The Middle Manager as Innovator," *Harvard Business Review* (July–August, 2004), 150–161; Hugo Uyterhoeven, "General Managers in the Middle," *Harvard Business Review* (September–October, 1989), 136–145.

13. Ronald L. Nichol, "Get Middle Managers Involved in the Planning Process," *The Journal of Business Strategy* (May 1, 1992), 26–33.

14. Bill Wooldridge, Torsten Schmid, and Steven W. Floyd, "The Middle Management Perspective on Strategy Process: Contributions, Synthesis, and Future Research," *Journal of Management* (December 2008), 1190–1221; and Rosemary Stewart, "Middle Managers: Their Jobs and Behavior," *Handbook of Organizational Behavior*, ed. Jay W. Lorsch (Englewood Cliffs, NJ: Prentice-Hall, 1987), 385–391.

15. Leonard A. Schlesinger and Janice A. Klein, "The First-Line Supervisor: Past, Present, and Future," *Handbook of Organizational Behavior*, ed. Jay W. Lorsch (Englewood Cliffs, NJ: Prentice-Hall, 1987), 370–384.

16. Carol Davenport, "America's Most Admired Corporations," *Fortune*, January 30, 1989, 68–94.

17. "CEO of the Year: Stanley Gault of Goodyear," *Financial World*, March 31, 1992, 26; Peter Nulty, "The Bounce is Back at Goodyear," *Fortune*, September 7, 1992, 70–72.

18. Thomas Bateman, Hugh O'Neill, and Amy Kenwothy-U'Ren, "A Hierarchical Taxonomy of Top Managers' Goals," *Journal of Applied Psychology* (2002), 1134–1148.

19. John C. Crotts, Dundan R. Dickson, and Robert C. Ford, "Aligning Organizational Processes with Mission: The Case of Service Excellence," *Academy of Management Executive* (2005), 54–68; Andrew Campbell, "The Power of Mission: Aligning Strategy and Culture," *Planning Review*, September 10, 1992, 10–15.

20. "Our Legacy," Citigroup website at www.citigroup.com (accessed July 22, 2010); Edward Boyer, "Citicorp: What the New Boss is Up To," *Fortune*, February 17, 1986, 40.

21. For more discussion of the goal-setting process, see John E. Gamble and Arthur A. Thompson, Jr. *Essentials of Strategic Management: The Quest for Competitive Advantage, 2nd ed.* (New York: McGraw-Hill, 2011).

22. Charles W. L. Hill and Gareth R. Jones, *Strategic Management, 8th ed.* (Cincinnati, OH: Cengage Learning, 2009); and Michael E. Porter, *Competitive Advantage* (New York: Free Press, 1985).

23. "The PepsiCo Family," PepsiCo website at www.pepsico.com (accessed July 24, 2010).

24. JP Mangalindan, "PepsiCo CEO: 'If All Consumers Exercised… Obesity Wouldn't Exist'," *Fortune* (April 27, 2010) at money.cnn .com (accessed July 24, 2010).

25. Anne-Sophie Poisot and Siobhán Casey, (2004) *Report of the FAO Internal Workshop on Good Agricultural Practices*; Rome, Italy: Food and Agricultural Organization of the United Nations; Amy Russell and Tony Battaglene, (2007) *Trends in Environmental Assurance in Key Australian Wine Export Markets*. Adelaide, Australia: Winemakers' Federation of Australia

26. U.S. Department of Health and Human Services, (1998) *Guide to Minimize Microbial Food Safety Hazards for Fresh Fruits and Vegetables*. Washington, DC: Food and Drug Administration, Center for Food Safety and Applied Nutrition; Jerry May, 2011 Updates to Michigan Siting GAAMPs available at http://news.msue.msu. edu/news/article/2011_updates_to_michigan_siting_gaamps.

27. Gary Hector, "Yes, You Can Manage Long Term," *Fortune*, November 21, 1988, 68.

28. G. Colvin, "Erik Fyrwald: The king of water," *Fortune*, June 24, 2010, at money.cnn.com (accessed July 10, 2010); S. Nunes, "Water: Act Now to Conserve the New Oil," *Fortune*, July 8, 2010 at tech.fortune.cnn.com (accessed July 12, 2010); "Our Company History," Nalco website at www.nalco.com (accessed July 11, 2010); D. Frykholm, "Finns tap Arctic water for Arab exports," April 10, 2004; "Dawn the Internet" at www.dawn.com (accessed July 12, 2010).

29. Michael T. Jacobs, "A Cure for America's Short-Termism," *Planning Review*, January 1992, 4–9; see also Grady D. Bruce and James W. Taylor, "The Short-Term Orientation of American Managers: Fact or Fantasy?" *Business Horizons*, May/June 1991, 10–15, who argue that the problem is not a short-term orientation.

30. Ricky W. Griffin, *Management, 10th ed.* (Mason, OH: Southwestern Cengage Learning, 2011); Michael Watkins and Max Bazerman, "Predictable Surprises: The Disasters You Should Have Seen Coming," *Harvard Business Review* (March 2003), 72–81.

31. "Analysts: Russia Chicken Decision to Help US Cos," *Business Week* (June 25, 2010) at www.businessweek.com (accessed July 2, 2010).

32. Donald C. Hambrick and David Lei, "Toward an Empirical Prioritization of Contingency Variables for Business Strategy," *Academy of Management Journal* (December 1985), 763–788.

33. Ari Ginsberg and N. Venkatraman, "Contingency Perspectives of Organizational Strategy: A Critical Review of the Empirical Research," *Academy of Management Journal* (July 1985), 421–434.

34. Gregory A. Baker and Joel K. Leidecker, "Does it Pay to Plan? Strategic Planning and Financial Performance," *Agribusiness* (2001), 355–364; George A. Steiner, *Strategic Planning: What Every Manager Must Know* (New York: Free Press, 1979).

35. Noel Tichy and Ram Charan, "Speed, Simplicity, and Self-Confidence: An Interview with Jack Welch," *Harvard Business Review* (September–October 1989), 112–120.

36. Robin Heath, *The Sun, Moon and Earth*. (NY: Walker and Company, 1999), 8; Thomas Friedman, *The Lexus and the Olive Tree*. (NY: Anchor Books, 2000), 46–60; Tom Polansek, "Wheat Goes Up, Prices to Follow," *The Wall Street Journal* (August 4, 2010) at www.wsj.com (accessed August 5, 2010); Anna Raff and Alex MacDonald, "Wheat Drops More Than 7%," *The Wall Street Journal* (August 7-8, 2010) at www.wsj.com (accessed August 5, 2010); Andrew Johnson, "Wheat Falls on a Supply Rethink," *The Wall Street Journal* (August 10, 2010), C5; Liam Pleven, Nours Malas, and Patrick Barta, "Decision Time Looms for Wheat Farmers," *The Wall Street Journal*, August 9, 2010, C1; Liam Denning, "When Wheat Withers, Corn Could Grow," *The Wall Street Journal* (August 10, 2010) at www.wsj.com (accessed August 5, 2010).

# Strategy and Strategic Planning

## LEARNING OBJECTIVES

After studying this chapter, you should be able to:

- Describe the nature of strategic planning, including the components and levels of strategy and strategy formulation and implementation.

- Understand the environmental forces important to strategic planning and how managers position their organizations within those forces.

- Identify major approaches to corporate strategy.

- Identify major approaches to business strategy.

- Identify the major functional strategies developed by most organizations.

- Describe the process of strategy implementation.

## Chockfinger's Chocolate Hedge

That innocent piece of chocolate—the dieter's "Waterloo." One bite and the dieter "morphs" or changes from virtuous to sinful because one bite is never enough. It seems that chocolate (or at least its commodity precursor, cocoa) has the same effect on commodity traders or speculators. And some folks are crying foul or at least not fair—not nice.

**FIGURE 9.1** Thanks to speculators, the cocoa bean's increase in value makes it attractive to smugglers as well as commodity traders and chocolate manufacturers.
© iStockphoto/Ewen Cameron

Anthony Ward, CEO of Armajaro, a hedge fund, has been labeled "Chocfinger" by the British press. This title is a passing reference to the James Bond film, "Goldfinger." What is Mr. Ward's transgression? His company strategically took possession of 240,100 tons of cocoa beans (Figure 9.1). These beans stored in temperature-controlled (refrigerated) warehouses represent 7 percent of the world's annual cocoa bean production, or 25 percent of the total stock in Western Europe in 2010. The $1 billion purchase (£650 million) can stay in this controlled environment for more than 20 years, although that was certainly not the intent of this purchase.

In commodity trading there are two types of strategic participants: users and speculators. Users are the food producers who use the market to acquire raw materials or to hedge against price increases for a key product ingredient, like cocoa beans in the manufacture of chocolate. Nestlé SA, one of the largest producers of cocoa-based products, uses the commodity market in both ways. Speculators are traders who "bet" on the upswing or downswing of commodity prices. They provide liquidity in the market. If all of the traders were producers or commodity users, the commodity exchange might become smaller or seasonal. The speculators provide someone to take the opposite position in a trade. Speculators rarely take physical possession of the product. This is not their goal. Their goal is to make money on the trade or close out the position before it expires, forcing them to take physical possession of the product.

Armajaro, on the other hand, believed that taking possession of the product was a gamble that the decrease in world production would make stockpiling of a product even more valuable to the producers. This hedge fund felt that cocoa bean production would decline in 2010 and that the price of cocoa beans would increase concomitantly. Cocoa production is down in West African countries, one of the effects of civil strife. These countries supply one-third of the world's supply. Some of the trees are older than 30 years—the trees in Africa were planted in the 1970s—and Bean production drops after the trees reach 30 years of age. Other countries may want to fill the gap created by this situation, but it takes four years for a cocoa tree to mature and produce "black pods." The supply-and-demand relationship complicates agribusiness decisions because of the delay in response time.

In addition to future decreases in production, Anthony Ward knows that demand has bee[n]
ing steadily. So the move by his hedge fund, Armajaro, was speculative with a "twist." He was bett[er]
both the supply and price. Normally, supply and demand set price. Ward knew, however, that even if o[ther]
producers come to the rescue of the world cocoa supply, it will not meet demand. In addition to the tree[s']
age problem, a huge supply gap will also occur if the average Chinese "chocaholic" aspires to European
standards. For example, the average Chinese person consumes about 3.5 ounces of chocolate a year, while
the average European consumes 22.2 pounds a year.

Whether the Chinese develop an appetite for chocolate is interesting, but the "gap" will lead to
increases in packaged food goods or at least a consideration to raise prices. This situation was acknowledged by Nestlé SA, Dannone SA, Unilever PLC, and Kraft Foods, Inc.

How do the producers feel? Sixteen cocoa companies complained to NYSE Liffe (New York Stock
Exchange and London International Financial Futures Exchange) about market manipulation and the need
for more regulation. Francisco Redruello, a senior food analyst at Euromonitor International, said that this
purchase and possession has already caused greater price volatility and attracted the interest of other
hedge funds.

The cleverness or questionable nature of this strategic speculation is in the eyes of the beholder. But
at least it was open and transparent. However, the pricing of chocolate is further complicated by the different and more difficult situation in Ghana, one of the nations that could make up some of the supply deficit.
Because the price of cocoa elsewhere is greater than offered by Ghana, up to 60 percent (one million tons)
of the cocoa produced in Ghana has been smuggled out of the country.

The "Chocfinger" situation accentuates the importance of strategic planning, particularly in high risk
situations—and gives a whole new meaning to "food fight."'

# INTRODUCTION

By most objective indicators, the strategy of Nestlé SA, one of the
largest producers of cocoa-based products, has enabled it to be a
very effective organization. Nevertheless, changing environmental
conditions such as cocoa supply and demand compel this highly
successful company to continue making a concerted effort to
understand its environment and take appropriate action. The
activities involved in doing these things are a part of strategy and
strategic planning.

This chapter is about the various elements of strategy and
strategic planning. First, it describes the nature of strategic planning.
It then focuses on environmental analysis and discusses corporate,
business, and functional strategies. The chapter concludes with a
description of strategy implementation.

...ng

...t strategic plans are the
... managers to guide the
...ization. To expand on
... the nature of strategic
... identify the components
... strategic planning, draw a distinction between
strategy formulation (including environmental monitoring) and strategy implementation, and note the
levels of strategy. Strategic planning is important to
all organizations, public and private.[2] Nevertheless,
some organizations neglect it. In recent years, Coca
Cola admitted that it had spent too much time and
energy on quantitative methods and not enough on
strategic planning.[3]

## The Components of Strategy

In general, strategy can be thought of as having four
basic components: scope, resource deployment, distinctive competency, and synergy.[4]

The **scope** of strategy specifies the position the
firm wants to have in relation to its environment.
More specifically, it details the markets or industries
in which the firm wants to compete—the range of
markets in which the organization will compete.[5]
The scope of an organization, then, helps determine
how it will accomplish its purpose and mission. For
example, the scope of Ford's strategy specifies that
the company wants to produce and sell automobiles
around the world, whereas Hershey Foods restricts
its scope to the confectionary business and related
food processing. While ethanol provides substantial
profits to some firms and Archer Daniels Midland
has done well with ethanol, this company nevertheless recognizes that ethanol performance can fluctuate and therefore it maintains its traditional strength
as a food processing and distribution business.[6]

**Resource deployment** indicates how
the organization intends strategically to allocate
resources.[7] General Electric (GE) wants each of its
businesses to be either No. 1 or No. 2 in its industry.
This suggests that GE might sell a business that is
No. 5 and losing ground, allocate enough resources
to a business firmly entrenched as No. 1 for it to stay
there, and provide new resources to a business that is
No. 3 and gaining ground on No. 2. Unilever divested
some less profitable units to acquire more closely
related ones such as Ben & Jerry's Homemade Ice
Cream and Slim-Fast.[8] In strategy implementation,

tactical resource deployment decisions will be made
to carry out the strategic one.

**Distinctive competency** is the specification of what advantage or advantages the firm holds
relative to its competitors. Also known as an organization's core competency, it is what enables an
organization to outperform its competitors. IBM's
distinctive competencies include its well-known
name and dominance of the computer market.
Likewise, Kodak used its strengths in name recognition and distribution channels to introduce a new
line of batteries.

**Synergy** is the extent to which various businesses within a firm can expect to draw from one
another.[9] Disney, for example, realizes considerable
synergy from its theme parks, movies, and merchandising businesses. Families familiar with Disney
characters from movies and books are motivated to
visit the theme parks. After an enjoyable experience,
they are subsequently motivated to buy merchandise
and see future movies. McDonald's uses synergy as
it considers adding pizza, Mexican foods, and other
related businesses.[10]

## Strategy Formulation and Implementation

Another important element of strategic planning is
the distinction between **strategy formulation** and
**strategy implementation**.[11] Actually, the words
themselves convey the meaning of the two terms: formulation is the set of processes involved in creating or
developing strategic plans, and implementation is the
set of processes involved in executing them, or putting them into effect.[12] Most of our attention in this
chapter is directed to formulation issues, although
the end of the chapter touches on implementation.

## The Levels of Strategy

A final important perspective to understanding the
nature of strategic planning is the level of strategy, of which there are three: corporate, business,
and functional.

The **corporate strategy** charts the course
for the entire organization and attempts to answer
the question, "What businesses should we be in?"
Walmart developed a corporate strategy that calls for
continued growth and emphasis on volume retailing.
As Summer Farms grows, it will need to take a closer
look at its corporate strategy, which appears to have
already changed greatly from the time it started as a
local vegetable grower.

A **business strategy** is charted for each individual business within a company. Managers at Walmart have set strategies for its discount store division, its Sam's Club division, and its Walmart .com division. Summer Farms has strategies for each of its six "fields." Although there may be some similarities in business strategies across divisions, there are also clear differences.

Similarly, **functional strategies** are developed to correspond to each of the basic functional areas within the organization. Common functional strategies include marketing, financial, and production. Walmart has a marketing strategy of low-cost and high-volume retailing, a financial strategy that calls for low debt, and a human resource strategy that emphasizes hiring college graduates as management trainees.

Clearly, each level of strategy is important. If any level is neglected, the entire organization can and will suffer. We shall therefore consider each one in more detail, but first we will look at another important dimension of strategic planning: environmental analysis.

### FOOD FOR THOUGHT 9.1

Worldwide demand for water is at least doubling every 21 years, thus placing a tremendous importance on strategic planning by large corporations and government entities in particular.

## Environmental Analysis

Environmental analysis is the starting point in strategic planning. It helps managers develop a thorough understanding of an organization and its environment. Agribusinesses have adopted many of the sophisticated strategic planning frameworks, such as environmental analysis, core competences, and SWOT analysis.[13]

### Organizational Position

Many of the key issues involved here were discussed earlier in Part Two. As shown in Table 9.1, from a strategic-management perspective, managers must

**TABLE 9.1** SWOT Analysis

| INTERNAL (SW) ANALYSIS | EXTERNAL (OT) ANALYSIS |
| --- | --- |
| **Strengths**<br>Identifying the organization's distinctive competencies perhaps using the VRIO framework.<br><br>• What does your organization have that makes you successful? Think about members of the organization, your family, labor, machinery, size, location, reputation, etc.<br>• What do you do better than anyone else does? Think about quality, marketing, branding, keeping costs low, meeting production targets, etc.<br>• What do your customers think about your organization? | **Opportunities**<br>Identifying aspects of the environment that the organization can use to achieve higher levels of performance.<br><br>• What economic or social trends loom that you can use to improve/expand the business?<br>• Are there markets in which you could expand or create a niche?<br>• Is there a new or emerging technology that you can exploit to improve/expand your business?<br>• Are there different packaging, distribution, or delivery methods that would give you an edge over your competitors? |
| **Weaknesses**<br>Identifying aspects of the organization that need improvement or that must be overcome for the organization to accomplish its objectives.<br><br>• What could be improved?<br>• Could operations or management be more efficient?<br>• What areas, products, services, or markets should you avoid?<br>• Have you conducted a thorough financial analysis of the business?<br>• In what ways are your competitors better than you? | **Threats**<br>Identifying aspects of the environment that impede or could impede the organization's ability to accomplish its objectives.<br><br>• What economic or social trends loom that could limit or obstruct your business?<br>• What are your competitors doing?<br>• Is changing technology becoming a problem?<br>• Is your financial position a problem? |

ask questions and think in terms of how to balance an organization's position in terms of its internal **strengths and weaknesses** [**SW**ot] against basic environmental **opportunities and threats** [sw**OT**] to the firm.

**Organizational strengths** are those aspects of an organization that let it compete effectively.[14] Different strengths or distinctive competencies call for different strategies.[15] For instance, Intel has tremendous strengths as an innovative computer chip manufacturer, but those strengths might not support a strategic move into the health-care industry. It is important for an organization to know which of its strengths are common to similar organizations and which are distinctive. The distinctive strengths or competencies are more important in shaping distinct strategies for the organization.

In evaluating strengths, one approach is to use the VRIO framework.[16] The **VRIO framework** consists of asking whether each identified strength is valuable, rare, difficult to imitate, and exploitable. The organization would, then, use the information to help in establishing its strategic plan.

**Organizational weaknesses** are those aspects of an organization that prevent or deter it from competing effectively. Weaknesses can be overcome through investing whatever resources are necessary, or they can be allowed to exist with corresponding changes in the mission and strategy of the organization. ConAgra's acquisition of Beatrice was an investment of resources to overcome a weakness in its grocery-store distribution system.[17] The March of Dimes, on the other hand, changed its mission from finding a cure for polio, which happened in the 1950s, to supporting research on birth defects because the weakness, a non-existent target, could not be overcome.[18]

**Environmental opportunities** refer to those aspects of an organization's environment that, if acted upon properly, would enable it to achieve higher than planned levels of performance. When the cost of entry into a market is high or if innovation or imitation is difficult, an organization already in that market has the opportunity to solidify and even expand its market share and establish brand loyalty. If the organization has numerous buyers and suppliers such that none of them has any significant power, that organization has an opportunity to influence product design and/or characteristics to lower costs, facilitate delivery, or in other ways establish itself as dominant in the market. Finally, if other companies are not actively engaged in attempting to increase market share or establish brand/product identity and loyalty, the organization has an opportunity to jump ahead of its competition.

**Environmental threats** refer to those aspects of an organization's environment (present or future) that if not countered in some way would impede its progress in achieving its goals. If entry and innovation are easy, the possibilities of new entrants and/or substitute products are clear threats. If there are few buyers or suppliers such that one or more of them could exercise power over the organizations, that would represent a threat. If other competitors are actively competing or jockeying strongly for market positions and brand identity, the organization is constantly threatened with a loss of its position unless it continually responds to such moves.

Of course, two of the biggest environmental threats that agribusinesses face are the weather and pests that can devastate entire crops. Consider, for example, recent floods in the Midwest, the prolonged droughts in Texas and other Southwestern states in 2012; storms on the Gulf Coast and in the Northeast; the Southern pine beetle (*Dendroctonus frontalis* Zimmermann) that killed thousands of acres of pine trees in the Southeast, causing more than $900 million of damage in the 1960s–2000s; and the kudzu beetle (*megacopta cribraria*) that is threatening soybean crops all over the United States since it was first identified in Georgia in 2009.

Consider how Summer Farms would perform just such a SWOT analysis. As it considers all aspects of its situation, it might come up with the results shown in Table 9.2.

## Critical Environmental Forces

The five critical environmental forces mentioned in Chapter 4 and illustrated in Figure 9.2 must be considered in strategic planning.[19] In general, these forces relate to the competitor dimension of the firm's task environment discussed in Part Two.

**New Entrants.** The threat of new entrants refers to the ease with which new competitors can enter a market (Figure 9.3). If entry and innovation are easy, the possibilities of new entrants and/or substitute products are clear threats. For example, it takes very little in the way of resources to enter the restaurant business, the dry-cleaning business, or the home video rental business, so in these environments the threat of new entrants is high. On the other hand, it takes

**TABLE 9.2** SWOT Analysis for Summer Farms

| INTERNAL (SW) ANALYSIS | EXTERNAL (OT) ANALYSIS |
|---|---|
| **Strengths**<br>• The Summers have been farming for more than 80 years.<br>• There are multiple generations involved in the farm.<br>• Family members actively engage in the farm's operations.<br>• Operations are balanced to minimize seasonal impacts.<br>• Some land and/or buildings can be converted to alternative uses, if necessary.<br>• The farm provides most of the family's own food.<br>• Family members get along with one another.<br>• Family members are open to exploring new opportunities and trying new technologies and crops.<br>• There is some off-farm income.<br>• Seasonal and other employees are available at competitive rates. | **Opportunities**<br>• Increasingly, family members are going to college.<br>• Sizable markets are within easy shipping distances.<br>• Truck, rail, and air transportation facilities are nearby.<br>• The state's university has an agribusiness program that may be interested in using the farm for class purposes.<br>• Agritourism is becoming popular in the area.<br>• Genetically modified crops show considerable promise.<br>• Organic crops are increasingly in demand and can sell at higher prices. |
| **Weaknesses**<br>• Insurance and other operating costs have risen dramatically.<br>• Sending family members to college has become expensive.<br>• The company has limited experience with genetically modified and organic crops.<br>• Must use non-family employees more frequently.<br>• Difficult to keep abreast of government requirements and regulations. | **Threats**<br>• Increased governmental regulation may raise costs and impede movement into new areas.<br>• Long-time employees retire, leaving gaps in expertise and loyalty.<br>• Dependable farm employees are difficult to locate.<br>• Other farms are moving rapidly into both organics and genetically modified crops.<br>• Increasingly, newer generations want to be in organizations other than the farm.<br>• Large, global agribusinesses take over many of Summer's smaller competitors. |

© Cengage Learning 2014

**FIGURE 9.2** Environmental forces affecting strategic planning.

© Cengage Learning 2014

a tremendous investment to enter the automobile manufacturing business or the petroleum-refining business, so in these instances the threat of new entrants is low. Summer Farms does not feel a strong threat of new entrants, as its business also is capital intensive. Green Things Landscaping, however, faces a constant threat of new competitors.

**Suppliers.** The power of suppliers refers to the extent to which suppliers can influence the organization. If Delta Air Lines wants to buy fifty jumbo jets, it can deal with only three suppliers—Boeing, McDonnell Douglas, and Airbus. Thus, each of these suppliers has considerable power over Delta and can take a hard line on price, delivery, and so forth. To offset the power of suppliers, an organization may try to have a relatively large number of suppliers. On the other hand, it may be better to have only a few, but

© iStockphoto/Mlenny Photography

**FIGURE 9.3** Staying ahead of the competition requires a better strategy than just running faster.

located nearby to provide faster service, volume discounts, and perhaps better relations. If Green Things purchases its plants from a single grower, that supplier may have substantial influence on the landscaper.

**Competitors.** Competitive rivalry or jockeying among contestants refers to the extent to which major competitors in a market are constantly trying to outmaneuver one another. In the fast food industry, for example, McDonald's, Burger King, and Wendy's are almost always running promotions aimed at one another. Burger King may be touting its method of cooking hamburgers while Wendy's is pushing its newest sandwich or price specials and McDonald's is running special promotions for Happy Meal toys. Each contestant is trying to get the upper hand.

**Substitute Products.** The threat of substitute products is the extent to which a new product might supplant demand for an existing product. For instance, calculators eliminated the need for slide rules, and personal computers have reduced the market for typewriters. Manufacturers of glass bottles saw their demand fall as plastic bottles became capable of holding more and more products, but some of that demand has been restored as a result of the worldwide conservation or "green" movement. Some products seem to have indefinite staying power, whereas others come and go quickly. Thus, most

businesses attempt to maintain a line of stable products and services while taking advantage of new market opportunities presented by fluctuating consumer tastes and preferences. Summer Farms, for instance, is maintaining its traditional nonorganic products as it cautiously enters the organic vegetable business.

**Buyers.** The power of buyers is the fifth environmental force. Organizations that rely on only one or just a few major customers for most of their sales are susceptible to this threat. This force may apply to Green Things, depending on how much it depends on one or two large developers for its landscape design business. Lockheed sells virtually all of the airplanes it makes to the U.S. military. Thus, the military and its various branches have considerable power over Lockheed and can negotiate on everything from prices and delivery dates to the colors of the planes it buy.

**FOOD FOR THOUGHT 9.2**

In 2011, about 1 percent of total world farmland (more than 91 million acres) was farmed organically. Leading countries were Australia, Argentina, China, and the United States. Organic food and textile market sales were estimated at $29 billion.

## The Organization–Environment Interface

Once managers develop a clear understanding of relevant environmental forces, they must come to grips with how they want to interact with those forces. Indeed, the purpose of strategy is to determine what position in the environment the firm wishes to take. As shown graphically in Figure 9.4, the key to developing effective strategy is to understand environmental opportunities and threats and organizational strengths and weaknesses.

For example, in 1989 Ford found itself with several billion dollars in surplus funds (an organizational strength). Managers at Ford also believed, however, that the company lacked a strong presence in the luxury car market (an organizational weakness). Meanwhile, Toyota and Nissan were introducing new luxury car divisions (both environmental threats). Managers at Ford learned that Jaguar, Ltd. was for sale (an environmental opportunity) but

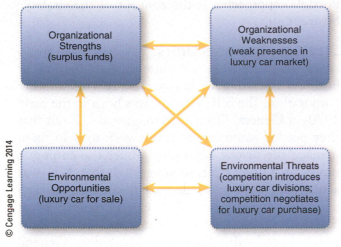

**FIGURE 9.4** The organization–environment interface.

© Cengage Learning 2014

that General Motors was already negotiating its purchase (still another threat). So Ford jumped in and paid a premium price to acquire Jaguar first. Ford hoped that the result would be a successful organization-environment interface achieved by matching environmental opportunities and threats with organizational strengths and weaknesses. Alas, that proved not to be the case.[20]

## Corporate Strategy

As defined earlier, corporate strategy involves determining in which businesses the firm expects to compete. The two most common approaches to corporate strategy are the development of a generic, or grand, strategy and the use of a portfolio approach.

### Alternative Generic Strategies

A **generic strategy**, also called a grand strategy, is an overall framework for action developed at the corporate level. It is generally used when the corporation competes in a single market or in a few highly related markets.[22] The three basic generic strategies that organizations adopt are growth, retrenchment, and stability.

### A FOCUS ON AGRIBUSINESS
#### Strategic Planning at Syngenta

Swiss-based Syngenta is a new firm with old roots. It was formed in 2000 when Novartis and Astra-Zeneca merged their agribusinesses, creating what may have been the first truly global group focusing exclusively on agribusiness. Novartis and AstraZeneca trace their origins back to 1758 and 1876. As of 2112, Syngenta had sales in excess of $13 million and employed over 26,000 people in 90 countries. Its products include herbicides, insecticides, fungicides, field crops, vegetable and flower seeds, seed-care products, and turf, garden, home care and public health products.

Syngenta's vision is to deliver "better food for a better world through outstanding crop solutions" while meeting commitments to stakeholders. Its goal is to be "the leading global provider of innovative solutions and brands to growers and to the food and feed chain." To achieve this vision, Syngenta has a series of business principles, including being future oriented and innovative, working in partnership with external stakeholders, having challenging and rewarding work, delivering on commitments, and striving for outstanding performance.

From this vision and business principles, then, Syngenta developed the following seven strategic goals:

- Maximize land productivity while conserving scarce resources, such as water, through innovation.
- Build leadership in plant performance by offering full crop programs.
- Expand sales of both genetically modified and conventional seeds to increase profitability.
- Expand in emerging markets.
- Create new businesses.
- Maintain cost efficiency.
- Outperform the industry.[21]

**Growth.** A **growth strategy** is adopted when the corporation wants to generate high levels of growth in one or more areas of its operations.[23] For example, Walmart is pursuing a growth strategy when it builds a distribution center and then opens a hundred new stores as it moves into new markets all across the United States.

**Retrenchment.** A **retrenchment strategy** is employed when managers want to shrink operations, cut back in some areas, or eliminate unprofitable operations altogether. Firestone pursued a retrenchment strategy when it cut its workforce from 107,000 to 55,000, closed eight plants, and sold several unrelated businesses. Many businesses, including Kmart, adopted this strategy during the recession that became obvious in 2009 or even earlier. Summer Farms may adopt this strategy temporarily with its timber business as a result of the national and international economic downturn (Figure 9.5).

**Stability.** Finally, an organization uses a **stability strategy** when it wants to maintain its status quo. Such an approach is often adopted immediately after a period of sharp growth or retrenchment. For example, following its recent shift in pricing strategy, Sears is attempting to maintain stability.

## Portfolio Approaches

A **portfolio approach** to corporate strategy views the corporation as a collection of different businesses. When the firm competes in several different markets simultaneously, it often uses one of several different portfolio approaches. The foundation of

**FIGURE 9.5** In bad economic times, many organizations adopt a retrenchment strategy by cutting their workforces.

portfolio approaches is the concept of the strategic business unit.

**Strategic Business Units.** A **strategic business unit (SBU)** is an autonomous division or business operating within the context of another corporation. The SBU concept was born in the early 1970s at General Electric. Managers at GE felt that they needed some sort of framework to help them manage the diverse businesses under the corporate umbrella. Close scrutiny suggested that GE was actually engaged in 43 distinct businesses, so each of these businesses was then clearly defined as an SBU for purposes of corporate strategy. In short order, several other large firms, including General Foods, began to realize that they, too, comprised a set of SBUs. In general, an SBU has the following characteristics:

- It has its own set of competitors.
- It is a single division or set of closely related divisions within the corporation.
- It has its own distinct mission.
- It has its own strategy that sets it apart from other SBUs within the organization.

These four characteristics are present, for example, in all of the strategic business units of the Mars candy company: its candy business, pet food business, packaged food business, and electronics business.

**The Portfolio Matrix.** Of course, the notion of SBUs by itself is of only marginal value to managers. This approach becomes truly significant only when managers can logically group the SBUs into meaningful categories. The best known of several different portfolio approaches is the portfolio matrix method. The **portfolio matrix** method classifies SBUs along two dimensions: market growth rate and relative market share. Market growth rate is the extent to which demand for the product in question is growing rapidly, at a modest pace, or not at all. Relative market share is the proportion of that market controlled by the product.

If we classify market growth rate as high or low and relative market share as high or low, we get the two-by-two matrix shown in Figure 9.6. Note that this creates four different categories of products. Classifying SBUs into the appropriate cells helps managers determine how to manage them better.

A **star** is a product which has a high share of a fast-growing market. When IBM personal computers were first introduced, they were clearly stars. The

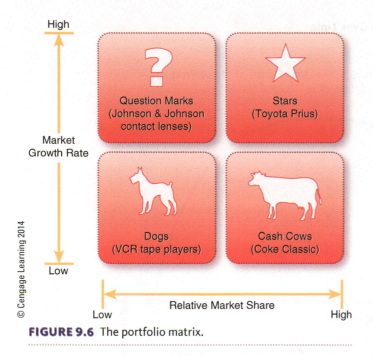

High

Market Growth Rate

Low

Relative Market Share

Low    High

© Cengage Learning 2014

**FIGURE 9.6** The portfolio matrix.

market was expanding dramatically, and the IBM PC controlled a large portion of it. Of course, since market growth rates must eventually stabilize, stars tend to be fairly short-lived.[24] Other recent stars have included the Toyota Prius and Apple's iPod and iPad. Stars need little investment to be sustained and generate large amounts of revenue. Understandably, managers like stars.

**Cash cows** are products that control a large share of a low-growth market. If market growth stabilizes and the product can still control a large portion of it, it has become a cash cow. Since the market is stable, there is little need to promote products aggressively, so these products generate large amounts of cash with relatively little support. Cash cows might include Crest toothpaste, Right Guard deodorant, and Coca-Cola Classic.

A **question mark** is a product with a small share of a growing market. When confronted with a question mark, a manager must decide whether to invest more resources in the hope of transforming it into a star, to simply maintain the status quo, or to drop the product or business from the portfolio. For example, Bausch & Lomb controls the lion's share of the market in the contact lens industry. Divisions of Revlon and Johnson & Johnson have been trying to gain a larger share of the market for the past few years but have made little headway. For them, the contact lens business is a question mark, whereas it is a star for Bausch & Lomb—at least until corrective

eye surgery (e.g., Lasik) drastically reduces the market for contact lenses.

A **dog** is a product with a small share of a stable market. Trying to salvage dogs is very difficult because growth has to come at the expense of competing products. Given their unappealing nature, dogs tend not to stay around long. An example might be a small division of an electronics division that still produces black-and-white televisions. Similarly, because of the growing popularity of compact discs and disc players, one-time stars such as vinyl records and turntables, VCR tapes and players have become dogs. CDs, DVDs, and non-HD products will eventually reach the same status. On the other hand, so long as a dog helps cover fixed costs, an organization may elect to retain it.

The key to using the portfolio matrix is to manage the portfolio effectively. For instance, cash cows can generate the resources needed to support stars and maintain question marks. Dogs are usually sold or dropped, although occasionally they are turned into viable products again.

On balance, the portfolio matrix is the dominant view of corporate strategy today. Managers are comfortable with it, and research generally supports its validity.[25] Nonetheless, it should be seen for what it is—a guiding framework to be used with caution and judgment.

## Product Life Cycle Approaches

The **product life cycle** refers to how sales volume for a product changes during the existence of the product. Analyzing the life cycles of different products can be useful in shaping corporate strategy. As shown in Figure 9.7, generally the product life cycle goes from development, growth, competitive shakeout, maturity, and saturation, to decline.[26] In the *development stage*, demand for the product may be high and the organizational response is focused on production. In the *growth stage*, more firms enter the market and the organizational response usually turns to quality, service, and delivery. Then competition intensifies and some firms are driven from the market during the *competitive shakeout*. As the market reaches *maturity*, demand slows down and the number of competitors drops off. The organizational response is usually to focus on low costs and product differentiation. As the market reaches *saturation* and begins to decline, organizations continue to lower costs and also to explore new products or services.

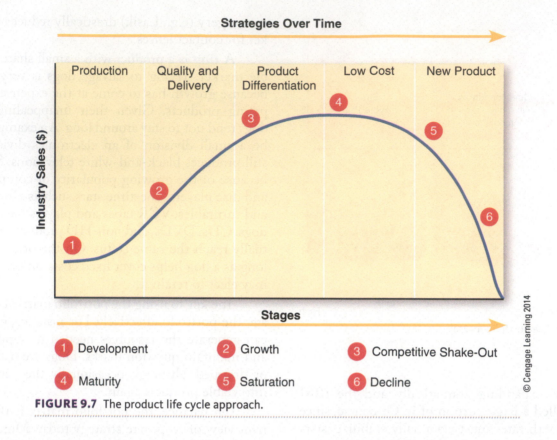

**Strategies Over Time**

| Production | Quality and Delivery | Product Differentiation | Low Cost | New Product |

Industry Sales ($)

**Stages**

① Development    ② Growth    ③ Competitive Shake-Out

④ Maturity    ⑤ Saturation    ⑥ Decline

© Cengage Learning 2014

**FIGURE 9.7** The product life cycle approach.

Most products, ranging from dolls to computer chips, have life cycles measured in a few years, although the actual time period involved may vary from extremely short to extremely long (Figure 9.8).[27] Fads like hula hoops and pet rocks may go through the cycle in a matter of months (although they may recur periodically), while a product like Levi's 501 jeans has a life measured in decades.

In assessing an organization's response during the life cycle, it is important to keep in mind the competitive position and the market share of the organization. Having a small market share, for

**FIGURE 9.8** The life cycle of cell phones will be extraordinarily long, but not for these particular models, which were pushed aside by various smart phones.

instance, in the development stage might not be so bad if the organization had a strong competitive position based on its other strengths. However, a small market share coupled with weak competitive positions is not encouraging and would probably suggest that the organization should drop that product from its line.

Each of these approaches (generic, portfolio, and product life cycle) should be used in strategic planning. One is not "better" than or an alternative to another. They are complementary ways of visualizing an organization's position and possibilities. The important thing is to analyze carefully and accurately the organization's SWOT situation to determine the strategy most likely to be effective.

## Business Strategy

As we saw earlier, business strategy is the strategy managers develop for a single business. A business strategy is developed for each SBU within a portfolio matrix and for single-product firms that are not broken down into SBUs. The following sections explore a useful conceptual framework for understanding business strategy and identify and discuss the major kinds of business strategy that firms adopt.

## The Adaptation Model

One popular approach to business strategy, the **adaptation model**, suggests that managers should focus on solving three basic managerial problems—entrepreneurial, engineering, and administrative—by adopting one of three forms of strategy.[28]

### Problems of Management.

The first of the three managerial hurdles is the **entrepreneurial problem**. This problem involves determining which business opportunities to undertake, which to ignore, and so forth. Decisions regarding the introduction of a new product or the purchase of another business relate to the entrepreneurial problem.

The **engineering problem** involves the production and distribution of goods and services. For example, Canon might elect to manufacture its own facsimile machines, to have them manufactured by someone else, or to combine the two methods in some way. Similarly, the plant that produces the facsimile machines might use traditional assembly lines staffed by employees, total automation, or some combination. Finally, the company might do its own distribution or subcontract distribution to a wholesaler. Each of these decisions relates to the engineering problem.

The **administrative problem** involves structuring the organization. Top managers at Canon might choose to give operating managers considerable power and autonomy to make important decisions or to retain most of that power at the top. It might also choose to have a large number of different divisions or to maintain only a handful of divisions. These are aspects of the administrative problem.

### Strategic Business Alternatives.

According to the adaptation model, firms can use a variety of strategies to address these problems. In general, managers usually choose one of three basic alternatives shown in Figure 9.9: defending, prospecting, and analyzing.

**Defending** is the most conservative approach to business strategy. Defenders attempt to carve out a clearly defined market niche for themselves and then work hard to protect that niche from competitors. They tend to ignore trends and remain within their chosen domains. They concentrate on efficiency and attempt to create and maintain loyal groups of customers. Mobil, Levi Strauss, many regional universities, and most local community hospitals use the defending mode of strategy.

**Prospecting** is the exact opposite of defending. Prospectors attempt to discover and explore new market opportunities including possible acquisitions. They prefer to avoid dependence on a narrow product or product group. Indeed, they attempt to shift frequently from one market to another. 3M is a good example of a prospector. One of its goals is to have 25 percent of its sales coming from products that did not exist five years ago.[29] Other well-known prospectors include Miller Brewing Company and Amazon.

**Analyzing** is a midrange approach that falls between defending and prospecting. Analyzers attempt to move into new market areas but at a deliberate and carefully planned pace. The analyzer keeps a core set of products that provide predictable revenues but at the same time systematically looks for new opportunities. Examples of analyzers are Procter & Gamble, Unilever, and Nestlé.

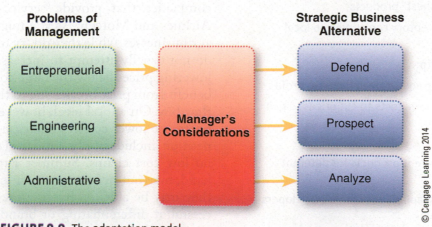

**FIGURE 9.9** The adaptation model.

## FOOD FOR THOUGHT 9.3

From 1999–2004, in the U.K. the market share for orange juice fell each year as apple juice and other blended juices increased.

Overall, the adaptation model is a useful framework for helping managers understand business strategy. It has been supported by research and is well known among practicing managers.[30]

## Competitive Strategies

In addition to the adaptation model, there are three competitive strategies, shown in Table 9.3, that are pursued by some businesses: differentiation, overall cost leadership, and targeting.

**Differentiation** is the process of setting the firm's products apart from those of other companies (Figure 9.10). Such differentiation might be in terms of quality (Volvo, Rolex), style (Ralph Lauren, Calvin Klein), or service (Maytag, Federal Express, American Express). Small- and medium-size pet food shops use this strategy to compete against larger firms like PetsMart and PetCo. Summer Farms has attempted to differentiate itself in the minds of individuals and business customers by claiming only "the freshest" or "straight from the farm to the table" or "you know where your food is coming from." Whitfill Nursery survives the aggressive marketing of other area nurseries by using a differentiation

**FIGURE 9.10** Adding flavors and labeling can differentiate tasteless, colorless water.

© iStockphoto/Media Photos

strategy when selecting product inventory that is native to the local area and therefore more likely to survive in that climate and soil conditions. Firms with products that involve delivery, including online businesses and other mail-order firms, sometimes differentiate on delivery availability, cost, or expediency. Mail-order plant nurseries have long used the strategies of targeting rural area customers, suggesting products that will grow in the customer's climate and soil, and delivering them at the appropriate time. It remains to be seen, however, whether those nurseries can survive the continued increases in shipping costs.

**Overall cost leadership** involves trying to keep costs as low as possible so that the firm is able to either charge low prices and thus increase sales volume and/or market share or charge competitive prices and earn greater profits. Examples of companies that use the cost leadership strategy are discount retailers like Kmart and Walmart, manufacturers like Bic Corp. and Black & Decker, and companies that provide services like Southwest Airlines and Motel 6. As you might expect, smaller agribusinesses, while they do try to hold down costs, do not usually attempt to increase sales volume by lowering prices below the large corporations, which benefit from price breaks due to the large quantities they buy. On a local level, however, one firm might try to use lower costs to decrease prices if its competition includes only similar size businesses and no low-cost chains. Summer Farms, on the other hand, may not need to use cost leadership—in fact, they may be able to charge more—if they are successful at differentiating or branding themselves as mentioned above.

**TABLE 9.3** Examples of Competitive Strategies for Small- or Medium-Size Agribusinesses

| Differentiation: |
| --- |
| **Summer Farm:** "Freshest" products |
| **Whitfill Nursery:** Inventory can survive in local conditions |
| **Overall Cost Leadership:** |
| **Summer Farms:** Keep costs low to increase profits |
| **Green Things:** Use lower costs to cut prices to gain new customers |
| **Targeting:** |
| **Summer Farms:** Different products for different markets |
| **Green Things:** Landscaping for builders/developers rather than individuals |

© Cengage Learning 2014

Finally, **targeting** occurs when a firm attempts to identify and focus on a clearly defined and often highly specialized market. Some companies produce cosmetics just for African Americans, for instance, or food for regional markets (chili in the Southwest, clam chowder in the Northeast). Others attempt to focus on special categories of consumers (the young, the upper class, and so forth). For a company like Summer Farms, the targeted markets differ for its various products. Green Things, although it will do landscaping for any customer, has chosen to target primarily homebuilders/developers rather than individual homeowners because of such factors as lower advertising costs and less time required to deal with builder/developers than with individual clients.

Table 9.3 suggests how some smaller agribusiness firms might employ these different competitive strategies. These are, of course, examples and many others could be developed for particular firms in specific markets.

## Functional Strategies

We saw earlier that the lowest level of strategy in the organization includes functional strategies. The six basic types of functional strategy are listed in Table 9.4, along with their major areas of concern.

### Marketing Strategy

The **marketing strategy** is the functional strategy that relates to the promotion, pricing, and distribution of products and services by the organization. For example, on its vegetable farms, Summer Farms may decide to add to its product line only if the demand for a new vegetable would justify planting a minimum of ten acres. In another example, Reebok is determined to avoid discount chains and sell only to fashionable retailers. The firm also limits the amount of apparel it produces, even though it could sell more.[31] These decisions are part of the firm's marketing strategy. Similarly, decisions about how many variations of each product to market (six versus nine sizes of toothpaste dispensers), desired market position (Walmart versus Kmart for the number one position in retailing), pricing policies (high prices with an emphasis on quality, versus lower prices with an emphasis on quantity), and distribution channels (Timex's early decision to sell only through drugstores) are all or part of the devel-

**TABLE 9.4**  Concerns of Basic Functional Strategies

| FUNCTIONAL AREA | MAJOR CONCERNS |
|---|---|
| Marketing | Product mix<br>Market position<br>Distribution channels<br>Sales promotions<br>Pricing issues<br>Public policy |
| Finance | Debt policies<br>Dividend policies<br>Asset management<br>Capitalization structure |
| Production | Productivity improvement<br>Production planning<br>Quality<br>Production<br>Plant location<br>Government regulations |
| Research and Development | Product development<br>Technological forecasting<br>Patents and licenses |
| Human Resources | Human resource policies<br>Labor relations<br>Executive development<br>Employee training<br>Government regulation |
| Organization Design | Degree of centralization<br>Methods of coordination<br>Bases of departmentalization |

© Cengage Learning 2014

opment of the marketing strategy. In any event, an emphasis on quality has become important in strategic planning.[32]

### Financial Strategy

The **financial strategy** of a firm is also important. Companies need to decide whether to pay out most of their profits to stockholders as dividends, retain most of the earnings for growth, or take some position between these extremes. They must also make decisions about the proper mix of common stock, preferred stock, and bonds; and they must establish policies regarding how to invest surplus funds and how much debt the organization can and is willing to support. For example, Disney's financial strategy calls for low debt. It managed to spend almost $1 billion from operating funds on Epcot Center without incurring any debt.[33] Family-owned Summer Farms wants to provide the participating family members with an acceptable, but not excessive income, to pay employee wages that will motivate them to remain with the company, and to earn enough profit (which it has defined as minimally $30,000 a year) to continue expanding and improving the company. Another part of Summer's financial strategy is to use proceeds from timber sales to purchase additional growing acreage for food products.

## Production Strategy

In many ways **production strategy** follows from marketing strategy. For example, if a company emphasizes quality, production costs may be of secondary importance. To emphasize price, on the other hand, low-cost production techniques may become critical. Like most companies, Summer Farms definitely emphasizes quality but also recognizes that financial and development or expansion goals are also important.

This strategy usually addresses several areas of concern. The location of factory sites (or crop sites at Summer Farms) is a production issue. So, too, are production planning and productivity improvement efforts. Production managers must also deal with governmental agencies such as the Environmental Protection Agency. This has always been true at Webb Dairy, and it will become increasingly more applicable to Summer Farms as it becomes increasingly involved with organic farming. In recent years, production strategies have become even more complex because more and more companies subcontract the manufacture and assembly of their products to other firms.

## Research and Development Strategy

The **research and development strategy** relates to the invention and development of new products and services as well as the exploration of new and better ways to produce and distribute existing ones.[34] Some firms, such as IBM, Rubbermaid, and Texas Instruments, spend large sums of money on research and development (R&D) whereas others spend less. Long-term gains from R&D investment can be impressive, as in the case of Bridgestone Tire Company, where a strong R&D program increased worker productivity by 10 percent each year in the 1970's and has helped Bridgestone remain highly competitive ever since.[35] Summer Farms expanded initially by buying adjacent property as fast as possible, and it was careful to keep various crops adjacent to the same types of crops (hay fields by hay fields,

corn fields by corn fields). Now the company faces the possibility of needing contiguous acreage and having to choose between paying for land that is not available at an agreeable price or buying lesser-priced land across the valley. It must formulate an investment strategy that concerns the financial and other pros and cons so it can make the best decisions when the need arises or when area property becomes available.

## Human Resource Strategy

Most organizations also develop a **human resource strategy**. This may deal with issues such as whether the firm plans to pay premium wages to get better-qualified workers, whether it will welcome unions, how it will attempt to develop executives more effectively, and how it will comply with federal regulations, such as the equal employment opportunity guidelines. Summer Farms and similar family-owned enterprises face unique problems relative to HR strategy. Which family members do you hire? Can the enterprise even consider the risk of family dissension by not hiring a family member? Non-family enterprises often deal with the related problem of nepotism—hiring a close relative of your own or of another employee. Nepotism is forbidden in many large corporations but is standard or automatic in many family businesses. Is forbidding it justifiable when the relative is actually more qualified than other applicants? Organizations need strategies for HR decisions like this.

## Organization Design Strategy

Finally, companies often develop an **organization design strategy**, which is concerned with how to arrange the various positions and divisions within the organization. For example, some organizations allow field managers to make fairly important decisions without consulting the home office (as is done at Summer Farms), whereas others require home office personnel to approve virtually all field decisions (the owner/designer of Green Things, for example). Determining which policy to follow is a part of the organization design strategy. (Organization design is discussed more fully in Part III.)

Integration, as well as development of the six major functional strategies, is crucial. Specifically, managers must ensure that all functional strategies follow logically from a unified corporate and business strategy and that they fit together logically.

For example, if the marketing strategy calls for a sales increase of 50 percent, the production strategy may be to build a new plant to come on line in five years. This strategy, in turn, means that the human resource strategy will include a provision for developing the necessary managerial talent to run the plant and that a financial strategy must be developed to pay for the plant. The organization design strategy specifies how the new plant will fit into the existing organizational structure.

---

### FOOD FOR THOUGHT 9.4

The U.S. Department of Energy (DOE) has set a goal of producing 5 percent of the nation's electricity from wind by 2020. DOE projects will provide $60 billion in capital investment to rural America, $1.2 billion in new income to farmers and rural landowners, and 80,000 new jobs during the next 20 years.

## Strategy Implementation and Control

The final part of strategic planning, and one we will review only briefly, is the implementation and control of strategic plans by managers of the organization.[36] Figure 9.11 summarizes the process of **strategy implementation**.[37] First, it must follow logically from strategy formulation; that is, managers must think strategically by first formulating and then systematically implementing strategies.[38] Implementation itself consists of three elements. Tactical planning, as detailed in Chapter 8, is the real way in which strategy is implemented. Contingency planning, also described in Chapter 8, is also important for the proper implementation of strategic plans. Finally, strategy and organization design must be properly integrated. A mismatch can result in numerous problems for organizations and can serve as a major barrier to the effective accomplishment of strategic plans.[39]

In reality, of course, strategy implementation is far more comprehensive and complex than this simple overview implies. Process improvement and other such techniques have been shown to be tools which can be useful in strategy implementation.[40] In addition, modeling and scenario generation have become widely used tools.[41] As a result, each strategy and its corresponding organizational context is unique, making each effort to implement a strategy unique as well. Much of the material in this book is in some way an attempt to show how to implement all or part of a strategic plan developed by top management.

**Strategic control** refers to the process whereby management assures that the strategic planning process itself is effective.[42] Such control involves evaluating the organization's progress with

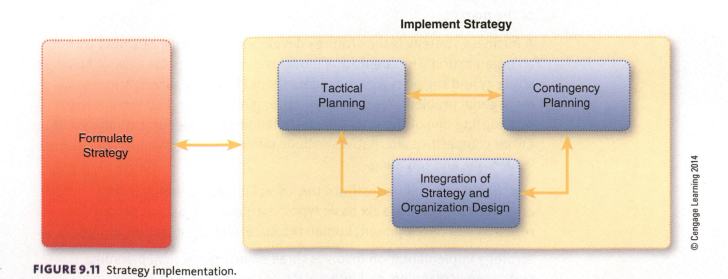

© Cengage Learning 2014

**FIGURE 9.11** Strategy implementation.

its strategy, its flexibility in meeting changing environmental conditions, and the resources actually consumed by the strategic planning process. This latter evaluation involves the strategic budget as opposed to the operations budget which is separate in many companies such as Texas Instruments and 3M.[43] Strategic control necessitates having indicators of the performance of the strategic planning process as well as mechanisms for corrective action should it not be accomplishing its objectives.

## CHAPTER SUMMARY

Strategic planning is the broad planning of top management to guide the general direction of an organization. The four basic components of strategy are scope, resource deployment, competitive advantage, and synergy. Strategy formulation is the set of processes involved in creating or developing strategic plans, and strategy implementation is the set of processes involved in executing those plans. Finally, there are three levels of strategy: corporate strategy, business strategy, and various functional strategies.

Environmental analysis is the specific study of the company's environment and how it affects the company. The five critical environmental forces are the threat of new entrants, the power of suppliers, competitive rivalry or jockeying among contestants, the threat of substitute products, and the power of buyers. Managers must understand these forces and use them as a framework to match environmental threats and opportunities with organizational strengths and weaknesses.

Corporate strategy means determining what businesses the firm expects to compete in. Three generic or grand strategies are growth, retrenchment, and stability. Some organizations also use a portfolio approach involving strategic business units, or SBUs. The portfolio matrix is a system for analyzing SBUs in terms of relative market share and the growth rate of the market.

A business strategy is the strategy developed for a single business within an organization. The adaptation model is the most popular view of business strategy and suggests that managers should focus on solving three basic managerial problems (entrepreneurial, engineering, and administrative) by adopting one of three strategies (defending, prospecting, or analyzing). Other competitive alternatives include differentiation, overall cost leadership, and targeting.

Functional strategies constitute the lowest level of strategy in an organization. There are six basic types: marketing, financial, production, research and development, human resource, and organization design.

Strategy implementation involves three elements: tactical planning, contingency planning, and integration of the strategy with the organization's design.

## CHAPTER ACTIVITIES

### REVIEW QUESTIONS

1. What are the three levels of strategy? Do all firms have all three levels? Why or why not?
2. What are the five critical environmental forces that organizations must consider when developing strategies? How do organizations position themselves relative to those forces?
3. Identify three grand strategies that organizations might choose to pursue.
4. What are the three problems of management and the three strategic business alternatives described in the adaptation model?
5. What are the six basic functional strategies most organizations develop?

### ANALYSIS QUESTIONS

1. Apply the concepts of corporate, business, and functional strategies to your university or college.
2. Which of the five environmental forces are more likely to exist together than others? Identify other examples of organizations that each of the five forces are likely to affect.
3. What are the risks involved in selling dogs quickly? Why would anyone want to buy a dog?
4. Identify examples beyond those noted in the chapter to illustrate defenders, prospectors, and analyzers.
5. What are the critical issues in implementing a new strategy within an organization?

### FILL IN THE BLANKS

1. The advantages or disadvantages that a firm holds relative to its competitors are known as _____ _____.
2. Strategies that chart the course for the basic functional areas within an organization (marketing, finance, production, etc.) are known as _____ strategies.
3. Those aspects of the organization that let the organization compete effectively are known as organizational _____.
4. The VRIO framework consists of asking for each strength whether it is _____, _____, _____, and _____.
5. Those aspects of the organization that prevent or deter it from competing effectively are known as organizational _____.
6. Those aspects of the organization's environment that, if not countered in some way, would impede the organization's progress to achieve its plans are known as organizational _____.
7. When managers want to shrink operations, cut back in some areas, or eliminate unprofitable operations altogether, the organization uses a _____ strategy.
8. The changing of a product's sales volume over the lifetime of the product is known as the _____ _____.
9. The process of setting the firm's products apart from those of other companies on some basis such as quality, style, or service is called _____.
10. When products have a high share of a market, they are known as _____ if the market is fast growing, or as _____ if the market is low growth.

# CHAPTER 9 CASE STUDY
## THE PRICE OF TEA IN CHINA

> "Seven steps up, you have to rest.
> Eight steps down, you have to rest.
> Eleven steps flat, you have to rest.
> You are stupid, if you don't rest."

This was the chant of male and female tea porters in China. They regularly carried loads of 150–200 pounds for 20 days over 140 miles through 17,000-foot mountain passes. They walked the "Tea Horse Road."

What was the price of tea in China? One horse for 130 pounds of brick tea. China had many people, but few horses. This is a remarkable and ancient example of David Ricardo's theory of comparative advantage, a source of competitive advantage. Simply put: Market forces will allocate a nation's resources to those industries where it is relatively most productive. The trade between Tibet and China was horse for tea.

The Chinese produced a strong bitter tea compressed into bricks that the Tibetans enjoyed. The Tibetans produced a unique horse, the Nangchen. It is about 4.5 feet tall, has fine legs and enlarged lungs, is sure-footed in snowy mountain passes, and is almost inexhaustible. This tea-horse commerce began in 641 and was terminated by Mao in 1949. By the thirteenth century millions of pounds of tea were traded for 25,000 horses annually. This trade expanded into other goods and services over the 1,300 years of activity.

Strategic trade between countries is the backbone and purpose of globalization. Countries often continue to trade with each other even when engaged in hostilities, although in this extreme case trade is often accomplished through third parties. When the economy of the world declined in 2008–2012 in the "Great Recession," an often spoken fear was that nations might resort to protectionist tariffs to preserve national economies. Such protectionism would stifle international trade. It was just such protectionism that prolonged the Great Depression in the 1930s. Many organizations began to develop strategic plans for just such an eventuality. However, the fall in trade that occurred in the Great Recession was predominantly due to a fall in demand and not due to the imposition of tariffs and anti-dumping campaigns. Both rich and poor countries kept their markets open to encourage trade and turn the global recession around.

The recession of 2008–2012 may be the first global economic downturn that was reversed by emerging-nation resilience and not the developed economies. Globalization has underscored the importance of trade and the necessity of good trade relations between countries regardless of economic strength. Trade is about developing economic potential.

The United States is the largest single country economy in the world, with a GDP (Gross Domestic Product) of almost $16 trillion (2012). It is larger than Japan, Germany, France, and the United Kingdom combined. All together they and China are the six largest economies in the world. All are dependent upon trade with each other. Yet it took a concerted effort by all countries to revitalize trade.

Historically, countries have gone to great lengths to initiate trade relations, as depicted by the opening story. It has always been a trade relationship based on an agribusiness industry or mining industry that opens doors between countries. All initial trade began as comparative-advantage activities centered around agribusiness, with some absolute-advantage trade in minerals or resources, and developed into competitive advantage trade as the relationships matured. Once the door had been opened, other products were discovered and traded.

The message of trade is beginning to have its effect on all nations. Myanmar (formerly Burma) has been in the grip of a trade-phobic military regime since the 1980s. Contact with the outside world has been minimized and controlled, yet in 2010 the military regime adopted policies that loosened controls on its rice farmers. The intent of these policies was to produce enough rice for Myanmar and for export. It took an economic downturn and a major cyclone to initiate change—change was a long time in coming. Myanmar produced as much rice in 2010 as it did in 1960. Its neighbors, Thailand and Vietnam, started at the same level as Myanmar but increased production by eight times in Thailand and six times in Vietnam. The vitality of Thailand and Vietnam reflects the result of trade. As they increased production of rice, they acquired money to obtain other resources, for example, seed stock, fertilizer, and equipment that further increased production for increased exports. Firms around the world are taking advantage of this opening up of strategic opportunities.

It was the unspoken hope of the first and second waves of globalization in the twentieth century that world trade would eliminate the need for war. It did not happen nor is it likely to happen, but the current wave of globalization could make war less appealing. That's worth all the tea in China.

### ▶ Case Study Questions

1. Why do you suppose that Chairman Mao terminated tea-horse commerce in 1949? Why are some countries like Myanmar still reluctant to embrace the idea of world trade?

2. Explain why international trade begins as comparative-advantage activities centered on agribusiness and then develops into competitive-advantage trading of other products.

3. Explain why world trade may or may not eliminate future wars.[44]

## REFERENCES

1. "Trading Cocoa: Sweet Dreams," *The Economist* (August 7, 2010) at www.economist.com (accessed August 19, 2010); Ben Bouckley, "Cocoa Bean Speculation is Causing Price Volatility, Analyst," (August 16, 2010) at Foodnavigator.com (accessed August 20, 2010); Paul Bulcke, "2009 Full Year Results Roadshow," (February 23, 2010) at Nestle.com (accessed August 19, 2010); Paul Sonne and Goran Mijuk, "Food Makers Chew Over Prices: Nestlé Profit Rises but Packaged Goods Firms See Raw Materials Costing More," *The Wall Street Journal*, August 12, 2010, B1; Mike Stones, "Cocoa Smuggling Undermines Ghana's Aim to Boost Production," (August 5, 2010) at www.confectionerynews.com (accessed August 20, 2010).

2. A. de Ferias Filho, M. L. D. Paez, and W. J. Goedert, "Strategic Planning in Public R&D Organizations for Agribusiness: Brazil and the United States of America," *Technological Forecasting and Social Change* (2002): 833–847.

3. Dean Foust, "Neville Isdell: Shaking Up Coke Abroad," *Business Week* (April 4, 2005) at www.businessweek.com (accessed March 2, 2006); Devon Jarvis, "Coke Is What?" *Fast Company* (September 2005) at www.fastcompany.com (accessed March 2, 2006).

4. For a review, see Charles W. L. Hill and Gareth R. Jones, *Strategic Management, 8th ed.* (Cincinnati, OH: Cengage Learning, 2009).

5. Richard P. Rumelt, "How Much Does Industry Matter?" *Strategic Management Journal* (1991): 167–186.

6. Ben Steverman, "Food, Not Ethanol, Fuels ADM's Big Quarter," *Business Week* (November 6, 2007) at www.businessweek.com (accessed June 30, 2010).

7. Jay Barney, "Firm Resources and Sustained Competitive Advantage," *Journal of Management* (1991): 99–120.

8. "For Unilever, It's Sweetness and Light," *The Wall Street Journal*, April 13, 2000, B1, B4. See also "Unprofitable Businesses Getting Axed More Often," *The Wall Street Journal*, February 17, 2009, B1, B2.

9. Kathleen M. Eisenhardt and D. Charles Galunic, "Coevolving— At Last, A Way to Make Synergies Work," *Harvard Business Review* (January-February 2000): 91–100.

10. "Did Somebody Say McBurrito?" *Business Week*, April 10, 2000, 166–170.

11. John E. Gamble and Arthur A. Thompson, Jr. *Essentials of Strategic Management: The Quest for Competitive Advantage, 2nd ed.* (New York: McGraw-Hill, 2011).

12. I. MacMillan and P. Jones, *Strategy Formulation* (St. Paul, MN: West Publishing, 1986).

13. Morgan P. Miles, John B. White, and Linda S. Munilla, "Strategic Planning and Agribusiness: An Exploratory Study of the Adoption of Strategic Planning Techniques by Co-operatives," *British Food Journal* (1997): 401–408.

14. Jay B. Barney and Ricky W. Griffin, *The Management of Organizations* (Boston: Houghton Mifflin, 1992), 216.

15. A. Tanzer, "We Do Not Take a Short-Term View," *Forbes*, July 13, 1987, 372–374.

16. Jay Barney, "Organizational Culture: Can It Be a Source of Sustained Competitive Advantage?" *Academy of Management Review* (1986): 656–665.

17. Russell Mitchell, Lois Therien, and Gregory L. Miles, "ConAgra: Out of the Freezer," *Business Week*, June 25, 1990, 24–25.

18. W. Olcott, "Taking Care of America: 50 Years of Philanthropy," *Direct Marketing*, May 1988, 98–102.

19. Michael E. Porter, *Competitive Strategy: Techniques for Analyzing Industries and Competitors* (New York: Free Press, 1980); for an update, see Michael E. Porter, "The Five Competitive Forces That Shape Strategy," *Harvard Business Review* (January 208): 79–90.

20. Heather Timmons, "Ford Closes Sale of Jaguar and Land Rover to Tata of India," *The New York Times* (March 26, 2008) at www.nytimes.com (accessed July 25, 2010).

21. A. Xydias, "Swiss Stocks Gain for Fourth Day; Syngenta, Transocean Advance," *Business Week* (July 9, 2010) at www.businessweek.com (accessed on July 9, 2010); "Annual Review 2009, Syngenta," at annualreport.syngenta.com (accessed July 7, 2010); J. Birger, "Hopeful Signs for Agri-Business," *Fortune* (March 2, 2009) at money.cnn.com (accesses July 9, 2010).

22. Colin Campbell-Hunt, "What Have We Learned About Generic Competitive Strategy? A Meta-Analysis," *Strategic Management Journal* (2000): 127–154.

23. Donald L. Laurie, Yves L. Doz, and Claude P. Sheer, "Creating New Growth Platforms," *Harvard Business Review* (2006): 80–91.

24. See "Personal Computers: And the Winner is IBM," *Business Week*, October 3, 1983, 76–79; see also "Mike Armstrong is Improving IBM's Game in Europe," *Business Week*, June 20, 1988, 96–101.

25. See, for example, Ian C. MacMillan, Donald C. Hambrick, and Diana L. Day, "The Product Portfolio and Profitability—A PMS-Based Analysis of Industrial Product Businesses," *Academy of Management Journal* (December 1982): 733–755.

26. Charles W. Hofer and Dan Schendel, *Strategy Formulation: Analytical Concepts* (St. Paul, MN: West Publishing, 1978).

27. Susan Benway, "Coleco: Out of the Cabbage Patch and Into the Fire," *Business Week*, March 30, 1987, 54; and Michael Porter, "Note on the Electronic Parts Distribution Industry," *Cases in Competitive Strategy* (New York: Free Press, 1979), 1–19.

28. Raymond E. Miles and Charles C. Snow, *Organizational Strategy, Structure, and Process* (New York: McGraw-Hill, 1978).

29. "Masters of Innovation," *Business Week*, April 10, 1989, 58–63.

30. Donald C. Hambrick, "Some Tests of the Effectiveness and Functional Attributes of Miles's and Snow's Strategic Types," *Academy of Management Journal* (March 1983): 5–26.

31. Stuart Gannes, "America's Fastest-Growing Companies," *Fortune*, May 23, 1988, 28–40.

32. Wayne Claycombe, "Building A Sound Management Foundation for Strategic Planning," *Industrial Management*, May 1, 1992, 17–19.

33. "Disney's Epcot Center, Big $1 Billion Gamble, Opens in Florida," *The Wall Street Journal*, September 16, 1982, 1 and 9.

34. Donald C. Hambrick and Ian C. MacMillan, "Efficiency of Product R&D in Business Units: The Role of Strategic Context," *Academy of Management Journal* (September 1985): 527–547.

35. Bernard Krisher, "A Different Kind of Tiremaker Rolls into Nashville," *Fortune*, March 22, 1982, 136–146.

36. Gamble and Thompson, *Essentials of Strategic Management*.

37. For a complete discussion of strategy implementation, see J. Galbraith and R. Kazanjian, *Strategy Implementation* (St. Paul, MN: West Publishing, 1986); and L. Hrebiniak and W. Joyce, *Implementing Strategy* (New York: Macmillan, 1984).

38. "The 'Art' of Taking the Long View," *Industry Week*, November 18, 1991, 12–23.

39. Lawrence G. Hrebiniak, "Obstacles to Effective Strategy Implementation," *Organizational Dynamics* (February 2006): 12–21.

40. Alan Brache, "Process Improvement and Management: A Tool for Strategy Implementation," *Planning Review* (September 1, 1992): 24–26.

41. Mason Tenaglia and Patrick Noonan, "Scenario-Based Strategic Planning: A Process for Building Top Management Consensus," *Planning Review* (March 1, 1992): 12–19.

42. P. Lorange, M. Morton, and S. Ghoshal, *Strategic Control* (St. Paul, MN: West Publishing, 1986), 10.

43. Lorange, Morton, and Ghoshal, *Strategic Control*, 35 and 53.

44. Mark Jenkins, "The Forgotten Road," *National Geographic* (May, 2010), 217:8, 102–119; Michael Porter, *The Competitive Advantage of Nations* (New York: Free Press, 1990), 11; "Defying Gravity and History," *The Economist* (August 7, 2010): 71–72; Kathy Chen and Brian Spegele, "U.S. Seeks Bigger Asia Role to Check Beijing," *The Wall Street Journal*, August 16, 2010, A8; Andrew Batson, Daisuke Wakabayashi, and Mark Whitehouse, "China Output Tops Japan," *The Wall Street Journal* (August 12, 2010) at online.wsj/com (accessed August 16, 2010); "Myanmar Loosens Yoke on Farmers," *The Wall Street Journal* (July 31, 2010) at online.wsj/com (accessed August 16, 2010); Thomas Friedman, *The Lexus and the Olive Tree: Understanding Globalization* (New York: Farrar Straus Giroux, 1999).

# Planning Tools and Techniques

## LEARNING OBJECTIVES

After studying this chapter, you should be able to:

- Describe the different organizational planning techniques businesses use today.
- Discuss when to use appropriate project planning tools.
- Identify personal planning techniques that managers must develop.

## Hershey Trusts Children

When John Schmalbach and Milton Hershey finished cleaning a huge copper vacuum kettle, they poured in the skim milk from the Holstein cows and added a large amount of sugar. Schmalbach took control, gradually raising the temperature of the kettle and gently cooking the contents. After a few hours of low-heat evaporation, he let the mixture cool. He then added cocoa powder, cocoa butter, and other ingredients to the warm, smooth, sweetened condensed milk. The mixture did not become lumpy. Milton exclaimed, "Look at that beautiful batch of milk. How come you didn't burn it? You didn't go to college." Schmalbach may not have been to college, but he knew how to carefully execute a plan for producing chocolate. And it was the plan that Hershey had been seeking. It created a sweet chocolate with just a single faint hint of sour. The Hershey chocolate bar was born.

**FIGURE 10.1** Kids love chocolate and Hershey's Chocolate Co. loves kids, as evidenced by the Milton Hershey School in Hershey, PA.
© iStockphoto/Fatihhoca

Milton Hershey sold his current business, the Lancaster Caramel Company, for $1 million in 1900 and put all of the resources into his new company, Hershey Chocolate Company. After failed candy business ventures in Denver and New York City, he had finally achieved success. The candy bar that would be included with C-rations in World War II and whose nickel bar was an affordable lunch during the Great Depression was on a path to become an American icon (Figure 10.1).

Hershey Chocolate Company was so successful that Hershey developed plans for a new factory in 1903 in nearby Derry township and for a new town for the employees, both management and labor. His planned location was perfectly situated next to shipping routes for raw materials, in a huge dairy area, and with an eager, hard-working populace. The name of the town was Hershey, Pennsylvania, of course.

This is an American success story with an unusual twist. "Milton oughtn't to be spending his time with printing and newspapers...he ought to be learning to make something—something that would bring him a good living." So Fanny Hershey paid Joseph A. Royer, a confectioner in Lancaster, Pennsylvania, to hire her son as an apprentice.

Milton never looked back and never forgot his origins. He learned from his mistakes and became an astute businessman and a utopian. He not only built the town around the plant but in 1909 he and his wife, Kitty (Catherine), established the Hershey Industrial School, a school for orphan boys. The farm boy and the poor immigrant's daughter never forgot their roots as they planned to provide the same opportunity for others.

In 1915, Catherine Hershey died of an incurable disease, locomotor ataxia. The only love of Milton's life was gone. In his eyes, his life was gone. Nevertheless, life went on; and he devoted his remaining 50 years to improving the plans for both the company and the school.

The astute businessman, former farm boy with little formal schooling, Hershey formed the Hershey Trust Company. To this company he gave 486 acres of land (Hershey, Pennsylvania) and $60 million in Hershey Company Stock. The trust's charge was to prudently use this endowment to maintain the school he and Kitty had started. Or, in his words, "to provide for the health, education, and welfare of orphaned boys."

The trust for the children controls the Hershey Chocolate Company. That was evident when Hershey was trying to buy Cadbury. In 2009, Kraft Food Company made a hostile bid for Cadbury. Hershey's did not make a counteroffer. It could not without running afoul of the articles of incorporation that gave the Milton Hershey Trust voting control despite owning just a third of the company. Milton's legacy, the children, said "No." Kraft acquired Cadbury and Hershey's long-ago plan continues to operate.

The Milton Hershey School began with ten students. Today it has 1,300 and it admits boys *and* girls. The trust company that oversees the school has assets exceeding $7 billion and the fiduciary responsibility not to jeopardize the endowment.[1]

## INTRODUCTION

Increasingly, a manager's task is not only to forecast the future but to create it and mold it to the organization's needs. Managers must know how to plan effectively and efficiently so they are in a position to act as well as react. Planning approaches are critical to the success of any project, regardless of the size of the company or the project. Summer Farms recognized the importance of planning from the first time it bought timberland and sold timber to raise capital to expand its farms. More recently it saw the growing demand for organic foods and began planning to take advantage of that market. Even international activities that at first seem unrelated to a company may have planning consequences for that organization, positively or negatively. Since the United States and Mexico signed their NAFTA agreement, for instance, the Hershey Chocolate Company has relocated some operations to Mexico.

This chapter addresses techniques and tools specifically designed to increase planning efficiency and effectiveness. The Milton Hershey School and the Hershey Trust Company require the use of just those sorts of tools. Planning tools are described for the macro level (strategic planning management), the mesolevel (monitoring within and between projects), and the micro level (planning tools for the individual manager).[2]

# Organizational Planning Techniques

Organizational members have the responsibility to plan for future events. To do a credible job in this function, managers must make assumptions about the future, using their experience and, increasingly, appropriate planning tools. Planning tools assist us with planning the future rather than merely reacting to it. "The objective of planning should be to design a desirable future and to invent ways to bring it about."[3] How we design the future has everything to do with planning.

## Environmental Scanning

Regardless of company size, there is an emerging group of organizational members who are referred to as *information rich* members. These individuals not only control the majority of the useful information about the environment and the competitors but also discern quickly what information is valuable and what information is not. The majority of information about the future of an organization resides external to the organization. To access this information members use the process of environmental scanning. **Environmental scanning** is a proactive approach to monitoring trends and anticipating changes that may affect an organization at the strategic and tactical levels of scanning.[4] The vast amounts of information available remain passive and useless until they are connected with active intelligence.[5]

That organic foods represents a growing market is of no use to Summer Farms unless the company is aware of and acts upon that information, deciding to accept or skip the opportunity to serve that market. In another example, knowing the smoking policy for another company is relatively useless information that may be shared among coworkers who play racquetball together. It might become very useful information, though, if the company decides to adopt a smoking policy and is looking for examples from other companies where such policies have already been implemented.

Scanning the environment does not always involve information provided by an information system (Figure 10.2). Information can be valuable without coming through a computerized method. For example, Philips Petroleum Company established an advantageous relationship with the highly technical retirement community in Bartlesville, OK. Many of the community members were avid ham radio operators who listened to news from all over

**FIGURE 10.2** Before computers, data had to be located within printed files, such as these, which was a tedious, time-consuming task.

© iStockphoto/Cunhek

the world late into the night. Philips's marketing department arranged for a security guard to pick up their "listenings" early each morning and then transcribed and placed the newsworthy items on the desks of department members before eight o'clock in the morning.[6] Organizations can gather intelligence in a variety of ways, but it is a unified approach to sharing the information that makes it valuable for leading to a competitive advantage.[7]

A more expansive approach to environmental scanning is the attempt to review relevant information for the company.[8] This can be a major task, given that the number of articles written each day is estimated at more than 20,000. Online retrieval systems facilitate access to this vast amount of information by allowing individuals to sort through the massive quantities of data and random facts. Also, more information is not necessarily better information, and it certainly is not a managerial asset until it has been scanned for relevance and quality. Consequently, better information is not a luxury but a critical tool to help deal with environmental uncertainties.[9] As you would expect, this technique is more relevant to larger companies that are supplied by or sell to wider geographic markets than would characterize most small businesses.

During the decade of the 1980s, companies gained access to external databases through communications networks. Previously, these huge databases were inaccessible to individual organizations because of their cost and storage problems. Access to these databases is often provided through connect-time costs. Using public access telecommunications systems, the databases charge users fees based on the amount of time they spend connected. This is similar to the way charges are computed for long-distance phone services.

**Online searching** is the expression used for computerized literature searches using public databases. All popular web browsers provide free searching, but there are organizations that can do much more sophisticated searches for a fee. All of these databases allow searching through the use of key words from the articles. Databases have become easier to use because they allow the stringing together of key words. For example, a person might ask a database for all of the citations on Fortune 500 companies *and* strategic planning in the future *or* forecasting. Green Things could search for updated information on the schedules and locations of new sites. Summer Farms could find locations of local Farmers' Markets or find market data on the growing demand for various farm products on a wider scale.

A major difficulty involved with environmental scanning is the wide range of available choices.

Systematic searching with consistent databases, rather than a scattered approach to information, reveals the greatest advantage. It is essential to understand this because one database (from over a total of 100 databases offered by one service) might include all of the United Press International (UPI) coverage since 1976.[10] Access to and appropriate use of these sources aids in keeping individuals and organizations headed toward the "information rich" philosophy.[11] In fact, it is fair to say that the information revolution is transforming the nature of competition.[12]

## Competitor Intelligence

One of the fastest-growing specific areas of environmental scanning is competitor intelligence. **Competitor intelligence** is an attempt to scan the information available publicly about competitors.[14] It is a form of environmental scanning that uses external sources to assure that a company has adequate information for competing effectively. Intelligence-gathering techniques are vital in maintaining

---

## A FOCUS ON AGRIBUSINESS
### Environment Scanning

Environmental scanning has proven its worth in agribusiness. In Thailand, for instance, it was demonstrated that managers in more turbulent environments were more likely to use environmental scanning. And the use of that scanning was shown to improve new product performance among small- and medium-sized enterprises.

To Kraft, environmental scanning revealed a new trend: Americans' preference for eating out (restaurant sales), which accounts for more than half of every dollar spent on food. In response, Kraft provides high-quality, convenience foods for in-home consumption in an effort to win back food dollars.

In New Zealand, farmers found that, even though they had to invest more capital in their operations and possibly rearrange their human resources, the use of environmental scanning paid off in terms of increased profitability. The farmers did not use formal scanning tools,

though. They used their downstream supply chain partners to scan their environments. They examined their current operations (including trying new products and markets) to determine which products and markets were more or less profitable and made production adjustments accordingly. Viewing the end consumer as the primary customer enabled them to evaluate their marketing mix better. Products and markets that were more profitable or showed potential were frequently expanded.

Environmental scanning does not have to involve expensive capital investments. It can be carried out through mailed surveys, meetings, focus groups, face-to-face interviews, trend data, data available from government agencies and groups, and, of course, the Internet. These sources can help identify current issues, competitive actions, and other forces in the environment that affect particular organizations.[13]

a competitive advantage. An organization keeps a competitive advantage by moving from identifying market needs to satisfying them faster than its competitors or the industry leader.[15] Like environmental scanning, competitive intelligence does not necessarily involve a computer.

For example, personnel can have a major impact on the information brought into an organization from external sources. Some organizations may select an applicant for a position in part because he or she has worked for a competitor recently (but there are laws that must be followed). To ward off such potential raids on their talented professionals, many companies are examining their hiring policies relating to nondisclosure after leaving the company. Another source of competitor intelligence for marketing is the sales force. Or a firm's advertising department may scan the environment for competitors' ads (including recordings of radio and TV advertisements) and obtain information in informal conversations at events that are or are not company-related. The purchasing department staff could work with vendors who also sell to the competition. A company's real estate department often gets information on competitors' expansion plans since many of the plans are filed with the local courthouse. The appropriate technical personnel may find useful information about future competition by keeping tabs on new patent applications filed with the U.S. Patent & Trademark Office.

It is important to separate the concept of competitor intelligence from industrial espionage; for example, selling information about the product plans of your company. Industrial espionage is illegal and unethical. On the other hand, the information retrieved from public databases such as ProQuest is considered legal competitor intelligence.[16] The information you obtain by simply listening and observing is legally yours. Your sales representatives may learn a significant amount by listening to nearby table conversations when eating in selected restaurants (Figure 10.3). The local manager of Denny's can learn a great deal by simply ordering breakfast at its competitors' establishments, watching what is ordered, observing the demographics of customers, and listening for approvals or disapprovals. Many businesses (e.g., sausage makers) carefully monitor their sellers (grocery stores) for changes in prices and the product lines of their competitors. And many stores (especially grocery stores), large or small, monitor local competing grocers for prices and for

**FIGURE 10.3** Company employees can pick up valuable information regarding their competitors by being observant and listening surreptitiously.

product quality and availability of the products that they both carry. Your local Safeway or Kroger store definitely collects data about the new Whole Foods or Sprouts stores that enter the neighborhood.

Human resource management professionals know which people in the organization have worked for competitors. In fact, companies have become so effective with these databases that they are aware that competitors are checking on them frequently. It is not unusual for a company to practice a form of counterintelligence by providing confusing or contradictory information in the public databases. For example, a pharmaceutical company may file a patent for compounds that its competitor assumes is a breakthrough in a drug on which they are both working. This patent sends a false signal to the competitor about the pharmaceutical company's actual status on the drug.

## Conceptual Modeling

Another tool for strategic planning is **conceptual modeling**, which helps a manager think of his or her organization as a whole with many interrelated parts by forming patterns and creating models. Conceptual modeling focuses on process: how we seek data, turn data into information, and use data to help reach conclusions.[17]

The most common way individuals reflect what is going on in their heads is through lists and hierarchies. Everyone jots down a list in a hurry to guide thinking. To give the list shape and significance, they may then sequence the list, relating and numbering items, shifting them around into an order that better reflects how they want to attack a particular problem.

Turning a list into a "map" instantly improves the thinking tool. The special feature of maps includes questions that are built into them. Maps provide structure and links that help people think up ways to handle tasks.

Rules and questions are exactly what expert systems attempt to capture from humans. **Expert systems** are computer-based programs that provide advice to users just like human experts do. The approach simulates the rules of thumb based on experience and education that we use to make decisions. This conceptual technique may aid us in recognizing that something essential in the planning is not known or is not right and then to work out the thinking that is needed. Organizations also use expert systems for activities other than planning, such as performance appraisals.[18]

Expert systems are built from the kind of questions that drive thinking and reflect mental patterns. These *patterns* are in fact *models*, but most people are unaware of them. For example, by your third year of college, how you select a course has become a simple procedure. You consider many things almost at once: the time that the classes meet, your work schedule, assignments listed in the syllabi, family obligations, and so on. You have become information rich in your environment. Though at one time it may have been an agonizing problem for you to decide which courses to select, now it is routine. If the model that we use remains unconscious, we will not have access to the patterns that drive similar questions. We will be trapped in routines of thinking over which we have limited control. On the other hand, if we are alert to our pattern-making capacity, we can use this effective skill. Once we are aware, we can manipulate the models in our mind and create and recreate them. We can use them to understand, predict, plan, map, and observe our own thinking.

The mental map becomes the essential link required for a task like making a decision or developing a plan. A good conceptual map is actually a blueprint for the future, generic enough in its design to be useful for many people. Wise managers are more likely to make plans for thinking ahead to avoid the dangers of being forced to react inadequately, starting fires that they must put out later. Managers who are too busy are often those who have failed to think things through in advance. They wind up spending an exorbitant amount of time because they have not given enough forethought to something that later became a problem.

## Quantitative Tools and Techniques

In this section we briefly look at a variety of quantitative tools and techniques that can assist managers in planning activities, including forecasting, linear programming, and break-even analysis.

**Forecasting.** The systematic development of predictions about the future is called **forecasting**. One of the most critical kinds of forecasting that managers must do is revenue forecasting. All organizations depend on revenues to remain in operation. For businesses, revenues come from the sales of products and services. For banks, revenues come from interest paid by borrowers. The government derives its revenues from taxes, and schools and universities get much of their revenues from the government and from student tuition.

It follows logically that managers need to know what their future revenues will be so they can plan effectively. For example, if a company wants to open a new factory, its managers need to know that they will have enough funds available to pay for it. Similarly, a university needs a budget for the next year so it can hire instructors, schedule classes, pay the staff, and so forth. It must also know the revenue from each college. Similarly, Summer Farms wants to know its expected revenue both in total and from each of its different farms or crops.

Thus, one of the first pieces of information most managers seek when developing plans is a projection of future revenues. This information is available through revenue forecasting, which involves statistical projections based on past earnings.

Numerous quantitative techniques are available to assist managers in developing forecasts. One technique is time-series forecasting, which involves plotting the subject of the forecast (sales, demand, or whatever) against time for a period of several years. Summer Farms may find this useful for planning

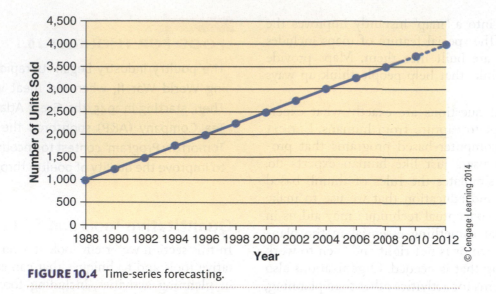

© Cengage Learning 2014

**FIGURE 10.4** Time-series forecasting.

when to plant new crops from the new land it buys with the revenue from its forestland. A "best-fit" line is then determined and extended into the future. An example is given in Figure 10.4, which plotted the number of units sold for the years 1988–2009. As the line moved beyond 2009 (dashed line), it forecasted a demand of just over 3,500 in 2010, 3,750 in 2011, and almost 4,000 by 2012.

Another common technique for planning is Delphi forecasting. **Delphi forecasting** is the systematic refinement of information that takes advantage of expert opinion. Under the Delphi method, experts are asked to make various predictions. Each individual then shares his or her response with the rest of the panel, and the process is repeated. After a few repetitions the experts fine-tune their opinions, and a consensus—the forecast—usually emerges.[19]

**Linear Programming.**   A useful method for determining the optimal combination of resources and activities is called **linear programming (LP)**. Consider a small manufacturer that produces sofas, chairs, and ottomans. Each product is made of a wood frame, fabric covering, and wooden legs. Further, each product goes through the same production and inspection system. Employees work on only one product line at a time. Since it is costly to change frequently from one line to another and since each product has a different profit margin, the question is how to schedule the work: How many sofas, chairs, and ottomans should be produced during a given period to optimize the efficient use of resources and simultaneously satisfy demand? Similarly, how many landscaping projects can Green Things undertake with its current designers and work crews?

LP quantifies the required raw materials and human resources, profit margins, and demand for each product into an equation. The entire set of equations is then solved, and the resulting solution suggests the best number of units of product to produce.

**Break-even Analysis.**   Another useful planning technique, **break-even analysis**, helps managers determine the points at which revenues and costs will be equal. It is especially useful for small- and medium-sized businesses who do not have the expertise on staff for more quantitative methods or business that are seeking less complicated solutions such as how much to charge or whether to add another product. For example, suppose a manager is trying to decide whether to produce a new product. Two kinds of costs are always associated with the product: fixed and variable. *Fixed costs* are incurred regardless of the level of output, such as rent or mortgage payments on the plant, taxes, guaranteed wages and salaries. *Variable costs* result only from producing the product, such as raw materials, direct labor, and shipping. Total costs, then, are the fixed costs plus variable costs. Because fixed costs always exist, the total cost line never begins at zero but at the minimum level of fixed costs if nothing is produced. Total costs rise from there in direct proportion to the volume of output.

To determine the break-even point, the manager plots total costs and total revenues on the same graph, as shown in Figure 10.5. Total revenue from a product is simply the projected selling price times the volume of output. The point at which the lines cross is a break-even point—the point at which the company covers all costs and begins to make a profit. If the company produces and sells less than that lower break-even point,

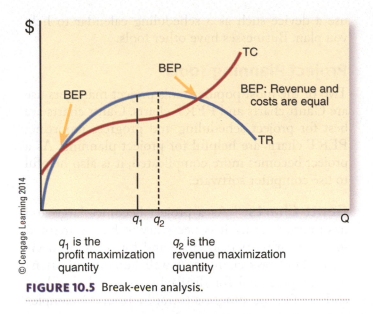

$q_1$ is the
profit maximization
quantity

$q_2$ is the
revenue maximization
quantity

**FIGURE 10.5** Break-even analysis.

it will have a loss because total costs exceed total revenues for the product at the given selling price. However, if the company produces and sells more than the upper break-even point, it will also have a loss.

You should note that the profit maximization quantity is not the same as the revenue maximization quantity in this example. Also, note that the break-even point would change if the unit selling price is raised or lowered.

## Using the Tools and Techniques.

In using various tools and techniques for planning, the manager needs to remember two things: the relative strengths and weaknesses of these aids and the increasingly important role of the computer. To use an analogy, most carpenters know *how* to use a handsaw, a table saw, a saber saw, and a ripsaw; good carpenters also know *when* to use each one. When the carpenter selects the right saw, he is using a conceptual skill. When the carpenter actually uses the saw, he is demonstrating a technical skill. In a similar fashion, managers should recognize that the various techniques described here are tools that provide technical skill. Some are useful in some situations; others are useful in other situations. When the manager has developed the skill to select the best planning tool for the situation, the manager has developed a conceptual skill for planning.

When choosing a technique, a manager must consider several points. On the positive side, these tools offer powerful ways to address certain kinds of problems. They help simplify and organize information, they make planning easier, and they are applicable in a wide variety of situations. On the negative side, they

may not reflect reality accurately, some factors may not be quantifiable, the tools may be costly to use, and the manager may use a technique too rigidly without giving enough credibility to either intuition or insight.

As virtually everyone knows, the last several years have seen a revolution in information technology. The foundation of this revolution has been the computer, in particular the personal computer. This machine greatly enhances the manager's ability to use quantitative techniques in a meaningful fashion. Today's managers have certainly added another powerful tool to their kit. Tomorrow's managers will be even better equipped to handle the complex and dynamic nature of the planning process. The computer also has opened the way to the Internet and the use of social media (e.g., Facebook, Twitter, LinkedIn, and the like). Social media allows companies to inform customers about "specials" and allows customers to inform organizations about their preferences.

While quantitative analysis and/or the application of tools in planning and decision making are important, one must always question and test his or her underlying assumptions. Sensitivity analysis is one way in which to do just that. **Sensitivity analysis** enables the decision maker to vary the values of critical assumptions and re-do an analysis. This, then, provides a clearer understanding of what is actually happening. In turn, this increases the confidence that the decision maker has in the results of the analysis.

## Project Planning Techniques

To best use the previous tools, it is important to schedule and plan who will be responsible for employing them and when. **Project planning tools** are designed to assist in the development of an acceptable solution to a problem within a reasonable time frame and at minimum cost. There are several reasons why projects require special techniques for their management. Projects are different than traditional tasks in that projects typically have specific beginning and ending points. A project is normally a one-time effort and tends to be complex in nature. Tools help break down the complexity into segments for analysis to make the project more approachable and amenable to planning.

Without techniques for scheduling and keeping track of milestones and key individuals, missed deadlines become what has been referred to as the mythical man-month.[20] As a project gets behind, the typical leader solution is to assign more people

© Cengage Learning 2014

to it to "catch up." But the problem does not work that logically. There is no linear relationship between time and number of personnel. The addition of personnel creates more communications and political interfaces. The result is that the project gets even further behind schedule. Project planning tools, such as PERT and Gantt charts (see below), help leaders plan and control projects. By mastering the ability to use these tools, which we will introduce shortly, you develop both conceptual and technical skills.

## FOOD FOR THOUGHT 10.2

Through composting operations at its North American beef and pork plants, Cargill has reduced organic waste by more than 70 percent.

## Project Planning and Scheduling

Because of the complexity of tasks and the number of individuals involved, scheduling competence becomes essential for managers. Scheduling in any environment becomes part of a typical day. For example, in your personal life, you have tailored your study habits around your work, courses, meals, and family obligations. The more complicated your individual schedule becomes, the more likely you are to use a device such as a scheduling calendar to help you plan. Businesses have other tools.

## Project Planning Tools

Two of the most popular tools project managers use are Gantt charts and PERT charts. Gantt charts are best for project scheduling and progress reporting; PERT charts are helpful for project planning. As a project becomes more complicated, it is also helpful to use computer software.

**Gantt Charts.** As a project becomes large and has multiple tasks, it is necessary to have a method for monitoring performance and for reaching decisions about task changes. A well-known and often-used graphic tool for displaying time relationships and monitoring progress toward a project's completion is the **Gantt chart**.

The Gantt chart is a chart that uses bars to represent project tasks. It depicts the overlap of scheduled tasks. Although developed in 1917, it is still a useful tool for time-scheduling projects that involve a graphic approach. The popularity of Gantt charts is based on their simplicity: they are easy to prepare, read, and use. When a task is complete, a shaded bar corresponds to that task. Figure 10.6 shows a Gantt chart created by a few staff members used to plan a picnic for their large organization.

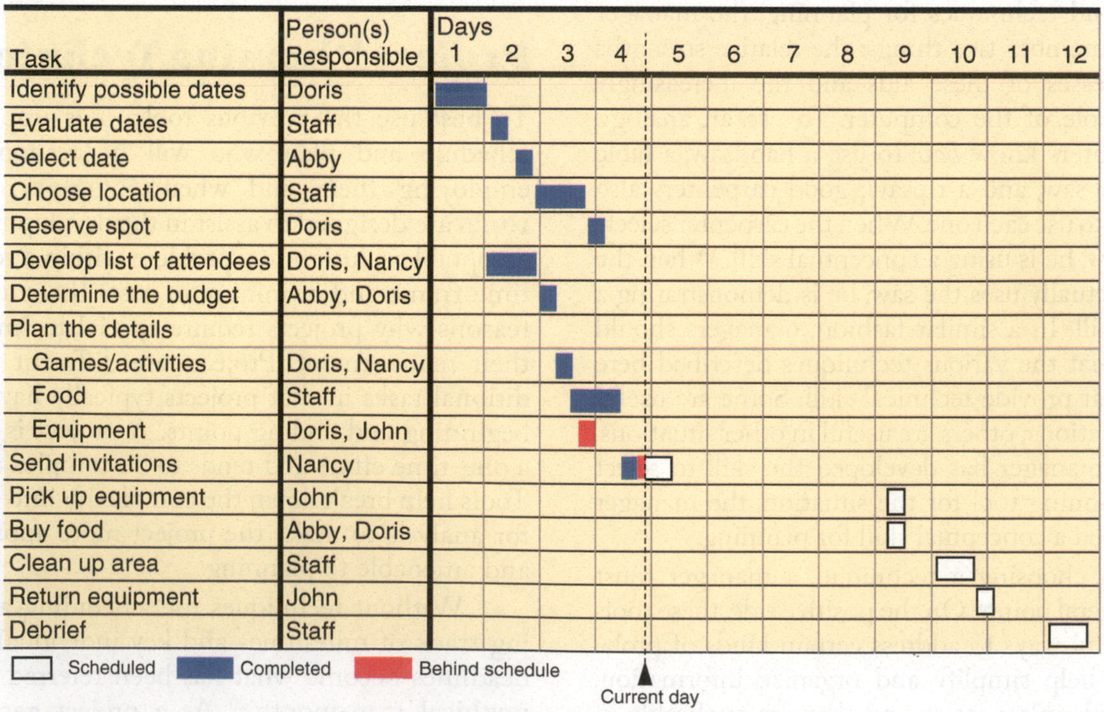

FIGURE 10.6 An example of a Gantt chart for planning a company-wide picnic.

**PERT Charts. Program Evaluation and Review Technique (PERT)** was developed in the late 1950s to assist the Navy in scheduling, coordinating, and controlling the Polaris submarine project. Thanks to PERT, the Navy saved two years in the development of the submarine. PERT is usually recommended for larger projects where the tasks are dependent on each other for completion. PERT and Gantt can be used in a complementary manner to plan, schedule, evaluate, and control systems development projects.

PERT involves identifying the various activities necessary in a project, developing a network that specifies the interrelationships among those activities, determining how much time each activity will take, and refining and controlling the implementation of the project using the network. For the planning process, PERT charts assist by determining the approximate time required to complete a given project, in deriving actual project dates, and in allocating the necessary resources to accomplish the task. PERT charts allow users to organize projects in terms of events and tasks.

Before drawing a PERT chart, events must be determined. **Events** represent points in time, including when the project begins, when tasks are completed, or when the project is completed. A variety of symbols are available to depict events on PERT charts. The PERT diagram in Figure 10.7

uses circles to represent the events. The numbers in the circles represent events that are specific, definable accomplishments. For example, in this exhibit Event 2 might be "Complete soil testing," and Event 6 might be "Complete clearing field."

The arrows identified by letters are activities or tasks necessary to complete the various events. The number in parentheses beside each activity letter is an estimate of how much time is necessary to complete the activity, so an estimate of the time needed for each task or activity must be calculated. Assume that the time estimates have been calculated as shown in Figure 10.7.

The direction of the arrow indicates the order in which events must be completed. Note that some activities can be worked on simultaneously, whereas others must be completed in sequence. In the PERT diagram in Figure 10.7, *Activities a, b,* and *c* can all be undertaken at the same time, but *Event 6* will not be completed nor *Activity h* started until *Activities g* and *e* are accomplished. For example, farmers cannot do anything toward sowing the seed until the field has been cleared and plowed. After that, however, one crew can be sowing while another tamps the earth and a third arranges irrigation equipment. Similarly, at the other end of the network, all other events must be completed before starting *Activity i,* so that *Event 8* can be attained.

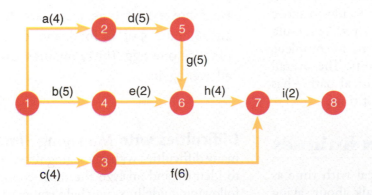

| Activities | Events |
|---|---|
| a—test the soil | 1—start preparing the field for planting |
| b—remove rocks | 2—complete soil testing |
| c—breakup compacted soil and/or hardpan | 3—complete breaking up the soil |
| d—install fencing to keep animals out | 4—complete rock removal |
| e—clear weeds | 5—complete securing field |
| f—install irrigation | 6—complete clearing field |
| g—add amendments (lime, sulfur, nitrogen, potassium, etc.) as needed | 7—complete plowing |
| h—plow | 8—finish planting |
| i—plant | |

© Cengage Learning 2014

**FIGURE 10.7** A PERT chart.

A specific advantage of PERT charts over Gantt charts is the determination of the **critical path**—the longest path through the entire project. The critical path is calculated by summing up the time estimates for each possible path through the network and then selecting the path with the longest time estimate. It is called the critical path because, if any task on the path gets behind schedule, the entire project is in jeopardy of being delayed.

In the example given, the critical path is 1-2-5-6-7-8. This combination of tasks should take 20 units of time to complete. A different path, and not critical to the overall project, is the path 1-4-6-7-8. This path takes 13 units of time. A final option of 1-3-7-8 takes 12 units of time to complete all tasks on the network. An understanding of the critical path helps managers in two ways. First, the manager recognizes that a delay in any activity along the critical path delays the entire project. If *Activity c* takes five units of time instead of four, no real harm has been done. On the other hand, if *Activity d* takes six units of time instead of five, the entire project has been delayed by one unit of time. Of course, the manager may be able to regain the lost time by working overtime, hiring extra help, or providing additional machinery.

Understanding the critical path allows managers to reallocate resources to shorten the overall project. For example, after seeing the PERT diagram in Figure 10.7, the manager might decide to move some workers from *Activity e* to *Activity d*. As a result, *Activity e* now takes three units of time to complete, but *Activity d* is finished in four units. The overall time of completion—that is, the critical path—has therefore been reduced by one unit of time.

## Personal Planning Techniques

Many of our current expressions deal with time as a scarce resource. For example, we talk about saving time, using time wisely, being short of time, and having a time crisis. Managers get the same amount of time as everyone else. Because it is definitely a finite resource, managers must plan the time they do have as effectively and efficiently as possible.

### Time Management

Time is a scarce resource that requires planning and monitoring. **Time management** is the act of setting priorities for how to use our time in achieving our needs and desires. A clear separation is beginning to develop between what is considered a healthy work attitude versus the attitude of a work addict. Workaholics are addicted to activity. **Peak performers**—individuals who excel at their performance—are committed to results. Job obsession and job addiction are not the same.[21]

Job-related stress is associated with job satisfaction or dissatisfaction.[22] Stress can be measured through time-related characteristics such as boredom (too much time with little of value to fill it) or procrastination (misuse of time resulting in crisis and anxiety). Too much stress has a multiplying effect, so that your judgment may be slowed down to the point that it will take you longer to reach the right decision, and it may prevent you from making a decision at all. Stress can cause you to exaggerate the importance of one class of problems at the expense of another to the point of obsession. In that situation you can neither settle the problems nor lay them aside long enough to consider the other ones that may be equally important.[23] For example, it may be difficult for a manager to give a negative performance appraisal to a cooperative employee. Consequently, the manager delays and worries about the appraisal to the point of missing a deadline for a proposal.

---

**FOOD FOR THOUGHT 10.3**

Hens are expert time managers. It takes them 24–26 hours, 5 oz. of food, and 10 oz. of water to produce one egg. Thirty minutes later they start all over again.

---

**Difficulties with Managing Time.** There are many difficulties with managing time, yet it is possible to identify and analyze the most common ones. The following guidelines may help you master your time.

First, *do not let being "busy" pass for being "productive,"* and *do not confuse visibility with productivity.* Everyone knows someone who is always in motion. For those individuals, the crisis and the urgent situation may crowd out the important tasks. For example, a manager on the *Apollo 11* project that put the first people on the moon impressed on his subordinates that his three main values were (1) arriving early, (2) staying late, and (3) looking busy—even though many of them were extremely bright young scientists hired to make creative contributions to the space program. He was interested in seeing people at their desks in case he and they were visited by senior managers. He worried about having people out of sight doing experiments or research in libraries. The staff knew he checked the coat rack when he came in to see who had arrived early each morning. One enterprising individual hired a janitor who started work at 6 A.M. to hang up his coat so that he could go to the library and work.[24]

Second, *have confidence in your ability to proceed.* Several personal symptoms force individuals into a cycle that is destined to delay decision making and induce procrastination. Underestimating oneself paralyzes an individual from action, for fear of not being able to handle the correct outcome. Henry Kissinger said that in the final analysis, the decision is yours, and after the decision is made there is a kind of calm that settles around you.[25] If you constantly distrust your own judgment, you may be so cautious that much elapses before you are able to act. Many managers, distrusting their experience or intuition, may request increasing amounts of information to shed new light on an issue. The assumption is that more information is better, and that more information will increase decision-making expertise. It does not indicate, however, that a manager knows how to use this additional information.

Third, *trust subordinates with decisions.* One type of time waster is the person who is not willing to delegate important tasks to others for fear that they will be unable to accomplish them correctly. Consequently, this person gathers the task back from the employee, very often after much time has already been spent on a project.

Fourth, *do not unnecessarily or needlessly prolong making a decision.* In today's approach toward participative management, many individuals are concerned that if a decision is made too quickly, or singularly, they may be perceived as misusing their authority rather than being decisive.[26]

Finally, *do not get sidetracked from making decisions.* It has often been stated that the less we understand something, the more words we require to explain it.[27] This tendency may be an avoidance of proceeding to closure and consequently is a time waster. If the next step is to venture into unpopular or difficult tasks, it is not unusual to stall at the stage where we are most comfortable. For example, term papers are often difficult to pull together. Therefore, you may be stuck at the information-gathering stage when you should be starting to write the paper.

**Improving Time-Management Skills.** There are several approaches to help improve your time management skills (Figure 10.8). First, *do not let adversity defeat you.* Individuals who manage time effectively become skilled jugglers, "able to change focus back and forth among a network of enterprises."[28] Barriers to your time, including unexpected crises, are not insurmountable for achieving sound time management accomplishments. Bertrand Russell was a mathematician and a political activist. When the government imprisoned him for his anti-war writings toward the end of World War I, he used the time in jail to write articles and the *Introduction to Mathematical Philosophy.*[29]

Second, *use technology to assist you.* One of the techniques to assist in time management is the environmental scanning approach to access the wealth of information available. To keep your stress in check, remember that you do not need to know everything. It is more advantageous to know where to find information.[30] Additionally, many good ideas have been left just sitting on the shelf. Simply begin. This approach may amount to doing the hardest thing first. Often we cannot find time for the really difficult tasks. We unconsciously let more task-oriented activities crowd out accomplishing long-term goals.

Third, *manage your own time rather than letting others manage it.* One of the more difficult problems is being selfish with your time to accomplish your

**FIGURE 10.8** Improving time-management skills.

goals. Set your patterns and get to know your tendencies. If you are most productive in the late afternoon for interpersonal communications, let people know that is when you will return their phone calls. Do not let the phone dictate your time management. You should not rationalize or blame someone else for all time wasters. Once you identify the time wasters, you can replace them with more productive activities. The key is to know exactly what you want to do with the time you free up when you eliminate a time waster. Otherwise, the time will quickly be dominated by another low-payoff activity.[31]

Fourth, *distinguish high-payoff activities from low-payoff activities*. Just as it is necessary to monitor the time you spend with others, it is necessary to determine how to spend your time. High-payoff activities are often long-range goals and consequently may not seem urgent. They are also not day-to-day tasks and may seem fuzzy or ill-defined. Therefore, low-payoff activities crowd into each day. Consciously evaluate yourself and these low-payoff activities. Do you continue to do them because you have always done them or because you feel like you have to do them to be valued?

Fifth, *self-correct when you determine a time-waste problem*. A common illusion in managing time is to think you are being effective (doing the right job right) when actually you are being efficient (doing the job right, but not doing the right job). A good example of this problem is studying very hard for an examination, but studying one of the chapters that will not be on the exam. Just as in project planning, path correction is an integral skill for peak performers. Rather than spinning in one place, a perceptive manager develops the skill to know when to alter the course. A critical path in a personal planning horizon is the most efficient or appropriate route to take toward a goal. Along the way, there is room for mistakes and corrections.[32]

## FOOD FOR THOUGHT 10.4

The average minutes per day spent on eating and drinking varies across cultures. France leads with 135 minutes, followed by Japan (117) and Italy (114), then Sweden (94), the United States (74), and Mexico (66).

Finally, *make a personal time study* not only to determine where the time goes but also to determine how you feel about the way it is used. One common illusion in managing time is that you think you know where your time goes. Managers have recognized and given credit to project planning techniques and scheduling for planning company projects and activities. If the purpose of scheduling is to ensure that the right things are finished at the right time with the right items and/or people to create them, it only makes good sense that we need the advantage of scheduling techniques for our hectic personal planning.

The tools for projects and overall organizational planning benefit large groups. What PERT and Gantt charts do for accomplishing a large project, action planning does for individual perspective on time scheduling.

## Action Planning

An **action plan** explains detailed implementation plans to organizational members.[33] Action plans specify decisions that call for actions; for example, to market new products, build new factories, or sell old machines.[34] Some of the proposed actions may be taken within single units, but others can cut across unit boundaries.

If a company's performance goal states that there should be an increase in sales by 10 percent a year, an action plan might state to do it by introducing blue widgets. Although the action plan may have different formats, it should answer who, what, when, where, and how, and consider the obstacles and aids for success of the goal.

An action plan should provide better estimates of the time needed to carry out a strategy, thereby resulting in more realistic deadlines for completing projects. This could be particularly important where producing the product takes considerable time (e.g., trees on a tree farm).[35] Action plans may help avoid delays caused by failure to carry out a critical action step or to start the action step early enough.

Action planning differs from strategic (or tactical or operational) planning in that it compels an individual to add specific steps to the ideas presented in the original plan. For example, if your goal is to achieve an overall grade point average of 3.6 upon graduation, an action plan can provide specific steps to ensure that your goal is met. Some of these action steps might include limiting course enrollment to 15 hours during the spring semester while working; taking only two courses in your major during any one semester; studying at least two hours an evening during the week; and not procrastinating on assignments.

# CHAPTER SUMMARY

All managers must spend some time planning for the future if their organizations are going to survive. Environmental scanning is an attempt to monitor the external factors that affect an organization so the manager can determine which objectives to pursue. Competitor intelligence is a specific form of environmental scanning that tracks competitors and what they are doing.

Conceptual models allow managers to develop mental maps of related topics. These conceptual models assist managers in scanning the environment and in making decisions. In addition, managers have specific planning tools for specific organizational situations.

Forecasting is the systematic development of predictions about the future. Linear programming and break-even analysis are other tools that managers can use in planning. These two tools focus on resource planning (linear programming) or costs (break-even analysis).

In addition to planning for organizational achievement, managers must plan for specific projects. There are many techniques for planning projects, such as PERT and Gantt charts. These techniques provide managers with technical skills to plan projects to achieve organizational objectives.

Personal planning tools aid managers in planning their time and actions. Time management focuses on a scarce resource called time. Action planning focuses on what actions should be accomplished and the necessary steps to accomplish them. Each of these personal planning tools assists the manager in achieving the overall organizational goals.

# CHAPTER ACTIVITIES

## REVIEW QUESTIONS

1. Why is environmental scanning a planning tool?
2. What is the reason for break-even analysis?
3. If you are sure of your resources, why would you want to spend the time on a PERT chart for a project?
4. What are some barriers to effective time management?
5. What is the difference between an action plan and strategic planning?

## ANALYSIS QUESTIONS

1. Review the assignments you have for the entire semester:
   (a) Develop a Gantt chart for the semester projects.
   (b) Develop an action plan for the semester.
   (c) Develop a time management schedule for the semester.
2. Comment on this statement: "Time management is just common sense. Either you have the ability to be organized or you don't."
3. Describe a situation where competitor intelligence borders on industrial espionage.

4. If you were planning to enter graduate school immediately after graduation, which environmental sources would be helpful to you in making your selection of a program?

5. Try to map your thought process for determining your major. Which steps did you take for granted and almost leave out of your notes?

## FILL IN THE BLANKS

1. A proactive approach to monitoring trends and anticipating changes that may affect an organization at the strategic and tactical levels of planning is called _____ _____.

2. The active approach of scanning information available publicly about competitors is known as _____ _____.

3. The systematic development of predictions about the future is called _____.

4. The systematic refinement of expert opinions to develop an expert-based prediction is known as _____ _____.

5. A method for determining the optional combination of resources and activities in making a product is called _____ _____.

6. A planning technique that determines the point at which revenues and costs will be equal is known as _____ _____.

7. A charting technique that involves identifying the various activities necessary in a project, developing a network that specifies the interrelationships among those activities, determining how much time each activity will take, and refining and controlling the implementation of the project using the network is known by its initials, _____.

8. A bar chart, with each bar representing a project task, for depicting the overlap of scheduled tasks is called a _____ _____.

9. The longest path of connecting tasks in a PERT chart as a project moves toward completion is the _____ _____.

10. The act of setting priorities for how using our time to achieve our needs and desires is known as _____ _____.

## ▼ CHAPTER 10 CASE STUDY
### VESTIS VIRDUM REDDIT

The Roman author Quintilian provided the quote still used in the fashion industry today: *vestis virdum reddit*—clothes make the man. Mention Roman clothing and one item of apparel springs to mind—the toga. The toga is as much an icon for the ancient Romans and their empire as it is today an occasion for a university fraternity party. The woolen toga worn over a linen tunic was a symbol of status and wealth. Interestingly, because it was not worn by Roman soldiers, the toga became a symbol for peace.

But time marches on, styles change, and the toga is no longer a part of everyday life. Similarly, the apparel and clothing industry is not often considered an important part of the agribusiness industry. Yet it is an important industry that developed with agriculture and the concomitant settling of communities. For the Romans, the shepherd guarding his flocks was as much a part of the countryside as wheat for bread and vineyards for wine. So it remains today. We have synthetic clothing materials, for example, nylon and other polyesters, and we also have natural materials. Synthetic material has its origin in petroleum derivatives; natural material is derived from plants and animals. The agribusiness linkage is obvious.

The agribusiness orientation of the apparel and clothing industry means that it is under the same economic pressure as food—it is a basic necessity—and planning is critically important. Ask people to list their basic needs and they will mention three things automatically: food, clothing, and shelter. Food and clothing are rather similar. We have common, everyday foods and gourmet foods and we expect to pay a premium for the latter. So it is with clothing. We have common, everyday clothes and fashion-designer clothes with the attendant premium.

Clothing and food share another attention-getter: price. We do not buy clothing as often as we purchase food; however, it is a far more frequent purchase than many other items. A consequence of this purchasing pattern is a constant awareness of price. However, unlike food, clothing production for many countries is an easier export item than food. This is especially true of emerging countries seeking their places in global trade. And like many foods, the price of cotton and wool are often set in large commodity exchanges located in Chicago, New York, and London (and to some extent a few lesser exchanges). Thus, the price of these materials is generally the same all over the globe.

So the distinguishing competitive characteristic of clothing production centers on the cost of production and its consequent consumer price. Textile and clothing production is very labor intensive and therefore favors countries with inexpensive labor. Because of this, China has used its large population to capture a significant share of world clothing production. This competitive advantage led to the establishment of ten million small businesses that accounted for 60 percent of China's economy and 80 percent of its jobs.

However, the success of this industry has led to pressure to increase wages to retain productive employees. Increased wages contributed to increased labor costs. This is a two-edged sword. It allows the employees to participate more actively in the Chinese economy as consumers; however, it also means that China's competitive advantage is now only a comparative advantage and it is shrinking.

Rising labor costs in China are forcing U.S. apparel and accessories retailers, such as Ann Taylor Stores Corp. and Coach Inc., to develop plans that consider relocating at least some of their production to countries with cheaper workforces. Suddenly Pakistan, Bangladesh, and Vietnam are becoming competitive. For example, the average monthly wage of a textile worker in China is $412.50. A similarly employed worker in Thailand earns $245.50; a Filipino, $169.80 per month; a Vietnamese worker, $136 per month; and in Indonesia her counterpart makes $128.90 each month.

This wage differential is beginning to attract global attention. Some countries, such as Malaysia or city-state Singapore, have had to focus on different high-end clothing segments because of their non-competitive wage structure, $666.10 and $2,832 per month, respectively. Other countries are trying a different tactic. Pakistan has formally requested that the United States lower its textile tariffs, but only for Pakistani products. The competition to garner or to retain these industries has become intense. China counters its neighbors' wage competitiveness by pointing out that their quality products are hard to match, and that they understand the U.S. companies and meet their expectations. Its neighbors say that they can learn about the expectations of U.S. companies as did the Chinese.

Competition is more than just a word in the business lexicon. It is real and personal. Just ask the 798 million working-age people in China, Thailand's and Philippines' 40 million each, Vietnam's 48 million, Indonesia's 118 million, or India's 88 million workers. These approximately 1.2 billion working-age people constitute four times the entire population of the United States. For countries it also is very important. China, beware!

Of course, for U.S. importers the process of planning for clothing sales at acceptable consumer prices in the U.S. economy goes beyond wages and similar costs of production. Transportation costs, the cost and development of new infrastructure, workforce and management development, and educational sophistication all must be considered.[36]

### ▶ Case Study Questions

1. What planning tools or techniques are important in dealing with global competitiveness?

2. Discuss some of the reasons clothing manufacturers shift production from one country to another.

3. If you were a consultant to a clothing manufacturer, what sort of computer information database would you recommend for scanning the global environment?

# REFERENCES

1. Michael D'Antonio, *Hershey: Milton S. Hershey's Extraordinary Life of Wealth, Empire and Utopian Dreams* (New York: Simon & Schuster, 2006), 107, 150–151; Kim Severson, "The Chocolate Wars," *The New York Times,* December 19, 2009, WK2; "Milton S. Hershey: The Man and His Legacy" (2010) at www.hersheys.com (accessed August 20, 2010); "The Hershey Heritage" (2010) at www.hersheytrust.com (accessed August 20, 2010); "About the Hershey Trust Company" (2010) at www.hersheytrust.com (accessed August 20, 2010); Rob Cox, Aliza Rosenbaum, and Hugo Dixon, "The Emotions of Hershey's Bid," *The New York Times,* November 24, 2009, B2; Aliza Rosenbaum, Rob Cox, and Anthony Currie, "Hershey Could Finance Purchase of Cadbury, but Shouldn't," *The New York Times,* November 26, 2009, B2.

2. John G. Bruhn and Howard M. Rebach, *Problem Solving at the Mesolevel, 2nd ed.* (New York: Springer, 2007); Hans Liljenström and Uno Svedin (eds.), *MICRO MESO MACRO: Addressing Complex Systems Couplings* (London: World Scientific Publishing Co., 2005).

3. Russell Ackoff, *The Art of Problem Solving* (New York: John Wiley & Sons, 1978).

4. L. E. Lanyon and C. W. Abdalla, "An Environmental Scanning Indicator Proposed for Strategic Agribusiness Management," *Agribusiness* (1997): 613–622; B. K. Boyd and J. Fulk. "Executive Scanning and Perceived Uncertainty: A Multidimensional Model," *Journal of Management* (1996): 1–21.

5. Donald A. Marchant and Forest W. Horton, Jr., *Infotrends, Profiting from Your Information Resources* (New York: John Wiley & Sons, 1986).

6. Bill Dausses, Manager, External Communications, Phillips Petroleum, April 13, 1993.

7. L. Jeen-Su, T. W. Sharkey, and K. I. Kim, "Competitive Environmental Scanning and Export Involvement: An Initial Inquiry," *International Marketing Review* (1996): 65–80.

8. M. Yasai-Ardekani and P.C. Nystrom, "Designs for Environmental Scanning Systems: Tests of a Contingency Theory," *Management Science* (1996): 187–204.

9. H. Skip Weitzen, *Infopreneurs* (New York: John Wiley & Sons, 1988), 9.

10. Peggy C. Smith, "Incorporating Public Information Systems into Systems Analysis," *Proceedings of the Computers and Business Schools* (Raleigh, NC: North Carolina State University, October 1989), 53–69.

11. M. Hudson, "Toward a Framework for Examining Agribusiness Competition," *Agribusiness* (1990): 181–189.

12. Michael E. Porter and V. E. Miller, "How Information Gives You Competitive Advantage," *Harvard Business Review,* Vol. 63.

13. C. Ngamkroeckjoti and M. Speece, "Technology Turbulence and Environmental Scanning in Thai Food New Product Development," *Asia Pacific Journal of Marketing and Logistics* (2008), 20(4): 413–432; R. K. Bowmar, *Farmer Level Marketing: Case Studies in the South Island, of New Zealand, Research Report No. 305* (Christchurch, New Zealand: Agribusiness and Economics Research Unit (AERU) of Lincoln University); L. E. Lanyon and C. W. Abdalla, "An Environmental Scanning Indicator Proposed for Strategic Agribusiness Management," *Agribusiness* (1998), 13(6): 613–622; M. P. Miles, J. B. White, and L. S. Munilla "Strategic Planning and Agribusiness: An Exploratory Study of the Adoption of Strategic Planning Techniques by Co-operatives," *British Food Journal* (1997), 99(11): 401–408.

14. Ben Gilad, "The Future of Competitive Intelligence: Contest for the Profession's Soul," *Competitive Intelligence Magazine* (2008): 22.

15. Stanley Davis, *Future Perfect* (New York: Addison-Wesley Publishing, 1987).

16. Craig S. Fleisher and Babette E. Bensoussan. *Business and Competitive Analysis: Effective Application of New and Classic Methods* (Upper Saddle River, NJ: FT Press, 2007).

17. Jerry Rhodes, *Conceptual Toolmaking: Expert Systems of the Mind* (Cambridge, MA: Basil Blackwell, Inc., 1991).

18. David D. Van Fleet, Tim O. Peterson, and Ella W. Van Fleet, "Closing the Performance Feedback Gap with Expert Systems," *Academy of Management Executive* (2005): 19:3, 38–53.

19. Andre L. Delbecq, Andrew H. Van de Ven, and David H. Gustafson, *Group Techniques for Program Planning* (Glenview, IL: Scott, Foresman, 1975); see also Ruth S. Raubitschek, "Multiple Scenario Analysis and Business Planning," in *Advances in Strategic Management*, eds. Robert Lamb and Paul Shrivastava (Greenwich, CT: JAI Press, 1988), V:181–205.

20. Frederick P. Brooks, *The Mythical Man-Month* (Reading, MA: Addison-Wesley, 1975).

21. Charles Garfield, *Peak Performers* (New York: Avon Books, 1986), 227.

22. Sherry E. Sullivan and Rabi S. Bhagat, "Organizational Stress, Job Satisfaction and Job Performance: Where Do We Go from Here?" *Journal of Management* (June 1992): 353–374.

23. Stuart D. Sidle, "Workplace Stress Management Interventions: What Works Best?" *Academy of Management Perspectives* (August 2008): 111–112.

24. Charles Garfield, *Second to None: How Our Smartest Companies Put People First* (Homewood, IL: Irwin, 1992).

25. Marc U. Porat and Michael R. Rubin, *The Information Economy*, U.S. Department of Commerce, Washington, DC, 1977.

26. Saul W. Gellerman, *Time Robbers: Worries and Tension*, unpublished manuscript.

27. Ackoff, *The Art of Problem Solving.*

28. William Oncken, Jr., *Managing Management Time* (Englewood Cliffs, NJ: Prentice-Hall, Inc., 1984).

29. "Bertrand Russell," *Stanford Encyclopedia of Philosophy* (first published December 7, 1995; substantive revision March 29, 2010) at plato.stanford.edu (accessed July 25, 2010).

30. Richard Saul Wurman, *Information Anxiety* (New York: Doubleday, 1989), 52.

31. Robert D. Rutherford, *Just in Time* (New York: John Wiley & Sons, 1981).

32. Garfield, *Peak Performers*, 199.

33. Daniel Hunt, *Quality in America: How to Implement a Competitive Quality Program* (Homewood, IL: Irwin, 1992).

34. Anja Schulze and Martin Hoegl, "Knowledge Creation in New Product Development Projects," *Journal of Management* (April 2006): 210–236.

35. C. McDougall, R. Prabuh, and Y. Kusumanto, "Participatory Action Research on Adaptive Collaborative Management of Community Forests: A Multi-Country Model," in *Managing Natural Resources for Sustainable Livelihoods—Uniting Science and Participation*, eds. Barry Pound, Sieglinde Snapp, Cynthia McDougall, and Ann Braun (London: Earthscan and IDRC, 2003), 189–191.

36. Andrew Batson, "Rising Wages Rattle China's Small Manufacturers," *The Wall Street Journal* (August 1, 2010) at online.wsj.com (accessed September 12, 2010); Elizabeth Holmes, "U.S. Apparel Retailers Turn Their Gaze Beyond China," *The Wall Street Journal* (June 15, 2010) at online.wsj.com (accessed September 12, 2010); Patrick Barta and Alex Frangos, "Southeast Asia Linking Up to Compete with China," *The Wall Street Journal* (August 23, 2010) at online.wsj.com (accessed September 12, 2010); Tom Wright, "Pakistan Calls for Lower Textile Tariffs," *The Wall Street Journal* (August 18, 2010) at online.wsj.com (accessed September 12, 2010).

# Managerial Problem Solving and Decision Making

## MANAGER'S VOCABULARY

- behavioral model
- bounded rationality
- brainstorming
- certainty
- creativity
- decision framing
- decision making
- decision tree
- distribution model
- escalation of commitment
- expected value
- inventory models
- nonprogrammed decisions
- payoff matrix
- probability
- problem solving
- programmed decisions
- queuing models
- rational model
- risk
- satisficing
- uncertainty

## LEARNING OBJECTIVES

After studying this chapter, you should be able to:

- Describe how managers solve problems and make decisions.
- Describe the problem-solving, decision-making process.
- Identify the conditions under which managers must make decisions and solve problems.
- Discuss the role of creativity in solving problems.
- Describe the rational and behavioral models of decision making.
- Describe decision-making techniques, including payoff matrices and decision trees.

## Tracking a Big Decision

**O**maha, Nebraska, was the site that transcontinental railroad chose in the 1860s to begin its westward trek in laying a rail line. Today, a century and a half later, Omaha is the home of Berkshire Hathaway Corp. and its legendary chief, Warren Buffett. And perhaps reflecting its history, Omaha became the partial home of the Burlington Northern-Santa Fe Railroad (BNSF) when Berkshire bought BNSF for $44 billion in 2010.

**FIGURE 11.1** Agribusiness has had a long relationship with railroads, shipping large quantities of heavy items, such as grain, cattle, and farm equipment.
© iStockphoto/R. Sherwood Veith

The Burlington Northern purchase took 15 minutes to consummate, according to BNSF's CEO, Matthew K. Rose. Prior to the deal Berkshire owned 25 percent of the business already. "Buying 100 shares of a stock is like buying a piece of a business. If I don't know enough about a company to buy the whole business, I don't know enough to buy 100 shares." This was the rationale for a decision to spend nearly $3 billion/minute.

"We admire Warren's leadership philosophy supporting long-term investment that will allow BNSF to focus on needs of our railroad, our customers, and the U.S. transportation infrastructure," said Mr. Rose.

Railroads and agribusiness have had a long and intimate relationship. From their beginnings in the nineteenth century to today, railroads have had two purposes: moving people and moving goods. Agriculture/agribusiness needed to ship large quantities of heavy items like grain, cattle, and farm equipment. Railroads were the answer when waterways were not close at hand and before the creation of freight trucks.

Railroads and agribusiness often were intertwined. The United Fruit Company established itself in Central America at the beginning of the twentieth century by building railroads and plantations simultaneously. This synergy provided the company with the clout to work with local governments and gain near monopolies in these areas. However, because of miscalculations this game plan did not work perfectly for United Fruit. In Cuba, the company did not treat a local, prominent plantation owner in a fair manner and his two sons never forgot the unfair treatment. The sons were Fidel and Raul Castro.

Regardless of historical miscues, railroads and agribusiness need each other. During the past decade, trucking has suffered from rising fuel prices, highway congestion, and price wars; and railroads have been re-emerging as a fuel- and cost-efficient means of moving goods, especially commodities such as coal, wheat, and lumber. In 2009, 27 percent of BNSF's business was transporting coal; 21 percent was from transporting farm equipment, lumber, and chemicals; and 20 percent from hauling agricultural products like corn, soybeans, and wheat. BNSF serves more of the nation's major grain-producing regions than any other railroad.

BNSF doesn't need the government to build new highways and airports to serve it. It has already invested heavily in infrastructure and technology, and it plans to invest more to keep up with the growing demand. The railroads invested $480 billion in infrastructure since 1980. This is equivalent of 40 percent of their revenue (Figure 11.1).

As in any business, efficiency and effectiveness are important in agribusiness. Railroads excel at both. For example, an average BNSF train hauls as much freight as does 280 trucks. Society has an enormous interest in using less oil to transport goods, which gives BNSF an important edge, as it can move an average of one ton of goods 470 miles on a single gallon of diesel fuel. But will this marriage of convenience between agribusiness and the freight railroads continue into the twenty-first century as it has through the last two centuries?

Early in railroad history, the railroads carried people and goods. In the 1960s and 1970s the missions were separated. People and goods moved on separate trains. In the United States, even though people and goods traveled on different trains, they used the same tracks. All trains traveled along freight ways. As a result, all "people carriers" had to travel under 80 mph to accommodate the trundling freight trains traveling at 50 mph. Commerce ruled over commuter needs for speed.

However, this is about to change with the proposal of high-speed rail. Making room for high-speed trains will likely require more regulations to limit freight activity. Trains traveling at 110 mph are not effective on freight rails. Rail lines will have to be altered, and freight may have to travel different routes as commuters may prevail over commerce.

Though only a few high-speed lines have been proposed, the freight lines are working with individual states to control this activity. What they have in common is an interest in who will pay for these trains. Who will carry the tax burden? As commerce and exports become more vital, rail sharing becomes an important issue. In fact, Europe is engaged in a similar debate among French and German lines about freight and passenger issues.

Progress has never been easy. Historical precedent is not always a guarantee. Prudent investment decisions might not be as prudent as they appear either. Did Buffet make a sound management decision? Is Berkshire's investment safe?[1]

## INTRODUCTION

The legendary Warren Buffett as well as executives at both Berkshire Hathaway Corp. and the Burlington Northern-Santa Fe Railroad (BNSF) must make many different types of decisions at different times to solve various types of problems. Indeed, decision making is a critical part of every manager's job and almost all organizational activities. Successful decision making requires a great deal of skill.[2] We cover it here because it is perhaps most closely linked to planning. However, the organizing, leading, and control functions also involve problem solving and decision making.

This chapter first explores the nature of problem solving and decision making with a focus on the individual. We will cover group decision making in Chapter 19. Then we'll outline the problem-solving and decision-making processes in detail, discuss various types and conditions of managerial problems, and identify major approaches to making decisions. We will conclude with a brief discussion of some useful problem-solving and decision-making techniques.[3]

# The Nature of Problem Solving and Decision Making

As noted earlier, problem solving and decision making are pervasive parts of all managerial activities. Virtually every action that managers take involves making one or more decisions. For example, a simple decision to raise prices must be made within the context of its probable effects on consumer and competitive behavior. A manager must also consider how much to raise prices, when to initiate the new prices, and a variety of other issues such as the product mix and the contribution of various products to revenue and profits. Decision making is most closely linked to planning, since all planning involves making decisions.[4]

## Managers as Problem Solvers

**Problem solving** occurs when a manager faces an unfamiliar situation for which there are no established procedures that specify how to handle the problem. A manager may receive information from his or her boss, subordinates, customers, or some other source that indicates something is not going as planned. Usually, the manager is aware of only some symptoms of the problem and must do additional research, data gathering, or fact finding to uncover the true cause of the problem. It can take considerable skill to separate symptoms from causes and properly diagnose a situation. Quite often this skill is affected by the way managers prefer to gather and evaluate information that they receive from the organization or the environment. The preference for how information is gathered and evaluated is called a problem-solving or decision-making style.[5] Rather than considering problem solving as a negative situation that a manager should fear, some optimists suggest that problems are really opportunities for making new and creative things happen. Whether problems are crises or opportunities, managers often face situations with which they have little or no experience, no decision rules, and little or no guidance about what to do.

Suppose that a plant manager for a manufacturer of farm trailers realizes that turnover in the plant has been increasing substantially over the past year. Having never encountered this problem before, the manager investigates and discovers that employees are leaving to take higher-paying jobs in other companies because wage rates in her plant have not kept up with the prevailing wage rate in the area. The manager must therefore address the wage issue. That problem involves a basic question: Which alternative will best solve the problem: raise wages, keep wages at their current level and live with higher turnover, or offer incentives other than higher wages? But what if the manager had found that the employees were leaving for reasons other than pay? The skills required include differentiating between symptoms and root causes of a problem, developing alternatives, and choosing the best alternative for solving the problem. Managers in large organizations increasingly find themselves with complex situations and difficult problems to solve.

## FOOD FOR THOUGHT 11.1

Problem solving in agribusiness often involves nature, such as pollination, over which managers have no control. Overall, pollinator-dependent crops were reported to make up an estimated 23 percent of total U.S. agricultural production in 2006, up from an estimated 14 percent in the 1960s.

## Managers as Decision Makers

**Decision making** is the process of choosing one alternative from among a set of alternatives. When managers make decisions, they identify a number of potentially feasible alternatives and choose what they believe to be the best alternatives for each situation. For example, before Harry Cunningham made the decision to open the first Kmart stores, he had several options: stay in variety retailing with the Kresge stores; move into another branch of retailing, such as food or specialty retailing; open only one or two Kmarts on a trial basis, or open several Kmart stores. Cunningham chose the fourth alternative and the rest, as they say, is history. Decision making and problem solving are slightly different processes, but they are also interrelated.[6]

## The Problem-Solving and Decision-Making Process

Many times every day managers face decisions to take advantage of new opportunities and solve problems. At each opportunity they must be prepared to make the best decision possible. Many managers seem to make decisions quite easily while others agonize over each and every one. Either way, managers must use

their best skills to make the appropriate decisions. The steps in the problem-solving and decision-making process, summarized in Figure 11.2, include recognizing and diagnosing the situation, generating alternatives, evaluating alternatives, selecting the best alternative, implementing the chosen alternative, and evaluating the results. Following these steps is the key skill that a manager needs for making the best decisions and solving problems effectively.[7]

## Recognizing and Diagnosing the Situation

The first step for a manager is to recognize the need for a decision and to define its parameters. In the earlier chapter on strategic planning (Chapter 9) the SWOT analysis was shown to precede many decisions about the strategy of the firm. In effect, the SWOT analysis leads to a strategy that then serves as the guide for all managerial decision-making. Sometimes the catalyst for a decision is the recognition that a problem exists. For example, a drop in the price of lumber or an increase in the price of local

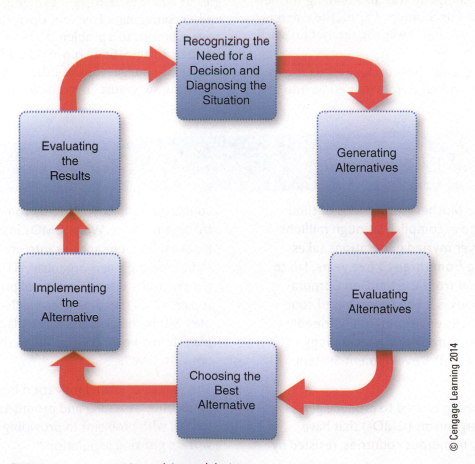

© Cengage Learning 2014

**FIGURE 11.2** The problem-solving and decision-making process.

farm acreage would require a decision by Summer Farms about whether it should modify its plans to sell timber and buy more land. Other examples of problem indicators include cases where employee turnover increases by 10 percent, profits unexpectedly drop by 15 percent, or a customer files a lawsuit against the company.

Positive developments can also prompt the need to make a decision. For example, the manager of an organization with surplus profits must decide how to use those profits. Likewise, a manager who makes offers to five outstanding engineers and then has all of them accept must decide about initial job assignments for them. A sudden surge in home building in one area may prompt Green Things to hire an additional landscape designer and planting crews.

A new business opportunity is another positive development that can act as a catalyst. For example, managers at Kodak reacted to an opportunity when they entered the battery market. Increased demand and cost-cutting technological breakthroughs combined to provide a perfect opportunity for Kodak, which had high name visibility and resources to back the venture. The company was also looking for new markets to enter.[8] For Summer Farms, the emerging popularity of organic foods was the catalyst for starting a new business line.

Recognizing and diagnosing a situation is a stage at which individual factors may come into play. An individual's predispositions and motives may influence how she or he sees the decision situation. For example, if a manager has a negative attitude about labor unions, she or he may define a situation prompted by a union-organizing campaign strictly in terms of how to avoid unionization. In any case, recognizing and diagnosing the situation are key skills in solving problems and making decisions.

## Generating Alternatives

The second step of the process is to generate alternative solutions. Since one of the characteristics of a decision-making situation is that the manager must choose from several alternatives, the identification of those alternatives is a very important part of the process. After all, if the "best" alternative is never considered, the "right" decision can never be made.

It is usually best to identify standard and obvious alternatives as well as innovative and unusual ones. Standard solutions are those that come to mind with little thought, such as things that the organization or the manager has done in the past. Innovative approaches may be developed through such strategies as **brainstorming**—bringing people together and encouraging a free and open discussion of creative solutions to a problem.[9] They are encouraged to let their imaginations run wild and not to mock or ridicule the suggestions of others. Although some of the ideas suggested in brainstorming sessions may

## A FOCUS ON AGRIBUSINESS
### Accelerating Mother Nature

"What may take Mother Nature tens of thousands of years to accomplish through millions of mutations over myriad generations takes Cibus anywhere from three to five years, lab to market." Spun off from the ValiGen Corporation in 2001, Cibus is a San Diego-based company that produces environmentally friendly crop traits using a proprietary technology called the Rapid Trait Development System (RTDS™).

Genetic engineering is used to produce genetically modified organisms (GMOs) that have been banned by numerous countries, resisted by consumers, and boycotted by some environmental organizations. While GMOs insert genetic material from one species into another, essentially creating a new organism, RTDS introduces genetic traits through a natural process of gene repair within the same species. Because it operates within the genome of the plant, like normal plant breeding, RTDS eliminates the issues present with GMOs.

This sort of creative innovation is what makes agribusiness exciting and promises to be able to deal with problems in providing food for the world's growing population."

be of little practical value, a surprisingly large number may have potential.[10] Family get-togethers at the Summers frequently become informal brainstorming sessions.

One consideration in choosing how much time to devote to generating alternative solutions is the significance of the decision. If the decision is extremely important, the organization is likely to conduct a lengthy and thorough search for alternatives.[12] If the decision is fairly minor, the search for solutions may be brief. When Union Carbide decided to build a new corporate headquarters, the firm spent more than two years looking for sites. On the other hand, when a company decides to buy new uniforms for employees, it will no doubt spend much less time and energy searching for alternatives.[13]

## Evaluating Alternatives

After an acceptable list of alternatives has been generated, the manager must evaluate each one on that list. Quite often this evaluation process takes place in two stages.

**Stage One.** First, evaluate each alternative in terms of its feasibility, satisfactoriness, and the acceptability of its consequences. Figure 11.3 provides a useful framework for this initial assessment.[14]

This assessment format involves subjecting each alternative to three questions. First, the manager asks if the alternative is feasible—whether or not it is even possible. For instance, if one alternative calls for a general layoff of operating employees but the firm has a labor contract that prohibits such layoffs, that alternative is not feasible. Nor is it always feasible for a food-producing business to grow organic foods on farmland

that lies too close to fields that have previously been contaminated by chemicals or currently sprayed by crop dusters (Figure 11.4). Similarly, if a company has limited capital, alternatives that require capital outlays are not feasible and may therefore be eliminated.

Second, the manager addresses the extent to which the alternative is satisfactory—whether it actually addresses the problem. Suppose the company wants to raise $1 million. One alternative is to sell some land the company owns. An appraisal suggests that the land will bring in $500,000, so the manager must either combine this alternative with another to raise the desired funds or drop the alternative from consideration and move to another solution such as using working capital. Certainly Webb Dairy could not address the loss of one cow to Mad Cow or Bang's Disease by simply deciding to buy a disease-free replacement cow or recall the milk it has sold since a specified date. Federal laws would dictate the dairy's decisions for this severe problem, in some cases starting with the destruction of the entire herd.

Third, the manager must consider the possible related consequences of the alternative. In some cases such consequences may render the alternative unacceptable. The degree to which each alternative is ethical and socially responsible may also be part of the evaluation of consequences. A plan to boost sales of a particular product by increasing advertising may work well within that narrow context. On the other hand, the boost in advertising may cut profits to an unacceptable level, may increase demand beyond the firm's production capacity, and may create resentment by managers of some products at having funds diverted to another manager's product. Thus, the alternative could be both feasible

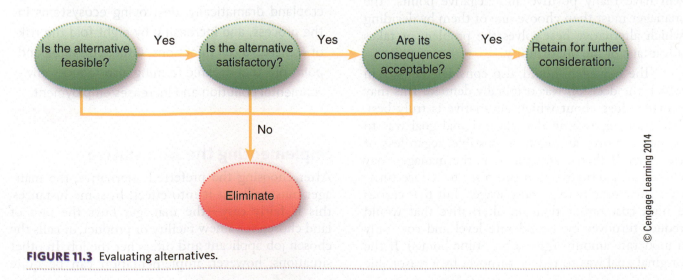

**FIGURE 11.3** Evaluating alternatives.

© iStockphoto/Phil Augustavo

**FIGURE 11.4** Evaluating alternatives to expanding or building new facilities includes considering federal and state ecological constraints, including pollution of streams.

and satisfactory but have such objectionable side effects that it is unacceptable.

**Stage Two.** The second stage of the evaluation phase is to continue gathering information and to analyze the remaining alternatives in terms of their potential for solving the problem or taking advantage of the opportunity. This may mean using expected value analysis, payoff matrices, various modeling and simulation techniques, and other tools that project which remaining alternative is most appropriate. The essence of this process is to try to predict the short- and long-run outcomes of each alternative in terms of the short- and long-run goals of the organization.

## Selecting the Best Alternative

The initial evaluation phase will probably eliminate some of the alternatives, and the few that remain will have many positive and negative points. The manager must then choose one of them by deciding which alternative best solves the problem or takes advantage of the opportunity.

The manager should also consider the way in which the decision was originally defined. This may provide clues about which alternative is truly best. For example, assume that the original goal was to reduce turnover as much as possible, regardless of the costs. If that is still the goal, the manager may choose an alternative that promises to reduce turnover substantially (e.g., raise wages) but that carries a high cost rather than an alternative that would reduce turnover by a moderate level and cost only a moderate amount (give a one-time bonus). If the original goal was to reduce turnover by a reasonable

amount, or if that goal is more desirable now, the second alternative may be better.

Finally, the manager may be able to choose more than one of the alternatives simultaneously by developing contingency plans. Contingency plans, described in Chapter 8, are alternative courses of action if certain conditions occur in the future. More specifically, a manager may select one preferred alternative but note that, if something unusual occurs in the future, then another alternative will be put into action. Thus, contingency planning can also be part of effective problem solving and decision making. Suppose the manager is hiring an assistant and has two strong candidates for the position. One strategy is to offer the position to one candidate and keep the other candidate on hold. If the first offer is refused, the manager still has an acceptable alternative. Knowing when to put contingency plans into effect is a key managerial skill.

Choosing alternatives is frequently an extremely difficult process. Consider the problem General Motors and IBM face in deciding whether to leave South Africa in the late 1980s. Both companies had substantial investments and employed hundreds of people there. Only after months of internal agonizing about the best course of action did they decide that the deteriorating political situation warranted a change in the company's policy. A related issue that occasionally arises pertains to the ethics of the various alternatives available.

### FOOD FOR THOUGHT 11.2

Some experts believe that organic farming can feed only 2.4–4.0 billion people even after expanding cropland dramatically, destroying ecosystems in the process, and increasing by eight fold the risk of *Escherichia coli* infections. Others disagree and point out that organic farming also reduces environmental pollution and increases employment.

## Implementing the Alternative

After choosing the preferred alternative, the manager must still put it into effect. In some instances, this is fairly easy—the manager buys the plot of land chosen for a new facility or product, or calls the chosen job applicant and offers her the job. In other situations, however, implementation can be quite complicated. Members of the organization may resist changes brought about by the decision to hire someone for a new position, for example. Similarly, even though it may be easy to buy the land for a new facility, it may be nearly impossible to convince the existing townspeople that the facility should be built because it will generate too much traffic, strain the put strains the current infrastructure, or create other logical or illogical issues.

The key to effective implementation is proper planning, including both contingency planning (Chapter 8) and strategic planning (Chapter 9). Changes take time, they are subject to unexpected pitfalls, and they do not always work as expected. Managers should exercise patience and understanding during this phase.

## Evaluating the Results

The final step of the problem-solving and decision-making process is to evaluate the results or consequences of the implementation of the chosen alternative. One big mistake that managers occasionally make is to implement an alternative and then assume that the problem has been corrected. Things seldom go this smoothly. It is necessary to follow up and evaluate the results of the alternative in light of the original situation.

One general way to handle this stage involves three steps. First, restate the desired consequences of the decision and estimate how long it will take to realize those consequences (Step 1). For example, suppose the catalyst for the decision was an unusually high absenteeism level. The manager may first investigate to see if the absenteeism is associated with a particular day of the week or is more general. Using this information, then, the manager may conclude that the desired consequence of the chosen alternative is to reduce absenteeism by 10 percent within one year. The chosen alternative is to pay a bonus to workers with a low absenteeism rate. This alternative is implemented as part of normal organizational procedures (Step 2). After a year, the manager measures absenteeism again (Step 3). If it has declined to the appropriate level, he assumes the problem has been solved. If not, more time, a different solution that may have been set up in contingency plans, or both might be needed to solve it.

One reason so many managers neglect the evaluation step is that they fear what may happen if their idea has been unsuccessful. In some organizations this kind of "failure" is considered a major black mark against the manager responsible for making the decision.[15]

## Managerial Problems: Types and Conditions

Managers must make many different types of decisions under many different conditions. A skilled manager must understand these differences and react accordingly.

### Routine Decisions and Nonroutine Problems

One of the important ways in which decision situations differ is the degree of routineness of the situation. Sometimes managers face factors that are familiar or have occurred in the past. In such situations, managers may be able to fall back on company policy, previously established procedures, or other decision rules to make the decision. A feed-seed-fertilizer business knows that its sales will drop substantially during certain weeks or seasons of the year so it asks employees to take their vacations during these times. Or the business could add a product line with a different seasonal demand (holiday decorations) and encourage employees to perform community service. These are **programmed decisions** and are quite common in organizations.

On the other hand, some situations are unique—they have never occurred before or they have such large consequences that managers cannot apply corporate procedures or some decision rule. Managers must go through extensive information gathering and alternative search and evaluation before making the decision (Figure 11.5). These are called **nonprogrammed decisions**. Crises call for nonprogrammed decisions although plans can still be made to prepare for them (see Chapter 8).

**FIGURE 11.5** Managers may need to gather and consider extensive information before making nonprogrammed decisions.

Managers should develop the skills to properly differentiate between programmed and nonprogrammed situations in order to be effective problem solvers and decision makers in organizations. If a problem is really nonprogrammed but the manager thinks it is programmed, there is danger that an inappropriate decision rule or procedure may be used to solve the problem. On the other hand, if the situation is really a programmed one and the manager thinks that it is a nonprogrammed one, then many hours and expense could be wasted to generate a solution that was already available. Either type of error results in the less than the best utilization of organizational resources.

## Certainty, Risk, and Uncertainty

Just as important as the type of situation, managers also need to understand the three conditions under which decisions are made: certainty, risk, or uncertainty.[16] These situations are illustrated in Figure 11.6.

**Certainty.** Decision making under a condition of **certainty** occurs when the manager knows exactly what the alternatives are and that each alternative is guaranteed. That is, the managers know that if Alternative 1 is chosen, it will result in certain outcomes. In reality, of course, managers encounter few situations of this nature. One example that approximates this condition occurs when American Airlines decides to buy a new jumbo jet. The company has exactly three alternative suppliers—Boeing, McDonnell Douglas, or Airbus—and knows the probable reliability, cost, delivery time, and so forth, for each.

**Risk.** Under a condition of **risk**, the manager has a basic understanding of the available options, and can estimate with some confidence the probabilities associated with each alternative.[17] That is, some element of risk is associated with each outcome. For example, suppose Oscar Meyer is considering two possible sites for a new plant. Except for taxes, the two sites are equal. Site 1 has a relatively high tax rate, but the rate is not likely to be increased for several years. Site 2 has a lower tax rate that will be increased next year. Managers at Oscar Meyer could reach several conclusions, including a 40 percent chance the new tax rate at Site 2 will be higher than that at Site 1; a 30 percent chance the new tax will be slightly lower; and a 30 percent chance the tax rate will increase only a little beyond its current level. Oscar Meyer's managers must consequently deal with a large element of risk in making their decision. Decision making under conditions of risk occurs frequently. Dr. Frank Baldino, Jr., the founder and head of Cephalon, is a decision maker and a risk taker. He has made the biotech firm a formidable competitor in a highly volatile industry.[18] The key to making effective decisions in these circumstances is estimating the probabilities correctly.[19]

**Uncertainty.** The most common decision-making condition that managers must confront is **uncertainty**. Not only are the probabilities hard to assess but the list of available alternatives is not clear; the manager may not even be able to identify all the feasible alternatives that should be considered. Agribusinesses, for example, never know for certain what the year's crops will yield locally or globally, how weather or pests or disease will affect them, the kind of demand there will be for them, etc.

Managers face their own unique uncertainties in today's changing scene in Eastern Europe and Russia. Developments in those areas toward a free market economy appear to hold considerable promise for businesses and managers who are able and willing to supply the right goods and services at the right price. Still, considerable uncertainty persists regarding a possible reversion to a command economy, competition from other countries, the spending power of consumers in those countries, and many other unknown factors. Therefore, a manager contemplating a decision to enter one or more of those markets faces considerable uncertainty.

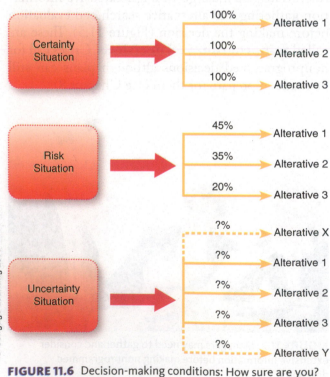

**FIGURE 11.6** Decision-making conditions: How sure are you?

# Creativity

Regardless of their types or conditions, problem solving and decision making often require that managers develop new ideas and approaches. A key ingredient in doing so is the creativity of the manager. This may be a slight adjustment to the old ways of doing things or it may be a radical departure from tradition. In either case, managers often need to be creative.

**Creativity** is a way of thinking that generates new ideas or concepts.[20] It is useful at every step of the decision-making process. It can help managers see problems or opportunities looming in the future before specific symptoms are evident. Similarly, creativity can help in the generation, evaluation, and selection of alternatives. Creative people, for instance, are more likely to think of novel alternatives, and creativity can also help managers figure out new ways of implementing and evaluating alternatives.

How does the creative process work? Some see it as a spontaneous event while others think of it as a much more complex process. One writer suggests that "sudden insights [are] then developed through an orderly and logical process."[21] As shown in Figure 11.7, most experts see creativity as a five-step process: preparation, frustration, incubation, illumination, and verification.

The first stage, *preparation*, relies heavily on intuition and often carries with it the germ of a new idea long before it has been used.[22] An agricultural student, for example, might develop an interest in a particular plant (cotton, sage, blueberry) while still in school. She may think of her interest as only a curiosity and do little formal work on it.

During the second stage, *frustration*, people more actively define the problems they are interested in, begin to manipulate ideas, and analyze and study previous experiences (Figure 11.8). The agriculture student may be drawn to a field of specialization (agriculture, botany, horticulture, plant pathology) that is especially pertinent to the plant she is interested in, without specific plans to pursue a treatment for the problem she has considered. At the same time, however, her subconscious may occasionally think about the problem and link it with her current studies. Quite often this stage is frustrating because the individual does not feel much success with solving the problem.

*Incubation*, the third stage of the creative process, is a period of inactivity and carries with it little active work on the idea. It still receives subconscious

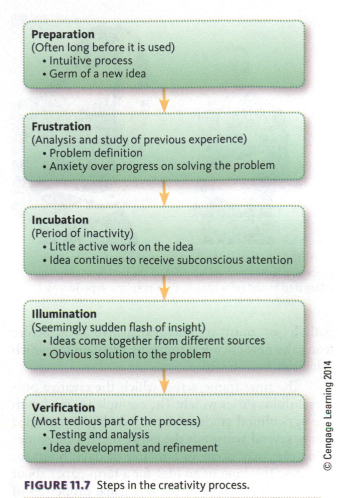

**FIGURE 11.7** Steps in the creativity process.

© Cengage Learning 2014

attention, however. During her graduate studies, the student may have little time or energy to work on her interests, but at the same time her thought process continues to work on it and develop new questions and complex insights.

Stage four is *illumination*. The original interests, preparation, and incubation all combine to yield what seems like a sudden flash of insight. Ideas come together from different sources and an obvious solution to the problem emerges. While examining a plant one day, or while reading a new research report in a research journal, the plant pathologist (former ag student) sees the potential for treating an old problem with a new chemical developed for a totally different situation.

In stage five, verification, the *veracity of the idea* must be confirmed. This may happen while still in school or later in the laboratory of her employing company. Testing and analysis are necessary to make sure that the idea will work, and additional development and refinement may be necessary. The plant

© Igor1308/www.Shutterstock.com

**FIGURE 11.8** During the frustration phase of the creative process, people more actively define the problems they are interested in and refer back to their previous experiences.

pathologist must conduct exhaustive tests before using the new chemical to treat a problem for which it has not been tested.

The time frame across which the creative process unfolds varies widely. The example we used with the agricultural student/plant pathologist extended across several years; but the entire process for something else may occur in a matter of days, hours, or minutes. Thinking of an idea for a term paper about insect-resistant crops or the medicinal properties of sage and determining its appropriateness, for example, may take only a few minutes.

Because of its significant role in organizational effectiveness, managers are keenly interested in creativity.[23] Firms often provide management training programs for their executives that are intended to help them become more creative. For decades Japanese managers have deemphasized creativity, preferring instead to identify ideas created in other countries and then figuring out how to exploit them. More recently, however, firms throughout Japan have started to place more value on creativity. Companies like Shiseido, Fuji, Omron, and Shimizu, for example, have executive training programs that emphasize creativity.[24] Media-Com, a cable company, also uses creativity training, and Marriott even has a chief creative officer.[25]

## Approaches to Decision Making

Managers approach decision making and problem solving in many different ways. Most managers like to think, of course, that they are completely rational in their decision making. However, the complexities and variations of problems and decision situations make it virtually impossible to use rational decision processes consistently.[26] There is a general consensus that most managerial decision making follows one of two models: the rational model or the behavioral model.[27]

### The Rational Model

The **rational model** of decision making, which many managers claim to follow, assumes that decision makers are objective, have complete information, and consider all alternatives and consequences when making decisions. Table 11.1 summarizes the basic premises of the rational model.[28]

First, this model assumes that managers have perfect information; that is, they have all the information that is relevant to the situation and the information is completely accurate. Second, the model assumes that the decision maker has an exhaustive list of alternatives from which to choose. If there are eight potential alternatives, the rational model assumes that the manager has complete knowledge and understanding of all eight.

Next, as the model's name implies, it holds that managers are always rational. It assumes that managers are capable of systematically and logically assessing each alternative and its associated probabilities and then making the decision that is best for the situation. Finally, the rational model assumes that managers always work in the best interests of the organization. Even if their decisions make them suffer, managers will still be motivated to make the decision that other managers in the same organization would make in that situation. Thus, if the clearest course of action will result in budget cuts for a manager's own department, the manager is expected to choose that alternative anyway.

Clearly, the rational model is not always realistic in its depiction of managerial behavior. A variety of forces often intervene in most decision situations, including emotions, fatigue, personal motives, individual preferences and biases, and organizational politics and reward systems. The behavioral model represents an attempt to incorporate these individual processes into managerial decision making.

**FOOD FOR THOUGHT 11.3**

If the price of cottonseed oil can improve by only a penny per pound, the added value to the cottonseed crop would be in the millions of dollars.

**TABLE 11.1**  Rational and Behavioral Models of Decision Making

| RATIONAL MODEL | BEHAVIORAL MODEL |
| --- | --- |
| The decision maker has perfect information (relevant and accurate). | The decision maker has imperfect information (incomplete and possibly inaccurate). |
| The decision maker has an exhaustive list of alternatives from which to choose. | The decision maker does not have a complete set of alternatives or does not completely understand those he or she does have. |
| The decision maker is rational. | The decision maker has bounded rationality and is constrained by values, experiences, habits, etc. |
| The decision maker always has the best interests of the organization at heart. | The decision maker will select the first minimally acceptable alternative (satisficing). |

© Cengage Learning 2014

## The Behavioral Model

The **behavioral model** of decision making recognizes that managers have incomplete information about the situation, alternatives, and their evaluation, thus limiting their potential for making the best possible decision. The behavior approach was first explained by Herbert Simon, who was subsequently awarded a Nobel Prize for his contributions.[29] The basic premises of the behavioral model are also summarized in Table 11.1.

First, this view assumes that managers have imperfect information. That is, the information may be incomplete and/or parts of it may be inaccurate. Second, the behavioral model assumes that managers also have an incomplete list of alternatives. There may be alternatives that managers simply do not know about, they may not completely understand some of the alternatives, and the probabilities associated with various alternatives may be difficult to predict.

Managers are also assumed to be characterized by **bounded rationality**. While they may attempt to be rational, it is constrained by their own values and experiences and by unconscious reflexes, skills, and habits. For example, if a manager has a history of making decisions in a certain fashion, she or he will probably continue to follow that same pattern, even when an objective observer might see the need for a new approach.

Finally, the behavioral model assumes that decision makers engage in what is called satisficing. **Satisficing** is selecting the first minimally acceptable alternative even though a more thorough search could uncover better ones. Suppose, for example, that a college student wants a job in marketing, preferably in marketing research, with a minimum salary of $35,000 and within 100 miles of his hometown. When he is offered a job in marketing (although it is in sales) for $35,500 at a company 80 miles from home, he may

be inclined to take it. A more comprehensive search, however, may have revealed a job in marketing research for $40,000 and only 25 miles from home.

## Other Behavioral Processes

Beyond the concepts of bounded rationality and satisficing, are other important behavioral processes that affect how decisions are made. One such process, called **escalation of commitment**, refers to the tendency of people to continue with a course of action when evidence indicates that the project is doomed to fail.[30] The idea behind escalation of commitment is that people sometimes become so committed to a decision that they fail to see that it was incorrect. For example, an investor buys stock in a company for $50 a share. As the price starts to drop, instead of selling at $40, he may doggedly hold on to the stock until the total investment is lost. Managers at ABC may have been guilty of this error in their 1988 production of a costly sequel to the highly successful miniseries "Winds of War." The company was expecting the sequel to also be a big hit. Halfway through the project everyone involved knew that the unexpectedly high production costs were dooming this "big hit" to be a profit loser. However, instead of "cutting their losses" by abandoning the project, managers at ABC continued production and probably lost about $20 million.[31]

Another behavioral process that affects managers when they make decisions is **decision framing**, which is the way the decision situation is perceived by the decision maker—either as a potential gain or as a potential loss.[32] When a decision is framed in terms of a loss, decision makers tend to choose more risky alternatives. On the other hand, when the decision is framed in terms of a gain, decision makers tend to choose less risky alternatives.

Disastrous consequences may occur when decision makers frame their choices between two losses and subsequently choose the more risky alternative. Examples include Coca-Cola's decision to change its original formula for Coke and introduce the new Coke; the decision by President Jimmy Carter to attempt the rescue of American hostages held in Iran; and the Iran-Contra affair in which arms were sent to Iran in exchange for hostages during the Reagan administration. In each situation existing evidence suggests that the ultimate decisions were framed as a choice between losses which could have led to the selection of a risky alternative. Suggestions on how to avoid the problems associated with decision framing include (1) training decision makers to be aware of the effects of framing a problem in certain ways, (2) approaching problems from multiple frames of reference, and (3) avoiding reacting with a first impulse.[33]

Other behavioral forces that affect decision making include power, political behavior, and coalitions. Power, discussed more fully in Chapter 17, is the ability to affect the behavior of others. Political behaviors are activities carried out for the specific purpose of acquiring, developing, and using power to obtain a preferred course of action. Finally, coalitions are informal alliances of people or groups formed to achieve common goals.[34] For example, suppose a ten-person board of directors is deciding on a new CEO. If the board members are split into factions of 3-3-4, no single faction can get its way. However, by forming a coalition, any two factions that can compromise on a candidate will control a majority vote. Thus, the coalition can now force the group to make a decision in the way that it wants.

# Tools for Improving Problem Solving and Decision Making

As with planning, there are several useful techniques that managers can use to enhance their decision-making skills. Two of the more popular techniques are the payoff matrix and the decision trees.

## The Payoff Matrix

The **payoff matrix** involves the calculation of expected values for two or more alternatives, each of which is associated with a probability estimate.[35] This technique is useful when the decision-making conditions have an element of risk such that the probability of occurrence of each outcome can be estimated. **Probability** is the likelihood, expressed as a percentage,

that an event will or will not occur. If something is certain to happen, its probability is 1.00. If it is certain not to happen, its probability is 0. If there is a 50–50 chance it will occur, its probability is .50.

The **expected value** of an alternative is the sum of all its possible outcomes multiplied by their respective probabilities. Thus, if there is a 50 percent chance that an investment will earn $100,000, a 25 percent chance that it will earn $10,000, and a 25 percent chance that it will lose $50,000, the expected value (EV) of the investment is

$$EV = .50(100,000) + .25(10,000)$$
$$+ .25(-50,000)$$
$$= 50,000 + 2,500 - 12,500$$
$$= \$40,000$$

The following example of an extended version of an investment decision illustrates how this concept relates to the payoff matrix. Suppose that we are considering buying either a restaurant or a plant nursery. We have determined that the success of each business is dependent on inflation. If inflation increases, we make $5 million from the restaurant or $3 million from the nursery. If inflation decreases, however, we lose $4 million in the restaurant but lose only $2 million in the nursery. We have also estimated that there is a 70 percent chance that inflation will increase and a 30 percent chance that it will decrease.

The resultant payoff matrix (Figure 11.9) shows that investing in the restaurant is likely to result in a higher profit than investing in the nursery. The

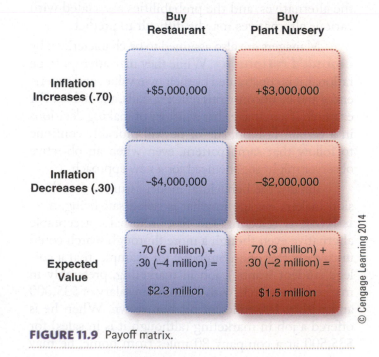

**FIGURE 11.9** Payoff matrix.

expected values are calculated as described above. For the restaurant, the expected value is

$$EV = .70(5 \text{ million}) + .30(-4 \text{ million})$$
$$= 3.5 \text{ million} - 1.2 \text{ million} = \$2.3 \text{ million}.$$

For the nursery, the expected value is

$$EV = .70(3 \text{ million}) + .30(-2 \text{ million})$$
$$= 2.1 \text{ million} - .6 \text{ million} = \$1.5 \text{ million}.$$

Payoff matrices have a number of applications in organizational settings, and the popularity of personal computers promises to make them even more pervasive. Of course, the manager must always remember that the estimates of expected value are only as good as the quality of the estimates for potential payoffs and their associated probabilities—garbage in, garbage out.

## Decision Trees

A **decision tree** is an extension of a payoff matrix that diagrams alternatives and includes second- and third-level outcomes that can result from the first outcome.[36] Consider a medium-sized manufacturing company that is thinking about building a new facility. It needs the plant because demand for the company's products is projected to increase. There is some chance that the increase will be large (or high) and some chance that it will be small (or low); so the manager is trying to decide whether to build a large

or a small facility. Figure 11.10 illustrates this decision scenario in decision tree format.

First, the manager must decide on the size of the new plant. Demand will be either high or low, regardless of the size chosen. If demand is high and a small plant is built, another decision will be necessary: build another small plant, sell the new small plant and build a larger one, or leave the demand unsatisfied. Of course, if demand is low and the manager builds a small plant no further action is needed to satisfy the demand. If demand is high and the manager builds a large facility, again no action is needed. On the other hand, if demand is low and a large plant is built, there are again new alternatives to consider. Should the plant produce at partial capacity, with the rest of the plant sitting idle? Should the plant produce at capacity and create an excess inventory in hopes of future increases in demand? Or should the manager again wait for more information?

To use the decision tree, the manager must estimate probabilities for each alternative on all branches of the tree. Then, working backward from the right to the left, the manager can estimate the expected values of building a large and a small plant.

As with payoff matrices, the key managerial skill necessary for using a decision tree effectively is to forecast potential outcomes and estimate the

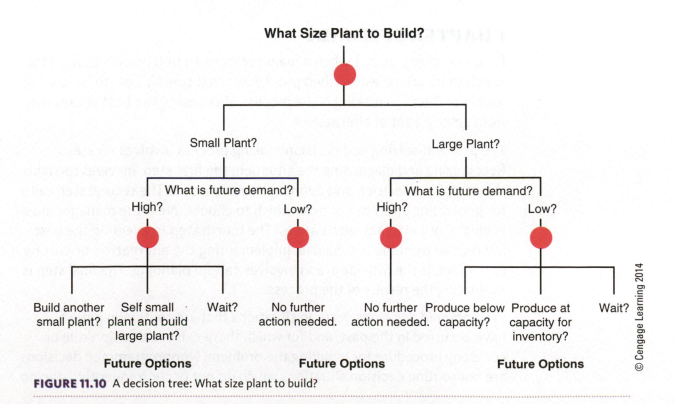

**FIGURE 11.10**  A decision tree: What size plant to build?

probabilities of each potential outcome accurately. Again, computers make decision trees much easier for managers to use.

## Other Techniques

Other fairly common quantitative techniques for decision making are inventory models, queuing models, and distribution models.

An **inventory model** is a decision-making technique that helps managers plan the optimal level of inventory to carry. For example, ordering large quantities of raw materials decreases the chance that the organization's supply will be depleted but it also increases storage costs and ties up money in excess inventory. Ordering smaller quantities reduces the storage costs but increases the chances of running out. An inventory model can help estimate the quantity, frequency, timing, and cost of materials ordered.

A **queuing model** is a decision-making technique that helps plan waiting lines, as in a Safeway grocery store, for instance. Having one check-out operator will reduce the labor cost but will increase waiting lines (Figure 11.11) and therefore customer dissatisfaction. Having 20 check-out operators on duty at all times will keep customers happy but will dramatically increase personnel costs. Queuing

**FIGURE 11.11** American customers in particular are impatient with queuing and will frequently forgo a purchase rather than wait in line.

models help determine the best number of operators to have on duty at various times of the day.

Finally, a **distribution model** helps managers plan routes for distributing products. Suppose a produce wholesaler must drop off shipments of products at twenty different businesses around the city. Left to their own devices, drivers might not proceed from one stop to another in the best sequence. A distribution model helps develop that sequence to minimize travel time and fuel expenses, and maximize employee productivity.

## CHAPTER SUMMARY

Problem solving occurs when a manager faces an unfamiliar situation for which there are no established procedures that specify how to handle the problem. Decision making is the process of choosing the best alternative from among a set of alternatives.

The problem-solving and decision-making process involves six steps. Recognizing and diagnosing the situation, the first step, involves recognizing the need for a decision and defining its parameters. The second step calls for generating alternatives from which to choose. Next the manager must evaluate, or judge, the alternatives. The fourth step is selecting the best alternative from those available. Implementing the alternative, or putting it into effect, is the fifth step, and involves careful planning. The final step is evaluating the results of the process.

Programmed decisions occur in situations that are routine in nature, have occurred in the past, and for which there exists a decision rule or company procedure for handling the problem. Nonprogrammed decisions are nonroutine decision situations which do not occur frequently, have no decision rule or company procedure, or have very significant consequences. In addition decisions are made under three conditions: certainty, risk, and uncertainty. Creativity is a way of thinking that generates new ideas or concepts and is a five-step process: insight, preparation, incubation, illumination, and verification.

There are two general approaches to decision making, the rational model and the behavioral model. The rational model assumes perfect information, an exhaustive list of alternatives, managers who are capable of systematically and logically assessing those alternatives, and managers who will always work in the best interests of the organization. The behavioral model is a modification of this approach and assumes imperfect information, an incomplete list of alternatives, bounded rationality, and self-satisficing behavior. Escalation of commitment, decision framing, power, political behaviors, and coalitions also influence how decisions are made.

Numerous techniques have been developed to aid managers in making decisions. Two general techniques are the payoff matrix and decision trees. Other techniques include inventory, queuing, and distribution models.

## CHAPTER ACTIVITIES

### REVIEW QUESTIONS

1. List and describe the steps in the decision-making process.
2. What is the difference between programmed and nonprogrammed decision situations?
3. Describe the three basic conditions under which managers make decisions.
4. What is the role of creativity in decision making?
5. Compare and contrast the rational and behavioral models of decision making.
6. Describe the differences and similarities between payoff matrices and decision trees.

### ANALYSIS QUESTIONS

1. Identify a decision-making circumstance that has elements of both certainty and uncertainty.
2. Which model of decision making do you think is most common? Is a given manager most likely to follow one of the models almost exclusively, or will the same manager go back and forth between models? Why?
3. Identifying all the potential alternatives to a decision situation may be too costly, if not impossible. Yet satisficing is a problem if too few alternatives are explored. How might a manager guard against these extremes?
4. Can creativity be practiced apart from decision making, or are the two always linked? Why?
5. Identify a recent situation in which you framed a decision in terms of a choice between losses. Did you eventually choose the riskier of the two alternatives? How could you have framed the decision differently so that your perspective of the choices might have been different?

## FILL IN THE BLANKS

1. The process of bringing people together and encouraging a free and open discussion of creative solutions to a problem is called _____.

2. Situations which occur frequently and therefore allow the decision maker to use a decision rule or company procedure are called _____ decisions.

3. Decisions that have significant or expensive consequences or that have not occurred in the past and have no established decision rules or procedures are called _____ decisions.

4. The process of identifying and using new, unusual, and/or creative solutions and alternatives to problems is known as _____

5. The model of decision making that assumes managers are objective, have perfect information, and understand all alternatives and consequences is called the _____ model.

6. The tendency of people to let the way a decision is stated—either as a potential gain or loss—affect their choice of risky or cautious alternatives is known as _____ _____.

7. The likelihood that an event will or will not occur is known as _____.

8. The sum of all possible outcomes of an alternative multiplied by their respective probabilities results in the _____ _____ of the alternative.

9. A decision-making technique that helps plan waiting lines is called a _____ model.

10. The three possible decision-making conditions managers may face are _____, _____, and _____.

## ▼ CHAPTER 11 CASE STUDY
### SOLVING A DRINKING PROBLEM

Ponce de Leon arrived in the New World with one quest in mind: to find the fountain of youth. In the fifteenth and sixteenth centuries the search for youth was paramount. It was important because most people died before they reached 50 years of age. The rigor of life and disease took its toll. Today the quest is for a magic potion, lotion, or fountain to stave off obesity or being overweight. We know that in most advanced economies we will live well into the eighth decade of our lives, but the specter of obesity casts a broad shadow over our existence. There is a caveat that makes this need a challenge: We want to make decisions that enable us to attain this state without giving up lifestyle and food.

Interestingly, perhaps magically, a "magic" potion does exist, and it can be obtained from a fountain. The name of this magic elixir is water. The most common, least expensive, and most important liquid on our planet seems to have almost magical properties. A 12-week randomized, controlled study with nearly 50 human volunteers from Virginia Tech demonstrated that drinking a half-liter of water (a little more than one pint) before each meal every day for three months led to weight loss. A year after the study ended, the participants could still eat whatever they wished and return to their normal lifestyles without gaining back their earlier weight. Most continued to drink water before meals because they liked it.

Water is most common because it is known by everyone, not because everyone possesses it. Common water, fresh water, is not that common; 97.5 percent of Earth's water is not fresh. It is salt water; only 2.5 percent is fresh, drinkable water. Of this 2.5 percent that is drinkable, only 0.8 percent is available in surface, atmosphere, and ground water. The remainder is locked in glaciers, ice caps, and permafrost. Water, it turns out, is not that common after all.

Water and agriculture/agribusiness are practically synonymous terms. Water is vital to agriculture. Soil viability and seed fertility are important, but they also are meaningless without the presence of water. Measured as consumptive use, that is, water unavailable for further use in the system, agriculture consumes 93 percent of all water, and domestic and industrial uses account for the remainder.

We know instinctively how important water is to our existence. As we become more sophisticated and travel through our solar system now and galaxies later, what commodity will designate a planet as a candidate for colonization? Water.

On Earth we have a problem. It is reported that dirty water and poor sanitation kill 5,000 children a day. By 2050, there will be 9 billion humans on the planet. The area under irrigation to feed them will double and the water drawn to irrigate the crops will triple. The problem is greater than it appears because it will not be sufficient to grow enough to sustain the population, while at the same time the people will demand food that is better and more interesting. In addition, the water must be clean—not just available.

The continuing movement toward increased urbanization only exacerbates the problem in every country. Even the United States is not immune. The fastest growing urban centers in the United States are in the western part of the country. Large urban areas and proximal agricultural activity to provide food will put pressure on the fresh water supply by 2025.

This is a situation that requires wise decisions and significant efforts to solve, and it is integrally tied to agribusiness. Agribusiness has even found ways to produce crops without soil, but not without water. There is no agribusiness without water. There can be agribusiness only by using water wisely and efficiently. There can be agribusiness only by using as little water as possible.

Adding to the challenge is another consideration—water is a basic human resource or need. Like air, it should be available to all and as cheaply as possible, preferably free. That makes for an interesting conundrum—the most valuable commodity on Earth should be free; but because it is free (absolutely or relatively), it is abused. Putting a price on water would increase its value and the respect it should get, but will it result in water being priced out of availability for all?

In the United States, water is available free from public fountains, but you pay for bottled water. Bottled water once represented one of the fastest growing segments of the beverage industry. Its value was its utility. Water in a bottle can be carried anywhere, and it is "the" diet drink since it is calorie free. It would seem to be the ultimate product for the soft-drink producers, and competition heated up accordingly.

At the same time, though, we are distracted by individuals who tell us that bottled water is a problem because of the bottle. We are not obese or overweight because we lack self-control but because of the food containers themselves. It is the chemicals in the plastic bottle, phthalates and Bisphenol A (BPA), which are making us what we are. We are the targets of "obsegens." Also, "bottled water is costly for the environment, our pocketbooks and our public water systems," said Gigi Kellett, national director of a "Think outside the bottle campaign." The source of the utility for bottles, that is, that which allows the industry to charge for the water, is under challenge.

Every country has a different context to view its water issues. Unlike food, no one sends water elsewhere. Awareness in the United States is very high about the plight of less fortunate individuals in other countries and their struggle to obtain good, clean water. There are about one billion people who would gladly trade problems with us. What decisions will organizations, especially bottled drink companies, make?[37]

### ▶ Case Study Questions

1. Is the transporting of water, by bottle or otherwise, a solution to the worldwide shortages of clean, potable water that already exist? Explain your answer.

2. What are some possible solutions to the shortage of water for agricultural use?

3. The availability of water differs from one state to another and one country to another, but are all people equally entitled to water? How can we ensure that people who don't have enough water can get access to water at an affordable price?

# REFERENCES

1. Dan Reed, "Warren Buffett Sees Strong Rail System as Key to U.S. Growth," *USA Today* (November 4, 2009) at www.usatoday.com (accessed August 29, 2010); Adam Shell, "Buffett: Railroad Business is 'In Tune with the Future,'" *USA Today* (November 4, 2009) at www.usatoday.com (accessed August 29, 2010); "American Railways: High-Speed Railroading," *The Economist* (July 22, 2010) at www.economist.com (accessed September 4, 2010); Peter Chapman, *Bananas: How the United Fruit Company Shaped the World* (Edinburgh, Scotland: Canongate Books, 2007); Scott Patterson and Douglas Blackmon, "Buffett Bets Big on Railroad," *The Wall Street Journal* (November 3, 2009), A1, A22; Josh Mitchell, "High Speed Rail Costs Irk States," *The Wall Street Journal* (August 21, 2010), A3; Steve Forbes, "Railroading the Taxpayer," *Forbes* (August 30, 2010) at www.forbes.com (accessed September 5, 2010); "High-Speed Rail in Europe: Trouble Ahead," *The Economist* (August 26, 2010) at www.economist.com (accessed September 5, 2010).

2. For a more complete discussion of decision making, see E. Frank Harrison, *The Managerial Decision Making Process, 5th ed.* (Boston: Houghton Mifflin, 1999).

3. Elke U. Weber and Erick J. Johnson, "Mindful Judgment and Decision-Making," in *Annual Review of Psychology 2009*, eds. Susan T. Fiske, Daniel L. Schacter, and Robert Sternberg (Palo Alto, CA: Annual Reviews, 2009), 53–86.

4. R. Duane Ireland and C. Chet Miller, "Decision-Making and Firm Success," Academy of Management Executive (2004): 8–12.

5. Kenneth Brousseau, Michael Driver, Rikard Larsson, and Gary Hourihan, "The Seasoned Executive's Decision-Making Style," *Harvard Business Review* (2006): 111–112.

6. G. Donaldson and J. Lorsch, *Decision Making at the Top* (New York: Basic Books, 1983).

7. M. W. McCall and R. E. Kaplan, *Whatever It Takes: Decision-Makers at Work* (Englewood Cliffs, NJ: Prentice Hall, 1985).

8. "History of Kodak," Kodak website at www.kodak.com (accessed July 24, 2010).

9. A. F. Osborn, *Applied Imagination* (New York: Charles Scribner & Sons, 1963).

10. Robert C. Litchfield, "Brainstorming Reconsidered: A Goal-Based View," *Academy of Management Review* (2008): 649–668; "The Art of Brainstorming," *Business Week*, August 26, 2002, 168–169.

11. M. Arndt, "An Alternative to Monsanto and Gene Splicing," *Business Week* (May 5, 2010) at www.businessweek.com (accessed July 13, 2010); "About Cibus," Cibus website at www.cibus.com (accessed July 13, 2010); J. M. O'Brien, "Ag-Tech Upstart Is Armed to Take on Monsanto," *Fortune* (September 21, 2009) at tech.fortune.cnn.com (accessed July 12, 2010).

12. Paul Nutt, "Expanding the Search for Alternative During Strategic Decision-Making," *Academy of Management Executive* (2004): 13–22.

13. Walter McQuade, "Union Carbide Takes to the Woods," *Fortune*, December 13, 1982, 164–174.

14. This section is based on Ricky W. Griffin, *Management, 10th ed.* (Mason, OH: South-Western Cengage Learning, 2011).

15. Glen Whyte, "Decision Failures: Why They Occur and How to Prevent Them," *Academy of Management Executive* (1991): 23–31; Jerry Useem, "Decisions, Decisions," *Fortune*, June 27, 2005, 55–56.

16. Kenneth MacCrimmon and Ronald Taylor, "Decision Making and Problem Solving," in *Handbook of Industrial and Organizational Psychology*, ed. Marvin Dunnette (Chicago: Rand McNally, 1976), 1397–1454.

17. Rene M. Stulz, "Six Ways Companies Mismanage Risk," *Harvard Business Review* (2009): 86–94.

18. "2009 Annual Report," Cephalon website at media.corporate-ir.net (accessed June 30, 2010); Andrew Pollack, "A Biotech Outcast Awakens," *New York Times*, October 20, 2002, BU1, BU13.

19. L. S. Baird and H. Thomas, "Toward a Contingency Model of Strategic Risk Taking," *Academy of Management Review* (April 1985): 230–243.

20. Watts S. Humphrey, *Managing for Innovation: Leading Technical People* (Englewood Cliffs, NJ: Prentice-Hall, 1987).

21. Ibid. 101.

22. Eugene Sadler-Smith and Erella Shefy, "The Intuitive Executive: Understanding and Applying 'Gut Feel' in Decision-Making," *Academy of Management Executive* (2004): 76–91.

23. Richard W. Woodman, John Sawyer, and Ricky W. Griffin, "Toward a Theory of Organizational Creativity," *Academy of Management Review* (April 1993): 293–326.

24. Emily Thornton, "Japan's Struggle to Be Creative," *Fortune*, April 19, 1993, 129–134.

25. Stephen Allan, "Guest Post: How to Inspire Your People," *Fortune* (August 19, 2009) at money.cnn.com (accessed July 24, 2010); Marc Gunther, "Marriott Gets a Wake-Up Call," *Fortune* (June 25, 2009) at money.cnn.com (accessed July 23, 2010).

26. David W. Miller and Martin K. Starr, *The Structure of Human Decisions* (Englewood Cliffs, NJ: Prentice-Hall, 1976).

27. Amitai Etzioni, "Humble Decision Making," *Harvard Business Review*, July–August 1989: 122–126.

28. Alvar Elbing, *Behavioral Decisions in Organizations, 2nd ed.* (Glenview, IL: Scott, Foresman, 1978).

29. Herbert A. Simon, *Administrative Behavior* (New York: Free Press, 1945).

30. Barry M. Staw and Jerry Ross, "Good Money After Bad," *Psychology Today* (February 1988): 30–33; and Jerry Ross and Barry M. Staw, "Expo 86: An Escalation Prototype," *Administrative Science Quarterly* (June 1986): 274–297.

31. "ABC's 'War and Remembrance' Fails to Deliver Audience, Big Losses Loom," *The Wall Street Journal*, November 28, 1988, B4; "'War' Miniseries Starts Strong, But Loss Is Expected," *The Wall Street Journal*, November 22, 1988, A6.

32. Glen Shyte, "Decision Failures: Why They Occur and How to Prevent Them," *Academy of Management Executive* (August 1991): 23–31.

33. Ibid.

34. Thomas A. Stewart, "New Ways to Exercise Power," *Fortune*, November 6, 1989, 52–64.

35. Robert Markland, *Topics in Management Science, 3rd ed.* (New York: John Wiley & Sons, 1989).

36. Everett Adam and Ronald Ebert, *Production and Operations Management, 4th ed.* (Englewood Cliffs, NJ: Prentice-Hall, 1989).

37. "Drink Till You Drop." *The Economist* (August 26, 2010) at www.economist.com (accessed September 6, 2010); "For Want of a Drink," *The Economist* (May 20, 2010) at www.economist.com (accessed September 6, 2010); "Enough Is Not Enough." *The Economist* (May 20, 2010) at www.economist.com (accessed September 6, 2010); Bureau of Reclamation, "Water 2025: Preventing Crises and Conflict in the West," *Water 2025 Status Report*; Valerie Bauerlien, "Pepsi to Pare Plastic for Bottled Water," *The Wall Street Journal* (March 25, 2009) at online.wsj.com (accessed September 6, 2010); Allysia Finley, "Are Plastics Making Us Fat?" *The Wall Street Journal* (August 13, 2010) at online.wsj.com (accessed September 6, 2010).

# ORGANIZING IN AGRIBUSINESS

4

# Organizing Concepts

## LEARNING OBJECTIVES

After studying this chapter, you should be able to:

- Discuss the nature of organizing, describe the organizing process, and identify key components and concepts involved in organizing.

- Discuss job design in organizations, including job specialization and alternatives to it.

- Indicate how jobs are grouped, particularly grouping by function, product, and location.

- Define and discuss authority, responsibility, delegation, and decentralization.

- Discuss the concept of the group effectiveness, its impact on the shape of organizations, and how a manager might determine what will make a group effective.

- Define line and staff positions and indicate their roles in organizational analysis.

## Speed Bumps on the Global Trade Route

The export and import of agricultural commodities, food products, and farm equipment is the life-blood of global agribusiness trade. The ability to export or import products is the cornerstone of any international business or global enterprise and, as such, the firm must be carefully organized. Simple exporting—shipping your products to another country/market—is one of the "easier" ways to organize for going global and exploring the potential of new markets. But there are "speed bumps" on the global trade route; the free market is not so free after all. Tariffs are one example.

**FIGURE 12.1** Like these figs, fruits and vegetables that could carry disease or insects are usually inspected in the exporting country or off shore, prior to reaching the United States.

© Ralf Siemieniec/www.Shutterstock.com

For all internationally organized companies, a tariff is another "cost of doing business." It is a tax or toll that must be paid to gain access to a particular country and its market. The same toll should apply to all international or global companies seeking access to a country's market. If it is applied to all, then for these firms the toll/tariff is a tax. They continue to compete against each other in whatever manner they have used to differentiate themselves; only their products have become more expensive by the same amount, the tariff.

This "tax" affects different companies differently, however, depending on their orientation. A domestic firm does not pay the tariff. In fact, tariffs are often instituted to provide domestic firms with advantages over global competitors. A firm organized as a multinational and located in the "buying" country is spared the tariff also. It is producing its product internally, so it is treated like a domestic firm. It has already "paid its dues" in the form of foreign direct investment (FDI). It built a factory in the target country and supplied jobs and paid internal taxes. These companies—multinationals—have decided that an organizational arrangement that incurs FDI costs would be less expensive than tariff costs over some long-time horizon. Agribusiness firms make these calculations and assessments on a regular basis.

Some companies and/or countries view and use a tariff as a "trade barrier," an impediment to global trade. When this occurs, countries sometimes engage in an international game of "chicken" as they try to boost the opportunities for the corporate national champions. The United States and Brazil engaged in a game of tariff "chicken" in 2010. Brazil objected to the U.S. support (subsidies) of its cotton farmers. The World Trade Organization (WTO) agreed with Brazil and said the U.S. subsidies violated global trade rules. To retaliate for something that Brazil could not control (American farmer subsidies), Brazil said it would charge $591 million in higher tariffs based on the WTO pronouncement. Half of these tariffs would be assessed on U.S. biotechnology firms selling products in Brazil. Each country hopes the other will yield. They usually compromise because trade is important.

A similar game of tariff "chicken" was played between the European Union (EU) and Latin American banana growers. The EU favored its former colonies' growers, Caribbean and Africa, over growers in Latin America by charging a lower tariff to its former colonies. Again the WTO intervened and the banana tariff was reduced from $176/ton to $148/ton. By 2017 the banana tariff will be reduced to $114/ton. This dispute is 15 years old.

Agribusiness firms are also subject to a form of "trade barrier" that cannot be negotiated away. It has to do with food safety. No country wants the integrity of its food supply compromised by the importation of unsafe food, especially pests that could infest local products (Figure 12.1). These barriers are called "sanitary" and "phytosanitary" measures. If the problem is a plant, fruit, or vegetable infected with a disease, the measure is phytosanitary. If the infestation is harmless, the commodity may be imported into a target country but at a discount. If the importing country deems the infestation a hazard, it may ban the import entirely; the importing country places a "sanitary restriction" on the commodity. The commodity cannot be imported until documented and substantiated proof is submitted that the infestation has been eliminated. The design of these measures is to protect the health of the importing country's crops and population. However, sometimes the infestation is more political than real, which makes dismissing the sanctions more difficult, less scientific, and more open to a political solution.

Other more serious kinds of "speed bumps" on the global trade route encourage illegal solutions, thus making organizing for international operations considerably more difficult. Some exporters and importers turn to the black market if they lose access to the legitimate market. The "dark side" of global trade is smuggling, which is illegal importation. Many people are familiar with smuggling and its interdiction because of newspaper stories about cocaine, marijuana, or other illicit drug smuggling.

But smuggling also occurs in agribusiness. For example, in some parts of the world "bushmeat" is considered a delicacy. "Bushmeat" is meat obtained from non-domesticated species (e.g., Nile crocodile or other exotic animals). This smuggled "bushmeat" sells for twice the price of standard species meat. Authorities regularly check food shipments in shipping containers from different countries, especially Africa, to insure that no illegal "bushmeat" is present. However, a 2010 study found that up to five tons of "bushmeat" passes through Paris airports in the luggage of passengers every week. Also as recently as 2010, authorities in Chicago broke up the biggest food smuggling ring in U.S. history. It involved the collusion of six German and Chinese companies, around 13 high-level executives, and at least $40 million of product. The 12 people arrested each faced significant fines and up to 20 years in prison. The food commodity they smuggled into the United States? Chinese honey.

Organizing for global trade is difficult not only from operational and marketing viewpoints but also from a legal perspective. Thus, a variety of organizational arrangements exist with which to deal with global trade because no product is trivial when it has a market.[1]

## INTRODUCTION

In Chapter 2 we noted that organizing was the second basic managerial function, following planning and decision making. **Organizing** is the process of grouping activities and resources in a logical and appropriate fashion. The pattern of organization which may have worked well in a company's early years must often give way in later years to one better suited to its newer, more complex environment and its expanded strategy. Its earlier organization may have been eminently logical at first but became obsolete over time. This is true whether a company is transformed from a domestic business to a multinational company, or a family business becomes a corporation or joins a cooperative, or a one- or two-person business hires a third employee.

This is the first of four chapters devoted to the organizing function and to helping you understand when different organization patterns are and are not logical and appropriate. This chapter introduces the five basic concepts of organizing. The next chapter focuses on how the components are put together to form organization designs. Chapter 14 addresses approaches to organizational change and innovation, and Chapter 15 covers the staffing of organizations.

# The Nature of Organizing

First, though, to develop a more comprehensive understanding of the critical organizational function, we need to elaborate on the definition of organizing by addressing the nature of organizing and identifying its key components.[2]

## The Organizing Process

Giant, complex organizations such as AT&T, DuPont, Monsanto, and Procter and Gamble were not created in a day. Nor were your neighborhood dry-cleaning establishments, plant nurseries, or pizza parlors. Instead, most organizations start out in one form and evolve into other forms as they grow, shrink, or otherwise change.[3] One such organization which had to change as indicated is EDS, Electronic Data Systems. This computerized data processing systems company was started by H. Ross Perot in Dallas, but its environment changed substantially when it was purchased by its largest customer, General Motors.[4]

Consider the case of Angela Price, who started breeding plants at home in her spare time just because she enjoyed this hobby that developed from a science fair project. Developing new varieties for changing conditions is a challenge for her "green thumb." When she started this "hobby business" 15 years ago, she prepared various plants and checked on their development in the morning hours, then opened her home-based business, Angel Plants, to area customers in the afternoons. On Saturday she updated her records, and on Sunday afternoons she worked on new developments. In the meantime, she also tried to find time to expand her plant selection and to tend her tiny homemade greenhouse. Sometime in between activities, Ms. Price had to consider and act upon decisions regarding product line expansion (perhaps offer some plants in decorative pots at higher prices?) and promotion (maybe print "tip" sheets for plant care or write a monthly article for a local or regional newspaper or magazine?), rather than relying only on word-of-mouth advertising.

Eventually, one of her new varieties was a real breakthrough and Ms. Price had trouble performing all of these activities. She hired a part-time salesperson and a part-time bookkeeper, which allowed her more time for production and experimentation. As sales continued to expand, she soon added a shop (Angel Plants) and a small greenhouse in her back yard. The salesperson and the bookkeeper became full-time employees, and Ms. Price also hired an assistant. In a few more years she may be in charge of two sales managers, each of whom oversees eight sales representatives. She may also have larger facilities and several dozen production workers in addition to a production manager, a full-time accountant, and a personnel manager. At that point the scope of Angel Plants' operation will have expanded from a true one-person effort to a full-fledged organization.

Each of the steps the plant breeder takes along the way—such as creating new jobs, grouping those jobs under new management positions, and delegating authority to those managers—is a part of the organizing process. Moreover, Ms. Price will never truly finish the organizing process. New circumstances, opportunities, and threats will always cause her to modify and adjust the organization of Angel Plants to meet competition more efficiently and more effectively.[5]

## Key Organizing Components and Concepts

Three basic concepts of organizing were noted in the above example—designing jobs, grouping jobs, and delegating authority. Our discussion here will include those three and two others—establishing group effectiveness and managing line and staff positions.[6]

**Designing jobs.**    As we will see, the process of designing jobs involves determining the best level of job specialization to use and grouping jobs into meaningful categories. This grouping, called *departmentalization*, is necessary to facilitate supervision and coordination.

**Defining authority and responsibility.** Between a manager and his or her subordinates, the process of defining authority and responsibility is accomplished through delegation. Across the entire organization, the process is called *decentralization*.

**Organizing for group effectiveness.**    Managers may coordinate the efforts of a few or many others in the organization. Characteristics of the manager, those reporting to him or her, and the situation determine what leads to group effectiveness.

**Managing line and staff positions.**    Line positions are usually thought of as positions in the direct chain of command that hold the responsibility for accomplishing organizational goals (Figure 12.2). Staff positions, in contrast, are generally thought to be advisory positions, primarily facilitating the work of line managers.[7]

© Shock/www.Shutterstock.com

**FIGURE 12.2**  Technical support employees may be considered staff personnel as they solve problems and give help and support to all divisions of the company.

## Designing Jobs

**Job design** is the process of determining the procedures and operations the employee in each position will perform. For example, a new employee at a Heinz plant or even a local travel agency does not simply sit down and start working. Instead the manager shows the new employee how to do the job. Moreover, this manager is not making up the job as he or she goes along. Other operating managers and managers from the human resource department at some time carefully decided how a job should be performed, based on the best design for the job. The basis for all job-design activities is **job specialization**. Thus, we will next investigate the nature of job specialization and identify several alternative approaches to designing jobs.

### Job Specialization

Job specialization was considered earlier in Chapter 2, during our discussion of scientific management. As you recall, Frederick W. Taylor, the chief proponent of scientific management, advocated extremely high levels of specialization and standardization as ways to increase the efficiency of organizations.[8]

The nature of job specialization at an abstract level is straightforward (Figure 12.3). Let's look at Agrium Inc. as an example. Agrium essentially has one "job": producing fertilizer. To execute this job, however, it must break this job down into smaller parts. Some of these smaller jobs include buying raw materials, transporting the raw materials to where they are needed, transforming the materials into fertilizer, and preparing it for distribution to dealers. And, of course, there are dozens more jobs along the way. So the "total job" of the organization gets broken down into many smaller "jobs" to accomplish the one big one. Further, most of these jobs require several people to perform them. As can be seen, specialization can involve managerial and professional jobs as well as those of non-managerial personnel.[9]

The smaller jobs that are created through specialization add back to the total. That is, if it were possible to add up the contributions of each of the individual jobs, the total would equal the original overall job of the organization, including managerial coordination.

Specialization has a number of advantages and disadvantages. On the plus side, it allows each employee to become an expert. If the job is simple and straightforward, people trained to do it should become very proficient. Specialization also allows managers to exercise greater control over workers, as they know

© Picsfive/www.Shutterstock.com

**FIGURE 12.3** This butcher specializes in one task in a meat-processing factory.

what each person should be doing at any given time and also can easily observe and monitor employees doing simple jobs. Specialization is also presumed to facilitate the development of equipment and tools that can increase the efficiency of the job holder.[10]

On the other hand, it is possible to overspecialize. If the job is simplified too far, employees may spend so much time passing the work from person to person that efficiency is actually decreased. Even more significant, if jobs are too simple and specialized, workers quickly get bored performing them. They become dissatisfied, and their performances may drop or they may consider looking for more exciting work elsewhere.[11]

To counter these problems, managers have begun to search for alternatives that will maintain the positive benefits that specialization can provide.

## Alternatives to Specialization

The three most common alternatives to job specialization are shown in Figure 12.4, which uses a single square to represent job specialization at the top. Such jobs can be thought of as narrowly defined and standardized. The three alternatives to job specialization are job rotation, job enlargement and job enrichment.

**Job Rotation.**  **Job rotation** involves systematically moving employees from one job to another. As noted in the exhibit, the jobs themselves are still narrowly defined and standardized. For instance, suppose an employee works in a sausage factory that has four sets of jobs it rotates among four employees. During the first week of the month, Employee #1 operates a machine that grinds the meat and dumps it into large tubs. The next week, Employee #2 operates that machine while Employee #1 moves to the machine that forms the ground meat into round patties. For the third week, Employee #1 moves to the machine that seals the sausage patties inside airtight, see-through wrapping. The fourth week of the rotation is spent on the machine that packs and seals the sausages in the box that consumers buy from the store. In the fifth week the Employee #1 rotates back to the first job again.

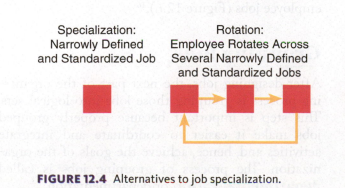

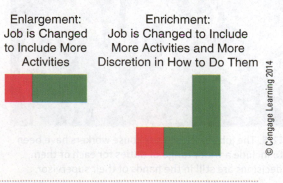

© Cengage Learning 2014

**FIGURE 12.4** Alternatives to job specialization.

Eli Lilly, Nokia, Prudential Insurance, and Bethlehem Steel have all experimented with job rotation, and many companies use this system in one form or another. Unfortunately, however, job rotation by itself is not entirely successful at decreasing the boredom associated with highly specialized jobs—each of the jobs is still fairly monotonous. In general, job rotation is used as a way to train employees in a variety of skills and/or as a part of a more comprehensive job design strategy.[12]

**Job Enlargement.**   In contrast to job rotation, **job enlargement** actually changes the nature of the job itself. As shown in Figure 12.4, this system involves adding more activities to the job. Job enlargement might be used in the sausage factory to decrease the total number of jobs from four to two. Two workers might now operate the machines that grind the meat and form it into patties, and the other two workers could then take responsibility for the two steps involved in wrapping and boxing the patties for distribution.

IBM, Maytag, AT&T, and Chrysler have all tried job enlargement. In general, the results have been somewhat more positive than those of job rotation, and workers report slightly less boredom with enlarged jobs. Still, adding more and more simple activities to a job that is already simple does not really change the nature of the work that much (Figure 12.5).[13]

**Job Enrichment.**   Another alternative to job specialization is **job enrichment**. Figure 12.4 points out the critical difference between job enlargement and job enrichment: Under job enrichment, workers are given not only more activities to perform but also more discretion as to how to perform them.[14]

**FIGURE 12.5** The jobs of these warehouse workers have been enlarged to include a wider range of duties for each of them, although decisions are still in the hands of their supervisor.

**FIGURE 12.6** This office worker's job has been enriched by assigning her a variety of duties, including computer entry, researching information, helping her supervisor prepare reports, and allowing her to schedule her own time.

Suppose, for example, that workers at a Kwik-brand toaster plant are told that they have to insert the heating element first and then attach the wires in a specified order using standard equipment. After enrichment they are given various options: They can attach the wires before inserting the element, use a different kind of tool to attach the wires, and so on. Similarly, the inspection operation may dictate that defective toasters be turned over to a supervisor, who then decides what to do. Enrichment can give the inspector more discretion: She might repair small problems herself, take the defective toaster back to the worker responsible for the defect, or give it to a supervisor. Enrichment combined with cross-training (training people in multiple skills so they can do more than one job) is frequently called re-engineering.[15]

Texas Instruments, General Foods, Texaco, and Volvo have all used job enrichment. Whereas this approach is far from being universally successful, it does frequently decrease employee boredom and dissatisfaction. Many organizations are continuing to experiment with new and innovative ways to design employee jobs (Figure 12.6).[16]

## Grouping Jobs

After designing jobs, the next part of the organizing process is grouping those jobs into logical sets. This step is important because properly grouped jobs make it easier to coordinate and integrate activities and, hence, achieve the goals of the organization. The process of grouping jobs is called **departmentalization** or departmentation.

The key word here is, of course, "logical." Managers do not just randomly pull together whatever jobs are at hand and call them a department. Instead, they must use a plan or a set of guidelines. These guidelines are the basis for departmentalization. The most common groupings are by function, by product, and by location. First, let's discuss each of these bases for departmentalization, and then we can briefly note others that are occasionally used.

## Departmentalization by Function

When organizations departmentalize by function, they group together employees who are involved in the same or very similar functions or broad activities.[17] An illustration of how a company that uses **functional departmentalization** is organized appears in Table 12.1. Note that this organization has a marketing department, a finance department, and a production department. Thus, marketing researchers, product managers, advertising managers, sales managers, and sales representatives are all included in the marketing department; and operations managers, distribution managers, plant managers, and quality control managers are all in the production department.

The key advantages of the functional approach are that, since each department is staffed by experts in that particular function or activity, (1) the managers in charge of each function can easily coordinate and control the activities within the department, and (2) areas of responsibility are clearly defined. On the other hand, functional departmentalization also has

certain disadvantages. Decision making tends to be slow. Employees may concentrate so much on their functional specialties that they lose sight of the total organization, and communication between departments can be difficult.[18]

Smaller organizations tend to use the functional approach to departmentalization; but as they grow, they often change bases. Frequently they go on to adopt departmentalization by product, as Summer Farms has done.

## Departmentalization by Product

When organizations use **product or process departmentalization**, they group together all the activities associated with individual products or closely related product groups. A simple example of this approach is also shown in Table 12.1. Product Group A may be a line of packaged foods, such as cereals, instant breakfast mixes, and so forth. All of the financial, marketing, and production activities associated with this line of products are grouped together in one department. Product Group B could consist of several different small retail chains. As with group A, all the marketing, financial, and operations functions for these chains are organized together in Group B. Summer Farms is another example of grouping by products—nonorganic and organic vegetables, corn and soybeans, and hay and timber.

There are several advantages of product/process departmentalization. For one thing, all of the activities associated with unique products are kept

**TABLE 12.1** Departmentalization by Function, Product, and Location

| STRUCTURE | DEPARTMENTS | POSITIONS |
|---|---|---|
| **Function** | Marketing | Marketing researchers, product managers, sales managers, etc. |
| | Finance | Budget analysts, accountants, financial planners, etc. |
| | Production | Operations managers, plant managers, quality control specialists, etc. |
| **Product** | Product Group A | Financial, marketing, and production managers for packaged foods. |
| | Product Group B | Financial, marketing, and production managers for small retail groceries. |
| | Product Group C | Financial, marketing, and production managers for national grocery chains. |
| **Location/Geographic** | North American Operations | Production, distribution, financial, marketing, and human resource managers for the United States, Canada, and Mexico. |
| | European Operations | Production, distribution, financial, marketing, and human resource managers for Europe. |
| | Southeast Asian Operations | Production, distribution, financial, marketing, and human resource managers for Southeast Asia. |

© Cengage Learning 2014

FIGURE 12.7 Grocery stores, department stores, and multi-product manufacturers are usually organized by product.

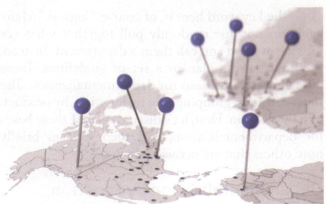

FIGURE 12.8 International businesses often organize by location as indicated by the "push pins" here.

together (Figure 12.7). Marketing cereal might be quite different in nature from marketing meat products. For another, decision making is faster because managers responsible for individual products are closer to those products. Finally, it is easier to monitor the performance of individual product groups under this arrangement.

As you might expect, of course, product departmentalization also has several disadvantages. First, administrative costs are higher because each department has its own marketing research team, its own financial analysis team, and so forth. Second, conflict or resentment occasionally arises between departments as each group thinks the other is getting more than its fair share of attention or resources.[19]

## FOOD FOR THOUGHT 12.2

Process or product departmentalization may be appropriate for whiskey manufacturing. The only whiskey distilled from at least 51 percent corn is bourbon. Rye whiskies must be made from at least 51 percent rye; other whiskies contain a mix of various grains.

## Departmentalization by Location

The third major form of departmentalization is **locational**. In this arrangement, jobs that are in the same or nearby locations are grouped together in a single department (Figure 12.8). Table 12.1 presents a simplified view of a company that uses this approach. All activities of the organization

that pertain to North America—production facilities, distribution systems, financial considerations, marketing activities, and human resource management activities—are grouped together in one department. Likewise, similar activities that relate to European operations are grouped together, as are those that relate to Southeast Asian operations.

The locational approach to departmentalization gives managers the basic advantage of being close to the location of their decision-making responsibilities. Managers in North America may not be fully in tune with cultural and social norms in Europe, nor understand the nature of European marketing, nor appreciate the financial difficulties involved in foreign exchange rates. By putting a manager with the requisite insights and authority directly on the scene, however, the company may achieve more effective operations. The locational or geographic approach may be advantageous for domestic organizations such as a large plant nursery or clothing manufacturer where demand varies by climate. A locational approach would probably not make sense for Summer Farms, even if their fields were spread over the entire state, because at least some degree of product specialization is essential.

On the other hand, this approach results in a duplication of staff, just as product departmentalization does. For instance, the company would need a marketing manager in North America, one in Europe, and one in Southeast Asia, or in the South, Southwest, East Coast, Northwest, etc. Nonetheless, this approach is becoming more widely used, especially by firms that decide to go multinational.[20]

## Other Considerations in Departmentalization

Some companies use other bases of departmentalization, such as departmentalization by *customer*. This approach allows the company to group activities associated with individual customers or customer groups. A Midwestern bank, for example, may have departments for consumer loans, business loans, and agricultural loans. *Time* also serves as a base for departmentalization. For instance, a plant might operate on three shifts, and the company might view each shift as a department. Departmentalization by *sequence* occurs when a sequence of numbers or other identifying characteristics defines the separation of activities. Customers who are picking up orders or buying tickets may be organized into different lines on the basis of their last names or the status of their order (place order here, pick up prepaid orders, reserved or will-call tickets, etc.).

Our discussion of departmentalization considers each base in its pure form, but in reality most organizations use *multiple bases* or combined bases; i.e., they mix bases of departmentalization within the same organization.

One way to mix approaches is by *level*. For example, a firm might use product departmentalization at the top of the organization but departmentalize each product group by function. Each marketing department could then be broken down by location or customer.

Similarly, bases of departmentalization might be mixed at the same level to suit individual circumstances. A firm doing business in Europe on a moderately small scale might decide to establish a marketing group in Paris but keep all production in the United States, at least until sales grow. Thus the marketing activities are broken down by location, but the production activities are not.

## Authority And Responsibility

Another important part of the organizing process is determining how to manage authority and responsibility.[22] At the level of an individual manager and his or her group, this is the delegation process. At the total organizational level, it is related to decentralization.

---

### A FOCUS ON AGRIBUSINESS
### Mixed Departmentalization

In small organizations regardless of the legal form of operation (sole proprietorship, partnership, cooperative, LLC) managers tend to be involved in all aspects of business operations. As the organization grows, functional and/or temporal structuring occurs (accounting, production, and sales; day shift and night shift). And in agribusiness with multiple products, some product segmentation also emerges.

The Texas Produce Association, for instance, has committees on citrus, onions, marketing, shipping, transportation, and trade. So products and functions are used to organize the work of this cooperative. Monsanto is structured similarly. It has product units (Seeds & Traits, Vegetable Business, Crop Protection) and functional units (Human Resources, Treasurer, Community Relations, Controller, Finance).

Agribusiness giants Nestlé S.A. and Bunge go one step further. Both of them have product and functional units as well as geographic units. Nestlé's product units include Nestlé Nutrition, Nestlé Waters, and Nestlé Professional. Its functional units cover Human Resources, Operations, Finance & Control, and Corporate Communication. Nestlé's geographic units are Zone Eur (Europe), Zone AOA (Asia/Oceania/Africa), and Zone AMS (Americas). Bunge has geographic units for Asia, North America, Brazil, Argentina, and Europe with two basic "product" units, Bunge Limited and Bunge Product Lines. Bunge's functional units are located within the product and geographic units.[21]

## Delegation

**Delegation** is the process through which a manager assigns a portion of his tasks to those reporting to him.[23] In discussing delegation, let's first identify the steps in the process and then address barriers to effective delegation.

**Steps in Delegation.** Delegation essentially involves the three steps shown in Figure 12.9. First, the manager must assign **responsibility**. For example, when a manager tells a subordinate to mow a field, prepare a sales projection, order additional raw materials, or hire a new assistant, he or she is assigning responsibility

Second, the manager must grant the necessary **authority** to the subordinate to carry out the task. Preparing a sales projection may call for the acquisition of sensitive sales reports, ordering raw materials may require negotiations on price and delivery dates, and hiring a new assistant may mean submitting a hiring notice to the human resource department. If these activities are not formal parts of a group member's job, the manager must give the group member the authority to complete the delegated assignment.

Finally, the manager must create **accountability**. This suggests that the group member incurs an obligation to carry out the job. If the field is not mowed, the sales report is never prepared, the raw materials are not ordered, or the assistant is never hired, the group member is accountable to his boss for failing to perform the task. An extreme focus on such accountability can be dysfunctional however. If the manager is not careful, some personnel will focus on satisfying short-term accountability standards and, as a result, may lose sight of his or her major, longer-term responsibilities.

Of course, these steps are not carried out in rigid, one-two-three fashion. Indeed, in most cases they are implied by past work behavior. When the manager assigns a project to a group member, for instance, the group member probably knows without asking that he has the authority necessary to do the job and that he is accountable for seeing that it does, indeed, get done.

**Barriers to Delegation.** Unfortunately, the ideal delegation process never materializes. Several factors can contribute to this failure. The manager may be too disorganized or too unsure to delegate systematically (Figure 12.10) or too insecure or afraid that the group member will do such a good job that he will look bad in comparison. Alternatively, the manager may be afraid that the group member is incapable of doing the job properly. Finally, the group member may be unwilling or unable to accept the job.[24]

## Decentralization

A closely related issue is **decentralization**, the result of maximum delegation throughout the organization. Johnson & Johnson, McDonald's, and Caterpillar are decentralized.[25] On the other hand, Sony centralized procurement in 2009 to counter losses, and Rubbermaid centralized its European operations in 2010.[26] IBM has recently moved from being centralized to becoming much more decentralized in an effort to speed decision making and respond more quickly to its customers.[27] Even Japanese companies, which have traditionally been centralized, have begun to become more decentralized.[28] Although he probably never thought about it in those terms, Joshua Summer began to decentralize when he turned over the fields bit-by-bit to his children.

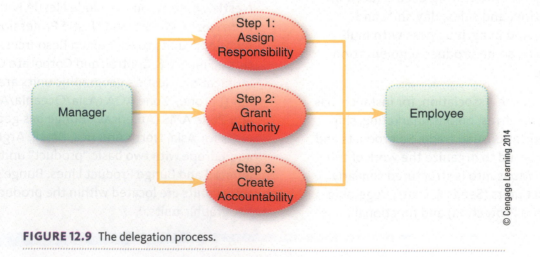

**FIGURE 12.9** The delegation process.

© iStockphoto/Otmar Winterleitner

**FIGURE 12.10** Managers can easily become frustrated if they do not learn to delegate.

Under conditions of decentralization, power and control are systematically delegated to lower levels in the organization. Under conditions of **centralization**, however, power and control are systematically kept at the top of the organization.

Some organizations choose to practice decentralization to keep managers who are close to problems responsible for making decisions about them. That is, managers who come into contact with customers, suppliers, and competitors on a daily basis may be in a better position to make decisions than managers who are isolated back at headquarters.

In general, decentralization is pursued when the environment is complex and uncertain, when lower-level managers are talented and want more say in decision making, and when the decisions are relatively minor. In contrast, centralization is often practiced when the environment is more stable, when the home office wants to maintain control, when lower-level managers are either not talented enough or do not want a stronger voice in decision making, and when decisions are more significant.[29]

Part of the reason behind decentralization is to speed up decision making, while another part is

---

**FOOD FOR THOUGHT 12.3**

Land O'Lakes does business in all 50 states plus 50 countries and handles 12 billion pounds of milk annually. Kraft Foods owns a variety of products, including Kraft cheese, Oscar Mayer lunch meats, Planters nuts, and Cadbury Confectioner.

---

to assure that the individual companies can respond quickly to customer wants and needs. These are precisely the kinds of reasons why General Motors reorganized in the early 1990s and why Johnson & Johnson and Caterpillar more recently decentralized.

## Group Effectiveness

Another important concept in organizing is group effectiveness. Group effectiveness traditionally was thought to be partly a function of the **span of management**, which is the number of people who directly report to a given manager. Companies must consider the relation of this number to group effectiveness and to overall organizational effectiveness. Let's look at the notions of the span, its effect on the "height" of the organization, and other factors that influence group effectiveness even more strongly than the span.

### Group Size and Span of Management

A manager who is responsible for a large group is said to have a wide span of management. Similarly, having a relatively small group defines a narrow span of management.

What difference does it make? At the request of an early management pioneer, a French mathematician, A. V. Graicunas, attempted to illustrate the impact of wide and narrow spans.[30] Graicunas noted that groups contain three basic kinds of relationships: *direct* (the manager's one-on-one relationship with each group member), *cross* (relationships among group members), and *group* (relationships between clusters of group members). Using a mathematical formula for these relationships, Graicunas showed that, as the number in a group increases linearly, the number of total relationships increases exponentially. That is, the number of interactions or relationships increases rapidly as more people are added to the group. This means that, if a manager has two people reporting to him, there are six relationships; for three people the number of relationships increases to 18; and five people in the group create 100 relationships. Adding one person to an existing group of five is quite different from adding one to an existing group of 17.

Although Graicunas's work is not based on business organizations, it does carry an important message: that groups involve a complex network of interrelationships, and the complexity of that network increases greatly as the group grows larger.

Other early writers sought to identify the ideal span for all managers, but they quickly realized that an ideal number did not exist. Indeed, the appropriate span varies considerably from one setting to another and may be much larger than these early writers speculated.[31] This fact relates both to the effects of the span of management on organizations and to the factors that influence a particular group's effectiveness.[32]

## Organizational Levels: Tall vs. Flat Organization

One key effect that span of management has on organizational structure concerns the "height" of the organization. A wide span of management results in an organization that has relatively few levels of management, or a **flat organization**. Smaller organizations may tend to be flatter than larger ones and more centralized. On the other hand, a narrow span of management adds more layers of management and therefore leads to a **tall organization**. Figure 12.11 illustrates these relationships.

Tall organizations tend to be more expensive because of the greater number of managers needed. They also may have more communication problems because of the larger number of people through which some communications must pass. In general, flat organizations tend to be characterized by greater communication between upper- and lower-level management, an increased capacity to respond to the environment, and lower total managerial costs than tall organizations. Many corporations—including Whole Foods, Zappos, CBS, Avon, and General Motors—have taken steps to eliminate layers of

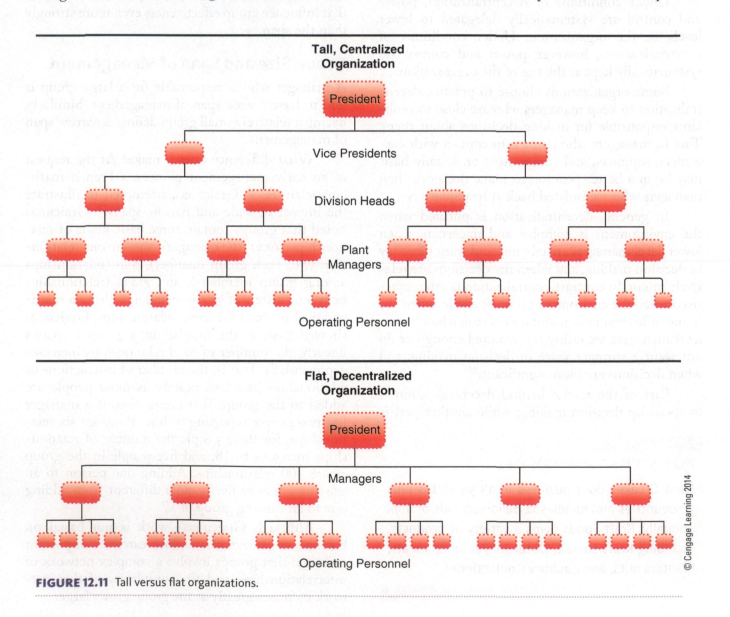

**FIGURE 12.11** Tall versus flat organizations.

© Cengage Learning 2014

management in an effort to create a more stream-lined and efficient organization.[33] Shortly before it became part of Kraft Foods, Cadbury PLC eliminated a whole layer between the CEO and the heads of its operating units.[34] Reducing the number of levels in an organization is one example of downsizing.[35] Improved organizational communication then enables substantially larger spans and correspondingly flatter organizations, thus making both the groups and the overall organization more effective.[36]

## Factors Influencing Group Effectiveness

Since a group's effectiveness and the impact of the span of management depend on the circumstances of the situation, what exactly are those circumstances? Figure 12.12 notes several of the more important ones.

The *competence* of both the manager and his group is one significant factor. If both are competent, a wide span is possible; but if either is less competent, a narrower span may be necessary. *Physical dispersion* is another important variable. In general, if the manager and the group are scattered throughout a building or territory, a narrow span is indicated. If, on the other hand, everyone works in close proximity, a wider span can be used. Small family businesses,

then, would generally be flat organizations, in large part because communication or sharing of information is less complicated.

*Preference* is also an important factor in some situations. If everyone wants a wider span, then that is what should probably be used. A narrow span may be desirable, though, if the manager prefers that. *Task similarity* is significant, too. The manager will probably want a narrower span if each group member is doing a different job, whereas a wider span can be adopted when everyone is performing similar tasks. As Summer Farms added products or crops, it became more difficult for Joshua to oversee the entire business, which led him to decentralize by adding a layer of management that consisted of his children. The degree to which the manager has *nonsupervisory work* to do is also an important consideration. If all the manager has to do is supervise, he or she can use a wider span. On the other hand, if the manager has a lot of other tasks to perform, a narrower span may be indicated.

A related factor is *required interaction* between the manager and the members of the group. Since interaction takes time, it follows that the more interacting a manager needs to do, the smaller the group will need to be. Similarly, less required interaction

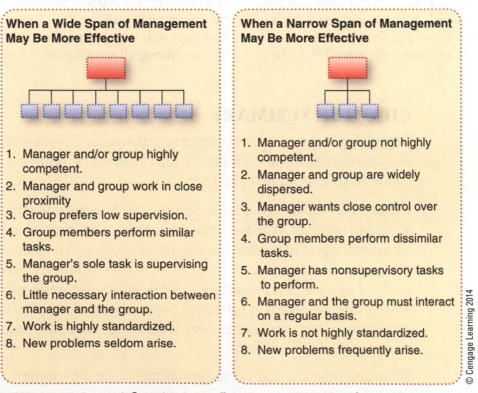

**When a Wide Span of Management May Be More Effective**

1. Manager and/or group highly competent.
2. Manager and group work in close proximity
3. Group prefers low supervision.
4. Group members perform similar tasks.
5. Manager's sole task is supervising the group.
6. Little necessary interaction between manager and the group.
7. Work is highly standardized.
8. New problems seldom arise.

**When a Narrow Span of Management May Be More Effective**

1. Manager and/or group not highly competent.
2. Manager and group are widely dispersed.
3. Manager wants close control over the group.
4. Group members perform dissimilar tasks.
5. Manager has nonsupervisory tasks to perform.
6. Manager and the group must interact on a regular basis.
7. Work is not highly standardized.
8. New problems frequently arise.

© Cengage Learning 2014

**FIGURE 12.12** Factors influencing group effectiveness and the span of management.

allows for a larger group. *Standardization* is another important variable. Highly standardized work, like task similarity, can accommodate a wider span than lack of standardization can. Finally, the *frequency of new problems* should be considered. If new problems arise often, a narrow span may be necessary, whereas few new problems will allow for a wider span.

In summary, the company must consider several factors when organizing a group and establishing the span of management.[37] The manager must properly assess and consider each of these factors, along with others which might be relevant, to achieve effective group performance.[38] Today, the trend is toward larger groups and correspondingly flatter organizations.

## Line and Staff Positions

The last element of the organizing process is the idea of line and staff positions. **Line positions** are traditionally in the direct chain of command with specific responsibility for accomplishing the goals of the organization. **Staff positions** are outside the direct chain of command and are primarily advisory or supportive in nature.

The roles of president and division head are line positions. Each has goals that derive from and contribute to those of the overall organization. The positions of special assistant and legal adviser are staff positions. These people perform specialized functions that are primarily intended to help line managers. For example, the legal adviser is not expected to contribute to corporate profits. Instead, she answers questions from and provides advice to the president about legal issues that confront the firm. Historically, staff managers tended to be better educated, younger, and more ambitious than their line counterparts. They were frequently hired directly out of school, whereas line managers usually worked up the corporate ladder. Moreover, there tended to be considerable conflict between line and staff managers, especially when staff managers were given the responsibility of finding shortcomings in the efforts of line managers.[39]

In recent years, however, this state of affairs has begun to change. The results of one survey show that top managers are redirecting the efforts of their staff managers in more cooperative and constructive directions.[40] Staff positions are also beginning to lose part of their glamour. Recent corporate cutbacks have often reduced the number of staff positions, and line managers have been given greater decision-making power and discretion. These jobs have therefore become more attractive to people graduating from college as well as to those managers already in staff positions.[41]

In the future, the distinction between line and staff managers is likely to become more vague. Effective managers are starting to see that everyone within the company is really on the same team and that the best approach is to promote participation and cooperation among all members of the organization.[42]

## CHAPTER SUMMARY

Organizing is the process of grouping activities and resources in a logical and appropriate fashion. Designing jobs involves determining the degree or extent of specialization in each job.

The basis for job design is job specialization. Specialization has advantages and disadvantages: it allows employees to become experts at their jobs, it allows managers to control workers more easily, and it makes it simpler to design equipment and tools that can increase efficiency. Overspecialization can lead to increased movement of work, which decreases efficiency. Much more important, it can cause boredom and dissatisfaction among employees, which can lead to negative consequences that outweigh the advantages of specialization.

Three common alternatives to specialization are job rotation, job enlargement, and job enrichment. Job rotation involves systematically moving employees from one job to another. Although useful in training, job rotation has not been very effective in reducing dysfunctional aspects of overspecialization. Job enlargement changes the nature of the job by adding activities; some short-term alleviation in boredom may result, but the impact is limited. Job enrichment has had more success. In job enrichment, employees are given more activities to perform and more discretion over how to perform them.

The process of grouping jobs into logical sets to make coordination and integration easier is known as departmentalization. Common bases are function, product, and location. Functional departmentalization refers to grouping together jobs that call for similar work. Product departmentalization refers to grouping together all activities associated with one product (project, process, program, etc.). Departmentalization by location involves recognizing the advantages of physical or geographical proximity as a basis for grouping jobs.

Authority is the power to get things done and responsibility refers to who is supposed to get them done. Delegation involves both of these, as managers assign portions of their tasks to group members. The manager assigns responsibility to do a task, grants the authority necessary to accomplish the task, and creates the obligation for the group member to carry out the job.

Decentralization is the result of delegation. Power and control are systematically delegated to lower levels of the organization. Decentralization enables organizations to respond more rapidly to their environments by keeping in close touch with customers, suppliers, and competitors.

Span of management refers to the number of people who directly report to a given manager. Large groups or wide spans result in fewer levels of management, or flat organizations, whereas narrow spans result in more levels, or tall organizations. Flat organizations tend to have better communication, an increased capacity to respond to the environment, and lower total managerial costs.

Eight factors influence the relation of span to group effectiveness. They are the competence of the manager and workers, physical dispersion, individual preferences, the degree of task similarity, the amount of nonsupervisory work done by the manager, the extent of necessary interaction among group members, the degree of standardization of the work, and the frequency with which new problems occur.

Traditionally, line positions carry specific responsibilities for accomplishing the goals of the organization. Staff positions are primarily advisory or supportive in nature. As line managers have become better educated, they have taken on certain staff functions; and as staff managers have become more experienced, they have acquired certain line functions. The distinctions are becoming less clear. Effective managers realize that everyone in the organization is part of the same team and that all are important to effectiveness.

## CHAPTER ACTIVITIES

### REVIEW QUESTIONS

1. What is job specialization? What are its benefits and limitations?
2. What is meant by departmentalization? What are the common bases for departmentalization?
3. Identify and describe the major parts of the delegation process.
4. What are centralization and decentralization?
5. What are the major differences between tall and flat organizations? Why are these differences important?
6. What are the major factors influencing group effectiveness?

### ANALYSIS QUESTIONS

1. Specialization has dominated organization design for centuries. Do you think it is likely to continue, or will companies move toward less specialization in the future? Why or why not?
2. How can organizations deal with the boredom and dissatisfaction created by specialization? Suggest some ways other than those mentioned in the chapter.
3. Comment on this statement: "You can delegate authority, but you can't get rid of your obligation."
4. Do companies that are heavily centralized have managerial processes similar to or different from those of companies that are heavily decentralized? Why?
5. Which kind of position is more important to an organization, line or staff? Why? Could an organization function with only one of these kinds of positions? Why or why not?

### FILL IN THE BLANKS

1. Narrowly defining and standardizing the tasks that set one job apart from others is called job _____ .
2. Alternating employees across several narrowly defined and standardized jobs to prevent them from becoming bored or dissatisfied is known as job _____ .
3. Modifying a job by adding more activities, hopefully making the job less boring without losing efficiency, is called job _____ .
4. Modifying a job to include more activities and giving the worker more discretion in how to perform the activities is called job _____ .
5. Grouping jobs according to similar functions or broad activities, such as marketing, production, etc. is called _____ departmentalization.
6. Assigning the manager's tasks to group members and giving the employee the responsibility for a job, the authority to perform it, and accountability for seeing that it gets done, is known as _____ .
7. Delegating power and control to lower levels rather than keeping it at the top of the organization is called _____ .
8. The number of people reporting to a manager is called _____ _____ _____ .
9. An organization that has relatively few levels of management so each manager has a wide span of management, is a _____ organization.
10. An organization that has many levels of management and each manager has only a narrow span, is a _____ organization.

Ever since Goldman Sachs global economic analyst Jim O'Neill introduced the acronym in a 2001 publication, "BRIC" has become a nickname for four fast-emerging economic powers: Brazil, Russia, India, and China. Few are surprised by the presence of India and China, some question Russia's presence in the list, but many others are surprised by Brazil's presence. They should not be surprised.

Approaching Brazil from the air, one can see what appears to be a magical desert of sand dunes that resemble bed sheets shaken by the wind. It is Lencois Maranhenses, "bed sheets of Maranhao" (a Brazilian state), its dunes ever shifting with the winds. So it appears, but this Brazilian desert is not a desert—it only looks like one. Actually, the area receives 40 inches of rain a year. This is four times the amount of rain that a desert receives. (The definition for a desert is an area of land that receives ten inches or less of rain per year.) Brazil says, "Enjoy the desert; do not worry about definitions."

Just as "the desert that is not a desert" surprises people, so does Brazil's emergence as an economic power. However, it is a fitting response because Brazil is truly unique and it is a country that does things, like free-market capitalism, in its own style—the Brazilian way. It begins in Rio, its former capital city, and its juxtaposition of wealth and poverty so visibly on display from its legendary beaches. Many countries would attempt to hide this disparity. Brazil hides nothing. In the case of Rio, it says "Enjoy the carnival; life is short." Brazil is emerging as an agribusiness giant, a key player in the international business arena. It was not always this way but changed with the presidency of Fernando Henrique Cardoso (1995–2003). He introduced a new currency, the real, and allowed it to float against world currencies. His economic reform was not readily apparent while he held office; but his successor, Luiz Inacio Lula de Silva (2003–2010) was smart enough to recognize a good thing and keep the reforms in place. The result: nothing short of an economic miracle. All of this happened on the watch of a "reluctant free-marketer, socialist, labor leader president." In Brazil even socialists are not what they appear to be.

Brazil has more than doubled its sugar cane crop in 20 years. This makes the country a leader in sugar and ethanol production. Brazil not only exports sugar and ethanol, it practices what it espouses. By 2017, 90 percent of the cars in Brazil will have flex fuel engines capable of using gasoline and ethanol. But unlike the United States with its 85-15 gasoline/ethanol mixture, Brazil uses a mixture of 20 percent gasoline and 80 percent ethanol. Cosan, Brazil's biggest sugar and ethanol producer, uses a co-generation scheme in ethanol production, allowing Cosan to generate and sell electricity to the Brazilian electricity grid by burning the waste product from sugar/ethanol production.

JBS, a Brazilian meat packer that began as a small meat-packing operation in 1953 is now the world's largest meat packer as a result of purchasing Swift (USA) and other companies around the world. JBS is a reflection of Brazil. The largest cattle herd outside of India is in Brazil. Brazil increased its beef exports ten times in ten years. It has displaced Australia as the world's largest meat exporter.

Then there is the miracle of the Cerrado. In northeastern Brazil there is a huge swath of land that is dry bush, not rain forest land. The Cerrado is Brazil's savannah and is ripe for development. Brazil has more spare farmland than any other country, around 400 million hectares (hectare = 2.5 acres), in addition to the 50 million hectares currently being farmed. Brazilian spare farmland equals the amount of spare farmland in the United States and Russia combined.

Agribusiness requires water resources, land access, and land availability, all of which Brazil has. Brazil has as much in water resources for its population (190 million) and agriculture as all of Asia has for its population of four billion people. This is not all the Amazon River basin. Piaui, a dry region but not a desert in Brazil, has 30 percent more water than the entire U.S. Corn Belt.

This development is not by chance but by design. Embrapa—Empresa Brasileira de Presquisa Agropecuaria (Brazilian Agricultural Research Corporation)—is responsible for this transformation. It is a public company established in 1973 to do what the Extension Service of the Department of Agriculture does in the United States. Embrapa took land that Nobel Prize winner Norman Borlaug once told *The New York Times*, "Nobody thought these soils were ever going to be productive," and made it productive. Embrapa's task is to make Brazil's agriculture and agribusinesses second to none.

Brazil is the first country to catch up to the "big five" traditional grain exporters—the United States, Canada, Australia, Argentina, and the European Union. It is the first tropical food giant; the "big five" are temperate zone producers. While 30 years ago Brazil was a net food importer, it now appears more than prepared to take its place as a global agribusiness heavyweight. And considering the large oil discovery found off the coast of Brazil in 2010, the country may emerge as an oil power as well.

Nothing is as it appears in Brazil. It is a "BRIC" that knows how to party. In Brazil's carnival spirit … the beat continues. Brazil has arrived … or so it appears.[43]

## ▶ Case Study Questions

1. What forms of organization might be used by JBS or other agribusiness firms in Brazil?

2. What are the advantages and disadvantages of using a public company to promote agribusiness?

3. What suggestions might you make to top managers of agribusinesses in Brazil about organizational designs? Why?

# REFERENCES

1. Sewell Chan, "U.S. and Brazil Reach Agreement on Cotton Dispute," *The New York Times,* April 7, 2010, B2; Stephen Castle, "Pact Ends Long Trade Fight Over Bananas," *The New York Times*, December 16, 2009, B3; "The WTO Agreement on the Application of Sanitary and Phytosanitary Measures," World Trade Organization (implemented January 1, 1995) at www.wto.org (accessed September 29, 2010); Anne-Lise Chaber, Sophie Allebone-Webb, Yves Lignereux, Andrew Cunningham, and J. Marcus Rowcliffe, "The Scale of Illegal Meat Importation from Africa to Europe via Paris," *Conservation Letters*, June 7, 2010, 1–7; Mira Oberman, "U.S. Cracks Down on Chinese Honey Smuggling Ring," AFP (*Agorce France-Presse*), September 1, 2010, at news.yahoo.com (accessed September 19, 2010).

2. David Lei and John Slocum, "Organization Designs to Review Competitive Advantage," *Organizational Dynamics* (2002): 1–18.

3. Robert H. Miles, *Macro-Organizational Behavior* (Santa Monica, CA: Goodyear, 1980).

4. Joe Mantone, "A Look Back at EDS: From Ross Perot to Cowboys Herding Cats," *The Wall Street Journal* (May 12, 2008) at blogs.wsj.com (accessed July 24, 2010).

5. Henry Mintzberg, *The Structuring of Organizations* (Englewood Cliffs, NJ: Prentice-Hall, 1979).

6. For a more complete listing of the components of organizing, see Gareth Jones, *Organization Theory, Design, and Change, 6th ed.* (Upper Saddle River, NJ: Pearson, 2010); Robert C. Ford, Barry R. Armandi, and Cherrill P. Heaton, *Organization Theory: An Integrative Approach* (New York: Harper & Row, 1988).

7. For an alternative view of organization design, see Michael W. Stebbins, "Organization Design: Beyond the Mafia Model," *Organizational Dynamics* (Winter 1989): 18–30.

8. Frederick W. Taylor, *Principles of Scientific Management* (New York: Harper and Brothers, 1911).

9. A. S. Miner, "Idiosyncratic Jobs in Formal Organizations," *Administrative Science Quarterly* (September 1987): 327–351.

10. For a classic treatment of the advantages of job specialization, see Adam Smith, *Wealth of Nations* (New York: Modern Library, 1937, originally published 1776).

11. Ricky Griffin, *Task Design—An Integrative Approach* (Glenview, IL: Scott, Foresman, 1982).

12. Geoff Colvin, "How Top Companies Breed Stars," *Fortune* (September 20, 2007) at money.cnn.com (accessed July 25, 2010); and Griffin, *Task Design.*

13. Ricky W. Griffin and Gary McMahan, "Motivation through Job Design," in *Organizational Behavior: The State of the Science*, ed. Jerald Greenberg (Hinsdale, NJ: Lawrence Erlbaum Associates, 1994), 23–44; and Griffin, *Task Design.*

14. Frederick Herzberg, *Work and the Nature of Man* (Cleveland World Press, 1966); and Robert Ford, "Job Enrichment Lessons from AT&T," *Harvard Business Review* (January–February 1973): 96–106.

15. A. Ehrbar, "'Re-Engineering' Gives Firms New Efficiency, Workers the Pink Slip," *The Wall Street Journal*, March 1, 1993, A1.

16. Recent analyses of job design issues may be found in Sharon K. Parker, Uta K. Bindl, and Karoline Strauss, "Making Things Happen: A Model of Proactive Motivation," *Journal of Management* (July 2010): 827–856; and Ricky W. Griffin, "A Long-Term Investigation of the Effects of Work Redesign on Employee Perceptions, Attitudes, and Behaviors," *Academy of Management Journal* (June 1991): 425–435; see also Ricky W. Griffin and David D. Van Fleet, "Task Characteristics, Performance, and Satisfaction," *International Journal of Management* (September 1986): 89–96.

17. Daniel Twomey, Frederick C. Stherr, and Walter S. Hunt, "Configuration of a Functional Department: A Study of Contextual and Structural Variables," *Journal of Organizational Behavior* (Vol. 9, 1988): 61–75.

18. Ford, Armandi, and Heaton, *Organization Theory.*

19. Miles, *Macro-Organizational Behavior.*

20. Richard L. Daft, *Organization Theory and Design, 10th ed.* (Mason, OH: South-Western Cengage Learning, 2008).

21. "Texas Produce Association's Organizational Structure" Texa-Sweet Citrus, Inc. at www.texasproduceassociation.com (accessed July 15, 2010); "Management Team," www.bunge.com (accessed July 15, 2010); "Monsanto Company Leadership," www.monsanto.com (accessed July 14, 2010); "Corporate Governance Report June 2010," www.nestle.com (accessed July 14, 2010).

22. Henry Mintzberg, *Power In and Around Organizations* (Englewood Cliffs, NJ: Prentice-Hall, 1983).

23. Carrie R. Leana, "Predictors and Consequences of Delegation," *Academy of Management Journal* (December 1986): 754–774.

24. Dale McConkey, *No Nonsense Delegation* (New York: AMACOM, 1974).

25. Geoff Colvin and Jessica Shambora, "J&J: Secrets of Success," *Fortune* (April 22, 2009) at money.cnn.com (accessed July 27, 2010); Peter Gumbel, "Big Mac's Local Flavor," *Fortune* (May 2, 2008) at money.cnn.com (accessed July 26, 2010); Alex Taylor, III, "Caterpillar: Big Trucks, Big Sales, Big Attitude," *Fortune* (August 13, 2007) at money.cnn.com (accessed July 26, 2010).

26. The Associated Press, "Newell Rubbermaid to Revamp European Ops," *Bloomberg Business Week* (June 17, 2010) at www.businessweek.com (accessed July 24, 2010); Kenji Hall, "Sony Sees a Second Year of Losses," *Bloomberg Business Week* (May 14, 2009) at www.businessweek.com (accessed July 24, 2010).

27. Michael W. Miller and Paul B. Carroll, "IBM Unveils a Sweeping Restructuring in Bid to Decentralize Decision Making," *The Wall Street Journal*, January 29, 1988, 3.

28. "Maverick Managers," *The Wall Street Journal*, January 29, 1988, 3.

29. Mintzberg, *Power In and Around Organizations*. V. Graicunas, "Relationships in Organizations," *Bulletin of the International Management Institute*, March 7, 1933, 39–42.

30. David D. Van Fleet and Arthur G. Bedeian, "A History of the Span of Management," *Academy of Management Review* (October 1977): 356–372.

31. David D. Van Fleet, "Span of Management Research and Issues," *Academy of Management Journal* (1983): 546–552.

32. For example, see Scott Cendrowsik, "Whole Foods Stock on the Rise," *Fortune* (July 9, 2009) at money.cnn.com (accessed July 23, 2010); Jeffrey M. O'Brien, "Zappos Knows How to Kick It," *Fortune* (January 22, 2009) at money.cnn.com (accessed July 23, 2010); Peter W. Barnes, "CBS Inc. is Said to be Planning More Dismissals," *The Wall Street Journal*, September 26, 1986, 4.

33. "Cadbury Gives its CEO More Control," *The Wall Street Journal*, October 15, 2008, p. B2.

34. Bernard Baumohl, "When Downsizing Becomes 'Dumbsizing'," *Time*, March 15, 1993, 55.

35. Brian Dumaine, "The Bureaucracy Busters," *Fortune*, June 17, 1991, 36–50.

36. For additional information on the span of management, see David D. Van Fleet, "Span of Management Research and Issues," *Academy of Management Journal* (September 1983): 546–552.

37. For a recent discussion of such factors, see Edward E. Lawler, III, "Substitutes for Hierarchy," *Organizational Dynamics* (Summer 1988): 4–15.

38. Vivian Nossiter, "A New Approach Toward Resolving the Line and Staff Dilemma," *Academy of Management Review* (January 1979): 103–106.

39. "Lean But Not Mean," *The Wall Street Journal*, October 28, 1986, 1.

40. Jeff Balley, "Where the Action Is: Executives in Staff Jobs Seek Line Positions," *The Wall Street Journal*, August 12, 1986, 29.

41. Jerry Useem, "What's That Spell? Teamwork," *Fortune*, June 12, 2006, 64–66.

42. Ian Bremmer, *The End of the Free Market* (New York: Porfolio, Penguin Group, 2010), 80; Ronaldo Ribeiro, "A Sea of Dunes," *National Geographic Magazine*, July, 2010, 218(1): 108-117; John Prideaux, "Getting it Together at Last," *The Economist: Special Report on Business and Finance in Brazil*, November 14, 2009, 1-3; "Ethanol's Mid-Life Crisis," *The Economist* (September 4, 2010) at www.economist.com (accessed September 20, 2010); "The Miracle of the Cerrado," *The Economist* (August 28, 2010) at www.economist.com (accessed September 20, 2010).

# CHAPTER 13

# Organization Design

## LEARNING OBJECTIVES

After studying this chapter you, should be able to:

- Discuss the meaning of organization design and the role of organization charts.

- Describe early approaches to organization design, including the bureaucratic design and System 4 design.

- Name and discuss several major contingencies that affect organization design.

- Name and discuss several major contemporary organization design alternatives.

- Define and discuss corporate culture, including its determinants, components, and consequences.

## Finding Our Comfort Zone

Globalization and world trade have opened doors for the gastronomically adventurous. Urbanites, 50 percent of the world population, have the opportunity to sample unique cuisine presentations in boutique restaurants, watch and learn from a variety of accomplished chefs on television, read about restaurant openings, and agree to meet friends to try "a new place." Reality television makes its contribution through programs that appear to "pit" the host against some immodest food dish at a local eatery famous for the immodesty. Of course, if the contest were really man versus food, the food would win—if it were a poison. However, even television can go only so far.

**FIGURE 13.1** More and more agribusinesses are devising foods resembling comfort foods that remind customers of happy times.
© Dmitriy Shironosov/www.Shutterstock.com

When families gather for a significant occasion, the centerpiece of the event is also food—food that reflects the family's origin and its history. It is food that is served when there is a need to "connect" with the past or to "reconnect" with roots. It offers memories, warmth, solace, and most of all, comfort. This is basic fare. This is home cooking.

Interestingly, the preceding paragraphs could be about the same type of food. "Markets that cater to migrants, whether from a different part of the country or from far-flung corners of the globe are not just great for gourmands," the *Economist* recently noted. "They are also a testament to the fact people often retain very strong preferences for the kinds of food they grew up eating." Organizations are developing to take advantage of just those preferences.

Indeed, a study by several economists revealed that a strong preference for "regional specialties" is retained even if it costs more to obtain them. Once again individuals who focus exclusively on price are "irrational" and are not willing to spend less money for more local food if the "home" food can be obtained somewhere at some price. They will do so even if it means more money and less food. In other words, they are willing to pay for "comfort" food. The price doesn't capture the full value of the product's value to consumers.

This effect is not lost on the food industry; its organizations reflect this bias or need. Kraft has designed its organization to include a huge research laboratory north of Chicago, where Kraft's food scientists test new foods and tinker with old recipes. Campbell's soup even has a vice-president of consumer insights,

Charles Vila. ConAgra has an organization unit that tinkers with items like Chef Boyardee Beefaroni, not a gourmet food but a convenient comfort food (Figure 13.1). All of this effort is to design and prepare a food that is novel, traditional, and healthy; and organizational designs are required to support these efforts, especially for Americans who prize taste, convenience, and value more than health and nutrition.

Organizations are being designed to meet this challenge to devise foods that resemble the food people want to eat but render them also healthy and nutritious. Unfortunately, this apparently simple mission does not always work. The product—healthy, nutritious, comfort food—can be produced, but the consumer does not wish to purchase it because it is "not the same" as what they are accustomed to. In a competitive environment like the food industry, a product that does not sell is removed from the shelves.

This is not just an American phenomenon; it is global. In South Korea, for example, the desire is for kimchi. Kimchi is a seasoned, pungent and fermented napa cabbage. Kimchi is Korean comfort food. In 2010 President Lee Myung-bak acknowledged the shortage of napa cabbage and said that he would eat Kimchi made from round cabbage (cole slaw cabbage in United States). The people listened and considered his words. Then they acted—they bid up the price of napa cabbage from $2.19 (2500 won) to $10.09 (11,500 won) rather than go without "real" kimchi. The message was clear: Regardless of what our leader says and does, we want the real thing; and we are willing to pay for it.

What does this mean? Because we are creatures of habit, we are willing to "pay" to eat familiar food, comfort food? Or because people are creatures of habit, will the effects of a given increase in income on people's well being be smaller than if people's tastes adjusted immediately? And, more importantly perhaps, are we willing to pay twice—first in money and later in health—just to stay in our comfort zones?

The final word from the *Economist:* "At least the world's cities have more interesting markets as a result." And they have more different organizations as well.[1]

## INTRODUCTION

Kraft, Campbell's, and ConAgra have organizational units designed just to develop foods for varying taste preferences, including foods that fit our comfort zones. They also have production units, financial units, and other particular organizational arrangements. One of the keys to understanding such large, complex organizations is to understand their **organization designs**—the overall configurations of positions and interrelationships among positions within the organizations.[2]

More specifically, we can think of the design of an organization as being like a puzzle, with a number of pieces that can be put together in certain ways. Some use functional departmentalization and others use product departmentalization; some have wide spans of management and others have narrow spans of management; some are decentralized and others are centralized. Thus, describing exactly how the various pieces of a specific organization are to be put together results in its organization design.[3]

In Chapter 12, we identified and discussed a number of basic organizing concepts. This chapter considers how these concepts integrate into one overall organization design. First, we discuss the nature of organization design and summarize some early approaches to it. Then we describe three important contingency factors that can affect organization design and several contemporary approaches to this design. To conclude, we consider the nature of organizational culture. Each of these subjects helps us better understand giant agribusinesses like Kraft, Campbell's, and ConAgra. Smaller organizations have simpler designs but the basic concepts still apply.

# The Role of Organization Charts

To many people, organization design is best represented by an **organization chart**. Throughout this book we use organization charts to present examples and to illustrate important points. These charts generally are composed of a series of boxes, each of which is connected to others with one or more lines. There are numerous conventions for developing such charts but the simplest is to have each box represent a position within the firm (Figure 13.2). The lines then represent the nature of the relationship between that position and other positions with support positions shown to the side. Several things about a company can be gleaned from an organization chart. The form of departmentalization used by the firm is shown as reporting relationships among the positions in the firm. Figures 13.3 and 13.4 show charts for Summer Farms at different points in time.

The organization chart can be thought of, then, as a picture or map of the organization. Such charts

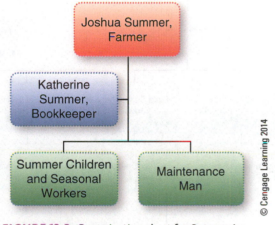

**FIGURE 13.3** Organization chart for Summer's Farm, circa 1975.

© Cengage Learning 2014

are especially useful to outsiders and newcomers to an organization. They help outsiders know whom to contact in the organization to get certain things done. They help newcomers better understand their place in the overall scheme of things, and they clarify reporting relationships between positions.

However, as an organization grows in size, organization charts become more difficult to use because of the large number of positions and the complex relationships that can exist among those positions. For example, in large, complex firms such as DuPont, Monsanto, and Archer Daniels Midland, some managers report to more than one higher-level manager. Further, in the contemporary world, organizations change so frequently that a chart drawn at any one point may not be accurate shortly thereafter. In such cases, organization charts may not be used at all. If they are used, they usually show only the major positions in the organization.[4]

Consider the changes that have taken place at Summer Farms. In its early years, it looked like the one shown in Figure 13.3. However, as it grew

**FIGURE 13.2** Organizing a company is somewhat like putting together a puzzle.

© iStockphoto/Studiovision

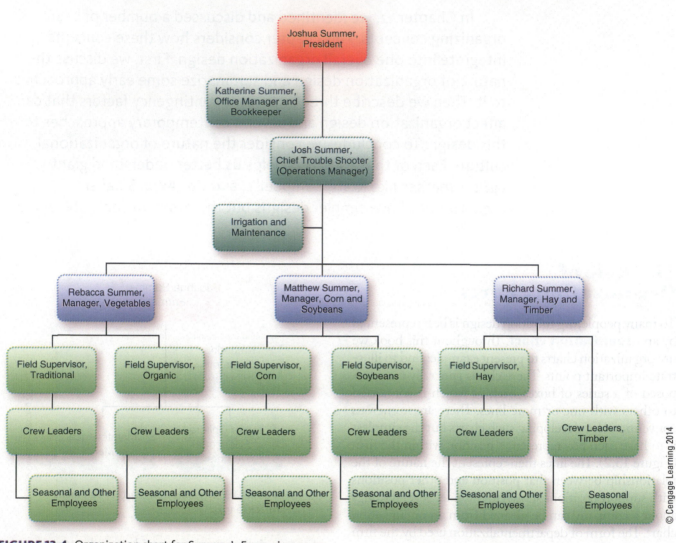

**FIGURE 13.4** Organization chart for Summer's Farm, circa 2000.

© Cengage Learning 2014

(and changed its name), it came to look like the one shown in Figure 13.4. More recently, both Joshua and Katherine have retired from active roles in the organization so the organization chart would have changed yet again. Josh has become the president (although he prefers to keep the title of Chief Trouble Shooter), and he has brought on board a CPA, unrelated to the family, to replace Katherine.

# Early Approaches to Organization Design

Back in Chapter 2 you were introduced to administrative management, the area of classical management theory concerned with organizational structure. The chapter briefly noted the contributions of the German sociologist Max Weber, one of the first people to describe how organizations should be designed to promote effectiveness.

## The Bureaucratic Design

Weber coined the term **bureaucracy** to describe what he saw as the ideal kind of organization design.[5] His goal was to identify and prescribe a set of guidelines that, if followed, would result in efficient and effective organizations. The foundation of his bureaucratic guidelines was the creation of a formal and legitimate system of authority. This system, he argued, would serve to guide rational and efficient organizational activities.

The guidelines Weber developed to create this system are summarized in Box 13.1. First, managers should strive for strict divisions of labor. Each position should be clearly defined and filled by an expert in that particular area to take advantage of the specialization of labor. Second, there should be a consistent set of rules that all employees must follow in performing their jobs. These rules should be impersonal and rigidly enforced so that family connections and political favoritism will be minimized. Third, to minimize confusion there should be a clear chain of command: Everyone should report to one, and only one, direct supervisor. Moreover, communication should always follow this chain and never bypass individuals. Fourth, business should be conducted in an impersonal manner. In particular, managers should maintain an appropriate social distance from the members of their groups and not play favorites. Fifth, advancement within the organization should be based on technical expertise and performance rather than on seniority or favoritism. Finally, careful records should be kept so that organizational learning can take place.[6] Weber expected this to enhance employee loyalty to the organization.

Over the years, however, the term bureaucracy has come to connote red tape and slow, hassle-ridden decision making. Many universities, hospitals, and governmental agencies have a bureaucratic flavor to them. In fact, the bureaucratic approach to organization design may be appropriate when the environment of the organization is stable and simple. However, since few organizations today have such environments, this approach should be used only with great care.[7]

## System 4 Design

As the human relations school of thought emerged, new perspectives on organization design naturally also emerged.[8] One of the more popular views has come to be known as the **System 4 approach**.[9] This view, which holds that the bureaucratic model has numerous drawbacks and deficiencies, advocates an entirely different way of designing organizations. Its basic premises are summarized in Table 13.1.

The proponents of the System 4 design argue that organization design can be described as a continuum. At one end of the continuum is a hierarchical design, largely bureaucratic in nature, called **System 1** in the table. At the other end is System 4, a design that has more openness, flexibility, communication, and participation. In between are other organization designs that show characteristics relatively similar to System 1 (called System 2) or relatively similar to System 4 (called System 3).

The premise is that most organizations start out as hierarchical bureaucracies like System 1 organizations. Theoretically, through a series of prescribed steps a manager can transform the organization first to a System 2, then to a System 3, and ultimately to a System 4 design.

Those who designed this approach helped demonstrate that the bureaucratic model is not the only way in which organizations can be designed, and one early study at General Motors found that a System 4 design was indeed more effective.[10] On the other hand, the System 4 model, like the bureaucratic model before it, was presented as a universal guideline that all managers should follow. As we see in the next section, research has shown that there are no universal guidelines. Instead, the appropriate form of organization design for any given company is contingent on a variety of critical factors.

## Contingency Factors Affecting Organization Design

In Chapter 2 we saw that the contingency approach to management suggests that no single method of management will always be successful. So, too, no

### BOX 13.1
### THE IDEAL BUREAUCRACY

## Weber's Organizational Guidelines

1. The division of labor should be clearly defined.
2. One consistent set of rules should apply to all employees.
3. A clear chain of command and communication should exist.
4. Business should be conducted in an impersonal manner.
5. Advancement should be based solely on expertise and performance.
6. Careful records should be kept for organizational learning.

**TABLE 13.1** System 1 and System 4 Organization Designs

| SYSTEM 1 (EXPLOITIVE) | SYSTEM 4 (PARTICIPATIVE) |
| --- | --- |
| **Leadership** includes a lack of confidence and trust in subordinates. Subordinates do not feel free to discuss their job problems with their superiors, who in turn do not solicit subordinates' ideas. | **Leadership** includes confidence and trust between superiors and subordinates. Subordinates discuss job problems with their superiors, who solicit subordinates' ideas. |
| **Motivational** processes tap only physical, security, and economic motives through the use of fear and sanctions. Unfavorable attitudes prevail among employees. | **Motivational** processes tap a full range of motives through participatory methods. Attitudes are favorable toward the organization and its goals. |
| **Communication** is top down and tends to be distorted, inaccurate, and viewed with suspicion by subordinates. | **Communication** flows freely in all directions. The information is accurate and undistorted. |
| **Interaction** processes are closed and restricted; subordinates have little effect on goals, methods, and activities. | **Interaction** processes are open and extensive; superiors and subordinates affect goals, methods, and activities. |
| **Decisions** occur only at the top of the organization; it is relatively centralized. | **Decisions** occur at all levels through group processes; it is relatively decentralized. |
| **Goal-setting** is at the top of the organization, a process which discourages group participation. | **Goal-setting** involves group participation in setting high, realistic objectives. |
| **Control** is centralized and emphasizes determining blame for mistakes. | **Control** is dispersed and emphasizes self control and problem solving. |
| **Performance goals** are low and passively sought by managers, who make no commitment to developing the human resources. | **Performance goals** are high and actively sought by superiors, who make a commitment to development through training, human resources. |

Source: Adapted from John H. Wilson, "Authority in the 21st Century: Likert's System 5 Theory," *Emerging Leadership Journeys* (Vol. 3, No. 1, 2010), 33–41; Rensis Likert, *The Human Organization* (NY: McGraw-Hill, 1967), 197–211; Rensis Likert, *New Patterns of Management* (NY: McGraw-Hill, 1961); and Marvin R. Weisbord, "For More Productive Workplaces," *Journal of Management Consulting* (Vol. 4, 1988), 7–14.

single organizational design is best. Three major situational elements or contingency factors that have been found to affect the appropriate design for an organization are size and life cycle, technology, and the environment.[11]

## Size and Life Cycle

The size of an organization can be assessed in any number of ways: by number of employees, assets, sales, and so forth. In most cases, however, these characteristics are closely related. Regardless of what criteria are used, Cargill, DuPont, General Mills, Monsanto, and Tenneco are large companies; and a small farm, a local sausage-maker, a neighborhood dry cleaner, and a locally owned pizza parlor are small organizations.

The effects of size on organization design are summarized in Figure 13.5. In particular, research has found that large and small organizations differ from one another in three important ways.[12] First, smaller organizations tend to be *less specialized* than larger organizations. It is not uncommon that every employee in a small company has the ability to do a number of different jobs, but as the

company grows each employee tends to stick to one well-defined job.

Second, smaller organizations tend to be somewhat *less standardized* than large ones. This means that they generally have fewer rules for how things should be done and more flexibility in how employees can confront problems. As an organization grows, however, it has a tendency to create more rules and to eliminate some of the individual flexibility in problem solving. It is interesting to note, however, that some small business owners seem to feel that having lots of rules is a good way to manage, while most large organizations are moving to reduce the number of rules and regulations they have.

Finally, organizations tend to be *more centralized* when they are small. This relates to the fact that the original owner or founder is probably still in charge and is accustomed to having the final say in all decision making. In larger organizations, decision making tends to become more and more decentralized. Indeed, larger companies often find that decentralization can aid in achieving high quality. Managers of growing organizations should recognize that alterations in the design of the company may be necessary as it grows.

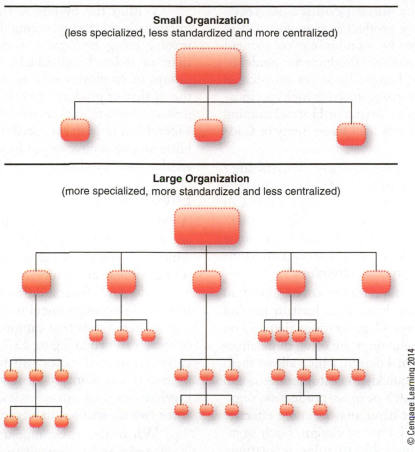

**Small Organization**
(less specialized, less standardized and more centralized)

**Large Organization**
(more specialized, more standardized and less centralized)

© Cengage Learning 2014

**FIGURE 13.5** The effect of size on organization design.

**Life cycle** is also important because it tends to be related to size. Most organizations begin small, grow and mature over a period of years, become stable as mature organizations, and then undergo change. That final stage may be one of decline or regrowth, depending upon the organization's response to its environment. Indeed many organizations are be cyclical—going through periods of decline only to reemerge in strong positions again. The organization design needed in each stage of its life cycle is different from that needed in other stages.

## Technology

**Technology** is the set of conversion processes used by an organization to transform inputs into outputs. Obviously, the job of processing food, preparing it for sale, and distributing it worldwide at Heinz is quite different from the job of displaying produce at a local farmers' market. It follows, then, that Heinz and the local farmers' market will have quite different management needs. One manifestation of those differences is the kind of organization design that is appropriate.

**FOOD FOR THOUGHT 13.2**

Through composting operations at its North American beef and pork plants, Cargill has reduced organic waste by more than 70 percent.

The most common approach to describing technology is to classify it as unit or small-batch, large-batch or mass-production, or continuous-process technology.[13]

**Unit** or **small-batch technology** is used when a product is made in small quantities, usually in response to customer orders. For example, Boeing doesn't keep an inventory of 747s on hand. Instead, it makes them to customer specifications after they have been ordered. John Deere provides only a limited inventory, particularly during some seasons. Tailor shops, printing shops, small farms, many food establishments, and sometimes small bakeries use a similar approach to production.

**Large-batch** or **mass-production technology** occurs when a product is manufactured in assembly-line fashion by combining component parts into a finished product. Products are made for inventory that will be bought by as yet unspecified buyers, instead of according to customer specifications. Examples include a Heinz or Hormel canning facility and bottling plants for Ocean Spray or Cadbury Schweppes.

Finally, **continuous-process technology** means that the composition of a raw materials, including foods, is changed through a series of mechanical or chemical processes (Figure 13.6). A Union Carbide chemical plant and many units within Bunge, Cargill, ConAgra, DuPont, and Monsanto use continuous-process technology.

In general, the appropriate form of organization design a company uses depends at least in part on its dominant technology. Organizations that rely on unit or small-batch technology are often more effective if they use a System 4 design, which allows them the flexibility to react quickly to customer needs and expectations. Large-batch or mass-production organizations, on the other hand, may be more effective if they use a System 1 type of design. Such organizations are more amenable to rules, regulations, and other formal practices. Continuous-process firms like unit or small-batch companies may be most effective if they adopt a System 4 organization design because they need its flexibility to oversee their complete technology as well as to enhance the introduction of automated production processes into the system.[14]

Within the System 4 framework, some new organizational arrangements have emerged. Most involve using employees working as teams rather than as isolated individuals. **Quality circles** are groups of employees who focus on how to improve the quality of products. **Semi-autonomous work groups** include workers who operate with no direct supervision to perform specific tasks, such as assembling an automobile or producing a circuit board for an electronic product.

## Environment

A third factor that can directly affect the appropriate design of an organization is the environment, which was discussed extensively in Part Two. Now we must relate environment more directly to organization design.[15] Several different perspectives on the impact of the environment have been developed, and a synthesized view that captures the essential points of each is shown in Figure 13.7. The basic idea is that **environmental uncertainty** can be captured by two dimensions: **environmental change** and **environmental complexity**.[16] The exhibit places these two dimensions in a graph.

When the environment changes frequently, is dynamic, and is difficult to predict, it has a high rate of change. In contrast, if the environment seldom changes, is fairly static, and is relatively easy to predict, it has a low rate of change. Similarly, if the environment contains many different elements, it can be considered to have a high level of complexity. When the number of elements is low, however, the complexity of the environment is also low.

**FIGURE 13.6** An example of continuous-process technology, these harvested brussels sprouts are moved toward the remaining steps of the process, including packaging and labeling.

© iStockphoto/David Gomez

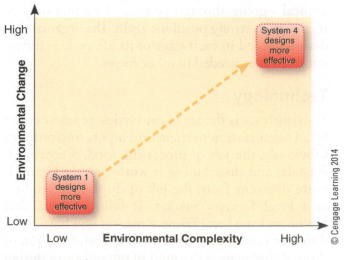

**FIGURE 13.7** The environment and organization design.

© Cengage Learning 2014

When both change and complexity are high, uncertainty is also high. In this instance, a System 4 or organic type of organization design will probably be most effective. However, when both change and complexity are low, so too is uncertainty. For this condition, a System 1 or bureaucratic or mechanistic design may work best.

For example, consider Intel Corporation versus a small liberal arts college. Intel competes in a complex and rapidly changing environment, the electronics industry. Faced with uncertainty, it uses a System 4 design, which allows it to respond quickly and relatively easily to shifts, threats, and opportunities in the environment. In contrast, the college functions in the relatively stable and simple environment of higher education. Confronted with little uncertainty, it uses a System 1 design, akin to a bureaucracy.

It is the manager's responsibility to analyze the three factors of size, technology, and environment effectively and then properly match the design of the organization to them. Now let us explore some of the more common design alternatives that many companies currently adopt.

# Contemporary Organization Design Alternatives

As we have seen repeatedly, there is no one best method for organization design. Managers must therefore carefully consider their circumstances and choose the one design that is most appropriate for those circumstances. Among the options that are being chosen with increasing regularity are the functional design, the conglomerate design, the divisional design, and the matrix design.[17]

## The Functional Design (U-Form)

Traditionally the most common design is the **functional design**, based on functional departmentalization as discussed in Chapter 12. It is also known as the **U-Form** because it uses a unitary or uniform approach to design.[18] This form of organization makes maximum use of functional specialization and therefore achieves the benefits of that specialization. However, it also requires considerable integration and coordination. Every part of the organization is dependent upon the rest; none can survive without the others. The operations unit, for example, depends on the marketing unit to sell the product or service, the finance unit to provide funds for equipment, and

the human resource unit to facilitate staffing. McIlhenny Company, maker of Tabasco sauce, uses the U-form organizational design.

The functional design is popular because it lends itself to centralized coordination. An individual CEO can integrate and coordinate the entire organization with this design at least up to some fairly large size. That is the reason this design is so common among smaller firms. However, as the organization grows in either size or complexity, it becomes increasingly difficult for a single person or even a single group of persons to perform such coordination.

## The Conglomerate Design (H-Form)

The **conglomerate design** is typically found in an organization that has grown through the development of new and perhaps relatively unrelated product lines. It is also called the **H-Form** because such conglomerates may be holding companies for groups of diverse products. This design uses the product/process form of departmentalization and takes advantage of specialization based on knowledge of production and marketing of specific products or services.

In the conglomerate design the units are separate businesses usually headed by general managers. These general managers are responsible for the profits and losses of their units and operate independently of one another. These units report to a central, corporate group. That group evaluates their performances, allocates resources, and guides the decision making of the general managers.

Duplication of costs and efforts, lack of communication and coordination, and the difficulties associated with trying to integrate highly diverse companies are among the limitations to this form of organization. Most companies using this design find that they achieve only average financial performance and are therefore moving away from the H-Form.[19]

## The Divisional Design (M-Form)

A third popular approach to organization design is the **divisional design**, called the **M-Form** for its multi-divisional characteristics. This form, popular with multinational organizations, combines a product approach to departmentalization with a strategic business unit strategy, as discussed in Chapter 9. An example of this kind of design is shown in Figure 13.8.

Each division of the organization is responsible for all aspects of the management of a given product

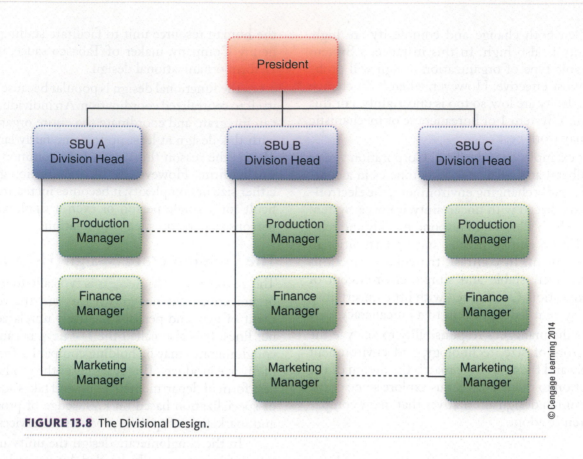

**FIGURE 13.8** The Divisional Design.

© Cengage Learning 2014

or product family. A company such as General Foods, for example, establishes a division for each of its major types of food products. Each division then takes care of its own suppliers and handles its own advertising campaigns. The head of the division is usually given the title of vice president, although titles such as division head and division manager are also common. Summer Farms is also an example of grouping by products—nonorganic and organic vegetables, corn and soybeans, and hay and timber. It makes sense for Summer Farms to organize and staff in this manner, as different crops require different marketing methods and contacts as well as different planting, tending, and harvesting knowledge.

Each division is also thought of as a strategic business unit, or SBU. That is, each has its own market and competitors. Moreover, a division might be a star, a cash cow, a question mark, or a dog (Chapter 9); so it may be sold, used to generate cash for other divisions, given extra cash, or put on a "wait and see" basis.

Note that positions in the different divisions in the exhibit are linked by dashed lines. These lines indicate that cooperation between divisions is encouraged, although usually not required. For instance, two divisions of General Foods may link up to negotiate

a better contract with a supplier, or two divisions of Purina may use the same advertising agency. Each division, however, is free to undertake and terminate such collaborative arrangements as it chooses.

Finally, we should note that even though the divisions are given considerable autonomy, certain functions are probably retained at a centralized level. At General Motors, for instance, all labor negotiations for all divisions are handled at the corporate level.[20] The divisional design generally is the most effective approach to organizational design.[21] However, it can have problems, which is one reason other designs are also common.

## The Matrix Design

Another important contemporary form of organization design, which is also used by international firms, is the **matrix design**.[22] It is created by superimposing a product-based form of departmentalization on an existing functional departmentalization.[23] An example is shown in Figure 13.9.

Note that in the exhibit four functional departments are arrayed across the top of the organization, and each is headed by a vice president. Down the side of the organization chart is a list

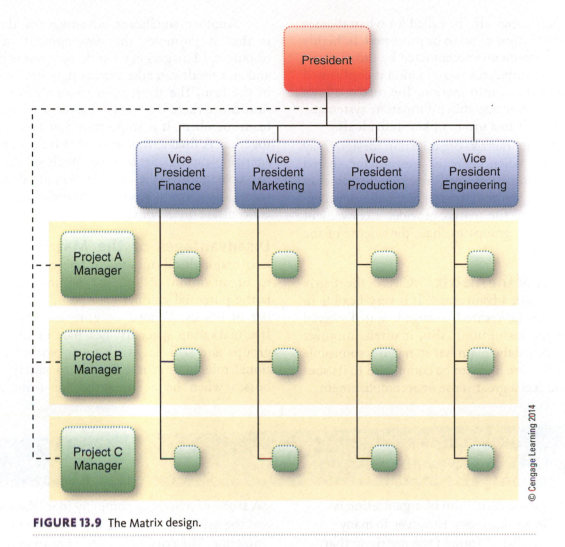

**FIGURE 13.9** The Matrix design.

© Cengage Learning 2014

of three project managers. Each of these project managers is similar to the head of a product-based department (as shown by the dashed lines). In the matrix, each manager heads up a project that cuts across the functional areas (shown by the shaded areas). Thus, the employees within the matrix are a part of two (or more) departments and report to two (or more) bosses at the same time, depending on the activities they are performing.

The rationale for a matrix is really quite simple. The functional departments allow the firm to develop and retain unified and competent functional specialists, and the product-based deparments direct special and focused attention to individual products or product groups. For example, suppose a firm using the matrix design wants to create and produce a new product. Specialists from each of the functional departments are brought together to form a team under the direction of a project manager instead of their usual managers. Thus, the new product gets the specialized attention it needs from each functional area, but each specialist can work on many projects at the same time and still have a functional "home." The matrix allows the firm to use the advantages of both forms simultaneously.

In recent years such well-known companies as Monsanto, Syngenta, and Unilever have adopted the matrix form of organization design. Although the design has disadvantages as well as advantages, there are generally clear indications of when to use a matrix.

**When to Use a Matrix.** In general, a matrix design is most likely to be effective in one of three situations.[24] First, it may be useful when the firm has a diverse set of products and a complex environment. Although the diverse set of products may suggest product departmentalization, the strength provided by the functional approach may be necessary to retain the requisite number of specialists.

The matrix may also be called for when there is a great deal of information to be processed. In highly uncertain environments accentuated by broad product lines, for example, the organization is confronted with a mountain of information. The matrix allows managers to categorize this information systematically and direct it to a group of key individuals.

Finally, the matrix organization design may be appropriate when there is pressure for shared resources. A company may need eight product groups, yet have the resources to hire only four marketing specialists. The matrix provides a convenient way for the eight groups to share the talents of the four specialists.

### Advantages of the Matrix.

Clearly, the matrix design has certain advantages.[25] It is very flexible in that teams can be created, changed, and dissolved without major disruption. Also, it often improves motivation. Since the team has so much responsibility, its members are likely to be committed to its success and will feel a great sense of accomplishment.

Another significant advantage of the matrix is that it promotes the development of human resources. Managers get a wide range of experiences and as a result can take increasingly important roles in the firm. The matrix can also enhance cooperation. Since there is so much interdependence among team members, it is important that they work well together. A final advantage is that it facilitates managerial planning. Because so much of the day-to-day operations of the organization are delegated to teams, top managers have considerably more time to concentrate on planning.

### Disadvantages of the Matrix.

There are also major disadvantages to the matrix approach to organization design. Paramount among these is the potential conflict created by having a number of bosses. Whose assignment takes priority? If a marketing specialist is a part of three project groups and still has work to do within his functional role, he may not be able to satisfy all of his bosses when he is pressured for time. Another

---

## A FOCUS ON AGRIBUSINESS
### Using the Matrix in Agribusiness

Increasingly the matrix form of organization is being used in agribusiness. However, in many instances it is a "soft" form of the matrix so that it is more of a team concept than a rigid multiple reporting structure. As Frank Braeken of Unilever explained, "A modern organization is no longer one where you can dictate decisions. In a matrix organization, the CFO needs to persuade on many levels and directions—senior finance people, the CEO, stakeholders, suppliers."

When Syngenta was in its early development, it needed to establish a new culture, the "Syngenta Way." It set up a matrix structure and created a program that spelled out the new organizational values and provided an overview of the company's history and future ambitions. "Creating a Matrix Mindset was valuable because it's something that we're doing more and more. Both employees and leaders have many people they're doing work for, not just their managers."

A study of a Serbian company identified some of the problems with the matrix form of organization. The company with more than 600 employees specializes in drugs and medical materials for nearly 40 percent of the medical institutions in Serbia. Some managers issued contradictory instructions and employees did not know whose instructions to follow. It was also unclear whether product managers or sales managers were responsible for key decisions. While the problems arose more from the inadequate implementation of the structure than with the structure itself, these sorts of problems have been frequently reported by organizations using the matrix organization.

Even though many firms use the matrix successfully, it is also known to have shortcomings, so caution should be exercised in its adoption.[27]

disadvantage is that coordination is difficult in a matrix. When two or more groups need the same information, each group may pay a market research firm to get the information without realizing that the other group or groups could use it too. A final drawback is the fact that group work tends to take longer than individual work. Each manager in a matrix is likely to spend considerable amounts of time meeting and talking with other managers and putting one set of activities aside to pick up others. Thus, the manager may have less individual time to devote to task accomplishment.[26]

## Cooperatives

Agribusinesses traditionally use the cooperative form of organization. **Cooperatives (co-ops)** have played an important role in the development of agribusiness, but their numbers have been decreasing. Farmer cooperatives went from more than 10,000 in 1950 to 3,346 in 2000 and less than 3,000 in 2009.[28]

There are three distinct types of cooperatives in agribusiness: input (supply/purchasing), output (marketing), and support (service) cooperatives. Input cooperatives, as the name implies, purchase feed, seed, fertilizer, fuel, and other inputs using the co-op size to obtain quantity discounts, thus reducing costs to members. Output cooperatives provide marketing and branding for members to increase sales (Figure 13.10). Support cooperatives provide services to members at reduced rates. Support cooperatives include banking and credit services, irrigation, or other specialized services.

Cooperatives (co-ops) are different from other forms of ownership—limited liability companies, partnerships, and proprietorships. Cooperatives are owned and controlled by their members, and their prime objective is to maximize the benefits they generate for the members. All members share the profits, which are usually distributed as patronage-based refunds at the end of the fiscal year. Pooling resources enables the co-op to secure the benefits of size and hence lowers the costs to the members. Users of a co-op's services are voting members and therefore exercise control of the co-op. However, just because members share profits and exercise control doesn't assure the co-op's success. Farmland Industries was the largest agricultural cooperative in the United States when it went bankrupt in 2002.[29] Agribusiness co-ops can range from small, local ones to large ones like Land O'Lakes, Inc.; Sunkist Growers, Inc.; Plains Cotton Cooperative Association; REI; Riceland Foods, Inc.; NORPAC Foods, Inc.; Swiss Valley Farms; Tennessee Farmers Cooperative; South Dakota Wheat Growers Association; or Sun-Maid Growers of California.

In the agribusiness global economy, farm cooperatives are important suppliers of food. Jack Gherty, president and CEO of Land O'Lakes, one of America's leading farmer-owned cooperatives, indicated that a unique aspect of leading a cooperative is educating its member-owners to understand the risks and rewards of growing a modern cooperatively organized company in the food economy.[30]

Cooperatives have been popular because they have numerous advantages. The owner-members have limited liability, have equal votes, share in the profits/benefits, and there may be tax benefits. And, of course, cooperatives tend to be member friendly at least so long as they are relatively small. However, there are disadvantages as well. As they grow larger, individual control becomes diluted and a small group of members may exert dominance. Cooperatives can also be more difficult and expensive to form and maintain. Thus, as with any form of business, the positives and negatives must be carefully evaluated.

**FIGURE 13.10**  A Midwestern U.S.A. grain co-op.

© iStockphoto/Akit

### FOOD FOR THOUGHT 13.3

Food co-ops have been innovators in the areas of unit pricing, consumer protection, organic and bulk foods, and nutritional labeling.

## Other Designs

Organizations that are becoming truly multinational in character have evolved the **global design**. This design refers to modifications of the functional, conglomerate, and divisional designs rather to another single specific design. A global design may be relatively decentralized like Nestlé, or relatively centralized like Matshushita.[31] The key to a successful global design is the development of a design which will provide the necessary coordination and integration for worldwide business while also enabling the flexibility and autonomy necessary to compete in regional and local markets.[32]

The **organic design** was developed in the early 1960s by two British researchers and is similar in some ways to the System 4 design.[33] Its designers' goal was to help managers better align their organizations with their environments. They found that some firms could effectively use what they called a **mechanistic design**, which is similar to System 1, that was found to work best in stable conditions. For example, the Singer Sewing Machine Company operates in an environment that is fairly stable, has few domestic competitors, and so forth, so Singer might be best designed along mechanistic lines. Wendy's also uses a mechanistic organizational design. In contrast, firms such as Adobe, Hewlett-Packard and Yum! Brands use an organic design, which is presumed to be most effective when the environment is fluid and when constant adjustments are necessary to respond to shifts and changes. The organic organization is based on open communication systems, a low level of specialization and standardization, and cooperation. These characteristics are summarized in Table 13.2.

The view that seems to be emerging in organization design is that there is no "one best way" to organize and that companies should use whatever design seems most appropriate for them to accomplish their objectives. As a result, numerous **hybrid designs** have appeared in ongoing organizations. **Hybrid** means that different parts of the same organization are designed along different lines. One part may be divisional, another may use a matrix, and still others may be more or less bureaucratic in design. Some of those designs that merit noting are new venture units, alternate ownership patterns, and front-end-back-end organizations, and network organizations.

**New venture units**, or "**skunkworks**" as they are called informally, are small, semi-autonomous, voluntary work units. They are protected from many normal day-to-day corporate activities and pressures so that they can concentrate on the generation of new ideas or the development of new products or ventures. Companies that have used this design include Lockheed, where the term "skunkworks" originated, Baskin-Robbins, Coca Cola, McDonald's, Boeing, Genentech, Monsanto, and 3M.[34]

Alternate ownership patterns have also emerged as a variety of designs is increasingly being used in an attempt to bring new ideas or new monies into the organization. For instance, **Employee Stock Ownership Plans (ESOPs)** transfer stock ownership to employees in an effort to increase their commitment, involvement, and motivation. This also means, however, that executives cannot set corporate policy and strategy without involving the employees.

**Research and Development Limited Partnerships (RDLPs)** are consortia, usually among high technology firms, that are designed to do basic research.[35] The resources come from participating firms' contributions, and those firms then share in

**TABLE 13.2** Mechanistic and Organic Designs

| MECHANISTIC | ORGANIC |
|---|---|
| Static, rigid structure | Flexible structure |
| Centralized | Decentralized |
| Communication is exclusively vertical | Open communication in all directions |
| Roles carefully defined in detail | Roles not rigidly defined |
| Boundary spanning (crossing organizational lines) uncommon | Boundary spanning common |
| Work relationships are vertical | Horizontal and diagonal work relationships occur |
| Reliance on methods, rules, and procedures | Reliance on group/team processes |

Source: Wesley D. Sine, Hitoshi Mitsuhashi, David A. Kirsch, Hitoshi Mitsuhashi, and David Kirsch (2006) "Revisiting Burns and Stalker: Formal structure and new venture performance in emerging market sectors," *Academy of Management Journal*; Gibson, J.L., Ivancevich, J.M., & Donnelly, J.H., Jr., *Organizations: Behavior, structure, processes,* 8th ed., (Boston, MA: Irwin, 1994); Tom Burns and G. M. Stalker, *The Management of Innovation* (NY: Oxford University Press, Revised edition, 1994): 119–122; Peter J. Frost, Experiencing Mechanistic Versus Organic Systems: Adding Affect To Students' Conceptual Grasp of "Abstract" Organizational Concepts (1990), *Journal of Management Education,* 14: 87–92.

the results of the research. **Joint ventures**, whereby two firms jointly form a third one to produce a new product, are also becoming more common. Examples include joint ventures between the AWB Group, one of Australia's largest agribusinesses, and Gavilon, formerly ConAgra Trade Group Inc.; between Mexico's Modelo and Cargill; and between Perdue and Agri-Recycle, the company that helped to develop the litter-pelletizing technology (now Perdue AgriRecycle).[36]

Finally, many firms are buying into other companies to establish **equity positions** through the purchase of significant portions of stock. This financial interdependency tends to lead to endeavors that benefit both firms.[37]

Many newer organizational designs are early forms of what has been called the **high involvement organization**.[38] These designs are based on a process orientation, open communications, a low level of functional specialization and standardization, and

cooperation. These designs have also been labeled as horizontal organizations as opposed to the traditional, vertical organization.[39] Figure 13.11 illustrates the nature of such an organization, where parallel processes consisting of self-managed teams coordinated by process owners (who may be team representatives) are focused by an executive team on customers' needs.[40] Versions of such high involvement organizations are the front-end-back-end and network organization as well as the new plant design.

**Front-end-back-end organizations** simultaneously employ product and customer departmentalization. The "back end" produces the goods and services, which are then marketed by the "front end." This form of organization seems to be increasingly used by organizations that are selling complex sets of products or services to numerous groups of customers.[41]

Still another emerging organizational design is the **network organization**. Also known

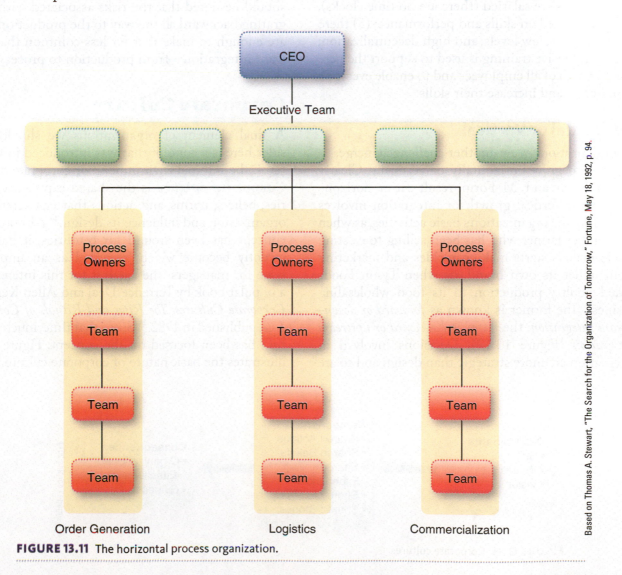

Order Generation    Logistics    Commercialization

**FIGURE 13.11** The horizontal process organization.

Based on Thomas A. Stewart, "The Search for the Organization of Tomorrow," *Fortune*, May 18, 1992, p. 94.

as a **value-added partnership** or a **hollow corporation**, network organizations usually engage in **outsourcing**, which is the contracting to other firms of many of their usual functions. Such organizations are the center of a network of companies that perform the activities of the business. Apple Computer, Reebok, Nike, and Benetton are examples of firms that have used this form of organization.[42]

Finally, Procter & Gamble, Mead, Cummins Engine, General Foods, and other companies have used a high-involvement approach in some of their newest plants. That has come to be called the new plant design and includes the following six characteristics:[43] (1) the selection process is long and realistic about the kind of management to be used to increase the fit between those employed and the organization design; (2) the physical layout is designed to suggest an egalitarian and team approach; (3) jobs are enriched and performed by autonomous teams; (4) all employees are salaried (there are no time clocks), and pay is based on skills and performance; (5) there are wide spans, few levels, and high decentralization; and (6) extensive training is used to support the personal growth of all employees and to enable everyone to maintain and increase their skills.

## Horizontal or Vertical

As organizations grow another challenge emerges—whether to grow horizontally or vertically. Clearly the H-Form and M-Form result from horizontal growth. Vertical growth or integration involves expanding the organization's basic activities, as when a cucumber farmer who has been selling to existing pickle makers starts making pickles and marketing them under its own brand, or when Tyson Foods added poultry production to its food wholesaling business. The former is known as *forward or downstream integration*; the latter is *backward or upstream integration* (Figure 13.12). Decisions involved in integration are more strategic than design and so are

**FIGURE 13.12**  A new example of upstream or backward integration is this solar farm that a large utility company built in the Southwest United States to provide more economical electricity for its customers.

beyond the purview of this basic book. However, it should be noted that the risks associated with integrating backward all the way to the production stage are enough to make that far less common than forward integration—from production to processing.

## Corporate Culture

A final element of organizations we should consider here is a concept that was introduced in Chapter 4: the internal environment, or corporate culture. **Corporate culture** is the shared experiences, stories, beliefs, norms, and actions that characterize an organization and influence its design.[44] Although the concept has been around for centuries, it has only recently become widely regarded as an important issue for managers. The catalyst for this interest was a popular book by Terrence Deal and Allen Kennedy, *Corporate Cultures: The Rites and Rituals of Corporate Life*, published in 1982.[45] Since that time, much attention has been focused on the concept. Figure 13.13 illustrates the basic nature of corporate culture.[46]

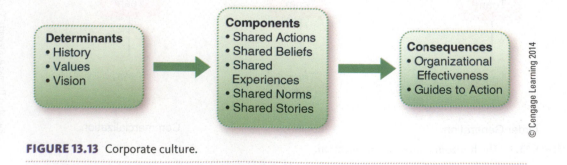

**FIGURE 13.13**  Corporate culture.

## Determinants of Culture

As noted in the exhibit, three basic factors determine the culture of an organization. One key determinant is the set of values held by top management. If top executives are antagonistic toward the government, if they want to stamp out all competition, or if they just want to earn fat profits, they set that tone for the firm. On the other hand, if they want to cooperate with the government, if they want to coexist peacefully with competitors, or if they want to treat customers honestly and fairly, a different atmosphere is prevalent.

The history of the organization also helps determine its culture. A company founded by a strong personality, one that leaves a mark on the firm, will follow the original model. For example, Steve Jobs left an indelible imprint on Apple Computer and Sam Walton did the same for Walmart. Les Wexner imbued Limited Inc. with an overwhelming self-confidence; those in the organization still believe that Limited can do no wrong. Even relatively old companies, such as Ford, still carry vestiges of their founders.

Finally, top management's vision for the firm also helps shape its culture. If the CEO decides that the company needs to undertake significant new ventures and aim for rapid growth and expansion, this vision will permeate the entire organization. On the other hand, if the CEO is content to maintain the status quo and take a defensive posture, this, too, will shape the culture. At Hewlett-Packard, the slogan "HP avoids bank debt" is derived from a story about the company's founders declining a project because it involved bank financing.[47]

## Components of Culture

How is an organization's design translated into culture? In general, it shapes five basic components or dimensions that we can use to characterize corporate culture.

1.  *Shared experiences* are the common events in which people participate and that become a part of their thinking. For instance, if a group of employees work closely together for an extended period, putting in 12-hour days and 7-day weeks to create a new product on schedule, the experience becomes a part of the group's culture. Even after the group breaks up, its members will always have this experience in common.

2.  *Shared stories* are the tales that have entered an organization's mythology. "Do you remember … ," "That was the time … ," and "This company has always …" are common beginnings to those stories.

3.  *Shared beliefs* are those things that all members of the organization accept as fact about the company. Employees at 3M, for example, believe the company will win any battle it chooses to undertake. Likewise, Walmart associates (employees) believe they can sell more this month than they sold last month.

4.  Similarly, *shared norms* are generally accepted ways of doing business. For example, having employees and managers jump in and help get the job done is a norm that helps eliminate back-ups in Southwest's passenger luggage at an airport or in the assembly line at the White Castle burger packing plant.

5.  Finally, *shared actions* are day-to-day behaviors that most people will perform. At Texas Instruments the "shirt-sleeve" culture suggests that men do not wear ties to work. In contrast, at IBM, the more formal culture dictates that all men wear ties. Other common actions involve work hours and social interactions.

## Consequences of Culture

There are two consequences of culture that we need to address: how it affects effectiveness and its usefulness as a guide to action.

**Culture Affects Effectiveness.** Although relevant research is scant, one consequence of culture is that it affects organizational effectiveness.[48] In particular, three aspects of culture can affect a company's success.

First, there may not be a single best culture, but it does seem that everyone in the company must understand what the culture is. Thus, it seems to be important for top management to create and transmit a clear, strong but not arrogant culture.[49] Common reasons cited for the success of firms such as Hewlett Packard, Disney, IBM, EDS, Southwest Airlines, and Walmart are that everyone in the organization understands the culture.[50] Second, the culture must fit the strategy. Effectiveness tends to be enhanced if the corporate culture is consistent with the organization's strategy. When the culture and strategy appear not to be in tune, effectiveness often suffers.[51] Third, top management must keep the culture adaptable to respond to changes in the organization's environment.[52] The organization then comes to value the importance of satisfying its numerous constituencies and balancing their various needs.

**Culture Provides a Guide to Action.** The second consequence of culture, which is actually a corollary of the first, is that it provides a guide to actions for newcomers. A new employee can look around at Texas Instruments or Campbell Soup headquarters and know immediately how he or she should dress the next day. Newcomers also quickly learn whether high performance is expected or not.

## CHAPTER SUMMARY

Organization design refers to the overall configuration of positions and interrelationships among positions within an organization. The ideal bureaucracy is to have a strict division of labor, consistent rules, a clear chain of command, impersonal decision making, advancement based on expertise and performance, and well-kept records. Such organizations possess some desirable features but are most appropriate only in relatively simple and stable environments. Another early design is the System 4 view, which uses eight characteristics to describe organizations along a continuum. The characteristics include leadership, motivation, communication, interaction, decision-making, goal-setting, and control processes, and the performance goals sought by managers.

Three major contingencies affect organization design: size, technology, and the environment. Technology refers to the set of conversion processes used by an organization to transform inputs into outputs. Environmental uncertainty is accounted for by two components, rate of change and degree of complexity; when these are high, uncertainty is high, and when they are low, uncertainty is low.

There is no one best organization design. Contemporary organizations seem to recognize this by having changeable, flexible designs. Major contemporary approaches to organization design are the functional, conglomerate, divisional, matrix, and cooperative designs.

New designs are emerging to deal with differing environments. New venture units, ESOPs, RDLPs, joint ventures, and equity positions are among those based on ownership. Early forms of high involvement organizations are also beginning to appear. Those include front-end–back-end and network organizations and new plant designs. Hence, designs based on a process orientation, open communications, a low level of functional specialization and standardization, and cooperation are increasingly used.

The major determinants of corporate culture are values held by the top management of the organization, the history of the firm, and the top managers' vision of the firm. These translate into culture through shared experiences, memories or stories, beliefs, norms (generally accepted ways of doing business), and actions. There is a link between corporate culture and organizational effectiveness. Effective organizations have strong, clear cultures that are consistent with their strategies. Further, the corporate culture provides a guide to action for newcomers and for new situations.

# CHAPTER ACTIVITIES

## REVIEW QUESTIONS

1. What is meant by organization design?
2. Describe two major early approaches to organization design.
3. What are three major contingency factors that affect organization design? How do they do so?
4. Briefly describe the major contemporary organization design alternatives.
5. What is corporate culture, and why is it important?

## ANALYSIS QUESTIONS

1. Comment on these statements: "The only real value of an organization chart is in doing the analysis necessary to draw it. Once it is drawn, the only value it has is covering cracks in plaster walls."
2. Can bureaucratic organizations avoid red tape and other problems usually associated with them? If so, how? If not, why not?
3. Rensis Likert, who developed the System 1–4 approach, said that when he asked managers to describe the best organization with which they had ever been associated, they invariably described a System 4 design; when asked to describe the worst, they described System 1. Yet when he asked, "What would you do if you took over a company in trouble?" they all described actions that characterize System 1 more than System 4. What reasons can you give for this? How could you prevent it?
4. Would you rather work in an organic or a mechanistic organization? Why? Which form of organization would be more likely to appeal to someone who believes "there is a place for everything and everything should be in its place"? Why?
5. What factors, besides those mentioned in this book, can influence corporate culture?

## FILL IN THE BLANKS

1. The overall configuration of positions and the interrelationships among positions within an organization is known as _____ _____.
2. Pictures or maps consisting of boxes connected in such as way as to show the overall arrangement of positions and clarify the interrelationships among positions are called _____ _____.
3. The manufacture of products in assembly-line fashion for inventory rather than to customer specifications is known as _____ _____ or _____ _____ technology.
4. Making products that require that the composition of the raw materials be changed mechanically or chemically is called _____ _____ technology.
5. Groups of workers who operate with no direct supervision to perform specific tasks are called _____ work groups.
6. The _____ or _____ design is a unitary or uniform approach to design that makes maximum use of functional specialization.
7. The _____ or _____ design uses the product form of departmentalization and takes advantage of specialization based on knowledge of specific products or services, their production, and marketing. General managers usually head the units.
8. The design that establishes fairly autonomous product departments that operate as strategic business units is the _____ or _____ design.
9. A new enterprise set up by two firms that are sharing in its control and ownership is called a _____ _____.
10. Ownership positions obtained through the purchase of significant portions of stock are called _____ positions.

## CHAPTER 13 CASE STUDY
### FISH INDUSTRY FINDS FRANKENFISH HARD TO SWALLOW

We humans are "hard wired." Our psyches say we must work, but we are also social animals. We work to live; we work to socialize. And both are accomplished in organizations.

Next to agriculture, the second oldest legal occupation on planet Earth is fishing. When man encountered bodies of water, he/she learned that water could be a source of food. Getting the food, however, was not an easy task; the prey was elusive and required a very different skill set than land-based hunting. So it was that since the dawn of civilization two different industries developed simultaneously. Agriculture required one set of skills—to grow and nurture food production, not just hunt for it. Fishing required a transfer of the hunter-gatherer instincts developed on land to a new environment—water. Each industry provided sustenance, commerce, and work. Each industry provided a livelihood. One represented "new technology"—farming; the other, an extension of the hunter-gatherer lifestyle to a new environment—water. Both required organizations.

This division between livelihoods has existed for thousands of years. Farmers have farmed and fishermen have fished, while other humans have reaped the benefit of their labors through a diversified and broad nutritional base, commerce and trade. At the beginning of the twenty-first century, however, this ancient dichotomy seems about to be broken. The irony of the rupture is that it emanates from the success of organizations in both industries.

The Food and Agriculture Organization (FAO) of the United Nations characterizes the fisheries industry as about to enter its "last stand." Among 600 marine fish species, only 4 percent are underexploited or recovering from depletion, i.e. are of viable commercial use. This means that 96 percent of marine fish species are characterized by a gradient from moderate exploitation to depleted stocks. Indeed, in 2006 newspaper headlines shouted, "All Seafood Will Run Out by 2050, say Scientists." Global fishermen have been doing their job too well. Attempts to establish moratoria on various species often have not succeeded since it takes only one country to ignore the stoppage before others cave in. Something must be done. Fish is far too important as a protein source for our species and as a means of preserving genetic diversity in the water.

Into this fray, agriculture entered, applying the techniques of land-based crop production to fish production. After a slow start the "aquaculture" industry has come into its own. As of 2010, aquaculture in the United States was a $1.5-billion dollar industry distributed over 4,800 farms. The largest segment of the industry, 43 percent, involved the production of food fish. This segment was followed by mollusk (oysters and clams) farms, crustacean farms, bait fish farms, and sport fish farms, respectively. Even with this assistance from aquaculture to the fishing industry, the United States remains a net seafood importer. The two industries coexist peacefully.

Unfortunately, organizations in the agricultural industry have not stopped tinkering with this "thorny" problem and the fishing industry is now getting upset. A U.S. company, AquaBounty Technologies, Inc., has developed an Atlantic salmon that grows twice as fast as conventionally farmed fish and its wild cousin. The Food and Drug Administration has tested the "new" fish and found that chemically and biologically it is identical to its conventional Atlantic salmon cousin. Many folks are upset, including Gloucester Fisherman's Wives Association of Massachusetts, Senator Lisa Murkowski (R-AK), Pacific Coast Trollers, Atlantic fishing companies and Stonyfield Farm (yogurt manufacturer). Is it smart to play with Mother Nature?

Two issues are involved: the intrusion of agriculture into the hunter-gatherer lifestyle of fishing (a removal of the separation between the two) and the imposition of a genetically modified organism as a solution for the fishery industry. The "new" salmon has added a gene from another type of salmon, the Chinook, that encourages a faster growth rate; and the gene is "turned on" by a gene from the eel-like ocean pout, another salmon. Land agriculture is already employing this technology so it can feed the expected nine billion people on this planet. The extension of gene technology seems very obvious to organizations in agriculture, but not to those in fisheries.

While agriculture adapted to its new environment several thousand years ago, hunter-gatherers did not; they simply employed new technologies to accomplish the hunter-gatherer task. Asking them to "catch up" in only a few years may be seen as unfair. Walking away or ignoring reality is not helpful, either. It is not just the Americas. Researchers at European-Union financed Selfdott announced they succeeded in spawning the Atlantic bluefin tuna in captivity without hormonal intervention. Now Japanese fisherman may be upset. Already, about half of the seafood consumed worldwide is farm-raised by different organizations.

Josh Ozersky, James Beard Award-winning writer, said it best: "There are no Black Angus cows grazing in the wild; they're the product of breeding for size, marbling, and fast growth, not unlike the genetically modified salmon." This argument while true is not going to win friends along the coastlines. This will be a prolonged debate. It will not end easily. But one fact is obvious: the fishery people are about to "have a cow" over this issue. And organizations must be designed to deal with both sides.[53]

### ▶ Case Study Questions

1. What form of organization might be best in each of these industries? Why?

2. Would the forms of organization likely to be different or similar in these two industries over time?

3. In organizations within these two industries, how would the corporate cultures differ and how would they be similar?

## REFERENCES

1. "The Marmite Effect," *The Economist* (September 23, 2010) at www.economist.com (accessed October 4, 2010); "Yuck: Making Healthy Food is Easy. Making People Eat It is Not," *The Economist* (September 16, 2010) at www.economist.com (accessed October 4, 2010); Jaeyeon Woo and Kanga Kong, "South Korea Faces Pinch in Kimchi Supply," *The Wall Street Journal*, October 4, 2010, A20.

2. Gareth Jones, *Organization Theory*, 6th ed. (Upper Saddle River, NJ: Pearson, 2010).

3. For a recent discussion, see N. Anand and Richard L. Daft, "What Is the Right Organization Design?" *Organizational Dynamics* (2007): 329–344.

4. A mentor for one of the authors of this book, noting that charts get out of date almost as fast as they are constructed, said that the best use of charts is to have them framed and use them to cover up cracks in walls.

5. Max Weber, *Theory of Social and Economic Organization*, trans. T. Parsons (New York: Free Press, 1947).

6. These records were to be kept in drawers in cabinets or bureaus, hence, the term bureaucracy.

7. Richard L. Daft, *Organization Theory and Design*, 10th ed. (Mason, OH: South-Western Cengage Learning, 2008).

8. Daniel A. Wren and Arthur G. Bedeian, *The Evolution of Management Thought*, 6th ed. (New York: Wiley, 2008).

9. Rensis Likert, *New Patterns of Management* (New York: McGraw-Hill, 1961); and *The Human Organization* (New York: McGraw-Hill, 1967).

10. William F. Dowling, "At General Motors: System 4 Builds Performance and Profits," *Organizational Dynamics* (Winter 1975): 23–28.

11. Robert C. Ford, Barry R. Armandi, and Cherrill P. Heaton, *Organization Theory: An Integrative Approach* (New York: Harper & Row, 1988).

12. Derek S. Pugh and David J. Hickson, *Organization Structure in Its Context: The Aston Programme* (Lexington, MA: D. C. Heath, 1976); see also "Is Your Organization Too Big?" *Business Week*, March 27, 1989, 84–94.

13. Joan Woodward, *Industrial Organization: Theory and Practice* (London: Oxford University Press, 1965).

14. Patricia L. Nemetz and Louis W. Fry, "Flexible Manufacturing Organizations: Implications for Strategy Formulation and Organization Design," *Academy of Management Review* (October 1988): 627–638.

15. Michael Russo and Niran Harrison, "Organizational Design and Environmental Performance: Clues from the Electronics Industry," *Academy of Management Journal* (2005): 582–593.

16. For example, see Tom Burns and G. M. Stalker, *The Management of Innovation* (London: Tavistock, 1961); and Paul R. Lawrence and Jay W. Lorsch, *Organization and Environment* (Homewood, IL: Richard D. Irwin, 1967); for a review, see Daft, *Organization Theory and Design*.

17. Henry Mintzberg, *The Structuring of Organizations: A Synthesis of the Research* (Englewood Cliffs, NJ: Prentice-Hall, 1979).

18. Oliver E. Williamson, *Markets and Hierarchies* (New York: Free Press, 1975).

19. Michael E. Porter, "The Five Competitive Forces That Shape Strategy," *Harvard Business Review* (January 2008): 79–90 and Michael E. Porter, "From Competitive Advantage to Corporate Strategy," *Harvard Business Review* (May–June 1987): 43–59.

20. Mintzberg, *The Structuring of Organizations*.

21. Robert E. Hoskisson, "Multidivisional Structure and Performance: The Contingency of Diversification Strategy," *Academy of Management Journal* (December 1987): 625–644.

22. For a more complete discussion of organizations in international contexts, see Ricky W. Griffin and Michael Pustay, *International Business: A Managerial Perspective*, 6th ed. (Upper Saddle River, NJ: Prentice-Hall, 2009).

23. Stanley M. Davis and Paul R. Lawrence, *Matrix* (Reading, MA: Addison-Wesley, 1977).

24. Harvey F. Koloday, "Managing in a Matrix," *Business Horizons*, March–April 1981, 17–24.

25. See, for example, Jeffrey Barker, Dean Tjosvold, and I. Robert Andrews, "Conflict Approaches of Effective and Ineffective Project Managers: A Field Study in a Matrix Organization," *Journal of Management Studies* (March 1988): 167–178.

26. James Owens, "Matrix Organization Structure," *Journal of Education for Business* (November 1988): 61–65; and Kenneth Knight, "Matrix Organization: A Review," *Journal of Management Studies* (May 1976): 111–130.

27. "Matrix Is the Ladder to Success," *Business Week* (August 11, 2009) at www.businessweek.com 2009 (accessed July 15, 2010); "Future Perfect: The CFO of Tomorrow," CFO Asia Research Services in Collaboration With KPMG (July 2008): 12; Nebojša Janicijevic & Ana Aleksic, "Complexity of Matrix Organisation and Problems Caused by its Inadequate Implementation," *Economic Annals* (2007): 28–44; "Client Success," Development Dimensions International, Inc. (2006) at www.ddiworld.com (accessed July 15, 2010); Quentin M. West, "Economic Research Trade-Offs Between Efficiency and Equity," *Southern Journal of Agricultural Economics* (July 1973): 9–12.

28. USDA/RBS/Cooperative Services, "Cooperative Historical Statistics," Cooperative Information Report 1, Section 26 (revised April 1998), p. 7; Charles A. Kraenzle et al., "Farmer Cooperative Statistics, 2000," USDA/Rural Business-Cooperative Service, Service Report 60 (December 2001), p. 2; *Understanding Cooperatives: Farmer Cooperative Statistics*, Cooperative Information Report 45, Section 13. Washington, DC: United States Department of Agriculture, June 2011 (original, December 1996), p. 1.

29. David Barboza, "Facing Huge Debt, Large Farm Co-op Is Closing Down," *The New York Times*, September 16, 2003 (available at www.nytimes.com).

30. Michael A. Boland and Jeffrey P. Katz, "Jack Gherty, President and CEO of Land O' Lakes, on Leading a Branded Food and Farm Supply Cooperative," *Academy of Management Executive* 17(3), 24–30.

31. "Matsushita Electric Industrial Company," *The Wall Street Journal*, June 29, 1988, 18–19; and G. Turner, "Inside Europe's Giant Companies: Nestlé Finds a Better Formula," *Long-Range Planning*, (June 1986): 12–19.

32. William G. Egelhoff, "Strategy and Structure in Multinational Corporations: A Revision of the Stopford and Wells Model," *Strategic Management Journal* (Vol. 9, 1988): 1–14.

33. Burns and Stalker, *The Management of Innovation*.

34. News Editor, "Whirl of Change Wins Popular Vote," *World Dairy Diary* (October 22, 2008) at www.wdexpo.org (accessed July 27, 2010); Patrick F. Coveney, Jeffrey J. Elton, Baiju R. Shah, and Bradley W. Whitehead, "Rebuilding Business Building," *McKinsey Quarterly* (June 2002) at www.mckinseyquarterly.com (accessed July 28, 2010); and Christopher K. Bart, "New Venture Units: Use Them Wisely to Manage Innovation," *Sloan Management Review* (Summer 1988): 35–43.

35. Gerardine DeSanctis, Jeffrey Glass, and Ingrid Morris Ensing, "Organizational Designs for R&D," *Academy of Management Executive* (2002): 55–64.

36. Sally White, "AWB, Gavilon Confirm Joint Venture," *The Land* (March 30, 2010) at theland.farmonline.com.au (accessed July 28, 2010); "Mexico's Modelo in Joint Venture with Cargill," *Reuters* (April 14, 2010) at www.reuters.com (accessed July 27, 2010); "About Perdue AgriRecycle," Perdue AgriRecycle website at www.perdueagrirecycle.com (accessed July 28, 2010).

37. D. Bruce Shine and Donald F. Mason, Jr., "ESOP: The American Workers' Leveraged Buy Out," *Case and Comment*, January 1, 1990, 24; "More Competitors Turn to Cooperation," *The Wall Street Journal*, June 6, 1989, B1; Tyzoon T. Tyebjee, "A Typology of Joint Ventures," *California Management Review* (Fall 1988): 75–86; and Howard Grindle, Charles W. Caldwell, and Caroline D. Strobel, "RDLP: A Tax Shelter That Provides Benefits for Everyone," *Management Accounting* (July 1985): 44–47.

38. Edward E. Lawler, III, *High Involvement Management: Participative Strategies for Improving Organizational Performance* (San Francisco, CA: Jossey-Bass, 1986).

39. Thomas A. Stewart, "The Search of the Organization of Tomorrow," *Fortune*, May 18, 1992, 92–98.

40. John Mathieu, M. Travis Maynard, Tammy Rapp, and Lucy Gilson, "Team Effectiveness 1997–2007: A Review of Recent Advancements and a Glimpse into the Future," *Journal of Management* (2008): 410–476.

41. Edward E. Lawler III, *The Ultimate Advantage* (San Francisco, CA: Jossey-Bass, 1992), 67–69.

42. Ibid., 69–71. See also R. Johnston and P. R. Lawrence, "Beyond Vertical Integration—The Rise of the Value-Adding Partnership," *Harvard Business Review* (1988): 94–101.

43. This section is based on Lawler, *The Ultimate Advantage*, 307–312.

44. Ricky W. Griffin and Gregory Moorhead, *Organizational Behavior, 9th ed.* (Mason, OH: South-Western, Cengage Learning, 2010); David D. Van Fleet, *Behavior in Organizations* (Boston: Houghton Mifflin, 1991); and W. Jack Duncan, "Organization Culture: 'Getting a Fix' on an Elusive Concept," *Academy of Management Executive* (August 1989): 229–235.

45. Terrence Deal and Allen Kennedy, *Corporate Cultures: The Rites and Rituals of Corporate Life* (Reading, MA: Addison-Wesley, 1982). For an excellent summary of that work, see Harrison M. Trice and Janice M. Beyer, *The Cultures of Work Organizations* (Englewood Cliffs, NJ: Prentice-Hall, 1993).

46. Deal and Kennedy, *Corporate Cultures*; and Vijay Sathe, "Implications of Corporate Culture: A Manager's Guide to Action," *Organizational Dynamics* (Autumn 1983): 5–23.

47. "Hewlett-Packard's Whip-Crackers," *Fortune*, February 13, 1989, 58–59.

48. Sathe, "Implications of Corporate Culture"; and Ralph H. Kilman, Mary Jane Saxton, and Roy Serpa, eds., *Gaining Control of Corporate Culture* (San Francisco: Jossey-Bass, 1985).

49. John P. Kotter and James L. Heskett, *Corporate Culture and Performance* (New York: Macmillan, 1992), 144–149.

50. Fathi El-Nadi, "Examples of Strong Corporate Cultures," Evan Carmichael at www.evancarmichael.com (accessed July 27, 2010); G. Sadri and B. Lees, "Developing Corporate Culture as a Competitive Advantage," *Journal of Management Development* (2001): 853.

51. Kotter and Heskett, *Corporate Culture and Performance*, Chapter 3.

52. Ibid., Chapter 11.

53. "General Situation of World Fish Stocks" (2007) at www.fao.org (accessed October 2, 2010); Charles Clover, "All Seafood Will Run Out in 2050, Say Scientists." *Telegraph* (November 3, 2006) at www.telegraph.co.uk (accessed September 29, 2010); "Research and Markets: Aquaculture in the U.S. Industry," includes about 4,800 Farms with Combined Annual Revenue of $1.5 Billion. *Business Wire* (October 2, 2010) at www.businesswire.com (accessed October 2, 2010); "Aquaculture Industry Facts" (2010) at www.jobmonkey.com/aquaculturejobs/aquaculture-industry-facts.html (accessed October 1, 2010); Gautam Naik, "Gene Altered Fish Closer to Approval," *The Wall Street Journal* (September 21, 2010) at online.wsj.com (accessed October 2, 2010); Alicia Mundy and Bill Tomson, "Industry Fights Altered Salmon," *The Wall Street Journal* (October 1, 2010) at online.wsj.com (accessed October 2, 2010); Paul Greenberg, "Taming the Wild Tuna," *The New York Times*, September 5, 2010, WK5; James C. Greenwood, "Don't Be Afraid of Frankenfish," *The Wall Street Journal* (September 23, 2010) at online.wsj.com (accessed September 30, 2010).

# Organization Change and Innovation

## LEARNING OBJECTIVES

After studying this chapter, you should be able to:

- Discuss the nature of change, including the reasons and need for change, planned organization change, and the steps in change.

- Explain why people resist change and how to overcome that resistance.

- Identify strategic, structural, technological, and people-focused approaches to change.

- Define and discuss the nature and techniques of organization development.

- Explain why and how organizations may need to undergo revitalization.

- Discuss the importance of innovation in organizations and describe how it can be facilitated and implemented.

# Trees

**T**rees are an important industry in agribusiness and have been for hundreds of years. Of course, the forestry industry is important for several contributions. The forest produces lumber; it is a significant control on "global warming gases"; it is a nice place to visit for ecotourism; the forests along the equator are a tremendous pool of biological diversity; and forestry is the first renewable industry (Figure 14.1). These are true but there are a couple far more important contributions from trees. They are so relevant to our existence that they are overlooked or ignored because they are so common in our everyday lives, yet they represent fiercely competitive industries. Think about competition, longevity, and innovation; think about the pencil and the aspirin tablet.

**FIGURE 14.1** Trees provide more than food and timber for people. This rubber tree provides milk (latex) for manufacturing rubber.
© Bunpot/www.Shutterstock.com

For example, two major German wood pencil manufacturers have been competing against each other for almost 200 years, officially. Unofficially they have been competing for more than 300 years. Staedtler Mars Gmbh celebrated its 175th anniversary in 2010, and Faber-Castell AG celebrated its 250th anniversary in 2011. A good trivia question would be which company is older? The answer: Staedtler. Nuremburg city records demonstrate that carpenter Friedrich Staedtler was listed as a "pencil craftsman" in 1662. However, Faber was the first to brand top-quality pencils and in 1870, he registered one of the first American trademarks. At one time Staedtler owned Faber-Eberhard but sold it back to the Faber-Castell family. It is hard to imagine eight generations of competition. It puts to shame current three- to five-year strategic plans.

For the record, the "lead" came first, then the wood enrobing of the lead. "Black lead" was discovered and then adapted for writing, but it was not enrobed in wood. Instead, those early editions used string wound around a chunk the lead. Also, "black lead" was not lead (Pb), but graphite (C)—the same substance as a diamond, but different molecular arrangement.

The point (sorry!):This simple tool has been around for hundreds of years, and its two major manufacturers have competed against each other just as long. Yet, it is possible that you could care less about the product; it is so ubiquitous that it is invisible. The pencil has achieved what many products aspire to—total and instant global recognition. Faber-Castell produced 2.2 billion pencils per year out of a 2009 global production of 15–20 billion. Faber-Castell is the world's biggest branded pencil manufacturer. China is the largest producer of unbranded or generic pencils, as it produces about half of the world supply of wood pencils.

After eight generations Faber-Castell is still innovating and looking for new designs. Count Anton Wolfgang von Faber-Castell, who bought an antique pencil made around 1890 or 1895, says, "It writes perfectly, even after all of these years. That's the fantastic thing with a pencil."

One thing is certain: No matter where the newest technology is under development and no matter how sophisticated the entrepreneur/innovator is, more than likely the initial designs/concepts will be drawn with pencils. We use them daily; we just never think about them. Think about the wooden pencil, the next time you "pencil in" the details of a deal.

In discussing the pencil, the wood was peripheral. However, it is interesting to note that while the wood of the willow tree is not suitable for wooden pencil production, its bark has been used for centuries in folk medicine for pain relief. Hippocrates recommended a preparation made from willow bark and leaves. The bark extract, taken internally, relieved pain and reduced inflammation and fever. The active agent in the tree bark is salicylic acid, sometimes called salicin. Bayer AG began marketing acetylsalicylic acid (ASA) in 1899 after its discovery in synthesis in 1897. It was marketed as "Aspirin."

Today aspirin (lowercase letters) is as ubiquitous as the wooden pencil. Its patent long since expired, aspirin is produced by almost any small pharmaceutical company. This common analgesic may have even more uncommon usage. It seems that a number of scientific studies demonstrated an efficacy of aspirin that no one suspected.

An article in *Lancet* (British medical journal) about a four- and eight-year randomized controlled study of 25,570 patients demonstrated that aspirin users were 21 percent less likely to die from cancer than placebo users. After five years of aspirin use, death rates from all cancers fell by 35 percent and for gastro-intestinal cancers by 54 percent. In fact, taking aspirin for five–ten years beat all initiatives to screen for breast and prostate cancer. The lowly aspirin tablet was demonstrated to have this efficacy.

A potential wonder drug, or as *The Economist* magazine put it, "…ask yourself what a pharmaceutical firm might charge for a drug that would reduce the chance of death by cancer by 20 percent—and then note that a 100-day supply of low dose aspirin can cost less than a dollar. By anyone's measure, that is a bargain."

The wooden pencil and aspirin have become common—so common as to be invisible—but they still have the potential to become uncommon. Their industries remain highly competitive, and product innovation continues. Indeed the lesson is simple: product innovation never stops…ever. Innovation does not always involve just the newest and latest and greatest … it also involves the steadfast and the durable.

And you thought this was going to be about trees … how boring![1]

# INTRODUCTION

Both pencil and aspirin manufacturers perform value-added activities to products that come from an agricultural production industry: forestry. The pencil manufacturers are an excellent example of how organizations must change if they are to maintain an effective alignment with their environments.[2] Indeed, an organization's ability to effectively manage change is a critical ingredient to its very survival in the long run. At the same time, however, if change is not properly managed, it may do as much harm as good. Given the changes in agriculture and agribusiness outlined in Chapter 1, the importance of an organization to effectively respond to change should be apparent.

This chapter explores organization change and three of its most important related areas. We first discuss the nature of change, then describe ways that change can be more effectively managed, highlighting why people resist it and ways to overcome that resistance. Then we identify the four general areas of organization change. The final three sections focus on organization development, organization revitalization, and innovation in organizations, each of which represents special forms of change that are often managed as distinct activities.

## The Nature of Organization Change

Johnson & Johnson announced the elimination of hundreds of jobs at its Fort Washington plant. AT&T expanded to deal with activation for iPhones and iPads. Goodyear altered virtually every facet of its organization design in the last several years. Tenneco is trying to motivate its employees to work harder in an effort to boost profitability. And Weyerhaeuser has installed new production equipment in many of its mills.[3] What do these examples have in common?

They all represent a fundamental change in some aspect of the organization—a form of **organization change**.[4] Johnson & Johnson is shrinking, while AT&T is growing. Tenneco is changing people's behavior. Goodyear is changing its organization design. And Weyerhaeuser is changing the technology it uses to make paper products. Why were these changes necessary? To answer this question, we must explore some of the reasons for organization change and look at the steps that managers usually follow in making such changes.

### Reasons for Change

An organization may find it necessary to change for a variety of reasons. As shown in Figure 14.2, the most common reasons stem from one or more forces in the organization's general and/or task environments, as discussed previously in Chapter 5.[5]

Consider, for example, how shifts in an organization's general environment might cause it to respond. Technological forces may provide new equipment or make existing technology obsolete. Newly automated work processes may need to be implemented to reduce labor costs and improve quality and productivity. Political-legal forces might introduce new legislation or regulations that affect the organization (Figure 14.3). For example, a new law that increases safety standards for the firm's

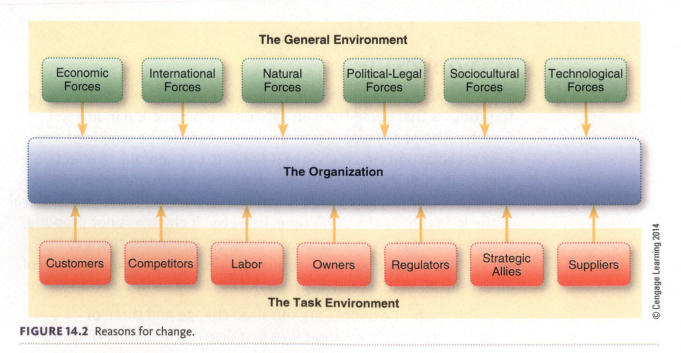

**FIGURE 14.2** Reasons for change.

**FIGURE 14.3** A reduced demand for cigarettes forced tobacco manufacturers to modify their product lines and find new uses for their excess manufacturing space.

products might alter the parts the firm buys to use in assembling the product. Similarly, increasing or decreasing the number of employees may subject smaller firms, such as Summer Farms and Green Things Landscaping, to additional regulations.

Economic forces could affect an organization in several different ways. For example, inflation or changes in unemployment might affect how it hires new employees. Similarly, shifts in interests might change a firm's ability or willingness to borrow money. International forces are also becoming increasingly important. New markets and/or competition may emerge in other countries. Finally, sociocultural forces are important in many different circumstances. The general shift in consumer values and preferences about tobacco and alcohol, for example, is changing how cigarette makers like Philip Morris and R.J. Reynolds and alcohol producers like Miller and Coors do business. The same is true for vegetables as public demand for organically-grown vegetables continues to accelerate. Natural forces—floods, drought, freezes, tornados, hurricanes, and the like—can lead to unexpected or at least unpredictable changes.

In similar fashion, the task environment can also bring about organization change. Suppliers can raise or lower prices, alter quality standards, or change delivery schedules. Large organizations like Walmart have enough buying power to force their suppliers to lower their prices or to dictate the size of packaged products. On the other hand, there is a limit to how much even a large organization can raise its prices unless it has a significant product edge or it can function as a monopoly. Customers may turn to alternative products or demand higher quality or lower prices, and new competitors may find it feasible to enter the market. Existing competitors may raise or lower prices, introduce new products or services, or adjust their advertising. Unions may demand higher wages or better working conditions or a greater voice in making decisions. Regulators could impose new restrictions on how the organization does business or modify the ways the organization can do certain things. Owners could exert pressure for higher profitability. And strategic allies may want to negotiate new joint-venture agreements or modify or cancel existing ones.

## Unexpected Change

Sometimes change occurs unexpectedly. **Organizational resilience** refers to an organization's ability to recover from such unexpected change.[6] Although important to all organizations, organizational resilience is particularly important in agriculture and agribusiness. Agricultural organizations face all sorts of natural disasters—floods, droughts, freezes, hail storms, hurricanes, tornados, and pests (Figure 14.4).

**FIGURE 14.4** The corn in this field is checked for radiation level after a nuclear leak was suspected at a reactor in another state. The corporation that owns the field has plans already in place for dealing with nuclear contamination of the crop.

In addition to those disasters, agribusinesses face accidents, recalls, product tampering, and terrorism.

Obviously an organization is not likely to be resilient unless its members are.[7] Individual members of an organization, however, take their cues from the organization's leaders so the leaders must set priorities and allocate resources that foster resilience. This means that the leaders strive to create and maintain organizations with workforces that are productive, efficient and effective, innovative and motivated, and have high morale and satisfaction.

As indicated in the previous chapter and in Chapter 4, corporate culture is vital. A resilient organization has a culture that supports flexibility, adaptability, innovation, and change.[8] It fosters empowerment of organizational members through delegation and participation. Experiences, memories, or stories about how the organization or other organizations responded to unexpected events in the past also serve to develop a resilient culture.

Environmental scanning (Chapter 10) and understanding risks (Chapter 11) are also important. Environmental scanning can help an organization to be aware of aspects in its environment that might lead to unexpected events and, hence, prepare for them. Examples include storing water for use in a drought, using floating row covers to protect crops from cold weather, and having food handlers wear protective gloves and masks to reduce possible product contamination. Understanding risks may involve the use of risk assessments.[9]

## Planned Organization Change

As just noted, one of the ways in which organizations can deal with the forces in the general and task environments is to anticipate and prepare for them.[10] In most cases, the company that senses the need for a change before the change is actually needed and then plans for that change in a careful and systematic fashion will be more effective than the company that waits to be forced to respond.[11]

An aspect of such change is growth. One way to grow is through acquisition. McDonald's bought Boston Market, Conato's, and Chipotle Mexican Grill while Wendy's obtained Baja Fresh Mexican Grill. Tricon Global Restaurants elected to grow through multibranding. It changed its name to Yum! and developed Pizza Hut, Taco Bell, KFC, Long John Silver's, and A&W.[12] Some acquisitions are made by individuals convinced that they can strengthen the acquired organization. Dean Metropoulos, known

for his success with Chef Boyardee and Vlasic Pickles, took over Pabst Brewing Co. in 2010 for just that reason.[13] In 2008, Bunge announced that it was acquiring Corn Products International, but those plans changed in 2010 after a successful stock buyback. That, in turn, paved the way for a takeover of Tate & Lyle, the maker of Splenda.[14]

The former approach is called **planned organizational change**. In general, planned organizational change involves the anticipation of possible changes in the environment to which the organization must respond and some consideration of how that response will occur ahead of time. Of course, managers cannot always accurately predict the future. Even an imperfect vision of what the future holds for a business, however, is almost certain to be better than no vision at all. Summer Farms probably did not see the demand for organic foods as early as it should have, but it did get started in this new product line at a relatively early time.

In contrast, an organization is likely to engage in **reactive change** if it pays little attention to anticipating environmental shifts and consequently, must allow its reaction to be dictated by what happens. For example, some critics argue that IBM struggled some years ago primarily because its managers did not anticipate the need for change. When worldwide computer sales slumped just as new competitors were emerging, IBM found itself with excess capacity and a heavy payroll. The changes it had to make in response were more painful than they might have been had the firm taken the time earlier to plan for them in an orderly fashion.

### FOOD FOR THOUGHT 14.1

Some things are not meant to be changed. Bourbon is the only whiskey distilled from at least 51 percent corn; rye whiskies must be made from at least 51 percent rye. Other whiskies contain a mix of various grains.

## Steps in Planned Change

When an organization is successful in its efforts to anticipate the need for change, it will be able to manage that change. Figure 14.5 summarizes the steps to follow in this circumstance.[15] Logically, the process must first start with the recognition of a need

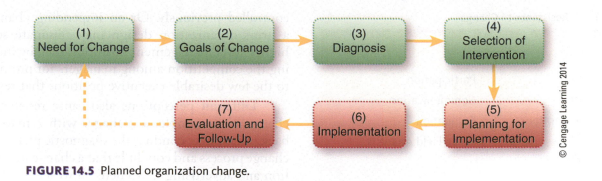

**FIGURE 14.5** Planned organization change.

for change. This recognition might come in the form of an anticipated event. For example, a manager may read about a new technological breakthrough that will soon be available and she believes it will be applicable to her organization. Or an employee might casually mention that workers in the organization would greatly appreciate an on-site child-care facility.

After recognizing the need for change, the manager must set goals for it. That is, she must consider why the change is being considered and what should be gained from it. For instance, she might decide that adopting the new technology should cut costs or that the on-site daycare facility should lower employee absenteeism. Obviously, if the change will not benefit the organization in some way, it may be advisable to delay making it until it has more potential.

The third step is diagnosis. This is where many managers fall short either by failing to do due diligence or by accepting fad dictums. Diagnosis means that the manager should look carefully at the organizational system to identify all of the possible effects of the change. Unfortunately, managers often err at this point by overlooking some of the requirements that will be necessary or by making assumptions that are unsustainable or just plain wrong. The new technology that is being considered may call for training of employees, a new performance appraisal system, and higher pay. The child-care facility may require the services of a full-time nurse and other additional employees.

Next, the manager chooses the actual intervention to use. For example, several forms of the technology may be available, and it may be possible either to buy it or to lease it. The child-care facility could be operated by the organization itself or managed by an outside company. Managers must consider each of these options, as well as others.

The dynamics associated with the intervention must also be carefully planned. The manager must ask herself when the technology will be introduced, whether it will be introduced in stages or all at once, and whether it will be installed in all plants or just as a pilot in one location. When will the training be done, who will do it, and who will be eligible for the training? When will the child-care center be opened, what are its potential hours of operation, how much will employees pay, and what are the insurance implications?

When these and a myriad of other questions have been answered, implementation can take place. Of course, some time will probably be necessary to get things fully operational, and some problems will have to be worked out. The manager should anticipate and expect these, so as not to abandon the change before it has had ample time to demonstrate its usefulness. The daycare center, for example, might need to be phased in gradually.

Finally, after the change has been fully installed and is operating smoothly, the manager should evaluate it carefully to ensure that it has met its original goals. After installing the new technology, productivity should go up and costs should go down. Absenteeism should also go down once the child-care facility is in place. Assuming that the goals have been reasonably met, the process is essentially complete, and the manager can wait for the next opportunity for change. If the goals are not being met (for example, if the child-care facility is not being used or absenteeism has not dropped), additional fine-tuning may be in order.

## Managing Organization Change

Many of these points relate to the management of organization change, but there are also other aspects of the process. Two aspects that are especially critical, include recognizing that people often resist change and subsequently understanding ways to overcome this resistance.[16]

**TABLE 14.1**   Resistance to Change

| COMMON REASONS FOR RESISTANCE | WAYS TO OVERCOME RESISTANCE |
|---|---|
| Uncertainty | Participation |
| Self-interests | Communication |
| Differing Perceptions | Facilitation |
| Feelings of loss | Force-Field Analysis |

© Cengage Learning 2014

## Resistance to Change

People in organizations resist change for a variety of reasons. As summarized in Table 14.1, the four most common reasons for resisting change are uncertainty, self-interests, perceptions, and loss.

First, for a variety of reasons, change breeds uncertainty. For instance, some people may fear for their place in the organization. They may worry that they cannot meet new job demands or that their job will be eliminated. Or they may dread the ambiguity that frequently accompanies change. As a result of this uncertainty, they may also feel anxious and nervous. They resist the change to cope with these feelings. Many U.S. government workers resisted efforts to automate their work because of their uncertainty about the new technology they were having to learn.[17]

People also resist change because it threatens their own self-interests (Figure 14.6). A plant manager may resist a change to automate his plant because he fears it threatens his control of the plant's human resources. Likewise, another manager may resist a new automated information network because it will give others access to information that he alone

controlled previously. Or an impending change in a firm's organization design may eliminate several higher-level management positions, thereby increasing the competition among managers for promotion to the few desirable executive positions that remain.

Different perceptions also cause resistance to change. For example, a manager with a marketing background may conduct the diagnostic phase of the change process and conclude that a change in promotion and advertising is needed. However, a manager with an operations background might see the problem in terms of quality and productivity. Thus, the second manager sees the situation in a different light and will likely conclude that a different intervention is needed. Consequently, he may resist the change as first proposed by the marketing manager because he perceives the situation differently. At Summer Farms the grandchildren with college educations have different perceptions as to what is important than do Joshua and Katherine.

Finally, an individual may resist change because of feelings of loss. Members of family-owned businesses like Summer Farms often experience a feeling of losing control when outsiders are hired in managerial positions. Many changes involve alterations in work assignments and work schedules, and thus informal groups and close working relationships among peers are broken up. For example, when J.C. Penney moved its corporate headquarters from New York to Dallas, many people resisted the change because, among other reasons, they did not want to leave their friends and existing working relationships. Loss of power, status, security, or familiarity with existing procedures can also be problems.

## Overcoming Resistance to Change

Fortunately, managers can at least partially overcome resistance to change in several ways. Table 14.1 notes four especially useful ways. One key approach is to encourage participation by those people who will be involved in the change. When people participate, they feel less threatened. They recognize that they have a say in what happens, and are less concerned about feelings of loss. Logically, then, they will tend to lower their resistance to the change.

Open communication also helps overcome resistance. Complete and accurate information helps remove the uncertainty that so often accompanies change, as we have noted. Managers should provide information that is relevant to the change as often as they can. This can be achieved by disseminating

**FIGURE 14.6** People are less fearful of change when they understand what they are doing and that their jobs are not threatened. Here, a manager explains an upcoming change to one of his employees.

© Andrew Taylor/www.Shutterstock.com

## A FOCUS ON AGRIBUSINESS
### Environmental Change = Organization Change

All organizations should assure that their structures fit their environments. In agribusiness, where family businesses are common, this is important to ensure that operations remain profitable from one generation to the next. Thus, when the environment of the organization changes, so too should its structure.

For example, in 2005 members of Diamond Walnut voted to convert the cooperative to a stockholder-owned corporation and merge it into its wholly owned subsidiary, Diamond Foods. While initial profits were expected to decrease, it was thought that the stock value would offset that decrease, although that would depend on the length of time that the members held their stock and/or continued to grow walnuts.

Other cooperatives are moving to a different structure known as New Generation Cooperatives (NGCs). Like traditional cooperatives NGCs have one-member, one-vote arrangements where earnings are based on member patronage. There are important differences, though. NGCs have the ability to trade equity shares and delivery rights; they require equity contributions by members; and there is an obligation to deliver products based on equity contributions.

NGCs are formed for a variety of reasons. In some cases they are a way for coop members to withdraw participation but regain their capital investments. In other cases, NGCs may be formed to protect local jobs, especially if a local processing plant closes. In yet other cases they are formed because individuals see them as a way to potentially increase income, to diversify their investments and thereby reduce their risks, or to increase the size of the market for their products.[18]

information as it becomes available and be open and receptive to questions and inquiries from employees about the impending change.

Facilitation can also reduce resistance. Facilitation simply means that managers recognize that resistance may be present and must therefore actively work to manage it. For example, introducing change gradually helps minimize its impact. Being sensitive to people's concerns and helping them to resolve those concerns rationally can also be effective.

Another management approach to overcoming resistance to change is to use **force-field analysis**. The first step in using force-field analysis is to look systematically at the pluses and minuses associated with the change from the standpoint of the employees. Next, the manager should make efforts to increase the pluses and decrease the minuses to tip the scales toward acceptance. For example, if a major barrier to an impending change is that a certain work group will be broken up, the manager might try to figure out a way to keep the group intact. If that can be arranged, one minus has been eliminated. Likewise, if one thing working in favor of the change is greater autonomy over how the job is done, providing even more autonomy and making sure employees see how this will benefit them will further increase acceptance of the change.

## Areas of Organization Change

Thus far we have talked a lot about organization change but have not really focused exactly on what the organization may actually be changing. In this section, we identify the four major areas of organization change. As shown in Table 14.2, these are strategy, structure, technology, and people.[19]

### Strategic Change

Whenever an organization modifies its strategy or adopts a new strategy, it has engaged in **strategic change**. At the corporate level, for example, a firm may move from a growth to a retrenchment strategy (strategies are discussed in Chapter 9), as Johnson & Johnson was doing in 2010. Or it could move from

**TABLE 14.2**  Areas of Organization Change

| STRATEGIC CHANGE | STRUCTURAL CHANGE | TECHNOLOGICAL CHANGE | PEOPLE-FOCUSED CHANGE |
|---|---|---|---|
| Corporate Strategy | Components of Structure | Equipment | Skills |
| Business Strategy | Organization Design | Work Processes | Performance |
| Functional Strategy | Reward System | Work Sequence | Attitudes |
| | Performance Appraisal System | Automation | Perceptions |
| | Control System | Information-Processing System | Behaviors |
| | | | Expectations |

© Cengage Learning 2014

a retrenchment strategy to one of stability. From a portfolio perspective, a corporation may eliminate one or more strategic business units (Chapter 9) and/or acquire new ones (as when PepsiCo divested its bottling group). At the business level, an organization may shift among the defending, prospecting, and analyzing approaches. Likewise, it may move from a differentiation strategy to one of cost leadership, or from cost leadership to targeting. Finally, the organization may also change one or more of its functional strategies. These changes might occur in any or all of the areas of marketing, finance, production, research and development, human resource, or organization design. For example, Sara Lee sold its Ambi Pur air care business to Procter & Gamble, sold its stake in a joint venture to Godrej Consumer Products Ltd., and sold its detergent business to Unilever to refocus its strategy.[20]

## Structural Change

**Structural change** is any change directed at a part of the formal organizational system.[21] In general, such changes relate to structural components, overall organization design, or other aspects of the organization.

In Chapter 12 we identified several basic components of organization structure. One or more of these components will need to be changed from time to time. That is, the organization may change its degree of decentralization, its span of management, or its bases of departmentalization. Other structural components can also change: The organization may adopt new methods of coordination, change the degree of job specialization, or modify working hours for employees.

On a larger scale, the company might need to change its overall design. It has been estimated that most companies need changes of this magnitude every five years or so.[22] In Chapter 13, we looked

at several forms of organization design. Whenever a firm changes its design, it is embarking on a major organizational change. In the late 1970s, for instance, Texas Instruments changed from a functionally organized firm to a matrix organization. In the early 1980s it made modifications in its matrix design.[23] And ten years later, it decided to alter its design once again to more effectively compete in the international arena.[24]

Structural changes may be made in other areas also, such as the organization's reward system, its performance appraisal system, use of participative management, and/or its control system.[25]

## Technological Change

A third major type of change is related to technology. **Technological change** has been increasingly widespread in the last few years.[26] Like strategic and structural change, it comes in many forms. One obvious area is new equipment. Agribusinesses continually look for innovations that will save crops from pests and from weather-related damage such as sustained freezes, ice storms, and hailstorms. As new and more efficient machines are introduced into the market, manufacturers occasionally need to upgrade to keep pace.

Some technological changes may make an otherwise unacceptable job more attractive. At Summer Farms, for example, the one-horse plow and cultivating machine have been happily swept aside by tractors, then tractors with air-conditioned cabs, radios and electronic sound equipment, and even computers. Even a Summer grandchild who otherwise is eager to leave the homestead may "sign on" until he or she has definitely decided on a different career.

Changes also may affect work processes or sequences. Tractors with air-conditioned cabs, for instance, enable their operators to increase the length of their workdays. A decision to use plastic rather than paper packaging necessitates major changes in the work processes. Similarly, the sequence of work can change as management considers new ways of arranging the workplace. This area will become increasingly important as flexible manufacturing systems come to replace traditional assembly lines.

Automation is a major form of technological change (Figure 14.7). Such changes as those made by California Dairies, Inc. in Turlock, California, which uses Motoman Robotics in its palletizing operations to greatly increase productivity, also represent a major capital investment.[27] Automatic irrigation equipment is an excellent example of the type of technological change that has become a necessity.

Finally, significant changes can also be made in the organization's information processing system. For example, the typewriter, a mainstay of many offices, has been widely replaced with word processors, and most managers now use personal computers. Each individual piece of equipment can be tied with the others in an integrated network to provide access to large blocks of data. Such data processing capabilities represent major changes.

**FIGURE 14.7** Modern food factory production line using automatically controlled machines.

© iStockphoto/Richard Clark

## People-Focused Change

Finally, people within an organization can be the focus of change. **People-focused change** can actually affect two distinct areas: (1) skills and performance, and (2) attitudes, perceptions, behaviors, and expectations.

In general, managers can take three approaches to upgrade skills and performance. First, they can replace current employees. This is clearly a difficult route to choose and usually should be taken only when no other options are available. Alternatively, managers can gradually upgrade selection standards. With this approach, existing employees are not dismissed; but when someone leaves for a better job, retires, or is fired, his or her replacement is selected with higher standards. A third option is to train existing employees to upgrade their performance-related skills. Change focused more on attitudes, perceptions, behaviors, and expectations is often undertaken from the perspective of organization development. We discuss this area in more detail in our next section. Many of the underlying causes of organization development are also discussed in depth in Chapters 15–20.

**FOOD FOR THOUGHT 14.2**

In 1996, the U.S. Agriculture Department proposed an unpopular change for school lunches: yogurt as a substitute for meat.

## Organization Development

A major approach to people-focused change is organization development. It focuses on changing attitudes, perceptions, behaviors, and/or expectations.[28]

### The Nature of Organization Development

**Organization development** can be defined as a planned, organization-wide effort to enhance organizational health and effectiveness through the systematic application of behavioral science techniques.[29] Rather than being a single, isolated change, organization development actually represents a complete philosophy of management. It assumes that people want to grow and develop, that they are capable of making useful contributions, and that one of the missions of the organization is to help facilitate personal growth as a way of enhancing employees' contributions.

Many large organizations practice organization development on a regular basis.[30] Some, such as Trimark Technologies in Chicago and Hanson Trust in England, have executive positions for organization development. American Airlines, Federated Department Stores, ITT, Polaroid, Procter & Gamble, and B. F. Goodrich have used organization development programs at times. In some cases, those trained in organization development serve as internal consultants, helping other managers who feel a need to use it.

## Organization Development Techniques

A number of change techniques fall under the organization development umbrella. One of the best known is the **Managerial Grid**, which is used to assess current leadership styles in an organization and then to train leaders to practice an ideal style of behavior.[31] The basic Grid has two axes—the vertical axis represents concern for people while the horizontal axis represents concern for production. Each leader is scored from one to nine along each dimension resulting in five main managerial styles Impoverished Management (1,1), Country Club Management (1,9), Authority-Compliance Management (9,1), Middle of the Road Management (5,5), and Team Management (9,9).

The Grid assumes that the 9,9 combination is the ideal style of behavior for leaders. Managers in the organizations go through a six-phase training program to move them in the general direction of the 9,9 coordinates. Although some people are critical of the Grid's effectiveness, others claim it has been quite beneficial.

In addition to the grid, of course, other organization development techniques are used by many organizations.[32] **Team building**, for example, is used to enhance the motivation and satisfaction of people in groups. It is a series of activities and exercises designed to foster mutual understanding acceptance and group cohesion. **Survey feedback**—surveying subordinates about their perceptions of their leader and then providing feedback to the entire group—helps increase communication between leaders and their subordinates.

## Organization Revitalization

A second related area of organization change that organizations occasionally must use is revitalization. **Organization revitalization** is a planned effort to bring new energy, vitality, and strength to an organization.

### Reasons for Revitalization

Why is revitalization necessary? For the most part, organizations do an imperfect job of maintaining an effective alignment with their environments. Given that environments change rapidly, the best that most organizations can do is to approximate the best strategy, design, and so forth that fits their particularly unique situations. Even such well-managed organizations as Disney and Southwest Airlines must occasionally step back and work to improve their alignment with their environments. Sometimes, however, an organization falls so far out of alignment with its environment that simple adjustments and shifts are not enough. When this happens, the organization must undergo a major change to get back in proper alignment with its environment.

### Stages in Revitalization

Figure 14.8 shows the general cycle that revitalization can follow. First of all, an organization has what might be called normal momentum. It is growing and effectively reaching its goals regularly. At some point, however, the organization falls so far out of alignment with its environment that its growth reaches a plateau or goes into a decline. For example, U.S. automakers were not aligned with their environment when they were turning out gas-guzzling, low-quality products whereas economy-minded consumers were concerned about quality and gas prices.

If ineffective alignment occurs and the firm cannot turn things around, it may have to enter a planned period of contraction. During this stage the organization cuts back on its operations, eliminates unnecessary facilities, and so forth. Next comes consolidation. During this phase, the organization learns to live with a leaner and tighter budget. Eventually, if things go well, the organization can start expanding and growing once again.

During these various stages, the company may undertake a number of initiatives to facilitate its revitalization. For example, it may bring in a new management team. It may also seek an infusion of new capital through extended bank loans or new investment. Some firms even go so far as to change their names altogether. After extensive programs of revitalization, International Harvester became Navistar, for example, and United States Steel became USX.

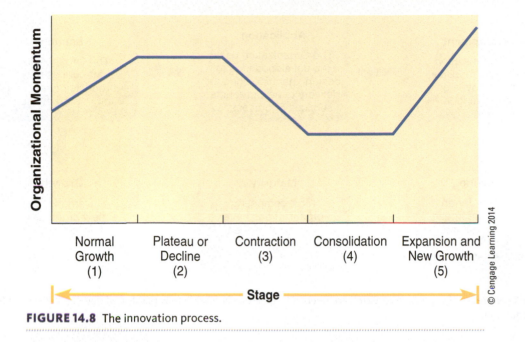

FIGURE 14.8 The innovation process.

© Cengage Learning 2014

## Innovation in Organizations

Finally, many organizations today are recognizing the critical importance of yet another aspect of change, innovation. **Innovation** is the managed effort of an organization to develop new products or services and/or new uses for existing products or services.

---

**FOOD FOR THOUGHT 14.3**

Pearl Milling Company developed Aunt Jemima, the first ready-made pancake mix, in 1889. A hundred years later, in 1989, the image of Aunt Jemima was updated by removing her headband and giving her pearl earrings and a lace collar.

---

### The Innovation Process

Organizations that emphasize innovation usually start with a culture that values innovation and then use that culture to attract creative employees and managers. To harness this creativity, however, the organization must keep itself focused on managing the innovation process itself. The process of developing, applying, launching, growing, and managing the maturity and decline of a creative idea is called the **organizational innovation process**.[33] Figure 14.9 is a depiction of this process.

**Innovation Development.** Once an idea for innovation is identified and approved, product prototypes are usually built or grown. Most original ideas are not ready to be instantly transformed into new products or services. **Innovation development** is the stage in which an organization evaluates, modifies, and improves on a potential innovation before turning that idea into a product or service to sell. Innovation development can transform a product or service with only modest potential into a product or service with significant potential. Parker Brothers, for example (manufacturers of the board games Monopoly, Risk, and Life), was developing plans for an indoor volleyball game. During innovation development, managers decided not to market the game itself but instead to sell separately the appealing little foam ball that designers had created for the game. That product, the Nerf ball, has generated millions of dollars in revenues for Parker Brothers.[34]

**Innovation Application.** Even after development, an idea has yet to be applied to real products or services. Innovation application is the stage in which an organization takes a developed idea and uses it in the design, manufacture, or delivery of new products, services, or processes. At this point the innovation emerges from the laboratory and is transformed into tangible goods or services. One example of innovation application is the use of radar-based focusing systems in Polaroid's instant

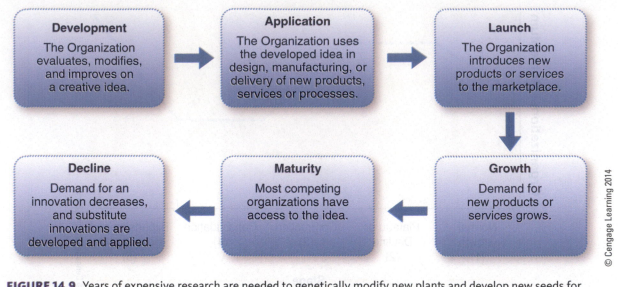

© Cengage Learning 2014

**FIGURE 14.9** Years of expensive research are needed to genetically modify new plants and develop new seeds for crop production.

cameras, starting in 1986. The idea of using radio waves to discover the location, speed, and direction of moving objects was first applied extensively by Allied forces during World War II. As radar technology developed over the following years, the electrical components needed became smaller and more streamlined. Researchers at Polaroid hit on radar as a creative idea and applied this well-developed technology in a new way.[35]

**Application Launch.** Application launch is the stage in which an organization introduces new products or services to the marketplace. The key question is not "Does the innovation work?" but "Will customers want to purchase the innovative product and service?" History is full of creative ideas that did not generate enough interest among customers to be successful. Some notable innovation failures include the Edsel automobile (with its notoriously homely front end), Polaroid's SX-70 instant camera (which cost $3 billion to develop, but never sold more than 100,000 units in a year), R.J. Reynolds' smokeless cigarette, Coca-Cola's New Coke and DuPont's Corfam. Thus, despite individual creativity, development, and application, it is still possible for new products and services to fail at the application launch phase of innovation.[36]

**Application Growth.** Once an innovation has been successfully launched, it then enters the stage of application growth, where demand for an innovation increases. This is a period of high economic performance for an organization because the demand for the product or service may exceed the supply, as has often been the case with electronics. Organizations that fail to anticipate this stage may unintentionally limit their growth, as Gillette did by not anticipating demand for its Sensor razor blades. At the same time, overestimating demand for a new product or service can be just as detrimental to performance. This occurred for many "western wear" retailers in the early 1980s after the popularity of the John Travolta movie "Urban Cowboy" suddenly, but very briefly, increased demand for cowboy boots, shirts, and other clothes. Many stores lost money anticipating higher demand than what actually existed and purchasing a great deal of western wear inventory that took years to sell.

**Innovation Maturity.** After a period of growing demand, an innovative product or service often enters a period of maturity. **Innovation maturity** is the stage in which most organizations in an industry have access to an innovation and are applying it in approximately the same way. The technological application of an innovation during this stage of the innovation process can be very sophisticated. However, because most firms have access to the innovation (either because they have developed the innovation on their own or have copied the innovation of others), it does not provide a competitive advantage to any one of them.

The time that elapses between innovation development and innovation maturity varies significantly, depending on the particular product or service. Whenever an innovation involves the use of complex skills (such as a complicated manufacturing process or highly sophisticated teamwork), it will take longer to move from the growth phase to the maturity phase. In addition, if the skills needed to implement these innovations are rare and difficult to imitate, then strategic imitation may be delayed and the organization may enjoy a period of sustained competitive advantage.

One innovation that has taken a very long time to move from growth to maturity is the user-friendliness of computers. When Apple first introduced the Macintosh, it immediately provided a level of user-friendliness that was unique in the personal computer industry. Over the years, however, Microsoft's Windows software begin to address user-friendliness effectively and Apple began offering Windows on its Macintosh models, too.

On the other hand, when the implementation of an innovation does not depend either on rare or difficult-to-imitate skills, then the time between the growth and maturity phases can be brief. In the market for computer memory devices, for example, technological innovation by one firm can be very quickly duplicated by other firms because the skills needed to design and manufacture these electronic devices are widespread. Computer memory devices thus move very rapidly from growth to maturity.

**Innovation Decline.**  Every successful innovation bears its own seeds of decline. **Innovation decline** is the stage during which demand for an innovation decreases and substitute innovations are developed and applied. Since an organization does not gain a competitive advantage from an innovation at maturity, it must encourage its creative scientists, engineers, and managers to begin looking for new innovations. This continual search for competitive advantage usually leads new products and services to move from the creative process through innovative maturity, and finally to innovative decline. It happens to any business regardless of size, from the entrepreneurial plant breeder mentioned in an earlier chapter to multinational corporations.

## Types of Innovation

Each creative idea that an organization develops poses a different challenge for the innovation process. Innovations can be radical or incremental, and technical or managerial.

### Radical Versus Incremental Innovations.

**Radical innovations** are new products or technologies that completely replace the existing products or technologies in an industry. For example, compact disk technology has replaced long-playing vinyl records in the recording industry, high definition television replaced regular television technology, and genetically engineered seeds are affecting farming. **Incremental innovations** are new products or processes that modify existing products or technologies. Whereas radical innovations tend to be visible and public, incremental innovations actually are more numerous but less obvious. In the meat industry, incremental innovations including meat processing automation machinery and engineering, meat quality enhancement and packaging innovation have improved productivity. Thus, it is an evolutionary change—one logical step following another.

> ### FOOD FOR THOUGHT 14.4
> Clarence Birdseye, the father of frozen food, says that he simply took Eskimo knowledge and scientists' theories when he used $7 worth of buckets of brine, cakes of ice, and an electric fan to freeze fish under high pressure.

### Technical versus Managerial Innovations.

**Technical innovations** are changes in the physical appearance or performance of a product or service, or the physical processes through which a product or service is manufactured. Many of the most important innovations over the last fifty years have been technical. For example, the serial replacement of the vacuum tube with the transistor, the transistor with the integrated circuit, and the integrated circuit with the microchip have greatly enhanced the power, ease of use, and speed of operation of a wide variety of electronic products. Foods that are processed and packaged for microwave cooking have become standard purchases for many consumers.

**Managerial innovations** are changes in the management process by which products and services are conceived, built, and delivered to customers. Managerial innovations do not necessarily affect the physical appearance or performance of products or services directly, but they can. Many Japanese firms have used managerial innovations to improve the quality of their products or services. One of the most important of these innovations developed in the last thirty years is called the quality circle.[37] It helped Oki Electronics become one of the premier electronics companies in the world. Organizations that have been able to incorporate quality circles and total quality management (in which small groups of concerned workers discuss how to improve product quality) have found that quality improves dramatically and costs of operations decrease.[38]

## Barriers to Innovation

To remain competitive in today's economy, it is necessary to be innovative. Yet many organizations that should be innovative are not successful at bringing out new products or services, or do so only after innovations created by others are mature. There are at least three reasons why organizations may fail to innovate.[39]

### Lack of Resources.

Implementing innovative change can be expensive in terms of dollars, time, and energy (Figure 14.9). If a firm does not have sufficient money to fund a program of innovation or does not currently employ the kinds of creative individuals it needs to be innovative, it may find itself lagging behind in innovation. Even highly innovative organizations cannot become involved in every new product or service its employees think up. For example, other commitments in the electronic instruments and computer industry kept Hewlett-Packard from investing in two young inventors who came to the firm with an idea for a personal computer. Those inventors, Steve Jobs and Steve Wozniak, eventually became frustrated and formed their own company, Apple Computer, instead. (Would things have turned out differently if Hewlett-Packard had not had a culture of avoiding bank financing?)

### Failure to Recognize Opportunities.

To obtain a competitive advantage, organizations usually must make investment decisions before the innovation process reaches the mature stage. Since they cannot pursue all innovations, they should develop the capability to evaluate innovations carefully and to select the ones that hold the greatest potential. If not skilled at recognizing and evaluating opportunities, they may be overly cautious and fail to invest in innovations that turn out later to be successful for their competitors.

### Resistance to Change.

As already noted, there is a tendency in many organizations to resist change. Innovation means giving up old products and old ways of doing things in favor of new products and new ways of doing things. These kinds of changes can be personally difficult for managers and other members of an organization.

## Facilitating Innovation

A wide variety of ideas for promoting innovation in organizations has been developed over the years. Three specific ways for promoting innovation are described below.

### The Reward System.

An organization may establish a reward system to encourage individuals and groups to develop innovative ideas. Key components of the reward system include either financial or nonfinancial rewards, such as salaries, bonuses, perquisites, and so forth. Using the reward system to promote innovation is a fairly mechanical but nevertheless effective management technique. Once the members of an organization understand that they will be rewarded for their innovations, they are more likely to work creatively. With this end in mind, Monsanto gives cash awards to the scientists and educators to further new developments and to strengthen science education.

Although it is important for organizations to reward innovative ideas, it is just as important to avoid punishing people when their innovative ideas are not successful. Fewer than 25 percent of new products are ever commercialized, and half of them don't become financially successful.[40] If innovative failure is due to incompetence, systematic errors, or managerial sloppiness, then an organization should respond with appropriate sanctions such as withholding raises or reducing promotion opportunities. However, people who act in good faith to develop an innovation that simply does not work out should not be punished for failure. If they are, they will probably not be as motivated to innovate in the future. To avoid punishing failure inappropriately, managers must thoroughly understand both the skills and capabilities of a creative individual and the goal the individual was attempting to accomplish. By developing this understanding, managers will be able to distinguish among activities that simply did not work out and other activities that reflect managerial incompetence, stupidity, or poor judgment.

**Intrapreneurship.** Intrapreneurship also helps organizations encourage innovation.[41] **Intrapreneurs** are similar to entrepreneurs except that they develop a new idea in the context of a larger organization.[42] There are three intrapreneurial roles in large organizations.[43] To use intrapreneurship successfully for encouraging creativity and innovation, the organization must find one or more individuals to perform these roles.

The *inventor* is the person who actually conceives of and develops the new idea, product, or service by means of the creative process. However, because the inventor may lack the expertise or motivation to oversee the transformation of the product or service from an idea into a marketable entity, a second role comes into play.

A *product champion* is usually a middle manager who learns about the project and becomes committed to it. He or she helps overcome organizational resistance and convinces others to take the innovation seriously. The product champion may have only limited understanding of the technological aspects of the innovation. However, product champions are skilled at knowing how the organization works, whose support is needed to push the project forward, and where to go to secure the resources necessary for successful development.

A *sponsor* is a top-level manager who approves of and supports a project. This person may fight for the budget needed to develop an idea, overcome arguments against a project, and use organizational politics to ensure the project's survival. With a sponsor in place, the inventor's idea has a much better chance of being successfully developed.

Several firms have embraced intrapreneurship as a way to encourage creativity and innovation. Colgate-Palmolive has created a separate unit, Colgate Venture Company, staffed with intrapreneurs who develop new products. General Foods developed Culinova Group as a unit to which employees can take their ideas for possible development. S.C. Johnson and Sons established a fund to support new product ideas, and Texas Instruments refuses to approve a new innovative project unless it has an acknowledged inventor, champion, and sponsor.

**Organizational Culture.** As we discussed in Chapter 4, an organization's culture is the set of values, beliefs, and symbols that help guide behavior. A strong, appropriately focused organizational culture can be used to support creative and innovative activity.[44] A well-managed culture can communicate a sense that innovation is valued and will be rewarded and that occasional failure in the pursuit of new ideas is not only acceptable but even expected. In addition to reward systems and intrapreneurial activities, firms such as 3M, Monsanto, Procter and Gamble, Johnson & Johnson, Apple, and Merck are all known to have strong, innovation-oriented cultures. These cultures value individual creativity, risk taking, and inventiveness. And, as noted in Chapter 5, an organization could pursue a Blue Ocean Strategy in which it attempts to innovate through simultaneous differentiation and low costs.

## CHAPTER SUMMARY

Organization change is a meaningful alteration in some part of the organization. Forces in the general and task environments can prompt the need for change. To the extent possible, organizations should plan for change that can be pursued through a series of rational and logical steps.

People resist change for a variety of reasons, including uncertainty, self-interests, different perceptions, and feelings of loss. There are four common methods for overcoming this resistance: participation, open communication, facilitation, and force-field analysis.

Organization change generally takes place in one or more of four areas—strategy, structure, technology, and people. Strategic change can occur at the corporate, business, or functional levels. Structural change involves changes in the formal organizational system. Technological change might involve new and different machines or work processes. People-focused change involves changes in the skills, performance, attitudes, perceptions, behaviors, and expectations of the people in the organization.

Organization development is a planned, organization-wide effort to enhance organizational health and effectiveness through the systematic application of behavioral science techniques.

Organization revitalization is a planned effort to infuse new energy, vitality, and strength into an organization. The general stages in revitalization include growth, plateau or decline, contraction, consolidation, and renewed expansion and new growth.

Innovation is a managed effort to develop new products or services or new uses for existing products or services. The innovation process includes development, launch, growth, maturity, and decline. Forms of innovation include radical or incremental and technical or managerial. While several barriers to innovation can be identified, managers can also actively facilitate innovation.

## CHAPTER ACTIVITIES

### REVIEW QUESTIONS

1. Distinguish between planned and reactive change.
2. What are the basic steps to follow in planned change? Identify the basic reasons people resist change and the most common methods for overcoming that resistance.
3. Provide at least two examples of each area of organization change.
4. What is organization development? What are some of its more common techniques?
5. What is innovation? Describe the innovation process.

### ANALYSIS QUESTIONS

1. Heraclitus once said, "There is nothing permanent except change." Is change, in fact, inevitable? Why or why not?
2. The text notes that people tend to resist change. Are there some people who like change? Is it possible for people to initiate too much change?
3. Think of a change you recently experienced. How was it managed? How did you feel about it?
4. Identify an organization that has never had to go through a period of revitalization?
5. Can creativity—as part of the innovation process—really be managed? Can an organization concentrate too much attention on managing innovation?

## FILL IN THE BLANKS

1. When an organization modifies its strategy or adopts a new strategy, it has engaged in _____ change.

2. Changes made in anticipation of possible changes in the environment and how the organization should most likely respond to those changes is called _____ _____ change.

3. An unplanned response to environmental changes as they occur is called _____ change.

4. A planned effort to bring new energy, vitality, and strength to an organization is known as _____ _____.

5. The managed effort by organizations to develop new products or services and/or new uses for existing products or services is called _____.

6. The planned, organization-wide effort to enhance organizational health and effectiveness through the systematic application of behavioral science techniques is known as _____ _____.

7. Modifications of existing products or technologies are called _____ innovations.

8. Products that involve changes in the physical appearance or performance of a product are known as _____ innovations.

9. Innovative people who work in the context of a large organization rather than for themselves are called _____.

10. The stage at which most organizations in an industry have access to an innovation and are applying it in approximately the same way is referred to as _____ _____.

## CHAPTER 14 CASE STUDY
### AGRIBUSINESS HAS ITS OWN "OIL SPILL"

Many years ago a popular show did a comedy skit that was a parody of television food advertising. The skit featured a husband and wife arguing whether the whipped-cream product they were featuring was a floor wax or a dessert topping. The absurdity of the juxtaposition of these two uses made for good comedy.

Fast forward to the present and such an apparent absurdity no longer seems so absurd. We have become accustomed to diverse and previously unthought-of uses for raw materials that were once limited to, or off-limits to, food products. It's all a matter of chemistry—food chemistry. Palm oil is one example. It can be used in baking and in making potato chips, margarine, peanut butter, ice cream, lipstick, shaving cream, shampoo, and biofuel. Palm oil is second only to soybean oil in vegetable oils used in food production, and vegetable oil is a huge agribusiness undertaking, amounting to 95 million tons per year.

The industry began in 1848 when it was discovered that the palm plant found in West Africa grew very well in Malaysia and Indonesia. Since then, it has become an economic mainstay for these two countries, which in 2009 produced almost 90 percent of the 35 million tons of palm oil exported to the world. At more than $800/ton, palm oil is the least expensive edible oil on the market. The oil palm is an efficient crop, yielding up to ten times more oil per hectare than soybeans.

For Malaysia and Indonesia palm oil has become an important export commodity—their comparative and competitive advantage in global trade. For the poorer farmers and plantation workers who bring down the giant bunches of red fruit by hand using a long-handled scythe, palm oil is an important source of calories and income. The versatility of palm oil has made it a mainstay in the products of Unilever PLC (Anglo-Dutch), Kraft Foods, Inc, and General Mills Inc.(United States), Danone SA (French), Nestle NV (Swiss), and Cargill (United States).

Because of the increased production of palm oil in Indonesia and Malaysia, it has also become the focus of worldwide attention. Indonesia alone has 11,550 square miles of palm oil plantations. Production has expanded through available grasslands, burning peat-lands and deforested land. Indeed, in the past the financing for the palm oil plantation arose from the sale of cleared timber. For the poorer farmers removing the other trees to make room for more palm trees was the only way for them to become producers in the industry. They viewed these unused lands as their future, but the world environmental community viewed their activities as wasteful and detrimental to the global environment. It was the classic clash of economic opportunity versus ecological consequences.

The large amount of palm production, the size of the multi-national corporations using palm oil, and the rising international concerns about climate have attracted the attention of many groups. A public pressure group, Friends of the Earth, published a report in 2006 entitled, "Oil for Ape Scandal." This study asserted that unless something is done, the habitat for the Orangutan in Indonesia would be destroyed in 12 years. The United Nations Environment Program also raised concern about deforestation. Greenpeace began issuing reports in 2009 alleging illegal rainforest clearing by PT Sinar Mas Agro Resources & Technology (SMART), an Indonesian palm oil company.

The effect of this publicity has been to prod the Roundtable on Sustainable Palm Oil (RSPO) into action. This organization was founded in 2004 by concerned growers, processors, food companies, and others to encourage sustainable palm oil, i.e. "certified" as not having resulted from a destruction of areas of high conservation value. As a result, "certified" palm oil has become an important commodity. International attention has also prompted other action as Nestlé removed SMART oil from its Kit-Kat bars.

Cargill—the largest privately owned corporation in the United States and the operator of two palm plantations in Indonesia and twelve palm oil refineries around the world—has gained certification for its palm oil. "Adoption of responsible and sustainable practices by small holders will lead to rising rural incomes as demand for certified palm oil continues to grow," said Angeline Ooi, CEO, Cargill's oil palm joint-venture company. Cargill's holdings involve 8,800 growers in 17 cooperatives who will become the first RSPO-certified group of small holders in the world.

Globalization provides many opportunities, but it also sets constraints. Economic development is an opportunity for all, especially in agribusiness; but awareness of the cost of this opportunity is essential. Cargill's approach may be more productive than engaging in endless debate as Sinar Mas intends to do. Competitive advantages are made, not just found.

Cargill's approach to this situation demonstrates how to turn lemons into lemonade. Or is it palm oil (certified) into candy bars? [45]

## ▶ Case Study Questions

1. All things considered, do agribusiness companies help or hurt countries and their citizens when they import products that enable people to rise from poverty, even if it is for only a few years? Explain your answer.

2. If citizens in countries like Malaysia and Indonesia are hurting themselves in the long-run (as in deforesting other trees or burning grassland) for a short-run boost in their standard of living, do the importers of the raw materials have a moral obligation to help the suppliers plan for their long-term economic survival?

3. Do you believe that "competitive advantages are made, not just found"? Explain your answer and how it applies in the palm oil case.

# REFERENCES

1. David Michaels, "As Pencil Makers Push the Envelope, Age-Old Rivalry Stays Sharp," *The Wall Street Journal* (September 29, 2010) at online.wsj.com (accessed December 12, 2010); "The Future of the Pencil," *The Economist* (September 16, 2010) at www.economist.com (accessed December 12, 2010); "Wonder Drug," *The Economist* (December 11, 2010) at www.economist.com (accessed December 12, 2010).

2. Michael A. Hitt, "The New Frontier: Transformation of Management for the New Millennium," *Organizational Dynamics* (Winter 2000): 7–15.

3. Parija Kavilanz, "Another Johnson & Johnson Drug Plant Gets Flagged," CNNMoney.com (July 19, 2010) at money.cnn.com (accessed July 28, 2010); "Leaders of Corporate Change," *Fortune*, December 14, 1992, 104–114; "IBM to cut 25,000 More Jobs, Spending," *USA Today*, December 16, 1992, 1B, 2B.

4. Achilles A. Armenakis and Arthur G. Bedeian, "Organizational Change: A Review of Theory and Research in the 1990s," *Journal of Management* (1999): 293–315.

5. Roy McLennan, *Managing Organizational Change* (Englewood Cliffs, NJ: Prentice-Hall, 1989).

6. Karl E. Weick and Kathleen M. Sutcliffe, *Managing the Unexpected: Resilient Performance in an Age of Uncertainty, 2nd ed.* (New York: Wiley, 2007); Cynthia A. Lengnick-Hall and Tammy E. Beck, "Adaptive Fit Versus Robust Transformation: How Organizations Respond to Environmental Change," *Journal of Management* (October 2005): 738–757.

7. Gary Hamel and Lisa Valikangas, "The Quest for Resilience," *Harvard Business Review* (September 2003): 62–75; Diane L. Coutu, "How Resilience Works," *Harvard Business Review* (May 2002): 46–55.

8. Dean Robb, "Building Resilient Organizations," *OD Practitioner* (2000): 27–32; Larry Mallak, "Putting Organizational Resilience to Work." *Industrial Management* (Nov/Dec 1998): 8–13.

9. Clive de W. Blackburn and Peter J. McClure (eds.) *Foodborne Pathogens: Hazards, Risk Analysis and Control* (Boca Raton, FL: Woodhead Publishing Limited and CRC Press, 2002); C. Haas, "The Role of Risk Analysis in Understanding Bioterrorism," *Risk Analysis* (2002): 671–677; Michel Crouhy, Robert Mark, and Dan Galai, *Risk Management* (New York: McGraw-Hill, 2001).

10. "To Maintain Success, Managers Must Learn to Direct Change," *The Wall Street Journal*, August 13, 2002, B1.

11. "Managing Change: Reinventing America," a special issue of *Business Week*, 1992, 59–74.

12. 2009 Annual Report, "The Power of Yum!" Yum! Brands, Inc., website, www.yum.com (accessed on June 30, 2010); Gerry Khermouch, "Tricon's Fast-Food Smorgasbord," *Business Week* (February 11, 2002) at www.businessweek.com (accessed January 16, 2003).

13. David Kesmodel, "Pabst Changes Hands," *The Wall Street Journal* (June 28, 2010) at online.wsj.com (accessed July 19, 2010).

14. Paul R. La Monica, "Betting the Farm on Agriculture Stocks," *Fortune* (June 23, 2008) at money.cnn.com (accessed July 19, 2010); Shruti Date Singh, "Bunge Buyback Plan Makes Bid Less Likely, BMO Says," *Business Week* (June 9, 2010) at www.businessweek.com (accessed June 30, 2010); and Thomas Biesheuvel, "Tate & Lyle May Be Takeover Target for Bunge, Evolution Says," *Business Week* at www.businessweek.com (accessed June 30, 2010).

15. Rosabeth Moss Kanter, "Change: Where to Begin," *Harvard Business Review* (July–August 1991): 8–9; Michael Beer, *Organization Change and Development: A System View* (Santa Monica, CA: Goodyear, 1980).

16. Paul R. Lawrence, "How to Deal with Resistance to Change," *Harvard Business Review* (March–April, 1979): 106–114.

17. "Revolt of Uncle Sam's Paper Pushers," *Business Week*, October 30, 1989, 156.

18. N. M. Manalili, D. M. Campilan, and P. G. Garcia, "Cooperative Enterprises: Losing Relevance or Still Responsive to the Challenges of Dynamic Markets?," *Stewart Postharvest Review*, 4(5): 1–7; J. G. Carlberg, C. E. Ward, and R. B. Holcomb, "Success Factors for New Generation Co-Operatives." *International Food & Agribusiness Management Review*, 9: 62–81; S. D. Hardesty, "The Bottom Line on the Conversion of Diamond Walnut Growers," *UPDATE: Agricultural and Resource Economics* (University of California Gianini Foundation), 8(6): 1–4 and 11.

19. John P. Kotter and Leonard A. Schlesinger, "Choosing Strategies for Change," *Harvard Business Review* (March–April, 1979): 106–114.

20. "Investor Relations," Sara Lee website at www.saralee.com (accessed July 27, 2010).

21. Harold J. Leavitt, "Applied Organizational Change in Industry: Structural, Technical, and Human Approaches," in *New Perspectives in Organization Research,* eds. W. W. Cooper, H. J. Leavitt, and M. W. Shelly (New York: Wiley, 1964), 55–71.

22. Kotter and Schlesinger, "Choosing Strategies for Change."

23. Bro Uttal, "Texas Instruments Regroups," *Fortune*, August 9, 1982, 40–45.

24. "U.S. Exporters That Aren't American," *Business Week*, February 20, 1988, 70–71; "What's Behind the Texas Instruments-Hitachi Deal," *Business Week*, January 16, 1989, 93–96.

25. Brian Dumaine, "The Bureaucracy Busters," *Fortune*, June 17, 1991, 36–50.

26. Dorothy Leonard-Barton and William A. Kraus, "Implementing New Technology," *Harvard Business Review* (November–December 1985): 102–110.

27. Amanda Nolz, "High Volume Dairy Implements Robotic Palletizing," *World Dairy Daily* (October 27, 2009) at www.wdexpo.org (accessed July 28, 2010).

28. Paul Bate, Raza Khan, and Annie Pye, "Towards a Culturally Sensitive Approach to Organization Structuring: Where Organization Design Meets Organization Development," *Organization Science* (2000): 197–211.

29. Richard Beckhard, *Organization Development: Strategies and Models* (Reading, MA: Addison-Wesley, 1969).

30. William Pasemore, "Organization Change and Development," *Journal of Management* (June 1992).

31. Robert R. Blake and Jane S. Mouton, *The Managerial Grid* (Houston: Gulf, 1964).

32. Wendell L. French and Cecil H. Bell, Jr. *Organization Development: Behavioral Science Interventions for Organization Improvement, 2nd ed.* (Englewood Cliffs, NJ: Prentice-Hall, 1978).

33. L.B. Mohr, "Determinants of Innovation in Organizations," *American Political Science Review* (1969): 111–126; G. A. Steiner, *The Creative Organization* (Chicago: University of Chicago Press, 1965); R. Duncan and A. Weiss, "Organizational Learning: Implications for Organizational Design," in *Research in Organizational Behavior, Volume 1, ed.* B. M. Staw (Greenwich, CT: JAI Press, 1979), 75–123; and J. E. Ettlie, "Adequacy of Stage Models for Decisions on Adoption of Innovation," *Psychological Reports* (1980): 991–995.

34. Beth Wolfensberger, "Trouble in Toyland," *New England Business*, September 1990, 28–36.

35. Alan Patz, "Managing Innovation in High Technology Industries," *New Management* (1986): 54–59.

36. An excellent guide to these kinds of management errors is Robert F. Hartley, *Management Mistakes and Successes, 3rd ed.* (New York: John Wiley, 1991).

37. David D. Van Fleet, Ricky W. Griffin, and Tom C. Head, "Quality Circle Effectiveness in High Technology Organizations," *Implementation Management in High Technology*, eds. L. R. Gomez-Mejia and M. W. Lawless (Greenwich, CT: JAI Press, 1995): 105–123.

38. For a discussion of quality circles at Oki Electric, see William G. Ouchi, *Theory Z* (Reading, MA: Addison-Wesley, 1980).

39. Clayton M. Christensen, Stephen P. Kaufman, and Willy C. Smith, "Innovation Killers," *Harvard Business Review* (January 2008): 98–107.

40. V. Kasturi Rangan, Rajiv Lal, and Ernie P. Maier, "Managing Marginal New Products," *Business Horizons* (September–October 1992): 35–42.

41. David A. Garvin and Lynne C. Levesque, "Meeting the Challenge of Corporate Entrepreneurship," *Harvard Business Review* (October 2006): 102–113.

42. Geoffrey Moore, "Innovating Within Established Enterprises," *Harvard Business Review* (2004): 87–96.

43. Gifford Pinchot III, *Intrapreneuring* (New York: Harper and Row, 1985).

44. Steven P. Feldman, "How Organizational Culture Can Affect Innovation," *Organizational Dynamics* (Summer 1988): 57–68.

45. Anthony Fletcher, "Palm Oil Research Targets Food Industry Benefits" (February 14, 2007) at www.foodnavigator.com (accessed September 5, 2010); "The Campaign against Palm Oil: The Other Oil Spill," *The Economist* (June 24, 2010) at www.economist.com (accessed September 3, 2010); Lorraine Heller, "Cargill Gets Palm Oil Sustainability Certification" (August 20, 2010) at www.foodnavigator.com (accessed September 3, 2010); Shie-Lynn Lim and Andreas Isma, "Palm Oil Firm Rebuts Greenpeace Claim," *The Wall Street Journal* (August 11, 2010) at online.wsj.com (accessed September 5, 2010); Laura Crowley, "Palm Oil Production May Trigger Climate Bomb" (November 8, 2007) at www.foodnavigator.com (accessed September 4, 2010); "A Commitment to Growth: 2010 Cargill Summary Annual Report" at www.cargill.com (accessed September 5, 2010); Yoga Rusmana, "Sinar Mas Says Greenpeace Deforestation Allegations 'Unfounded,'" *Bloomberg Businessweek* (August 10, 2010) at www.businessweek.com (accessed September 5, 2010).

# Staffing and Human Resources

## LEARNING OBJECTIVES

After studying this chapter, you should be able to:

- Discuss the nature of staffing, including the staffing process and legal constraints.

- Describe human resource planning and indicate how to use job analysis and forecasts to match supply and demand.

- Discuss the selection of human resources, including recruitment, selection, and orientation.

- Describe the assessment of training and development needs, various training and development techniques, and the importance of evaluating those techniques.

- Define performance appraisal and discuss objective and judgmental methods, management by objectives, and feedback.

- Discuss compensation decisions regarding wages and salaries, as well as various kinds of benefits.

- Discuss labor relations, including how unions are formed and the nature of collective bargaining.

## The Glass Ceiling: A Window to the Stars

Late in the twentieth century many business schools had a common problem. It was already an established fact that more women than men were pursuing baccalaureate degrees, a subtle shift that occurred in the 1990s. Now it was becoming apparent that more and more women were gaining competitive admission into the best MBA programs in the country. While this achievement was being hailed at the admissions gate, it was a cause for concern in the hallways of academia. The problem was not that these women were not qualified; unquestionably they were. The question was what would these women do when they encountered the "glass ceiling" (Figure 15.1)?

**FIGURE 15.1** The glass ceiling is still high and not easy to break through.
© Auremar/www.Shutterstock.com

The "glass ceiling" is a euphemism for a barrier that prevents women from gaining the chief executive officer post at major corporations. Countless business articles have attempted to identify the problem, citing boardroom bias, inappropriate training, different career goals, and distractions along the career path between job and family. Each of these probably contained a grain of truth. The factor that was really at work was missed by all business journal writers—time.

Over time a critical mass of dedicated, bright, and visionary women began to form. These were individuals with impressive track records, leadership skills, and experience. After all, experience is just another word for time. Change began to occur.

The 2012 Fortune 500 listed 19 female CEOs; the 2012 Fortune 1000 listed another twenty-one female CEOs. This was not all: a milestone occurred at Xerox, not an agribusiness company, when a woman replaced another woman as CEO. Ursula Burns replaced Anne Mulcahy as chief executive officer. While it might be argued that 40 out of 1,000 positions, or 4.0 percent, is hardly parity, it is a great and visible start.

Among agribusiness firms in 2012, a disproportionate number of women are CEOs: Indra Nooyi (Pepsico Inc.), Ellen J. Kullman (DuPont), Carol Meyrowitz (The TJX Companies), Irene Rosenfeld (Mondelēz International, Inc.), Sheri McCoy (Avon), Heather Bresch (Mylan), Patricia Woertz (Archer Daniels Midland Company), Ilene Gordon (Ingredion Incorporated), Linda Lang (Jack in the Box, Inc.), Kay Krill (ANN, Inc.), Gretchen McClain (Xylem), Diane M. Sullivan (Brown Shoe Company), Helen McCluskey (Warnaco Inc.), Sandra Cochran (Cracker Barrel), and Denise Morrison (Campbell Soup). Among the women CEOs mentioned in *Fortune* magazine, 37.5 percent of them were in agribusiness.

These are necessary and transformational steps. A country is not ready to assume its place in global competition unless all of its population can participate regardless of gender. A mind is still a terrible thing to waste.

So maybe the glass ceiling is not a barrier but a "window" to the stars. It is important to see where you are going before you leap.[1]

## INTRODUCTION

Most organizations must change the way they do business to remain competitive. One major aspect of such changes that we have already discussed involves developing new business strategies. Another may involve an entirely new approach to managing human resources, which includes developing a new philosophy about staffing—hiring, developing, and rewarding its employees.[2] An excellent example of this new approach is the hiring and promotion of women and minorities, especially allowing qualified and experienced individuals to break through the glass ceiling to become the leaders of major organizations.

## The Nature of Staffing

**Staffing** is the process of procuring and managing the human resources an organization needs to accomplish its goals. Staffing, however, is more than just hiring new employees. When Avon hires new direct marketers it is engaging in staffing. But if General Mills gives one hundred managers early retirement, it is also engaging in the staffing process. And when Welch sends 450 employees to a training seminar, it is engaging in staffing. CF Industries is involved in staffing when its employees get a new health insurance option. If Quaker Oats has to close a plant because of safety hazards, it is also exercising part of the staffing process. And while Outback Steakhouse seems to be doing things right (it has a culture consisting of good food, good quality, good value, good hospitality, and good customer service), things were less than perfect just a few years ago. Outback agreed to a $19 million dollar gender bias settlement in 2009.[3]

We have already defined staffing, so now let's develop a more complete framework for understanding staffing and for organizing our discussion. Then we can discuss the legal environment of the staffing process.[4] First, we provide more background information about the nature of staffing. We then examine human resource planning. Next we look at the selection process and training and development. Performance appraisal and compensation are covered next. Finally, we explore labor relations.

### FOOD FOR THOUGHT 15.1

In the U.S. in 2012, the agriculture industry employed over twice as many people as the automotive industry.

## The Staffing Process

Figure 15.2 presents the basic staffing process in detail. First, as indicated along the left side of the framework, the staffing process must take place within a series of legal constraints that restrict what a firm can do. Within this legal environment, the first actual step in staffing is human resource planning. This involves assessing the current situation through developing a complete understanding of the various tasks within the organization. Then the number of employees needed to perform those tasks are compared with current employees to determine staffing requirements. That will involve forecasting how many people are needed and will be available to perform the tasks.

Several steps follow the initial planning. If Ghirardelli Chocolate needs more employees, it must recruit qualified applicants to consider employment with the firm and then choose the ones best suited for the available jobs. After joining the company, the new employees must be trained and developed. They must be compensated from the time they begin work and also have the opportunity to participate

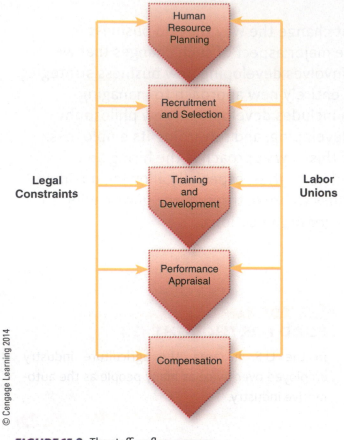

© Cengage Learning 2014

**FIGURE 15.2** The staffing flow process.

in relevant benefit programs. Managers must evaluate the employees while they are working in order to provide appropriate rewards or remediation, and the employees' base-level compensations may be adjusted as a result of the performance appraisal process. Finally, as the exhibit indicates, labor unions may affect and be affected by these various activities.

Who performs these functions? In very small organizations, the owner/manager must perform these functions; but as the organization grows, a personnel manager or HR manager, handles them. (**Personnel** is the traditional name but the term **human resources** is increasingly used instead.) As an organization grows, a department of personnel or human resources is created. Line managers also assist in performing many of the human resource functions in most organizations. In the past line managers and human resource managers did not always work well together, but this is changing, in large part because of the increased time, technical knowledge, and paperwork that is required to keep abreast of personnel legislation. Line managers and the human resource department need to act jointly in the best interests of the organization.

## Legal Constraints

One factor that has contributed to the increased importance of human resource managers is the number and complexity of legal constraints organizations face. These constraints affect selection, compensation and benefits, labor relations, and working conditions. Table 15.1 lists the most significant of these.

Among constraints on the selection of employees is *Title VII of the 1964 Civil Rights Act*, which prohibits discrimination on the basis of sex, race, color, religion, or national origin in all areas of employment, including hiring, layoff, compensation, access to promotion, and training. *The Age Discrimination Act* prohibits discrimination against people over the age of 40. In addition, various executive orders prohibit discrimination in organizations that do business with the government, provide extra protection for Vietnam-era veterans, and so forth.[5]

Wages and salaries are affected by the *Fair Labor Standards Act*, which sets minimum wages that must be paid to employees. Over the years, the federal government has raised the minimum wage for non-farm jobs several times; in 2012 it was $7.25 per hour although several states have opted to set higher minimums. Another important law in this area is the *Equal Pay Act*, which prohibits wage discrimination on the basis of sex.

Labor relations are strongly influenced by legal constraints as well. For example, the *National Labor Relations Act* governs the collective bargaining process between companies and organized labor unions. Another important law, the *Labor-Management Relations Act*, provides additional guidelines for dealing with labor unions.

**TABLE 15.1** Legal Constraints that Affect Staffing

| STAFFING CONSIDERATIONS | LEGISLATION |
|---|---|
| Employee Selection | Title VII of the 1964 Civil Rights Act<br>Age Discrimination Act<br>Executive Orders |
| Compensation and Benefits | Fair Labor Standards Act<br>Equal Pay Act |
| Labor Relations | National Labor Relations Act<br>Labor-Management Relations Act |
| Working Conditions | Occupational Safety and Health Act<br>Americans with Disabilities Act |

© Cengage Learning 2014

Legal constraints also affect working conditions. For example, the *Occupational Safety and Health Act* requires organizations to provide safe, nonhazardous working conditions for employees. Another more recent law (1990) is the *Americans with Disabilities Act*, which prohibits discrimination on the basis of physical handicaps and requires that all employers make their workplaces accessible to the disabled.[6]

Because of these and other laws, the staffing process is perhaps more affected by the legal environment than any other area of management. It is little wonder, then, that human resource managers have come to be vital members of organizations.

### FOOD FOR THOUGHT 15.2

More than half of all Cargill employees live and work in developing countries. More than 100 different nationalities work at Nestlé.

# Human Resource Planning

As we have seen, the first actual phase of the staffing process is human resource planning. This consists of three steps: job analysis, forecasting human resource supply and demand, and matching supply and demand. Figure 15.3 illustrates these steps.

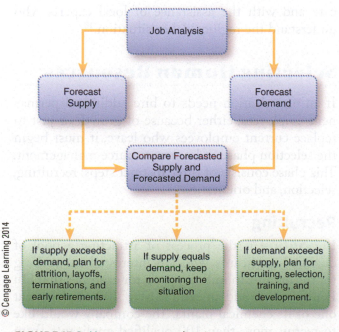

**FIGURE 15.3** Human resource planning.

© Cengage Learning 2014

## Job Analysis

**Job analysis** is the systematic collection and recording of information about jobs in an organization.[7] It actually consists of two different activities: developing job descriptions and listing job specifications. A **job description** summarizes the duties encompassed by the job; the working conditions where the job is performed; and the tools, materials, and equipment used on the job. A **job specification** lists the skills, abilities, and other credentials necessary to perform the job. Taken together, the job description and the job specification provide the human resource manager with the information he or she needs to forecast the supply and demand of labor within the organization. The information provided by a job analysis is also used for a variety of other purposes, including compensation and performance appraisal.

## Forecasting Supply and Demand

Forecasting the supply and demand for various kinds of employees involves knowing about the availability of potential employees in your industry. In addition, any number of sophisticated statistical techniques may be used.[8] While a discussion of those techniques is beyond the scope of this book, we will focus on the forecasting process at a general and descriptive level.

Forecasting demand involves determining the numbers and kinds of employees that the organization will need at some point in the future. If General Mills plans to open three new food-processing plants in five years, its human resource managers must begin planning now to staff those plants. Likewise, if the company intends to close a plant, there will be less demand for employees. Demand, then, is based partly on the projected overall growth of the organization and partly on where in the organization that growth is expected to occur.

Forecasting supply involves determining what human resources will be available, both inside and outside of the organization., Employees currently working for the firm, for example, will fill many upper-level management positions, whereas technical employees like engineers and programmers are usually brought in from the outside. It may be necessary to identify highly skilled specialists months or even years in advance, whereas the identification of lower-level and unskilled workers may be done according to a much more immediate timetable. If Summer Farms plans to increase its organic farming in time for next year's harvesting, someone must

## A FOCUS ON AGRIBUSINESS
### Agricultural Workers

According to the Bureau of Labor Statistics, in 2010 the "agriculture, forestry, and fishing" industry employed approximately 1.3 million wage and salary workers plus over 800,000 self-employed and unpaid family workers, making it one of the largest industries in the nation. Nearly 90 percent of those workers are employed in crop and animal production. Almost 80 percent of the establishments in this industry employ fewer than ten workers.

Agricultural workers, a subset of the industry, are those whose labor gets food, plants, and other agricultural products to market. These individuals work mostly on farms or ranches, but also in nurseries and slaughterhouses as well as other businesses. Their activities include planting and harvesting crops, installing irrigation,

and delivering animals. In 2010 almost 1.5 million individuals were employed in animal production. Approximately one-fourth of all crop workers are in Arizona, California, Colorado, New Mexico, and Texas, while California, Florida, and Oregon employ the most nursery workers.

A substantial number of these workers are seasonal and migrant workers, many of whom are illegal immigrants. The role of these workers has always been a political issue and became a heated issue again in 2010. In June of that year the United Farm Workers union—in a tongue-and-cheek campaign—suggested that all U.S. citizens become farm workers. Their view was that this would create a greater understanding of the issues surrounding these workers as the political debate progressed.[9]

---

decide how many new workers with what specific skills will be needed to plant, tend, harvest, and sell the additional food.

## Matching Supply and Demand

After the appropriate forecasts have been prepared, the results must be compared and actions must be planned. As shown earlier in Figure 15.3, managers can select from among three alternatives. If the supply of labor is projected to exceed the demand for labor, management must plan for normal attrition, layoffs, terminations, and early retirement. But if the demand for labor exceeds its projected supply, management must plan to recruit, select, train, and develop new employees. Finally, if supply and demand are roughly the same, no immediate action is necessary, although the situation should be monitored in case either supply or demand changes.

The human resource planning process can be especially difficult in international business. For example, international firms must engage in human resource planning within each country where they do business. Given the high costs of moving employees

between countries, different patterns of supply and demand within each country, and tremendous variations in both cultural and legal contexts, planning within each country must be undertaken with great care and with the assistance of local experts who understand the regional labor situation.[10]

## Selecting Human Resources

If an organization needs to hire additional permanent employees, either because of growth or just to replace current employees who leave, it must begin the selection phase of human resource management. This phase consists of three distinct steps: recruiting, selection, and orientation.

### Recruiting

**Recruiting** is the process of attracting a pool of qualified applicants who are interested in working for the organization.[11] Suppose, for example, that Du Pont wants to add an extra shift of 100 employees at one of its chemical plants. The company would like to recruit more than 100 qualified applicants from which to select. Fewer than one hundred does not give the company ample choice, but several thousand applicants would pose a big logistical problem. The key is to attract enough recruits, but not too many.[12] As long as Summer Farms is looking for local employees, it may be able to recruit by word-of-mouth, classified ads in area newspapers, and internal recruiting.

**Internal recruiting** involves identifying existing employees who want to be transferred and/or promoted. Often called **job posting**, this method may increase worker motivation because employees see that they have opportunities to advance within the organization. On the other hand, it may also lead to numerous job changes as each internal recruit vacates his or her position.[13] External recruiting is advertising for and soliciting applicants from outside the organization, often through classified ads (Figure 15.4). This approach brings in new talent and perspective but may also upset existing workers who would like to have been considered for the openings.

Each of these sources, internal and external, has distinct advantages. However, the advantages of internal recruiting tend to be the disadvantages for external recruiting, and vice versa. For instance, an advantage for internal recruiting is that work histories are readily obtained, whereas a disadvantage for external recruiting is that work histories are not so easily obtained. Most organizations use a combination of internal and external recruiting to try to obtain the advantages of both. Table 15.2 summarizes some of the advantages and disadvantages of internal recruiting from the standpoint of both the organization and the employees.

**TABLE 15.2** Advantages and Disadvantages of Internal Recruiting

**Advantages**—From the *organization's standpoint*:
- Current employees' work history easily obtained and evaluated
- No time delays for visits, interviews, etc.
- Current employees already socialized

**Advantages**—From the *current employees' standpoint*:
- May serve to motivate if contingent on performance
- Already know the organization so the process may be easier

**Disadvantages**—From the *organization's standpoint*:
- Limits the pool, which may lead to inferior decisions
- Limits the ability to bring in "new blood" that could speed up innovation and lead to rapid change

**Disadvantages**—From the *current employees' standpoint*:
- Resultant competitive pressures could make working together cooperatively more difficult.
- Inability to get desired positions could lead to decrements in work performance.

© Cengage Learning 2014

**Other Forms of Recruiting.** Many organizations use external placement firms and private employment agencies (so-called "headhunters") to do their recruiting. While farms may use this strategy to obtain seasonal labor, it is especially common when the organization is recruiting for managerial and/or professional positions. As technology develops, human resource professionals are increasingly using computerized databases to assist them in the recruitment task.[14]

Although organizations do not always admit it, some prefer to find new employees through recommendations of their high-performing current employees. In fact, some companies reward current employees for recommending new individuals if the company actually hires them. While this may or may not be the ideal way to staff an organization, in a tight labor market this form of networking is highly recommended for applicants, as it may be the only way to find a new job.

**Internships.** **Internships** are a combination of both internal and external recruiting. They provide work experience for students who are potential employees, and they establish contacts between potential employees and organizations. The organization benefits by getting some work done and, more

**FIGURE 15.4** Advertising job openings is a common form of external recruiting.

importantly, getting a first-hand look at the performance of potential employees. The student benefits from getting some first-hand experience in the type of work performed by college graduates and from establishing contacts within the organization.

The organization may need to be proactive to get students to apply for internships, however. All too often students do not seek internships because they are unaware that the company hires interns or they are already temporarily employed to finance their undergraduate educations. Such temporary jobs are obtained on the basis of a student's school skill set. Internships are more directed, as the organization agrees to employ the student in a position similar to one held by a college graduate. The organization "anticipates" a very different skill set—one that the student is in the process of acquiring. Thus, while a temporary job and an internship may look similar, they have different expectations and outcomes. Both are important additions to a job seeker's resume, but each is interpreted differently by future employers. The job illustrates current and past skills; the internship focuses on future ability.

Consider this analogy. From a student's perspective, among the more important purchases (and often the first) he or she will make is a car. Although he or she may prefer a big, fast, fancy car, the reality of personal finance usually limits the scope of the purchase. Regardless of financial status, the purchase is significant, especially since once a car is bought, the purchaser is "stuck" with it at least for a while. So no one would think of buying a car without first checking it out—test-driving it. After all, such a big, important purchase and commitment should not be made impulsively. Similarly, important and expensive commitments are made when a company hires a new graduate and when a new graduate commits to a particular job or company. Doesn't it make sense to "test drive" these? An internship accomplishes a two-way test drive for the hiring company and the potential new graduate. Both the student and the organization get a feel for how well their expectations and skills will match and therefore lead to a happy and productive relationship.

## Selection

Once an organization recruits applicants, it must attempt to hire the ones that best fit its needs and opportunities. It makes this choice through the **selection process**—a systematic attempt to determine how well the skills, abilities, and aspirations of a job applicant match the needs, requirements, and opportunities within the organization. Managers can use a variety of techniques to do this. However, each of these qualifications must be job related and have no discriminatory effects.

**Application Forms.**   The typical first step in selection is to have prospective employees complete application forms. An **application form** is a standardized form for collecting information from job applicants about their background, education, experience, and so forth. Figure 15.5 is an example of an application form.

An application form generally serves as a screen or filter to eliminate some applicants from the process and to retain others for further consideration. For example, if a particular job requires a college degree, application form data can easily separate those who meet this requirement from those who do not. Today, using computers to scan resumés and application forms serves the same purpose: to eliminate individuals who do not meet some minimum criteria. Some firms use more sophisticated application forms that allow managers to weight some items mathematically and then combine various scores to provide an overall predictor of performance.[15]

**Tests.**   Organizations frequently use tests to select employees. Common **tests** include ability, skill, aptitude, and knowledge tests.[16] Computer tests can be given to applicants for office and many other jobs, and swimming tests to a potential lifeguard. Physical ability tests such as these are relatively easy to administer and interpret. Physical examinations are also used by some firms as part of their test program.

On the other hand, organizations that use aptitude tests or personality assessments in selection must be much more careful in their administration and interpretation. These characteristics are harder to accurately measure and are also more prone to have bias and/or to discriminate in ways that are unrelated to job performance.

**Interviews.**   The most common selection technique is the interview. In its simplest format an **interview** is a simple conversation between the job applicant and a representative of an organization (Figure 15.6). While evaluating the applicant, the interviewer can use this occasion also to promote the company. Many organizations use multiple interviews. Interviews may be relatively structured (with a set of prescribed questions to be asked and

## DN
**Desert Nuseries**

### APPLICATION FOR EMPLOYMENT

*TYPE OR PRINT CLEARLY*

**PERSONAL INFORMATION**

DATE _____     SOCIAL SECURITY
                          NUMBER _____

NAME _____
           LAST                    FIRST                    MIDDLE

PRESENT ADDRESS _____
                    STREET              CITY        STATE      ZIP CODE

PERMANENT ADDRESS _____
                    STREET              CITY        STATE      ZIP CODE

PHONE NO. _____

CAN YOU, AFTER EMPLOYMENT, SUBMIT
VERIFICATION OF YOUR LEGAL RIGHT. TO
WORK IN UNITED STATES?  CIRCLE ONE:        YES      NO       (IF YES, VERIFICATION
                                                              WILL BE REQUIRED)

**EMPLOYMENT DESIRED**

POSITION _____     DATE YOU            SALARY
                                      CAN START _____ DESIRED _____

ARE YOU EMPLOYED NOW? _____     IF SO MAY WE INQUIRE
                                           OF YOUR PRESENT EMPLOYER _____

| **EDUCATION** | NAME AND LOCATION OF SCHOOL | DEGREE OR CERTIFICATE | SUBJECTS STUDIED |
|---|---|---|---|
| HIGH SCHOOL | | | |
| UNIVERSITY OR COLLEGE | | | |
| TRADE, BUSINESS, OR CORRESPONDENCE SCHOOL | | | |

DO YOU HAVE ANY EXPERIENCES, SKILLS, OR QUALIFICATIONS THAT WILL BE OF SPECIAL BENEFIT IN THE POSITION FOR WHICH

YOU ARE APPLYING? _____

WHAT FOREIGN LANGUAGES DO YOU SPEAK FLUENTLY? _____

READ? _____     WRITE? _____

WHAT PROFESSIONAL ORGANIZATIONS DO YOU BELONG TO? _____

_____

*(LAST / FIRST / MIDDLE — vertical side tabs)*

CONTINUED ON OTHER SIDE

**FIGURE 15.5**  A job application form.

| EMPLOYED | LIST LAST EMPLOYER FIRST | JOB DESCRIPTION AND TITLE | RATE OF PAY | REASON FOR LEAVING |
|---|---|---|---|---|
| FROM<br>Month   Year | NAME OF EMPLOYER | TITLE | START | |
| | STREET ADDRESS | DUTIES | | |
| | CITY, STATE, ZIP | | LAST | |
| TO<br>Month   Year | TELEPHONE NUMBER | | | |
| | NAME OF SUPERVISOR | | | |
| FROM<br>Month   Year | NAME OF EMPLOYER | TITLE | START | |
| | STREET ADDRESS | DUTIES | | |
| | CITY, STATE, ZIP | | LAST | |
| TO<br>Month   Year | TELEPHONE NUMBER | | | |
| | NAME OF SUPERVISOR | | | |
| FROM<br>Month   Year | NAME OF EMPLOYER | TITLE | START | |
| | STREET ADDRESS | DUTIES | | |
| | CITY, STATE, ZIP | | LAST | |
| TO<br>Month   Year | TELEPHONE NUMBER | | | |
| | NAME OF SUPERVISOR | | | |
| FROM<br>Month   Year | NAME OF EMPLOYER | TITLE | START | |
| | STREET ADDRESS | DUTIES | | |
| | CITY, STATE, ZIP | | LAST | |
| TO<br>Month   Year | TELEPHONE NUMBER | | | |
| | NAME OF SUPERVISOR | | | |

(Left vertical label: BUSINESS EXPERIENCE)

If presently employed, may DN contact your employer for further information? _____ YES _____ NO

| Have you ever worked under another name? | ☐ YES  ☐ NO | If yes, what name(s) | Which co. or organization |
|---|---|---|---|

Machines you can operate

**MEDICAL STATUS**

This employer is a government contractor to section subject 503 of the Rehabilitation Act of 1973, which requires government contractors to take affirmative action to employ and advance in employment qualified handicapped individuals. If you have such a handicap, and would like to be considered under the affirmative action program, please tell us. This information is voluntary and refusal to provide it will not subject you to rejection for employment, discharge or other disciplinary treatment. However, in order to assure proper placement of all employees, we do request that you answer the following questions:

Do you have or have you had a mental or physical disability which would create a hazard to you or to others at the worksite?          ☐ YES          ☐ NO

If yes, please describe _____

**REFERENCES**

NAME 3 PERSONS NOT RELATED TO YOU WHO CAN ATTEST TO YOUR EXPERIENCE AND QUALIFICATIONS

| NAME | ADDRESS | CITY | STATE | ZIP CODE | PHONE NO. | OCCUPATION |
|---|---|---|---|---|---|---|
| | | | | | | |
| | | | | | | |
| | | | | | | |

THE ABOVE INFORMATION IS TRUE AND CORRECT TO THE BEST OF MY KNOWLEDGE. MISREPRESENTATION OR OMISSION OF MATERIAL FACTS (I.E., FACTS RELATED TO MY QUALIFICATIONS FOR THE POSITION FOR WHICH I AM APPLYING) IS CAUSE FOR SEPARATION FROM THE EMPLOYER, I FURTHER AUTHORIZE ALL FORMER EMPLOYERS, SCHOOLS (PROFESSIONAL AND VOCATIONAL). AND PERSONS NAMED ABOVE TO FURNISH REFERENCES AND ANY FACTS WHICH MAY BE PERTINENT TO MY EMPLOYMENT. I UNDERSTAND THAT EMPLOYMENT IS ALSO CONTINGENT UPON PASSING A PHYSICAL EXAMINATION.

Employment is voluntarily entered into and the employee is free to resign at any time.
Similarly, DN may, at any time, conclude the employment relationship where it believes it is in the company's best interest

_____          _____
(SIGNATURE)                                                      (DATE SIGNED)

**FIGURE 15.5** (continued)

**FIGURE 15.6** A job applicant shows her resume to a panel of interviewers.

answered) or unstructured (open-ended with no prescribed questions).

Unfortunately, despite their widespread use, interviews are rather poor predictors of job success.[17] Judgments of interviewers tend to have low or zero correlation with the later job performance of employees. One reason for this may be that many people are relatively unskilled in asking probing questions and interpreting responses. Another is that job applicants may not be completely truthful during the interview to avoid making a bad impression.

**Assessment Centers.** Finally, **assessment centers** are specially designed techniques used to select managerial employees. During an assessment center prospective managers spend an extended period of time (perhaps two or three days) undergoing a battery of tests, interviews, and simulated work experiences, which attempt to imitate various parts of the managerial job, such as decision making, time management, giving feedback to subordinates, and so forth. Each potential manager performs these tasks under the observation of skilled human resource managers. Assessment centers are an excellent way to predict an individual's managerial potential. On the other hand, they take a lot of time (for both applicants and human resource managers) and are expensive to operate.

Regardless of which selection techniques it uses, an organization must be able to demonstrate that it is not discriminating on the basis of irrelevant factors. For example, suppose an organization uses a selection test with a maximum potential score of 100. During the development of the test a manager asks 30 current employees to answer the test questions.

It turns out that the ten highest job performers all score between 90 and 100, the next ten job performers all score between 70 and 80 and the ten lowest job performers all score below 60. The manager now has evidence that the test is a valid predictor of performance. If the test proves to be unrelated to subsequent performance, however, its use is discriminatory and could result in legal problems for the organization.

Such discriminatory practices were the catalyst for much of the legislation that so tightly controls the human resource area today.[18] Most organizations now work hard not to discriminate in their employment practices.

## Orientation

After a new employee accepts an offer to join the organization, he or she must go through an **orientation** procedure. Such procedures vary widely from company to company and from job to job. Orientation for operating employees might simply be telling them when to come to work, when they get paid, and who to see if they have a problem. Orientation for managerial and professional employees tends to be more involved. It may take several weeks or months of work with other employees for the newcomer to become totally acquainted with all phases of the organization. (See Figure 15.7.)

## Training And Development

After new employees have been recruited, selected, and oriented, the next logical step is to train and develop them. **Training** typically involves specific job skills and applies more to operating employees.

**FIGURE 15.7** A company official briefs three newly hired employees as part of their orientation process.

**Development** usually refers to training for managers. Development is more general in nature and focuses on a wider array of skills, including conceptual, problem solving, and interpersonal skills.[19]

## Assessing Training and Development Needs

Before a manager can properly plan training and development activities, he must ascertain the training and development needs of both the employees and the organization. For example, training needs will be great if labor market conditions are such that a restaurant must hire individuals who are untrained and inexperienced (e.g., good short-order cooks are scarce) or a nursery cannot find employees who already know about trees and shrubs. On the other hand, if qualified employees are readily available, an organization's training needs are considerably less.

Regardless of how much formal training is necessary, employees will need at least some basic work to learn exactly how the organization requires them to perform their tasks. And management development is a long-term, ongoing process that never really stops.

In recent years organizations have intensified their training and development efforts in a number of specific areas. For one thing, with the growth of international business many companies now provide extensive language and/or cross-cultural training for their employees. For another, increased sensitivity regarding workforce diversity within organizations has also prompted training in the area of multiculturalism.[20] Many companies have had problems adjusting to a highly diverse workforce. As a result of legal action against it, Denny's became highly proactive and became known as one of the best companies for minorities.[21] Pro's Ranch Markets even uses diversity of its employees to create its distinctive competency.[22]

## Popular Training and Development Techniques

Table 15.3 notes several common techniques for training and development. The key, of course, is to match the technique with the goals of the training and development effort. For example, if the goal for employees is to learn about new company procedures, assigned reading may be an effective approach. If Food Lion has a goal to teach people how to relate better to others or how to make decisions more effectively, it may use role playing or case discussion groups. Training supervisors in conducting performance reviews might involve behavior modeling.[23] If the idea is to teach a physical skill such as operating a new kind of machine, vestibule or on-the-job training could be the most appropriate option. Whichever technique you use, however, it is important to get the trainees to accept the training.[24]

## Evaluating the Effectiveness of Training

The final component of a well-managed training and development program is evaluation. Considering how much time, effort, and money companies invest in training and development, they should make sure that the goals of the program are met.[25]

If an employee training program is designed to increase the proficiency of word-processing operators, for instance, the operators' performance should be measured both before and after the training program. If the program is effective, their performance should improve. Lack of improvement may suggest that the training should be revised.

Evaluation is more difficult in the case of management development, but it is not impossible. Managers who participate in many development activities and get high marks in those activities should subsequently have a good record of promotions, performance, and so forth. If this is indeed the case, then the organization has evidence that its development efforts are paying off. But again, if involvement and performance in management development activities are unrelated to future job performance, the organization should reassess its approach to developing managers.

## Performance Appraisal

After employees have been trained and have adjusted to their jobs, managers usually begin to evaluate their performances. There are several purposes behind this evaluation, or **performance appraisal**. First, the organization needs evidence to justify, or validate, the selection techniques it used to hire the person in the first place. Second, since performance is frequently a basis for rewards, it is important to evaluate performance to provide rewards fairly. In addition, the individual must know how well he or she is performing in

**TABLE 15.3** Common Training and Development Techniques

| METHOD | COMMENTS |
| --- | --- |
| Assigned Readings | Readings may or may not be specially prepared for training purposes. |
| Behavior Modeling Training | Use of a videotaped model displaying the correct behavior, then trainee role playing and discussion of the correct behavior. Used extensively for supervisor training in human relations. |
| Business Simulation | Both paper simulations (such as in-basket exercises) and computer-based business "games" to teach management skills. |
| Case Discussion | Real or fictitious cases or incidents discussed in small groups. |
| Conference | Small-group discussion of selected topics, usually with the trainer as leader. |
| Lecture | Oral presentation of material by the trainer, with limited or no audience participation. |
| On the Job | Ranges from no instruction, to casual coaching by more experience employees, to carefully structured explanation, demonstration, and supervised practice by a qualified trainer. |
| Programmed Instruction | Self-paced method using text or computer followed by questions and answers. Expensive to develop. |
| Role Playing | Trainees act out roles with other trainees, such as "boss giving performance appraisal" and "subordinate reacting to appraisal" to gain experience in human relations. Also used in international training. |
| Sensitivity Training | Also called T-group and laboratory training, this is an intensive experience in a small group wherein individuals give each other feedback and try out new behaviors. It is said to promote trust, open communication, and understanding of group dynamics. |
| Vestibule Training | Supervised practice on manual tasks in a separate work area where the emphasis is on safety, learning, and feedback rather than productivity. |
| Interactive Video | Newly emerged technique using computers and video technology. |

© Cengage Learning 2014

order to improve.[26] And finally, performance appraisals help determine what additional training the employee may need. Managers can use several different kinds of techniques for performance appraisal.[27]

## Objective Measures

**Objective measures of performance appraisal** are quantifiable indicators of how well an employee is doing. For instance, it may be possible to count how many units of a product an employee assembles, adjust this number for quality, and arrive at an objective index of performance. Similarly, the number of sales dollars generated by a sales representative reflects performance objectively.

Unfortunately, objective measures are often unavailable or, worse still, misleading. Assembly-line workers have little control over how many units they produce, and a sales representative with a lot of major customers in his territory should have more sales than one with only a few large customers. For these reasons, managers may need to adjust objective indicators of performance in order to have a valid representation of actual performance.

## Judgmental Methods

Another common approach to performance appraisal that many organizations use is **judgmental methods.** These methods involve having someone, usually the employee's immediate supervisor, subjectively evaluate that person's performance via a ranking or a rating procedure (Figure 15.8).

**FIGURE 15.8** Performance appraisal is an important part of a manager's job. This evaluation form is not a good one as it relies too heavily on the manager's judgment.

© Ragma Images/www.Shutterstock.com

**Ranking**, as the term implies, means that the supervisor ranks her subordinates in a continuum from high to low performance. Such a procedure forces the manager to differentiate among high, moderate, and low performers. At the same time, however, it also makes feedback more difficult to deliver as, unless ties are permitted, each rank must be dependent, and the last person on the list may still be a solid performer.

**Rating** is comparing each employee with one or more absolute standards and then placing the employee somewhere in relation to that standard. These scales rate the individual's level of conscientiousness and degree of initiative on the job. The manager considers the questions, judges how well the person stacks up, and then circles the appropriate numbers along the scales. Managers usually sum or average the various ratings to arrive at an overall index of performance.

Because they are flexible and relatively easy to use and to interpret, rating scales such as these are probably the most common kinds of performance appraisal devices currently in use. Unfortunately, however, they also suffer from a number of problems. For one thing, managers are sometimes inclined to give everyone the same relative rating (i.e., all high, all average, or all low). This makes it impossible to differentiate among the employees. Also, many people tend to be influenced by an employee's most recent behavior rather than by her overall level of performance over longer periods of time.

To help overcome these shortcomings some organizations develop guidelines to improve the process used in rating systems.[28] Some organizations also develop more sophisticated and intricate rating scales to make them more accurate. Two examples of these are Behaviorally Anchored Rating Scales (BARS) and Behavior Observation Scales (BOS). While each is complicated and takes considerable time and effort to develop, they are also a vast improvement over traditional rating scales.[29] Figure 15.9 shows two example rating scales.

## Management by Objectives

**Management by objectives (MBO)**, a popular vehicle for managing the goal-setting process, also serves as a useful method for evaluating the performance of those managers who set the goals to begin with. For example, suppose a sales manager for Cargill sets a goal of increasing sales in his territory next year by 15 percent. At the end of the year, this goal

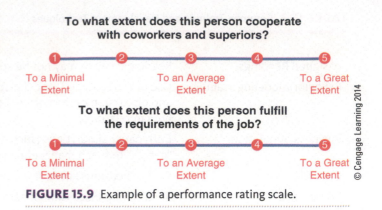

**FIGURE 15.9** Example of a performance rating scale.

© Cengage Learning 2014

provides an effective framework for performance appraisal. If sales have indeed increased by 15 percent or more, a positive performance appraisal may be in order. But if sales have increased by only 4 percent, and if the manager is directly responsible for the disappointing results, a more negative evaluation may be forthcoming.

## Feedback

The final part of performance appraisal, and often the most difficult, is providing **feedback** to the employee by telling him or her the results of the appraisal.[30] Because it is often so difficult, software has even been written to assist managers with the feedback task.[31]

In most instances, feedback is given in a private meeting between the superior and the subordinate. The superior typically begins by summarizing the results of the appraisal, then answers any questions, suggests ways to improve, and explains the immediate consequences of the appraisal. A poor evaluation might result in no salary increase, a cutback in authority, or even a warning that the employee will be fired if things aren't turned around. In contrast, a good evaluation can lead to a raise, a bonus, a promotion, or increased responsibilities.

The employee being evaluated is usually given an opportunity to respond to the results of the evaluation. For example, the employee may argue that the manager was unfair or that the evaluation itself did not tell the complete story. The manager may or may not choose to adjust the evaluation, but the employee can usually document her or his perceptions and add it to the evaluation before it becomes official.

One other approach to improving a performance appraisal is 360-degree feedback.[32] The basic idea, which has been around for some time, is to obtain evaluations and feedback from everyone around the individual—supervisor, subordinates, peers, and, in some cases, clients.[33] Obtaining many differing points of view provides extremely rich information for both the individual and the organization. These systems, however, do take more time and the information must be handled carefully to assure privacy and proper use.[34]

# Compensation and Benefits

The management of compensation and benefits is another important part of the human resource process. Employees must be paid **compensation**—wages and salaries—and they usually expect to receive various kinds of benefits. The organization also often uses financial incentives to increase motivation and reward past performance of productive workers.[35] The management of compensation and benefits can be a controversial process.

---

**FOOD FOR THOUGHT 15.4**

During the War of 1812 the U. S. Government paid soldiers in salt brine (a necessity for food preservation) because it was too poor to pay them with money.

---

## Wages and Salaries

The central part of compensation management involves determining wages and salaries for employees. This determination, in turn, consists of three parts: wage level decisions, wage structure decisions, and individual wage decisions.

**Wage Level Decisions.** Management's **wage level** decision is whether the organization wants to pay higher wages, the same wages, or lower wages

than the prevailing rate in the industry or geographic area. If Ralston Purina decides to attract the best possible food scientists, it would set a policy of paying recent college graduates starting salaries that are several percentage points higher than those of other companies hiring food scientists. Similarly, a small manufacturer might decide to pay the same rates as other local companies rather than attracting people on the basis of a high starting wage or losing employees to higher-paying jobs.[36]

**Wage Structure Decisions.** Another important decision pertains to the **wage structure** within the organization. The essential issue here is how much people performing one job should be paid relative to people performing another job. That is, all else equal, is Job A worth a higher salary than Job B, is Job B worth a higher salary than Job A, or are Jobs A and B worth exactly the same salary? The wage structure is usually determined through job evaluation. **Job evaluation** is the process of determining the relative value of jobs within the organization.[37]

Probably the most widespread approach to job evaluation is the **point system**. The point system starts with a committee of workers and managers determining what factors are most appropriately used to differentiate and characterize jobs within the organization. As shown in Table 15.4, these factors might include such things as education, responsibility, skill, and physical demand.

After the factors have been identified, each factor is assigned points based on its perceived importance to the organization. For example, points awarded for education required to perform the job might range from 20–100. Table 15.5 shows how these points may then be allocated for three jobs: secretary, office manager, and janitor. In the table the job of office manager warrants 80 points for education needed, 70 for responsibility, 60 for skill, and

**TABLE 15.4**  Sample Point System for Wage Determination

| COMPENSABLE FACTORS | POINTS ASSOCIATED WITH DEGREES OF THE FACTORS | | | | |
|---|---|---|---|---|---|
| | Very little | Low | Moderate | High | Very High* |
| Education | 20 | 40 | 60 | 80 | 100 |
| Responsibility | 20 | 40 | 70 | 110 | 160 |
| Skill | 20 | 40 | 60 | 80 | 100 |
| Physical Demand | 10 | 20 | 30 | 45 | 60 |

*The job evaluation committee that constructed the system believed that responsibility should be the most heavily weighted factor and physical demand should be least. That is why the maximum points for these factors are different.

© Cengage Learning 2014

**TABLE 15.5** An Application of the Point System

| COMPENSABLE FACTORS | JOB | | |
|---|---|---|---|
| | SECRETARY II | OFFICE MANAGER | JANITOR |
| Education | Moderate = 60 | High = 80 | Very Low = 20 |
| Responsibility | Low = 40 | Moderate = 70 | Low = 40 |
| Skill | High = 80 | Moderate = 60 | Low = 40 |
| Physical Demand | Low = 20 | Low = 20 | High = 45 |
| Total Points | 200 | 230 | 145 |

The job analysis committee carefully reviews the content of each job and decides what degree of each factor best describes the job.
© Cengage Learning 2014

only 20 for physical demand. Its total, then, is 230 points. This job should therefore be worth more than the secretarial job, which totals 200 points. In turn it is worth more than the janitorial job, with 145 points.

**Individual Wage Decisions.**    Finally, the manager must address individual wage decisions. These decisions involve deciding how much to pay each person within a job classification. Most organizations set wage ranges for jobs within certain point ranges. For instance, a company might decide to pay people $15–$20 an hour for jobs worth 350–375 points. Initial wages are then set according to the employee's experience. A person who is just starting her career might be paid $15, whereas a more experienced person receives, say, $17.50.

Later, wages are adjusted according to seniority and/or performance. (See Figure 15.10.) The new employee might get a 20-cents-an-hour raise after six months and another 30-cent raise for very good performance, bringing his total to $15.50 an hour. The other employee, meanwhile, might be given the same 15-cent raise after six months but no additional increase for performance.

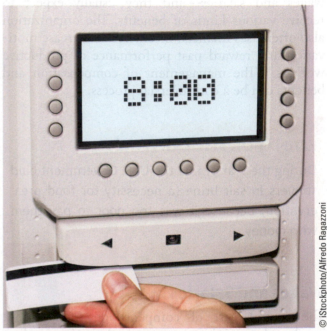

**FIGURE 15.10** Hourly workers usually clock in and out so the computer can calculate their wages for the pay period. Salaried workers normally go directly to their work stations and keep the expected number of work hours without "punching in."

## FOOD FOR THOUGHT 15.5

Ben & Jerry's had trouble finding good people to fill the upper ranks of the highly successful company until it abandoned its goal that no one in the little ice cream company would ever be paid more than five times what the lowest-paid employee earned.

## Benefits

Another important part of compensation is the employee benefit package. **Benefits** are payments other than wages or salaries. Benefits add substantial

costs to the total compensation received by employers, averaging more than 40 percent beyond the cost of wages and salaries.[38] The most common benefits include health, dental, disability, and life insurance coverage for the employee (and sometimes the family). Costs for these may be borne entirely by the organization or shared with the employee. Employees also usually receive some pay for time when they do not work, such as vacations, sick days, and holidays. Retirement programs are also common benefits. Not as prevalent but still provided by some organizations are benefits such as counseling programs, physical fitness programs, credit unions, and tuition reimbursement for educational expenses related to the job. On-site daycare facilities are also becoming more common, especially in larger firms.

An organization's benefit package is clearly significant for several reasons. In addition to representing a major cost to the organization, it is an important factor in attracting and retaining employees. Some organizations have experimented with "cafeteria benefits package"—an arrangement whereby each benefit is priced and employees can choose those they want within a total price. Therefore, a married worker with several children can concentrate his or her benefits on insurance programs, a single employee can choose more vacation time, and an older worker can put more into retirement. Such cafeteria programs are expensive to administer but are an attractive feature to many employees.[39]

## Labor Relations

The final aspect of human resource management to be discussed here is **labor relations**. This generally refers to dealing with employees when they are organized in labor unions.[40] In this section we first describe how unions are formed and then address collective bargaining.

### How Unions are Formed

Given the turbulent history of union-management relations, it should come as no surprise that government regulation closely defines the processes involved in forming a union.[41] The National Labor Relations Board (NLRB) was created to oversee these processes.

Figure 15.11 summarizes the actual steps in forming a union. First, someone must generate interest among employees. For example, either disgruntled employees or representatives of large national unions might initiate this action within a particular organization. Next, employees must collect signatures on authorization cards. These cards simply indicate that individuals who sign them believe that an election should be held to determine whether employees are interested in unionization. If fewer than 30 percent of the eligible employees sign the cards, the process ends. But if 30 percent or more employees sign authorization cards, the NLRB conducts a certification election.

A simple majority of those who vote—as opposed to a majority of all eligible employees in the organization—determines the outcome of the election. Of those casting a ballot, if the majority

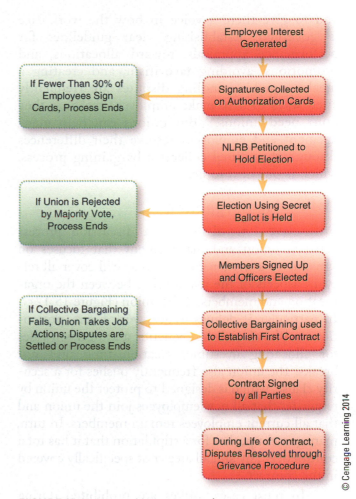

**FIGURE 15.11** How unions are created.

© Cengage Learning 2014

votes against the certification of a union, the process ends. But if a majority vote in favor of unionization, the union is officially certified by the NLRB and becomes the official bargaining representative for all eligible employees.

Immediately following certification, the union recruits members and elects officers. Members pay dues (to cover administrative costs) and expect to gain improved employment conditions as a result. After the membership is signed up, the union sets out to negotiate a labor contract with management. Either management and the union agree on a contract or the union takes various job actions, which might include strikes, work slowdowns, or similar activities. The contract also specifies a grievance procedure that will be used to settle disputes during the term of the contract.[42]

Obviously, management prefers that employees not belong to unions. In general, the best way to avoid unionization is to treat employees fairly and to give them a voice in how the workplace is governed. Establishing clear guidelines for performance appraisals, reward allocations, and promotions, avoiding favoritism, and creating a mechanism for handling disputes are frequently cited ways to help make employees feel that they don't need unions.[43] But unions can help management and employees resolve their differences through the formal collective bargaining process, which is our next topic.

## Collective Bargaining

**Collective bargaining** is a discussion process between union and management that focuses on agreeing to a written contract that will cover all relevant aspects of the relationship between the organization and members of the union (Figure 15.12).[44] In particular, it defines wages, work hours, promotion and layoff policies, benefits, and decision rules for allocating overtime, vacation time, and rest breaks. The union also frequently pushes for a security clause, which is designed to protect the union by requiring that all new employees join the union and that all current employees remain members. In turn, management pushes for a stipulation that it has total control over any and all areas not specifically covered in the contract.

In most cases, strikes are prohibited during the term of the contract. Any strikes that do occur are called *wildcat strikes* and do not have the official endorsement of the union. Of course, the union can call strikes after the existing contract expires and when the union and management representatives cannot agree on a new one.

**FIGURE 15.12** When collective bargaining fails, union members use influential tactics, such as going on strike or organizing a boycott to boost their bargaining power.

**Discipline.** Labor contracts frequently devote considerable attention to discipline. Everyone seems to agree that management can discipline employees for just cause, but disputes frequently arise over the meaning of "just cause." In general, management and union representatives make an effort to define work rules very clearly and to predefine penalties for violating those rules. For example, the first time an employee is late he may receive a reminder about work hours; the second time a note might go in his personnel file; a third offense could lead to a short suspension without pay. If everyone knows the rules and the penalties for breaking them, there will be few legitimate complaints.[45]

**Grievance Procedure.** Of course, no contract is perfect, and differences of opinion are inevitable. Thus the contract also specifies procedures that everyone will follow to resolve disputes. An employee who feels mistreated often files a written **grievance** with the union and discusses it with her supervisor. If the problem cannot be resolved, union officials and higher-level managers become involved. Ultimately, it may be necessary to make use of an arbitrator. An **arbitrator** is a labor law specialist jointly paid by the union and the organization. The arbitrator listens to both sides of the argument, studies the contract, and makes a decision as to how the dispute will be settled. Both the union and management agree to abide by his or her ruling.

## Human Resource Records

It should be clear that the human resource function generates a need for record keeping. While it already makes sense to maintain accurate information about employees for payroll and other purposes, state and federal governments also require that you keep certain records. On occasion, the Human Resource department may be asked to produce documentation about some aspect of an employee's performance or work history.

Smaller organizations are likely to keep paper records in a file cabinet although many are starting to use electronic records stored on computers. Larger organizations may develop some sort of computerized system that integrates human resources information and payroll information. Indeed, some organizations employ a standalone HRIS (Human Resource Information System).

Generally, two sets of records should be maintained—one for human resources or personnel and another for payroll and medical. The records in human resources are generally referred to as personnel or employee files. As the name implies, they contain information pertaining to the individual's employment—initial employment forms, performance evaluations, disciplinary actions, awards, attendance, training, and the like. Those records should not have medical, payroll, or tax information. Due to varying company rules and governmental regulations, managers should consult with the proper company, state, and federal officials concerning what should and should not be in these records, what information can be released, and how long the records must be maintained.

## CHAPTER SUMMARY

Staffing is the process of procuring and managing the human resources an organization needs to accomplish its goals. Staffing is vitally important to all organizations and is usually the shared responsibility of human resource specialists and line managers.

Human resource planning is the first phase of the staffing process and involves job analysis, forecasting human resource supply and demand, and matching supply with demand. Job analysis is the systematic collection and recording of information about jobs in the organization resulting in descriptions of jobs and specifications that explain the kind of person needed for them. With this information forecasts of the demand for and supply of human resources are made. If supply exceeds demand, the human resource manager must make plans to reduce employment through attrition, layoffs, terminations, and early retirements. If demand exceeds supply, the manager must engage in the next process, selection.

The selection phase of human resource management consists of recruiting, selecting, and orienting employees. Recruiting is the process of attracting a pool of qualified applicants interested in working for the organization. Selection is choosing which applicants to hire, using techniques such as application forms, tests, interviews, and assessment centers. Orientation is the procedure whereby new employees are brought into and informed about the company, its purpose, and their jobs within it.

Training is usually job specific, whereas development is more general. Before engaging in such activities, the organization should determine how much and what kinds of training are needed. Training and development involve a wide range of techniques, including role playing, reading, and on-the-job training, and all training and development activities should be evaluated in terms of effectiveness and efficiency.

Performance appraisal—the evaluation of each employee's performance—is used to justify the selection techniques used, to provide a basis for reward distribution, and to give feedback to the employee. Where employees exert considerable control over their work, the best measures are objective, quantifiable indicators such as units assembled or dollars sold. In other cases judgmental methods such as ranking or rating tend to be used. One approach that has some value when individuals have a fair amount of control over their jobs is management by objectives (MBO), whereby performance is appraised by how well the objectives have been accomplished.

The determination of employee compensation (wages and salaries) for employees is central to management and involves three types of decisions: wage level, wage structure, and individual wage decisions. Wage level decisions are external; wage structure decisions are internal. Job evaluation is the process used to determine the relative value of jobs within an organization. Benefits are yet another part of compensation management. Health care, dental care, disability and/or life insurance, vacations, sick leave, and holidays are all part of benefits that must be managed.

Labor relations refers to dealing with employees when they are organized into labor unions. The best way to avoid unions is by making them unnecessary. This means treating employees fairly, having them participate in setting clear guidelines for performance appraisal, reward allocation, and personnel decisions, and creating a mechanism for handling disputes. Negotiating a contract—collective bargaining—may be a complex process, since all relevant aspects of the relationship between the organization and members of the union are covered in such contracts.

## CHAPTER ACTIVITIES

### REVIEW QUESTIONS

1. What are the major components to human resource planning? How are they related?
2. What is selection? What are some of the techniques used in it?
3. What are popular training and development techniques? Where might each be appropriate to use?
4. What are some methods used in performance appraisal? What are the advantages and disadvantages of each?
5. Describe the general process through which unions are formed.

### ANALYSIS QUESTIONS

1. What are the advantages and disadvantages of internal and external recruiting? Which do you feel is best in the long term? Why?
2. How can you determine whether a particular technique is valid for selection? What are the costs and benefits of using invalid techniques?
3. An objective measure of performance for a research chemist may be "number of patents obtained." Why may this be a poor method for evaluating the chemist's performance? What might be a better approach?
4. Do you think wages or salaries are more important than benefits? Why or why not?
5. Respond to this statement: "Unions would not exist if it weren't for poor management."

## FILL IN THE BLANKS

1. The systematic investigation that involves collecting and recording information about jobs in the organization is called _____ _____.

2. A summary of the duties and the working conditions of a job, plus the tools, materials, and equipment used on the job is called _____ _____.

3. Finding current employees who would like to change jobs is called _____ _____, whereas finding qualified applicants outside the company is called _____ _____.

4. Preparing employees for a job by teaching them job skills is generally referred to as _____ _____, whereas _____ involves teaching more general abilities.

5. A system that measures employees against one another is a _____ system, whereas a system that compares each employee with a standard of performance is a _____ system.

6. Telling employees the results of their performance appraisal is called _____.

7. The comparison of a company's wages relative to others in the local economy is known as the _____ _____, whereas the comparison of wages for different jobs within the company is called _____ _____.

8. Payments to employees other than wages or salaries are known as _____.

9. Management and union officials reach agreement on wages, layoff policies, and the like through a process known as _____ _____.

10. A list of the skills, abilities, and other credentials necessary to perform a job is known as a _____ _____.

## ▼ CHAPTER 15 CASE STUDY
### THEY JUST WON'T GO AWAY

Though they appear to be dissimilar and involved in quite different industries, pharmaceutical companies are like oil companies—both industries must have new products in the pipeline. This is the reason an American firm, Genzyme Corporation, a Cambridge, Massachusetts, biotechnology company, and a French firm, Sanofi-Aventis SA, a pharmaceutical giant, began to dance with each other in 2010. It was a dance with financial and staffing implications.

It was a long and important dance, with both parties interested in more than a one-night stand. Sanofi-Aventis was in trouble. Five of its eight best-selling drugs faced generic competition in 2012. To Chris Viehbacher, Sanofi-Aventis CEO, Genzyme looked like a solution to his problems. Genzyme had a portfolio of drugs to treat rare diseases, many of which were complex and difficult for generic rivals to mimic.

Henri Termeer's company, Genzyme, had its problems, too. Manufacturing facilities were under scrutiny for production lapses that caused major shortages of key drugs. The resolution of its problems was being carefully monitored by U.S. regulators. These problems caused a drop of Genzyme's stock price and turnover among its employees.

So Sanofi-Aventis offered to buy Genzyme, initially for $69/share. This offer in a private letter from Viehbacher to Termeer valued Genzyme at $18.4 billion. Termeer rebuffed the offer stating, "The Genzyme board is not prepared to engage in merger negotiations with Sanofi based upon an opportunistic proposal with an unrealistic starting price that dramatically undervalues our company."

Termeer indicated that the problem was the price. In fact, the directors were not opposed in principle to a sale of the company. They indicated that they wanted a fair value if they were to do a strategic transaction. Sanofi countered with a "bear hug" letter. This was an open letter to Genzyme stockholders explaining the offer and its rationalization. The "bear hug" letter is one step removed from a "hostile takeover bid."

Part of the reason for this "gentle" approach was that Viehbacher had his own board problems. Two of his largest shareholders, Total SA (French oil firm) and L'Oreal (French cosmetics firm), were determined that Sanofi not overpay. These companies had four of the 13 board seats.

Meanwhile, Genzyme sold a genetic-testing business to Laboratory Corporation of America Holdings for $925 million. Genzyme wanted to concentrate on its core businesses, recover its manufacturing ability, and stabilize employment. It also hoped to fend off Sanofi-Aventis. The proceeds of the sale would go to an announced stock buyback in 2011. Sanofi-Aventis was not idle either. It approached Citigroup Inc., Bank of America Corporation, J.P. Morgan Chase & Company, BNP Paribas SA, and Societe General SA for financing to make a higher offer for Genzyme, should a "bidding war" erupt.

Sanofi-Aventis' strategy was very simple. It knew that Genzyme had to remedy its manufacturing problems because of regulator scrutiny and that these difficulties depressed Genzyme's stock. In addition, the slow approach by Sanofi-Aventis did not seem to attract a more eager bidder for Genzyme. In fact the casual nature of the process seemed to create an impression less of desperation and more that resistance is futile. Genzyme's hope, on the other hand, was for another bidder to enter the fray.

Eventually, tired of waiting, Sanofi-Aventis launched a "hostile bid" for Genzyme. It maintained the same price but hinted at the possibility of a higher price, as perhaps somewhere between its initial offer of $69/share and a top offer of $80/share. Genzyme sought a "white knight" (another bidder). According to the company it was not actively seeking a "white knight" but simply attempting to determine the value of Genzyme. However, it certainly looked like Genzyme was seeking a "white knight." While this was occurring, Genzyme reported soaring profits of $16 million and said that it was on track to resume full production of its top-selling biotech drugs. The stock exchange was unimpressed.

Finally, Genzyme proposed a different approach, a "contingent value right," or CVR. This approach gave shareholders an additional benefit once an acquired company hit a future benchmark. CVRs are typically used when buyers and sellers cannot agree on a price. In addition, Sanofi-Aventis offered $79/share or $21 billion.

Why did this endless "slow dance" occur?

For Viehbacher, it was a strategic deal to improve Sanofi-Aventis and acquire an "ailing" but solid company. For Termeer, it was personal. Genzyme began in 1981. He joined the company in 1983 and regarded it as "his" company.

So we have a tale of two businessmen working with the same issue from different sides of a deal; they were apparently in similar situations. However, this was not true. For one it was a business decision, whereas for the other it was a personal decision. Only from a distance do they look similar. Meanwhile both shareholders and employees faced profound uncertainty.[46]

### ▶ Case Study Questions

1. How do corporate struggles over ownership affect employees? Why?

2. What might be the advantages and disadvantages of changing corporate ownership?

3. If you were an employee in an organization that was a possible takeover target, what might you do to protect your personal future?

## REFERENCES

1. "Women CEOs of the Fortune 1000," *Catalyst: Expanding Opportunities for Women and Business* (September 2010) at www.catalyst.org (accessed October 1, 2010); "Women CEOs," *CNNMoney.com* (May 3, 2010) at money.cnn.com (accessed October 2, 2010); Anjali Cordeiro and Joann S. Lublin, "New Chief for Campbell Soup," *The Wall Street Journal* (September 29, 2010) at online.wsj.com (accessed October 3, 2010); Andrew Cleary, Susan Berfield, and Michael Arndt, "Cadbury Purchase May Top Nabisco in Kraft's Legacy." *Business Week* (June 20, 2010) at www.businessweek.com (accessed October 4, 2010. Leanne Atwater and David D. Van Fleet, "Another Ceiling? Can Males Compete for Traditionally Female Jobs?" *Journal of Management*, 23(5), 1997, 603–626.

2. For a complete discussion of human resource management, see Angelo S. DeNisi and Ricky W. Griffin, *Human Resource Management, 4th ed.* (Cincinnati, OH: Cengage, 2011).

3. Gene G. Marcial, "Outback: Set to Sizzle," *Business Week* (January 23, 2006) at www.businessweek.com (accessed June 30, 2010);

Joe Schneider, "Outback Steakhouse to Pay $19 million to End Gender Bias Suit," *Business Week* (December 30, 2009) at www.businessweek.com (accessed June 30, 2010).

4. For a more detailed treatment of the staffing process, see Cynthia Fisher, Lyle Schoenfeldt, and Ben Shaw, *Human Resource Management, 6th ed.* (Boston: Houghton Mifflin, 2006).

5. David P. Twomey, *Employment Discrimination Law: A Manager's Guide, 6th ed.* (Mason, OH: Thomson/Southwestern, 2005).

6. "The Disabilities Act is a Godsend—For Lawyers," *Business Week*, August 17, 1992, 29.

7. Benjamin Schneider, "Strategic Job Analysis," *Human Resource Management* (Spring 1989): 51–60.

8. Peter Cappelli, "A Supply Chain Approach to Workforce Planning," *Organizational Dynamics* (2009): 8–15; Norman Scarborough and Thomas W. Zimmerer, "Human Resource Forecasting: Why and Where to Begin," *Personnel Administrator*, (May 1982): 5–61.

9. "Agricultural Workers," *Occupational Outlook Handbook, 2010–11 ed.*, May 17, 2010, at www.bls.gov/oco on June 30, 2010; "Agriculture, Forestry, and Fishing," *Occupational Outlook Handbook, 2010–11 ed.*, December 17, 2009, at www.bls.gov/oco on June 30, 2010; Robert Rodriguez, "UFW union encourages citizens to become farmworkers, 'Take Our Jobs'," at www.cleveland.com on July 6, 2010.

10. Nancy Adler and Susan Bartholomew, "Managing Globally Competent People," *Academy of Management Executive* (August 1992): 52–60.

11. Karen Holcolme Ehrhart and Jonathan Ziegert, "Why Are Individuals Attracted to Organizations?" *Journal of Management* (2005): 901–919.

12. Claudio Fernandez-Araoz, Boris Groysberg, and Nitin Nohria, "The Definitive Guide to Recruiting in Good Times and Bad," *Harvard Business Review* (2009): 74–85.

13. Fisher, Schoenfeldt, and Shaw, *Human Resource Management*.

14. Laura M. Herren, "The Right Technology for Recruiting in the '90s," *Personnel Administrator*, (April 1989): 48–53.

15. James J. Asher, "The Biographical Item: Can It Be Improved?" *Personnel Psychology* (Summer 1972): 251–269.

14. Frank L. Schmidt and John E. Hunter, "Employment Testing: Old Theories and New Research Findings," *American Psychologist* (October 1981): 1128–1137.

17. Neal Schmidt, "Social and Situational Determinants of Interview Decisions: Implications for the Employment Interview," *Personnel Psychology* (Spring 1976): 79–102.

18. For an interesting look at some of the legal cases regarding selection, see Kathryn E. Buckner, Hubert S. Field, and William Holley, Jr., "The Relationship of Legal Case Characteristics with the Outcomes of Personnel Selection Court Cases," *Labor Law Journal* (January 1990): 31–40; for a review of recent research on selection, see Edwin A. Fleishman, "Some New Frontiers in Personnel Selection Research," *Personnel Psychology* (Winter 1988): 679–702.

19. Kenneth N. Wexley and Gary P. Latham, *Developing and Training Human Resources in Organizations, 3rd ed.* (Upper Saddle River, NJ: Prentice-Hall, 2002).

20. "Companies Use Cross-Cultural Training to Help Their Employees Adjust Abroad," *The Wall Street Journal*, August 4, 1992, B1, B9; Paul Vanderbroeck, "Long-Term Human Resource Development in Multinational Organizations," *Sloan Management Review* (Fall 1992): 95–104.

21. Jonathan Hickman, "America's 50 Best Companies for Minorities," *Fortune*, July 7, 2003, pp. 103–120.

22. Lisa Schnebly Heidinger, "Pro's Ranch Market: The Melting Pot," *Arizona Food Industry Journal* (April 2011): 12–14.

23. William M. Fox, "Getting the Most from Behavior Modeling Training," *National Productivity Review* (Summer 1988): 238–245.

24. "Training 101: How to Win Closure and Influence People," *Training and Development Journal*, (January 1990): 31–35.

25. Charles D. Pringle and Peter Wright, "An Empirical Examination of the Relative Effectiveness of Supervisory Training Programs," *American Business Review* (January 1990): 1–7, and Stewart J. Black and Mark Mehdenhall, "Cross-Cultural Training Effectiveness: A Review and a Theoretical Framework for Future Research," *Academy of Management Review* (January 1990): 113–136.

26. Amy DelPo, *The Performance Appraisal Handbook: Legal & Practical Rules for Managers, 2nd ed.* (Berkeley, CA: Nolo, 2007); Paul Levy and Jane Williams, "The Social Context of Performance Appraisal: a Review and Framework for the Future," *Journal of Management* (2004): 881–905.

27. Jeanette N. Cleveland, Kevin R. Murphy, and Richard E. Williams, "Multiple Uses of Performance Appraisal: Prevalence and Correlates," *Journal of Applied Psychology* (February 1989): 130–135.

28. William M. Fox, "Improving Performance Appraisal Systems," *National Productivity Review* (Winter 1987–1988): 20–27.

29. Gary P. Latham and Kenneth N. Wexley, *Increasing Productivity Through Performance Appraisal, 2nd ed.* (Reading, MA: Addison-Wesley, 1994).

30. Timothy M. Downs, "Predictors of Communication Satisfaction During Performance Appraisal Interviews," *Management Communication Quarterly* (February 1990): 334–354; and James R. Larson, Jr., "The Dynamic Interplay Between Employees' Feedback-Seeking Strategies and Supervisors' Delivery of Performance Feedback," *Academy of Management Review* (July 1989): 408–422.

31. David D. Van Fleet, Tim O. Peterson, and Ella W. Van Fleet, "Closing the Performance Feedback Gap With Expert Systems," *Academy of Management Executive* (2005): 38–53; Peter H. Lewis, "I'm Sorry; My Machine Doesn't Like Your Work," *The New York Times*, February 4, 1990, F27.

32. David Waldman and Leanne Atwater, *The Power of 360 Degree Feedback* (Houston: Gulf Publishing, 1998); Richard Lepsinger and Anntoinette E. Lucia, *The Art and Science of 360 Degree Feedback.* (San Francisco: Jossey-Bass/Pfeiffer, 1997); Mark R. Edwards and Ann J. Ewen, *360 Degree Feedback: The Powerful New Model for Employee Assessment & Performance Improvement.* (New York: AMACOM, 1996).

33. For a very early discussion of the use of multi-rater approaches, see Edward E. Lawler III, "The Multitrait-Multirater Approach to Measuring Managerial Job Performance," *Journal of Applied Psychology* (1967): 369–381.

34. Angelo S. DeNisi and Avraham N. Kluger, "Feedback Effectiveness: Can 360-Degree Appraisals Be Improved?" *Academy of Management Executive* (2000): 129–139.

35. Richard E. Kopelman, Janet L. Rovenpor, and Mo Cayer, "Merit Pay and Organizational Performance: Is There an Effect on the Bottom Line?" *National Productivity Review* (Summer 1991): 299–307.

36. Richard I. Henderson, *Compensation Management in a Knowledge-Based World*, 10th ed. (Upper Saddle River, NJ: Prentice-Hall, 2006). See also, Edward E. Lawler, *Rewarding Excellence: Pay Strategies for the New Economy.* (San Francisco: Jossey-Bass, 2000).

37. "Executive Pay," *Business Week*, May 1, 1989, 46-47; and Thomas H. Patten, Jr., *Employee Compensation and Incentive Plans* (New York: Free Press, 1977).

38. U.S. Chamber of Commerce, *Employee Benefits Study* (Washington, DC: U.S. Government Printing Office, 2008).

39. Karen L. Frost, Dale L. Gifford, Christine A. Seltz, and Kenneth Sperling, eds. *Fundamentals of Flexible Compensation 1994 Supplement* (New York: Wiley, 1997).

40. For a general treatment, see Arthur A. Sloane and Fred Witney, *Labor Relations, 13th ed.* (Upper Saddle River, NJ: Prentice-Hall,

2009). For a view of what may be expected in the future, see "Peter Drucker Looks at Unions' Future," *Industry Week*, March 23, 1989, 16–20.

41. Casey Ichniowski and Jeffrey S. Zax, "Today's Associations, Tomorrow's Unions," *Industrial and Labor Relations Review* (January 1990): 191–208; and Wendell French, *Human Resources Management, 6th ed.* (Mason, OH: Cengage, 2006).
42. Sloane and Witney, *Labor Relations*.
43. James Rand, "Preventive Maintenance Techniques for Staying Union Free," *Personnel Journal* (June 1980): 497–508.
44. Sloane and Witney, *Labor Relations*.
45. Ibid.
46. Hester Plumridge and John Jannarone, "Sanofi Seeks Genzyme Treatment," *The Wall Street Journal* (July 27, 2010) at online.wsj.com (accessed December 19, 2010); Dana Cimilluca, "Genzyme, Sanofi Are At Odds Over Price," *The Wall Street Journal* (August 24, 2010) at online.wsj.com (accessed December 19, 2010); Thomas Gryta, "Genzyme Rejects Sanofi's Overture," *The Wall Street Journal* (August 31, 2010) at online.wsj.com (accessed December 19, 2010); Gina Chon and Anupreeta Das, "Genzyme CEO Will Look for a Higher Offer," *The Wall Street Journal* (September 1, 2010) at online.wsj.com (accessed December 19, 2010); Jeanne Whalen and Dana Cimilluca, "Sanofi's Board Members Differ on Bid for Genzyme," *The Wall Street Journal* (September 1, 2010) at online.wsj.com (accessed December 19, 2010); Thomas Gryta, "Genzyme Sets Sale of Unit to LabCorp as it Refocuses," *The Wall Street Journal* (September 14, 2010) at online.wsj.com (accessed December 19, 2010); Dana Cimilluca and Gina Chon, "Sanofi Lining Up Genzyme Financing," *The Wall Street Journal* (September 27, 2010) at online.wsj.com (accessed December 19, 2010); John Jannarone, "Genzyme's Time Out with Sanofi," *The Wall Street Journal* (September 28, 2010), C10; Jeanne Whalen and Dana Cimilluca, "Sanofi Makes Hostile Genzyme Bid," *The Wall Street Journal* (October 5, 2010) at online.wsj.com (accessed December 19, 2010); Gina Chon and Joann W. Lublin, "Genzyme Mulls 'White Knight,'" *The Wall Street Journal* (October 8, 2010), B3; Jonathon D. Rockoff and Thomas Gryta, "Genzyme's Profits Soars Amid Fight," *The Wall Street Journal* (October 21, 2010) at online.wsj.com (accessed December 19, 2010); Gina Chon and Anupreeta Das, "Genzyme Sees Fix for Impasse," *The Wall Street Journal* (November 22, 2010), B1.

# CHAPTER 18
- Employee Motivation

# CHAPTER 19
- Groups and Teams

# CHAPTER 20
- Managerial Communication

# Individual and Interpersonal Processes

## LEARNING OBJECTIVES

After studying this chapter, you should be able to:

- Describe the relationship between individuals and the organization.
- Identify and discuss basic individual differences.
- Identify and discuss performance-based differences at work.
- Describe the role of stress in organizations.
- Discuss basic interpersonal processes at work.

## Can Women Smell Their Way to Success?

**F**ermented beverages are almost as old as humanity. Beer and wine are beverages that trace their origins far back into human history. They were alcoholic beverages that required the least amount of technology to master, although of the two beverages, wine was the most demanding alcoholic beverage. It required the identification and cultivation of vines and the nurturing of grapes before it could be made into wine. Beer, however, required only a fermentable sugar in a cereal crop, yeast, and water. Other additives, like hops, etc., came later.

**FIGURE 16.1** Both men and women are hired to taste or smell products (a standard part of the marketing process), but women appear to have a keener sense of smell than do men.
© Karuka/www.Shutterstock.com

Beer has been traced back in antiquity to the Egyptians; however, it was the Germans who made the most of their dual prowess for both producing beer and quaffing it. At one time beer was consumed by everyone—men, women and children—because it was a means of making bad water palatable. This was necessary since animals, people, and their waste lived in close proximity to each other and their water supplies (wells). Thanks to the monasteries in Europe, beer "morphed" into a brew enjoyed—and perhaps all too often *over* enjoyed—by males. Today, males account for 72.8 percent of the world's beer sales.

Ah … but the times they are a-changin'.

The 177th Munchen Oktoberfest (Munich) witnessed fewer celebrants, though they did consume a record seven million liters. One would think that German brewers would be pleased with this level of consumption since Germany remains the fifth largest beer-consuming nation, but alas they were not. The source of this Teutonic angst was that since 1991 Germans have been consuming less beer each year. By 2009 Austria, the Czech Republic, and Ireland each consumed more beer than Germany. Germany's 1300 breweries were locked in fierce competition and the smaller ones were struggling. This competitive environment is enough to make a brewer cry in his … —well, you know.

But this is not the only change occurring in the beer world.

Rhonda Dannenberg, a Milwaukee-suburb mother of three, stuck her nose in six glasses of beer at the MillerCoors brewery and swished a bit of each in her mouth (Figure 16.1). Then she delivered the kind of frank verdict that's shaking up the men's club world of beer tasting. "I got a strong bruised fruit," she said, drawing nods from three other women and two men sitting on the tasting panel table; "Slight cardboard taste. Oxidized. Unacceptable."

Only about one of every five people—male or female—who try out for tasting positions at breweries ascend to the level of corporate panelist, says Bill Simpson of Cara Technology Ltd., in Britain. And an important individual difference has been noted: it seems that women have a better sense of smell then men. SABMiller, maker of Pilsner Urquell, Peroni, and Grolsch (premium European beers), in addition to Miller and Coors, says its empirical evidence shows that females are the superior sex when it comes to identifying chemicals such as 3-methyl-2-butene-1-thiol, which makes beer "skunky."

If the United States Congress passes a bill to reduce the excise tax paid per barrel by small brewers (from $7 to $3.50) for the first 60,000 barrels they produce and lowers the tax on production beyond

60,000 barrels from $18 to $16, it may be a boon for more jobs in the United States. A Harvard University study indicated that another 2,700 jobs would be added to the already 100,000 jobs among small brewers.

"These breweries do a good job of pumping life back into the regions and creating tens of thousands of jobs. They also make some really good beer," said a Massachusetts Congressman, somewhat hyperbolically. The better news is both men and women can apply, of course, but the "the best jobs in the universe" may go to the women. Who would have suspected that a beer glass would chip the glass ceiling? Maybe Kaiser Wilhelm of Germany suspected this tectonic industry change when he said, "Give me a woman who truly loves beer, and I will conquer the world."[1]

## INTRODUCTION

Like it or not, admit it or not, physical characteristics and ability are among the individual differences that help determine how healthy or dysfunctional various interactions between people will prove to be. In the beer-sampling job discussed above, one form of physical ability (sense of smell and hence taste) may help determine whether you get the job and whether you can advance in the job. Other examples are individuals who are sight or hearing-challenged. To offset their challenges, these individuals often develop other sensory skills that are highly superior to individuals who feel no need to compensate.

Humor also plays an important role in most organizations, as do other fundamental individual and interpersonal processes, especially in dealing with conflict. People's attitudes, for instance, shape how they feel about the organization; and their motivation to perform clearly helps determine the effectiveness of the organization.

Individual differences should be recognized for what they add, though, rather than being used as negative discrimination factors. Discrimination by a company or its employees on the basis of physical features or ability can result in quick and serious legal action. This chapter does not deal with physical differences; rather, it introduces and discusses several basic individual and interpersonal processes in organizations. We first discuss the basic relationship between individuals and organizations. Then we examine such psychological individual differences as personality, attitudes, and perception, and discuss performance-based differences, followed by an examination of stress. Finally, we introduce and discuss several basic interpersonal processes. Other critical individual and interpersonal processes—leadership, employee motivation, group dynamics, and managerial communication—are covered in the next four chapters.

# Individuals and Organizations

Organizations are comprised of a collection of individual people—managers, clerical workers, administrators, custodial workers, scientists, and dozens of others. Each individual brings a unique set of contributions to the organization, and each expects to receive certain things from the organization. We begin by examining psychological contracts and the person-job fit.

## Psychological Contracts

A **psychological contract** is the set of expectations held by an individual about what he or she will contribute to the organization and what the organization will provide in return.[2] The nature of a psychological contract is illustrated in Figure 16.2. The individual makes a variety of contributions to the organization: effort, skill, ability, time, loyalty, and so forth. These contributions satisfy various needs and requirements of the organization.

In return for these contributions, the organization provides inducements to the individual. Some inducements, like pay, are tangible. Others, like status, are intangible. Just as the contributions from the individual must satisfy some of the needs of the organization, the inducements offered by the organization must serve some of the needs of the individual (Figure 16.3). That is, if a person accepts employment with an organization because he thinks he will earn an attractive salary and have an opportunity to advance, he will subsequently expect that those rewards will actually be forthcoming.

**FIGURE 16.3** Employees have come to expect not only their take-home pay but also benefits, such as health insurance.

If both the individual and the organization perceive that the psychological contract is fair and equitable, they will be satisfied with the relationship and continue it. But if either party sees an imbalance or inequity in the contract, it may initiate a change.[3] For example, the individual may request a pay raise or promotion, decrease her efforts, or look for a better job elsewhere. The organization can also initiate change by requesting that the individual improve his skills through training, transferring the person to another job, or terminating the person's employment altogether.

A basic challenge the firm faces, then, is managing psychological contracts. An organization must ensure that it is not only getting value from its employees but also providing employees with appropriate inducements. If a firm is underpaying its employees for their contributions, for example, the employees may perform poorly or leave for better jobs elsewhere. On the other hand, if they are being overpaid relative to their contributions, the firm is incurring unnecessary costs.

## The Person-Job Fit

One specific aspect of managing psychological contracts is managing the **person-job fit**—the extent to which the contributions made by the individual match the inducements offered by the organization. In theory, each employee has a specific set of needs that he or she wants fulfilled and a set of job-related behaviors and abilities to contribute. Thus, if the organization can take perfect advantage of those behaviors and abilities and exactly fulfill the employee's needs, it will have achieved a perfect person-job fit.[5] (Chapter 18 includes a discussion regarding needs and motivation.)

Of course, such a precise level of person-job fit is seldom achieved. For example, because organizational

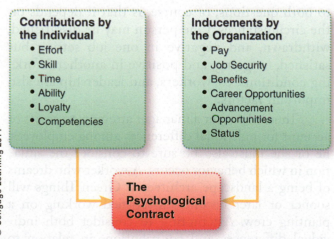

**FIGURE 16.2** The psychological contract between individuals and organizations.

## A FOCUS ON AGRIBUSINESS
### Psychological Contracts

Why would family members stay in the family business if they could make more money elsewhere? Because the longer term, intangible aspects of the psychological contract are stronger than the lack of direct financial incentives. Strong psychological contracts play a major role in reducing withdrawal behavior. The psychological contract is critical in any organizational setting, but perhaps especially so in many agribusiness firms.

The psychological contract doesn't exist just between employers and employees; it also exists between organizations, such as a grower and a distributor or wholesaler, or between a wholesaler and a retailer. Green Things Landscaping, for example, has a psychological contract with some of the developers who build neighborhoods or shopping centers; and Summer Farms maintains this relationship with a regional chain of grocery stores.

When one organization forms a strategic alliance with another organization, a psychological contract exists there as well. Such an alliance might include, for example, a producer's cooperative and a company that will handle one or more functions for the cooperative, such as human resources or marketing or accounting. Agricultural marketing cooperatives have been facilitating the transfer of the marketing function from the producers to the cooperative in varying degrees for a long time.

A study of cotton producers outsourcing the marketing of their fiber found that many producers are willing to place the marketing function in the hands of a cooperative. However, they will do so only if they both trust the marketing cooperative and feel that it possesses the expertise necessary to perform the activity effectively.[4]

selection procedures are imperfect, firms can only estimate employee skill levels when making hiring decisions. Similarly, both people and organizations change. An organization's adoption of new technology, for example, may change the skills the company needs from its employees. Or an individual who finds a new job exciting may find the same job monotonous after a few years of performing it. Summer Farms experienced this very problem with one of the grandsons, who was neither a fast learner nor a dedicated worker after he concluded that he would not be fired since he was "family." Finally, because each individual is unique, assessing individual differences simply cannot be done with complete precision.

### FOOD FOR THOUGHT 16.1

Before Cadbury became part of Kraft Foods in 2010, the company made an unpopular change when it began adding palm oil from New Zealand as a replacement for some cocoa butter. The move cost the company about 12 million in sales.

## The Nature of Individual Differences

**Individual differences** are personal attributes that vary from one person to another (Figure 16.4). Are specific differences that characterize a given individual good or bad? Do they contribute to or detract from performance? The answer to both questions, of course, is that it depends on the circumstances. One person may be dissatisfied, withdrawn, and negative in one job setting but satisfied, outgoing, and positive in another. Working conditions, co-workers, and leadership are also important factors.

Thus, whenever a manager attempts to assess or account for individual differences among employees, he or she must also be sure to consider the situation in which behavior occurs. A worker who dreams of being a landscape architect at Green Things will sooner or later become dissatisfied working on a planting crew. Attempting to consider both individual differences and contributions in relation to inducements and contexts, then, is a major challenge for managers as they attempt to establish effective

FIGURE 16.4 This group of employees demonstrates the wide variety of individual differences found in today's organizations.

**TABLE 16.1** Key Personality Attributes

| | |
| --- | --- |
| *Locus of Control* | The degree to which a person believes that behavior has a direct impact on its consequences. |
| *Authoritarianism* | The extent to which a person believes that power and status differences are appropriate within organizations. |
| *Dogmatism* | The rigidity of a person's beliefs and his or her openness to other viewpoints. |
| *Self-Esteem* | The extent to which an individual believes that he or she is a worthwhile and deserving person. |
| *Risk Propensity* | The degree to which an individual is willing to take chances and make risky decisions. |

© Cengage Learning 2014

psychological contracts with their employees and achieve optimal fits between people and jobs.

## Personality and Work

**Personality** is the relatively stable set of psychological and behavioral attributes that distinguishes one person from another.[6] Understanding basic personality attributes is important because they affect people's behavior in organizational situations and their perceptions of and attitudes toward the organization.[7]

**Personality Formation.** An individual's basic personality is formed before becoming a member of an organization, but personality can still change as a result of organizational experiences. For example, suppose a manager is subjected to prolonged periods of stress or conflict at work. As a result, he may become more withdrawn, anxious, and irritable. While removal of the stressful circumstances may eventually temper these characteristics, the individual's personality may also reflect permanent changes. From a more positive perspective, continued success and accomplishment at work may cause an individual to become more self-confident and outgoing, of course. Still, managers should recognize that they can do little to change the basic personalities of their subordinates. Instead, they should work to understand the basic nature of their subordinates' personalities and how attributes of those personalities affect the subordinates' work behavior.

## Personality Attributes in Organizations.

Table 16.1 defines several important personality attributes that are particularly relevant to organizations.

**Locus of control** is the degree to which a person believes that behavior has a direct impact on its consequences.[8] Some people believe that they will succeed if they work hard. They also believe that people who fail do so because they lack ability or motivation. Because these people believe that each person is in control of his or her life, they have an *internal locus of control*. Other people think that what happens to them is a result of fate, luck, or the behavior of other people. An employee who fails to get a promotion may attribute that failure to a politically motivated boss or to bad luck rather than to his own lack of skills or poor performance. Because these people think that forces beyond their control dictate what happens to them, they have an *external locus of control* (Figure 16.5).

FIGURE 16.5 This manager, who appears to have difficulty making a decision, may have an external locus of control and would therefore prefer assistance.

Locus of control has several implications for managers.[9] Individuals with an internal locus of control may have a strong desire to participate in the governance of the firm and have a voice in how they do their jobs. Thus, they may prefer a decentralized organization and a leader who gives them freedom and autonomy. They may also be most comfortable under a reward system that recognizes individual performance. External locus of control people, on the other hand, may prefer a centralized organization where others make decisions or gravitate to structured jobs where standard procedures are defined for them. They may also prefer leaders who make most of the decisions and reward systems that put premiums on seniority. We might expect, therefore, that the Green Things planting crew has an external locus of control whereas the landscape architect is an internal locus person.

Another important personality characteristic is **authoritarianism**—the extent to which a person believes that power and status differences are appropriate within organizations.[10] A person who is highly authoritarian may accept orders, without question, from someone with more authority purely because the other person is "the boss." A person who is not highly authoritarian may still carry out appropriate directives from the boss but is more likely to question things or express disagreement with the boss. Managers who are highly authoritarian may be autocratic and demanding, and subordinates who are highly authoritarian are more likely to accept this behavior from their leaders. On the other hand, a manager who is less authoritarian may allow subordinates bigger roles in making decisions, and less authoritarian subordinates will prefer this behavior.[11]

**Dogmatism**, another important attribute, is the rigidity of a person's beliefs and his or her openness to other viewpoints.[12] Popular terms for dogmatism are *close-minded* and *open-minded*. For example, suppose a manager has such strong beliefs about how to carry out procedures that he or she is unwilling to even listen to a new idea for performing it more efficiently. This person is close-minded, or highly dogmatic. A manager who is receptive to listening to and trying new ideas is said to be more open-minded, or less dogmatic.

**Self-esteem** is the extent to which an individual believes that he or she is a worthwhile and deserving person.[13] A person with high self-esteem is more likely to seek higher status jobs, be more confident in his or her ability to achieve high levels of performance, and derive greater intrinsic satisfaction from her accomplishments. In contrast, a person with less self-esteem may be more content to remain in a lower-level job, be less confident of her ability, and focus more on extrinsic rewards.[14]

**Risk propensity** is the degree to which an individual is willing to take chances and make risky decisions.[15] A manager with a high-risk propensity, such as Joshua Summer, may be expected to experiment with new ideas and gamble on new products. He may also lead the organization in new and different directions and be a catalyst for innovation. Of course, the same individual may also jeopardize the continued well being of the organization if the risky decisions prove to be bad ones. A manager with low-risk propensity might lead to a stagnant and overly conservative organization; on the other hand, he may help the organization successfully weather turbulent and unpredictable times by maintaining stability and calm. Thus, the potential consequences of risk propensity to a firm are heavily dependent on that firm's environment.

## Attitudes and Work

Another aspect of individuals in organizations consists of their attitudes. **Attitudes** are sets of beliefs and feelings that individuals have about specific ideas, situations, or other people.[16] Attitudes are important because they are the mechanism through which most people express their feelings. An employee's statement that he feels underpaid by the organization reflects his feelings about his pay. Similarly, when a manager says that she likes the firm's new advertising campaign, she is expressing her feelings about the organization's marketing efforts.

Attitudes have three components:

1. The *affective component* reflects feelings and emotions that an individual has toward something.

2. The *cognitive component* is derived from knowledge that an individual has about something. It is important to note that cognition is subject to individual perceptions (something we discuss more fully later). Thus, one person may "know" that a certain political candidate is better than another, while someone else may "know" just the opposite.

3. Finally, the *intentional component* reflects how an individual expects to behave toward or in the situation.

To illustrate these three components, consider the case of a bottled water company manager who placed an equipment parts and supply order with a new company. Two of the items he ordered were out of stock, one was over-priced, and another arrived damaged. When he called someone at the supply firm for help, he felt that he was treated rudely and was even disconnected before his claim was resolved. When asked how he felt about the new supplier, he responded: "I don't like that company (affective component). They are the worst supply firm I've ever dealt with (cognitive component). I'll never do business with them again (intentional component)."

People try to maintain consistency among the three components of their attitudes. However, circumstances sometimes arise that lead to conflicts called **cognitive dissonance**.[17] For example, the water-bottler manager may be forced either to buy again from the disappointing company or to replace his current equipment with a different brand, which would be quite costly. An individual who has vowed never to work for a big, impersonal corporation intends instead to open her own business. Unfortunately, a series of financial setbacks causes her to have to take a job with a large company. Thus, cognitive dissonance occurs—the affective and cognitive components of the individual's attitude conflict with intended behavior. She needs the job and is pleased to have this option, but she does not want the job and is not looking forward to working there. To reduce dissonance, which is usually uncomfortable, the individual may tell herself the situation is only temporary. Or she might revise her cognitions and decide that working for a large company is more pleasant than she had expected.

**Attitude Change.** Attitudes are not as stable as personality attributes. For example, attitudes may change as a result of new information. A manager may have a negative attitude about a new colleague because of the person's lack of job-related experience. After working with the new person, however, he may realize that the colleague is actually very talented and learns quickly. The manager may subsequently develop a more positive attitude toward that person. Attitudes can also change due to changes in the object of the attitude. For example, if employees feel underpaid and have negative attitudes about their pay, big salary increases may result in more positive attitudes about pay.[18]

Attitude change can also occur when the object of the attitude becomes less important or less relevant to the person. For example, suppose an employee has a negative attitude about his firm's health insurance. When the employee's spouse gets a new job with an organization that has outstanding health insurance benefits, his attitude toward his own insurance may become more moderate simply because it is no longer a worry. Finally, as noted earlier, individuals may change their attitudes to reduce cognitive dissonance.[19]

**Work-Related Attitudes.** People form attitudes about many different things. For example, employees may have attitudes about their salaries, promotion possibilities, their bosses, employee benefits, the food in the company cafeteria, and the color of the company softball team uniforms. But some attitudes are especially critical.

**Job satisfaction or dissatisfaction** is an attitude that reflects the extent to which a person is gratified by or fulfilled in his work. Extensive research conducted on job satisfaction suggests that factors such as individual needs and aspirations determine this attitude, along with factors such as relationships with co-workers and supervisors and working conditions, work policies, and compensation.[20] A satisfied employee tends to be absent less often, to make positive contributions, and to stay with the firm. But a dissatisfied employee may be absent more often, may experience stress that disrupts co-workers, and may be continually looking for another job. Contrary to what many managers believe, however, high levels of job satisfaction do not necessarily lead to higher levels of performance.

**Organizational commitment**, another important attitude, reflects an individual's identification with and attachment to the firm.[21] People with high commitment are likely to see themselves as true members of the organization (e.g., referring to the organization in personal terms, like "We make high quality products …"), to overlook minor sources of dissatisfaction, and to see themselves remaining members of the firm. In contrast, people with less commitment are more likely to see themselves as outsiders, using less personal terms like "They don't pay employees very well," to express more dissatisfaction about things, and not see themselves as long-term members of the organization.

## Perception and Work

As noted earlier, an important element of an attitude is the individual's perception of the object about which the attitude is formed. **Perception** is the set of processes by which an individual recognizes and interprets information about the environment. As Figure 16.6 illustrates, perception can be deceiving. Since perception plays a role in other workplace behaviors, managers need a general understanding of basic perceptual processes. Two basic perceptual processes that are particularly relevant to managers are selective perception and stereotyping.

**Selective perception** is the process of screening out information that we are uncomfortable with or which contradicts our beliefs. Suppose a manager has a positive attitude about a worker because he or she is a top performer. One day the manager sees that the worker appears to be loafing. Selective perception may cause the manager to ignore the undesirable behavior. Similarly, suppose the manager has a negative image of a worker and thinks he is a poor performer. When the manager sees that person working hard, he may ignore that good behavior. Selective perception allows us to disregard minor bits of information. On the other hand, if selective perception causes us to ignore important information, it can become a problem.

**Stereotyping** is the process of categorizing people on the basis of a single attribute. Common attributes from which people often stereotype are age, race, and sex.[22] Of course, stereotypes along these lines are inaccurate and can be harmful. For example, suppose a manager holds the stereotype that women can only perform certain tasks and that men are best suited for other tasks. To the extent that this affects the manager's hiring practices, he or she is (1) costing the organization valuable talent, (2) violating the law, and (3) behaving unethically. Another example of stereotyping is having a Summer Farms manager categorize job candidates with the Summer family name as either the hardest working, most trustworthy employees or as lazy, uncommitted employees who expect everything to be given to them.

But certain forms of stereotyping can be useful and efficient. Suppose, for example, that a manager believes that communication skills are important for a job and that speech majors tend to have exceptionally good communication skills. As a result, whenever interviewing job candidates, that manager pays close attention to speech majors. To the extent that communication skills do predict performance and that majoring in speech does indeed provide those skills, this form of stereotyping can be beneficial.

## FOOD FOR THOUGHT 16.2

According to the USDA, brown eggs are the same as white eggs in taste and nutrition but brown eggs usually sell at a higher price because some customers perceive them to taste better.

**Perception Can Be Deceiving.**
Is "A" a goblet or two faces?
Is the hat in "B" taller or wider?
Can the object in "C" exist?
Is "D" a young or old woman?

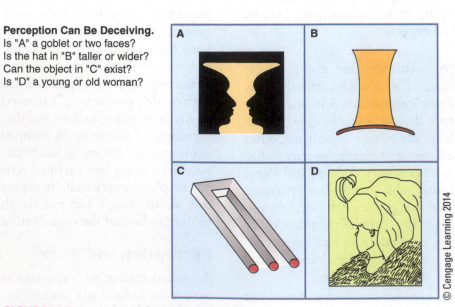

© Cengage Learning 2014

**FIGURE 16.6** Examples of deceptive perception.

# Performance-Based Differences and Work

Performance includes productivity, but it also includes all of an employee's job-related behaviors. Focusing only on productivity can be misleading as a valuation of an employee's overall worth to a company. An employee may have value quite apart from just the number of units he or she produces.

## Performance Behaviors

**Performance behaviors**, derived from the psychological contract, are the set of work-related behaviors that a firm expects people to display. Performance behaviors can be narrowly defined and easily measured for some jobs. An assembly line worker who sits by a moving conveyor and attaches parts to a product as it passes by has relatively few performance behaviors. He or she is expected to remain at the workstation and correctly attach the parts. Performance can often be assessed quantitatively by counting the percentage of parts correctly attached.

For many other jobs, performance behaviors are more diverse and difficult to assess. For example, consider the case of a research scientist at Monsanto. The scientist works in a lab trying to find new breakthroughs that have commercial potential. The scientist must apply knowledge with experience. Intuition and creativity are also important elements. The desired breakthrough may take months or even years to accomplish. As we discussed in Chapter 15, organizations rely on a number of different methods for evaluating performance. The key, of course, is to match the evaluation mechanism with the job being performed.

## Withdrawal Behaviors

Another work-related behavior results in withdrawal from being a committed and contributing member of the organization. **Absenteeism**, for example, occurs when an individual does not come to work. The cause may be legitimate (e.g., illness or jury duty) or feigned (reported as legitimate but actually just an excuse to stay home). When an employee is absent, his work does not get done or a substitute must do it. In either case, the quantity or quality of actual output may suffer.[23] Obviously, organizations can expect some absenteeism. The concern of managers is to minimize feigned absenteeism and reduce legitimate absences as much as possible. High absenteeism may also be a symptom of other problems, such as dissatisfaction and low morale.

**Turnover** occurs when people quit their jobs. A firm incurs costs in replacing individuals who have quit, especially if it must replace productive people. Turnover results from aspects of the job, the firm, the individual, the labor market, family influences, and other factors. Of course, some turnover is inevitable and may even be desirable. For example, if the organization is trying to cut costs by reducing its workforce, having people choose to leave is preferable to terminating them. If the people who choose to leave are low performers or express high levels of job dissatisfaction, the organization may also benefit from turnover.

## Organizational Citizenship

**Organizational citizenship** is the behavior of individuals that makes a positive overall contribution to the organization.[24] Consider, for example, an employee whose work is acceptable in terms of both quantity and quality. However, she refuses to work overtime, will not help newcomers learn the ropes, and is generally unwilling to make any contribution to the organization beyond the strict performance of her job. While this person may be seen as a good performer, she is not likely to be seen as a good organizational citizen. Another employee may exhibit the same level of performance and also work late when the boss asks, take time to help newcomers, and be helpful and committed to the organization's success. Although his level of performance may be seen as equal to that of the first worker, he is also likely to be seen as a better organizational citizen.

The determinant of organizational citizenship behaviors is a complex mosaic of individual, social, and organizational variables. For example, the personality and attitudes of the individual must be consistent with citizenship behaviors. Similarly, the group in which the individual works must facilitate and promote such behaviors. And the organization itself, especially its culture, must be capable of promoting, recognizing, and rewarding these types of behaviors to maintain them.[25]

## Stress at Work

Another important individual process in organizations is stress. **Stress** is an individual's response to a strong stimulus.[26] This stimulus is called a **stressor**. Stress generally follows a cycle referred to as the **General Adaptation Syndrome (GAS)**, shown in Figure 16.7.[27] According to this view, an individual's first encounter with a stressor initiates the GAS

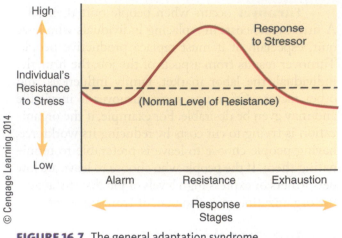

FIGURE 16.7  The general adaptation syndrome.

© Cengage Learning 2014

© Lichtmeister/www.Shutterstock.com

**FIGURE 16.8**  This highly stressed manager appears to be in Stage 3 of GAS exhaustion.

and activates the first stage: alarm. He or she may feel panic, may wonder how to cope, and may feel helpless. For example, suppose a manager is told to prepare a detailed evaluation of a plan by his firm to buy one of its competitors. His first reaction may be, "How will I ever get this done by tomorrow?"

If the stressor is too intense, the individual may feel unable to cope and never really try to respond to its demands. In most cases, however, after a short period of alarm, the individual gathers some strength and starts to resist the negative effects of the stressor. The manager with the evaluation to write, for example, may calm down, call home to say he's working late, roll up his sleeves, order out for coffee, and get to work. Thus, at stage 2 of the GAS, the person is resisting the effects of the stressor. In many cases, the resistance phase may end the GAS. If the manager can complete the evaluation earlier than expected, he may drop it in his briefcase, smile to himself, and head home tired but satisfied.

On the other hand, prolonged exposure to a stressor without resolution may bring on stage 3 of the GAS: exhaustion. At this stage, the individual literally gives up and can no longer resist the stressor. The manager, for example, may fall asleep at his desk at 3 a.m. and never finish the evaluation (Figure 16.8).

We should note that not all stress is bad.[28] Just as too much stress can have negative consequences, the absence of stress can lead to lethargy and stagnation. An optimal level of stress, on the other hand, can result in motivation and excitement. It is also important to understand that "good" as well as "bad" things can cause stress. Excessive pressure, unreasonable demands on our time, and bad news can all cause stress. Even receiving a bonus and then having

to decide what to do with the money can be stressful. So, too, can receiving a promotion, gaining recognition, and similar "good" things.

One important line of thinking about stress focuses on Type A and Type B personalities.[29] **Type A** individuals are extremely competitive, devoted to work, and have a strong sense of time urgency. They are likely to be aggressive, impatient, and work-oriented. They have a lot of drive and want to accomplish as much as possible as quickly as possible. **Type B** individuals are less competitive, less devoted to work, and have a weaker sense of time urgency. Such individuals are less likely to experience conflict with other people and more likely to have a balanced, relaxed approach to life. They are able to work at a constant pace without time urgency. Type B people are not necessarily more or less successful than are Type A people. But Type B people are less likely to experience stress.

### FOOD FOR THOUGHT 16.3

Did Pat "Deep Dish" Bertoletti experience stress when he ate 275 whole pickled jalapeños? In doing so, he set a new world record and beat out runner-up Sonya "The Black Widow" Thomas by only one jalapeño in a special double overtime individual challenge.

## Causes of Stress

Stress is obviously not a simple phenomenon. Table 16.2 lists several different things that can cause stress.[30]

**TABLE 16.2**  Common Causes of Stress

| |
|---|
| ***Organizational Stressors***—stressors derived primarily from the organizational context in which the person works: |
| Task demands associated with the task itself.<br>Physical demands associated with the job setting.<br>Role demands associated with position in group or organization.<br>Interpersonal demands associated with working relationships. |
| ***Life Change***—stressors derived from the rate of change in a person's life. |

© Cengage Learning 2014

**FIGURE 16.9**  Role demands create stress for managers, as in this scene where the manager fears that his team will be unable to meet the deadline set by his boss.

## Organizational Stressors.

Organizational stressors fall into one of four categories: task, physical, role, and interpersonal demands.[31] *Task demands* are associated with the task itself. Some occupations are inherently more stressful than others. Having to make fast decisions, decisions with less than complete information, or decisions that have relatively serious consequences are some of the things that can make some jobs stressful. The jobs of surgeon, airline pilot, and stock broker are relatively more stressful than the jobs of general practitioner, tractor driver, airplane baggage loader, and office receptionist. While a general practitioner makes important decisions, he is also likely to have time to make a considered diagnosis and fully explore a number of different treatments. But the surgeon must make decisions quickly during surgery while realizing that the wrong one may endanger her patient's life.

*Physical demands* are stressors associated with the job setting. Working outdoors in extremely hot or cold temperatures, or even in an improperly heated or cooled office, can lead to stress. A poorly designed office that makes it difficult for people to have privacy or does not provide for social interaction can result in stress, as can poor lighting and inadequate work surfaces. More severe stressors are actual threats to health. Examples include jobs like underground mining, toxic waste handling, and law enforcement.[32]

*Role demands* can also cause stress. (Roles are discussed more fully in Chapter 19.) A role is a set of expected behaviors associated with a position in a group or organization. Stress can result from either role ambiguity or role conflict that people can experience in groups (Figure 16.9).[33] For example, an employee who is feeling pressure from her boss to work longer hours while also being asked by her

family for more time at home will almost certainly experience stress. Similarly, a new employee experiencing role ambiguity because of poor orientation and training practices by the organization will suffer from stress.

*Interpersonal demands* are stressors associated with relationships that confront people in organizations. For example, group pressures regarding restriction of output (producing less than people are capable of producing) and norm conformity (meeting the expectations of others) can lead to stress. Leadership style may also cause stress. An employee who feels a strong need to participate in decision making may feel stress if his boss refuses to allow participation. And individuals with conflicting personalities may experience stress if required to work too closely together. A person with an internal locus of control may be frustrated when working with someone who prefers to wait and just let things happen.

## Life Change.

Stressors may also lie in events that are less connected to people's daily work lives. Some of these may have nothing to do with a person's work at all, while others may only be work-related in a general way. One early perspective was called *life change*—any meaningful change in a person's personal or work situation.[34] According to this view, major changes in a person's life can lead to stress. For example, when people finish college they have more responsibilities, usually have more money, may get married, and so forth. Each of these events represents a significant life change.

Table 16.3 summarizes the relative effects of different types of life change. Some of these changes,

**TABLE 16.3** Life Change and Stress

| RANK | LIFE EVENT | MEAN VALUE |
|:---:|---|:---:|
| 1 | Death of spouse | 100 |
| 2 | Divorce | 73 |
| 3 | Marital separation | 65 |
| 4 | Jail term | 63 |
| 5 | Death of close family member | 63 |
| 6 | Personal injury or illness | 53 |
| 7 | Marriage | 50 |
| 8 | Fired at work | 47 |
| 9 | Marital reconciliation | 45 |
| 10 | Retirement | 45 |
| 11 | Change in health of family member | 44 |
| 12 | Pregnancy | 40 |
| 13 | Sex difficulties | 39 |
| 14 | Gain of new family member | 39 |
| 15 | Business readjustment | 39 |
| 16 | Change in financial state | 38 |
| 17 | Death of close friend | 37 |
| 18 | Change to different line of work | 36 |
| 19 | Change in number of arguments with spouse | 35 |
| 20 | Mortgage over $10,000 | 31 |
| 21 | Foreclosure of mortgage or loan | 30 |
| 22 | Change in responsibilities at work | 29 |
| 23 | Son or daughter leaving home | 29 |
| 24 | Trouble with in-laws | 29 |
| 25 | Outstanding personal achievement | 28 |
| 26 | Spouse beginning or stopping work | 26 |
| 27 | Beginning or ending school | 26 |
| 28 | Change in living conditions | 25 |
| 29 | Revision of personal habits | 24 |
| 30 | Trouble with boss | 23 |
| 31 | Change in work hours or conditions | 20 |
| 32 | Change in residence | 20 |
| 33 | Change in schools | 20 |
| 34 | Change in recreation | 19 |
| 35 | Change in church activities | 19 |
| 36 | Change in social activities | 18 |
| 37 | Mortgage or loan less than $10,000 | 17 |
| 38 | Change in sleeping habits | 16 |
| 39 | Change in number of family get-togethers | 15 |
| 40 | Change in eating habits | 15 |
| 41 | Vacation | 13 |
| 42 | Christmas | 12 |
| 43 | Minor violations of the law | 11 |

The amount of life stress that a person has experienced in a given period of time, say one year, is measured by the total number of life change units (LCUs). These units result from the addition of the values (right-hand column) associated with events that the person has experienced during the target time period.

Source: Reprinted from Thomas H. Holmes and Richard H. Rahe, "The Social Adjustment Rating Scale," *Journal of Psychosomatic Research* (1967), 11:213–218, with kind permission from Pergamon Press, Ltd., Oxford, U.K.

such as marriage and outstanding personal achievement, are positive while others, such as the death of a spouse and serving a jail term, are negative. Early research suggests that higher total point values were likely to lead to health problems. For example, a total of 150 life change units within a one-year period was associated with a 50 percent chance of major illness the following year.[35] While this view provides some instructive viewpoints on stress, current thinking has advanced beyond these relatively simplistic ideas.

## Consequences of Stress

As noted earlier, the results of stress may be positive or negative. The negative consequences may be behavioral, psychological, or medical.[36] Behaviorally, for example, stress may lead to detrimental or harmful actions such as smoking, alcoholism, overeating, and drug abuse.[37] Other stress-induced behaviors are accident proneness, violence toward self or others, and appetite disorders.[38]

Psychological consequences of stress interfere with an individual's mental health and well being. These outcomes include sleep disturbances, depression, family problems, and sexual dysfunction. Managers are especially prone to sleep disturbances when they experience stress at work.[39] Medical consequences of stress affect an individual's physiological well being. Heart disease and stroke have been linked to stress, as have headaches, backaches, ulcers and related disorders, and skin conditions, such as acne and hives.[40]

Individual stress also has direct consequences for businesses. For an operating employee, stress may translate into poor quality work and lower productivity. For a manager, it may mean faulty decision making and disruptions in working relationships. Withdrawal behaviors can also result from stress. People who are having difficulties with stress in their jobs are more likely to call in sick or to leave the organization. More subtle forms of withdrawal may also occur, such as a manager missing deadlines or taking longer lunch breaks. Employees may also withdraw by developing feelings of indifference.[41] The irritation displayed by people under great stress can make them difficult to get along with. Job satisfaction, morale, and commitment can all suffer as a result of excessive levels of stress. So, too, can motivation to perform.

Another consequence of stress is **burnout**, a feeling of exhaustion that may develop when someone experiences too much stress for an extended period of time.[42] Burnout results in constant fatigue, frustration, and helplessness. Increased rigidity follows, as does a loss of self-confidence and psychological withdrawal. The individual dreads going to work, often puts in longer hours but gets less accomplished than before, and exhibits mental and physical exhaustion. Because of the damaging effects of burnout, some firms take steps to help avoid it. For example, British Airways provides all of its employees with training designed to help them recognize the symptoms of burnout and develop strategies for avoiding it.[43]

## Managing Stress

Given the potential consequences of stress, it follows that both people and organizations should be concerned about how to limit its more damaging effects. Numerous ideas and approaches have been developed to help manage stress, some for individuals and some for organizations.[44]

**Exercise.** One way people manage stress is through exercise. People who exercise regularly feel less tension and stress and are more self-confident and optimistic. Their better physical condition also makes them less susceptible to many common illnesses. People who don't exercise regularly, on the other hand, tend to feel more stress and are more likely to be depressed.[45] They are also more likely to have heart attacks.

**Relaxation.** Another method people use to manage stress is relaxation. Relaxation allows individuals to adapt to, and therefore better deal with, their stress. Relaxation comes in many forms, such as taking regular vacations. A recent study found that people's attitudes toward a variety of workplace characteristics improved significantly following a vacation.[46] People can also learn to relax while on their jobs, too. For example, some experts recommend that a person take regular rest breaks during his or her normal workdays.

**Time Management.** People can also use time management to control stress. The idea behind time management is to reduce or eliminate many daily pressures by doing a better job of managing time. One approach to time management is to make a list every morning of the things to be done that day. The items on the list are then grouped into three categories: critical activities that must be performed, important activities that should be performed, and optional or trivial things that can be delegated or postponed. The individual performs the items on the list in order of their importance.

**Support Groups.**   Finally, people can manage stress through support groups. A support group can be as simple as a group of family members or friends with whom to enjoy leisure time. Going with a couple of co-workers to a basketball game or a movie after work, for example, can help relieve stress built up during the day. Family and friends can help people cope with stress on an ongoing basis and during times of crisis. For example, an employee who has just learned that she did not get the promotion she has been working toward for months may find it helpful to have a good friend to lean on, to talk to, or to yell at.[47] People also may make use of more elaborate and formal support groups. Community centers or churches, for example, may sponsor support groups for people who have recently gone through a divorce, the death of a loved one, or some other tragedy.

**Organizational Stress Programs.**   Organizations are also beginning to realize that they should be involved in helping employees cope with stress. One argument for this is that, since the business is at least partially responsible for stress, it should also help relieve it. Another argument is that stress-related insurance claims by employees can cost the organization considerable sums of money. Still another is that workers experiencing lower levels of detrimental stress will be able to function more effectively for the company. AT&T initiated a series of seminars and workshops to help its employees cope with the stress they face in their jobs. The firm was prompted to develop these seminars for all three of the reasons noted above.[48]

A wellness stress program is a special part of the organization created specifically to help employees deal with stress. Organizations have adopted stress management programs, health promotion programs, and other kinds of programs for this purpose. Stress seminars are similar to this idea, but true wellness programs are on-going activities that have a number of different components. They commonly include exercise-related activities as well as classroom instruction programs dealing with smoking cessation, weight reduction, and general stress management.

Some companies are developing their own programs or using existing programs of this type.[49] Johns-Manville, for example, has a gym at its corporate headquarters. Other firms negotiate discounted health club membership rates with local establishments. For the instructional part of the program, the organization can again either sponsor its own training or perhaps jointly sponsor seminars with a local YMCA, civic organization, or church. Organization-based fitness programs facilitate employee exercise, a very positive consideration; but such programs are also quite costly. Still, more and more companies are developing fitness programs for employees.[50]

## Interpersonal Processes at Work

Thus far, we have focused primarily on characteristics of individuals in organizations. But work in organizations is done by people working together; therefore, we must also examine some fundamental interpersonal processes at work.

### The Nature of Working Relationships

The nature of working relationships in an organization is as varied as the individual members themselves. Personality and attitudes, as well as numerous other factors, affect working relationships in many different ways. At one extreme these relationships can be personal and positive. This occurs when the parties know each other, share mutual respect and friendship, and enjoy interacting with one another. Two managers who have known each other for years and are close personal friends will likely interact at work in a relaxed and effective fashion.

At the other extreme, relationships at work can have quite a negative cast. This is most likely the case when the parties dislike one another, do not have mutual respect, and do not enjoy interacting with one another. Suppose, for example, one of the lower-level managers at Summer Farms has sought a promotion for about three years. If another person is promoted or hired into that position, their relationship will most likely be strained and difficult.

Most working relationships fall between these extremes as members of the organization interact in a professional way focused on goal accomplishment. A professional relationship is more-or-less formal and structured and highly task directed. Two managers may respect each other's work and recognize the professional competence that each brings to the job. The fact that they may have few common interests and little to talk about other than their jobs will probably not be a factor in the quality of their work.

Working relationships, whether positive and professional or negative, exist among individuals, among groups, and among individuals and groups. They can also change over time. Two managers with a negative relationship, for example, might eventually resolve their disputes and develop a more positive and professional relationship. Similarly, people may also expand mutual professional respect into real friendship.

## Collaboration and Cooperation

From a pure performance standpoint, effective working relationships promote collaboration and cooperation—people working together toward the best interests of the organization. Good working relationships throughout a firm can be a tremendous source of synergy. People who support one another and who work well together can accomplish much more than people who do not. They focus more fully on meeting their goals and accomplishing their tasks, for example; and they are more inclined to help each other. Even when people are working individually, the presence of good relations minimize emotional distractions.

Positive working relationships can also serve as a solid basis for social support in a firm. When people offer reassurance and encouragement to one another, they are providing social support. Suppose an employee receives a poor performance review or is denied a promotion. Others in the firm can commiserate because they share a common frame of reference: the organization's culture, an understanding of the causes and consequences of the event, and so forth.[51]

## Competition and Conflict

On the other side of the coin, poor interpersonal relations at work can result in excessive competition and conflict. Like stress, competition and conflict may be either good or bad.[52] When there is absolutely no competition or conflict, performance often tends to be very low. This stems primarily from people who are often motivated by competition and spurred to action because they think their way of doing something is better than someone else's. Thus, as conflict increases, so too may performance. At some point, however, conflict reaches its maximum point of effectiveness. Additional conflict then begins to hurt performance.

**Causes of Conflict.**  Several things can cause conflict, one of which is the interdependency that can exists between people or groups within the organization.[53] For example, consider the case of a cotton farm. Work is performed in large blocks by teams of workers and then is passed from one team to another. If the picking team doesn't do its jobs properly or gets too far ahead or too far behind in its work, it causes problems for the seed-removal team, the baling team, and others.

Competition among people or groups can also cause conflict, especially when the stakes are high. For example, if a company establishes a sales contest among sales groups and plans to award a two-week vacation in Hawaii to the winning group, conflict is likely because the stakes are high and when one group wins the others all lose. Differences in goals and activities can also lead to conflict. For example, if the marketing department wants to increase product lines to boost sales and the production department wants to reduce product lines to cut costs, conflict may result. Finally, personalities may also come into play. Two people may be unable to work together, or simply not be able to get along.

**Consequences of Conflict.**  Just as there are several causes of conflict, there are also several consequences, including hostility or even withdrawal. The withdrawal may be confined to activities such as refusing to socialize with others or it may extend to actually leaving the organization. On a more positive note, conflict may also increase motivation. People who have a mild disagreement may each become more motivated to prove that the other is wrong. Finally, conflict may performance. For example, consider the case of two plant managers within the same company who disagree over the best way to improve productivity. Each may be allowed to pursue his own ideas, and each may develop new techniques that do actually improve productivity. Thus, the overall performance of the company increases, and each manager can feel like a "winner."

**Controlling Conflict.** Given the importance of competition and conflict to managers, it follows that managers should know how to control them. One approach is to increase resources. For example, if two departments are competing for access to a new computer network, it may be wise to expand the network to give each department full access. Rules and standard operating procedures can be established to better manage interdependencies. Setting overall goals can also help prevent conflict. We earlier used an example of potential conflict between the marketing department's goal to increase sales and the production department's goal to cut costs. The manager may get the two departments to agree that the best overall goal is to optimize sales increases and cut costs. Finally, interpersonal relations may also be manageable. For example, if a manager knows that one employee likes to eat tuna sandwiches at his desk and another person nearly gags when he smells tuna, the manager can ensure that their offices are as far apart as possible to minimize their interactions.

**Resolving Conflict.** Of course, regardless of the manager's best intentions, conflict is still likely to occur. Fortunately, there are several ways to resolve conflict. One technique is *avoidance*. As the term implies, avoidance involves ignoring the problem and hoping it will go away. If the conflict is minimal, avoidance may work; however, it should not be used simply because the manager doesn't want to deal with the problem.

**FIGURE 16.10** The conflict over this team's work appears to have been settled by compromise.

*Smoothing* is similar to avoidance. Here, the manager acknowledges the existence of the conflict but also downplays its importance. Like avoidance, smoothing may be effective if used wisely, but sometimes the conflict is simply too great to go away on its own.

*Compromise* involves reaching a point of agreement between what each of the conflicting parties initially wanted (Figure 16.10). Of course, like other strategies, compromise must be used with care. It is possible that both parties will end up feeling like they lost.

Finally, the conflict may be resolved through confrontation. *Confrontation* is the direct approach of addressing the conflict and working together to resolve it.

## CHAPTER SUMMARY

The basic relationship between an individual and an organization is defined by a psychological contract—the expectation of an individual about what he or she will contribute to the organization and what the organization will provide to them in return. One key aspect of managing psychological contracts is enhancing the person-job fit.

Individual differences are personal attributes that vary from one person to another. Key personality attributes include locus of control, authoritarianism, dogmatism, self-esteem, and risk propensity. Important job-related attitudes include job satisfaction or dissatisfaction and organizational commitment. Important perceptual processes are selective perception and stereotyping.

Actual performance is an important performance behavior. So, too, are withdrawal behaviors like absenteeism and turnover. Organizational citizenship is also important.

Organizational stressors, such as task, physical, role, and interpersonal demands cause stress. Life change also causes stress. Consequences of stress include behavioral, psychological, and medical outcomes; lower performance; and burnout. Individual stress usually follows the General Adaptation Syndrome. Type A and B individuals react to stress quite differently. Both individuals and organizations can help manage stress.

Working relationships vary significantly. Collaboration and cooperation are most effective. Competition has some benefits but can also lead to difficulties. Conflict is an important outcome that must be carefully managed.

## CHAPTER ACTIVITIES

### REVIEW QUESTIONS

1. What is a psychological contract?
2. Identify and define several basic personality attributes relevant to organizations.
3. What are the three components of an attitude? What does an individual experience if there is a conflict or inconsistency among these components?
4. Identify several factors that cause stress. Identify several consequences of stress.
5. Name several things that can cause conflict.

### ANALYSIS QUESTIONS

1. What can an individual do if he/she wants to change the psychological contract she/he has with his/her organization?
2. Characterize yourself on each of the personality traits discussed in the chapter.
3. Select an important attitude that reflects how you feel about something. Try to recall how this attitude was formed and break it down into its three components.
4. What causes stress for you? How do you deal with it?
5. Suppose a manager thinks there is too little conflict in the organization. Would it be ethical to stimulate conflict? What are the inherent dangers in doing this?

### FILL IN THE BLANKS

1. The set of expectations held by an individual about what he or she will contribute to the organization and what the organization will provide in return is a _____ _____.
2. The relatively stable set of psychological and behavioral attributes that distinguish one person from another is known as _____.
3. The rigidity of a person's beliefs and his/her openness to other viewpoints is known as _____.
4. The extent to which an individual believes that he is a worthwhile and deserving person is called _____ _____.
5. A conflict among the three parts of an attitude is known as _____ _____.
6. The process of screening out information that we are uncomfortable with or which contradicts our beliefs is called _____ _____.

7. The process of categorizing people on the basis of a single attribute is _____ .

8. The set of work-related behaviors that a firm expects people to display are called _____ _____ .

9. The condition that occurs when an individual is subjected to unusual situations, difficult demands, or extreme pressures is called _____ .

10. The extent to which the contributions made by an individual match the inducements offered by the organization is known as _____ _____ _____ .

## ▼ CHAPTER 16 CASE STUDY
### A TALE OF TWO BEERS

Carlos Brito transformed a Brazilian beer maker, Brahma, into one of the largest brewers in the world. This 50-year-old Brazilian with an MBA from Stanford merged Brahma with the maker of Stella Artois into InBev NV, a Brazilian-Belgium hybrid brewer. It was InBev that made a $52 billion bid for Anheuser Busch of St. Louis, an iconic American brand. The company is Anheuser Busch-InBev, or AB-InBev NV, based in Belgium. Normally, individuals involved in such high-profile deals have a swashbuckling attitude. This does not describe Mr. Brito, or as he prefers to be addressed, "Brito."

He instilled a no-frills culture at InBev in which executives gave up individual secretaries and company cars. It was this frugal environment that allowed InBev to purchase Anheuser-Busch. Why did he instill this culture at AB-InBev and how did he get other executives to buy into this concept? His answer is simple: Think like your customer. Or in Mr. Brito's own words: "If you are doing anything that you think a consumer would not be willing to pay a premium for—think twice before doing it." He uses his own frugality as an example: "I come to work by train. I tell the guys, 'Being efficient is what our consumers would do.' When I travel with my family, I don't go for five-course meals, five-star hotels."

"I also have to be humble and admit that we have learned a lot (through acquisitions)," he continues. "Our marketing and 'people' tool kit evolved a lot. Now we want to start growing the right brands."

Mr. Brito's challenge is to revive Budweiser—that American icon. It labeled itself the "king of beer" and after many years began to believe it. But even kings must pass away, so Budweiser is now the number two selling beer in the United States after Bud Light. This is the equivalent of Diet Coke outselling Classic Coke. They are produced by the same brewer, but Budweiser is the original beer and its market share has been slipping for 21 years.

When InBev took over Anheuser-Busch, it carried forward the directive of Mr. Brito, recognizing individual differences. InBev eliminated 1,500 jobs, redesigned the company's pay system, and reduced expenses. Business travelers on the road slept two to a room. Salesmen worked from a central office and used telephones rather than travel. All of these activities saved $1.67 billion and … Budweiser raised its prices. Only a beer with a premium image can do what Budweiser accomplished under Mr. Brito. Again in his words, "Brand people have been able to convince people to pay more for beer."

But operational efficiency alone is not going to win market share. Corporate frugality does not eliminate the occasional "big bet." In fact, Mr. Brito would say it is because of its frugal and efficient culture that it can make the "big bet." The "big bet" was a heavy investment in the World Cup (August 2010). AB-InBev increased its spending on sales and marketing by 10 percent. In the words of the company, "Budweiser activated the FIFA World Cup asset for the seventh time."

Sales of Budweiser at the games outpaced soft drinks, sports drinks, and bottled water combined. But it was not South African sales that mattered to AB-InBev—it was the television exposure in the company's three largest markets: Europe, Brazil, and the United States. Sales were up 2.1 percent in the quarter. This may not seem like a huge difference, but in the terrible economic environment of 2010 it was a huge victory. This was a "measured big bet," that is, it was a calculated move by Mr. Brito, not an impulsive one.

While a quiet, frugal Brazilian is working hard to revive an iconic American beer brand through operational efficiency, the oldest brewer in the United States continues to grow the old fashioned way—organically or through old fashion internal growth. D.G. Yuengling & Son has been brewing beer since 1829, well before Budweiser labeled itself the "king of beers." The company is under its fifth generation brewer, Dick Yuengling. It sells its products on the East coast but now is seeking to expand.

Yuengling is the seventh largest beer supplier in the United States. It is among the few remaining domestic, non-global beer brands. Yuengling has long taken a conservative approach to growth. Its brands have a cult-like appeal that have prompted consumers to send letters and place phone calls urging Yuengling executives to expand distribution. While it distributes in 15 states, the company has said that it "doesn't have a definitive timetable to introduce its beer in additional states." According to Yuengling.com, "It simply is not logistically feasible to transport beer outside of our current footprint."

Corporate culture, intra-company communication and leadership can take many forms to arrive at the same place. Both Anheuser Busch-InBev and Yuengling have frugal, conservative approaches to growth. One was re-instilled; the other never changed.

It is worth a footnote that the bottling plant Yuengling is purchasing was owned by Hardy Bottling Co, a local bottler, who bought the plant from Molson Coors Brewing Company. Molson Coors Brewing Company of Denver and Montreal resulted from the merger of two beer companies, Molson and Coors, which once produced local brews with an avid cult following.[54]

### ▶ Case Study Questions

1. How may individual differences help you understand the approaches of Brito and Dick Yuengling?

2. What may be the implications of these two approaches for other organizations?

3. Will Bud be revived in the United States? Will Yuengling successfully expand beyond the East coast? What could happen in the industry that might make Yuengling's expansion come too late?

## REFERENCES

1. David Kesmodel, "No Glass Ceiling for the Best Job in the Universe." *The Wall Street Journal* (June 29, 2010) at online.wsj.com (accessed November 18, 2010); John Porter, *All About Beer* (Garden City, NJ: Doubleday and Company, 1975), p. vii; "Oktobergloom," *The Economist* (October 9, 2010) at www.economist.com (accessed November 17, 2010); Kasey Wehrum, "A Craft-Beer Stimulus Plan," *Inc. Magazine* (October 1, 2010) at www.inc.com (accessed November 18, 2010).

2. Jacqueline Coyle-Shapiro and Neil Conway, "Exchange Relationships: Examining Psychological Contracts and Perceived Organizational Support," *Journal of Applied Psychology* (2005): 774–781.

3. Zhen Xiong Chen, Anne Tsui, and Lifeng Zhong, "Reactions to Psychological Contract Breach: A Dual Perspective," *Journal of Organizational Behavior* (2008): 527–548.

4. O. J. Ladebo (2005). "Relationship Between Citizenship Behaviors and Tendencies to Withdraw Among Nigerian Agribusiness Employees," *Swiss Journal of Psychology*, 64: 41–50; Mark H. Hansen and J. L. Morrow, Jr., "Trust and the Decision to Outsource: Affective Responses and Cognitive Processes," *International Food and Agribusiness Management Review*, 6(3): 40–69; Edwin J. McClenahan and Robert A. Milligan, "Profile of the Work Force on Dairy Farms in New York and Wisconsin," Cornell University at aem.cornell.edu (accessed July 15, 2010); Jeffrey Burkhardt, "Agribusiness Ethics: Specifying the Terms of the Contract," *Journal of Business Ethics* (1986), 5(4):333–345.

5. Arne Kalleberg, "The Mismatched Worker: When People Don't Fit Their Jobs," *Academy of Management Perspectives* (2008): 24–40.

6. Lawrence Pervin, "Personality," *Annual Review of Psychology*, eds. Mark Rosenzweig and Lyman Porter (Palo Alto, CA: Annual Reviews, 1985): 36: 83–114; and S. R. Maddi, *Personality Theories: A Comparative Analysis, 4th ed.* (Homewood, IL: Dorsey, 1980).

7. Lawrence Pervin, *Current Controversies and Issues in Personality, 3rd ed.* (New York: Wiley, 2002).

8. J. B. Rotter, "Generalized Expectancies for Internal vs. External Control of Reinforcement," *Psychological Monographs* (1966), 80: 1–8.

9. Simon S. K. Lam and John Schaubroeck, "The Role of Locus of Control in Reactions to Being Promoted and to Being Passed Over: A Quasi Experiment," *Academy of Management Journal* (2000): 66–78.

10. T. W. Adorno, Else Frenkel-Brunswik, and Daniel J. Levinson, *The Authoritarian Personality* (New York: Harper & Row, 1950).

11. "Who Becomes an Authoritarian?" *Psychology Today* (March 1989), 66–70.

12. Jason J. Dahling, Brian G. Whitaker, and Paul E. Levy, "The Development and Validation of a New Machiavellianism Scale," *Journal of Management* (2009): 219–257.

13. Jon L. Pierce and Donald G. Gardner, "Self-Esteem Within the Work and Organizational Context: A Review of the Organization-Based Self-Esteem Literature," *Journal of Management* (2004): 591–622; Barbara Foley Meeker, "Cooperation, Competition, and Self-Esteem: Aspects of Winning and Losing," *Human Relations* (1990), 43: 205–220.

14. "Hey, I'm Terrific," *Newsweek*, February 17, 1992, 46–51.

15. Hao Zhao, Scott E. Seibert, and G. T. Lumpkin, "The Relationship of Personality to Entrepreneurial Intentions and Performance: A Meta-Analytic Review,'" *Journal of Management* (2010): 381–404.

16. Michael G. Aamodt, *Industrial/Organizational Psychology, 6th ed.* (Belmont, CA: Wadsworth/Cengage Learning, 2010).

17. Leon Festinger, *A Theory of Cognitive Dissonance* (Palo Alto, CA: Stanford University Press, 1957).

18. Steven C. Currall, Annette J. Towler, Timothy A. Judge, and Laura Kohn, "Pay Satisfaction and Organizational Outcomes," *Personnel Psychology* (2005): 613–640.

19. For an example of an attitude change effort, see Deborah A. Byrnes and Gary Kiger, "The Effect of a Prejudice-Reduction Simulation on Attitude Change," *Journal of Applied Social Psychology* (1990): 20: 341–353.

20. Patricia C. Smith, L. M. Kendall, and Charles Hulin, *The Measurement of Satisfaction in Work and Behavior* (Chicago: Rand-McNally, 1969).

21. Omar N. Solinger, Woody van Olffen, and Robert A. Roe, "Beyond the Three-Component Model of Organizational Commitment," *Journal of Applied Psychology* (2008): 70–83; Steven M. Elias, "Employee Commitment in Times of Change: Assessing the Importance of Attitudes Toward Organizational Change," *Journal of Management* (2009): 37–55.

22. Richard A. Posthuma and Michael A. Campion, "Age Stereotypes in the Workplace: Common Stereotypes, Moderators, and Future Research Directions," *Journal of Management* (2009): 158–188.

23. John P. Hausknecht, Nathan J. Hiller, and Robert J. Vance, "Work-Unit Absenteeism: Effects of Satisfaction, Commitment, Labor Market Conditions, and Time," *Academy of Management Journal* (December 2008): 1223–1245.

24. For recent findings regarding this behavior, see Philip M. Podsakoff, et al., "Organizational Citizenship Behaviors: A Critical Review of the Theoretical and Empirical Literature and Suggestions for Future Research," *Journal of Management* (2000): 513–563.

25. Dennis W. Organ and Mary Konovsky, "Cognitive Versus Affective Determinants of Organizational Citizenship Behavior," *Journal of Applied Psychology* (1989): 157–164.

26. James L. Gibson, John M. Ivancevich, and James H. Donnelly Jr., *Organizations—Behavior, Structure, Processes, 13th ed.* (New York: McGraw-Hill, 2009).

27. Hans Selye, *The Stress of Life* (New York: McGraw-Hill, 1976).

28. Ibid.

29. M. Friedman and R. H. Rosenman, *Type A Behavior and Your Heart* (New York: Alfred A. Knopf, 1974).

30. James Campbell Quick, Jonathan D. Quick, Debra L. Nelson, and Joseph J. Hurrell Jr., *Preventive Stress Management in Organizations* (Washington, DC: American Psychological Association, 1997); see also Stephan J. Motowidlo, John S. Packard, and Michael R. Manning, "Occupational Stress: Its Causes and Consequences for Job Performance," *Journal of Applied Psychology* (August 1986): 618–629.

31. Anne B. Fisher, "Welcome to the Age of Overwork," *Fortune*, November 30, 1992, 64–71.

32. John M. Jermier, Jeannie Gaines, and Nancy J. McIntosh, "Reactions to Physically Dangerous Work: A Conceptual and Empirical Analysis," *Journal of Organizational Behavior* (January 1989): 15–33.

33. John Schaubroeck, John L. Cotton, and Kenneth R. Jennings, "Antecedents and Consequences of Role Stress: A Covariance Structure Analysis," *Journal of Organizational Behavior* (January 1989): 35–58.

34. T. H. Holmes and R. H. Rahe, "Social Readjustment Rating Scale," *Journal of Psychosomatic Research* (1967): 29: 213–218.

35. Ibid.

36. James Campbell Quick and Jonathan D. Quick, *Organizational Stress and Preventive Management* (New York: McGraw-Hill, 1984).

37. Ricky W. Griffin and Yvette Lopez, "'Bad Behavior' in Organizations: A Review and Typology for Future Research," *Journal of Management* (2005): 988–1005.

38. David D. Van Fleet and Ella W. Van Fleet, *The Violence Volcano: Reducing the Threat of Workplace Violence* (Charlotte, NC: Information Age Publishing, 2010); Ella W. Van Fleet and David D. Van Fleet, *Workplace Survival: Dealing with Bad Bosses, Bad Workers, Bad Jobs* (Frederick, MD: PublishAmerica, 2007).

39. Walter Kiechal III, "The Executive Insomniac," *Fortune*, October 8, 1990, 183–184.

40. Quick and Quick, *Organizational Stress and Preventive Management*; see also Brian D. Steffy and John W. Jones, "Workplace Stress and Indicators of Coronary-Disease Risk," *Academy of Management Journal* (September 1988): 686–698.

41. Ibid.

42. Leonard Moss, *Management Stress* (Reading, MA: Addison-Wesley, 1981).

43. Thomas A. Stewart, "Do You Push Your People Too Hard?" *Fortune*, October 22, 1990, 124–128.

44. Alan Farnham, "Who Beats Stress—And How," *Fortune*, October 7, 1991, 71–86.

45. C. Folkins, "Effects of Physical Training on Mood," *Journal of Clinical Psychology* (April 1976): 385–390.

46. John W. Lounsbury and Linda L. Hoopes, "A Vacation From Work: Changes in Work and Nonwork Outcomes," *Journal of Applied Psychology* (May 1986): 392–401.

47. Daniel C. Ganster, Marcelline R. Fusilier, and Bronston T. Mayes, "Role of Social Support in the Experiences of Stress at Work," *Journal of Applied Psychology* (February 1986): 102–110.

48. Quick and Quick, *Organizational Stress and Preventive Management*.

49. Ibid.

50. Richard A. Wolfe, David O. Ulrich, and Donald F. Parker, "Employee Health Management Programs: Review, Critique, and Research Agenda," *Journal of Management* (Winter 1987): 603–615.

51. Marcelline R. Fisilier, Daniel C. Ganster, and Bronston T. Mayes, "Effects of Social Support, Role Stress, and Locus of Control on Health," *Journal of Management* (Fall 1987): 517–528.

52. David B. Lipsky, Ronald L. Seeber, and Richard Fincher, *Emerging Systems for Managing Workplace Conflict: Lessons from American Corporations for Managers and Dispute Resolution Professionals* (San Francisco: Jossey-Bass, 2003).

53. James Thompson, *Organizations in Action* (New York: McGraw-Hill, 1967).

54. David Kesmodel, "Brazil's Brito Aims to Revive Bud in the U.S.," *The Wall Street Journal* (June 27, 2010) at online.wsj.com (accessed November 15, 2010); John W. Miller, "Anheuser Gets Boost from World Cup," *The Wall Street Journal* (August 13, 2010) at online.wsj.com (accessed November 15, 2010); David Kesmodel, "Yuengling Seeks Memphis Brewery in Bid to Expand Beyond the East," *The Wall Street Journal* (October 15, 2010): B4; David Kesmodel, "MillerCoors Grooms No. 2 Executive," *The Wall Street Journal* (September 13, 2010) at online.wsj.com (accessed November 15, 2010).

# Leadership

## LEARNING OBJECTIVES

After studying this chapter, you should be able to:

- Define leadership, indicate the difference between leadership and management, and identify the challenges of leadership.

- Name and describe several types of power, including their uses, limits, and outcomes.

- Briefly discuss the trait approach to the study of leadership.

- Discuss leadership behaviors, and compare and contrast the Michigan and Ohio State studies.

- Describe several situational approaches to leadership, including the LPC model, the path-goal model, the participation model, and an integrative framework for these models.

- Discuss such contemporary perspectives on leadership as charismatic, transformational, and symbolic leadership; and substitutes and neutralizers of leadership.

## *Nothing Runs Like a . . . Buffalo*

"**T**he fundamental impulse that sets and keeps the capitalist engine in motion comes from the new consumer goods, the new methods of production or transportation, the new markets, the new forms of industrial organization that capitalist enterprise creates." With these words Joseph Schumpeter described the renewal activity of capitalism, a process he called "creative destruction."

**FIGURE 17.1** The market in capitalistic economic systems demands change, but some farm owners prefer to keep their less efficient animals rather than adapt to equipment that can do the work 20 times faster.
© iStockphoto/Josef Muellek

The vitality of capitalism is in market forces: the constant need to change, to develop the "better mouse trap." The market demands change. Sometimes change is prompted by the development of better products; sometimes it is prompted by the need to change the need.

So it is for Jared Hayes, product manager, who develops and markets for Deere & Company products. A leader at Deere, Hayes works with John Deere "Green-Star™" products. "I grew up on a farm, and my dad would never let me plant corn because I couldn't drive as straight as he could." Now with Deere & Company's new AutoTrac™ system he is able to steer a 500-horsepower tractor pulling a 120-foot implement within 0.5 inch of where it needs to be, automatically. And while he is performing this task he can be watching Tosh™—a computer link in the cab of the tractor that monitors real time grain prices. Plant your crop, perfectly, while monitoring the futures market on your onboard computer.

The world's largest maker of farm equipment has its fortune linked to agricultural commodity prices, just like the computer link in the cab of its tractors. As farmers achieve more return for their crops, they reinvest in new and more efficient machinery to enhance productivity and yield. In the United States, where a mere 1 percent of the population feeds the entire population and exports additional products, the key to prosperity is productivity and efficiency.

Agricultural efficiency is a driver of economic development. "The lesson is clear. For a country to move up the economic ladder, more workers need to be freed to do other things. To do this, agriculture must become more efficient." Every country striving to grow economically follows this trend.

This evolution or "creative destruction" means farmers must change also. So it is with Somsak Baitalum, as he guides his Nagano NT 350 four-wheel drive, twelve-speed, three-reverse-speed tractor with a canopy to protect from the sun over his rice field. "Such machines never get tired," says Pornipa Wasusatien, manager of a local tractor shop.

In Thailand, the forces of Schumpeter's creative destruction have a picturesque, faithful, historic, if slow companion of the rice farmer in its crosshairs—the water buffalo. It takes a water buffalo 20 days to accomplish a one-day task for the tractor. However, in Bua Yai, Thailand (four hours north of Bangkok), one man, Komin Mongkolpanya, is not about to throw in the towel and concede victory to the tractor. Mr. Mongkolpanya is a government livestock officer and a leader in trying to get farmers to continue using animals. He has even established a school for water buffalos in the hope of improving their efficiency. In fact the government even lends water buffalos to farmers. In 2010, 33,000 were on loan.

However, even the indomitable Mr. Mongkolpanya admits that it is a difficult task. "We are going to train you—you should pay attention," he whispers to a four-year-old water buffalo named Kam. "If you don't make a good plow, people will kill you and eat you. We love you."

It is not likely that "Kam the Buffalo" knows ... or cares ... who Joseph Schumpeter is. Mr. Mongkolpanya should know, but change is not easy and economic development demands change. Business or agribusiness decisions affect not only economics but also culture.[1]

## INTRODUCTION

Just as agricultural efficiency is a driver of economic development, managerial and leadership ability are two key ingredients for building and sustaining a company that is successful in achieving that efficiency. Whereas managerial ability relies heavily on relatively objective talents and skills, leadership is much less tangible. The importance of leadership in family businesses is particularly critical. Thus, most observers agree that it is very difficult to understand, much less practice, effective leadership.

We begin this chapter explaining the nature of leadership and then define power and its relationship to leadership. After looking briefly at early trait models of leadership, we examine in more detail two behavioral approaches and focus much attention on situational approaches. The chapter concludes with a discussion of other contemporary perspectives on leadership.

## The Nature of Leadership

We can define **leadership** as an influence process directed at shaping the behavior of others.[2] Various tactics can be used when attempting to influence others.[3] When Vince Lombardi exhorted the members of his football teams to play harder, he was leading with one set of influence tactics. When Bill Gates as CEO of Microsoft encouraged his managers to work harder, he was leading with another tactic. A friend who convinces you to try a new restaurant you have been avoiding is also leading. While many leaders come from within one organization, others come from outside, as was the case with Patricia A. Woertz who came from Chevron to become CEO of Archer Daniels Midland.[4] Not all leaders are equally effective, though—Scott Livengood was the CEO of Krispy Kreme when its stock dropped precipitously and came under investigation by the SEC.[5]

The study of leadership covers more than a century and the number of books, articles, and papers on it literally number in the thousands.[6] This chapter is not intended to be a comprehensive review of that literature, but it does reflect major consensus views supported by extensive research.[7] Thus, popularized views of leadership for which supporting research does not exist are not included. This does not imply that to follow their prescriptions would be in error; it does imply that sufficient research to support their claims does not now exist.

Leadership occurs in a variety of settings and in a variety of ways. Before discussing the various forms of leadership, however, let us first make a clearer distinction between leadership and management, and then explore some of the challenges of leadership.

## Leadership Versus Management

Leadership and management are in some ways similar but in more ways different.[8] People can be leaders without being managers, or managers without being leaders, or both managers and leaders at the same time. Nevertheless, there do appear to be some differences between the two roles, as Table 17.1 indicates.

In creating an action agenda, managers are more likely to emphasize planning and budgeting while leaders tend to focus more on direction. In focusing on the human element necessary to achieve that agenda, managers tend to think in terms of organizing and staffing while leaders seem more concerned with communication and cooperation. In carrying out that agenda, managers tend to focus on problem solving and control whereas leaders emphasize motivation.

The bases of power used by managers and leaders also tend to differ. For one thing, managers can direct the efforts of others because of their formal organizational power and control of resources. If a department head tells a member of the department to do three things and the person does exactly what was dictated but nothing else, the department head is probably being a manager but not a leader. If Joshua Summer instructed a field manager to prepare for planting on a specified date, he was acting as a manager. A leader, on the other hand, does not have to rely on his formal position to influence someone but may rely more on expertise, personality, charisma, or competence that is respected. Thus, if Mr. Summer asks the field manager the status of the fields and is told when the plowing is scheduled to start, when the crops will be planted, etc., he has been acting as a leader whose employees get the work done without direct supervision. If a secretary organizes a group effort to help a coworker who has personal problems, she is acting as a leader but not as a manager.

---

**FOOD FOR THOUGHT 17.1**

The Zulu people of South Africa use their knowledge of animal behavior to help train people for leadership positions.

---

From the standpoint of organizational effectiveness, people who are both leaders and managers are valuable resources.[9] Such individuals are able to carry out their managerial responsibilities effectively while also commanding the loyalty and respect of those they lead. They are usually quite successful at doing almost anything they set out to do. They are also quite scarce.

## The Challenges of Leadership

To fulfill others' expectations of them, leaders must confront numerous challenges. In large measure, the success of any given leader depends on his or her ability to address these challenges in a way that people will accept. Although any number of challenges are inherent in a given situation, three are relatively constant—multiple constituencies, unpopular decisions, and diversity.

**Multiple Constituencies.** Satisfying **multiple constituencies** means that the leader must attempt to deal with several different people and groups at the same time and in a way that is relatively acceptable to every party. This concern is compounded by the fact that the different constituencies

**TABLE 17.1** Differences Between Management and Leadership

| | MANAGEMENT | LEADERSHIP |
|---|---|---|
| **Creating an Action Agenda** | Focuses on planning and budgeting | Concentrates on establishing direction |
| **Achieving That Agenda** | Thinks in terms of organizing and staffing | More concerned with communication and cooperation |
| **Carrying Out That Agenda** | Focuses on problem solving and control | Emphasizes motivation |
| **Bases of Power** | Formal organizational position and control of resources | Expertise and personality |

Sources: Cliff Ricketts and John C. Ricketts, *Leadership: Personal Development and Career Success*, 3rd ed. (Clifton Park, NY: Delmar, Cengage Learning, 2011); Alan Murray, *The Wall Street Journal Essential Guide to Management* (New York: HarperBusiness, 2010); John P. Kotter, *A Force for Change: How Leadership Differs from Management* (New York: The Free Press, 1990).

often desire conflicting things from the organization. Employees may demand higher wages while stockholders desire bigger dividends. Consider the case of Richard T. Clark. While at Merck, Clark had dealings with the government, union officials, suppliers, creditors, competitors, and employees. One of his major challenges was to see that each party was in basic agreement with what he felt necessary to build the company's health. Even Summer Farms must answer not only to customers and employees but also state and federal government agencies that regulate input, production, and value-added agricultural industries.

**Unpopular Decisions.**   Hand in hand with the notion of multiple constituencies is the simple fact that leaders must occasionally make decisions that are unpopular, at least among some of their constituents. When Clark closed Merck plants, employees at those plants were unhappy.[10] Bill Marriott encountered considerable resistance when he announced the decision to sell some of the company's restaurant operations. The Summer heirs can expect to face unpopular decisions in the future as the organization moves from its original family leader to a new generation and perhaps even a family outsider. The mark of a good leader is the ability to recognize when such decisions must be made and the perseverance to see them through (Figure 17.2).[11]

**Diversity.**   Both managers and leaders are becoming more diverse as a group and are having to deal with groups composed of more diverse members than in the past. Organizations that employ only a few members of a different race, sex, age, or culture—

or even non-relatives in a business that is heavily staffed by family members—may take advantage of the skills such diverse groups possess; but if not careful, the organization could also alienate members of these diverse groups and suffer rather than gain from the increasing diverse labor force worldwide.

Organizational members from different nations bring rich and valuable perspectives to bear on problems. They may also bring communication problems and misunderstandings. Developing organizational members versed in international experience and capable of communicating in more than one language is vital to all organizations attempting to do business internationally, but it can also be useful to any organization.

Demographic issues—race, gender, age, ethnic origins, region of the country—are also becoming significant to leadership and organizations. There is some evidence that effective leaders display the same behaviors regardless of sex.[12] On the other hand, there is also some evidence which suggests that they are different.[13] Race has likewise been a subject with mixed results.[14] Age issues have received less attention, but there is some research suggesting that managers who lead groups that are older than they are face considerable difficulties.[15] Despite the difficulties inherent in highly diverse organizations, the potential benefits certainly seem to outweigh the costs.

In addition to these three critical challenges, leaders have others. They must set good examples for their followers, they must continually monitor situations so that new actions can be taken as needed, and they must develop the potential of employees in the organization. Leaders must use their powers wisely and without infringing on the rights and privileges of others. Finally, leaders must be ethical in all of their dealings. In the next section we turn our attention to a more detailed consideration of power in organizational settings.

## Power and Leadership

The foundation of leadership is power. Leaders have power over their followers and wield this power to exert their influence. The various kinds of power can be used in several different ways.[16]

### Types of Power

Most people agree that there are five basic types of power: legitimate, reward, coercive, expert, and referent power.[17]

**FIGURE 17.2** Managers and leaders must satisfy various constituents, make unpopular decisions, and adapt to diverse groups.

© iStockphoto/Jacob Wackerhausen

**Legitimate power** is power created and conveyed by the organization. It is the same as authority. The person formally in charge of a group can generally tell group members how they should be doing their jobs, how they should allocate their time at work, and so forth. Legitimate power alone does not make someone a leader; all managers have legitimate power, but only some of them are leaders. As we have seen, orders and requests from someone with legitimate power may be carried out by others but only to the minimum extent needed to satisfy the person in charge.[18]

A second type of power is **reward power**—the power to grant and withhold various kinds of rewards. Typical rewards in organizations include pay increases, promotions, praise, recognition, and interesting job assignments. The greater the number of rewards a manager controls and the more important they are to others, the more reward power the manager has.

**Coercive power** is the power to force compliance through psychological, emotional, or physical threat. In some settings, such as the military and prisons, coercion may take the form of physical force. In most settings today, though, coercion is practiced more subtly, through verbal reprimands, disciplinary layoffs, fines, demotions, the loss of privileges, and excessive public criticism. As with reward power, the more punitive elements a manager can bring to bear and the more important they are to those reporting to that manager, the more coercive power he or she has. However, the use of coercion also tends to increase hostility and resentment.

**Expert power** is power based on knowledge and expertise. A manager who knows the best way to deal with a difficult customer, a dairy manager who understands what increases or decreases milk production or a secretary who knows the ins and outs of the organization's bureaucracy all have expert power. The more important the knowledge is and the fewer people who are aware of it, the more expert power the person has.

The fifth type of power is **referent power**—the power that generally sets leaders apart from nonleaders. This power is based on personal identification, imitation, and charisma. If a child dresses and talks like his favorite rock singer, the rock singer has referent power over the child. If an ambitious middle manager starts to emulate a successful top manager (dressing like her, going to the same restaurants for lunch, playing the same sports, and so on), the top manager has referent power. We cover this aspect of leadership again later, when we look at charisma.

**FIGURE 17.3** This industrial manager has expert power which he gets from his knowledge of the manufacturing process and his expertise with the equipment.

Of course, most leaders use several different bases of power at the same time. For example, no matter how effective an individual is as a leader, he or she will sometimes find it necessary to rely on legitimate power. Indeed, many managers who lack leadership characteristics are still somewhat effective by using legitimate and reward power together. Likewise, leaders are often successful by combining expert and referent power.

## Uses, Limits, and Outcomes of Power

So all leaders use power. The question then becomes how do they use that power and what are the results?

**Uses of Power.** There are numerous ways to use power.[19] The manager can make a legitimate request—that is, simply ask someone to do something that falls within the normal scope of his job, for example. The manager may also try to gain instrumental compliance—that is, use reward power by letting the person know that a reward will be forthcoming if he does what is needed.

Coercion is the use of coercive power to get one's way. In the business context, it involves threatening a group member. For example, a manager might threaten to fire a member of his or her group if he does not perform a specific action. A more reasonable approach might be rational persuasion—convincing the group member that compliance is in everyone's best interest. Suppose a manager is trying to initiate a wage reduction. Most employees will not be enthusiastic, but if the manager can convince everyone that the cuts are necessary and that they will be temporary, people may be more receptive.

Sometimes managers also use personal identification and inspirational appeals. These approaches derive from referent power. The idea is that the leader will strive to set a good example and attempt to inspire others to follow it. For example, while Sam Walton, founder of Walmart, was one of the richest men in America, he lived an unpretentious life.

Finally, managers occasionally distort information to get their ways. This misuse of expert power can be dangerous and frequently backfires. Consequently, it should seldom, if ever, be used. Distorting information was precisely what Enron's Kenneth Lay, Jeffrey Skilling, and Andrew Fastow were accused of.

**Limits and Outcomes of Power.** Regardless of the manager's skill, power always has its limits. As a general rule, people can be influenced only up to a point. Moreover, their willingness to follow someone may be quite short-lived. Few leaders can maintain long-term support for their ideas and programs when mistakes are made and faulty decisions are implemented.

When a person in charge attempts to influence group members, the members usually respond in one of three ways: commitment, compliance, or resistance. **Commitment** is the outcome when the manager is also a leader. People are committed to the person and therefore respond favorably to his or her attempt to influence them. **Compliance** occurs when the person in charge is strictly a manager but has little leadership quality. Employees go along with the request but do not have any stake in the result.

Finally, **resistance** occurs when the manager's power base is weak or inconsistent with the situation. In this case, employees actively resist attempts to influence them.

Table 17.2 summarizes the outcomes of different kinds of power uses. In particular, it shows when commitment, compliance, and resistance are likely or possible in different power bases and situations.

In sum, various kinds, uses, and outcomes of power are relevant in organizational settings. We are now ready to look more closely at leadership itself. Through the years, writers have focused on three distinct approaches to studying and describing leadership, called the trait, behavioral, and contingency approaches.[20]

## FOOD FOR THOUGHT 17.2

Dolphins don't respond positively to threats or punishments of any kind. Rather, if you want the dolphin to do your bidding, you must coax, cajole, and praise it.

## Leadership Traits

One early systematic approach to the study of leadership is the trait approach, whose adherents assumed that great leaders such as Napoleon, Lincoln, and Gandhi possessed a set of stable and enduring **leadership traits** or characteristics that set them

**TABLE 17.2** Outcomes of the Uses of Power

| SOURCE OF LEADER INFLUENCE | TYPE OF OUTCOME | | |
|---|---|---|---|
| | COMMITMENT | COMPLIANCE | RESISTANCE |
| **Legitimate Power** | *Possible*—if request is polite and appropriate | *Likely*—if request or order is seen as legitimate | *Possible*—if arrogant demands are made or request does not appear proper |
| **Reward Power** | *Possible*—if used in a subtle, personal way | *Likely*—if used in a mechanical, impersonal way | *Possible*—if used in a manipulative, arrogant way |
| **Coercive Power** | *Very unlikely* | *Possible*—if used in a helpful, nonpunitive way | *Likely*—if used in a hostile or manipulative way |
| **Expert Power** | *Likely*—if request is persuasive and subordinates share leader's task goals | *Possible*—if request is persuasive but subordinates are apathetic about task goals | *Possible*—if leader is arrogant and insulting. or subordinates oppose task goals |
| **Referent Power** | *Likely*—if request is believed to be important to leader | *Possible*—if request is perceived to be unimportant to leader | *Possible*—if request is for something that will bring harm to leader |

Source: Table adapted by Gary A. Yukl from information in John R. P. French Jr. and Bertram Raven, "The Bases of Social Power," *Studies in Social Power*, ed. Dorwin P. Cartwright (Ann Arbor, MI: Institute for Social Research, the University of Michigan, 1959), 150–167. Copyright 1955. Data used by permission of the institute for social research.

apart from followers. Their goal was to identify these traits for use as a basis for selecting managers.

A great deal of attention was focused on the search for traits, with researchers studying common traits, such as intelligence, height, self-confidence, attractiveness, and vocabulary.[21] Unfortunately, traits proved to be ineffective predictors of leadership. For one thing, the list of characteristics soon grew to unmanageable lengths. For another, the list of exceptions was almost as long as the list of leaders who possessed each trait. For instance, it has been suggested that leaders are taller than nonleaders, but many historical leaders (such as Napoleon and Hitler) and contemporary leaders (like H. Ross Perot) are of slight build.

Scholars soon realized that the search for leadership traits was interesting but of limited scientific merit. Consequently, they started focusing attention on other areas instead. However, many people cling to the trait notion despite the failure of researchers to identify a replicable set of traits that can predict leadership.[22] Scholars, on the other hand, next turned to the study of leadership through what we now call the behavioral approach.

## Leadership Behaviors

Whereas the trait approach attempts to identify characteristics that differentiate leaders from nonleaders, the behavioral approach seeks to define behaviors that set effective leaders apart from ineffective leaders. One of the first studies to note the complicated nature of leadership was conducted at International Harvester (now Navistar International).[24] Although numerous **leadership behaviors** have been found, those identified by two major sets of studies have received special attention.[25]

### The Michigan Studies

Beginning in the late 1950s, researchers at the University of Michigan identified two critical leadership behaviors: job-centered behavior and employee-centered behavior.[26] A leader who practices

## A FOCUS ON AGRIBUSINESS
### A Top Leader

Born in Morocco, raised in France, and with an MBA from the European Business School, Michel Landel is a top leader in agribusiness. After completing his studies he went first to Chase Manhattan Bank and then to the Poliet group.

He joined Sodexo in 1984 as Chief Operating Manager for Eastern and North Africa; two years later he was President of all African operations; and three years after that, he headed up the North American operations. He founded the STOP Hunger Sodexo Foundation in 1996, and in 1998 he oversaw the successful merger with Marriott Management Services and watched Sodexo become a top food and facilities organization.

Landel's career continued its upward trajectory when in 1999 he became President and CEO of Sodexo's North America operations. The next year he was named Vice President of Group Sodexo. And in 2005 he was appointed Group CEO, taking over from Pierre Bellow who, in 1966,

had founded Sodexo (formerly Sodexho) in Marseilles, France. Landel became the first non-family CEO at Sodexo, which had been an exclusively family-run business up to that time. In 2009 he also became a member of the Board of Directors.

In an interview in 2010, Landel indicated that to be a successful leader in a family-run business you must respect the family's values and what the family has done to build the organization. In addition, he noted that because it is difficult for another company to take over the business, leaders can "have a really long-term vision to work, invest, and do what is needed to create a sustainable company."

Under Landel's leadership Sodexo has been named one of the "Worlds' Most Ethical Companies," one of the world's best outsourcing companies, a DiversityInc Top 50 Company, a "Most Admired Company" by *Fortune* magazine, and is recognized for its commitment to sustainable development.[23]

job-centered behavior engages in close supervision to monitor and control the performance of those reporting to him or her. The manager's interests are primarily in getting the job done, and he or she takes an active role in explaining this task.

In contrast, employee-centered behavior focuses on reaching high levels of performance by building a sense of team spirit through the human element of the workplace. An employee-centered leader is concerned with developing a desire to achieve high levels of performance among group members by paying attention to them. The leader is willing to let employees have a voice in how they do their jobs and tries to develop job satisfaction and group cohesion.

The Michigan researchers felt that job-centered and employee-centered behaviors represent a single dimension, with one of the two basic behaviors at each end. That is, they believed that if leaders become more job centered, they simultaneously become less employee centered, and vice versa. They also felt that leaders who were employee-centered would generally be more effective as managers than leaders who were primarily job centered. That is, employees would perform at a higher level and also be more satisfied under employee-centered leaders.

## The Ohio State Studies

Starting in the late 1940s, researchers at Ohio State University identified many of the same concepts as those developed in Michigan but also extended and refined them.[27] After studying managers at

International Harvester, they agreed that there were two critical leadership behaviors, which they called initiating structure behavior and consideration behavior. *Initiating structure behavior*, which is similar but not identical to job-centered behavior, focuses on getting the job done. *Consideration behavior* involves employee satisfaction and friendliness. It is similar to only a part of employee-centered behavior since it does not have the high performance focus found in employee-centeredness.

The basic difference between the Ohio State and Michigan findings is shown in Figure 17.4. Again, the Michigan researchers argued that leaders could be job centered or employee centered, but not both, while the Ohio State researchers found that the two forms of leader behavior they identified were independent. Therefore, as the exhibit shows, a leader can use initiating structure behavior and consideration at the same time.

As noted earlier, much of the early research of Ohio State was conducted with managers from International Harvester (now known as Navistar). In general, the researchers found that high initiating structure behavior resulted in higher performance but also led to lower levels of job satisfaction. High levels of consideration behavior caused higher levels of job satisfaction but lower levels of performance.[28] However, they also found that, depending upon the situation, there were numerous other types of leader behavior that could be important. The dynamics of leader behavior are complex and anything but straightforward.

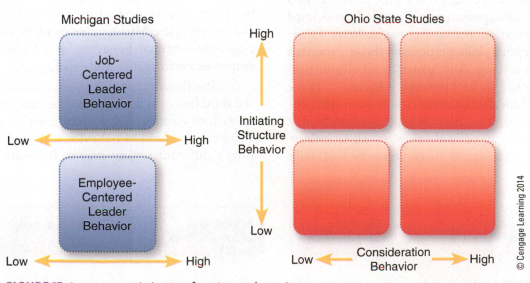

**FIGURE 17.4** Leadership behaviors from two early studies.

## A Contemporary View

As behavioral research continued, numerous categories of leader behavior were identified. As each researcher used his or her own labels, a bewildering array of lists of leader behaviors emerged. However, only one approach to integrating them seems defensible in terms of both research and being understandable and useful to practicing managers.[29] This approach suggests that the two previously identified categories of leader behavior can readily be seen to consist of four major categories: Job Centered and Initiating Structure become Giving-Seeking Information and Making Decisions; while Employee Centered and Consideration become Influencing People and Building Relationships.[30]

Giving-Seeking Information includes monitoring and informing behaviors. Making Decisions includes planning, organizing, delegating, and problem solving. Influencing People includes motivating and rewarding behaviors. And Building Relationships includes team building, networking, and managing conflict. Each of those could be refined still further to fit particular circumstances. These more specific categories appear to be more useful because they are readily understandable and learnable. However, just as had been found in the Ohio State studies, research has shown that effective leaders use different combinations of these behaviors in different situations. Thus, there is no one best set of leader behaviors.

Further, numerous skills have been found to be associated with effective leadership.[31] While considerably more research is necessary to clearly identify those skills, some are rather obvious. Leaders need interpersonal skills, particularly those skills associated with communication, persuasiveness, and tact. Leaders need conceptual skills, particularly those associated with problem solving. And leaders need administrative and technical skills. However, it is clear that the particular technical skill needed will vary with the situation. Indeed, it is clear from this research that no one type of skill or leadership fits all situations. **Leadership style** refers to combinations of skills and behaviors. This means, then, that there is no one best style but rather style must fit circumstance.

The notion that one style of leadership will always be appropriate has been unacceptable for many years. Thus, researchers have shifted their efforts to the development of contingency models of leadership, which attempt to define those circumstances in which one style of leadership is best and those in which an alternative style will be more appropriate. The next section introduces several interesting contingency models of leadership.

**FOOD FOR THOUGHT 17.3**

The Cattle Feeders Hall of Fame presents its Industry Leadership Award to individuals who have been instrumental in shaping modern beef production and ensuring the success and growth of the industry outside of the feed yard.

## Situational Approaches

Earlier in the book we investigated the nature of **contingency**, or **situational**, **approaches** to management. Leadership was one of the very first areas in which situational theories were developed. Figure 17.5 illustrates the differences between behavioral approaches and situational approaches. The behavioral approach would simply go from leader behavior to subordinate response, whereas the situational approach would involve the whole process. The basic premise is that appropriate leadership behaviors lead to desired group member responses. For instance, the Michigan researchers assumed that employee-centered behavior would always lead to employee performance and satisfaction.

Situational approaches, in contrast, introduce the third box into the exhibit. They suggest that situational, or contingency, factors must be considered. Whereas one kind of behavior will work in one setting, a different setting may well dictate a different

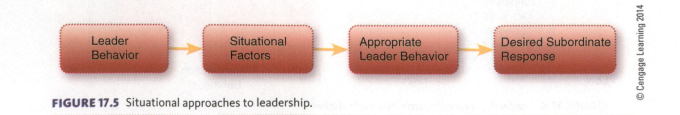

**FIGURE 17.5** Situational approaches to leadership.

© Cengage Learning 2014

form of behavior. The goal of situational approaches is to define the situational variables that managers should consider in assessing how different forms of leadership will be received.

Although a few early researchers noted the potential importance of situational factors, these factors did not receive widespread attention until the 1960s. Since then, however, virtually all approaches to leadership have adopted a situational view. Next, we describe the three most widely known situational theories—the LPC Model, the path-goal model, and the participation model.[32]

## The LPC Model

The first of these contingency models of leadership, called the **LPC** (for least preferred coworker) **model**, was developed by Fred E. Fiedler and is shown in Figure 17.6.[33] Fiedler suggested that appropriate forms of leadership style varied as a function of the favorableness of the situation.

**Leadership Styles.**  The LPC model includes two basic forms of leadership style: task oriented and relationship oriented. The *task-oriented style* is similar to

the earlier job-centered and initiating structure behaviors, and the *relationship-oriented style* is like employee-centered and consideration behaviors. The name of the model is derived from a questionnaire developed to measure task-oriented and relationship-oriented style. People who complete the questionnaire do so in reference to the employee with whom they least prefer to work, that is, their least preferred coworker.

One interesting aspect of the LPC model is that it assumes leadership style to be a stable personality trait of the leader. That is, some leaders use one style and others use a different style, and these styles are basically constant. Any given leader is unable to change her or his behavior.

**Favorableness of the Situation.**  Again, the LPC model sees appropriate leadership behavior as a function of the favorableness of the situation. Favorableness is defined by three elements—leader-member relations, task structure, and position power.

**Leader-member relations** defines the nature of the relationship between the leader and the members of the group. If the relationship is characterized by confidence, trust, liking, and respect, it is defined as

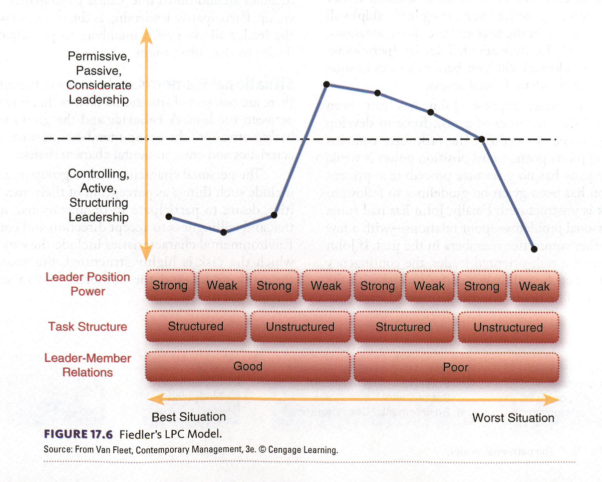

**FIGURE 17.6** Fiedler's LPC Model.

Source: From Van Fleet, Contemporary Management, 3e. © Cengage Learning.

good and is favorable for the leader. In contrast, if the relationship lacks confidence, trust, and respect, and if the leader and the followers do not like each other, the relationship is bad and unfavorable for the leader.

**Task structure** is the degree to which the group's task is well defined and understood by everyone. If the task is highly structured, the situation is probably more favorable.

Finally, **position power** is the power vested in the leader's position. Strong power is favorable for the leader and weak power is unfavorable. Thus, the best possible situation is good leader-member relations, a structured task, and strong position power. The worst situation is poor relations, an unstructured task, and weak power.

### Combining Styles and Situations. 
Figure 17.6 also illustrates how leadership styles combine with the situation to determine group effectiveness. Note that the situation can be defined in eight unique ways. The left side of the chart represents the best situation. As the situation progress to the right, however, it gets gradually worse. The line above the situations predicts which style of leadership will be most effective in each situation. A task-oriented leader (using controlling, active, structuring leadership) will be most effective in the best and the worst situations, whereas a relationship-oriented leader (permissive, passive, considerate) will have better chances in situations of intermediate favorableness.

For example, suppose John has just been appointed the chairman of a task force to develop a new grievance procedure. The task force contains several of John's peers, so his position power is weak. The company has no grievance procedure at present and John has been given no guidelines to follow, so the task is unstructured. Finally, John has had some interpersonal problems—poor relations—with a few of the other committee members in the past. If John is by nature a task-oriented leader, the contingency model predicts he will have greater success with the group than if he is a relationship-oriented leader.

While the LPC model was the first major approach to attract our attention to the situation as an important part of leadership, on balance, it has received mixed research support.[34] For example, people have questioned its assumptions about the flexibility of leadership style and how it defines situations. There are also major questions about the questionnaire that is used to measure leadership behavior.

### The Path-Goal Model

The **path-goal model** also provides interesting insights into the situational nature of leadership. This model essentially suggests that the purpose of leadership in organizational settings is to clarify for members of the organization the paths to desired goals. That is, leaders should determine what employees want from their jobs and then show them how to acquire those things through their work.[35] The basic framework of the path-goal model is shown in Figure 17.7.

### Leadership Styles. 
Like the other models discussed so far, the path-goal model includes one task-oriented style and one employee-oriented style, called directive and supportive leadership. It also includes an additional one, called participative leadership. Participative leadership is the extent to which the leader allows group members to participate in decisions that affect them.

### Situational Elements. 
As shown in the exhibit, there are two sets of situational factors that intervene between the leader's behavior and the group member's motivation: the group members' personal characteristics and environmental characteristics.

The personal characteristics of group members include such things as perception of their own abilities, desire to participate in organizational activities, and willingness to accept direction and control. Environmental characteristics include the extent to which the task is highly structured, the nature of the work group, and the authority system within the organization.

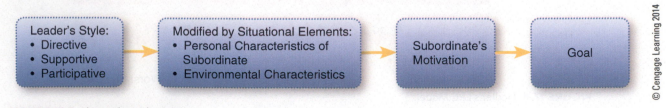

**FIGURE 17.7** The path-goal model.

**Combining Styles and Situations.** Unlike the precise LPC model, the path-goal model is general and suggests that the leader needs to use a lot of common sense. The path-goal model also assumes that a leader's style is flexible and can be changed as needed.

In using the model, then, a manager should assess the relevant dimensions of the situation and choose an appropriate combination of behaviors that will complement that situation. For instance, suppose that a group member lacks confidence in her abilities and is assigned to a new task that is highly unstructured. It may be appropriate in this situation for the group leader to be highly directive in order to clarify the member's task demands and reduce her anxiety about her ability to get the job done. Similarly, if group members want to participate, the leader should consider allowing them to do so, whereas the leader may not even have to think about it if the group has no desire to participate.

The path-goal model is still in the early stages of development, so several variations exist. It has received generally favorable support from research, however, and will probably continue to be developed for future use.[36]

## The Participation Model

The final situational model for discussion here is the **participation model**, which involves a much narrower aspect of leadership than do the preceding models. In particular, it addresses the specific question of how much group members should be allowed to participate in decision making.[37] As with the other models, the participation model includes alternative styles and situational factors to consider.[38]

**Leadership Styles.** As shown in Table 17.3, the participative model includes five different degrees of participation or leader behaviors:

Decide: The manager makes the decision alone with no involvement of group members.

Consult (individual): The manager asks individual group members for information that he or she needs to make the decision, but still makes the decision alone. Group members may or may not be informed of the decision.

Consult (group): The manager shares the situation with group members and asks for information and advice. The manager still makes the decision, but keeps the group actively informed.

Facilitate: The manager meets together with the whole group. Information is freely shared, although the manager still makes the decision.

Delegate: The manager and the group meet and freely share information, and the group makes the decision.

**Situational Elements.** The participation model suggests that a manager needs to ask several questions before choosing a degree of participation. Moreover, different circumstances call for different questions. In some cases, the manager will have a group-related problem to address, whereas in other cases the problem will relate more to an individual. In addition, in some cases the goal will be to make a decision as quickly as possible, whereas in other situations the manager will be striving to help an individual or group improve its decision-making skills.

**Combining Styles and Situations.** Styles and situations are combined in a very structured and precise fashion in the participation model, as illustrated in Table 17.3. This particular model is suggested for a group problem when time is important. The manager asks the questions and answers high or low, yes or no. Depending on the answer to any given question, the manager identifies the suggested style. This style is the one that optimizes group members' acceptance of the decision, the quality of the decision (from the organization's standpoint), and the demands on the manager's time.

In general, the participation model has been supported by research and accepted by managers.[39] Of course, it should not be followed too rigidly. Managers should recognize that the model provides a set of guidelines rather than a set of rules that should always be followed.

## An Integrative Framework

Recently, a framework to integrate these diverse results has been devised.[40] An extension of that framework is shown in Figure 17.8, the framework includes all of the major factors discussed so far—power, traits, behavior, and situations.

This framework suggests that leadership is a complex phenomenon rather than a simple one and that all the approaches to leadership have merit and should be seen as complementary rather than contradictory. All other things being equal, traits do make a difference in leadership. Given the careful selection process in many organizations, however,

**TABLE 17.3**  The Participation Model of Leadership

| LEADER PARTICIPATION BEHAVIOR | WHEN LEADER PARTICIPATION BEHAVIOR IS MOST APPROPRIATE TO USE | | | | | | | |
|---|---|---|---|---|---|---|---|---|
| | DECISION QUALITY IS IMPORTANT | FOLLOWER COMMITMENT IS IMPORTANT | LEADER HAS SUFFICIENT INFORMATION | PROBLEM STRUCTURED | LIKELIHOOD OF FOLLOWER COMMITMENT | FOLLOWERS SHARE GOALS | FOLLOWER CONFLICT | FOLLOWERS HAVE SUFFICIENT INFORMATION |
| **Decide** | Yes | Yes | Yes | | High | | | |
| | Yes | No | Yes | | | | | |
| | No | Yes | | | High | | | |
| | No | No | | | | | | |
| **Consult (individually)** | Yes | Yes | No | Yes | High | No | | Yes |
| | Yes | Yes | No | Yes | High | Yes | No | Yes |
| **Consult (group)** | Yes | Yes | No | Yes | High | Yes | Yes | Yes |
| **Facilitate** | Yes | No | No | No | | | | |
| | Yes | Yes | No | Yes | Low | No | | |
| | Yes | Yes | No | Yes | Low | Yes | No | No |
| | Yes | Yes | No | No | Low | No | | |
| | Yes | Yes | No | No | Low | | | No |
| | Yes | Yes | No | No | High | | | |
| | Yes | Yes | Yes | | Low | Yes | | No |
| | Yes | Yes | Yes | | Low | No | | |
| **Delegate** | No | Yes | | | Low | | | |
| | Yes | Yes | Yes | | Low | Yes | | Yes |
| | Yes | Yes | No | Yes | Low | Yes | Yes | |
| | Yes | Yes | No | Yes | Low | Yes | No | Yes |
| | Yes | Yes | No | No | Low | Yes | | Yes |

Sources: Adapted from Ricky W. Griffin, *Management*, 10th ed., Mason, OH: South-Western Cengage Learning, 2011; Victor H. Vroom and Arthur G. Jago, *The New Leadership: Managing Participation in Organizations*, Englewood Cliffs, NJ: Prentice-Hall, 1988; Victor H. Vroom and Philip W. Yetton, *Leadership and Decision-Making*, Pittsburgh, PA: The University of Pittsburgh Press, 1973; V. H. Vroom and A. G. Jago, "The Role of the Situation in Leadership," *American Psychologist*, 2007, Vol. 62, No. 1, 17–24; and V. H. Vroom, "Decision Making: The Vroom/Yetton/Jago Models." In G. Goethals, G. Sorenson, and J. M. Burns, *The Encyclopedia of Leadership*, McGraw-Hill, 2004, Vol. 1, 322–325.

managers will have such similar traits that power or behavior are more likely to make the difference in terms of performance.

The situation surrounds the other components of the framework to convey the idea that its impact is general rather than particular. That is, it affects everything, not just one or a few aspects. The situation can influence which behavior is appropriate for leaders, as in the situational models; and it can influence which forms of power are available to the leader. Also, of course, it can directly influence results beyond the leader's impact. For example, a shortage of materials may hinder performance in ways the leader

cannot control. Understanding the situation, then, is the key to being an effective managerial leader.[41]

This framework also introduces the issue of criteria. As indicated, there are numerous criteria for assessing the impact of leadership. Stated another way, different combinations of the leadership factors can lead to different outcomes.[42] It is essential, therefore, for the organization to have its objectives clearly in mind.

Contingency theories suffer from their reality. Reality is complex and so are these theories. They do not translate into simple, behavioral guides for practicing managers. Nevertheless, these theories are

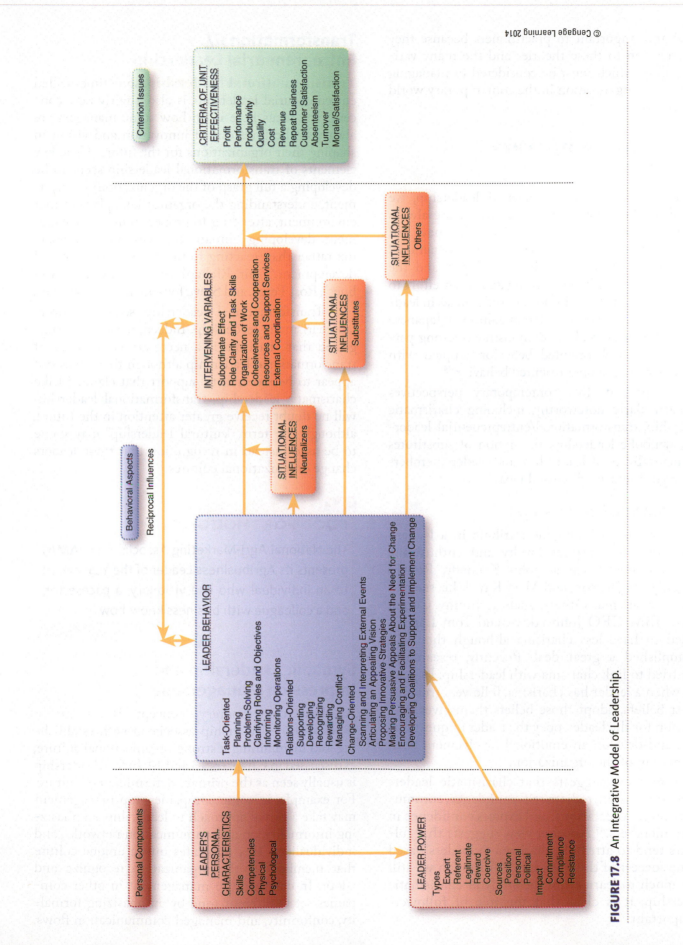

**FIGURE 17.8** An Integrative Model of Leadership.

useful and important to practitioners because they call attention to these theories and the many variables, all of which must be considered in managing an effective organization in the contemporary world of business.

## Other Contemporary Perspectives

Given the obvious importance of leadership, we should not be surprised that researchers are always devoting a lot of attention to identifying new and even more insightful perspectives. Cross-national differences, such as those between Japan and the United States, are being investigated in an effort to determine what, if any, important differences in leadership styles exist across different cultures.[43] Japanese leaders, for instance, have demonstrated strong performance or task-oriented behavior coupled with maintenance or people-oriented behavior.[44]

At present, five contemporary perspectives are particularly noteworthy, including charismatic leadership, transformational/entrepreneurial leadership, symbolic leadership, the notion of substitutes and neutralizers of leadership, and leader-members exchange. These are discussed next.

### Charismatic Leadership

**Charisma** is an intangible attribute in a leader's personality that inspires loyalty and enthusiasm. Certain leaders, such as John Kennedy, Ronald Reagan, Jesse Jackson, and Mary Kay Ashe, seemed to have charisma. Others, such as Jimmy Carter, former IBM CEO John Akers, and Tom Landry, tended to have less charisma although they still accomplished a great deal. Recently, researchers have tried to link charisma with leadership. In general, when a leader has charisma, followers trust his or her beliefs, adopt those beliefs themselves, feel affection for the leader, obey the leader unquestioningly, and develop an emotional involvement with the mission of the organization.

One view suggests that charismatic leaders articulate a vision, radiate self-confidence, communicate high expectations, and express confidence in group members.[45] Another view suggests that followers tend to attribute charisma to leaders based on the success of those leaders.[46] Although we still have much to learn from the study of charismatic leadership, it is clear that the role of followers is important.[47]

## Transformational/ Entrepreneurial Leadership

**Transformational leadership**, sometimes called entrepreneurial leadership, is also a fairly new concept. The focus here is on how some managers are always at the forefront of innovation and vision in shaping their organizations for the future.[48] The key elements of transformational leadership seem to be developing a full grasp of the organization's environment, understanding the organization's place in that environment, attending to proper strategic management, developing human resources, and anticipating rather than reacting to the need for change and development.[49] Sir Richard Branson (Virgin Airlines), Ross Perot, and Steve Jobs are recent examples.

Transformational leadership seems to overlap charismatic leadership. Indeed, it has been suggested that charisma is a necessary component of transformational leadership although there does not appear to be research to support that claim.[50] Like charismatic leadership, transformational leadership will no doubt receive greater attention in the future, although the term "cultural leadership" may come to be used instead in recognition that these leaders change organizational cultures.[51]

### FOOD FOR THOUGHT 17.4

The National Agri-Marketing Association (NAMA) presents its Agribusiness Leader of the Year award to an individual who is a visionary, a pacesetter, and a colleague with business know-how.

## Symbolic Leadership and Impression Management

A third contemporary concept is **symbolic leadership**—leadership associated with establishing and maintaining a strong organizational culture, as discussed in Chapters 13 and 14. In fact, leadership is usually seen as the primary determinant of culture. For example, in some companies, top management may take a casual approach to leadership, emphasizing informality, open communication networks, and individualism. This translates into a unique culture that members of the organization recognize and adopt. In contrast, top management in other companies set a different tone by emphasizing formality, conformity, and managed communication flows.

The wrong symbols can lead people into unethical behavior. Once again, the notion of symbolic leadership is a relatively new one. Managers are becoming aware of the importance of symbolism in their work.

Closely related to symbolic leadership is impression management.[52] Impression management refers to direct and intentional efforts on the part of one person to enhance his or her image in the eyes of others. Leaders use symbols and other techniques in impression management to enhance their power with subordinates. Followers use impression management to enhance their relationships with superiors. "Looking good" may improve their chances for advancement or boost their self-esteem.[53]

## Substitutes and Neutralizers

The notion of substitutes and neutralizers for leadership arose to account for situations in which leadership does not seem to be necessary.[54] Many situations exist where people go about doing their jobs without any specific direction from either managers or leaders (Figure 17.9). They know their jobs so well or the situation is so specific that they act without the presence of a leader. Thus, group member expertise can substitute for leadership as can the nature of the task itself (emergency conditions, for instance).

Indeed, in situations where group members have considerable expertise, efforts by an individual to influence them may be rendered ineffectual, that is, be neutralized. The efforts of the individual may also be inappropriate in, say, attempting to offer rewards which group members do not need or want.

**FIGURE 17.9** These data analysts work without specific directions from managers or leaders.

Here, too, the effort is neutralized by aspects of the situation or characteristics of the group.

Finally, characteristics of the organization can also substitute for or neutralize leadership. Formal and inflexible policies and procedures can be developed such that leadership is not needed. A rigid reward system may neutralize a leader's attempts at using reward power to influence group members. Group solidarity may also resist influence attempts.[55]

To repeat, leadership is a complex phenomenon. There is no one style, no one basis of power, no one set of behaviors, and no one use of skills that fits every situation and so can be recommended for managers. Rather, contemporary managers must understand different styles, bases of power, and combinations of behaviors and skills, and shape them to fit the situations in which they are trying to effectively manage.

## The Leader-Member Exchange Approach

The Leader-Member Exchange (LMX) approach recognizes superior-subordinate pairs or vertical-dyads as crucial in leadership.[56] Even though a leader may have the same formal relationship with his or her subordinates, the actual relationships may differ in important ways. Managers usually develop special trusting relationships with only a few subordinates— the "in-group." When a new member joins the organization, the manager often decides early in the relationship whether that new employee will become one of this select group. While this may depend on a variety of factors, personal compatibility and the new employee's competence will be important. Over time the vertical-dyad linkage model evolved into the leader-member exchange approach.

People who are not selected for special relationships—the "out-group"—receive less of the manager's time and attention. People in the "in-group" tend to be more productive and have higher levels of job satisfaction than those in the "out-group."[57] It is, however, unclear whether "in-group" status leads employees to work harder, get more rewards, and hence be more satisfied or whether managers simply pick hardworking, confident people for "in-group" membership.[58]

## Leadership Development

Given the complexity of leadership, how is it developed? How does one learn to be a leader? In many organizations a senior manager may act as a mentor for a junior person to help them learn about leadership. Joshua Summer has served as a mentor for all of his children to assure that they will be effective both as managers and as leaders. Most people learn by doing and observing, but other approaches have become available.

Books about leaders and leadership can provide useful insights. Professional organizations and consulting groups provide leadership training programs. Colleges and universities have formal courses and extension or executive education programs covering leadership. And more recently a computer simulation has been shown to be quite helpful in developing an individual's leadership skills.[59]

## CHAPTER SUMMARY

Leadership is an influence process directed at shaping the behavior of others. People can be managers but not leaders, leaders but not managers, or both. The most valuable individuals are those who combine the two qualities, but they are scarce.

Leadership depends on the use of one of the five types of power: legitimate, reward, coercive, expert, and referent. Leaders may use each of these forms to influence people to pursue certain goals. When people are confronted with someone's attempt to influence them, their response can be commitment, compliance, or resistance.

Leadership traits are those characteristics of leaders that set them apart from followers. Unfortunately, no one has ever established that any single set of traits can reliably distinguish leaders from nonleaders in a wide variety of situations.

The behavioral approach attempted to differentiate successful or effective leaders from unsuccessful or ineffective ones on the basis of their behaviors. The Michigan studies identified two major leadership behaviors termed job-centered and employee-centered behaviors. They regarded these behaviors as mutually exclusive and felt that employee-centeredness was more effective. The Ohio State researchers also tended to emphasize two major forms of behavior, which they called initiating structure and consideration behaviors. Unlike the Michigan group, the Ohio State researchers felt that these two forms of behavior could and should be combined for most effective leadership.

Situational approaches to leadership attempt to define circumstances in which one form of leadership is more appropriate than another. The LPC model assumes that the behavior of managers is not changeable and is either task oriented or relationship oriented. The path-goal model suggests that the purpose of leadership is to clarify for followers the paths to desired goals. The participation model focuses on a narrow but important aspect of leadership: determining how much group members should be allowed to participate in decision making.

A framework to integrate these diverse models has recently been developed. It suggests that leadership is a complex phenomenon rather than a simple one and that all approaches to leadership have merit and should be seen as complementary rather than contradictory. Three new and insightful perspectives on leadership focus on charismatic, transformational, and symbolic leadership.

Senior managers can help some employees become leaders, as can leadership training provided by professional organizations, consulting groups, and university courses.

## CHAPTER ACTIVITIES

### REVIEW QUESTIONS

1. Explain how someone could be a manager but not a leader, a leader but not a manager, or both a leader and a manager.
2. List and define the five types of power.
3. What are the major forms of leadership behavior identified in the Michigan and Ohio State studies?
4. Summarize the LPC, path-goal, and participation models.
5. Describe three contemporary perspectives on leadership.

### ANALYSIS QUESTIONS

1. Many people have heard the saying "Power corrupts, but absolute power corrupts absolutely." What does this mean? What implications does it have for organizations and for managers?
2. Compare and contrast the three situational approaches to leadership.
3. What leadership traits do you think are most important?
4. What forms of leadership behavior can you identify beyond those discussed in the chapter?
5. Who do you think are today's charismatic leaders?

### FILL IN THE BLANKS

1. An influence process that is directed at shaping the behavior of others is called _____ .
2. A leadership perspective that focuses on innovation and vision to shape organizations for the future is called _____ _____ leadership.
3. The leader's power that is based on personal identification, imitation, and charisma is known as _____ power.
4. The power to force compliance through psychological, emotional, or physical threat is known as _____ power.
5. The combination of leadership skills and behaviors is called _____ _____ .
6. The _____ or _____ approach to leadership attempts to specify circumstances under which different kinds of leadership behavior are appropriate.
7. The leadership model that suggests that appropriate leadership behavior is a function of the favorableness of the situation is the _____ model.
8. The _____ _____ model of leadership suggests that leaders should attempt to determine their group members' goals and then clarify the paths to achieving them.

9. The _____ model of leadership helps managers determine how much participation employees should be allowed in making various kinds of decisions.

10. The type of leadership that relies on the personality and other attributes of the leader to inspire loyalty and enthusiasm is called _____ leadership.

## ▼ CHAPTER 17 CASE STUDY
### THE PEPSI CHALLENGES

**M**ore than a decade ago Coca-Cola and Pepsi-Cola set the standard for a fierce, competitive, industry-wide rivalry. Each company purchased huge swaths of time to "hawk" their beverages. It made for entertaining television, boosted the coffers of broadcast networks, and provided a nice living for advertising executives. PepsiCo's tag line for this activity was, "Take the Pepsi Challenge." Coca Cola lost the challenge but won the cola war. Time passed and each multinational company evolved differently. PepsiCo diversified beyond beverages into other foods like snacks (Frito-Lay) and breakfast food (Quaker Co.) and divested itself of fast-food restaurants (Pizza Hut, Kentucky Fried Chicken, and Taco Bell).

PepsiCo is a company that thrives on challenges, or at the very least it does not shy away from them. When Vice President Richard Nixon was showing Soviet Premier Nikita Krushchev around the America National Figureion in Moscow, they stopped at the Pepsi Cola kiosk. An enterprising young PepsiCo executive, Donald Kendall, seized the photo opportunity and thrust a cup of Pepsi Cola into the Soviet leader's hands. It was great PR and lead to a big opportunity for PepsiCo. Pepsi Cola in 1974 became the first western consumer product produced and sold in the USSR. Mr. Kendall went on to become a CEO of PepsiCo.

The 107 million viewers of the 2010 Super Bowl saw many engaging and entertaining commercials. There were some who believed the advertising competition outrivaled the football engagement; however, the 2010 viewers did not see any ads for Pepsi Cola. Rather than spend $20 million dollars for a few 30-second spots, PepsiCo decided to give the money away. It engaged in "cause marketing." Under the banner of "Refresh Everything," the Pepsi Refresh campaign asked the public to vote online for charities and community groups to receive substantial contributions from PepsiCo. It worked. People participated enthusiastically—it is always fun to give away someone else's money—and the campaign attracted a "ton" of publicity. PepsiCo challenged its customers and its rivals—and won.

Indra Nooyi, PepsiCo's CEO, wants her company to be "seen as one of the defining companies of the first half of the 21st century, a model of how to conduct business in the modern world." What makes this a challenge is the nutritional dichotomy in PepsiCo's stable of foods. It has Frito-Lay snacks as well as its eponymous soft drink at one end of the nutritional spectrum and Quaker Oats cereal at the other end. It will not be easy selling one end of the spectrum without affecting sales at the other end. A Pepsi with a bowl of cereal, for example, is not a nutritionist's ideal and therefore could be a real challenge to implement.

PepsiCo's 2009 revenue from international markets was 48 percent while Coca Cola's was 74 percent. However, PepsiCo is betting that its special relationships in Russia will lead to greater long-term profits. In 2008 PepsiCo purchased Russia's largest juice maker, Lebedyansky, for $2 billion, and in 2010 it purchased Wimm-Bill-Dann juice and dairy group for $5.4 billion. Wimm-Bill-Dann introduced yogurt to the Russian market. Apparently as Russians emulated western-style tastes, yogurt sales skyrocketed. PepsiCo admitted that the WBD purchase was its largest-ever international acquisition and that it was part of its strategic plan to build a $30 billion nutrition business.

The challenge did not stop with those purchases. With WBD PepsiCo will compete with the other large dairy goods company in Russia, Unimilk. Unimilk is controlled by Groupe Danone SA, a French company and is number three in juice sales and number one in baby food in Russia. For PepsiCo to acquire WBD, Danone had to sell its 18 percent share on the open market to acquire money to purchase Unimilk. In case you were wondering about Coca Cola, in 2005 it bought juice-maker Multon Co. for $500 million. Challenges seem to follow PepsiCo.

Competitive juices—pardon the pun—never stopped flowing. When Indra Nooyi flew to Russia to receive Vladimir Putin's approval for the sale, she brought with her a retired veteran, Donald Kendall, the fellow that started it all. Nice touch.

A side bar on competition: If you pronounce the name Wimm-Bill-Dann quickly it will sound like "Wimbledon." No coincidence, as one of the founders of WBD loved tennis and nothing symbolized competitive tennis to him more than Wimbledon. Also, in 2010, Indra Nooyi joined the board of directors of the U.S. Soccer Federation to help bolster its bid to host the 2018 World Cup. The United States lost its bid to host the FIFA World Cup in 2018 … to Russia.[60]

## ▶ Case Study Questions

1. Which approach to leadership seems most nearly to fit Donald Kendall? Indra Nooyi?
2. How would you describe the style of each leader?
3. Do any of the contemporary perspectives seem applicable to either of them?

## REFERENCES

1. Joseph Schumpeter, *Capitalism, Socialism and Democracy* (New York: Harper Books, 1942) 83; Stephanie Schomer, "Farm Fresh," *Fast Company Magazine*, September, 2010, 52; Robert Tita, "Deere Garners Robust Earnings," *The Wall Street Journal*, August 19, 2010, B3; George Seperich, *Food Science and Safety. 2nd ed.* (Upper Saddle River, NJ: Prentice-Hall, 2004), 375; Patrick Barta and Wilawan Watcharasakwet, "It's Back-to-School Season for the Water Buffalo, Too," *The Wall Street Journal* (September 10, 2010) at online.wsj.com (accessed September 20, 2010).

2. For a thorough review of various definitions of leadership, see Bernard M. Bass with Ruth Bass, *The Bass Handbook of Leadership, 4th ed.* (NY: Free Press, 2008) and Gary A. Yukl, *Leadership in Organizations, 7th ed.* (Upper Saddle River, NJ: Prentice-Hall, 2010).

3. Chad Higgins, Timothy Judge, and Gerald Ferris, "Influence Tactics and Work Outcomes: A Meta-Analysis," *Journal of Organizational Behavior* (2003): 89–106; G. Yukl and C. M. Falbe, "Influence Tactics and Objectives in Upward, Downward, and Lateral Influence Attempts," *Journal of Applied Psychology* (1990): 75: 132–140.

4. Joseph Weber, "The Downside of ADM's Focus on Biofuels." *Business Week* (December 31, 2008) at www.businessweek.com (accessed July 20, 2010).

5. "The Best & Worst Managers of the Year," *Business Week*, January 19, 2005, 55–84.

6. David D. Van Fleet and Gary Yukl, "A Century of Leadership Research," in *Papers Dedicated to the Development of Modern Management*, eds. D. A. Wren and J. A. Pearce II, *Academy of Management*, 1986: 12–23.

7. For a more complete review of the literature, see Bruce J. Avolio, Fred O. Walumba, and Todd J. Weber, "Leadership: Current Theories, Research, and Future Directions," in *Annual Review of Psychology 2009*, eds. Susan T. Fiske, Daniel L. Schacter, and Robert Sternberg (Palo Alto, CA: Annual Review, 2009); Gary Yukl and David D. Van Fleet, "Theory and Research on Leadership in Organizations," in *Handbook of Industrial & Organizational Psychology. 2nd ed., volume 3*, eds. Marvin D. Dunnette and Leaetta M. Hough (Palo Alto, CA: Consulting Psychologists Press, Inc.), 147–197.

8. John P. Kotter, "What Leaders Really Do," *Harvard Business Review* (May–June 1990): 103–111.

9. John P. Kotter, *A Force for Change* (New York: Free Press, 1990).

10. "Merck Closing 8 Plants, 8 Research Sites," CBS News (July 8, 2010) at www.cbsnews.com (accessed July 31, 2010).

11. Kenneth Labich, "The Seven Keys to Business Leadership," *Fortune* (October 24, 1988), 58–66.

12. Cynthia Epstein, "Ways Men and Women Lead," *Harvard Business Review* (January–February 1991): 150–160; David D. Van Fleet and Julie Saurage, "Recent Research on Women in Leadership and Management," *Akron Business and Economic Review* (Summer 1984), 15:15–24.

13. Judy B. Rosener, "Ways Women Lead," *Harvard Business Review* (November–December 1990): 119–125.

14. David A. Thomas, "The Impact of Race on Managers' Experiences of Developmental Relationships (Mentoring and Sponsorship): An Intra-Organizational Study," *Journal of Organizational Behavior* (November 1990): 479–492; E. M. Van Fleet and David D. Van Fleet, "Entrepreneurship and Black Capitalism," *American Journal of Small Business* (Fall 1985): 31–40.

15. "Older Workers Chafe Under Young Managers," *The Wall Street Journal*, February 26, 1990, B1.

16. George J. Seperich and Russell W. McCalley, *Managing Power and People* (London: M.E. Sharpe, 2006).

17. John R. P. French and Bertram Raven, "The Bases of Social Power," in *Studies in Social Power*, ed. Dorwin Cartwright (Ann Arbor, MI: University of Michigan Press, 1959), 150–167.

18. Henry Mintzberg, *Power In and Around Organizations* (Englewood Cliffs, NJ: Prentice-Hall, 1983) see also Thomas A. Stewart, "New Ways to Exercise Power," *Fortune*, November 6, 1989, 52–64.

19. Gary A. Yukl, *Leadership in Organizations, 2nd ed.* (Englewood Cliffs, NJ: Prentice-Hall, 1989).

20. Bass, *Bass & Stogdill's Handbook of Leadership*; Yukl, *Leadership in Organizations*; see also John W. Gardner, *On Leadership* (New York: Free Press, 1989).

21. Bass, *Bass & Stogdill's Handbook of Leadership*.

22. Shelley A. Kirkpatrick and Edwin A. Locke, "Leadership: Do Traits Matter?" *Academy of Management Executive* (May 1991): 48–60.

23. "Executive Committee members," Sodexo website at www.sodexo.com (accessed July 5, 2010); Toddi Gutner, "Leading a Food Management Empire," *The Wall Street Journal* (June 18, 2010) at online.wsj.com (accessed June 30, 2010); Bret Thorn, "Michel Landel," *Nation's Restaurant News* (January 2000) at findarticles.com (accessed July 5, 2010).

24. Edwin A. Fleishman, E. E. Harris, and H. E. Burt, *Leadership and Supervision in Industry* (Columbus, OH: Bureau of Business Research, Ohio State University, 1955).

25. Bass, *Bass & Stogdill's Handbook of Leadership*; and Yukl, *Leadership in Organizations*.

26. Rensis Likert, *New Patterns of Management* (New York: McGraw-Hill, 1961); and *The Human Organization* (New York: McGraw-Hill, 1967).

27. Ralph M. Stogdill and A. E. Coons, eds., *Leader Behavior: Its Description and Measurement* (Columbus, OH: Columbus Bureau of Business Research, Ohio State University, 1957); see also Bass and Bass, *The Bass Handbook of Leadership*.

28. Timothy Judge, Ronald Piccolo, and Remus Ilires, "The Forgotten One? The Validity of Consideration and Initiating Structure in Leadership Research," *Journal of Applied Psychology* (2004): 36–51.

29. Gary Yukl, Steve Wall, and Richard Lepsinger, "Preliminary Report on Validation of The Managerial Practices Survey," *Measures of Leadership*, eds. Kenneth E. Clark and Miriam B. Clark (Greensboro, NC: Center for Creative Leadership, 1990), 223–237; and Yukl, *Leadership in Organizations*, Chapter 7.

30. Yukl, *Leadership in Organizations*, loc. cit.

31. D. Hosking and I. E. Morley, "The Skills of Leadership," in *Emerging Leadership Vistas*, eds. James Gerald Hunt, B. Rajaram Baliga, H. Peter Dachler, and Chester A. Schriesheim (Lexington, MA: Heath, 1988), 89–106.

32. For a description of other situational approaches, see Bass, *Bass & Stogdill's Handbook of Leadership*.

33. Fred E. Fiedler, *A Theory of Leadership Effectiveness* (New York: McGraw-Hill, 1967).

34. For a review, see Yukl, *Leadership in Organizations*.

35. Robert J. House and Terrence R. Mitchell, "Path-Goal Theory and Leadership," *Journal of Contemporary Business* (Autumn 1974): 81–98.

36. Bass, *Bass & Stogdill's Handbook of Leadership*.

37. Victor H. Vroom and Philip H. Yetton, *Leadership and Decision-Making* (Pittsburgh, PA: University of Pittsburgh Press, 1973); and Victor H. Vroom and Arthur G. Jago, *The New Leadership* (Englewood Cliffs, NJ: Prentice-Hall, 1988).

38. Victor Vroom, "Leadership and the Decision-Making Process," *Organizational Dynamics* (2000): 2–94.

39. R. H. G. Field and R. J. House, "A Test of the Vroom-Yetton Model Using Manager and Subordinate Reports," *Journal of Applied Psychology* (1990): 75: 362–366.

40. Yukl, *Leadership in Organizations*.

41. P. B. Smith, J. Misumi, M. Tayeb, M. Peterson, and M. Bond, "On the Generality of Leadership Style Measures Across Cultures," *Journal of Occupational Psychology* (1989), 62: 97–107.

42. David D. Van Fleet and David Rubinstein, "The Criterion Problem in Leadership Research: Convergence Through an Alternative Literature Review," *Arab Journal of Administrative Sciences,* Vol. 1, No. 1 (November 1993): 190–211.

43. A. B. Shani and M. Tom Basuray, "Organization Development and Comparative Management: Action Research as an Interpretive Framework," *Leadership and Organizational Development Journal* (1988): 9: 3–10; J. Misumi and M. Peterson, "The Performance-Maintenance (PM) Theory of Leadership: Review of a Japanese Research Program," *Administrative Science Quarterly* (1985): 30:198–223; J. Misumi, *The Behavioral Science of Leadership: An Interdisciplinary Japanese Research Program* (Ann Arbor, MI: The University of Michigan Press, 1985).

44. Abraham Zaleznik, "The Leadership Gap," *The Executive*, February 1990, pp. 7–22.

45. Robert J. House, "A 1976 Theory of Charismatic Leadership," in J. G. Hunt and L. L. Larson (eds.), *Leadership: The Cutting Edge* (Carbondale, IL: Southern Illinois University Press, 1977).

46. J. A. Conger, *The Charismatic Leader: Behind the Mystique of Exceptional Leadership* (San Francisco: Jossey-Bass, 1989).

47. Jane Howell and Boas Shamir, "The Role of Followers in the Charismatic Leadership Process: Relationships and Their Consequences," *Academy of Management Review* (2005): 96–112.

48. James M. Burns, *Leadership* (New York: Harper & Row, 1978).

49. Cynthia A. Montgomery, "Putting Leadership Back into Strategy," *Harvard Business Review* (2008): 54–63.

50. Bernard M. Bass, *Leadership and Performance Beyond Expectations* (New York: Free Press, 1985); B. J. Avolio and F. J. Yammario, "Operationalizing Charismatic Leadership Using a Levels-of-Analysis Framework," *Leadership Quarterly* (1990): 1:193–208.

51. Harrison M. Trice and Janice M. Beyer, *The Cultures of Work Organizations* (Englewood Cliffs, NJ: Prentice-Hall, 1993): Chapter 7.

52. William L. Gardner, "Lessons in Organizational Dramaturgy: The Art of Impression Management," *Organizational Dynamics* (1992): 51–63.

53. Mark C. Bolino, K. Michele Kacmar, William H. Turnley, and J. Bruce Gilstrap, "A Multi-Level Review of Impression Management Motives and Behaviors," *Journal of Management* (2008): 1080–1109.

54. Steven Kerr and John M. Jermier, "Substitutes for Leadership: Their Meaning and Measurement," *Organizational Behavior and Human Performance* (December 1978): 375-403.

55. Jon P. Howell, David E. Bowen, Peter W. Dorfman, Steven Kerr, and Philip M. Podsakoff, "Substitutes for Leadership: Effective Alternatives to Ineffective Leadership," *Organizational Dynamics* (Summer 1990): 20–38.

56. Fred Dansereau, Jr., George Graen, and William J. Haga, "A Vertical Dyad Linkage Approach to Leadership Within Formal Organizations: A Longitudinal Investigation of the Role Making Process," *Organizational Behavior and Human Performance* (1975): 46–78.

57. Robert P. Vecchio and Bruce C. Gobdel, "The Vertical Dyad Linkage Model of Leadership: Problems and Prospects," *Organizational Behavior and Human Performance* (1984): 5–20.

58. Kathryn Sherony and Stephen Green, "Coworker Exchange: Relationships Between Coworkers, Leader-Member Exchange, and Work Attitudes," *Journal of Applied Psychology* (2002): 542–548.

59. www.simulearn.net.

60. "Pepsi's Russian Challenge," *The Economist* (December 9, 2010) at www.economist.com (accessed December 10, 2010); PepsiCo Inc. 2009 Annual Report at www.pepsico.com (accessed December 9, 2010); Guy Chazan, Dana Cimilluca, and Betsy McKay, "Pepsi Juices Up in Russia," *The Wall Street Journal* (December 3, 2010) at online.wsj.com (accessed December 9, 2010); "Give and Take: Will Pepsi Profit by Enlisting the Public in Its Philanthropic Efforts?" *The Economist* (February 11, 2010) at www.economist.com (accessed December 10, 2010); "Pepsi Gets a Makeover: Taking the Challenge," *The Economist* (March 25, 2010) at www.economist.com (accessed December 10, 2010); Jason Bush, Christopher Hughes, and Martin Hutchinson, "PepsiCo's Big Bet on Changing Russian Tastes," *The New York Times* (December 3, 2010), B2.

# CHAPTER 18

# Employee Motivation

## LEARNING OBJECTIVES

After studying this chapter, you should be able to:

- Discuss the nature of human motivation and explain the basic motivational process.

- Identify important human needs and discuss two theories that attempt to outline the way in which those needs motivate people.

- Describe employee motivation from the perspectives of expectancy, satisfaction, equity, goal setting, and participation.

- Discuss reinforcement processes, including kinds of reinforcement and schedules of reinforcement.

- Identify several kinds of rewards, indicate how reward systems can be effective in motivation, and describe several new reward systems.

## Necessity, Innovation, and Motivation

Innovate is to introduce new methods or devices. Motivate is to impel, incite. Somehow in the lexicon of business these two words became separated. Books were written on the need for innovation and its importance in maintaining national leadership. Similar books were written about motivation and the need to look within ourselves, or how to incite others into action. So the word *innovation* became the purview of engineers, scientists, designers and inventors (Figure 18.1). Similarly, the word *motivation* found itself in management, philosophy, sociology, and psychology. Innovation was hard science; motivation was soft science.

**FIGURE 18.1** The need to develop sources of renewable energy motivates individuals and companies to innovate.
© Pixel 4 Images/www.Shutterstock.com

This separation is unfortunate. The two words belong together. It is the motivation to be successful that drives innovation.

The Great Recession (2008–2010) affected all businesses, but it did so in different ways. The restaurant industry witnessed dramatic decreases in "foot traffic." Some restaurants were affected more than others, e.g. fine-dining establishments more than fast-food restaurants. Most restaurants tried to regain customers with a "tried and true" technique—advertising, coupons and specials—with mixed success. Others did something truly innovative. They turned the "foot traffic" around. Their innovation: If the customer will not come to us, we will go to them.

In Los Angeles, the venerable Canter's Deli, a fixture on Fairfax Avenue since 1931, began serving potato pancakes and matzo-ball soup from a truck. The upscale Border Grill restaurant with its celeb chef, has had two trucks serving gourmet tamales in paper cups so they're convenient for pedestrians to eat.

"Ten percent of the top 200 chains will have trucks on the road within the next 24 months," predicted Aaron Noveshen, a restaurant industry consultant.

"I became a food-truck-crazy maniac like everybody else because I couldn't figure out what in the world was making people stand in line for 45 minutes. I was thinking these people should be going to Sizzler," suggested Kerry Kramp, CEO, Sizzler USA. Now Sizzler is outfitting a truck so Sizzler can come to the customers.

Many brick-and-mortar eateries have added mobile units in recent years, and more are expected to do the same, including national brands. By the end of 2011, "you'll begin to see some big brands rolling this out," said Robert Stidham, president, Franchise Dynamics LLC, a company that helps businesses develop franchises.

In 2012, the National Restaurant Association devoted 1,500 square feet of convention floor show space for truck exhibitors. While the trucks are priced between $80,000–$150,000, the investment is far less than a traditional storefront restaurant. Now some startups are literally "starting up." Van Leeuwen

Artisan Ice Cream started with one truck and expanded to three trucks. In two years of operation the company had sales of $900,000 and a profit of $300,000.

Lack of mobility should not impede innovation. Try a new pricing model as an innovation. "You can pay what you like. We will tell you the recommended price for your meal, but it is up to you if you want to pay that, a bit more or less," said two female greeters at the door of a Panera restaurant in St. Louis. The goal is to let customers who are feeling the strain of the weak economy dine with dignity among regular customers, with none of the stigma of the soup kitchen.

About 4,000 people a week visited the restaurant, operated as a non-profit entity under the brand "Panera Cares." Sixty-five percent pay the recommended amount, and the remainder split evenly between over-payers and those who pay less or nothing. There is a one meal per person limit. Panera, a company formerly under Au Bon Pain, has seen double-digit growth in same-store sales.

Panera has developed under Ron Shaich, who founded Au Bon Pain (ABP). He stepped down as CEO of ABP to operate Panera, which at the time was a smaller company. In 2009, there were 1,400 stores, and he plans to open 80 more at a time when everyone else is forming the wagons in a circle because of the economy. Earnings per store? They earn around $2,000,000 per store.

What this illustrates is that motivation and innovation are inseparable . . . and soft. They are soft because each is a function of interpretation—or is it imagination? But that's another story. As customers' motivation to eat out changes, restaurants' motivation to provide only one form of service must also change.[1]

## INTRODUCTION

Many contemporary managers are just beginning to learn that the success of any business is largely dependent on a manager's ability and motivation to innovate, plus the employees' motivation and willingness to work together in the best interests of the organization. After Hurricane Katrina flooded the Domino Sugar Refinery in Chalmette, LA, its employees helped to get it going again. Domino's vice president, Mickey Seither, stated, "We can fix anything; we can rebuild anything, but if we don't have employees, it's for naught."[2] John Mackey, founder and CEO of Whole Foods, suggests that managers should take care of employees who in turn take care of customers and, by doing so, take care of shareholders.[3] For employees to do this, though, three conditions must be met: (1) the employees must know how to do their jobs; (2) they must have the proper tools, materials, and equipment to do their jobs; and (3) they must want to do their jobs well. That third factor is motivation. In a tight labor market, keeping a job may be enough to properly motivate employees to help their bosses innovate.

In this chapter, we explore the topic of employee motivation in detail.[4] First we examine the nature of motivation, then identify important human needs that are relevant to the workplace and investigate various complexities of human motivation. We continue by discussing reinforcement processes and conclude with a summary of how reward systems affect motivation.

# The Nature of Human Motivation

Let us define **motivation** as the set of processes that determine behavioral choices.[5] Note the word *choices*. You can choose whether to study a lot, study a little, or not study at all. Managers can choose whether to innovate and keep their companies operating or choose not to change as customers' purchasing habits change during bad economic times. Our concern here is how to create conditions which motivate essentially all employees to choose to help the manager make the necessary changes that enable the company, and hence the employees, to succeed.

## FOOD FOR THOUGHT 18.1

You will never plow a field if you only turn it over in your mind.

*—Irish Proverb*

## Historical Perspectives

Managers have been aware of the importance of employee motivation for decades. In general, their thinking about motivation has progressed through three distinct stages.

The *traditional view*, popular during the era of scientific management (1880s–1930s), was very simplistic. The dominant opinion in those days was that employees worked only for economic reasons.

Presumably, people found work unpleasant and did it only for money; so the more people were paid, the harder they would work. Although the importance of money should not be underestimated, managers soon recognized that money was only one of several factors that led to motivation.

The *human relations view*, which was part of the human relations school of thought, held that social forces were the primary determinants of motivation. In particular, the adherents of this view believed that the more satisfied people were with their jobs, the harder they would work. As we see later in this chapter, this assumption is also extremely simplistic and not often true.

The *human resources view* is reflective of most contemporary thinking and takes the most positive attitude toward employee motivation. This philosophy argues that people are actually resources that can benefit the organization that they want to help, and that managers should look on them as assets. These notions relate to current interests in employee participation, workplace democracy, and so forth.[6]

## The Motivational Process

Exactly how does motivation occur? Although the complete set of processes is quite complex and not totally understood, we can devise a general framework for the motivational process, which is illustrated in Figure 18.2.

The starting point in the process is **need**—the drive or force that initiates behavior. People need

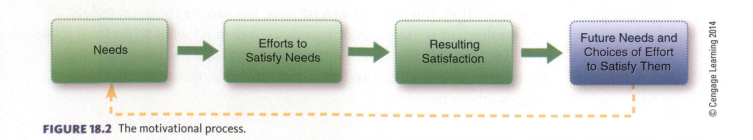

**FIGURE 18.2** The motivational process.

© Cengage Learning 2014

recognition, feelings of accomplishment, food, affection, and so forth. When our needs become strong enough, we engage in efforts to fulfill them. For instance, suppose you began to experience pangs of hunger at ten o'clock this morning. By noon the pangs became too great to ignore, so you went looking for a food source, probably a restaurant.

As a result of such efforts, people experience various levels of need satisfaction. If you had a good meal, you were no longer hungry; if only a candy bar, you will soon need to satisfy the hunger pangs again. The extent to which people find their needs satisfied then influences their future efforts to satisfy the same needs. For example, if the meal was filling but not particularly tasty, you may look for a different restaurant the next time you get hungry.

Obviously, then, the motivational process is a dynamic one. We always have a number of needs to satisfy, and we are always at different places in the process of satisfying them. Likewise, different time frames are involved. Satisfying your hunger may take only a couple of hours, but satisfying the need to accomplish meaningful work could take months or years. At any rate, the starting point is always the same: need. Now let us explore this concept in more detail.

## Important Human Needs

Needs are the starting point in all motivated behavior. Both our biological craving for food and water and our emotional longing for companionship are needs. All people have many needs, even in the workplace; and for many people work itself is an important need.[7] Several different theories exist that describe human needs in the workplace. We will examine two of the more popular ones.

### The Need Hierarchy

Although there are several different theories about needs, the one most familiar to managers is Abraham Maslow's **need hierarchy**.[8] Maslow argued that humans have a variety of different needs which he classified into five specific groups and then arranged in a hierarchy of importance. Figure 18.3 illustrates the basic framework of that hierarchy.

At the bottom are the **physiological needs**— the things we need to survive, such as food, air, and sufficient warmth. In the workplace, adequate wages for food and clothing, reasonable working conditions, and so forth are generally thought to satisfy these needs.

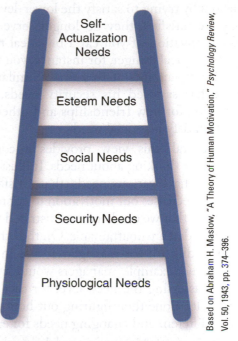

**FIGURE 18.3** Maslow's need hierarchy.

Next are **security needs**, which reflect the desire to have a safe physical and emotional environment. Job security, grievance procedures, and health insurance and retirement plans can satisfy security needs.

Third in the hierarchy are **social needs**—the need for belongingness. These include the desire for love and affection and the need to be accepted by our peers. Making friends at work and being a part of the team are common ways in which people satisfy these needs in the workplace.

**Esteem needs** come next. These actually comprise two different sets: (1) the needs for recognition and respect from others and (2) the needs for self-respect and a positive self-image. Job titles, spacious offices, awards, and other symbols of success help satisfy the externally focused needs, whereas accomplishing goals and doing a good job help satisfy the internally focused needs.

Finally, at the top of the hierarchy are the **self-actualization needs**—the needs to continue growing, developing, and expanding our capabilities. Opportunities to participate, to take on increasingly important tasks, and to learn new skills may all lead to satisfaction of these needs. Many people who break away from jobs in large corporations to start their own businesses may be looking for a way to satisfy their self-actualization needs.

The manner in which the hierarchy is presumed to work is really quite simple. Maslow suggests that

Based on Abraham H. Maslow, "A Theory of Human Motivation," *Psychology Review*, Vol. 50, 1943, pp. 374–396.

we start out by trying to satisfy the lower-level needs. As they are satisfied, they no longer serve as catalysts for motivation. If you eat a big meal to satisfy your physiological hunger, for instance, you will stop looking for restaurants at this time. Similarly, if an employee has satisfied his security needs, he will begin to look for new friendships and other opportunities to satisfy his social needs.

In general, this view provides a convenient framework for thinking about needs. It illustrates the idea that we have various needs, that satisfaction of those needs decreases our motivation to get more satisfaction, and that we can become frustrated trying to satisfy needs that are unattainable. On the other hand, this view is difficult for managers to use in a clear-cut fashion. For example, managers would have a difficult time assessing the need level for each of their employees, let alone then figuring out how to satisfy the many different and changing needs for each.

Criticisms of Maslow's need-hierarchy led to the development of ERG Theory.[9] This theory collapses Maslow's needs into three levels—existence, relatedness, and growth—and suggests that more than one can be active at any given time. (Maslow had insisted that only one could motivate at a time.) Research tends to be more supportive of this conceptualization, but it, too, is of limited usefulness to practicing managers.[10]

## The Two-Factor View

Another popular way to describe employee needs is by using the **two-factor view**, which was developed by Frederick Herzberg in the 1960s.[11] This model grew from a study of 200 accountants and engineers in Pittsburgh. Prior to the study, it was commonly believed that employee satisfaction and dissatisfaction, and thus motivation and lack of motivation, were at opposite ends of the same dimension. That is, people were satisfied, dissatisfied, or something in between. However, the Pittsburgh study uncovered evidence that satisfaction and dissatisfaction are considerably more complex than this. The researchers found that one set of factors influenced satisfaction

and an entirely different set of factors influenced dissatisfaction, hence the term *two-factor*.

As shown in the two-factor model in Figure 18.4, at the top is the dissatisfaction dimension and some of the factors found to affect it. For example, when pay and security, supervision, working conditions, and so forth are deficient, employees tend to be "dissatisfied." When these factors are adequate, however, employees are not necessarily satisfied. Instead, they are simply "not dissatisfied."

The bottom of the exhibit shows the other dimension, satisfaction. Factors such as achievement, recognition, and responsibility influence this dimension. When these factors are present, employees should be "satisfied." When they are deficient, though, employees are not necessarily dissatisfied but merely "not satisfied."

The two-factor theory carries some very clear messages for managers. The first step in motivation is to eliminate dissatisfaction, so managers are advised to make sure that pay, working conditions, company policies, and so forth are appropriate and reasonable. Then they can address motivation itself. But additional pay, improvements in working conditions, and so forth will not totally accomplish this. Instead, managers should strive to provide opportunities for achievement, growth, and responsibility. The theory predicts that these things in turn will enhance employee motivation.

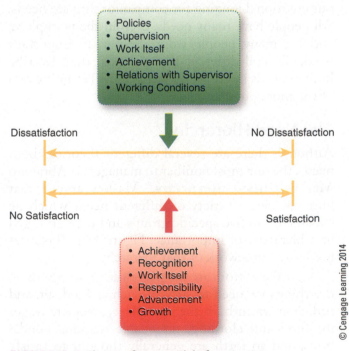

**FIGURE 18.4** The two-factor model of motivation.

© Cengage Learning 2014

The two-factor model has been the source of considerable debate. On the one hand, it has not always been supported by research and is somewhat arbitrary in its classification of factors with some items appearing on both dimensions. On the other hand, it does provide a useful and applicable framework for managers to use.[12]

## Affiliation, Achievement, and Power

Three specific needs incorporated in the need perspective warrant additional discussion: affiliation, achievement, and power.

**Affiliation Need.**  Affiliation refers to the need that most people have to work with others, to make friends in the workplace, and to socialize. Work settings that deprive people of social interaction may lead to dissatisfaction and low morale. The need for affiliation is similar to Maslow's social needs and Herzberg's interpersonal relationships.

**Achievement Need.**  Another important employee need is the need for **achievement** or the desire that some people have to excel or to accomplish some goal or task more effectively than they did in the past.[13] This need parallels Maslow's need for self-actualization and Herzberg's achievement factor. Research has indicated that people with a high need for achievement tend to have four common characteristics.

First, they set moderately difficult goals. A sales representative for Sara Lee, for instance, may set a sales goal of 115 percent of last year's sales. The goal is moderately difficult to reach but can be accomplished with hard work. Josh Summer, CEO of Summer Farms, and his siblings may set a goal of buying two existing farms in the area as a rapid means of expanding the business.

Second, people with a high need for achievement want immediate feedback. A new department manager at a Winn-Dixie store who calls the store manager every morning to learn how her department did the day before may well have a strong need to achieve. Josh Summer will quickly approach other family members with his idea, seek financial information and legal counsel, and determine if the other farm enterprises would be willing to sell.

Third, people with high needs to achieve tend to assume personal responsibility. Suppose a number of managers at Welch Foods are formed into a task force to study ways to improve productivity. If one of them continually volunteers to do the work for the entire group, he may be a high achiever. Josh, the eldest Summer son, showed signs of being a high achiever and soon moved to the top of the family business.

Fourth, such people are often preoccupied with their tasks. An engineer who thinks about her job at Potash Corporation while taking a shower, eating breakfast, and driving to work every day may also have a high need to achieve. The founder of Hartco and the landscape architect (designer) at Green Things may fit this model.

Researchers have estimated that only about 10 percent of the American population has this need, but there is evidence that the desire to achieve can be taught to people.[14]

**Power Need.**  The third need, power, is the desire to control a situation and the behavior of others. An organization with lots of managers with high-power needs would not be very effective. People with high-power needs tend to be high performers and have good attendance records. Supervisors tend to have high-power needs.[15] Further, there is some evidence that more successful managers have high-power needs than do less successful ones.[16]

## Complex Models of Employee Motivation

Although an understanding of basic human needs is a necessary starting point for enhancing motivation, managers also need to have a more complete perspective on the complexities of employee motivation. They must understand why different people have different needs, why individuals' needs change, that need frustration can lead to possible dysfunctional behavior, and that employees choose to try to satisfy needs in different ways. There are several useful theories for understanding these complexities.

### The Expectancy Model

The **expectancy model** is perhaps the most comprehensive model of employee motivation, but its basic notion is simple: Motivation is a function of how much we want something and how likely we think we are to get it.[17] As an example, consider a new college graduate looking for his first job in business. First he hears about an executive position with Diamond Foods with a starting salary of $200,000 per year. He wants the job, but he doesn't apply for it because he "knows" he has no chance of getting it. He then hears about a job with a local Safeway store, carrying bags of groceries for customers. He thinks he could get the job, but again he doesn't apply, this time because he doesn't want it. Finally, he hears about a management trainee position with Oscar Meyer. He will probably choose to apply for this job because it is similar to what he wants and he thinks he stands a good chance of getting it.

The problem, of course, is that in many situations we have various outcomes—some bad and some good—to consider. Suppose a recent graduate ends up with two reasonable job offers in a big city. His choice is more difficult, as one pays a little more and is in a more modern facility but also is a 30-minute longer drive each way. Is the higher pay and newer facility sufficiently motivating to offset the extra driving time, the cost of gas plus wear and tear on his car, and the increased probability of experiencing an automobile accident?

Figure 18.5 illustrates the basic expectancy model.[18] The theory holds that motivation leads to effort, which in conjunction with ability and environmental forces such as the availability of materials and equipment, leads to performance. Performance, in turn, has multiple outcomes. For example, high employee performance can result in a pay increase, a promotion, and better job assignments. However, it can also lead to stress and to the resentment of less successful colleagues. Therefore, the employee must choose how much effort to exert. She may weigh the potential outcomes and decide that the raise, promotion, and better assignments are more important to her than putting up with the stress and resentment, so she then exerts maximum effort to achieve those things. Another employee

in the same situation, though, might put a higher premium on reducing stress and avoiding resentment, so he would consequently exert less effort.

A real example of how the expectancy model works can be drawn from the Chaparral Steel Company, where pay is tied directly to output, promotions are based on merit rather than seniority, all employees participate in a lucrative profit-sharing plan, and all employees have a voice in decision making. Thus, Chaparral is making it easy for workers to figure out what outcomes are available to them and how best to achieve those outcomes.[19]

In summary, the expectancy model implies that managers should (1) recognize that employees have different needs and preferences, (2) try to understand each employee's key needs, and (3) help employees determine how to satisfy each of their needs through performance.

## Performance and Satisfaction

Managers should also recognize the complexity of the relationship between performance and satisfaction. In Chapter 2 we noted the human relations theory viewpoint that employee attitudes, such as satisfaction, would lead to changes in employee behaviors, such as performance. We also noted that this thinking is now considered inaccurate and overly simplistic.

Researchers now believe that performance properly rewarded leads to satisfaction. Thus performance is presumed to occur before satisfaction.[20] At first this seems unbelievable, but consider how you evaluate your classes each semester. When someone asks you how you

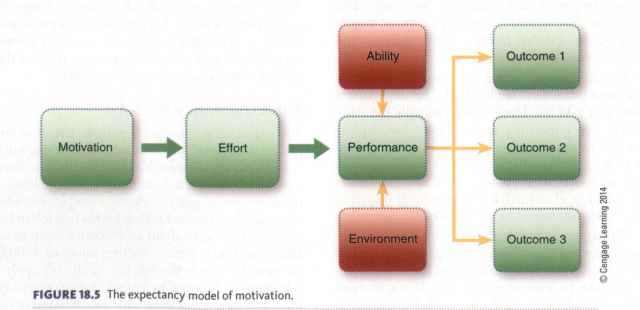

**FIGURE 18.5** The expectancy model of motivation.

© Cengage Learning 2014

feel about a certain class after the first week, you usually have a fairly neutral reaction. After you have taken the first exam, though, your attitude is somewhat more intense. If you get an A on the exam, you are likely to say the class is great. Under other circumstances, you may come up with a less favorable evaluation.

The same process occurs in work settings. During the early stages of employment, people tend to have fairly neutral attitudes toward the organization and their jobs. After they have worked a while and received various rewards (both extrinsic, like salary increases, and intrinsic, like a feeling of accomplishment), however, their attitudes become more extreme. For example, if Lionel works hard and subsequently gets praise and a pay raise, he is likely to express favorable attitudes toward the organization. But if another employee feels she worked just as hard but received only limited recognition and a small raise, she will be inclined to have less favorable attitudes.

Thus, it is more accurate to argue that performance, through the reward system, leads to satisfaction rather than the reverse. This raises the notion of equity. Much of how the two employees feel has to do with equity. That subject is discussed next.

### FOOD FOR THOUGHT 18.3

"People become really quite remarkable when they start thinking that they can do things."

—*Norman Vincent Peale*

## Equity in the Workplace

Another complex perspective on employee motivation is the role of **equity**, or fairness, in the workplace. Equity has been found to be a major factor in determining employee motivation.[21] Its power is demonstrated visibly in the sports arena. For example, until 1989, there were no baseball players being paid $3 million a year. However, when Will Clark, one of the game's top players, negotiated a new contract calling for such a salary, others followed suit almost immediately. As a result, by the end of that year, there were ten players earning a salary of more than $3 million.[22]

Although equity in the workplace is perhaps less visible, it is also important. Figure 18.6 illustrates how this works. First, each employee contributes to and gets things from the workplace. We contribute our education, experience, expertise, and time and effort. In return we get pay, security, recognition, and so forth. Given the social nature of human beings, it should come as no surprise that we compare our contributions and rewards with those of others. As a result of this comparison, we may feel equity or inequity; that is, we may feel that we are treated fairly or that we are not. Today, strong feelings of inequity are expressed on the part of major-league sports players not only toward one another but also toward their "employers" or owners, as evidenced by the 2011 NBA lock out and other similar game-canceling disputes. Do these feelings result, however, from the dissatisfaction of successful but lower-paid players toward the extremely high-paid super players rather than the expressed dissatisfaction with profit splits between owners and players?

We should note, of course, that everyone's contributions and rewards do not have to be the same for equity to exist. If one employee has a college degree, ten years on the job, and is a good performer, whereas another employee has only a high school diploma, little experience, and is only an average performer, the second employee should expect to be paid less. For equity to exist, people must perceive that the relative proportion of all rewards and all contributions

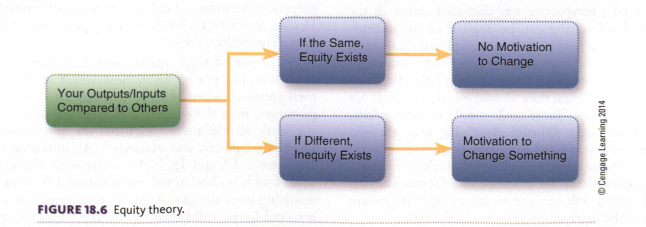

**FIGURE 18.6** Equity theory.

is equal. And if equity is present, people are generally motivated to keep everything as it is.

In contrast, if people experience inequity, they are generally motivated to change something. If someone believes that she is underpaid relative to a colleague, her only option is to decrease her contributions since she does not control her rewards. Working harder would only exacerbate the perceived inequity for, while it might lead to more rewards, it would also increase even further her contributions. Thus, the employee will decrease her efforts. She may, of course, try to exercise some control over rewards by asking for a raise or for her colleague's pay to be cut. She could also try to convince the colleague that he should work harder to justify his pay, or she can rationalize the problem away in some manner. And, of course, she also can quit so that the issue no longer exists. Regardless of which option she chooses, she will be uncomfortable with inequity and will try to do something about it.

This viewpoint also conveys several clear messages to managers. First, people should be rewarded according to their contributions, which includes their performance as well as their knowledge and skills. Second, managers should try to ensure that employees feel equity by assuring that the reward allocation system is linked to performance and clearly understood. Finally, managers should be aware that feelings of inequity are almost bound to arise. When they do, managers must be patient and either correct the problem if it is real or help people recognize that things are not as inequitable as they may at first appear.[23] Relatively recent modifications of equity theory may render it even more useful to managers in the future.[24]

## Goal Setting Theory

Still another useful perspective on employee motivation is **goal-setting theory**.[25] Goal setting from a planning perspective was discussed earlier in the book. Goal setting can be applied on an individual level as a way to increase employee motivation (Figure 18.7).[26] The starting point is for managers and their subordinates to meet regularly. As a part of these meetings, they should jointly set goals for the subordinate. These goals should be very specific and moderately difficult. Assuming they are also goals that the subordinate will accept and be committed to, the employee is likely to work very hard to accomplish them. The evidence thus far suggests that goal setting will become an increasingly important part of the motivational process in the future.[27]

**FIGURE 18.7**  Vision/mission/goals are linked together to motivate.

## High-Involvement Management

Finally, managers are using **high-involvement management** more and more as a way to enhance motivation in the workplace. They are finding that when employees are given a greater voice in how things are done, they become more committed to the goals of the organization and are willing to make ever-greater contributions to the success of the business. A first step in achieving high involvement is getting each employee involved in his or her job. At Summer Farms low-level managers are involved in such decisions as whether the crop is ready to harvest, whether the employees should work weekend days to save a crop, which equipment needs to be replaced, and so on. And all employees are encouraged (and rewarded for) safety suggestions that are adopted. Quality circles, discussed later in the book, are one popular method of increasing employee participation. This approach solicits employee volunteers who meet regularly in an attempt to first identify and then recommend solutions for quality-related problems in the workplace. Many businesses are attempting to encourage participation as a means of enhancing competitiveness.[28] Honda, for instance, has been using participation to gain a competitive edge.[29]

The use of high-involvement management is more than just participation, however. It involves a total approach whereby the organization's structure, processes, reward systems, and methods of doing the work all are altered to focus on information, knowledge, power, and rewards.[30] As indicated in Chapters 2, 12, and 13, high-involvement management builds on horizontal organizational structures involving work designs. A lot of group processes are involved (more will be said about the use of groups

## A FOCUS ON AGRIBUSINESS
### Motivation in Agribusiness

In an effort to improve conditions in the production and marketing of lamb, a cooperative was formed. Called American Lamb Producers, Inc. (ALPI), it was set up as an investor-owned firm with most of the stock controlled by lamb producers. It was a rather dismal failure. Among the reasons for that failure were conflict among employees, poor coordination, and lack of employee motivation. If employees aren't motivated, little gets accomplished.

Dhanuka Agritech is well aware of the importance of motivation. The company strives to understand its employees and their needs. Using a vigorous feedback system to develop a satisfied workforce, it constantly evaluates parameters to monitor employee motivation and commitment.

Keeping a workforce motivated is even more important in hard times to avoid experiencing "Zombie Turnover"—a condition where demotivated employees want to quit but cannot because of family or labor market conditions. They become "zombies"—they come to work, but they are not really there.[31]

in the next chapter) and reward systems are based on skills and performance. High-involvement management holds tremendous promise as a way of tapping the enormous potential in diverse organizations.

## Reinforcement Processes

A final question about motivation concerns how and why behaviors stay the same or change. Consider the case of two new workers: one starts out as an average performer and continually gets better; the other starts out as a top performer but later slacks off or even becomes a poor performer.[32] What has happened? The answer probably involves reinforcement processes.

The idea of **reinforcement** suggests that future behavior is shaped by the consequences of current behavior. If their current behavior leads to rewards, individuals are likely to engage in the same behaviors again. But if current behavior does not lead to a reward, or if it leads to unpleasant outcomes, individuals are more likely to follow different behavior patterns in the future.

Much of what we know about reinforcement can be traced to psychologists who have studied human learning processes,[33] but more and more people have come to see how clearly the concept relates to organizational settings.[34] The following sections describe kinds of reinforcement and schedules that managers can use to provide them.

## Kinds of Reinforcement

As shown in Figure 18.8, there are four basic kinds of reinforcement: positive reinforcement, avoidance, extinction, and punishment.[35]

**Positive reinforcement** is a reward, or desirable outcome, that is given after a particular behavior.[36] For instance, suppose that a supervisor notices a worker doing an extraordinarily good job. He stops and tells the worker what a good job he is doing and then recommends to his boss that the worker should get a small pay raise. The praise and the pay raise are positive reinforcements and as a consequence the worker is likely to continue to work hard.

**Avoidance** or **negative reinforcement** also increases the likelihood that someone will repeat a desirable behavior, but it uses a different perspective. In this case, the employee is allowed to avoid an unpleasant situation because of good performance. If a company has a policy that employees who are late for work get penalized and if all employees come to work on time, no penalties are imposed. As long as the threat continues, employees will be motivated to be on time every day.

**Extinction** is used to weaken behavior, especially behavior that has previously been reinforced. Consider the manager of a small office who in the past allowed employees to come by whenever they wanted to "shoot the breeze." Now the office staff has grown so large that she must curtail this practice. Rather than close her door, respond as in the past, or correct the employee for interrupting, the

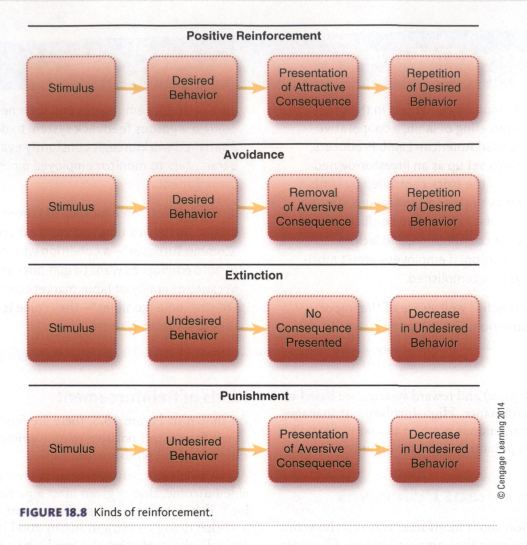

**Positive Reinforcement**

| Stimulus | → | Desired Behavior | → | Presentation of Attractive Consequence | → | Repetition of Desired Behavior |

**Avoidance**

| Stimulus | → | Desired Behavior | → | Removal of Aversive Consequence | → | Repetition of Desired Behavior |

**Extinction**

| Stimulus | → | Undesired Behavior | → | No Consequence Presented | → | Decrease in Undesired Behavior |

**Punishment**

| Stimulus | → | Undesired Behavior | → | Presentation of Aversive Consequence | → | Decrease in Undesired Behavior |

© Cengage Learning 2014

**FIGURE 18.8**  Kinds of reinforcement.

manager just ignores the undesired behavior by remaining cordial but still continuing to work at her desk until employees get the message that she does not have time to "shoot the breeze." Of course, she must also work to guard against resentment and a loss in communication.

**Punishment** is also used to change behavior. Common forms of punishment in organizations include reprimands, discipline, and fines. Since punishment usually engenders resentment and hostility, managers should usually use it only as a last resort. Suppose that an employee has been late for work three times in the last week with no valid excuses. His boss might choose to reprimand him, explaining that another absence within the next six months will lead to a suspension.

## Schedules of Reinforcement

For managers to use reinforcement effectively to enhance motivation, they must know when to provide it. Five basic schedules of reinforcement are available.[37]

Under the **continuous reinforcement schedule**, the manager provides reinforcement after every occurrence of the behavior; that is, the supervisor praises her subordinate every time she sees him doing a good job. Obviously, in this type of situation the power of the praise as reinforcement will rapidly diminish since it is so common and easy to get.

Under a **fixed-interval schedule**, the manager provides reinforcement on a periodic basis, regardless of performance. An example is the Friday paycheck that many employees get. The checks are obviously important to them, but they really do not affect their performances directly since they receive the checks regardless of how hard they work. Therefore, this schedule is also of limited value as a way to enhance motivation.

The **variable-interval schedule** also uses time as a basis for reinforcement, but the time intervals between reinforcements vary. If praise follows this non-regular schedule or is tied to random office visits by the manager, it tends to be more powerful.

Under the **fixed-ratio schedule**, the manager provides reinforcement on the basis of number of behaviors rather than on the basis of time. However, the number of behaviors an employee must display to get the reinforcement is constant. For example, suppose H.E. Butt Grocery decides to get more customers to use its discount cards. Each sales clerk is asked to solicit new applicants, and the clerks are given 50 cents for every five applications that are completed. The idea is that each clerk will be highly motivated to get new applicants because each one brings him or her closer to the fifth application needed for another reward. Many retailers use this schedule by rewarding employees for the number of applicants they sign up for the store's credit card.

Finally, there is the **variable-ratio schedule**, which is generally the most powerful one for enhancing motivation. Under this arrangement, the manager again gives reinforcement on the basis of behaviors, but the number of behaviors an employee needs to display to get the reinforcement varies so the employee cannot predict when reinforcement will occur. A variable ratio of 1:3 means that *on average*, one out of every three behaviors will be rewarded. It might be the first; it might be the third; it might even be the fourth—as long as it averages out to one in three. Thus, the subordinate is motivated to continue to work hard, because each incident raises the probability (though not the certainty) that the next will bring praise. (Slot machines in Las Vegas use this schedule.)

## Reward Systems and Motivation

Regardless of what motivational model or perspective a manager uses, it is typically made operational through the organization's reward system.[38] In this section we consider what kinds of rewards are usually available, the characteristics of effective reward systems, and some interesting reward systems currently in development.

### Kinds of Rewards

From the standpoint of the employee, a **reward** is anything the organization provides in exchange for services. Clearly, however, outcomes vary in terms of their potency as rewards. One category of reward includes base pay, benefits, holidays, and so forth. These rewards are not tied to performance. A second category includes pay increases, incentives, bonuses, promotions, status symbols (bigger offices, reserved parking spaces), and attractive job assignments. These

**FIGURE 18.9** A senior manager congratulates a colleague on receiving an award for a cost-saving suggestion that will save the company about $400,000 annually.

© iStockphoto/Daniel Laflor

are rewards in the truest sense; they represent significant forms of positive reinforcement and satisfy many of the basic needs of most employees (Figure 18.9).

Some fairly inexpensive, simple reward systems have had good success. One of your textbook authors helped achieve outstanding performance from individuals involved in a three-year project by handing out stars periodically. Everyone involved received a gold star once a month, but those who made some sort of special effort during a month would get a silver star. Those who got three silver stars in a row would then get larger red stars. People kept their stars visible at their workstations and took considerable pride in showing them to others.

The Holiday Inn at Briley Parkway in Nashville, Tennessee, evolved The Apple Award for its personnel to reward outstanding service. Guests or fellow employees could nominate any employee for an award. A committee would then review the nominations to assure that they were indeed for outstanding service. If so, the employee received a red apple pin. When more than one person in a department earned a recognition, the whole department might receive a basket of real apples to share. Personnel displayed their pins with great pride.[39]

## Effective Reward Systems

If reward systems are to serve their intended purpose, they must be effective. Effective reward systems tend to have four basic characteristics, as summarized in Table 18.1. First, they must satisfy the basic needs of the employees. Pay must be adequate, benefits reasonable, holidays appropriate, and so forth. Second, the rewards must be comparable to those that other organizations offer in the immediate area. If a Dow Chemical plant is paying its workers $10 an hour and a DuPont plant down the road is paying $12, employees at the Dow plant will always be looking for openings at the DuPont facility. If the major film studios in Hollywood give their employees more holidays and also close shop a half day before each holiday, other firms in that industry may experience a decrease in worker morale, excessive absences, a decline in productivity around each holiday, and more difficulty in retaining employees.

Rewards also must be distributed in a fair and equitable fashion. As we have discussed, people have a need to be treated fairly. If rewards are not distributed in an equitable fashion, employees will be resentful. Finally, the reward system must be multifaceted, which means that it must acknowledge that different people have different needs. A range of rewards must be provided, and people need to be able to attain rewards in different ways. For example, a marketing manager and a financial manager must each have an opportunity for promotion into the ranks of middle management.

## New Reward Systems

In an effort to compete with other firms for good employees, many organizations are experimenting with new kinds of rewards and new ways to achieve them.[40] One kind of contemporary system, for example, ties pay more directly to performance than was done in the past. And Du Pont has worked at developing a new reward system as a way to boost productivity.[41]

Another innovation is the all-salaried workforce. Under this arrangement, all workers are paid by the month rather than by the hour. People monitor their own work hours, which eliminates the need for time clocks. Both Gillette and Dow Chemical have used this plan. However, federal regulations regarding changing hourly personnel to salaried personnel to avoid paying overtime must be carefully examined to assure that this system will reduce costs and improve productivity.

Skill-based job evaluation systems are also becoming popular. In this case people are paid according to their levels of proficiency or skills. Teachers with master's degrees, for instance, are paid more than teachers with only undergraduate degrees, even though they do the same jobs. General Foods and Texas Instruments have also used this approach, increasing an employee's pay whenever he or she masters a new job or skill.

### FOOD FOR THOUGHT 18.4

The reasons for working are as individual as the person working. But we all work because we obtain something that we need from work.

## An Integrated View

The numerous models discussed here suggest alternative ways of approaching an understanding of motivation. They are, however, not competing models. Each one tends to focus on one or a few aspects of motivation to obtain a fuller understanding. As shown in Figure 18.10, all of those views can be integrated.

**TABLE 18.1** Characteristics of Effective Reward Systems

| CHARACTERISTIC | EXAMPLES |
|---|---|
| Rewards satisfy the basic needs of employees. | Adequate pay, reasonable benefits, appropriate holidays. |
| Rewards are comparable to those that similar organizations in the area offer. | Pay rates of nearby companies are equal; employees receive the same holiday time as employees in comparable positions in other organizations. |
| Rewards are distributed fairly and equitably. | Employees who work overtime for a special project receive extra pay or compensatory time off; employees in comparable positions receive equal rewards for similar work. |
| The reward system is multifaceted. | A range of rewards are given: pay, benefits, promotion, privileges and so on; rewards may be earned in different ways. |

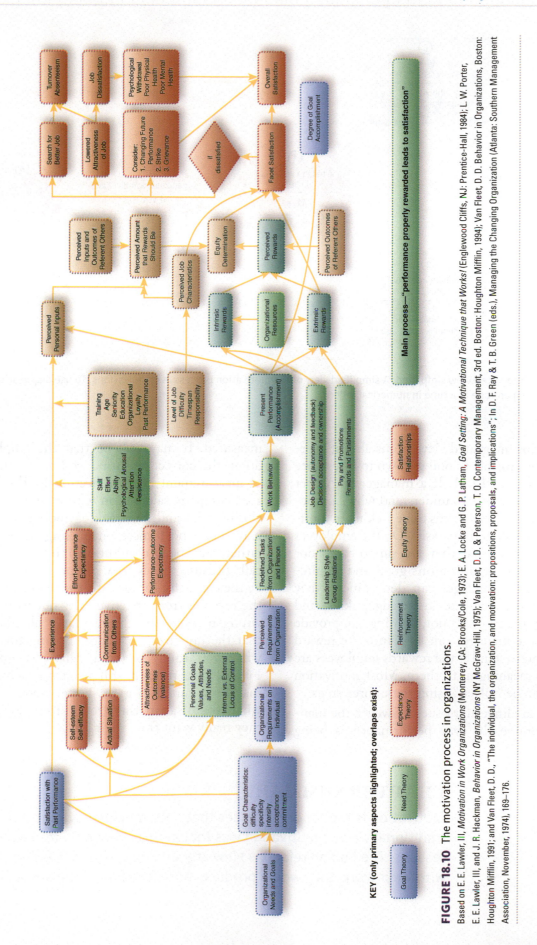

**FIGURE 18.10** The motivation process in organizations.

Based on E. E. Lawler, III, *Motivation in Work Organizations* (Monterey, CA: Brooks/Cole, 1973); E. A. Locke and G. P. Latham, *Goal Setting: A Motivational Technique that Works!* (Englewood Cliffs, NJ: Prentice-Hall, 1984); L. W. Porter, E. E. Lawler, III, and J. R. Hackman, *Behavior in Organizations* (NY McGraw-Hill, 1975); Van Fleet, D. D. & Peterson, T. O. Contemporary Management, 3rd ed. Boston: Houghton Mifflin, 1994); Van Fleet, D. D. Behavior in Organizations, Boston: Houghton Mifflin, 1991; and Van Fleet, D. D., "The individual, the organization, and motivation: propositions, proposals, and implications"; in D. F. Ray & T. B. Green (eds.), Managing the Changing Organization (Atlanta: Southern Management Association, November, 1974), 169–176.

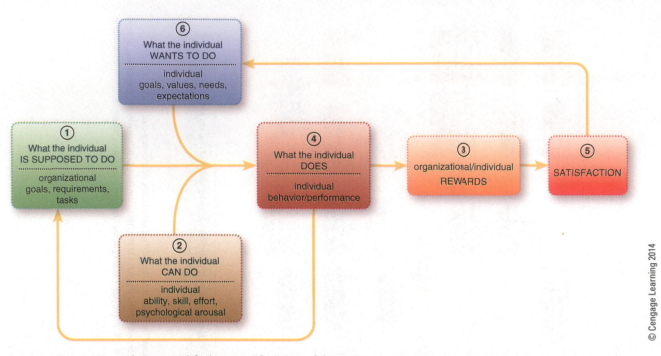

© Cengage Learning 2014

**FIGURE 18.11** Integrated view simplified. A simplified view of the motivation process in organizations. To use diagnostically, examine each element one at a time in numerical order.

While this integrated view seems almost overwhelming, it is possible to boil it down to its essence as shown in Figure 18.11. If a manager at Archer Daniels Midland wants to understand why a particular group of employees seems to be lacking in motivation, he or she should first ask: "Do the employees understand exactly what they are supposed to do? Do they know the organization's goals and requirements?" If so, the manager can determine if they actually have the requisite knowledge, skills, and abilities to do their jobs. If not, training can provide those. Next, the manager can look into the reward structure to understand if the rewards employees are receiving are meaningful to them. Most motivation "problems" can be "solved" using these three steps.

If there are lingering issues, however, the manager can examine the way performance is being measured. It may well be that the employees are doing the correct performance but that the system of measurement is not reflecting it. Think about a seed company salesperson who is highly capable of closing sales and thus making money for the company, but who focuses on making more cold calls rather than spending additional time to "close" fewer calls because that is how performance is measured by the company. In this case, the organization wants the revenue from effective sales but focuses on the wrong measure to achieve that goal. Making sure that the organization is not "hoping for A (sales), while rewarding B (cold calls)" is important.[42] Finally, if the problem still persists, expert assistance may be required to get a better understanding of what would satisfy this group of employees—what they really want from their jobs.

## CHAPTER SUMMARY

Motivation is the set of processes that determine behavioral choices. The traditional view of motivation was that people work only for economic reasons. The human relations view argued that if people were satisfied with their work, they would produce more. The human resources view suggests that people are actually resources that can benefit organizations and that managers should maintain and develop them like other assets for maximum productivity. The process of motivation begins with needs, drives, or forces that initiate behavior. The existence of needs leads to efforts to satisfy those needs. The extent to which needs are satisfied leads in turn to the levels of satisfaction that people experience, which influence their choices of future efforts to satisfy needs. Hence the process begins again.

Early theories emphasized needs and the need hierarchy. These theories, while important as starting points, are of limited value to practicing managers.

The dominant theory today is the expectancy model, which holds that motivation is a function of how much we want something and how likely we think we are to get it. Many early writers felt that if managers could satisfy the needs of employees, those employees would perform better. However, research shows that satisfaction does not cause performance. It is more accurate to say that performance properly rewarded leads to satisfaction, a key element of which is equity. Whenever someone perceives an inequity in a situation, he or she will try to do something about it; that is, the person will be motivated to reduce the perceived inequity. Goal setting and high-involvement management are also important ingredients in motivation.

Reinforcement processes are similar to the process described by the expectancy model. The theory suggests that future behavior is shaped by the consequences of current behavior. There are four basic kinds of reinforcement: positive reinforcement, avoidance, extinction, and punishment. Positive reinforcement is a reward that follows desired behavior, increasing the chances it will be repeated. Avoidance behavior is aimed at avoiding negative consequences. In extinction, nothing happens following a behavior; the behavior is ignored. Punishment, on the other hand, means that the behavior results in an undesirable consequence. The impact of these kinds of reinforcements on behavior is complicated, however, by their schedules, or the frequency with which the reinforcement occurs.

There are numerous kinds of rewards, both extrinsic and intrinsic. These include base pay, benefits, holidays, pay increases, incentives, bonuses, promotions, status symbols, praise, recognition, work assignments, and so on. Effective reward systems tend to have four characteristics: They must satisfy basic needs, be comparable to those used elsewhere, be distributed equitably, and be multifaceted.

## CHAPTER ACTIVITIES

### REVIEW QUESTIONS

1. What are the three basic historical perspectives through which motivational theory has passed?
2. What are the five basic need levels in Maslow's hierarchy of needs?
3. Summarize the basic premises of expectancy theory.
4. What is the relationship between performance and satisfaction?
5. What are the four basic types of reinforcement? What are the five schedules of reinforcement?

### ANALYSIS QUESTIONS

1. How important do you think money is in motivation? Explain.
2. Do you agree or disagree with the basic premises of the two-factor theory? Why or why not?
3. How do you think performance and satisfaction are related?
4. Explain how you personally might form equity perceptions in your role as a student.
5. Which new reward system do you think holds the most potential? Why? Why might workers oppose it?

### FILL IN THE BLANKS

1. The drives or forces that initiate behavior are called _____.
2. The set of processes that determine the choices that people make about their behaviors is called _____.
3. The need to work with others, to interact, and to have friends, is the _____ need.
4. Shaping behavior by rewarding desirable behavior but not rewarding undesirable behavior is known as _____.
5. The desire to excel or to accomplish some goal more effectively than in the past is the need for _____.
6. The model that suggests that motivation is determined by how much we want something and how likely we think we are to get it is called the _____ model.
7. The need hierarchy has five levels of needs: _____, _____, _____, _____, and _____.
8. The theory that specific and moderately difficult goals may increase motivation is called the _____ theory.
9. Four kinds of reinforcement are _____ reinforcement, _____, _____, and _____.
10. A view of motivation that suggests employee satisfaction and dissatisfaction are two distinct dimensions affected by different sets of factors is the _____ _____ view.

## ▼ CHAPTER 18 CASE STUDY
### LEAVING BEFORE THE GAME IS OVER

Most business schools spend a significant amount of instruction time teaching students how to think strategically, how to motivate a company and its employees, and how to reach (or at least strive for) the corner office. Much time is spent on "staying in the game" and setting achievable goals as well as "stretch" goals—all aimed at becoming "numero uno": the boss. It appears that 2010 may have been the year that introduced a new concept: the end game or exit strategy—or how to get off the field before the game is over.

Often, so much instruction time is spent on personal motivation and developing self-discipline—staying focused—that little or no time is spent on recognizing when the "game" is or should be over. It is just as important to recognize when the "game" has become too much and is exacting a toll far more personal and less business-oriented than was expected. Jeffrey Kindler, CEO of Pfizer, Inc, the world's largest drug company, abruptly quit as CEO, admitting that the job wore him out. He was only 55 years old. He had tried for months to convince his board of directors that he needed someone to lighten his load. The board did not act fast enough. Mr. Kindler did.

Compared with a few years ago, pressures on CEOs "are substantially different—especially in certain industries," said Steve Reinemund, retired CEO of PepsiCo. Jeffrey Sonnenfeld of the Yale University School of Management added that many CEOs now look at the corporate throne "… as a position with a limited term of office. They rarely seek to stay a minute more than a dignified decade." Sonnenfeld may be optimistic with his decade timeline.

Some CEOs wear themselves out. Sometimes enough is enough. Kellogg's CEO David Mackay announced to his board in December 2010 that he is retiring at the start of 2011. He was tired after facing two of the largest product recalls in Kellogg's history, being slapped by federal regulators for crossing the line on nutritional profiles on its packaging, failing to anticipate an aggressive campaign by its closest competitor (General Mills Inc.), watching cereal sales get battered by supermarket house brands, and coping with the flooding of a waffle-making plant. At least the board got a two-week notice.

Adding to this pressure-cooker environment is an acknowledgement that the corner office has plenty of windows. Everyone is watching and they are monitoring and measuring. The motivation to assume and stay in the office grows smaller. In 2010, Kenexa Research Institute conducted a survey to measure the effectiveness of senior corporate management based on several attributes, for example, people management skills and commitment to quality products. Of the 29,000 respondents, globally 55 percent of employees rated their senior managers highly. In the United States—the country that practically invented the "B-School"—only 56 percent of employees gave their managers high ratings.

Who finished in the top five spots? The top five scores in the leadership-effectiveness index were India with employee ratings of 72 percent, China with 71 percent, Switzerland with 63 percent, Mexico with 62 percent, and Russia with 59 percent Japan, once the icon of global competitiveness and internationally centric management, finished "dead last" with a 39 percent rating.

Sometimes the pressure prompts strange actions or the need to employ "gamesmanship" more along the lines of "brinksmanship." When the Prime Minister of Fiji, Voreqe Bainimarama, announced a 15-cent/liter tax on bottled water, Fiji Water, which bottles 3.5 million liter/month, announced it was closing the plant. This move eliminated the largest company from Fiji tax rolls. It was also an act of hari-kari for Fiji Water; it could not produce "fiji" water elsewhere. Everyone realized the "lose-lose" positions in this game and reversed themselves a day later. Newton's law—every action fosters an equal and opposite reaction—seemed to be in play.

This was not a one-time use of gamesmanship or brinksmanship. Lynda Resnick, owner/CEO of POM Wonderful LLC, launched a $10 million dollar advertising campaign against the Federal Trade Commission and the U.S. Food and Drug Administration for criticizing her company for its pomegranate juice product. The FTC and the FDA were not amused. What is the connection? Lynda Resnick and her husband, Stewart Resnick, through a holding company, Roll International, own Fiji Water. Sometime the best response to pressure is to return in kind.

Of course, this pressure-filled environment can induce forgetfulness. Mickey Drexler, CEO of J.Crew Group, the "preppy" clothing company, negotiated for almost two months with an outside company to sell J. Crew before he remembered to tell the J.Crew board of directors what he was doing. When you are under a great amount of stress, apparently it is hard to remember that you are selling the company for $3 billion dollars or at least you forget to mention it for two months.

Who bought the company? TPG Capital, which had a member on the J.Crew board of directors, and Leonard Green & Partners—and, oh yes, did we mention Mr. Drexler himself?

All of this indicates that a prize can be highly motivational and cause one to be singularly focused. So much so that once the prize is gained, a response may be, "Is that all there is?" Less philosophically, it might prompt a reality television program, "CEOs Behaving Selfishly." But for more and more CEOs, it may be that the star player just wants or needs to go to the locker room early. Maybe the media networks will offer him an on-air analyst job.[43]

### ▶ Case Study Questions

1. What theories of motivation seem useful in understanding why one might want to become the CEO of a corporation?

2. Based on the information in this case, how has the motivation to be CEO changed and how might it change in the future? See if you can find examples other than the ones mentioned in the case.

3. What might an organization do to motivate high-quality individuals to seek the CEO position?

# REFERENCES

1. Sharon Bernstein, "Food Trucks Are Rolling into the Mainstream," *Los Angeles Times* (September 9, 2010) at articles.latimes.com (accessed December 9, 2010); Sarah E. Needleman, "Restaurant Franchises Try Truckin' as a Way to Grow," *The Wall Street Journal* (October 28, 2010) at online.wsj.com (accessed December 9, 2010); Kimberly Weisel, "Spurred by a Passion for Gourmet Grub, Ben Van Leeuwen Launched a Fleet of Trucks Offering a 'Taste of Creamy Ecstasy,'" *Inc. Magazine* (October, 2010), 66; "Dough Rising," *The Economist* (October 7, 2010) at www.economist.com (accessed December 8, 2010).

2. "Domino Sugar Plant Reopens," *PBS NewsHour* (March 8, 2006) at www.pbs.org (accessed June 29, 2010).

3. Seth Lubove, "Food Porn," *Forbes*, February 14, 2005, 102.

4. Linda-Eling Lee, "Employee Motivation—A Powerful New Model," *Harvard Business Review* (2008): 78–89.

5. Lyman W. Porter, Gregory A. Bigley, and Richard M. Steers, eds., *Motivation and Work Behavior, 7th ed.* (New York: McGraw-Hill, 2003).

6. Edwin Locke and Gary Latham, "What Should We Do About Motivation Theory? Six Recommendations for the Twenty-First Century," *Academy of Management Review* (2004): 388–403.

7. Walter Kiechel III, "The Workaholic Generation," *Fortune*, April 10, 1989, 50–62.

8. Abraham H. Maslow, "A Theory of Human Motivation," *Psychological Review* (1943), 50: 370–396.

9. Clayton P. Alderfer, "An Empirical Test of a New Theory of Human Needs," *Organizational Behavior and Human Performance* (April 1969): 142–175.

10. Clayton P., *Alderfer, Existence, Relatedness, and Growth* (New York: Free Press, 1972).

11. Frederick Herzberg, "One More Time: How Do You Motivate Employees?" *Harvard Business Review* (January–February 1968): 53–62 (reprinted in Harvard Business Review, January 2003, 87–98).

12. Craig C. Pinder, *Work Motivation in Organizational Behavior* (Englewood Cliffs, NJ: Prentice-Hall, 1997). Pinder, Work Motivation.

13. David McClelland, "That Urge to Achieve," *Think*, November–December 1966, 22.

14. David McClelland, *The Achieving Society* (Princeton, NJ: Van Nostrand, 1961).

15. E. Cornelius and F. Lane, "The Power Motive and Managerial Success in a Professionally Oriented Service Company," *Journal of Applied Psychology* (January 1984): 32–40.

16. David McClelland and David H. Burnham, "Power Is the Great Motivator," *Harvard Business Review* (March–April 1976): 100–110.

17. Victor Vroom, *Work and Motivation* (San Francisco: Jossey-Bass, 1995).

18. See also David A. Nadler and Edward E. Lawler, III, "Motivation: A Diagnostic Approach," in Perspectives on Behavior in Organizations, *2nd ed.*, eds J. Richad Hackman, Edward E. Lawler, and Lyman W. Porter (New York: McGraw-Hill, 1983), 67–78.

19. Kurt Eichenwald, "America's Successful Steel Industry," *Washington Monthly*, February 1985, 42.

20. Lyman W. Porter and Edward E. Lawler III, Managerial Attitudes and Performance (Homewood, Ill.: Dorsey, 1968).

21. Richard T. Mowday, "Equity Theory Predictions of Behavior in Organizations," in Motivation and Work Behavior, eds. Steers and Porter, 91–113: see also J. Stacey Adams, "Toward an Understanding of Inequity," *Journal of Abnormal and Social Psychology* (November 1963): 422–436.

22. "Will Clack Package Zooms to $15 Million," *The New York Times* (January 23, 1990) at www.nytimes.com (accessed June 30, 2010).

23. Jerald Greenberg, "Equity and Workplace Status: A Field Experiment," *Journal of Applied Psychology* (November 1989): 606–613; Edward W. Miles, John D. Hatfield, and Richard C. Huseman, "The Equity Sensitivity Construct: Potential Implications for Work Performance," *Journal of Management* (December 1989): 581–588.

24. R. A. Cosier and D. R. Dalton, "Equity Theory and Time: A Reformulation," *Academy of Management Review* (April 1983): 311–319; and R. C. Huseman, J. D. Hatfield, and E. W. Miles, "A New Perspective on Equity Theory," *Academy of Management Review* (April 1987): 222–234.

25. "Paying Workers to Meet Goals Spreads, But Gauging Performance Proves Tough," *The Wall Street Journal*, September 10, 1991, B1, B8.

26. Yitzhak Fried and Linda Haynes Slowik, "Enriching Goal-Setting Theory with Time: An Integrated Approach," *Academy of Management Review* (2004): 404–422.

27. Gary P. Latham and Edwin Locke, "Goal Setting—A Motivational Technique That Works," *Organizational Dynamics* (Autumn 1979): 68–80.

28. Peter Österberg and Jerker Nilsson, "Members' Perception of their Participation in the Governance of Cooperatives: The Key to Trust and Commitment in agricultural cooperatives," *Agribusiness* (2009): 181–197; David J. Glew, Anne M. O'Leary-Kelly, Ricky W. Griffin, and David D. Van Fleet, "Participation in Organizations: A Preview of the Issues and Proposed Framework for Future Analysis," *Journal of Management* (1995): 395–421.

29. Louis Kraar, "Japan's Gung-Ho U.S. Car Plants," Fortune, January 30, 1989, 98–108; "The Americanization of Honda," *Business Week*, April 25, 1988, 90–96; "Honda Wins USA's Heartland," *USA Today*, December 2, 1987, 1B, 2B.

30. E. E. Lawler III, The Ultimate Advantage (San Francisco, CA: Jossey-Bass, 1992).

31. Wayne H. Howard, Kenneth A. McEwan, George L. Brinkman, and Julia M. Christensen Assistant, "Human Resource Management on the Farm: Attracting, Keeping, and Motivating Labor," *Agribusiness* (2006), 7(1): 11–26; M. Visser, "Zombie Turnover: The New Threat," *Leading Edge, Advantage Business, Inc.*, Volume 14, n.d. at www.advantagebusiness.co.nz (accessed July 15, 2010); "People," *The Dhanuka Group* at www.dhanuka.com (accessed July 13, 2010); R. D. Smith, E. G. Smith, E. E. Davis, R. A. Edwards, and G. Molina, "Contemporary Producer-Owned Lamb Processing Ventures: Lessons Learned," *RBS Research Report 167, United States Department of Agriculture, Rural Development, Rural Business-Cooperative Service* (1999).

32. Nigel Nicholson, "How to Motivate Your Problem People," *Harvard Business Review* (2003): 57–67.

33. B. F. Skinner, *Beyond Freedom and Dignity* (New York: Knopf, 1971).

34. Fred Luthans and Robert Kreitner, *Organizational Behavior Modification and Beyond: An Operant and Social Learning Approach* (Glenview, Ill.: Scott, Foresman, 1985).

35. Ibid. Luthans and Kreitner, *Organizational Behavior Modification and Beyond: An Operant and Social Learning Approach.*

36. "At Emery Air Freight: Positive Reinforcement Boosts Performance," *Organizational Dynamics* (Winter 1973): 41–50.

37. C. B. Ferster and B. F. Skinner, *Schedules of Reinforcement* (New York: Appleton-Century-Crofts, 1957).

38. Edward E. Lawler III, *Pay and Organizational Development* (Reading, MA: Addison-Wesley, 1981).

39. Personal interviews with employees conducted in December 1992 and the nomination form.

40. "Pay Raise Demands Appear to be Modest," *The Wall Street Journal*, February 28, 1989, A2.

41. Brian Dumaine, "Creating a New Company Culture," *Fortune*, January 15, 1990, 127–131; "All Eyes on Du Pont's Incentive Pay Plan," *The Wall Street Journal*, December 15, 1988, B1.

42. Steven Kerr, "On the Folly of Rewarding A, While Hoping for B," *Academy of Management Journal* (1975): 769–783. Reprinted in *The Academy of Management Executive* (1995): 7–14.

43. Joann S. Lublin and Jonathan D. Rockoff, "CEO's Stress Worried Pfizer," *The Wall Street Journal* (December 7, 2010) at online.wsj.com (accessed December 9, 2010); Joann S. Lublin, "Corner Office Turned Pressure Cooker," *The Wall Street Journal* (December 7, 2010) at online.wsj.com (accessed December 9, 2010); Ilan Brat and Paul Ziobro, "Kellogg CEO Mackay to Retire, Capping Tough Times at Company," *The Wall Street Journal* (December 7, 2010), B3; Joe Light, "Chinese, Indian Workers Give Bosses Top Marks," *The Wall Street Journal* (November 22, 2010), 12; Lauren A. E. Schuker, "Island's Tax Increase Gives Fiji Water a Bitter Taste," *The Wall Street Journal* (November 30, 2010) at online.wsj.com (accessed December 10, 2010); Lauren A. E. Schuker, "Fiji Water Reopens its Plant," *The Wall Street Journal* (December 1, 2010), B7; Alicia Mundy, "POM Lambastes FTC," *The Wall Street Journal* (October 5, 2010) at online.wsj.com (accessed December 10, 2010); Anupreeta Das and Gina Chon, "J.Crew CEO Waited Weeks to Tell Board of Deal Talks," *The Wall Street Journal* (December 7, 2010), C1.

# Groups and Teams

## LEARNING OBJECTIVES

After studying this chapter, you should be able to:

- Characterize organizations in terms of interpersonal processes.

- Define a group and elaborate on different kinds of groups.

- Discuss the psychological character of groups, including why people join groups, the stages of development through which groups tend to move, and the nature of the informal organization.

- Identify important group dimensions—role dynamics, cohesiveness, and norms—and the relationships among them.

- Discuss the management of functional groups, task forces and committees, work teams, and quality circles.

- Describe the advantages, disadvantages, and techniques of group decision making.

## The Promise of Biotechnology—Return of Dr. Jekyll and Mr. Hyde

"By all accounts, I'm a medical miracle," says Ozzy Osbourne. He is getting his entire genome sequenced by Knome, an American genetic-testing firm, for clues as to how his body coped with such prolonged abuse. Genetic sequencing is "in." A company, 23andMe, has held "spit parties" for celebrities—rich and famous—to compare genomes. They compared the genomes of Warren Buffett and Jimmie Buffet to see if they are related (they are not).

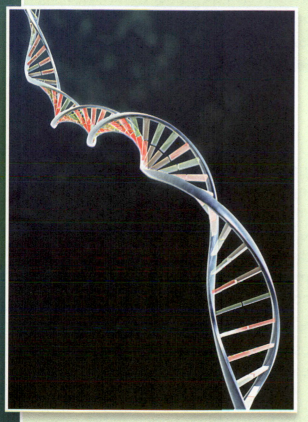

**FIGURE 19.1** The pharmaceutical industry is undergoing major changes as the emphasis shifts to drugs based on personal genetics.
© iStockphoto/George Paul

The pharmaceutical industry is approaching a fork in the road. It can continue along its usual pathway—having teams of scientists identifying potential drugs produced through the application of organic or biochemistry—or take the road less traveled by having teams focus on genomic testing and the design of genetic specific biologics. The former is slow but the means have been tested; the latter is new but with huge potential.

In 2007 Knome charged $350,000 to sequence a human genome. In 2010 that charge was reduced to $40,000/genome. By 2015 the price may be $1,000 per sequence. A competitor, Pacific Biosciences, has attracted $369 million in financing from the likes of the Blackstone Group, a hedge fund, and Kleiner, Perkins, Caufield and Byers, a venture capital group.

The reason for the interest is the potential contained in the human genome for personal genetics/pharmaceuticals (Figure 19.1). No longer will the pharmaceutical industry depend upon drugs aimed at large groups, as much as drugs designed for a specific individual. James Watson, Nobel laureate biochemist, once claimed that sequencing genomes was something even a monkey could do. Now the field is open to interpretation and exploration. In the words of the *Economist,* "Reading the human genome in the first place may, indeed, have been the work of mechanical monkeys. Interpreting the result will require the finest minds that humanity can muster."

Biotechnology is a field filled with potential but it is also completely fragmented. While giant companies with numerous teams of scientists dominate the pharmaceutical industry, the biotechnology industry is filled with many smaller companies with few teams exploring many avenues of expression with limited or restricted funding. The industry has tremendous potential, but for "big pharma" (the big firms in the pharmaceutical industry) the question is which ones have the key to success and which hold a key to pandora's box.

For example, scientists uncovered a gene in the "junk portion" of the human genome (around 90 percent of the genome) that is the culprit for one of the common forms of muscular dystrophy, facioscapulohumeral muscular dystrophy. An MIT geneticist, David Housman, stated, "As soon as you understand something that was staring you in the face and leaving you clueless, the first thing you ask is, 'Where else is this happening?'" On the other hand, rapid advances in bioscience are raising alarms, "If students can order (genetic sequences) online, somebody could try to make the Ebola virus," said Craig Venter, a pioneer geneticist.

So big pharma has gone on a buying spree to capture the potential being created by teams of scientists in smaller firms. Actelion Ltd, a Swiss biotechnology firm, is pursued by potential buyers: Bristol-Myers Squibb Co., Roche Holding AG, GlaxoSmith-Kline PLC, and Amgen Inc. Bristol-Myers Squibb Co. acquired Zymogenetics for $735 million to add hepatitis drugs to its pipeline. All of this activity is an attempt to improve productivity to generate new products. The internal rate of return on research and development in the pharmaceutical industry is around 7.5 percent below the cost of capital. Acquiring firms with ready-made drugs improves productivity by a factor of four.

The field of bioscience and biotechnology has become downright Darwinian.[1]

## INTRODUCTION

Interpersonal relations and groups are a ubiquitous part of organizational life and, increasingly, so is the use of teams.[2] Two of the oldest maxims known to mankind are that "two heads are better than one" and "the whole is greater than the sum of its parts." Nowhere is this more obvious than in research companies like the pharmaceuticals, where problem-solving minds must collaborate. Teamwork and groupthink are not new to other industries, however.

By definition, organizations involve networks of interactions among virtually everyone in those organizations.[3] Therefore, a manager should understand the dynamics of group and team activities within the organization.

As indicated in Chapter 16, the very nature of organizations means that they are composed of people working together. Heads of groups discuss work assignments and performance issues with members of their groups; peers at all levels conduct regular meetings to develop plans and solve problems; group members go to their group leaders with problems and questions; sales representatives talk to customers; and secretaries take telephone calls. Indeed, research has suggested that the average manager spends around three-quarters of his or her time interacting with others.[4]

However, the nature of any given interaction can vary considerably. Two managers talking at the water cooler may be reaching agreement on how to solve a problem, setting up a meeting for later in the day, discussing the latest football polls, deciding whether or not to fire someone, or arguing about how to resolve a point of disagreement. People can interact as individuals or as members of groups and teams. Whereas Chapter 16 focused primarily on individual processes, this chapter focuses primarily on group processes. Regardless of the purpose or the consequences of interpersonal activity, much of it occurs within the context of groups, and the use of groups is growing.[5] In this chapter, we first describe the interpersonal character of organizations, then we explore the nature of groups, establishing their psychological character and examining three critical group dimensions. Finally, we investigate some guidelines for using groups and teams in organizations.

# The Interpersonal Character of Organizations

Amgen, a leading human therapeutics company in the biotechnology industry, has developed top performance and effectiveness by capitalizing on the energy and strength provided by one of the most underused resources in organizations today: groups. After Litel Telecommunications restructured to a high-involvement, teamwork approach in 1989, within a month its order-processing time dropped from 14 days to one day and the error rate went from 40 percent to less than 5 percent. Within a year the company's revenues nearly doubled. A Corning plant in Blacksburg, VA, is organized around self-managed teams with only three managers for a 150-person plant. Harley-Davidson, Xerox, Procter & Gamble, and Digital Equipment are all using some form of teams as well.[6]

# The Nature of Groups

What is a group? First, let's examine the concept of a group so we can then identify the major kinds of groups found in organizational settings.

## Definition of a Group

A **group** is two or more people who interact regularly to accomplish a common goal.[7] There are three basic elements that are necessary for a group to exist.

As the definition states, a group is comprised of at least two people. Although there is no precise upper limit, a group that gets too large usually ceases to function as a group. Second, the members must interact regularly. This is one reason for setting an upper limit on group size. Once a group reaches a certain size—say, 20 people—it becomes difficult for everyone to interact regularly. At that point smaller groups tend to emerge from within the larger one.[8] Finally, group members must have a common purpose. Managers at CF Industries may create a group to develop a new plan or product; a group of workers may band together to try to change a company policy; a group of friends may go out together for dinner and a movie. In each case members of the group are working toward a common goal.

**FOOD FOR THOUGHT 19.1**

"Wearing the same shirts doesn't make you a team."

—*Buchholz and Roth*

## Kinds of Groups

There are many different kinds of groups, but we are most concerned about groups that exist in organizations. These can be classified as functional, task, or informal, as shown in Figure 19.2.[9]

A **functional group** is created by the organization to accomplish a range of goals with an indefinite time horizon. The operations division at McCormick, the marketing division of Seagram, the management department at the University of Notre Dame, the nursing staff at Scottsdale Healthcare in Arizona, and the field crews at Green Things Landscaping and Summer Farms are all functional groups. Each of these was formed by the organization, each has a number of goals, and each has an indefinite time horizon—that is, it is not slated to disappear at a certain time in the future. As the exhibit illustrates, functional groups generally conform to departmental boundaries on an organization chart.

A **task group** is created by the organization to accomplish a limited number of goals within a stated or implied time (Figure 19.3). Tootsie Roll Industries, for example, may appoint a design team to develop a new advertising campaign. The group

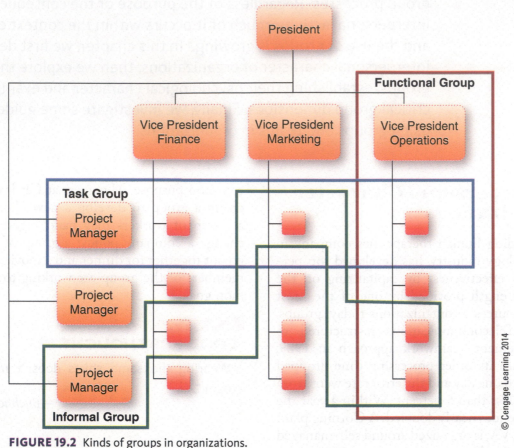

FIGURE 19.2 Kinds of groups in organizations.

© Cengage Learning 2014

is created by the organization, it has only one goal, and it has an implied time horizon (after the campaign has been developed and approved, the group will dissolve). Joshua Summer and his daughter used a task group that they referred to as "an exploratory committee" to research the possibility of entering the organic food business.

FIGURE 19.3 This functional group will be disbanded only after they finish the assigned project.

© Deklofenak/www.Shutterstock.com

There are actually several different forms of task groups in most organizations. As shown in the exhibit, a matrix design, as discussed in Chapters 12–13, places task groups under the direction of a project manager. Task forces, most committees, and many decision-making groups are also task groups.

An **informal group**, also called an *interest group*, is created by the members of the group itself for purposes that may or may not be related to the organization, and it has an unspecified time horizon. Five coworkers who have lunch together frequently, twelve employees who form a softball team, and three secretaries who take their afternoon coffee breaks together are examples of informal groups. Each person in the group chooses to participate and can stop whenever he or she wants.

When at lunch, the first group of workers may discuss how to solve an organizational problem (relevant to and in the best interests of the organization), how to steal a machine (relevant to but not in the best interests of the organization), or local politics and sports (not relevant to the organization). As you might guess, informal groups are extremely important to managers and can be a powerful force in determining organizational effectiveness. It is therefore important to understand the psychological character of groups.[10]

# The Psychological Character of Groups

Much of what is known about group processes comes from research in the field of psychology.[11] A substantial portion of this research relates to why people join groups, stages of group development, and the informal organization.

## Why People Join Groups

Sometimes people have no choice as to whether to join a group. Students may have to take a certain class, or employees may have to accept a specific job assignment that involves working with a designated group of other people. In many instances, though, people can choose whether to join a particular group. The four most common reasons for doing so are set forth in Table 19.1.[12]

**TABLE 19.1** Why People Join Groups

| REASON | EXAMPLE |
|---|---|
| Interpersonal Attraction | An employee joins three colleagues for lunch because they share his interest in local politics. |
| Group Activities | An employee joins the company bowling team because she loves to bowl. |
| Group Goals | An employee joins a union because she believes employees should negotiate for higher wages. |
| Instrumental Benefits | A manager joins a golf club because most of his business associates are members and he will therefore make useful contacts. |

© Cengage Learning 2014

One powerful reason for joining a group is interpersonal attraction. For instance, a new employee might find three other people who work in his department to be especially pleasant and to have interests and attitudes similar to his own, so he might start joining them for lunch most days. He thus joins the group because he is attracted to the group's members.

Another reason for joining a group is the group's activities. Suppose that another new employee is an avid bowler. She might inquire about and subsequently be invited to join the company bowling team (Figure 19.4). Of course, she will probably not remain on the team if she dislikes all the other members, but it is not her attraction to them that prompts her to join to begin with. Rather, it is a specific activity that she wants to pursue and group membership facilitates it.

A third reason people choose to join groups is that they identify with and want to pursue the goals of the group. This is a common reason for joining unions. Interpersonal attraction is irrelevant, and few people enjoy tedious contract negotiations or strikes. But employees may subscribe to the goals the union has set for its members, such as high wages, better working conditions, and so forth. Similar motives cause people to join organizations outside the company like the Sierra Club, charitable groups like the United Fund and the American Cancer Society, or clubs within the company such as the bowling or softball teams.

A final reason for joining groups is the instrumental benefits that may accompany group membership. For example, it is fairly common for college students entering their senior year to join one or more professional associations so they can list them

**FIGURE 19.4** These five computer programmers have formed an informal group for bowling after work.

© iStockphoto/Ozgurcankaya

on their resumes. Similarly, a manager may join one particular golf club not because he likes the other members (although he may) or because he likes to play golf (although, again, he may) but to make some useful business contacts. A former colleague of one of the authors of this textbook joined a particular church because of the potential business contacts it provides. Another joined the same church because it provides better childcare facilities than were available elsewhere near the worksite.

## Stages of Group Development

Regardless of the reasons that people join groups, the groups themselves typically go through periods of evolution or development. Although there is no rigid pattern that all groups follow, they usually go through the stages portrayed in Figure 19.5.[13]

The first stage, **forming**, is also known as mutual acceptance. As shown by the image on the left side of the exhibit and as the term itself suggests, forming involves the members of the group actually coming together to create it. During this stage the members acquaint themselves with each other, introduce newcomers, and begin testing which behaviors are acceptable and which are unacceptable to other members of the group.[14]

In the **storming** stage, the members may begin to pull apart if they see that the group may not meet their expectations. They may disagree over what needs to be done and how best to do it. Key elements in this stage are communication and decision making to offset any conflict and hostility that may emerge. Patterns of interaction may be uneven, as suggested by the exhibit, and informal leaders often begin to emerge.

**Norming**, a common third phase, is characterized by the resolution of conflict and the development of roles (discussed later). People either have left the group because the conflict is too great (it

didn't meet their expectations) or have accepted the group for what it is. The remaining members take on certain responsibilities and everyone develops a common vision of how the group will function. Motivation and productivity begin to emerge as a sense of unity evolves.

Finally, the group begins **performing**—moving toward accomplishing its goals, whether deciding which movie to see, developing a major planning document for the firm, or recommending that management spend $10,000 for new equipment. Members enact their respective roles and direct their efforts toward the goal(s). Control and organization evolve as the group becomes more stable and structured.

For groups or group projects that have definite times of existence, yet another stage exists. **Adjourning** is the completion of the project or break-up of the group. Group members who have bonded closely may feel a sense of insecurity or threat from this change. There may need to be some sort of "ceremony" to minimize the negative feelings associated with breaking up the group. As noted in Chapter 13, these sorts of ceremonies help bring about strong organizational cultures.

Of course, as noted earlier, every group may not follow these exact stages in a distinct and observable sequence. All groups, however, do generally deal with the kinds of issues associated with each stage as they mature. Mature groups, then, tend to exhibit the four characteristics that are discussed later in the chapter: role structure, behavioral norms, cohesiveness, and leadership.

---

**FOOD FOR THOUGHT 19.3**

Teamwork, simply stated, is less me and more we.

---

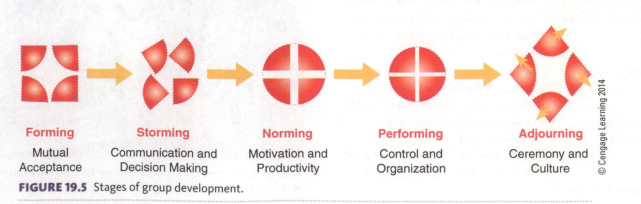

| Forming | Storming | Norming | Performing | Adjourning |
| Mutual Acceptance | Communication and Decision Making | Motivation and Productivity | Control and Organization | Ceremony and Culture |

**FIGURE 19.5** Stages of group development.

© Cengage Learning 2014

## The Informal Organization

It is critical for managers to recognize the existence and importance of the **informal organization**—the overall pattern of influence and interaction defined by the total set of informal groups within the organization.[15] As suggested earlier in Figure 19.2, the formal organizational structure is overlaid with informal groups. These groups actually get much of the organization's work done; that is, informal telephone calls, chance meetings at the coffee machine, and impromptu lunch gatherings go a long way toward defining the organization's goals and helping achieve them. Thus, managers should not ignore the power of the informal organization as they go about their business.

## Important Group Dimensions

Groups in and of themselves are of considerable importance and interest. However, managers can gain even greater insights into groups by considering their role dynamics, their levels of cohesiveness, and their norms.

### Role Dynamics

What is a **role**? In a movie or play, a role is a part played by an actor. People in groups also play one or more roles.[16] For example, some people play the role of "task specialists" to help the group accomplish its goals. Others, called "social specialists," may work to keep everyone happy. A few serve as leaders. Still others, called "free riders," may do very little.

As we have seen, everyone belongs to several different groups. Many people are part of a formal work group, one or more task groups, several informal groups, and a family. And in each group there are various roles to play. A given individual might be a task specialist in one group, the leader in another, and a free rider in yet another.

Figure 19.6 illustrates the way in which role dynamics occur in a group. **Role dynamics** are the process whereby a person's **expected role**—the one that others in the group expect that person to play—is transformed to his or her **enacted role**—how the person actually behaves in the group. The members transmit these expectations in the form of the **sent role**. (As we will see in the next chapter, however, communication breakdowns frequently occur, so there may be differences between the expected and sent roles.)

Perceptual factors can affect the role process, too. Therefore, still more differences may creep in as the sent role is translated into the **perceived role**—how the individual comes to think he or she should behave in the group. Finally, the **enacted role** is how the person actually behaves; and here, too, differences can arise. For example, the person may not be capable of executing the perceived role or may simply choose not to execute it in the way that others expect.

Figure 19.6 also introduces the concepts of role ambiguity and role conflict, two facets of group dynamics that warrant additional explanation.

**Role ambiguity** occurs when the sent role is unclear.[17] For instance, suppose a supervisor at Summer Farms asks a new employee to prepare a sales forecast for the next period. It is quite possible

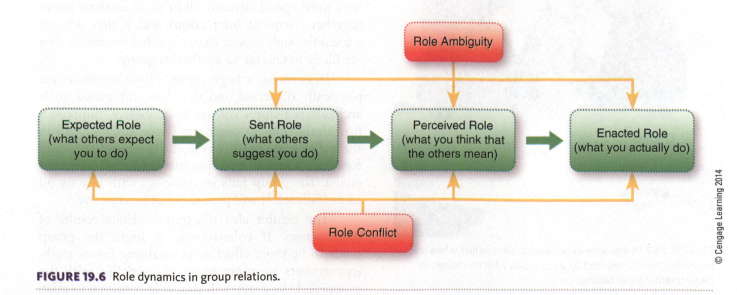

**FIGURE 19.6** Role dynamics in group relations.

© Cengage Learning 2014

that the employee does not know where to get the data, how many products to include in the forecast, what form to use for the report, or even what time period is involved. Thus, she will probably suffer from role ambiguity.

Of perhaps even greater concern is **role conflict**, which occurs when messages about a role are clear but involve some degree of inconsistency or contradiction.[18] There are several forms of role conflict.[19]

Interrole conflict occurs between two or more roles. For example, suppose an employee's leader tells him that he must work more overtime when the employee was already feeling guilty about working so much and had just resolved to spend more time with his family. Consequently, the employee will now experience *interrole* conflict. In contrast, *intrarole* conflict occurs when one person receives two or more conflicting messages. If a marketing vice president tells a sales manager to have the sales force travel more just when the controller is telling the sales manager to cut travel costs, the sales manager will experience intrarole conflict (Figure 19.7).

**FIGURE 19.7** A manager experiences role conflict when told his budget will be reduced by 27 percent, with no change in departmental expectations.

*Intrasender* role conflict arises if one person transmits conflicting expectations. Suppose that on Monday a manager tells her assistant that she may dress more casually, but on Thursday she reprimands the assistant for not looking professional. The obvious contradiction will result in intrasender role conflict for the assistant.

Finally, **person-role conflict** arises when the demands of the role are incongruent with the person's preferences or values. An employee experiences this conflict if his or her job demands a great deal of travel, for instance, but he or she prefers to remain at home. Similarly, an employee who has strong feelings against military buildups and the arms race may feel uncomfortable when the company gets several defense contracts.

## Cohesiveness

Another important dimension of groups is **cohesiveness**, or the extent to which the members of the group are motivated to remain together.[20] A highly cohesive group is one in which the members pull together, enjoy being together, perform well together, and are not looking for opportunities to get out of the group. In contrast, a group with a low level of cohesiveness is one in which the members do not like to be together, do not work well together, and would casually leave the group if an opportunity arose.

Figure 19.8 shows the determinants and consequences of cohesiveness.[21] As shown, small size, frequent interaction, clear goals, and success tend to foster cohesiveness. For example, if five people (small size) are assigned to a crash project for developing a new product by the end of the year (clear goals), if they must spend virtually all of their working hours together (frequent interaction), and if they achieve a breakthrough sooner than expected (success), they are likely to emerge as a cohesive group.

In contrast, a large group whose members are physically dispersed and that has ambiguous goals and suffers failures will be less cohesive. Suppose a firm designates 85 engineers (large size) located in three different plants (physically dispersed) as a task force to "explore new product ideas" (ambiguous goals). The group fails to come up with any useful ideas (failure). It will probably not be cohesive.

The exhibit also illustrates various results of cohesiveness. If cohesiveness is high, the group tends to be more effective in attaining future goals, its members are more personally satisfied with the group, and the group will probably continue to exist.

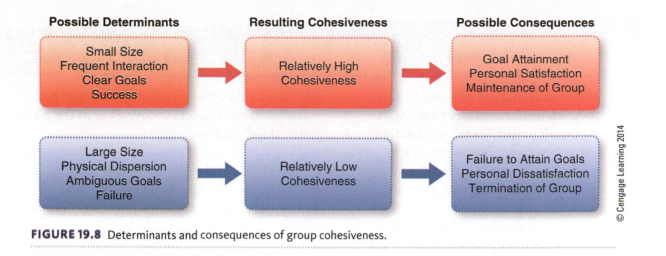

**FIGURE 19.8** Determinants and consequences of group cohesiveness.

On the other hand, if cohesiveness is low, the group is less likely to attain its future goals, members will express more dissatisfaction with the group, and the group is more likely to dissolve or fall apart.

## Norms

The importance of cohesiveness is accentuated when we consider it in the context of group norms. A **norm** is a standard of behavior that the group develops for its members.[22] For instance, a group may have a norm against talking to the head of the group too much. People who violate this norm may be "punished" with unpleasant looks, snide remarks, and the like. The norms themselves are, of course, socially defined and exist only in the minds of the group members.

Norms can arise for any number of work-related behaviors. In a typical work group, there may be norms that define how people dress, the upper and lower limits on acceptable productivity, things that can be told to the head of the group and things that should remain secret, and so forth.[23] How employees and others perceive the ethical norms of organizations is becoming increasingly important in the success of those organizations.

### FOOD FOR THOUGHT 19.4

"Individual commitment to a group effort is what makes a team, a company, a society, or a civilization work."

—*Vince Lombardi*

When confronted with a set of norms in an established group, an individual may do several dif-ferent things. One common reaction is to *accept and conform* to the group norm. A less frequent reaction is *total rebellion*, in which case the person completely rejects the norm. (Of course, if the norm is important to the rest of the group, the newcomer will probably be ostracized.) There is also *creative individuality*. In this situation, the person accepts some of the norms, especially the most important ones, but follows his or her own preferences within their limits. For example, if the norm is that male employees will not wear jeans to work, a newcomer who prefers to dress casually may be able to get away with wearing jeans once every two or three weeks, as opposed to never wearing them (conformity) or wearing them every day (rebellion).

The norm that governs the acceptable level of performance in the group becomes especially critical when linked with cohesiveness. Figure 19.9 shows

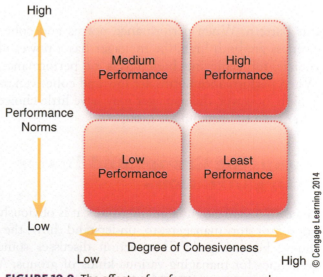

**FIGURE 19.9** The effects of performance norms and cohesiveness on group performance.

## A FOCUS ON AGRIBUSINESS
### Agribusiness Development Teams

A unique and controversial use of teams in agribusiness has developed as a result of war. ADTs (Agribusiness Development Teams) exist within National Guard Units to provide assistance in Afghanistan. The basic premise is that if they can help improve Afghan agriculture, they will improve the Afghan people's standard of living and the stability of the provincial and central governments. It will also decrease the likelihood that locals will support insurgents or terrorists.

As of August 2010 the teams, which started out with one team from Missouri, included teams from Arkansas, California, Illinois, Indiana, Iowa, Kansas, Kentucky, Missouri, Nebraska, Nevada, Oklahoma, South Carolina, Tennessee, Texas, and Wisconsin. These teams not only include technical experts in agriculture and livestock but they also have links to their states' university agricultural programs for training and advice.

ADTs are involved with numerous projects dealing with animal husbandry, horticulture, irrigation, storage and distribution, and agribusiness education. Immediate successes have been obtained dealing with row cropping and crop rotation, drip versus flood irrigation, and pest management. Projects underway or in the planning stages include:

- Building grain mills
- Introducing new wheat seed
- Developing canning and juicing factories for harvested vegetables and fruits
- Building cool storage facilities to store harvested crops operated by solar panels
- Overseeing micro-slaughter facilities to increase sanitization of livestock meat
- Launching vet clinics focused on de-worming the livestock
- Advising on reforestation projects
- Increasing the crop yield for commercial use
- Operating cold- and warm-water fish hatcheries

While this use of teams involves the National Guard in a foreign country, which is highly controversial, the basic approach could also be employed by agribusiness firms to help local communities in the United States.[24]

---

this pattern. When performance norms and cohesiveness are both high, the manager has a powerful vehicle for achieving high levels of performance. When performance norms are low and cohesiveness is high, however, the manager may have little choice other than to break up the group.

## Managing Groups and Teams in Organizations

Since all organizations contain groups, it is obviously important for managers to understand how they should be managed.[25] This section discusses some guidelines for managing various kinds of groups. A special case of managing groups, group decision making, is considered in the section following this one.

## Managing Functional Groups

Virtually all aspects of interaction between a formal leader and his or her work group are related to the management of functional groups, so all of the areas of management covered in the other chapters in this book are pertinent. However, several specific implications for managing functional groups can be drawn from earlier discussions of group dimensions.[26]

For one thing, the manager should be cognizant of the importance of role dynamics. In particular, he or she should recognize the potential problems associated with role ambiguity and role conflict and strive to avoid these problems whenever possible. For another thing, the manager should realize the importance of group cohesiveness. As we have noted, cohesiveness can be a powerful force in organizations, especially when combined with different levels of performance norms. The manager should therefore work to enhance cohesiveness when it is in the best interests of the organization. Finally, the manager should work to establish high performance norms by adding hard workers to the group, consistently rewarding high performance, and so forth. Such actions should both reinforce current levels of performance norms and simultaneously push them higher.

## Managing Task Forces and Committees

Organizations often use task forces and committees to perform various kinds of tasks. Since both of these are task groups, they can generally be managed in similar ways. There are, however, a few subtle differences.

**Managing Task Forces.** Several guidelines exist that relate specifically to task forces.[27] A **task force** is a temporary group within an organization created to accomplish a specific purpose (task) by integrating existing functional areas. First, the majority of the group should be line managers. Since line managers are ultimately responsible for implementing the group's ideas, they should be well represented in the group. Second, the group must have all relevant information. It will be more effective if it has access to information that affects its work.

Group members need to have the legitimate power necessary to translate group needs back to their respective functional groups. Within the group, though, the emphasis should be on expert power; that is, group members with expertise in specialized areas should be in charge of those areas. The task force should also be properly integrated with relevant functional groups. Ideally, this follows from the composition of the group, as suggested above; but special attention may be required to maintain this integration over time. A last point is to establish group membership with the goal of optimizing technical and interpersonal skills. Members should know how to do their respective jobs, but they also need to get along with one another.

**Managing Committees.** Another important part of almost every manager's job at one time or another is managing a committee. A **committee** is a special kind of task group. It may have only a few members or many members. A committee's purpose may be broad and short term, or narrow and long term. It often has a name, such as the Grievance Committee, Steering Committee, or President's Advisory Committee. Committees may appear at the bottom of the organization, or they may occur at the highest levels of the firm and have primary responsibility for its management.

Like other kinds of groups, committees can be made more effective if managers follow specific guidelines. For one thing, the goals and limits of the committee's authority need to be clearly specified to help keep the committee's activities focused and directed toward its intended purpose. For another, the committee should usually have a specific agenda. Care should also be taken to see that the committee does not interfere too much with its members' normal responsibilities; that is, people should not devote so much time to their committee duties that they neglect their regular jobs.

Finally, it is useful to specify the output of the committee. For example, suppose a manager creates a committee to locate a site for a new factory. He may want the committee to submit a list of three acceptable sites, either rank-ordered or unranked. He may need a lot of supporting materials for each site, or he may want only the recommendations. Obviously, he should clearly communicate each of these expectations to the committee before it begins to work.[28]

## Managing Work Teams

A fairly recent innovation in the use of groups in organizations is the establishment of work teams. A **work team** is a small group of employees that is responsible for a set of tasks previously performed by its individual members and that takes primary responsibility for managing itself. Many firms such as A. O. Smith have adopted the work team concept, and research in agribusiness has found them effective.[29]

There are several reasons for the use of work teams. Perhaps the most important is that work teams provide a natural vehicle for giving employees more say in their jobs—that is, for increasing participation. Given the general trend toward more participation, work teams are likely to be used more and more often. Moreover, they often result in increased productivity, higher quality, and better employee attitudes.

Organizations that want to use work teams need to consider several guidelines. First, it is often necessary to provide some initial training. People who previously did one simple task on their own may need to learn how to perform other tasks and work with others. Perhaps most importantly, organizations that use work teams must be willing to give work team members more responsibility and control over their tasks. If a first-line supervisor is always watching over his or her shoulders, the work team may actually be a detriment. The team members will succeed only if they have some of the responsibility previously held by their managers. Finally, selecting new employees may require a different approach. They may need heightened interpersonal skills and the ability to work with existing group members. Considerable care should be used in building work teams.[30]

High performance teams have been shown to have both a clear understanding of their goals as well as a belief that the goal is worthwhile.[31] Further, the structure of effective teams embodies (1) clear roles and accountability for each of those roles, (2) effective communications, (3) methods for monitoring performance and for providing feedback, and (4) objective evaluations.[32] As such, it is important that team performance be carefully evaluated.[33]

## Managing Quality Circles

A final type of group which many managers are beginning to contend with is the **quality circle**, or **QC**. QCs are groups of operating employees formed for the express purpose of helping the organization identify and solve quality-related problems.[34] They are made up of volunteers from the same or related work areas who meet regularly, usually for an hour or so each week, to talk about quality issues facing the company. Each member is free to suggest problems the group might tackle. Theoretically, the employees are so close to their tasks that they can suggest ideas and opportunities for improvement that might escape the attention of managers.

QCs are quite controversial. Many firms, such as Procter & Gamble and General Foods, have successfully adopted QCs. But others firms have had disappointing experiences with them. If an organization decides to try QCs, it can at least increase the potential for success by considering some of the hints given by successful companies. First, rely totally on volunteers. People who feel coerced to join will be of little value. Second, provide enough time and resources to allow the QC to do its intended job. Third, provide feedback and recognition to the QC regarding its suggestions. For instance, suppose a QC recommends a new way of doing something that results in substantial cost savings for the company. Managers should, of course, tell the QC of the success of its idea, but they should also communicate that fact to others through the company newsletter, special announcements, and other means. This allows the group members to develop feelings of pride and accomplishment. If an idea is not acceptable or does not really do any good, managers should communicate this fact to the group, albeit in a more subtle fashion.

We will discuss QCs more fully in Chapter 22.

## Mistakes to Avoid in Using Groups and Teams

There are four common mistakes that managers should avoid while using groups and teams.[35] Calling a work unit a team but also managing the members as individuals is the easiest mistake to make. The manager wants the benefits of a team without relinquishing the authority to the group. Closely related to this is the simple failure to really delegate authority to the team. Managers should set the direction and the constraints but delegate the means for accomplishment to the team. The third mistake is vague delegation—telling a group in general terms what is needed and letting them "work out the details." Teams need clear structures to function effectively. Finally, setting up a team properly but providing it with no organizational support will also doom it to failure.

In addition, as discussed in Chapter 16, conflict is frequently an aspect of interpersonal relations. Such conflict may be either good or bad, so understanding it is important.[36] Remembering that interdependencies such as those that exist in teams frequently can lead to conflict suggests another reason for the careful management of teams.[37] Since conflict is quite probable when using groups and teams, it is important that managers be aware of the possibility and how to deal with it.[38]

## Group Decision Making

As we have already seen, a great deal of decision making in organizations is done by groups. Executive committees make critical decisions about the company's future, project teams make decisions about which new products to introduce, and grievance committees make decisions about who is right and wrong in organizational disputes. Arnott Duncan of Duncan Family Farms tries to involve everyone in his relatively small business in decision making.[39] This section outlines the advantages and disadvantages of allowing groups to make decisions and then summarizes three techniques to promote better group decisions.[40]

## Advantages of Group Decision Making

Obviously, there must be certain advantages to group decision making. Why else would it ever be done? Table 19.2 summarizes the four most general advantages.[41] These factors tend to lead to a higher-quality decision than a single individual working alone might have obtained.

One advantage is the fact that more information is available to the group than is available to an individual. Each member is able to draw on his or her unique education, experience, insights, and other resources and (it is hoped) contribute these to the group. Similarly, a group is likely to generate more alternatives to consider than an individual can. That is, some individuals have ideas that escape others, so the total set of alternatives should be greater in the group than for any single individual.

A third advantage is that acceptance of the decision will probably be greater than it would be if an individual made the decision alone. Since more people participated in making the decision, more people will understand its origins. Those who did not participate may still feel that the decision was reached in a democratic fashion. Finally, groups just tend to make better decisions than individuals do. The extra information and additional alternatives, when properly considered and processed, promote a better outcome.

## Disadvantages of Group Decision Making

Unfortunately, there are also four disadvantages to group decision making. If these disadvantages did not exist, all decisions would automatically be assigned to groups. These factors, also listed in Table 19.2, serve as barriers to high-quality decisions.

One major disadvantage is that a group tends to take longer to reach a decision, as all members may want to discuss every aspect of it. Although this may be a plus, it adds a lot of time to the process. The group may also try too hard to compromise. Some degree of compromise may be necessary and perhaps even desirable, but compromise may be sought to the exclusion of a better decision that the group could have reached with more effort.

It is also possible that a single individual will dominate the process. If this happens, the decision has the appearance of having been made by the group and therefore may be widely accepted. Besides, allowing one member to make the decision sets aside all of the potential advantages of group decision making. Groups involved in making decisions may also succumb to a phenomenon known as **groupthink**.[42] Groupthink is what happens when group members become so interested in maintaining cohesiveness and good feelings toward one another that the group's original goals become lost (Figure 19.10). In this case, the group makes decisions that protect its members as individuals and the group as a whole rather than decisions that are in the best interests of the overall organization.

**TABLE 19.2** Advantages and Disadvantages of Group Decision Making

| ADVANTAGES | DISADVANTAGES |
|---|---|
| Availability of more information | Longer decision-making process |
| Generation of more ideas | Too much emphasis on compromise |
| Easier acceptance of decision | Dominance by an individual |
| Better decisions through collaboration | Possibility of groupthink |

© Cengage Learning 2014

**FIGURE 19.10** A group of individuals with similar characteristics and interests may succumb to groupthink.

**FOOD FOR THOUGHT 19.5**

"No one can whistle a symphony. It takes a whole orchestra to play it."

—*H.E. Luccock, theologian*

## Techniques for Group Decision Making

Managers have come up with several techniques for capitalizing on the advantages of group decision making while simultaneously minimizing the potential harm from the disadvantages.[43] As we saw in Chapter 11, **Delphi forecasting** uses experts to make predictions about future events. These predictions are systematically refined through feedback until a consensus emerges.

The **nominal group technique** is a structured process whereby members individually suggest alternatives, which are then listed on a chart for all to see. All the alternatives are discussed, and each member is asked to rank-order them. The average rankings are listed and the process is repeated until everyone agrees.

The **devil's advocate strategy** is to assign one member the role of devil's advocate. This person is expected to challenge and take issue with the actions of the rest of the group. The devil's advocate strategy is especially valuable in preventing groupthink.

## CHAPTER SUMMARY

Organizations are by their very nature highly interpersonal. Virtually all of the work of most organizations is accomplished by people working together increasingly in the context of groups. A group is two or more people who interact regularly to accomplish a common goal. Groups in organizations can be classified as formal, task, or informal.

People usually choose to join a group because of interpersonal attraction, group activities, group goals, and instrumental benefits. Regardless of why a group is formed, it normally goes through a four-stage developmental process consisting of forming, norming, storming, and performing. The informal organization is the overall pattern of influence and interaction defined by the total set of informal groups within the organization.

In all groups, people play certain parts, or roles. Role ambiguity occurs when the sent role is unclear and the person is not sure what she or he is supposed to do. Role conflict occurs when role messages are clear but inconsistent or contradictory. Another group dimension is cohesiveness, the extent to which the members of the group are motivated to stay together. This is connected to norms, which are the standards of behavior developed by the group. Norms may relate to performance (such as the quantity and quality of goods produced) or to nonperformance (such as how to dress or what to keep secret).

Managing groups is not easy. Fortunately, several guidelines exist that can facilitate the management of functional groups, task forces and committees, work teams, and quality circles. Group decision making, which is common to American corporations, has four basic advantages and four disadvantages. Managers can use several techniques to enhance the quality of group decision making.

> Conflict can be a consequence of many interpersonal processes in organizations. It has both positive and negative attributes and can be stimulated, reduced, or resolved through a variety of techniques.

## CHAPTER ACTIVITIES

### REVIEW QUESTIONS

1. Identify and define the three basic types of groups.
2. What are the four stages of group development?
3. Describe the various kinds of role conflict.
4. How do performance norms interact with cohesiveness to determine performance?
5. What are the advantages and disadvantages of group decision making?

### ANALYSIS QUESTIONS

1. Why is there an upper limit on the number of people that can reasonably constitute a group?
2. Have you ever experienced any of the forms of role conflict described in the chapter? Describe as many as possible.
3. If you were a manager in charge of a group that had high cohesiveness but very low performance norms, what would you do?
4. What are the common elements inherent in managing the different types of groups? What are the differences?
5. Have you ever been involved in a group decision-making process? If so, how did that process compare to those described in the text?

### FILL IN THE BLANKS

1. A small group of employees that is responsible for a set of tasks previously performed by individual members and that takes responsibility for managing itself is called a/an _____ _____.
2. A group created to accomplish a limited number of goals within a stated or implied time is called a _____ group.
3. The phenomenon that happens when group members begin making decisions that protect its members as individuals and the group as a whole rather than decisions that are in the best interests of the overall organization is known as _____ _____.
4. A group created by the members themselves for purposes that may or may not be related to the organization constitute a/an _____ group.
5. Standards of behavior that the group develops for its members are called _____.
6. The extent to which members of the group are motivated to remain together is known as _____.
7. Three common types of groups are _____ groups, _____ groups, and _____ groups.
8. Groups usually progress through these four stages as they develop: _____, _____, _____, and _____.
9. A lack of clarity about a person's role (incomplete information, for example), can cause _____ _____.
10. Some degree of inconsistency or contradiction about the role a person is supposed to play leads to _____ _____.

# ▼ CHAPTER 19 CASE STUDY
## IS THE GOVERNMENT ADDICTED TO TOBACCO?

Smoking originated in the Americas where tobacco grew wild. Around 46 million Americans are smokers, with men (24.8 million) slightly outnumbering women (21.1 million); and in the United Kingdom there are 20 million smokers. As a result of legislative action, however, smoking is down among minors.

In response to the U.S. government's efforts to reduce smoking, in November 1998 the nation's leading cigarette manufacturer's signed a contract called the Master Settlement Agreement (MSA) with the attorneys general of 46 states, five U.S. territories, and the District of Columbia. The original tobacco signatories on this agreement were agribusiness firms: Philip Morris Inc., R.J. Reynolds Tobacco Company, Brown & Williamson Tobacco Company, Lorillard Tobacco Company, Liggett Group Inc., and Commonwealth Brands Inc.

In the words of Philip Morris Inc., "The tobacco settlement agreements created fundamental changes in how tobacco products are advertised, marketed, and sold in the United States. The agreements include a variety of restrictions on the sale and marketing of cigarettes, including prohibiting … any action, directly or indirectly, targeting youth in the advertising, marketing, and promotion of tobacco products."

It is estimated that this agreement, also known as the multi-state tobacco settlement, will provide about $246 billion dollars to the states, territories, and District of Columbia over the first 25 years of the agreement. (Note: The agreement requires that the tobacco signatories will continue to make payments to the states in perpetuity.)

Pause for a moment. This one agribusiness industry will pay to the United States, and others close to a quarter of a trillion dollars until 2023 and then continue to pay even after that. This is a significant sum of money and surely must have a great effect on the industry. It will be instructive to see how the industry and government have fared over the past decade. Customers are fewer as the proportion of smokers has shrunk. Smoking is being banned in eating establishments, government offices, and by many other organizations. Employees frequently have to exit buildings where they work in order to smoke.

So it seems that the original tobacco signers on the agreement have followed the tenets of the agreement, but others are getting creative about trying to capture a share of the "smoker market." In the United States, for example, a federal tax loophole offers deep discounts to people who roll their own cigarettes. However, rolling your own cigarette is self-limiting. It requires skill and experience. Enter technology: some tobacco retailers have installed "Roll-Your-Own Cigarette" machines. These machines produce a carton of cigarettes in eight minutes, and the carton could cost half the price of a branded manufacturer's carton of cigarettes.

The reason this technology can deliver a carton of cigarettes at half the price has little to do with technology and everything to do with tax regulations. These cigarettes are made with "pipe" tobacco, not cigarette tobacco. Cigarette tobacco carries a federal excise tax of $24.78/pound versus pipe tobacco with a federal excise tax of $2.83/pound. How acceptable is the substitute? One retailer reported a one-hour wait by customers for their turn at the machine.

Not to be outdone by the retailers, the U.S. federal government ruled that retailers featuring these machines must obtain "manufacturing permits and pay applicable federal taxes." Now these 120 stores must obtain permits from the Treasury's Alcohol and Tobacco Tax and Trade Bureau, comply with bookkeeping rules, and pay $10.07 federal excise tax per manufactured carton.

There seems to be a "Darwinian" struggle between tobacco retailers, their customers, and the federal tax officials. It does not stop with machines … sometimes definitions work. Sometimes a cigar is a cigar, but at other times it may be little more than a cigarette. So, in 2009 Congress raised the federal excise tax on these "little" cigars, which are filtered, generally sweetly flavored products that are similar in size and shape to a cigarette. These little "cigarette cigars" were taxed at a rate of $10.07/carton. However, some manufacturers then responded by increasing the weight of their little cigars to qualify as "conventional" cigars. Cigars heavier than three pounds per thousand are taxed only $2–$4/carton. These products would be priced less than cigarettes also. The beat continues…

These are not exactly shining examples of compliance and restraint being demonstrated by the tobacco industry in a broader context than just the "settlement" signers. However, before we vilify the industry, let's look at the other side of the "settlement."

It seems that the "settlement" states were recipients of record amounts of tobacco revenue funds, $25.1 billion for 2009 (December 9, 2009). At the same time they were spending less to prevent kids from smoking and to help smokers quit. Only nine states funded tobacco prevention at half the level recommended by the Center for Disease Control; the others funded prevention programs at one-quarter the recommended level. Only North Dakota funds a tobacco prevention program at the original level.

So what was the purpose of the settlement, if not for health? Could it be taxes? Maybe nicotine is not the only addictive substance involved in the tobacco industry. Could it be that tax revenues are also addictive—to governments?[44]

### ▶ Case Study Questions

1. What is the impact of anti-tobacco actions (legislation, education, etc.) on organizations other than those in the tobacco industry?

2. Visit a variety of local businesses and determine what their smoking policies are for employees. What do your findings suggest?

3. Can, and should, smoking/non-smoking be used as a criterion in hiring? Why or why not?

## REFERENCES

1. "What Lies Within," *The Economist*, (August 12, 2010) at www.economist.com (accessed December 20, 2010); "Methylated Spirits," *The Economist* (October 17, 2009) at www.economist.com (accessed December 18, 2010); Gina Kolata, "Reanimated 'Junk' DNA is Found to Cause Disease," *The New York Times* (August 20, 2010), A1; Keith Johnson, "Bioscience Gains Cause Terror Fears," *The Wall Street Journal* (August 11, 2010) at online.wsj.com (accessed December 19, 2010); Anupreeta Das and Dana Cimilluca, "Suitors Knock, Swiss Biotech Weighs Options," *The Wall Street Journal* (October 8, 2010), B1; Peter Loftus and Jonathon Rockoff, "Bristol-Myers Agrees to Acquire Biotech Company Zymogenetics," *The Wall Street Journal* (September 8, 2010) at online.wsj.com (accessed December 20, 2010); Hester Plumridge, "Big Pharma is Winning War on Drugs," *The Wall Street Journal* (October 16, 2010), B16.

2. Bradley L. Kirkman and Benson Rosen, "Powering Up Teams," *Organizational Dynamics* (2000): 48–58; Richard A. Guzzo and Gregory P. Shea, "Group Performance and Intergroup Relations in Organizations," in *Handbook of Industrial and Organizational Psychology, Volume 3, 2nd ed.*, eds. Marvin D. Dunnette And Leaetta M. Hough (Palo Alto, CA: Consulting Psychologists Press, Inc., 1992), 269–313.

3. Rob Cross, Nitin Nohria, and Andrew Parker, "Six Myths About Informal Networks—and How to Overcome Them," *Sloan Management Review* (2002): 67–77.

4. Henry Mintzberg, *The Nature of Managerial Work* (New York: Harper & Row, 1973).

5. John Mathieu, M. Travis Maynard, Tammy Rapp, and Lucy Gibson, "Team Effectiveness 1997–2007: A Review of Recent Advancements and a Glimpse into the Future," *Journal of Management* (2008): 410–476; "A Braver New World?" *Industry Week*, August 3, 1992, 48–54; and Colin Coulson-Thomas, "The Responsive Organization," *Journal of General Management* (Summer 1990): 21–31.

6. This paragraph is based on Clay Carr, *Team-Power* (Englewood Cliffs, NJ: Prentice-Hall, 1992); see also Jon R. Katzenbach and Douglas K. Smith, *The Discipline of Teams* (New York: Wiley, 2001).

7. Gregory Moorhead and Ricky W. Griffin, *Organizational Behavior, 9th ed.* (Boston: Houghton Mifflin, 2010); Linda M. Jewell, *Contemporary Industrial/Organizational Psychology, 3rd ed.* (Pacific Grove, CA: Brooks/Cole Publishing, 1998).

8. T. Kameda, M. F. Stasson, and J. H. Davis, "Social Dilemmas, Subgroups, and Motivation Loss in Task-Oriented Groups: In Search of an 'Optimal' Team Size in Work Division," *Social Psychology Quarterly* (March 1, 1992): 47–56.

9. Marvin E. Shaw, *Group Dynamics—The Psychology of Small Group Behavior, 4th ed.* (New York: McGraw-Hill, 1985); see also Donelson R. Forsyth, *Group Dynamics, 5th ed.* (Belmont, CA: Wadsworth, Cengage, 2010).

10. Ibid.

11. Dorwin Cartwright and Alvin Zander, eds., *Group Dynamics: Research and Theory, 3rd ed.* (New York: Harper & Row, 1968); see also Alvin Zander, *Making Groups Effective, 2nd ed.* (San Francisco: Jossey-Bass, 1994).

12. Shaw, *Group Dynamics*.

13. B. W. Tuckman, "Developmental Sequence in Small Groups," *Psychological Bulletin* (1965), 63: 383–399; for a more recent treatment of group development stages, see Connie J. G. Gersick, "Marking Time: Predictable Transitions In Task Groups," *Academy of Management Journal* (June 1989): 274–309.

14. Gilad Chen, "Newcomer Adaptation in Teams: Multilevel Antecedents and Outcomes," *Academy of Management Journal* (2005): 101–116.

15. George Homans, *The Human Group* (New York: Harcourt, 1950).

16. David Katz and Robert L. Kahn, *The Social Psychology of Organizations, 2nd ed.* (New York: Wiley, 1978).

17. Ibid.

18. Robert L. Kahn, D. M. Wolfe, R. P. Quinn, J. D. Snoek, and R. A. Rosenthal, *Organizational Stress: Studies in Role Conflict and Role Ambiguity* (New York: Wiley, 1964).

19. Katz and Kahn, *The Social Psychology of Organizations*.

20. Shaw, *Group Dynamics*.

21. For how to increase cohesiveness, see P. F. Buller and C. H. Bell, Jr., "Effects of Team Building and Goal Setting on Productivity: A Field Experiment," *Academy of Management Journal* (June 1986): 305–328.

22. Shaw, *Group Dynamics*; see also Monika Henderson and Michael Argyle, "The Informal Rules of Working Relationships," *Journal of Occupational Behavior* (1986), 7:259–275.

23. Daniel C. Feldman, "The Development and Enforcement of Group Norms," *Academy of Management Review* (January 1984): 47–53.

24. Staff Army Sgt. 1st Class Jon Soucy, "Agricultural Development Teams," *National Guard* (October 6, 2009) at www.ng.mil (accessed August 1, 2010); Kentucky National Guard Public Affairs Office, "Kentucky Agriculture Commissioner Writes About Agribusiness Teams," *Unbridled Service* (June 30, 2010) at kentuckyguard.wordpress.com (accessed August 1, 2010); Staff Sgt. Stacia Zachary, "Agri-Business Development Team Plants Seeds of Hope for Afghan People," *WWW.ARMY.MIL* (July 6, 2009) at www..army.mil (accessed August 1, 2010); *Agribusiness Development Teams in Afghanistan, Handbook No. 10-10* (November 2009) U.S. Army Combined Arms Center.

25. In addition to the views presented in this section, see Kim B. Clark and Steven C. Wheelwright, "Organizing and Leading 'Heavyweight' Development Teams," *California Management Review* (Spring 1992): 9-20; and "Blueprint for a Successful Team," *Supervisory Management*, May 1, 1992, 2–3.

26. Bernard A. Nijstad, *Group Performance, 2nd ed.* (NY: Psychology Press, 2009).

27. Jay Galbraith, *Organization Design* (Reading, MA: Addison-Wesley, 1977).

28. Cyril O'donnell, "Ground Rules for Using Committees," *Management Review* (October 1961): 63–67.

29. György Szabados, "The Study of Group and Team Management in Agribusiness Companies," *Applied Studies in Agribusiness and Commerce* (2009): 75–77.

30. "The Payoff from Teamwork," *Business Week*, July 10, 1989, 56–62.

31. Frank LaFasto and Carl Larson, *When Teams Work Best* (Thousand Oaks, CA: Sage, 2001).

32. Ibid.

33. Mark R. Edwards and Michael W. Woolverton, "Appraising Work Team Performance: New Productivity Solutions for Agribusiness Management," *Agribusiness* (1986): 43–53.

34. George Munchus, "Employer-Employee Based Quality Circles in Japan: Human Resource Implications for American Firms," *Academy of Management Review* (April 1983): 255–261; see also Ricky W. Griffin, "Consequences of Quality Circles in an Industrial Setting: A Long-Term Field Experiment," *Academy Of Management Journal* (June 1988): 280–304.

35. This section is based on J. Richard Hackman, ed., *Groups That Work (And Those That Don't)* (San Francisco, CA: Jossey-Bass, 1990); see also J. Richard Hackman, *Leading Teams: Setting the Stage for Great Performances* (Boston: Harvard Business School Press, 2002).

36. Kristin J. Behfar, Randall S. Peterson, Elizabeth A. Mannix, and William M. K. Trochim, "The Critical Role of Conflict Resolution in Teams: A Close Look at the Links Between Conflict, Conflict Management Strategies, and Team Outcomes," *Journal of Applied Psychology* (2008): 170–188.

37. James Thompson, *Organizations In Action* (New York: McGraw-Hill, 1967).

38. Patrick Nugent, "Managing Conflict: Third-Party Interventions for Managers," *Academy of Management Executive* (2002): 139–148; and Danny Ertel, "How to Design a Conflict Management Procedure that Fits Your Dispute," *Sloan Management Review* (Summer 1991): 29–39.

39. "Duncan Family Farms," *Arizona Food Industry Journal* (June 2011): 12–14.

40. Shaw, *Group Dynamics*.

41. Davis, *Group Performance*; see also John P. Wanous and Margaret A. Youtz, "Solution Diversity and the Quality of Group Decisions," *Academy of Management Journal* (March 1986): 149–159.

42. Irving L. Janis, *Groupthink, 2nd ed.* (Boston: Houghton Mifflin, 1982).

43. Andre L. Delbecq, Andrew H. Van De Ven, and David H. Gustafson, *Group Techniques for Program Planning* (Glenview, IL: Scott, Foresman, 1975); see also David M. Schweiger, William Sandberg, and James W. Ragan, "Group Approaches for Improving Strategic Decision Making: A Comparative Analysis of Dialectical Inequity, Devil's Advocacy and Consensus," *Academy Of Management Journal* (March 1986): 51–71.

44. "Tobacco Settlement Agreements—Philip Morris USA,(n.d.) at www.philipmorrisusa.com (accessed October 2, 2010); "A Broken Promise To Our Children: The 1998 State Tobacco Settlement 11 Years Later," (December 12, 2009) at www.tobaccofreekids.org (accessed October 2, 2010); "Globalization of Cigarette Smoking," (May 17, 2010) at www.tobaccoworld.org (accessed October 2, 2010); David Kesmodel, "Roll-Your-Own Cigarette Machines Help Evade Steep Tax," *The Wall Street Journal* (August 30, 2010) at online.wsj.com (accessed October 3, 2010); David Kesmodel, "Cigarette Ruling Deals a Blow to 'Roll-Your-Own' Retail Shops," *The Wall Street Journal* (October 3, 2010): A5; David Kesmodel, "Close, and It Is a Cigar," *The Wall Street Journal* (September 23, 2010) at online.wsj.com (accessed October 3, 2010).

# Managerial Communication

## LEARNING OBJECTIVES

After studying this chapter, you should be able to:

- Define managerial communication and discuss its importance and pervasiveness.
- Describe the communication process and relevant behavioral processes.
- Discuss barriers to and skills for effective communication.
- Discuss oral, nonverbal, and written communication, and effective listening.
- Describe formal communication in organizations, including vertical and horizontal communication, information systems, and the chief information officer.
- Describe the grapevine and discuss its advantages and disadvantages.

## *Making Scents of It All*

**A**mong the species that populate our planet, we humans are among the worst at using our noses for sniffing, even though we may be inclined to poke them into places they do not belong. Other species are far more competent at recognizing scents and detecting the presence of hidden substances, e.g., dogs use their superior sense of smell to locate hidden drugs or bodies, and pigs use theirs to sniff out truffles in French forests. Yet humans are the only species that use the scents of other biologicals (plants and animals) to enhance our food and to communicate our personal auras (Figure 20.1).

**FIGURE 20.1** This field of lavender plants in full bloom will be harvested to manufacture expensive perfume products intended to communicate and influence feelings.
© iStockphoto/Carmen Martinez Banus

This often-overlooked corner of the agribusiness industry is a significant global participant with around $21 billion per year in revenue. Because of its significance many companies are searching for a niche. This is important because if you can describe an aroma or a scent, you may be able to duplicate it. But unlike the "taste" of a food product, which can be characterized with some degree of accuracy, the aroma or scent of a product may often defy description. Dr. Alirio Rodrigues, University of Porto, Portugal, and his colleagues have compiled a list of scent descriptions. They found that eight general terms could work as descriptors to more than 2,000 specific scents. The general terms were citrus, floral, green, fruity, herbaceous, musk, oriental, and woody. Dr. Rodrigues noted that one reason why people describe smells differently is a weakness of the nose. Despite attempts of scientists to define perfumes and fragrances, the discipline remains more art than science, but everyday art and science compete for the upper hand in this exotic corner of agribusiness.

Art is represented by Dominique Durbrana, a 54-year-old Frenchman living in Italy, who is a Sufi (Muslim) convert and a perfume genius. Durbrana is a perfumer and he invents his own fragrances. "Smells talk to people in three ways," he says. "First, through personal experiences, which are mostly from childhood, e.g., the smell of your vacations—a pine forest. Secondly, through culture, the smell of tobacco or cumin for Arabs." Finally, "there are genetic memories, the smell of burning wood, the sea, or hay. They affect even people who have never smelled them before."

Science is represented by a 52-year-old American living in Lexington, Kentucky. She is the CEO of Allylix, a San Diego-based biotechnology firm, and works with 16 biologists and chemists. Allylix has spent five years engineering baker's yeast to produce a variety of smells in a 6,000 gallon biofermentation tank. "Long term we're looking at producing these (smells and flavors) for a fifth or tenth of the cost. We see opening up new markets in personal care and household cleaning." The CEO continues by explaining that Allylix is well aware of the future profits that can be realized: "The price of grapefruit essence, nootkatone, is $2,000/pound; and orange essence, valencene, is $600/pound. The incentive for the work is obvious."

The art/science duality has stopped no one from looking for his or her niche in this industry. It is dominated by a few global brands, but it is a fragmented industry with many forms of expression. "We are tracking the trends that define American consumers," says Steve Tanner, president of Arylessence, a leading U.S. fragrance and flavor company. "When you understand the perceptions and preferences shared by millions of Americans, the consumer landscape comes sharply into focus. Our insights help marketers develop innovative fragrances and flavors that not only connect where consumers are today, but where they will be tomorrow."

This marketing venue is open to all, the large and small. Fragrances are used in all sorts of feel-good ways, including hospital rooms for cancer patients and Japanese office buildings to pep up workers, but their efficacy is largely anecdotal. "We don't make perfume, we blend for therapy," said Trygve Harris, owner of Enfleurage, a New York aromatheraphy store. "We have headache helper, a blend for sleep, one for comfort, if you're feeling stressed and you need to withdraw. We can't say 'This is good for that.' We have to say, 'This has been shown to…' or 'Some people use this for…'"

But do not confuse feeling good about yourself with feeling good. "People do not understand how much of an impact the indoor air has upon their health," said Angel De Fazio, president of the National Toxic Encephalopathy Foundation, a Las Vegas-based organization that lobbies on behalf of people with brain injuries caused by chemicals.

Or, with feeling… Dr. Noam Sobel, at the Weizmann Institute in Israel, has a far different approach to our noses. He is treating people suffering from "locked-in syndrome—mentally alert individuals who are physically paralyzed." It seems that sniffing is regulated by the soft palate, the flap of tissue in the back of your mouth. It is controlled by cranial nerves and not connected with the spinal column. Spinal damage does not affect these nerves. Dr. Sobel and his team have developed a device to communicate by dictating to a computer to steer wheelchairs and type messages using coded patterns of sniffing. Their first volunteer, a 51-year-old woman suffering for seven months with locked-in syndrome, was able to write a letter to her family after three weeks of training. Communication by reverse engineering one of our senses has very serious potential.

Obviously this segment of the agribusiness industry has unexplored potential. Maybe some of it should remain unexplored. In 2003, Pat Thomas, general curator at the Bronx Zoo conducted an experiment with 24 fragrances and two cheetahs. He measured how much time these "big cats" spent interacting with the scents. No contest. Estee Lauder's "Beautiful" occupied the cheetahs for an average of two seconds; Revlon's "Charlie," 15 seconds, Nina Ricci's "L'Air du Temps," 10 minutes. But for Calvin Klein's "Obsession for Men" it was an average of more than 11 minutes. A spokesperson for Coty, makers of Calvin Klein's scent, declined to comment. The perfume industry does not mind associating a good time with "wild" but perhaps not wildlife.

This corner of agribusiness has much potential for many forms of expression. It represents a novel form of communication and is certainly worth the time to sniff around.[1]

## INTRODUCTION

As the preceding vignette suggests, new companies will soon appear to manufacture and sell fragrances that are designed to help communicate a message or feeling. Those companies will be taking advantage of the huge potential for building a company around scents that millions of other companies and individuals will use to enhance their communication skills or messages.

This chapter explores managerial communication in the more established ways. First, communication is defined and its importance in organizations is discussed. Then the communication process is more fully described in terms of a comprehensive model, and the influence of important behavioral processes is discussed. Several barriers to effective communication are presented, as well as several skills that result in more effective communication. The chapter then describes three important types of communication—oral, nonverbal, and written—as well as the important skills of effective listening. Finally, the chapter presents formal and informal organizational networks and ways in which they may be managed effectively.

### FOOD FOR THOUGHT 20.1

The blue Chiquita stickers, which have become an iconic symbol for high-quality fruit, are placed on each banana by hand to avoid bruising the delicate fruit.

## The Nature of Communication

Communication has been studied from many different perspectives: as shared meaning between individuals, as the "glue" that binds an organization together, as both a one-way and a two-way process, and as the means for coordinated action in organizations. One of the major keys to an organization's success is the open and candid communication that occurs among everyone in the company. As we examine the role of communication in organizations, it is important to define communication and discuss the pervasiveness of communication in the manager's job.

### The Definition of Communication

In the broadest sense, **communication** is the process of transmitting information. Thus, communication can take place between organizations, between organizational units, between computer systems, and between nations. However, when communication occurs between people, it is **interpersonal communication**.[2] When two people are talking on the telephone, when a speaker is addressing a large group, and when someone is reading a letter or

e-mail, interpersonal communication is taking place. (See Figure 20.2.)

Of course, the message being received may be quite different from the one that was transmitted. So it is useful to differentiate between simple communication and effective communication. *Simple communication* is merely the transmission of information from one person to another; for example, "We need 2,000 pounds of cheese shrink-wrapped quickly." *Effective communication*, in contrast, occurs when the message that is received has essentially the same meaning as the message that was sent. The above request, for example, should have indicated what time frame the message sender had in mind when he said "quickly"—in 30 minutes? within a couple of hours? by noon tomorrow? Clearly, then, it is important for managers to have the skills to communicate effectively.

**FIGURE 20.2** Communication in organizations occurs in a variety of forms.

© Dmitriy Shironosov/www.Shutterstock.com

## The Pervasiveness of Communication

Communication is one of the major ingredients of the manager's job.[3] This is illustrated in Table 20.1, which outlines a day in the life of a typical, although hypothetical, manager. Virtually all of the activities listed involve communication in one form or another. The manager writes something that others will read, reads what others have written, talks, and listens.

Research has clearly documented the pervasiveness of communication in management. One early study, for example, revealed that the average manager spends 59 percent of his or her time in scheduled meetings, 22 percent on desk work (such as writing and reading), 10 percent in unscheduled meetings, 6 percent on the telephone, and 3 percent walking around the company premises.[4] All of these activities involve communication.

**TABLE 20.1** Communication in the Daily Activities of a Summer Farms' Manager

| TIME OF DAY | ACTIVITY |
| --- | --- |
| Breakfast | Read *The Wall Street Journal* |
| 8:00–8:30 | Read and answer e-mails |
| 8:30–8:45 | Receive telephone call from field supervisor |
| 8:45–9:30 | Meet with other farm managers |
| 9:30–10:00 | Read report from marketing |
| 10:00–11:00 | Meet with foreman at corn farm #3 |
| 11:00–11:30 | Review the day's mail |
| 11:30–12:00 | Meet with CEO/Operations Manager |
| Lunch | Lunch with family or friends |
| 1:00–2:00 | Draft report for Josh Summers & other farm managers |
| 2:00–2:30 | Return three telephone calls |
| 2:30–3:00 | Interview prospective crew leader |
| 3:00–4:00 | Meet with organic foods task force |
| 4:00–4:30 | Check and send e-mails |
| 4:30–5:00 | Read report from legal adviser re: organic farm |
| 5:00–5:30 | Attend grandson's soccer game while going over some issues with 2 field hands |
| 5:30–6:00 | Stop by the potato harvesting site to visit with Vegetable Farms managers |
| After dinner | Browse *Progressive Farmer* magazine |

© Cengage Learning 2014

## The Communication Process

Now that we have defined communication and discussed the pervasiveness of communication for managers, we can explore the communication process itself in more detail. First, we present a complete model that explains all of the dynamics of communication. We then note important behavioral processes that influence how communication occurs.

### The Communication Model

As noted earlier, communication involves at least two people. These are represented in Figure 20.3 as the **sender** (the person who transmits the message) and the **receiver** (the person who receives the message).

The starting point in the communication process is an idea. This idea may be a fact, an opinion, an observation, or anything else the sender needs to transmit. The idea is then translated into a message to be sent to someone else. This message may be all or part of the idea. The sender may feel that it is best to send only part of the idea in the first attempt at communication, expecting to send the remainder of the idea later. In the **encoding** process, the message is put into the exact mix of words, phrases, sentences, pictures, or other symbols that best reflect the content of the message.

The message is then transmitted through one or more **channels**: a face-to-face meeting, a letter, a telephone call, a facial expression, or any combination of these. The message is received and retranslated by the process of **decoding** into a message. The receiver then combines the message with other ideas and may send a return message to the sender in the form of feedback, a response, or a new message.

The difference between simple communication and effective communication manifests itself when the symbols are decoded into a message and combined with other ideas the receiver may have. If the idea formed by the receiver is similar to the idea that the sender originally formulated, effective communication has occurred. On the other hand, if the ideas are different in one or more important ways, the communication was not effective.

As shown in Figure 20.3, the process may continue under conditions of two-way communication. That is, receivers may respond to the original message with a message of their own. Thus, the receiver becomes the sender, transmitting a new message to the original sender, who is now playing the role of the receiver.

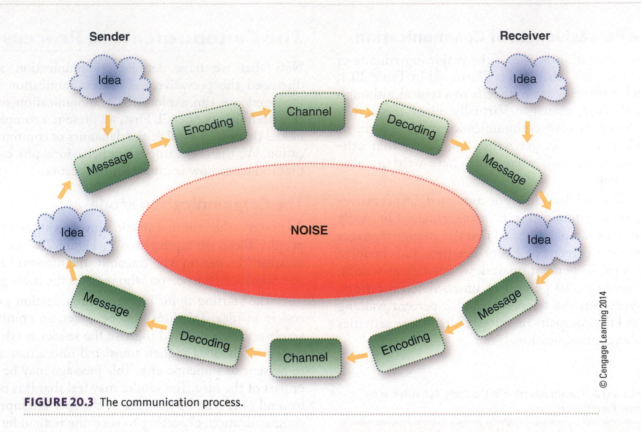

**FIGURE 20.3** The communication process.

The final part of the process is noise. **Noise** is anything that disrupts the communication process. It may be true noise, such as someone in another room or at another table talking so loudly that two people cannot hear each other speak, or a radio playing so loudly that the receiver cannot hear the sender's voice over the telephone. It can also be other things, however. A letter getting lost in the mail, a telephone call being disconnected, or a typographical error in a report can all reduce communication effectiveness.

---

**FOOD FOR THOUGHT 20.2**

The colors of the twist ties or plastic tabs on bread packaging indicate on what day of the week the bread was baked. This communication code was developed not to aid the customer but to tell the person stocking the shelves which loaves to remove from the shelves.

---

## Behavioral Processes and Communication

The communication process is also influenced by a number of important behavioral processes. Two of these are attitudes and perception, which were first introduced in Chapter 16.

**Attitudes.** **Attitudes** are sets of beliefs and feelings that individuals have about specific ideas, situations, or other people.[5] Each of us has attitudes toward school, jobs, other people, movies, politicians, sports teams, and almost everything else that may be a part of our lives.

Attitudes affect how we communicate with others in a variety of ways. For example, suppose you have two subordinates. You like one subordinate a great deal and have a high regard for her capabilities and dedication to the organization. In short, you have a positive attitude toward her. You strongly dislike the other subordinate, however, and question her capabilities and dedication. Your attitude toward her is negative. If the first subordinate asks for extra time off to visit a sick friend, you may respond in a favorable way and may convey feelings of concern. But if the second subordinate asks for time off for the same reason, you may deny the request and may even question her truthfulness. In other words, attitudes about the receiver affect the sender's encoding processes, and attitudes about the sender affect the receiver's decoding processes.

**Perception.** The other behavioral process that strongly affects communication is perception.[6] **Perception** refers to the processes by which we receive and interpret information from our environment. Most people are familiar with real incidents or stories in which two or more witnesses observe the same accident but report different details. Such differences are attributable to perception, which in general affects communication through familiarity. People tend to perceive things from a frame of reference with which they are comfortable. In one classic study, for instance, executives were asked to read a case about problems at a steel mill and then to describe the nature of those problems. Five out of six sales managers said the problems were related to sales, but four out of five production managers saw the problems as being related primarily to production.[7]

One of the ways that perception affects communication is through stereotyping. **Stereotyping** is the process of categorizing people into groups on the basis of certain presumed traits or qualities. A lot of research has been conducted on the process of stereotyping in organizations. One study found that business students tended to stereotype older workers as less creative, more resistant to change, and less interested in learning new skills.[8] Stereotyping affects communication most in the encoding and decoding phases. When a sender encodes a message into symbols, the sender is making assumptions about what symbols the receiver will understand. The same process occurs in reverse when the receiver decodes the symbols into a message. When making assumptions about other people, we often err in our perceptions due to stereotyping the other person's ability to understand the symbols.

It follows that perception affects communication in other ways as well. If managers see things from a biased perspective, they will respond accordingly. A manager who perceives a problem as falling into a certain area will communicate in a certain way with other managers in that area. If the manager views the problem in a different way, the communication will be different.[9] For example, if a Summer Farms manager says, "We need to reduce the loss of plants during their first month," others do not know whether (1) he has no idea as to the cause of the problem or (2) he perceives a problem with seed quality, planting depth, irrigation, or something else. In short, perceptions of the situation and the receiver affect the sender's encoding of the message into symbols, and perceptions of the sender and the situation affect the receiver's decoding of the symbols into the received message.

# Barriers to and Skills for Effective Communication

As we have indicated, the message sent is not always the message received. Whether the manager is the sender or the receiver, he or she must be very skilled at trying to make both the sent message and the received message the same. In general, these skills center on understanding the barriers to effective communication and knowing how to overcome those barriers.[10]

## Recognizing Barriers to Effective Communication

Of the many different kinds of barriers to effective communication shown in Figure 20.4, some are associated with the sender, some with the receiver, and some with both.[11]

**Sender Barriers.** From the standpoint of the sender, problems can arise due to inconsistency, credibility, and reluctance. *Inconsistency* occurs when the person sends conflicting messages. *Credibility* problems occur when the individual is considered to be unreliable. For example, when a public official makes statements that are later found to be untrue, the official will encounter credibility problems. Finally, people are sometimes simply *reluctant* to communicate. This may be the case when the news is bad or unpleasant.

**Receiver Barriers.** From the standpoint of the receiver, selective attention, receiver attitudes, and value judgments can be barriers. Not concentrating, trying to think of a response, letting attention wander, looking around when someone is talking, and daydreaming can all impede effective listening. Specifically, selective attention occurs when the receiver pays attention to only part of the message being sent. Sometimes people have already made up their minds about what a speaker is saying, so their attitudes get in the way of listening to the speaker's points and arguments. Making value judgments based on personal beliefs may also affect the way that the receiver hears the message.

Noise, as we have seen, also poses difficulties with effective communication. So can **communication overload**, which occurs when the sender is transmitting too much information for the receiver to process adequately.[12] Overload occurs, for example, when a lecturer talks too fast for students to take

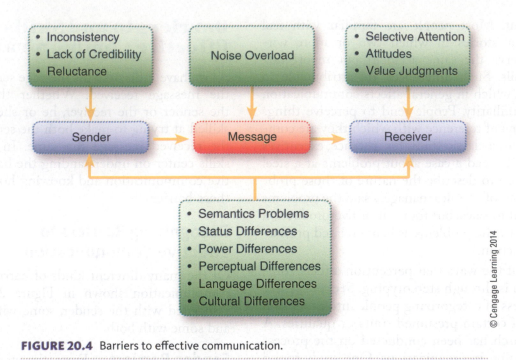

FIGURE 20.4  Barriers to effective communication.

notes as well as when someone is trying to watch television while a roommate listens to the stereo. See Figure 20.5.

FIGURE 20.5  Mixed messages make communication difficult.

**Sender and Receiver Barriers.**  Other barriers can be attributed to both the sender and the receiver. Obvious communication problems can occur when operating in a culturally diverse business situation or in the international business arena. When the receiver and the sender speak *different languages*, the encoding and decoding processes can be quite difficult. Anthropologists suggest that it is very difficult to understand a culture without knowing the language. The reverse is also true; it is difficult to understand the language without taking into account its cultural context. When doing business internationally, it is essential to understand the language and culture of the host country for communication efficiencies as well as for the benefit of the business.[13]

**Semantics problems** (problems with word meanings) involve both parties. For example, when an instructor says the course is "challenging and rigorous," the student may hear this as "hard and picky." One specific type of semantics problem arises due to professional **jargon**, which is the use of words that have specific meaning within a profession.[14] The computer industry is well known for the use of jargon particular to its industry. *Status* and *power differences* can also disrupt effective communication. For instance, a janitor may not be able to communicate effectively with a top manager, or a low-level manager may have problems communicating with a high-level manager even though all of them are in the same

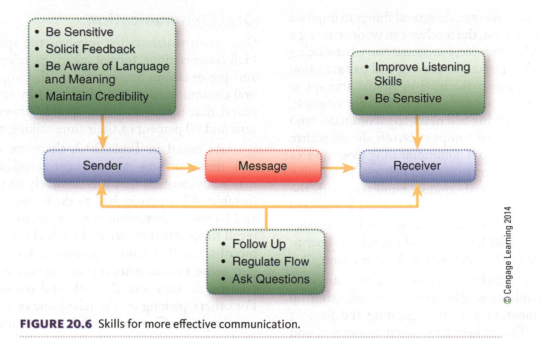

**FIGURE 20.6** Skills for more effective communication.

division. *Perceptual differences*, as described earlier in the chapter, also can disrupt communication. If one manager perceives an employee's sloppy work habits as laziness and another manager thinks the same habits indicate creativity, the two managers could have a difficult time discussing the worker's performance.

In today's increasingly diverse workforce, both domestically and internationally, *language and cultural differences* can present barriers for senders and receivers of communications. Does the lack of response of a Native American worker or the nodding of the head of a Japanese colleague indicate agreement with something that you said, or do they merely acknowledge that they are listening to you? Are highly participative processes equally espoused by all cultural groups? Does a pat on the back and a hug indicate the support of someone or does it suggest sexual harassment? Communication may be even more complex in the future than it is today.

## FOOD FOR THOUGHT 20.3

The newest kind of labels for fresh meat products are printed using a special ink to communicate the freshness of the meat. When the label becomes dark (smeared) enough to prevent the bar-code from scanning, the meat has surpassed the date of guaranteed freshness.

## Building the Skills for Effective Communication

Fortunately, there are things that managers can do to overcome some or all of these problems. The communication skills shown in Figure 20.6 can be used by the sender, the receiver, or both.[15]

The manager needs to be sensitive to different reactions that employees may have to communication they receive. For example, some workers might be happy to hear that their work schedule is being changed or their hours are being reduced. However, the manager should also expect that some may become upset and therefore temporarily hostile. Such sensitivity can keep the manager from getting upset if the subordinate says something in anger. The sender should also solicit feedback as a way to facilitate two-way communication. Asking the receiver if the message is understood, asking for opinions, and ensuring clarity in other ways enhance communication effectiveness.

Managers should also be aware of language and meaning. Employees almost always get concerned when they hear about major changes. Hence, a manager should not talk about an impending change as being "big" or "major" if it is in fact relatively routine or minor or if the employee will not be affected by it. The sender should also always attempt to maintain credibility. There is nothing wrong with admitting that you do not know something. Of course, it also helps to check facts and stay as up-to-date as possible.

The receiver also can do several things to improve communication. First, the receiver can work on being a good listener; for example, concentrate on what is being said, look at the speaker, be patient, and pay attention to meaning. Likewise, the receiver should attempt to be sensitive to the sender's perspective. For example, few people enjoy giving bad news, so subordinates who are receiving notice of company layoffs should realize that the manager who is communicating those notices probably is also feeling upset and helpless. In a multicultural or international context, familiarity with the culture and the language of host nations is essential for effective communication. This means that managers should study the language and the culture *prior* to initiating business in a different culture or country.[16]

Both the sender and the receiver can promote communication effectiveness by following up the communication and by regulating the flow of information. The sender can solicit questions, and the receiver can ask for clarification and demonstrate understanding. Suppose, for instance, that a sales executive calls a finance executive to schedule a meeting. The finance manager can send a note back to the sales executive confirming the meeting time and location. Both parties can also regulate information flows by working to prevent communication overload, the sender by making sure that he is not speaking too fast, and the receiver by interrupting and asking the sender to slow down when necessary.

# Forms of Interpersonal Communication

Interpersonal communication occurs in many different forms. In this section we will consider four types of communication skills that are necessary for managers: two forms of verbal communication (oral and written), nonverbal communication, and listening.

## Oral Communication

**Oral communication** involves the spoken word. Hall conversations, formal meetings, telephone calls, and presentations are examples. The importance of oral communication is underscored by research that found that managers often spend between 50 percent and 90 percent of their time talking to people.[17]

As noted in Table 20.2, there are advantages and disadvantages to oral communication. One major advantage is that it is relatively easy and comfortable. All a person has to do is open his mouth and let words come out. (Of course, many of us can recall times when we wished we had not opened our mouths, but that issue is addressed later.) There are also some people who, out of shyness or for other reasons, do have trouble with oral communication. For others, picking up the telephone or walking into a colleague's office to schedule a meeting is a relatively simple operation when compared to doing the same thing in writing. It has been suggested that more than half of all managers do not have confidence in their ability to write, so they feel more comfortable with the spoken word. Accordingly, they use oral communication whenever they can.[18]

The second major advantage of oral communication is that it facilitates immediate feedback. **Feedback** is the response from the receiver of a message to the sender of that message. If the sender wants an answer to a question or a verification that the listener understands what was said, all the sender has to do is ask. Similarly, the listener can interrupt to respond to the message or to seek clarification. An astute speaker can even tell how well the message is being received by looking at the facial expression of the listener.

Of course, there are major disadvantages to oral communication as well. For one thing, it can be quite inaccurate. If we simply talk "off the top of our heads," we can confuse the facts, omit important

**TABLE 20.2** Advantages and Disadvantages of Oral, Nonverbal, and Written Communication

| ORAL COMMUNICATION | | NONVERBAL COMMUNICATION | | WRITTEN COMMUNICATION | |
|---|---|---|---|---|---|
| ADVANTAGES | DISADVANTAGES | ADVANTAGES | DISADVANTAGES | ADVANTAGES | DISADVANTAGES |
| Easy to use | Causes inaccuracies | Gives completeness to communication | Can conflict with verbal communication | Is fairly accurate | Hinders feedback |
| Facilitates feedback | Provides no record | Can convey images without verbalizing | Can give unintended messages | Provides a record | More time consuming |

© Cengage Learning 2014

points, distort things, and so forth. We might also say some things that should not be said or divulge information inappropriately. Similarly, the listener may not hear everything accurately or may misunderstand or forget important details.

Finally, oral communication provides no permanent record of what has been communicated. After a conversation has taken place, the parties may have to recall details of what they said. Because memory can be quite faulty, the lack of a permanent record of the conversation can cause major problems.

## Written Communication

**Written communication** is the transmission of a message through the use of written words, including memos, letters, reports, notes, e-mail, and text-messaging. The advantages and disadvantages of this kind of communication are essentially the opposite of those of oral communication, as Table 20.2 points out.

One key advantage is improved accuracy. If a manager chooses to use written symbols to communicate with someone, the manager can dictate the letter, proofread it, revise it, check the facts, and have the letter retyped before it is mailed. Likewise, written communication can provide a relatively permanent record of the communication. The sender and the receiver can talk about the details of their exchange over the phone several months later and still agree about the contents of the letter. This, of course, means that managers must be careful about what they put into writing, particularly through electronic media.[19]

A major disadvantage of written communication is that it hinders feedback. After the manager decides to write a letter, several days can go by before it is typed, mailed, delivered, and read. If timing is critical, major problems can result from such delays. Of course, electronic media, overnight delivery services, and facsimile machines have reduced this factor to a great extent. Overnight physical delivery is expensive for routine correspondence, and many legal documents still require original signatures.

Written communication is generally more time-consuming than oral communication. It takes only a few seconds to pick up the phone and call someone, but it may take days or weeks to get a written response using traditional mail. And most managers generally do not prefer to use traditional written communication, preferring instead to use electronic means. We have noted their lack of confidence in their writing skills. The same study found

that most managers regard only about 13 percent of their mail as valuable to them, and they find almost 80 percent of it to be poorly done.[20]

## Nonverbal Communication

Another form of interpersonal communication to consider is **nonverbal communication**—communication that uses facial expressions, body movements, and gestures rather than words to convey a message.[21] It has been suggested that 55 percent of a message is transmitted through facial expression and body movement and that another 38 percent is conveyed by inflection and tone.[22]

In general, people communicate nonverbally in three ways. First, they use the setting, which is where the communication takes place, and the nature of its surroundings. The manager who sits behind a huge desk in a big chair in front of a wall covered with awards, honors, and accolades is clearly in a position of power and authority, which will usually influence the communication process. A visitor who sits in a small chair in front of the desk will be in a very different position in the communication process.

Second is body language. One element of body language is the distance we stand from someone with whom we are talking. Close contact can connote intimacy or hostility, and eye contact can convey positive or negative feelings (Figure 20.7). For example, looking the sender or receiver in the eye connotes honesty to most Americans but may make those from other countries feel uncomfortable. Body and arm movement, pauses in speech, and style of dress are also important parts of body language.

**FIGURE 20.7** Understanding facial expressions, body movements, and other forms of nonverbal communication is important to managers.

Another aspect of nonverbal communication that has become increasingly important in recent years relates to the cultural differences between people of different nations. We have seen that distance between people affects communication. Appropriate distances vary among cultures: the English and Germans stand farther apart than Americans do when talking, whereas the Japanese, Mexicans, and Arabs stand closer together. Looking the sender or receiver in the eye is considered rude to people of some other cultures whereas, as mentioned above, it generally indicates trustworthiness to most Americans.

The third aspect of nonverbal communication is the imagery conjured up by language.

Former NBA player Charles Barkley, who uses colorful language and catchy phrases when he talks, conveys a different image and meaning than does the quieter Michael Jordan. Such a person can transmit messages of confidence, boldness, or aggression, or of foolhardiness and recklessness. More mundane, bland language may convey images of cautiousness and thoroughness or of timidity and indecisiveness.

Nonverbal communication also has advantages and disadvantages. Nonverbal communication can provide confirming images to verbal (oral and/or written) communication. Certain facial expressions and body movements can indicate that the sender is confident in the message, thereby indicating to the receiver that the sender is an expert and can be trusted. On the other hand, other body movements or gestures can send a conflicting message, as when an expert speaks too softly and hesitatingly, and hence is perceived as an individual who is unable to explain or provide details. When this happens the receiver can be confused about which message to believe. Another example is the competent but "pushy" individual whose assertiveness causes the receiver to wonder whether the rush for action suggests that the product may not hold up under closer inspection.

**FOOD FOR THOUGHT 20.4**

More than half of the information conveyed in a conversation is probably communicated through nonverbal channels.

## Effective Listening

Listening is one type of communication that many people overlook; yet it must occur if effective communication is to take place. **Listening** is the process of receiving encoded symbols via the ear from a sender and decoding them into a message to be interpreted. Most education and training programs focus only on what to say, not on how to listen. Talking is more straightforward and easier to measure, control, and learn. Listening, however, is harder to predict, to control, and to practice.[23]

Most errors in listening occur because the receiver is not actively listening to what the sender is sending. This may be due to several reasons. The receiver may be discounting what the sender is sending because it goes against what the receiver already believes. The receiver may be thinking about what to say next rather than really listening to what the sender is saying. The receiver may not hear what the sender is saying because of noise in the system or because the symbols sent by the sender may not mean the same thing to the receiver that they do to the sender. This confusion occurs more often when the sender sends symbols that have specific meanings in a technical jargon that is not familiar to the receiver. Whatever the reason, the receiver must listen very carefully to the symbols sent by the sender and ask for clarification whenever the symbols are unclear. In addition, the sender can assist the receiver in listening better by speaking slowly, distinctly, using symbols (words) that are familiar to the receiver; and stopping periodically to ask if the receiver understands the message so far.

Skills in listening can be improved by asking others (spouse, children, manager, employees, peers) how they view you as a listener. Encourage them to be honest, listen carefully to what they have to say, and be ready to make changes in your listening habits (Figure 20.8). Second, you can tape your

**FIGURE 20.8** Effective communication requires the receiver to listen effectively while the sender communicates clearly.

## A FOCUS ON AGRIBUSINESS
### Talk and Listen

Agribusiness managers must communicate with employees, suppliers, processors, merchandisers, lenders, landlords, banks, and others. In addition, many agribusiness managers also must communicate with family members because either directly or indirectly they are involved with the business.

Communication, especially in a family business, is not always easy. Not only do individual differences in personality, communication styles, skills, and expectations come into play, but also family loyalty and culture. When an issue or a problem arises, it is tempting to let it rest a bit before talking about it with family members. On the other hand, many times a discussion cannot

wait even if it may "ruffle feathers." If the family is communicating, business communication will flow naturally.

But there are non-family employees in family-owned businesses, too. Frequently they will not speak up in groups or with other family members around, so managers should use the "management by walking around" (MBWA) technique. They should walk around the work area three to four times a day and informally chat with employees. Employees are more likely to mention concerns in this context if you are willing to listen. And, after all, listening is an important part of communication.[25]

conversations with others (with their permission, of course), listen to the tapes critically, evaluate how much of the time you spent talking versus listening, and see how much of what others said that you missed the first time.[24]

## Managerial Communication

Managerial communication is a part of the formal organizational hierarchy: that is, it occurs between people in various positions. Three elements of managerial communication are communication in the formal organization, management information systems, and the position of chief information officer.

### Communication in the Formal Organization

As we noted earlier, the manager's job is filled with activities that involve communication (see Figure 20.2). In Chapter 3 we discussed the roles that managers must fill within the organization. Recall Table 3.1 in which we showed the interpersonal roles, the decisional roles, and the informational roles. Interpersonal roles involve interacting with supervisors, subordinates, peers, and others outside the organization. Decisional roles require managers to seek out information to use in making decisions

and then communicate those decisions to others. Informational roles focus specifically on acquiring and disseminating information both inside and outside the organization. Each of the ten roles discussed in Table 3.1 would be impossible to fill without communication. The new roles identified as part of the integrated framework also required extensive use of communication.

Within the organization managerial communication is usually experienced only between managers and employees whereas external communication involves interacting with those outside the organization. This external communication is done by many different people. The sales force is often the group that does the most external communication. In addition to directing communication to potential buyers of the products or services of the organization, the sales force can also be a valuable receiver of external communication, which in turn it communicates back to others in the company as a form of internal communication. Listening to the sales force can be the best way an organization can stay close to customers and their needs. U.S. Surgical Corporation, for example, utilizes its sales force to bring back valuable information from surgeons for the company to use in developing new products that solve operating-room problems.[26] Summer Farms receives feedback sometimes directly from consumers but also indirectly from the wholesalers and retailers when they communicate their level of satisfaction by reducing or increasing their purchases.

## Vertical Communication.

**Vertical communication** takes place between managers and their subordinates. It can flow both down and up the organization. *Downward* communication includes the assignment of new job responsibilities, information that will assist subordinates in performing their job duties, or simple information about the organization. Such communication helps subordinates know about aspects of the organization that affect them. Unfortunately, managers do not always do a good job of keeping their subordinates informed.

*Upward* communication is also a vital part of organizational functioning. Information from employees keeps top management in touch with day-to-day operations of the company, significant successes and failures, and potential difficulties. Upward communication is also often associated with the practice of whistle blowing, as when Jerome LiCari discovered that his employer, Beech-Nut, was claiming its apple juice for babies was pure apple juice when in fact it was chemically modified. He went to his manager, who did not listen. Then he went to the president of the company and received the same non-response. Finally, he resigned and told his story to the media.[27] One of the reasons that Walmart executives spend so much time traveling to stores is to get upward communication from those people who are in closest contact with the customers.

## Horizontal Communication.

**Horizontal communication** takes place between two or more colleagues or peers at the same level in the organization. It is critical when there are high demands for coordination and integration. For instance, if the marketing manager is planning a new advertising campaign that will probably increase product demand by 10 percent, the manufacturing manager needs to be aware of this increase so that plans can be made for additional production. Similarly, if a plant manager locates a new supplier who will deliver an important raw material more reliably and at a lower price than current suppliers, it is important that this information be passed along to other plant managers in the company.

## Information Systems

Another important aspect of managerial communication is the various kinds of information systems the organization creates to manage the official flow of information within the business. In recent years management information systems have changed dramatically, mainly because of breakthroughs in electronic communication capabilities. Electronic typewriters and photocopiers were early breakthroughs, but the personal computer, electronic networks, and facsimile machines accelerated the process. Cellular technology and similar equipment for transmitting data make it possible to be in communication with almost everyone in the world from almost anywhere in the world, instantly.[28] Information systems will be covered extensively in Chapter 24. We need only remember here that they are an important part of many organizational communications activities and should be as fully institutionalized as possible if they are to be effective.

### FOOD FOR THOUGHT 20.5

The USDA communicates information about maple syrup by assigning grades that are based on lightness of color and strength of flavor, which are dependent mainly upon when the maple syrup is harvested.

## The Chief Information Officer

A final element of managerial communication is the position of **chief information officer (CIO)**. This person is the executive who oversees all aspects of information technology, such as computing, office systems, and telecommunications. Many people have never heard of the CIO because it is a new executive position that is just being created in many organizations.[29] The CIO position is becoming important because top managers are increasingly recognizing the value of information to the organization and of having a qualified individual responsible for managing it.

Many managers do not have the title but still fill the role of CIO; others, of course, actually are given the title. As the manager of all aspects of information technology, the CIO reports to the CEO or chairperson and usually concentrates on long-term strategic issues, leaving the nuts-and-bolts part of the communication job to technicians. Numerous companies have begun to use the CIO concept, including American Airlines, Bunge, Firestone, Heinz, Pillsbury, and Wells Fargo. Smaller companies also need to designate a company spokesperson who is trained well enough to at least do no harm when speaking with the press. In times of crisis, such as a food contamination outbreak or a case of workplace violence, this spokesperson can give the public the information it deserves while enabling upper management to concentrate its time on the problem at hand.

# Informal Communication: The Grapevine

The final element of organizational communication to be addressed in this chapter is the informal communication network that exists in all organizations: the **grapevine**.[30] The nature of organizational grapevines is illustrated in Figure 20.9, which shows a hypothetical organization and three messages that have wound themselves through the organization.

These paths reveal several interesting aspects of the grapevine. First, the grapevine can start anywhere; that is, any individual can start the process simply by telling someone something. Second, some people are included in virtually all of the messages. These people serve as focal points along the grapevine, receiving most messages and passing most of them on to others. Third, the grapevine flows in all directions. Messages can go up, down, or laterally in the organization.

Finally, not everyone is included; some people neither receive nor pass on informal news.

The grapevine exists because being social is one characteristic of human nature (see Figure 20.10). People like to interact with others. Because much of this interaction involves talking, information will naturally be passed among many different people. For another, the grapevine is often used as a way to get power. Controlling information, regardless of its source, makes anyone a more powerful person. And grapevines emerge in response to deficiencies in the formal communication network in the organization. If people are curious about something and do not hear about it officially, they are likely to solicit the information from others, through the grapevine.[31]

## Advantages and Disadvantages of Grapevines

The grapevine has many attributes, some of them good and some bad. On the plus side, the grapevine can be used to transmit information quickly, it builds a sense of togetherness and a feeling of being part of the same team, managers can use it to try out ideas or get informal reactions to potential decisions, and employees may find it useful when managers fail to communicate openly.

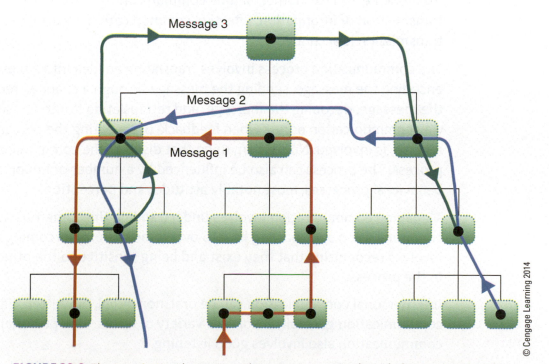

© Cengage Learning 2014

**FIGURE 20.9** The organizational grapevine. Three messages move through the organization using different routes.

**FIGURE 20.10** Information spreads through the grapevine anytime workers are together, as during this office break.

On the other hand, the grapevine can also be detrimental to the organization. The information carried along the grapevine can be inaccurate, or it may be information that the manager would prefer to keep confidential. Apple Computer had a problem with rumors and the grapevine. In the personal computer industry, where knowing "secrets" has become a status symbol, important product information was leaking out to other companies and to the media. Apple hired a manager of information security, who created an internal campaign to stop the leaks. The campaign included posters and lapel buttons reading, "I know a lot but I can keep a secret." Some insiders have reported that the campaign did not work and may have made matters worse. In this case, the grapevine may have extended beyond the organizational boundaries and may have implications for organizational success or failure.[32]

Managers should be fully aware of the potential benefits and pitfalls of the grapevine. Perhaps the best advice is to maintain open communication with employees at all levels and to respond quickly to inaccurate information. If people can come to their manager and get straight answers, they are less likely to pay attention to gossip and rumors.[33]

## CHAPTER SUMMARY

Communication is one of the major ingredients of the manager's job. Interpersonal communication is the process of transmitting information from one person to another. Simple communication is the mere transmission of information. Effective communication occurs when the transmission is accurate.

The communication process involves translating an idea into a message, encoding the message, sending the message through a channel, receiving the message, decoding the message, and retranslating it into an idea. Two-way communication occurs when feedback, or repeating the process in reverse, is involved. Noise is anything that disrupts the communication process. The process can also be influenced by a number of important behavioral processes, most notably attitudes and perception.

Managing communication involves understanding the numerous barriers to communication and knowing how to overcome them. Overcoming barriers involves recognizing that they exist and being sensitive to the other person in the process.

Interpersonal communication can be oral, nonverbal, or written. Effective communication generally involves a variety of forms. Interpersonal communication also involves good listening.

Managerial communication is communication that occurs as a part of the formal organization. Vertical communication occurs between bosses and subordinates and can flow downward or upward. Horizontal communication takes place between two or more colleagues or peers at the same level in the organization. The person who is in charge of information technology in major organizations is known as the chief information officer, or CIO.

The grapevine is the informal communication network that exists in all organizations. Managers sometimes use the grapevine to try out ideas or to get informal reactions to potential decisions. However, the grapevine also has several disadvantages.

## CHAPTER ACTIVITIES

### REVIEW QUESTIONS

1. What is the difference between simple communication and effective communication?
2. Summarize the communication model. How do attitudes and perception affect communication?
3. Note four major barriers to effective communication and four important ways to overcome those barriers.
4. What are the relative advantages and disadvantages of oral, nonverbal, and written communication?
5. What is a CIO? Why is a CIO important?

### ANALYSIS QUESTIONS

1. Relate an incident in which your attitude affected communication.
2. Which communication form (oral or written) do you prefer? Why?
3. Describe where you have used or observed nonverbal communication taking place.
4. If nonverbal messages contradict oral statements, which do you believe? Why?
5. Think of the last "message" you picked up from the grapevine. As far as you know, was it accurate or inaccurate?

### FILL IN THE BLANKS

1. The means of transmitting an idea, such as a face-to-face meeting, an e-mail, a telephone call, or a facial expression is a communication _____ .
2. The transmission of messages between bosses and subordinates is called _____ communication.
3. Transmitting too much information for the receiver to process adequately results in _____ _____ .
4. Anything that disrupts the communication process is called _____ .
5. Translating a message into a mix of words, phrases, sentences, pictures or other symbols is known as _____ .
6. An informal communication network is called _____ .
7. Transmitting a message between two or more colleagues or peers at the same level in the organization is called _____ communication.
8. The use of words that have specific meanings within a profession or group of people is referred to as _____ .
9. Transmitting messages through body movements, facial expressions, and gestures is called _____ communication.
10. A response from the receiver of a message to the sender of that message is called _____ .

## CHAPTER 20 CASE STUDY
### TRASH TALK

The agribusiness industry is remarkably productive; however, this productivity comes with a price. The United States produces about 591 billion pounds of food each year, and up to half goes to waste. This does not occur in one place or venue; everyone contributes: harvesting leaves 10 to 15 percent of a crop in the field, the transit of food loses another 10 to 15 percent, supermarkets dispose of 30 million pounds of food a day, the average shopper throws away 15 to 25 percent of the food she buys, commercial kitchens (hospitals, schools, etc.) dispose of another 4 to 10 percent, and restaurants discard around 10 percent of the food they purchase. Most of this wasted food ends up in landfills; only about 2.5 percent is composted. You might think the business to pursue is trash collecting. After all, someone must haul away all of this food.

However, David Steiner, CEO, Waste Management Inc., would disagree. In fact, he thinks the future of a traditional garbage-collection company looks bleak. "Picking up and disposing of people's waste is not going to be the way this company survives long term." Everywhere he looks he sees threats to his business: reusable grocery bags, concentrated laundry detergent in small packages, and Walmart embracing the idea of zero waste.

Walmart? As the world's largest retailer, Walmart is on a mission to determine the social and environmental impact of every item it puts on its shelves. It wants to create a universal rating system that scores products based on how environmentally and socially sustainable the products are over the course of their lives. This is the green equivalent to a food product's nutrition label. The company is looking to academics for help. People such as Jon Johnson at Sam M. Walton College of Business at the University of Arkansas and Jay Golden at Arizona State University have joined the fray.

"Because of Walmart's leadership ... they were able to set a standard for the entire industry," said Len Saucers, vice-president, customer development at Unilever. "Walmart has invited the Targets, the Costco's and the Tesco's of the world to come up with a solution ..." said Tim Marin, associate director of external relations, Procter & Gamble.

"We know we need to get ready for a world in which energy will only be more expensive," explained Michael T. Duke, CEO, Walmart. Its goal is to cut 20 million metric tons of greenhouse gas emissions from its supply chain by the end of 2015. This is the equivalent to removing 3.8 million cars from the road for a year.

This is a major reason why Waste Management is looking for ways to extract value—energy or materials—from the waste stream. "Think Green," is Waste Management's new tag line. It is more than a line. The company is trying to communicate a new direction to its stakeholders. It has invested in or acquired nearly 25 companies that capture materials or energy from stuff that's thrown away. Already Waste Management's waste-to-energy plants generate enough electricity to power 1.1 million homes—more than the entire solar energy industry can generate in the United States.

While most yard clippings are recycled, more than 95 percent of food waste goes to landfills. The cost of transporting this heavier and wetter waste exceeds whatever revenues can be realized from turning it into fertilizer or compost. "Organics, for us, is difficult, but it's also a huge opportunity," CEO Steiner says. Turning a ton of food waste into compost generates about $40 to $50 in revenue. Generating transportation fuels, like gasoline, from a ton of food waste could generate $200 to $250 in revenues. "If we can figure out a way to process and convert organic material better than anybody else, we're going to own that material," challenged Steiner.

Speaking of trash, there is some noise in the system. Frito-Lay, the maker of Sun Chips, thought it was onto a great idea—a healthy snack food in a 100 percent biodegradable chip package. The biodegradable bag, polylactic acid to chemists, will disappear in 14 weeks while a conventional potato chip bag takes over 100 years to degrade in a landfill. Great idea, great product, great ... noise?

"It is louder than the 'cockpit of my jet,'" said J. Scot Heathman, a U.S. Air Force pilot on a blog. Indeed, in a sound test that involved opening, closing, and reaching inside to get the chips, the bag recorded

95 decibels versus a Frito-Lay Tostitos Scoops bag at 77 decibels. He was not alone; over 40,000 people signed up on a Facebook page to complain. It seems even when you "go green," it must be quiet green or good green.

Realizing there is no escaping the noise, Frito-Lay attempted to communicate the positive aspects of the new bag. The company featured the noisy package in some of its marketing. In stores, the company attached signs to shelves that read: "Yes, the bag is loud, that's what change sounds like." However, there is noise and then there is noise. Customers also made a lot of noise. Consumer outcry won out and PepsiCo Inc. pulled the noisy packaging from five of its six flavors. Sun Chips Original kept the new packaging.

Apparently, "Going Green" is not enough. We need to "Go Quiet Green." [34]

### ▶ Case Study Questions

1. What elements of the communication model are most evident in this case?
2. What forms of communication are evident in this case?
3. How should organizations communicate their efforts to "go green" to stakeholders?

## REFERENCES

1. "Making Sense of Scents," *The Economist* (December 9, 2010) at www.economist.com (accessed December 10, 2010); "Smell-bound," *The New York Times* (August 13, 2010) at www.nytimes.com (accessed December 11, 2010); Kerry A. Dolan, "Allylix Sniffs Out Biotech For New Fragrances," *Forbes Magazine* (November 8, 2010) at www.forbes.com (accessed December 10, 2010); "New report by fragrance company Arylessence describes nine 'deep trends' that define U.S. consumer attitudes and predict future buying behaviors," *PR Newswire (United Business Media)* (October 22, 2010) at www.prnewswire.com (accessed December 9, 2010); Jennifer A. Kingson, "In Competition for Your Nose," *The New York Times* (July 28, 2010), E3; Jennifer Steinhauer, "Breath of Fresh Air, Indoors, in Las Vegas," *The New York Times* (May 25, 2010), A16; "Not to be Sniffed At," *The Economist* (December 9, 2010) at www.economist.com (accessed December 11, 2010); Ellen Byron, "Big Cats Obsess Over Calvin Klein's 'Obsession for Men,'" *The Wall Street Journal* (June 8, 2010) at online.wsj.com (accessed December 10, 2010).

2. Pamela S. Shockley-Zalabak, *Fundamentals of Organizational Communication, 7th ed.* (Boston: Allyn & Bacon, 2008). Norman B. Sigband and Arthur H. Bell, *Communication for Managers, 6th ed.* (Cincinnati, OH: South-Western, 1994).

3. Henry Mintzberg, *The Nature of Managerial Work* (New York: Harper & Row, 1973).

4. Henry Mintzberg, "The Manager's Job: Folklore and Fact," *Harvard Business Review* (July–August 1975): 49–61.

5. Martin Fishbein and I. Ajzen, *Belief, Attitude, and Behavior: An Introduction to Theory and Research* (Reading, MA: Addison-Wesley, 1975).

6. E. E. Jones and R. E. Nisbett, *The Actor and the Observer: Divergent Perceptions of the Causes of Behavior* (Morristown, NJ: General Learning Press, 1971).

7. D. C. Dearborn and H. A. Simon, "Selective Perception: A Note on the Departmental Identification of Executives," *Sociometry* (1985): 21–143.

8. B. Rosen and T. H. Jerdee, "The Influence of Age Stereotypes on Managerial Decisions," *Journal of Applied Psychology* (61, 1976): 428–432.

9. James P. Walsh, "Selectivity and Selective Perception: An Investigation of Managers' Belief Structures and Information Processing," *Academy of Management Journal* (December 1988): 873–896.

10. Courtland L. Bovee, John V. Thill, and Mukesh Chaturvedi, *Business Communication Today, 9th ed.* (Upper Saddle River, NJ: Prentice-Hall, 2008).

11. Jerry Wofford, Edwin Gerloff, and Robert Cummins, *Organizational Communication* (New York: McGraw-Hill, 1977).

12. "Information Overload is Here," *USA Today*, February 20, 1989, 1B–2B.

13. Gary P. Ferraro, *The Cultural Dimension of International Business, 5th ed.* (Upper Saddle River, NJ: Prentice-Hall, 2005).

14. Jane Whitney Gibson and Richard M. Hodgetts, *Organizational Communication: A Managerial Perspective, 2nd ed.* (New York: HarperCollins, 1991).

15. Wofford, Gerloff, and Cummins, *Organizational Communication*.

16. Ferraro, *The Cultural Dimension of International Business*.

17. Mintzberg, *The Nature of Managerial Work*.

18. Walter Kiechel III, "The Big Presentation," *Fortune*, July 26, 1982, 98–100; Michael T. Motley, "Taking the Terror out of Talk," *Psychology Today* (January 1988): 46–49.

19. "Watch What You Put in That Office E-mail," *Business Week*, September 30, 2002, 114–115; and Nicholas Varchaver, "The Perils of E-mail," *Fortune*, February 17, 2003, 96–102.

20. Kiechel, "The Big Presentation."

21. Michael B. McCaskey, "The Hidden Messages Managers Send," *Harvard Business Review* (November–December 1979): 135–148.

22. Ibid.

23. Theodore Kurtz, "Dynamic Listening: Unlocking Your Communication Potential," *Supervisory Management*, September 1990, 7.

24. Kurtz, "Dynamic Listening: Unlocking Your Communication Potential."

25. Aadron Rausch, "Communication in the Family," *Progressive Forage Grower* at www.progressiveforage.com (accessed July 15, 2010); Vera Bitsch and Elaine K. Yakura, "Middle Management in Agriculture: Roles, Functions, and Practices," *International Food and Agribusiness Management Review* (2007): 1–28; Tom Peters and Nancy Austin, *A Passion for Excellence* (New York: Random House, 1985).

26. Jennifer Reese, "Getting Hot Ideas from Customers," *Fortune*, May 18, 1992, 86–87

27. Chris Welles, "What Led Beech-Nut Down the Road to Disgrace," *Business Week*, February 22, 1988, 124–128.

28. Janet Guyon, "The World Is Your Office—What Will It Be Like When You Can Be at Work or On the Net or Trading Stocks Wherever You Are, Whenever You Want? Here's a Glimpse," *Fortune* (June 12, 2000) at money.cnn.com (accessed July 28, 2010).

29. Gordon Bock, Kimberly Carpenter, and Jo Ellen Davis, "Managements Newest Star," *Business Week*, October 13, 1986, 160–172.

30. Keith Davis, "Management Communication and the Grapevine," *Harvard Business Review* (September–October 1953): 43–49.

31. Nancy B. Kurland and Lisa Hope Pelled, "Passing the Word: Toward a Model of Gossip and Power in the Workplace," *Academy of Management Review* (2000): 428–438.

32. "At Apple Computer Proper Office Attire Includes a Muzzle," *The Wall Street Journal*, October 6, 1989, A1, A5; Brian Dumaine, "Corporate Spies Snoop to Conquer," *Fortune*, November 7, 1988, 68–76; "Mind What You Say: They're Listening," *The Wall Street Journal*, October 25, 1989, B1.

33. "Job Fears Make Offices All Ears," *The Wall Street Journal*, January 20, 2009, B7.

34. "Throwing Away Our Food," *The Wall Street Journal* (October 16, 2010), C12; Marc Gunther, "Waste Management's New Direction," *Fortune Magazine* (December 6, 2010) at money.cnn.com (accessed December 8, 2010); Stephanie Rosenbloom, "At Walmart, Labeling to Reflect Green Intent," *The New York Times* (July 16, 2009), B1; Stephanie Rosenbloom, "Walmart Unveils Plan to Make Supply Chain Greener," *The New York Times* (February 26, 2010), B3; Caroline Scott, "Let's Hear It for the Sound of Compostable Packaging," *Nutra* (August 23, 2010) at www.nutraingredients-usa.com (accessed December 10, 2010); Suzanne Vranica, "Snack Attack: Chip Eaters Make Noise About a Crunchy Bag," *The Wall Street Journal* (August 18, 2010) at online.wsj.com (accessed December 10, 2010); Suzanne Vranica, "Sun Chips Bag to Lose Its Crunch," *The Wall Street Journal* (October 6, 2010) at online.wsj.com (accessed December 11, 2010).

# CONTROLLING IN AGRIBUSINESS

**6**

# CHAPTER

# 21

# Organizational Control

## LEARNING OBJECTIVES

After studying this chapter, you should be able to:

- Describe the nature of control and why it is necessary, as well as some areas of control and the planning-control link.

- Identify and discuss various approaches to control, such as steering, concurrent, postaction, and multiple controls.

- Identify and discuss the steps in establishing a control system.

- Describe the characteristics of effective control.

- Discuss reasons for resistance to control and how to overcome this resistance.

- Describe the responsibility for control that lies with line managers and the controller.

- Identify the importance of food safety and the regulations supporting it.

# Heavy Metal

**S**odium. It is a metal that is so explosive that, to control it, it must be stored in liquid to avoid contact with air. Another element, chlorine, which normally exists in a gaseous state, also must be carefully controlled. Indeed, chlorine is so dangerous that it was once used on the battlefields of World War I. When combined, these two make a compound that is among the more corrosive materials on our planet. It is so corrosive it will destroy steel pipes and the hulls of massive war ships. Its simple chemical formula insures its deep penetrability and ability to ionize.

What do you do with a dangerous compound like this? You eat it. The two chemicals together taste great on French fries. We call it common table salt, Sodium Chloride (NaCl) (Figure 21.1).

**FIGURE 21.1** Salt being prepared for packaging and shipment.
© Olaf Speier/www.Shutterstock.com

The importance of salt in agribusiness has long been recognized. The Romans loved salt; they used 25 grams a day to flavor vegetables, to preserve fish, and to extend the utility of hams and sausages. Cato, whose second-century BCE (Before Common Era) book, *De Agricultura*, is the oldest complete book of Latin prose, was enamored of salted pork products. His family name said it all. It was Marcus Porcius—Porky Mark. Speaking of salt and pork, Poland is another ancient country famous for its salt mines and its pork products. The Chinese were among the first to mine salt, with the oldest salt mine dating back to 800 BCE, but they did not apply salt directly to food. Soy sauce, a fermented derivative of soybeans using salt, was the condiment of choice.

The control of salt resources was important. Wars were fought over salt mines, salt marshes, and other sources of salt. This compound was essential for savory cooking. The addition of salt to a recipe enhanced the flavor of the food. Salt's major attribute is the unique effect it has on taste buds. "Salt is a pretty amazing compound," says Alton Brown, Food Network star, in a Cargill video. "You might be surprised by what foods are enhanced by its briny kiss." Because salt is an essential element of cooking and food preparation, its use is ubiquitous. Its use typically is not controlled, however, and as such exceeds moderation from a health standpoint. "Salt is very addicting," claims Sidney Alexander, a cardiologist at the Lahey Clinic Medical Center (Boston, MA). "Even though there are good salt substitutes and other spices (my patients) can use, they have a hard time giving it up."

The stuff is everywhere. In the United States and the United Kingdom, 40 percent of salt is used in the chemical industry, 40 percent on slippery roads, and 20 percent in food and agriculture. This disparity in usage is reflected in a variety of prices: refined salt for the chemical industry, $150/ton; the lowest grade salt for de-icing roads, $50/ton; and for "foodies" and gourmets, *fleur de sel,* French sea salt, $70,000/ton.

What started around 5,000 years ago as a simple preservative has become omnipresent in processed foods. Three Oreo cookies account for 11 percent of the recommended daily intake (RDI) of salt. A serving of low-fat cottage cheese equals more than 25 percent of salt RDI. The U.S. Department of Agriculture reported that processed foods, along with restaurant meals, account for 80 percent of the salt in the American diet.

Even the so-called "recommended daily salt intake" is subject to interpretation. The U.S. Food and Drug Administration recommends a teaspoon of salt a day, or 2.3 grams for healthy adults, below middle age. However, for older individuals, children, younger adults, and people with hypertension, the FDA recommendation is only 1.5 grams per day.

In 2010, the City of New York decided to act to control this salt tide. Mayor Michael Bloomberg, unveiled a health initiative to encourage food manufacturers and restaurant chains across the country to curtail the amount of salt in their products. The goal is laudable, but it will be difficult to enforce. New York is asking for national cooperation in its initiative. While the goals are probably not achievable, bringing national attention to the issue is an achievable goal.

But control isn't just a governmental or corporate issue. We are battling ourselves. Salt is essential, but too much salt becomes a serious problem. The food industry is responding to this threat, albeit sometimes not directly. For instance, ConAgra submitted a study that asserted that more healthcare savings could be achieved ($58 billion) by people controlling their diets and eliminating 100 calories daily than from all salt reduction plans ($24 billion).

The issue is complicated, personal, and historical. Perhaps Michael Alderman, a professor at the Albert Einstein College of Medicine, summed it up best when he said, "I'm always worried about unintended consequences."

Heavy metal just gets heavier.[1]

# INTRODUCTION

Unintended consequences must be controlled in some manner and to some degree, whether dealing with chemicals or humans. Control was introduced in Chapter 2 as one of the four basic functions that managers must perform. Organizations that neglect control are likely to face severe consequences.[2] On the other hand, they can also make a mistake by concentrating too much on control—by over controlling.[3]

This chapter—the first of four devoted to the control function—provides an overview of control. We begin by examining the nature of control in more detail. We discuss approaches to control and how managers establish control systems. Then we distinguish between effective and less effective control. Behavioral issues in managing control are discussed next. Finally, we briefly describe who is responsible for control. Subsequent chapters in this part discuss three particularly important areas of control: total quality management, operations and technology management, and information systems management.

# The Nature of Control

**Controlling** is the process of monitoring and adjusting organizational activities toward goal attainment.[4] In some ways control is like the rudder of a ship. Without a rudder, a ship's captain could still cause the ship to move; but it would be impossible to steer the ship in any particular direction, and it might end up going in circles and eventually run aground. Likewise, a manager can get things done without control but will eventually run into serious difficulties. Control, then, like a rudder, helps steer and guide the organization in the direction its managers set.

## Reasons for Control

PepsiCo underwent restructuring because its managers realized that consumer buying habits were changing, shifting toward juices and teas and away from soft drinks.[5] Similarly, Summer Farms added organic vegetables to its product line and looked at changing its mix of organic and non-organic vegetables in 3 to 5 years according to whether the national demand for organic foods shows signs of continued growth. American Greetings instituted financial controls because it had years and years of ups and downs while its major competitor, Hallmark, was almost always profitable.[6] The parent organizations for IHOP, Chili's, and the Cheesecake Factory saw strong performance around 2008 through careful cost control.[7] Each of these examples reflects a company's response to a need for greater control. PepsiCo responded to changing consumer tastes; American Greetings responded to financial fluctuations; and the restaurants controlled costs. Not unlike the restaurants, the consumer staples sector (including firms such as Tyson Foods, Dr Pepper Snapple Group, and PepsiCo) is one of the stronger agribusiness sectors because the companies have a global presence and are not tied to just U.S. consumers.

In general, there are three basic reasons for control. Foremost among these is the contemporary *environment*. We have seen several times in previous chapters how contemporary environments change rapidly and how organizations need to respond to these changes. L.L. Bean struggled in recent years because of environmental shifts. IBM was forced to respond because domestic competitors like Apple, Dell, and dozens of clone manufacturers were continually increasing the quality of their products while reducing their prices. At the same time, demand for mainframe computers, an IBM staple, plummeted. In 2009 problems with pork led to bans, causing some people to change their eating habits. That, in turn, caused a drop in the stock prices of companies like Smithfield Foods, Hormel, and Tyson.[8] Control is one of the primary channels for recognizing the need for such responses.

A second reason for control is *organizational complexity*. Contemporary organizations are so complicated that a single manager cannot hope to grasp all of their inner workings. Thus, control is necessary to help the manager monitor internal operations. A properly designed control system can provide data on raw materials inventory, work-in-process inventory, and finished goods inventory. Without such a system, the manager can never get a true fix on what the company's inventory actually is. For example, a few years ago Emery Air Freight bought one of its competitors, Purolator Courier Corporation. Emery then tried to merge Purolator into its existing operations without changing its existing control systems. The organizational complexity created by the merger almost destroyed the firm because managers were not able to effectively control operations.[9] Years later Emery did go out of business, and its successor business, Menlo Worldwide Forwarding, was then bought by UPS in 2004.[10]

A final reason for control is the way that *small errors and problems, left unchecked, can grow rapidly into much bigger ones*. A satellite enroute to Jupiter that is only a little off course at launch time, for example, may ultimately miss the planet by millions of miles if the error is not identified and corrected. Similarly, a small deviation in costs in a manufacturing plant can grow significantly if uncorrected, or a small increase in crop damage from insects may translate into hundreds or thousands of barrels of soybeans and result in a price increase. Whistler Corporation, a manufacturer of radar detectors, once confronted a small manufacturing error by deciding to simply fix the defects after assembly was complete. Defects escalated rapidly from 4 percent to 9 percent to more than 25 percent. One day a manager realized that 40 percent of the plant's workforce was involved in fixing defective products.[11] Control can help the manager detect and correct small problems before they grow into bigger ones.

## FOOD FOR THOUGHT 21.1

If the price of cottonseed oil can increase by only a penny per pound, the value of cottonseed crop would increase $24 million in an average year.

## Areas of Control

There are four basic areas of control in most organizations: physical resources (equipment, physical plant, product quality), human resources (personnel, stakeholders), information resources (public perceptions, "good will," brand integrity), and financial resources (assets, cash, liabilities). As shown in Figure 21.2, financial resources are usually at the center of the control process.

**Physical resources control** deals with such areas as inventory control (having neither too much nor too little inventory), quality control (ensuring that products are being made to appropriate quality standards), and equipment control (having the proper equipment to do the job). Quality control is particularly important. For that reason, Chapter 22 addresses that topic more fully.

**Human resources control** focuses on such activities as employee selection and placement (hiring the right kinds of employees and assigning them to appropriate jobs within the organization), training (upgrading employee skills), performance appraisal (assessing employee performance), and compensation (paying neither too much nor too little). As we noted in Chapter 16, enhancing the fit between people and their jobs benefits both the organization and its employees. For example, if employees are underskilled, the firm will have to train them, supplement their performance, or accept less or inferior work. Overskilled people may become bored and unhappy and eventually leave.

**Information resources control** involves making sure that various forecasts and projections are prepared accurately and on a timely basis, that managers have access to the information they need to make decisions effectively, and that the proper

**FIGURE 21.3** Accountants provide financial control information for management.

© iStockphoto/Steve Cole

image of the organization is projected to the environment. Without the right information, managers cannot make decisions or they may make poorly conceived decisions.[12]

Finally, as already noted, **financial resources control** is all-important. First, financial resources themselves must be controlled (Figure 21.3). For instance, the organization needs to have enough cash on hand to be able to function but not so much that resources are used inefficiently. Second, many of the other areas of control relate to financial resources. Improper inventory management costs money, as do poor employee selection, inaccurate forecasts, and so forth. More than any other resource, financial resources are needed to maintain the other resources of the firm and to keep it on sound footing.

## The Planning-Control Link

In many respects, controlling is the other side of the planning coin. This **planning-control link** is shown in Figure 21.4. Note in particular that planning and control continually cycle into one another, with organizing and leading serving as ways to get the actual work of the organization done. Gantt charts were introduced in Chapter 10 as a planning tool. Once a project or activity is undertaken, the Gantt chart becomes a control tool as well, monitoring progress of the project.

The normal approach in establishing this linkage is for management to determine plans for the future and simultaneously specify control conditions to keep the organization moving toward achieving those plans. That is, managers simultaneously specify

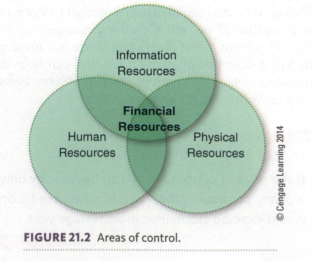

**FIGURE 21.2** Areas of control.

© Cengage Learning 2014

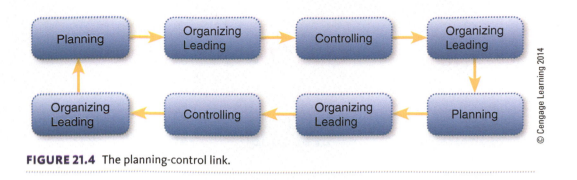

**FIGURE 21.4** The planning-control link.

© Cengage Learning 2014

where they want the firm to go and how they will know that it is headed in that direction. Organizing and leading activities also come into play as the organization implements the plans.

Control helps management determine whether to adjust plans. Suppose a firm plans to increase sales by 20 percent over the next 10 years. At the end of the first year, an increase of 2 percent suggests that things are on track. No increase in sales during the second year, however, suggests that modifications to plans may be necessary, such as increasing advertising or lowering the original projection. Subway used this approach as it expanded in the 1990s. If in any given year this target is not met, a reassessment of current expansion plans takes place, and in some cases renewed efforts to make up the shortfall the next year. In a later section we explore how to adjust things if plans are not being achieved as expected.

## Approaches to Control

Another important perspective that we need to consider is approaches to control. Most managers agree that there are three basic approaches: steering, concurrent, and postaction control.[14] As shown in Figure 21.5, steering control deals with inputs from the environment, concurrent control focuses on transformation processes, and postaction control is concerned with outputs to the environment.

### Steering Control

**Steering control** (also called *preliminary control* and *feed-forward control*) monitors the quality and/or quantity of various kinds of resources before they enter the system.[15] Firms like General Foods and Procter & Gamble, for instance, pay extra attention to what kinds of people they hire for future management positions.

### A FOCUS ON AGRIBUSINESS
#### Cost Control

Green Plains Renewable Energy, Inc., operates in three market segments: ethanol production, agribusiness, and marketing and distribution. It tries to maintain an environment of continuous improvement so that it can achieve its goal of increasing efficiency and effectiveness as a low-cost producer of ethanol. To do that it employs an extensive cost-control system at each of its plants. This way it is able to continuously monitor each plant's performance and then use that data to develop strategies for cost reduction and efficiency that can be applied across the whole organization.

Gary Townshend, CEO of Grasslands Consultants LLC, says that dealing with five different dairy markets—California, the upper midwest, the northeast, the southeast, and the southwest—presents unique opportunities. He is aggressively pursuing a low-cost agribusiness model. He said, "You have to be disciplined to adjust to price increases, inflation, and farming inputs."

But it is not just big firms that need to engage in careful cost control. Even small businesses—including restaurants, cafeterias, and "mom-and-pop" cafes—must monitor costs.[13]

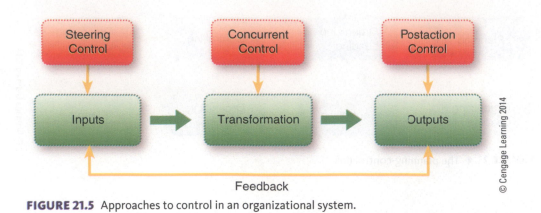

FIGURE 21.5  Approaches to control in an organizational system.

When Sears orders merchandise to be sold under one of its store brands, it specifies rigid standards to ensure appropriate levels of quality for the prices it will charge. Television networks such as CBS and NBC monitor the commercials that potential sponsors intend to run to be sure that appropriate standards are being met. Financial inputs are thus monitored to the extent that some sponsors are rejected. And some organizations control information inputs by contracting only with the best market research firms and paying attention to only the most valid economic forecasts.

Following the lead of successful Japanese firms, more and more companies are emphasizing steering control as part of a strategy to boost product quality.[16] By requiring suppliers to provide higher quality parts, for instance, manufacturers can enhance the quality of their own products. Each major U.S. automaker, for example, has increased numerous quality standards it demands of its suppliers. Ford alone has implemented more than 1,000 tougher standards since 1988.[17] To keep costs down, many automakers have turned to Korea to produce the same part; 41 of the world's 50 largest auto parts suppliers were located in Korea as of 2008.[18]

## Concurrent Control

**Concurrent control** (also referred to as *yes/no control* and *screening control*) focuses on activities that occur as inputs are being transformed into outputs (Figure 21.6). For example, a company might design various inspection stages during a production process to catch problems before too much damage is done. Green Things Landscaping crews inform their managers before planting when the nursery stock that the company has purchased appears to be inferior. Likewise, the performance of employees is usually assessed at regular intervals. Hopefully, care is taken to ensure that the information about ongoing operations that is

FIGURE 21.6  Worker using computers to control production lines in an insecticide factory is an example of concurrent control.

provided to managers is accurate. Financial resources are also carefully monitored through periodic audits.

Concurrent control is also becoming more widespread. As part of most employee involvement or participation programs, for example, workers are given new responsibility for halting production when problems arise. And when they point out flaws and inefficiencies, managers are more willing to listen to them.

## Postaction Control

**Postaction control** deals with the quality and/or quantity of an organization's outputs. When ConAgra inspects finished products before they are shipped, it is using postaction control. Rewarding employees after they have done a good job is an example of postaction control of human resources. Top management commonly screens news bulletins and press releases before these items leave the organization; this represents postaction control of information. And payments of dividends to stockholders and investments in the stock market are checked in order to provide postaction control of financial resources.

As steering and concurrent controls increase in usage, postaction control is perhaps becoming a bit less important. If higher quality inputs are going into production, for example, and if more problems are caught and corrected during production, it follows that fewer problems will exist in the finished product. While some sort of final inspection is likely always to be useful, firms may be able to test somewhat fewer units or to test only certain aspects of them in order to assure acceptable overall quality.

## Multiple Controls

Each of the various approaches to control is useful in particular circumstances. Indeed, most large organizations find it necessary to establish integrated control systems using multiple approaches. These systems include each of the three approaches to control noted above, and approach is integrated with the others for maximum effectiveness.

For example, consider a large manufacturing firm such as Boeing. Boeing carefully screens the engineers it hires, the materials it buys, the smaller firms it subcontracts with, and the financial solvency of airlines that place orders for new planes (steering control). Boeing also carefully monitors each stage of the construction process as each new plane is assembled. Information about new technology and the overall health of the airline industry is also tracked on an ongoing basis to assure that the firm's procedures and decisions are kept current. And the performance of individual engineers and managers is assessed regularly. Cash flow is also monitored continuously to make sure the firm has sufficient capital on hand.

After each plane is finished, it is subjected to numerous inspections and checks to make sure that it is flight worthy. To assure production control, managers get bonuses for completing work ahead of schedule. New information gleaned from each project is added to the firm's information system to assist in future projects.

Such integrated and comprehensive systems are necessary to provide adequate control for large, complex organizations.[19] This is especially true for multinational corporations with operations spread around the globe. Many smaller firms also find it necessary to create and maintain multiple control systems.

### FOOD FOR THOUGHT 21.2

During the packing process, eggs are separated by size, according to minimum weight per dozen: jumbo (30 oz.), extra large (27 oz.), large (24 oz.), medium (21 oz.), small (18 oz.).

## Establishing a Control System

A **control system** is a mechanism used to ensure that the organization is achieving its objectives. Regardless of which approach to control a manager is taking, he or she must follow four basic steps in establishing a control system or framework.[20] The four steps in control systems are illustrated in Figure 21.7.

### Setting Standards

The first step in control is to set **standards** or targets against which performance will be compared. For example, a fast-food outlet may set the following standards:

1. Customers will be served within four minutes of their entrance into the restaurant.

2. Drive-through customers will have their orders filled within five minutes of the time they enter the drive-through queue.

3. All tables will be wiped clean within three minutes after a customer leaves.

Note that each standard is stated in objective, measureable terms. Moreover, each is very clear and specific and also has a specific time frame.

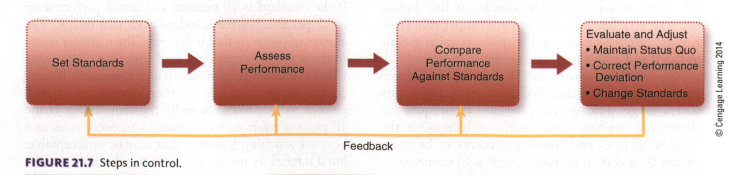

**FIGURE 21.7** Steps in control.

© Cengage Learning 2014

Standards are usually derived from, and therefore consistent with, the goals of the organization. The standards given above are appropriate for an organization whose primary goals are related to customer service and satisfaction. If growth were of greater importance, however, the standards might instead reflect increases in sales or number of customers served. In similar fashion, a university coming under criticism for its overemphasis on athletic programs may attempt to offset this criticism by shifting its emphasis back to academics. It may set a standard that 75 percent of all student athletes will graduate within five years of their initial enrollment, for example.

Of course, the specificity of standards will vary according to the level of the organization to which they apply. For example, a single Burger King restaurant has many different standards, all very specific and somewhat narrowly focused on areas related to customer service, cleanliness, and efficiency. At the corporate level, on the other hand, Burger King's standards are likely to be fewer and less narrowly focused. The corporation probably has performance standards for sales growth, growth in the number of outlets, menu adjustments, and so forth.

## Assessing Performance

The second step in the control process is to assess performance. This step relates specifically to those things the organization is attempting to control. That is, while assessing performance for control purposes is similar to and must obviously be consistent with the overall performance evaluation process, it must also be clearly focused on control-related standards. Thus, individual performance assessment at this stage may be a little less comprehensive than normal ongoing performance evaluations.

When managers are establishing standards, they should also specify how progress toward those standards will be assessed. Suppose that Safeway has determined that the average customer spends five minutes waiting in the check-out line before being served. Because of increased competition from Kroger, Safeway may want to reduce waiting time to three and a half minutes, by eliminating one half minute every 6 months for 18 months. In a given store, then, the manager will need to measure current average waiting time, develop techniques for shortening waiting time (such as increasing the number of lines and training checkers to be more efficient), and then monitor progress to ensure continued improvement.

In many situations measuring performance is fairly easy, especially when the standards are objective and specific. In other settings, however, performance assessment is considerably more difficult. A manager taking over a struggling company may need an extended time to turn things around. Or a research and development scientist may not be able to produce consistent breakthroughs that can be evaluated immediately; instead, his or her contributions may come at irregular intervals, and their value may take some time to assess.

Finally, we should note that the appropriate time intervals for assessing performance also vary a great deal. In some areas, such as the strategic goals of Caterpillar or John Deere, performance measurement may be appropriate every six months, or even longer. In others, such as a grocery store, a gambling casino or a clothing store, it is necessary to assess performance every day along several different dimensions.[21]

---

**FOOD FOR THOUGHT 21.3**

Archer Daniels Midland (ADM) has 43 wheat flour mills with the combined capacity of a million bushels a day.

---

## Comparing Performance with Standards

After setting standards and measuring subsequent relevant performance indicators, the manager must compare the two. Although this sounds fairly easy, it is actually somewhat difficult in many cases. Some standards are relatively difficult to quantify. For instance, a standard of increasing customer or employee satisfaction is hard to measure accurately. Similarly, it may be difficult to assess a standard of achieving technological innovation.

Another difficulty relates to the fact that performance and standards are seldom precisely the same. If the standard is 10 percent and actual performance is 9.7 percent, has the standard been met or not? The answer, of course, is that it depends. The manager needs to draw on his experience and insight to determine whether the company needs to attain precisely 10 percent to be successful or whether a range from, say, 9 percent to 11 percent is really the same thing. If the 10 percent refers to an increase in recovery rates in a hospital, anything less than that may be unacceptable; but if it refers to improvements in the cleanliness of the company parking lot, some slippage may be acceptable.

## Evaluating and Adjusting

The final step in control is to evaluate and adjust standards and performance. Depending upon the circumstances, three different courses of action may be appropriate.

One response is to do nothing—to keep things just as they are. This action is clearly most appropriate when the assessed performance meets the targeted standard for performance. For example, if a manager's standard is to increase sales in her department by 12 percent this year and at the end of six months she has achieved an increase of 6.05 percent, she probably does not need to make any changes in what people in the department are doing.

More likely some action will be needed to correct a deviation in performance from the desired standard. For example, if the standard for another manager also is to increase sales by 12 percent this year and his department only has a 3 percent increase after six months, the department is clearly not on track to meet the standard. He may need to increase advertising, plan more promotional activities, and/or motivate his sales group. Correcting a deviation first requires giving performance feedback.

Of course, we also have problems if we are exceeding our standards by too much. Suppose our standard is to hire 100 highly qualified new employees this year. If we end up hiring 75 individuals in the first six months, we may need to cut back on recruiting and hiring. An example of exceeding standards is the introduction of iPad, for which sales took off immediately, leaving many dealers with insufficient inventories and waiting lists of customers. While this was not a bad position to be in, Apple's advertising dollars did not produce as much current revenue as might have been expected because there were too few units to sell. In addition, some customers were annoyed by having to wait.[22]

In some situations it may be appropriate to change the standards against which performance is being assessed. Unexpectedly strong competition, for instance, may necessitate lowering an organization's expectations for growth. On the other hand, if all employees are exceeding their standards easily, the standards may have been set too low to begin with. Unfortunately, when this happens and the standards are subsequently raised, employees will probably be resentful and angry toward the organization. Therefore, it is important to do a good job in the beginning.

## Effective Control

The manager's job would be greatly simplified if establishing organizational controls were easy. However, as we have seen, managing the control process is actually quite difficult and demanding.[23] What can a manager do to enhance the effectiveness of the organization's control system? In general, effective control has five attributes, as summarized in Table 21.1.[24]

## Integration

First, and perhaps foremost, control systems must be integrated into the overall organizational system. This is most critical in terms of planning. Given the cyclical nature of planning and control, as described earlier, it is logical and necessary that the planning and control systems be properly coordinated and integrated with one another for them to work smoothly.

To see how this works, consider the situation confronted by Knight-Ridder, one of the largest media corporations in the United States. Although Knight-Ridder published some of this country's most prestigious newspapers, problems caused the company to take strong action to enhance profitability. Two interdependent actions were used: Top management at all of the company's newspapers had to prepare five-year plans aimed at boosting profit margins to 20 percent or higher, and tight controls were implemented to ensure that these targets were met.[25]

The mechanics involved in achieving proper integration are fairly straightforward. Managers need to consider relevant control elements as they develop plans, simultaneously using goals, strategies, and tactics to establish complementary dimensions of

**TABLE 21.1** Characteristics of Effective Control

| CHARACTERISTIC | EXPLANATION |
|---|---|
| *Integration* | Establishing control systems that take into account organizational plans |
| *Objectivity* | Supplying detailed, verifiable information |
| *Accuracy* | Providing complete and correct information |
| *Timeliness* | Providing information when it is needed |
| *Flexibility* | Establishing control systems that accommodate changes in the organization or the environment |

© Cengage Learning 2014

the control system. Similarly, results provided by the ongoing control system make very useful resources for future planning cycles.[26] A well-integrated control system will permeate the entire organization.[27]

## Objectivity

A second characteristic of effective control systems is objectivity. This simply means that, to the greatest extent possible, the control system should use and provide detailed information that can be verified and understood.

For instance, suppose a sales manager asks two sales representatives to assess how their clients feel about the company and its products. One salesperson reports that he talked to 15 customers and that 10 of them liked what the company was doing, three were indifferent, and two had complaints. He also reports on the exact nature of the clients' likes and dislikes and provides an estimate of how much each intends to order next quarter. The other sales representative reports that she talked to a few people, that some were happy and some were unhappy with the products (although she isn't too sure of the reasons for their attitudes), and that sales will be okay next quarter. Clearly, the data the first sales representative provided will be more useful than that of the second.

Of course, the manager needs to look beyond simple numbers. A plant manager may appear to be doing a great job of cutting costs, but closer inspection might reveal that the manager is using substandard materials, pushing workers too hard, and padding reports. On balance, the control system should be as objective as possible, but not so dependent on figures that managers lose contact with what is actually going on behind the scenes.[28]

## Accuracy

Obviously, the control system must be accurate in order to be effective. If it is providing erroneous information, it may be doing more harm than good. In reality, of course, any number of things can allow inaccuracies to creep into the system. A plant manager may be providing incomplete cost figures to make herself look better, or a sales representative may be padding his expense account and collecting more reimbursement than he is owed.[29] At another level, a human resources manager may overestimate the company's minority recruiting prospects in order to diminish short-term pressure to meet affirmative action goals.

The critical nature of such inaccuracies becomes apparent when we consider how managers use the control system. If a manager signs a contract to

**FIGURE 21.8** This warehouse manager knows that an inventory inaccuracy could slow or shut down production.

provide merchandise for a figure below what the true production costs are, the firm will lose money (Figure 21.8). Hence, managers need to take every precaution to ensure the accuracy of the information they receive from the control system.

## Timeliness

It is also important that the information provided by the control system be timely. Timeliness does not necessarily mean speed, but it does mean that information is in the manager's hands when it is needed. The manager of a Walmart, for example, will want and need to know precise sales figures on a daily basis, but she may need inventory figures only every two or three months. And the corporate office will need only weekly or monthly, not daily, sales information.

In general, the need for timeliness is related to uncertainty; that is, the more uncertain the situation, the greater the need for timely information. When a new product is introduced, the manager may desire daily sales reports, but for an older, more established product the manager may need them only every week or every month.

## Flexibility

Finally, effective control systems tend to be flexible. They are able to accommodate adjustments and change in the organization or the environment.[30] Suppose a control system is designed to manage data about two hundred raw materials that go into producing the company's products. A new technological breakthrough allows the company to produce the same products with only half as many materials. If the control system is not flexible, the managers will have to scrap the entire system and develop a new one. On the other hand, a flexible system will be able to accommodate the changes.

In summary, effective control systems generally have five basic attributes: they are integrated with other organizational systems, they are objective, they are accurate, they are timely, and they are flexible. In the next section we explore other ways to enhance control system effectiveness.

## Managing Control

In addition to making control systems effective by promoting the attributes described above, managers must deal with resistance issues. Some people tend to resist control in particular, so managers need to understand why resistance occurs and what they can do to overcome it.

### Understanding Resistance to Control

The most common factors underlying resistance and the best ways to deal with them are summarized in Table 21.2.[31] Four of the most common reasons that people resist control are overcontrol, inappropriate focus, rewards for inefficiency, and accountability.

Organizations sometimes make the mistake of practicing **overcontrol**, or too much control. This can be particularly problematic when the control relates to employees. Employees may require a certain degree of control in the workplace, but they also want a reasonable degree of autonomy and freedom. For example, an organization may specify normal working hours and work-related expectations for its employees, but it will probably not be successful in trying to dictate personal behavior such as mannerisms, recreational preferences, and so forth. Attention is increasingly being paid to ways in which managers can get employees to exert personal control while simultaneously accommodating both individual and organizational goals.[32] Personal control issues can be important to new and growing businesses.

Another reason for resistance to control is **inappropriate focus**, which occurs when the control is too narrowly focused or does not provide a reasonable balance between different outcomes that are important. For instance, if a sales manager concentrates so much on sales increases that nothing else really matters, sales representatives may come to ignore other parts of their jobs. Likewise, if a university encourages and rewards publication and provides few incentives for professors to be good teachers, its faculty will gradually devote more and more time to research and less and less time to teaching.

In other cases, organizations end up rewarding inefficiency and perhaps not rewarding efficiency. Many departments rush to spend any of their budget that is going to remain at the end of the year regardless of whether the expenses are necessary. They do this because they feel that if they have money left over, management will assume they need less money next year. But if they spend all of their money, and perhaps even report a small loss, their budget is likely to be increased next year because they ran out of funds. Obviously, this situation rewards inefficiency.

A final reason for resistance to control is that effective control creates **accountability**. That is, a properly designed control system will allow a manager to determine how each department, and in many cases how each individual manager, is performing. As a consequence, managers become more accountable for their actions, decisions, and performance. Obviously, some managers will not object to such accountability, but others—especially those who are not doing a good job—will resist.

> ### FOOD FOR THOUGHT 21.4
> Annually, The Andersons, Inc., formulates, stores, and distributes nearly two million tons of dry and liquid agricultural nutrients.

### Overcoming Resistance to Control

As also shown in Table 21.2, managers can at least partially overcome employee resistance to control in a number of ways. One obvious method is to make sure that the control system is properly designed. In particular, if it is designed to have the attributes of effective control already discussed (integration, objectivity, accuracy, timeliness, and flexibility), employees will be less likely to object to it.

Second, employee participation can reduce resistance to control. If employees have a voice in designing the parts of the control system that

**TABLE 21.2** Managing Control

| REASONS FOR RESISTANCE | WAYS TO COMBAT RESISTANCE |
| --- | --- |
| Overcontrol | Design the system well. |
| Inappropriate focus | Encourage employee participation. |
| Rewards for inefficiency | Use Management By Objectives (MBO) |
| Accountability | Provide checks and balances |

© Cengage Learning 2014

directly affect them and also have avenues for suggesting modifications, they are more likely to accept the system as fair and reasonable. They will have a better understanding of how the control system was developed and how it contributes to overall organizational performance. And because they know it can be modified, they may be more willing to try it and see how it works, rather than resisting it out of hand.

A third approach to overcoming resistance to control is to use management by objectives, or MBO, which is described in Chapter 8 as a collaborative goal-setting technique between managers and their subordinates. Employees in a well-conceived and well-managed MBO system know exactly what is expected of them, how they should attempt to achieve those goals, and what their rewards will be if they succeed.

Finally, the control system should have a built-in provision for checks and balances. This simply means that it should provide a mechanism for checking and potentially adjusting for discrepancies. For example, suppose an employee who is dismissed for frequent tardiness denies that he was late very often. If the human resource manager can prove via thorough documentation that the employee was actually late many times, he (and other employees) will be more likely to see the control system as fair and equitable. Or suppose that a sales representative claims that she did not meet her quota because of unexpectedly high levels of competition from other companies. If the sales manager has adequate information to assess the validity of her claim, he can more easily accept or refute that claim.

## Responsibility for Control

A final issue to consider is the responsibility for control. Exactly who needs to be concerned with control? As shown in Figure 21.9, this responsibility is shared between line managers and specialized managers called controllers.

### Line Managers

In a very real sense, all managers are responsible for the control function in an organization. They all help design the system, are responsible for implementing and using it, and are at least partially governed by it.

As Figure 21.9 illustrates, the CEO is responsible for the overall control of the total organization. Each division head is responsible for control within his or her division. In general, such managers have some degree of autonomy in adjusting the control system to fit their own preferences and views on how control should be executed.[33] No matter what variations they implement, however, the system within each division must be consistent with the overall system of the organization.

### Controllers

In most organizations, control is also the specific responsibility of one or more managers who have the title of **controller**. As shown in the exhibit, a large organization may have a corporate controller (probably a staff position with the title of Comptroller) as well as other controllers in each division.[34] The controller's job

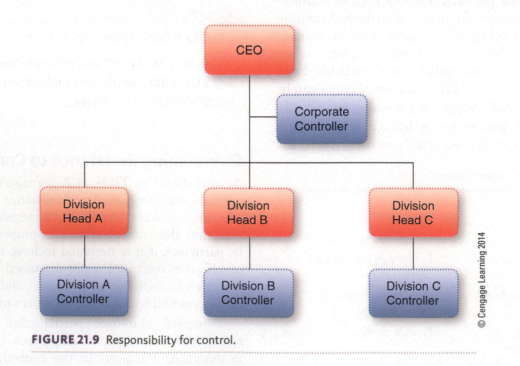

**FIGURE 21.9** Responsibility for control.

© Cengage Learning 2014

is to help line managers with their control activities, to coordinate the overall control system within the organization, and to gather relevant information and report it to all managers. Monsanto developed a program to train employees in controllership that it claims graduates broad-thinking, line-oriented managers.[35]

Controllers are particularly involved in the control of financial resources.[36] This is consistent with the pervasiveness of financial concerns in control. Because of the increased importance of the control function in recent years, the position of controller has taken on added stature in many organizations. For example, controllers at Coors, UPS, and Hyundai have all been given more responsibility in recent years.[37]

## Food Safety

One area of control that is of particular importance in agribusiness is food safety.[38] Most people do not think seriously about food safety until they or a loved one gets sick from unknowingly eating contaminated food. While the food supply in the United States is one of the safest in the world, millions of people get sick, thousands are hospitalized, and hundreds die each year from food-borne illnesses.

Food safety covers the handling, preparation, and storage of food from seed to consumption, so a lot of individuals and organizations are involved. While all manner of government agencies—local, state, and federal—attempt to regulate food safety, it is ultimately up to the industry and consumers to assure that safety. The industry must see that the food provided to consumers is safe, and consumers must store and prepare the food properly before using it.

### FOOD FOR THOUGHT 21.5

In 1984 members of an Oregon religious commune tried to influence a local election by poisoning salad bars with salmonella bacteria to sicken voters 751 people became ill. The attack took place near the end of the food-distribution chain but could occur at any point between farm and table.

When food-borne illnesses occur, both government and the private sector respond. In the mid-1990s, President Clinton announced the Produce and Imported Food Safety Initiative, which called for additional resources to improve domestic food safety

standards and to ensure that imports were equally safe.[39] The FDA then developed a guide entitled "Guidance for the Industry: Guide to Minimize Microbial Food Safety Hazards for Fresh Fruits and Vegetables" to assist firms in responding to the directive. That was later updated and followed by a voluntary produce action plan.[40] Each of these plans outlines so-called "Good Agricultural Practices" (**GAP**), "Good Handling Practices" (**GHP**), and "Good Manufacturing Practices" (**GMP**) that are intended to reduce contamination from irrigation water, the use of manure as fertilizer, the presence of animals in fields or packing areas, and the health and hygiene of workers handling produce during production, packing, processing, transportation, and distribution or preparation (Figure 21.10).

The FDA may inspect both imports and domestically produced foods. Foods found to be contaminated may be rejected for commercial sale. The current practice of inspecting for microbial contamination is difficult, as the source may be isolated within a small amount of a shipment and may not be found during random sampling. Handling the product after an inspection may spread the pathogen to other produce in the same shipment or other shipments. Thus, the most current effective method of reducing the risk of microbial contamination is through vigilant company adherence to the recommended guidelines for GAP, GHP, and GMP and possibly implementation of a Hazard Analysis Critical Control Point (HACCP) management program.

Retailers have a strong interest in requiring that their grower-suppliers meet these guidelines. Beginning in 1999 most U.S. grocery retailers began requiring their suppliers to have their production,

**FIGURE 21.10** Agribusinesses must comply with U.S. government regulations when spreading or spraying pesticide onto fields.

©iStockphoto/Esemelwe

packing, distribution, and logistics systems audited to ensure that GAP/GHP/GMP practices were being followed. Safeway was the first U.S. retailer to require suppliers to employ the services of third-party food safety auditing firms. At that time, Safeway required its suppliers to use the services of either the American Institute of Baking, PrimusLabs.com, or Scientific Certification Systems.[41] Many of these firms began their auditing services in testing for pesticide residues on crops or in conducting audits of Hazard Analysis and Critical Control Point (HACCP) programs established by food processors. Most other major retailers followed Safeway's lead. Indeed, in a survey of select fruit and vegetable growers and shippers in 1999 it was found that about half were required by their retail buyer to use third-party services.[42]

Recently one survey indicated that many Americans feel that the food industry is not doing enough to assure food safety. Another suggested that, while Americans are confident regarding food safety, their confidence in the government's ability to maintain a safe food supply had dropped. Yet another survey suggested that highly educated consumers have concerns about food safety. In that survey, more than 60 percent of the respondents had experienced illness from food and half of those then changed their eating habits and were more likely to view the government as needing to do more to assure food safety.[43]

But assuring food safety is no easy matter. Who sets the standard for what is safe? Who determines the procedures to be followed? What systems should be in place to trace the origins of contaminated foods? How should producers prevent cross-contamination of organic with non-organic products? Answering these questions is well beyond the scope of this book, but the very nature of the questions suggests the growing importance of the control of food safety by managers in all agribusiness organizations.

## CHAPTER SUMMARY

Controlling is the process of monitoring and adjusting organizational activities toward goal attainment. Most organizations have controls to deal with four areas: financial resources, physical resources, human resources, and information resources.

The planning-control link means that control is the other side of the planning coin; it helps management ensure that performance conforms to plans. Control-generated information can also facilitate future planning activities.

The three basic approaches to control are steering control, concurrent control, and postaction control. Steering control monitors the quality and/or quantity of various kinds of resources before they enter the system. Concurrent control focuses on activities that occur as inputs are being transformed into outputs. Postaction control deals with the quality and/or quantity of an organization's outputs. These approaches to control are usually used together as multiple controls.

The four basic steps to establishing a control system are setting standards, assessing performance, comparing performance with standards, and evaluating and adjusting.

Effective control has five attributes: integration, objectivity, accuracy, timeliness, and flexibility. Integration means that the control system must fit into the overall organizational system. Objectivity means that everyone who is affected by the control system must be able to understand it. Accuracy means that the control system must be reporting correct information that is pertinent to the company's goals. Timeliness means that the information provided by the control system must be available when it can be used. Information that arrives too late is of no value. Finally, flexibility means that the control system must be able to accommodate adjustments and change and should not be applied rigidly.

Managing control also involves dealing with resistance to control. There are four basic things a manager can do to overcome such resistance: (1) make sure that the control system is properly designed and has the characteristics of effective controls; (2) encourage participation, which tends to increase acceptance; (3) use management by objectives to establish collaborative goals and to enhance acceptance of controls; and (4) ensure that the control system has a built-in provision for checks and balances so that discrepancies do not occur or are corrected if they do.

The ultimate responsibility for control must be shared by everyone in the organization. In more practical terms, the responsibility is shared by line managers and specialized managers called controllers. All managers contribute to the planning process and so also share the responsibility for control. However, control is the specific responsibility of one or more managers who have the title of controller. The controller's job is to assist line managers with their control activities, coordinate the overall control system, and gather relevant information and report it to all managers.

One area of control that is of particular importance in agribusiness is food safety, which covers the handling, preparation, and storage of food from seed to consumption. Local, state, and federal agencies attempt to regulate food safety, but it is ultimately up to the industry and consumers to assure that safety. Consumer surveys have indicated that many Americans are confident regarding food safety but their confidence in the government's ability to maintain a safe food supply has decreased.

## CHAPTER ACTIVITIES

### REVIEW QUESTIONS

1. What is control, and why is it important to organizations?
2. What are the basic approaches to control?
3. Discuss each of the steps in establishing a control system.
4. What are the characteristics of effective control?
5. Why do people resist control, and how do managers overcome this resistance?

### ANALYSIS QUESTIONS

1. Do you ever feel that certain aspects of your college or university are out of control? If so, describe some reasons why they might need some control, and how the discussion in the text would apply.
2. Which step in the control system is likely to be the most difficult to carry out? Why? Which is likely to meet with the most resistance? Why?
3. Do all of the characteristics of effective controls fit together? Why or why not?
4. Which characteristic of effective control do you feel is most important? Why?
5. Why would it be unwise for an organization to have just one person be totally responsible for control? What persons should be responsible for control? Why?

## FILL IN THE BLANKS

1. The type of control that monitors the quality and/or quantity of resources before they enter a company's system is _____ control.

2. The type of control that involves the process of transforming resources into products is _____ control.

3. The type of control that monitors the quality and/or quantity of products as they leave a company's system is _____ control.

4. Standards should be consistent with the organization's _____.

5. The five characteristics of effective control are _____, _____, _____, _____, and _____.

6. So much control that employees' independence and autonomy are limited is labeled _____.

7. The manager's answerability for actions, decisions, and performance is called _____.

8. The company official who is directly responsible for helping line managers, coordinating the overall control system, and gathering important information to relay to all managers is the _____.

9. The way in which planning and control are integrated within the business cycle is called the _____ _____ _____.

10. Control that is too narrowly concentrated or fails to balance essential factors is known as _____ _____.

---

### CHAPTER 21 CASE STUDY
### HOLES IN THE GROUND PRESENT NEW CONCERN FOR AGRIBUSINESS FUTURE

It is well established that agribusiness involves the production and marketing of perishable commodities. Agribusiness appears to need only land, sunlight, and water. This would be a false impression. Ignoring other "soft" inputs like seeds and human capital, agribusiness also needs resources that are mined and manufactured. These are "hard" inputs. Among the most important is potash—a unique entity with a unique history.

On July 31, 1790, George Washington, first President of the United States, signed the first patent issued by the United States of America. The patent, assigned to Samuel Hopkins of Pittsford, Vermont, was for the making of "Pot Ash and Pearl Ash" by a new method. In the early days of the United States, potash was made from hardwood trees and other plants. It was used to make soap, glass, and salt peter for gunpowder. It was one of the earliest and largest exports of a young country.

Today this humble mineral derived from the earth remains an essential crop fertilizer. The significance of potash for agriculture is simple. It is the prime source of the element *potassium*. Potassium is needed to activate at least 80 important enzymes in plants. In fact, without potassium, crops cannot reproduce and make grain, fruit, or vegetables. Potassium, along with phosphorus and nitrogen, are the "big three" of any fertilizer mix. They are necessary elements for plant growth and reproduction.

Approximately 13 countries are major producers of potash for world agribusinesses. Five countries (Canada, Belarus, Russia, China, and Germany, in descending order of production), account for 75 percent of world production. The global reserves of this mineral are estimated at approximately 8.3 billion tons. At current rates of production this should last 250 years.

As the need for food production grows, a concomitant need is created for potash. The price of potash will reflect this need. Though plentiful, it must be mined, and this process uses a substantial amount of energy. The presence of potash is not evenly distributed on Earth; therefore, control of potash means control of production, market, and price.

In 2001, Billiton, a South African mining firm, merged with Broken Hill Proprietary (BHP), an Australian mining company. The combined South African-Australian company, BHP Billiton, is the world's ninth largest company with a market capitalization approaching $190 billion and the largest mining company in the world.

Mining companies need products—commodities—in the pipeline to market. In August 2010, BHP Billiton offered to purchase the Potash Corporation of Saskatchewan, a Canadian firm, for $38.6 billion, a 32 percent premium on the average price of Potash shares. The tender offer was rejected as "grossly inadequate." The rejection by Potash Corporation and the continued pursuit by BHP Billiton made this offer a hostile takeover.

Mining companies acquire mines. Mines are large natural resource extraction entities. Natural resources are integral to the countries in which they are located. Acquisition of one company by another is a standard form of corporate accretive growth. However, issues are interpreted differently when the acquired company has sole access to a country's mineral reserves. As a result, the Canadian government expressed concern about the impact of this acquisition. The Province of Saskatchewan liked the attention but worried about the impact on its revenues—taxes.

In the nomenclature of "M&A" (Mergers and Acquisitions), when a company is tendered a takeover bid, whether it is accepted, ignored, or rejected, the target company is said to be "in play," which means that other suitors may appear. Hopu Investment Management Company, a consortium of Canadian, U.S., China, and Asian investors, publicly considered a bid. Sinochem, a state-owned Chinese potash importer, also expressed interest. Their interest was a reflection of the Chinese government's interest in potash.

Suddenly Canadian *blasé* was aroused. A Canadian asset, Potash Corporation, was about to fall into foreign hands. BHP Billiton sweetened its offer to Potash Corporation to $40 billion. Government officials of Saskatchewan suggested to the Canadian government that the BHP Billiton bid be rejected because, "it doesn't benefit the province," even though BHP offered an additional $3 billion to the provincial government as payment in lieu of taxes.

All of these financial and political machinations became more complicated when farmers gave Potash Corporation its best financial quarter in history as they made early orders of fertilizer in anticipation of a possible price boost. The very customers everyone was seeking responded to the financial cacophony, as if to announce "remember us?"

What does all of this mean?

Globalization has changed much; but resources, especially limited natural resources, remain and will be tied to the country of placement. Globalization will simply be more complicated. Even in agribusiness, a mine is more than just another hole in the ground, and control is important.[44]

## ▶ Case Study Questions

1. What form of control over potash should be used by the government of the country where it is located and by the corporations who purchase it?

2. What measures of performance would be appropriate for each of these?

3. What problems can you imagine on the horizon for agricultural businesses who buy potash and for the countries and the companies who control potash?

## REFERENCES

1. Mark Kurlanksky, *Salt: A World History* (NY: Walker and Company, 2002), 66; Michael Moss, "The Hard Sell on Salt," *The New York Times* (May 30, 2010), A1; Bill Marsh, "Stealth Salt in the Pantry," *The New York Times* (April 25, 2010), WK2; "Salt Sellers," *The Economist* (January 14, 2010) at www.economist.com (accessed November 8, 2010); William Neuman, "Citing Hazard, New York Says Hold the Salt," *The New York Times* (January 11, 2010), A1.

2. Robert Anthony, John Deardon, and Norton M. Bedford, *Management Control Systems, 12th ed.* (New York: Irwin/McGraw-Hill, 2007).

3. Karynne Turner and Mona Makhija, "The Role of Organizational Controls in Managing Knowledge," *Academy of Management Review* (2006): 197–217.

4. William Newman, *Constructive Control* (Englewood Cliffs, NJ: Prentice-Hall, 1975).

5. "PepsiCo 3Q Profit Climbs on Cost Cuts; Sales Fall," *USA Today* (October 8, 2009) at www.usatoday.com (accessed August 1, 2010).

6. "American Greetings Is Carding Gains," *USA Today*, August 24, 1988, 3B; "Flounder," *Forbes*, April 25, 1988.

7. Matt Krantz, "Stocks of Restaurants, Financial Services, Retailers Bounce Back," *USA Today* (May 26, 2009) at www.usatoday.com (accessed August 1, 2010).

8. Paul R. La Monica, "It's Still Safe to Bring Home the Bacon" (May 1, 2009) at www.money.cnn.com (accessed July 25, 2010).

9. "Why Emery is Biting Its Nails," *Business Week*, August 29, 1988, 34.

10. "History," Menlo Worldwide Logistics at www.con-way.com (accessed August 1, 2010).

11. Joel Dreyfuss, "Victories in the Quality Crusade," *Fortune*, October 10, 1988, 80–88.

12. Myron Magnet, "Who's Winning the Information Revolution," *Fortune*, November 30, 1992, 110–117.

13. Natasha Holland, "A Bit of New Zealand in US Dairy Farms," *The Southland Times* (January 9, 2009) at www.stuff.co.nz (accessed July 6, 2010); "Overview," at www.gpreinc.com on July 5, 2010; Robert Bryant, "Eateries Need to Watch the Figure," *SmartCompany* (March 20, 2008) at www.smartcompany.com (accessed July 6, 2010).

14. Newman, *Constructive Control*.

15. Harold Koontz and Robert W. Bradspies, "Managing Through Feedforward Control," *Business Horizons*, June 1972, 25–36.

16. Thomas A. Stewart, "Brace for Japan's Hot New Strategy," *Fortune*, September 21, 1992, 62–74.

17. Charles W. L. Hill, "Establishing a Standard: Comparative Strategy and Technological Standards in Winner-Take-All Industries," *Academy of Management Executive* (1997): 7–16; and Alex Taylor III, "Do You Know Where Your Car Was Made?" *Fortune*, June 17, 1991, 52–56.

18. "Korea's Automobile & Auto Parts Industry 2009," *Scribd* (February 23, 2010) at www.scribd.com (accessed August 1, 2010).

19. Edward E. Lawler III and John G. Rhode, *Information and Control in Organizations* (Pacific Palisades, CA: Goodyear, 1976).

20. Robert N. Anthony, *The Management Control Function* (Boston: Harvard Business School Press, 1988).

21. Daniel Seligman, "Turmoil Time in the Casino Business," *Fortune*, March 2, 1987, 102–116.

22. "iPad Shortage Forces Apple to Delay International Launch Until May," *AppleInsider* (April 14, 2010) at forums.appleinsider.com (accessed August 1, 2010).

23. P. Rajan Varadarajan, Terry Clark, and William M. Pride, "Controlling the Uncontrollable: Managing Your Market Environment," *Sloan Management Review* (Winter 1992): 39–48.

24. William G. Ouchi, "The Transmission of Control Through Organizational Hierarchies," *Academy of Management Journal* (June 1978): 173–192.

25. "Knight-Ridder Acts to Boost Bottom Line," *USA Today*, November 11, 1986, 1B, 2B.

26. J. M. Horovitz, "Strategic Control: A New Task for Top Management," *Long Range Planning*, June 1979, 28–37.

27. Paul G. Makosz and Bruce W. McQuaig, "Is Everything Under Control? A New Approach to Corporate Governance," *Financial Executive*, January 1, 1990, 24–29.

28. Ronald Henkoff, "Cost Cutting: How to Do It Right," *Fortune*, April 9, 1990, 40–48.

29. Walter Kiechel III, "Managing Expense Accounts," *Fortune*, September 16, 1985, 205–208.

30. Stratford Sherman, "How to Prosper in the Value Decade," *Fortune*, November 30, 1992, 90–103.

31. For a thorough explanation of resistance to control, see Lawler and Rhode, *Information and Control in Organizations*.

32. David B. Greenberger and Stephen Strasser, "Development and Application of a Model of Personal Control in Organizations," *Academy of Management Review* (January 1988): 164–177.

33. Cortlandt Cammann and David A. Nadler, "Fit Control Systems to Your Management Style," *Harvard Business Review* (January–February 1976): 65–72.

34. Vijay Sathe, "Who Should Control Division Controllers?" *Harvard Business Review* (September–October 1978): 99–104.

35. Michael A. Robinson and Donald T. Hughes, "Controllership Training: A Competitive Weapon," *Management Accounting* (May 1, 1989): 20–24.

36. Robert N. Anthony and Leslie K. Breitner, *Essentials of Accounting, 10th ed.* (Upper Saddle River, NJ: Prentice-Hall, 2010).

37. Christopher Carr, Cyril Tomkins, and Brian Bayliss, "Strategic Controllership—A Case Study Approach," *Management Accounting Research* (1991): 89–197; Al Pipkin, "The 21st Century Controller," *Management Accounting* (February 1, 1989): 21–26.

38. George J. Seperich, *Food Science and Safety* (Upper Saddle River, NJ: Pearson Prentice-Hall, 2004).

39. Stephen R. Crutchfield and Jane Allshouse, "The Economics of Improving Food Safety," *Farm Foundation* (1998): 53–64.

40. Office of Food Safety, U.S. Food and Drug Administration (February 2008) at www.fda.gov (accessed July 28, 2010).

41. Maki Hatanaka, Carmen Bain, and Lawrence Busch, "Third-Party Certification in the Global Agrifood System," *Food Policy* (2005): 354–369.

42. Linda Calvin, Roberta Cook, Mark Denbaly, Carolyn Dimitri, Lewrene Glaser, Charles Handy, Mark Jekanowski, Phil Kaufman, Barry Krissoff, Gary Thompson, and Suzanne Thornsbury, "U.S. Fresh Fruit and Vegetable Marketing: Emerging Trade Practices, Trends, and Issues," U.S. Department of Agriculture, Economic Research Service (January 2001) at www.ers.usda.gov (accessed July 31, 1010).

43. David D. Van Fleet and Ella W. Van Fleet, "Attitudes of Well-Educated Consumers Toward the US Government's Food Safety Efforts," *Internet Journal of Food Safety* (2010): 45–52; "Food Safety: Majority of Americans Feel Industry Doesn't Do Enough," *American Society for Quality* (2009) at www.asq.org (accessed June 30, 2010); David D. Van Fleet and Ella W. Van Fleet, "Food Safety Attitudes among Well-Educated Consumers," *Internet Journal of Food Safety* (2009): 88–97; L. Morales, "Despite Salmonella Cases, Americans Confident in Food Safety" (2008) at www.gallup.com (accessed July 6, 2010).

44. Gautam Naik, "Humble Mineral Gleams as Farm Demand Climbs," *The Wall Street Journal* (August 19, 2010), B1; Phred Dvorak, "Potash Bid Stirs Hope—and Fear," *The Wall Street Journal* (August 23, 2010) at online.wsj.com (accessed November 6, 2010); Liam Denning, "Savior for Potash Comes From Left Field," *The Wall Street Journal* (October 30–31, 2010) at online.wsj.com (accessed November 6, 2010); Phred Dvorak and Dennis K. Berman, "BHP's Bid Faces Regional Setback," *The Wall Street Journal* (October 20, 2010) at online.wsj.com (accessed November 7, 2010); Phred Dvorak, Alison Tudor, and Edward Welsch, "BHP's Potash Offer Gets Modest Boost," *The Wall Street Journal* (October 5, 2010), B1; Edward Welsch, Dinny McMahon, and Phred Dvorak, "China Eyes Potash Corp," *The Wall Street Journal* (September 1, 2010) at online.wsj.com (accessed November 8, 2010); Robert Guy Matthews and Anupreeta Das, " BHP Billiton Goes Hostile With Potash Bid," *The Wall Street Journal* (August 19, 2010) at online.wsj.com (accessed November 8, 2010); Dinny Mcmahon, Phred Dvorak, and Gina Chon, "Chinese Investors Mull Bid for Potash," *The Wall Street Journal* (August 24, 2010) at online.wsj.com (accessed November 8, 2010); "Making the Earth Move," *The Economist*. (August 19, 2010) at www.economist.com (accessed November 6, 2010).

# CHAPTER

# 22

# Managing Quality

## LEARNING OBJECTIVES

After reading this chapter, you should be able to:

- Describe the nature of quality and productivity.
- Discuss the importance of quality and its relation to competition, costs, and effectiveness.
- Describe strategic initiatives firms use to improve quality.
- Describe operational techniques firms use to improve quality.

## *More Than Just a Game*

"**H**ey ump! What game are you watching?"

Those words hurled at a referee or umpire at a sporting event are not meant to be helpful, but to point out the official's shortcomings. All sports require umpires or referees to enforce rules and make on-the-spot judgments. They are the *sine qua non* of every sport. However, their necessity does not mean that they are embraced or encouraged by sports fans or even the participants. It is a sports axiom that the umpires and referees do their job well when they remain invisible. Few appreciate a great umpire or referee, but everyone reacts to a "bad call" or a "bad decision."

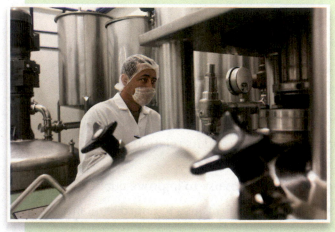

**FIGURE 22.1** Quality-control inspectors are the umpires in food-processing plants.
© iStockphoto/Miguel Malo

In industry the quality assurance department plays the role of on-site umpire or referee. The responsibility of this unit is to ensure the safety and quality of the final product. This function is essential in every industry, but it is crucial in the agribusiness industry. Consumers do more than just "use" the products of the food industry; the products become an integral part of us. We do not merely hope our food is safe—we expect our food to be safe. As Rich Nuheisel, UCLA Bruins football coach, said, "Hope is not a function of strategy; implementation is."

Some of the original authors in the quality field never spoke of quality control or quality assurance; they spoke of the "quality function," referring to physical and performance characteristics of products. To the early proponents, quality was a function of the product. Later on this function acquired another dimension: safety. A food product could be high quality but totally unsafe for consumption because of contamination. To reinforce food safety, the industry began employing HACCP or Hazard Analysis Critical Control Point principles and applications. Both food and pharmaceutical companies spent a great amount of time and money trying to prevent food contamination. The quality/safety function was performed competently, confidently, and silently. No one noticed because there was nothing to notice.

But when something goes awry, it becomes very bad and very visible very quickly. "The information I've seen during the course of our investigation," said the chairman of the House Committee on Oversight and Government Reform about Johnson & Johnson, "raises questions about the integrity of the company. It paints a picture of a company that is deceptive, dishonest, and has risked the health of many of our children." Yet J&J is a company that prides itself on its safety record. It manufactures products for the health industry, where extremely high standards are expected. J&J is not accustomed to this kind of negative exposure. Consumer trust is very hard to earn and so easy to lose.

What caused J&J's standards to diminish was too much attention to the bottom line necessitated by a queasy economic environment, distraction by a corporate acquisition, and an inability to blend divergent corporate cultures. During mergers and acquisitions these distractions occur and quality slips may occur. In the food and pharmaceutical industry, however, "slips" have the potential to injure or kill the consumer, to hurt or destroy corporate reputations, and to lead to millions of dollars in litigation. Like the famous bumper sticker "Don't Mess with Texas," likewise don't mess with food and pharmaceuticals. Mistakes like these are career-ending errors, as the J&J Consumer Chief experienced first-hand. When the umpire becomes visible, she has to go.

The food and pharmaceutical industry produces millions and millions of safe and effective products every day. All of this production is accomplished quietly and competently. While constant vigilance assures the quality/safety function (Figure 22.1), vigilance can be thwarted by the boredom of repetition and lulled to sleep by the paucity of errors. These two factors conspire to dull vigilance and to instill complacency … until something happens.

What many people do not understand is that product safety is the responsibility of the manufacturer, not the government. The Food and Drug Administration (FDA) and U.S. Department of Agriculture (USDA) monitor the manufacturer's vigilance. In other words, the FDA and the USDA review the umpire's decisions. While the USDA has on-site inspectors to grade the umpires, the FDA does not. And it's even more complicated than that as there are "sub" agencies within the USDA and multiple other agencies that can be involved, such as the Department of Commerce, the Centers for Disease Control, the Department of Homeland Security, the Environmental Protection Agency, to name just a few.

Sometimes it is only the hint of the possibility of a problem that earns the umpire visibility. This happened to Abbott Laboratories. It was in the process of merging with another pharmaceutical company in late 2010 when it announced the recall of Similac™, its powdered baby formula, because insects (beetles) were found in the production area—not in the formula, but in the area. The *possibility* of contamination is what determined the voluntary recall. HACCP principles say, "Stop," because the distance from the floor to the product is too short; it is always too short. The umpire must not only prevent errors but anticipate them as well.

To prevent the preventable, the umpire must monitor not only the product for the quality/safety function but also the means of production, For example, combustible dust from sugar production exploded and killed 14 people in an Imperial Sugar Company plant. Companies that produce fine powders like sugar and flour are aware that a static electric spark or any spark among the particles suspended in the air could lead to an explosion. "These are well recognized hazards and have been for decades. This was a terribly tragic and completely preventable tragedy," an OSHA deputy assistant secretary said. OSHA is the Occupational Safety and Health Administration, another umpire-watcher.

Suddenly the quality/safety function looks questionable and lax. It is not. It is a statistical occurrence: Given enough manufacturing on a continual basis, something will happen. The ultimate goal is to eliminate the occasion for disaster; the realistic goal is to minimize its appearance.

The situation is no better elsewhere. In fact, the Communist Party in Beijing China reported that an "average of 187 workers die daily in mishaps, as focus on growth often trumps safety concerns." This happens when the umpire is overwhelmed, looks the other way, or is not present.

No one says that an umpire's or referee's job is easy, but it is essential. Remember this next time something goes awry and extend a little respect; be thankful that the umpire is vigilant. After all, it's more than just a game.[1]

## INTRODUCTION

Quality and productivity are also "more than just a game." Many businesses today are facing a serious product quality problem that has resulted in lost sales and poor images. The solution? More vigilance by the on-site referees or management umpires—a renewed dedication to making product quality a cornerstone for managerial decisions and activities.

This chapter explores the concepts of quality and productivity in detail. We first describe the nature of quality and productivity. We then discuss the importance of quality and its relation to competition, costs, and long-term effectiveness. Then we introduce and discuss strategic initiatives for improving quality. Finally, we conclude by discussing operational techniques for improving quality.

## The Nature of Quality and Productivity

Most people have a general sense of what the word *quality* means. Nevertheless, given both its importance and its complexities, we will start by examining the meaning of quality in detail. We then relate quality to productivity, identify levels of productivity, productivity trends, and ways to improve productivity. With this basic understanding of productivity, we can then turn back to a more detailed analysis of quality.

### The Meaning of Quality

Suppose someone buys two pens, one of which writes for 40 hours before running out of ink while the other writes for 80 hours. Is the latter pen clearly of higher quality than the former-pen? Not necessarily. For example, if the first pen cost the consumer 69 cents and has a stated manufacturer's expectation of being able to write 35 hours, it actually comes out looking pretty good. And if the second pen cost $5 and was intended to write 95 hours, it wasn't such a good deal after all. Thus, it is important to recognize that quality has both an absolute and a relative meaning.[2]

It also important to note that quality can be used in reference to products and/or to services. The example above clearly pertains to product quality, with the product being the pen. **Product quality** refers to the quality of a real item. Common products for which quality is particularly relevant include mechanical equipment, electronic equipment, trucks and cars, and consumer products.

**Service quality** refers to the quality of an intangible service provided by an organization.[3] Suppose a customer wants to order merchandise from a catalog. She will likely feel that she has experienced high service quality if she is given an 800 number to call, the call is answered quickly, she is treated courteously, her order is taken by a trained professional, and the shipment arrives as promised. But if she has to pay for a toll call, is put on hold, is treated brusquely by someone who has little product knowledge, and the shipment arrives much later than promised, service quality is low.[4]

Following published guidelines established by the American Society for Quality Control, we define **quality** as the total set of features and characteristics of a product or service that bear on its ability to satisfy stated or implied needs.[5] Thus, assuming that the two pens noted above are equivalent in all other respects, the cheaper pen that writes only 40 hours is of higher quality than the pen that cost more and wrote for a while longer. Similarly, the first catalog ordering experience described above had higher service quality than did the second.

Table 22.1 lists eight specific factors or characteristics for assessing or evaluating product or service quality. For example, all else being equal, the more desirable features that a product or service has, the higher its quality relative to a product or service with fewer desirable features. Likewise, all else being equal, if a particular product or service is aesthetically pleasing, it is of higher quality relative to another product or service that is less aesthetically pleasing.

**TABLE 22.1**  Eight Dimensions of Quality

| DISCUSSION | EXPLANATION |
|---|---|
| 1. Performance | A product's primary operating characteristic, such as an automobile's acceleration and a television set's picture clarity. |
| 2. Features | Supplement to a product's basic functioning characteristics, such as power windows on a car. |
| 3. Reliability | A probability of not malfunctioning during a specified period. |
| 4. Conformance | The degree to which a product's design and operating characteristics meet established standards. |
| 5. Durability | A measure of product life. |
| 6. Serviceability | The speed and ease of repair. |
| 7. Aesthetics | How a product looks, feels, tastes, and smells. |
| 8. Perceived quality | As seen by a customer. |

Source: Based on data in David A. Garvin, "Competing on the Eight Dimensions of Quality," *Harvard Business Review* (November–December 1987).

It is important to recognize that the caveat "all else being equal" plays an important role in these quality assessments, however. For example, if a product has an abundance of features and looks beautiful but seldom works the way it is supposed to, its quality obviously suffers. Similarly, if the same product has few features and is relatively unattractive, but works reliably and dependably, its quality is enhanced.

Also, we must relate these ideas back to the previously noted dimensions of absolute quality and relative quality. **Absolute quality** is the generally understood level of quality that a product or system needs to be capable of fulfilling its intended purpose, rather than its quality relative to other alternatives. Suppose you want to buy a new sound system. You go to an electronics store and look at a Cowon MP3 player for $200; a Sony Blu-ray Disc™ Home Theater system for $500; and a Yamaha digital receiver with high-resolution audio for $500 but you will also have to buy a DVD/CD player and speakers for another $1,000. Regardless of which system you buy, you expect it to play your favorite music. You also expect it to do so for a reasonable period of time (longer than a few days, buy not necessarily twenty years). Thus, you evaluate each system in terms of some generally understood absolute level. A product or system must be capable of fulfilling its intended purpose.

At the same time, you also are likely to find that the $200 system will not perform as well in several ways as will the other systems. The portable system may not reproduce the bass tracks as well, does not have the same volume capabilities, may not have remote control, and may be expected to last only two or three years. On the other hand, the more expensive systems have louder and better sound, have remote controls, last much longer, and so forth.

Thus, you must interpret the relative component of quality in comparison to other alternatives in several ways. All three systems may be high quality relative to the standards they are expected to meet. The question becomes how much you want to pay in terms of what you will get. And you would also compare each system against "comparable" systems available from other manufacturers. Despite these comparative findings, however, you may choose the lesser expensive item regardless of quality if you decide that the technology is changing so rapidly that you will want to buy a replacement product relatively soon.

Then there is the brand itself. People develop brand loyalties and, associated with them, brand identifications and prejudices—Ford truck owners are different from Chevy truck owners, white Stetson wearers are different from those with black hats, etc. So if you have a brand preference for a Bose sound system you will not choose a Cowon, a Sony, or a Yamaha. You will continue shopping until you find a satisfactory Bose system.

## FOOD FOR THOUGHT 22.1

According to USDA regulations, the term "fresh" on a poultry label means that the raw poultry product has never been refrigerated below 26°F. Raw poultry held at 0°F or below must be labeled frozen or previously frozen.

## Quality and Productivity

From a managerial perspective, product or service quality is highly related to productivity.[6] Thus, before proceeding further with our discussion of quality, we will introduce the productivity construct and establish its basic principles.

In a general sense, productivity is a measure of efficiency. More specifically, **productivity** is an economic index of the value or amount of what is created relative to the value or the amount of resources necessary to create it. For example, assume that two workers in the same factory are being paid the same wage to spend eight hours assembling toasters. One assembles 100 toasters that meet standard quality tests while the other produces 125 toasters of the identical quality. Clearly, the worker who makes 125 toasters is more productive than the one who makes only 100 toasters in the same period of time.

Similarly, assume two manufacturing plants are of the same size and have identical production equipment as well as the same size and type of workforce. One plant, however, is able to manufacture 5,000 electrical instruments per day whereas the other can make only 4,500 comparable instruments. The former plant is more productive than the latter.

In the past, managers often assumed—incorrectly—that productivity and quality were inversely related. That is, they thought they could increase output only by lowering quality. More recently, however, as shown in Figure 22.2, many managers have come to realize that productivity and quality are actually related in a positive way. Increased productivity often means higher quality, and higher quality results in higher productivity.

For example, if a firm can enhance its quality, several things happen. First, since defects are almost certain to decrease, there will be fewer returns from unhappy customers. Second, because defects are lower, fewer resources need to be dedicated to reworking or repairing defective products. And finally, since fewer defective units are being produced and since operating employees are now involved in quality enhancement, fewer quality control inspectors are necessary. Overall, then, since producing products and services requires fewer resources, productivity (by definition) increases.

## Levels and Forms of Productivity

As we have seen, there are different levels and forms of productivity.[7] Figure 22.3 illustrates some of the more general levels of productivity that are of interest and concern to managers.

**Levels of Productivity.** The most basic level of productivity is **individual productivity** or the amount produced or created by a single individual relative to his or her costs to the organization. For example, Glen May works in a Walmart distribution center in Buckeye, Arizona. His productivity reflects the value of the orders he fills each day relative to his wages.

One step higher on the productivity chain is unit productivity. **Unit productivity** may be the productivity of a manufacturing plant within a firm (as shown in the Figure 22.3), a single restaurant within a chain of restaurants, or a group of workers within a single facility of an organization. Walmart, for example, determines the unit productivity of each of its distribution centers.

**Company productivity**, an even higher level, is the total level of productivity achieved by all employees and/or units of an entire organization. All of Walmart's distributions centers, retail stores, and other operations combine to determine the corporation's overall productivity. Likewise, managers at General Foods, Bunge, ConAgra, PepsiCo, and even Summer Farms and Green Things are all concerned about their respective level of company productivity.

**Industry productivity** is that level of productivity achieved by all companies in a single industry. Kmart, Sears, Walmart, Target, and other retailers contribute to the retailing industry's aggregate level of productivity. Ford, General Motors, and Chrysler combine to determine the productivity of the U.S. automobile industry. Likewise, Archer Daniels

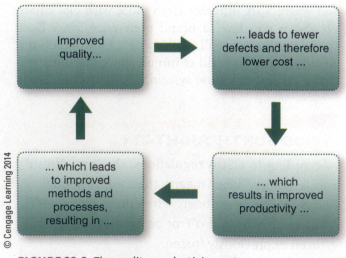

Improved quality...

... leads to fewer defects and therefore lower cost ...

... which results in improved productivity ...

... which leads to improved methods and processes, resulting in ...

© Cengage Learning 2014

**FIGURE 22.2** The quality-productivity cycle.

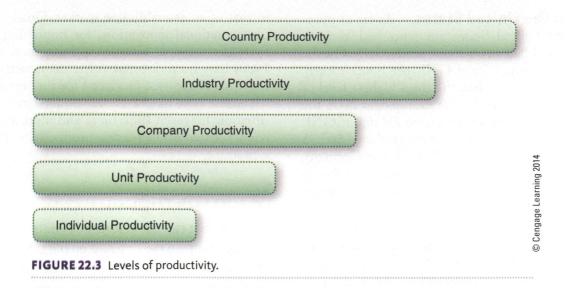

**FIGURE 22.3** Levels of productivity.

© Cengage Learning 2014

Midland, Cargill, ConAgra, Heinz, Hormel, and others determine the productivity of the food industry.

Finally, **country productivity** refers to the productivity levels achieved by entire nations. The retailing industry in the United States contributes to this country's total productivity. Industrialized countries like the United States, Japan, France, and Germany all have relatively high levels of productivity, whereas less industrialized countries like Poland, Vietnam, Cuba, and Kenya have relatively lower levels of productivity.

**Forms of Productivity.**  In many instances it also makes sense to consider different forms of productivity. For example, **overall productivity**—technically called **total factor productivity**—includes all the inputs that an organization uses. Dividing outputs by the sum of labor, capital, materials, information, and energy costs determines overall productivity.

In many instances, however, it makes sense to evaluate productivity in terms of only some of the resources an organization uses. For example, **labor productivity** is determined by dividing outputs by direct labor. Milk productivity is measured in terms of pounds per year, corn in bushels per acre, wheat in tons per hectare, and so on. Such a partial productivity index has several advantages. For one thing, resources can be expressed in different terms (i.e., hours of labor, units of raw materials, etc.), rather than having to transform them all into a common base, such as dollars. It also allows managers to focus their efforts to enhance productivity on specific areas and to more directly assess their effects.

Productivity is important for both obvious and not so obvious reasons. For example, the productivity of a single individual will contribute to the productivity of his or her unit. Likewise, unit productivity can help the organization determine where resources are being used wisely and where they are being used less wisely. Similarly, a firm's level of productivity will contribute directly to its profitability and thus to its ability to survive. The overall productivity of an industry will serve to encourage or discourage foreign competition. For example, if productivity is low for all the firms in one country in a particular industry, foreign competitors will feel that they have significant market opportunities and will enter the market aggressively. But if industry productivity is high, foreign competitors will find it much more difficult to enter the market.

Less obvious, but also very important, is the fact that productivity contributes to our overall quality of life and standard of living. For example, the more goods and services a country can create with its resources, the more goods and services will be available to its citizens. And if those goods and services are produced efficiently enough, they can also be shipped to foreign markets. The money that flows back will also add to the country's overall standard of living. Thus, the citizens of a highly productive country will have a better standard of living than will the people of a less productive country.

## Productivity Trends

In the preceding section we described the clear importance of productivity. Given that importance, it is equally critical that we understand trends and patterns in how productivity is changing.[8]

**Trends in the United States.**    As far as managers in the United States are concerned, there is both good and bad news about productivity. On the plus side, in terms of real GNP per hour worked, the United States has had the third highest level of productivity in the world. For example, in 2008 the typical American worker produced $50.69 of GNP per hour while a worker in Norway produced $63.35 and Belgium $51.71. France, $48.99, and Germany, $47.89 were fourth and fifth.[9] Moreover, there is little chance that U.S. productivity will drop anytime in the near future.[10]

On the other hand, productivity growth rates in the United States have been slow in recent years. For example, from 1948 through 2009 U.S. labor productivity increased at an annual rate of only 3.6 percent. And the level of U.S. farm output in 2009, while 170 percent above its level in 1948, was only growing at an average annual rate of 1.53 percent.[11] As we will see later, our major industrial competitors have been increasing their productivity at much higher rates.

Why did overall U.S. productivity slow down to begin with? Several factors contributed to this pattern. For one thing, because of booming production levels and relatively little foreign competition, U.S. manufacturing facilities deteriorated badly during the 1950s and 1960s. This resulted in less efficient production in the 1970s. Also, more and more relatively unskilled and/or inexperienced workers (e.g., minorities and women) entered the workforce for the first time during the 1960s and 1970s. Finally, to boost short-term profits many U.S. businesses are spending less and less on long-term research and development. As a consequence, they are achieving fewer breakthroughs in new technology.

Another consideration is the service sector, which has grown tremendously over the last several years while its productivity has remained essentially flat. Because the service sector has become such an important part of the U.S. economy, the stagnation in productivity in this sector has kept overall productivity from growing as much as it might have otherwise.

There are several different explanations as to why productivity levels in the service sector have lagged. For one thing, the rapid growth in services has carried with it some inefficiencies due to rapid start-up, lack of training, poorly designed operations, and so forth. Also, it is difficult to measure output in many service areas. Assessing the output of a lawyer or accountant, for example, is difficult relative to that of an assembly line worker. Finally, many of the operational innovations for improving productivity (for example, robotics and automation) of the 1980s have been directed at the manufacturing sector while fewer were produced for the service sector. Managers now have a clearer understanding of the situation, however. Thus, more and more attention is being focused on increasing productivity in the service sector.[12]

---

## FOOD FOR THOUGHT 22.2

A single bat can eat half its bodyweight in insects in one night, so the one million bats that died of a fungal disease in 2010 could have consumed about 694 tons of insects—many of them agricultural pests—in one year.

---

**International Trends.**    A major reason that people are concerned about productivity growth rates in the United States is because workers in many other countries are increasingly more productive. Indeed, over the last several decades, productivity growth in several industrial countries—Korea, Sweden, Taiwan, Singapore—has outstripped productivity growth in the United States. From 1978 through 2008, labor productivity increased 8.8 percent per year in Korea, 5.8 percent in Taiwan, 4.4 percent per year in Sweden, and 4.1 percent in Singapore. The growth rate for the United States in that same period was 3.8 percent.[13] And by 2010 China had replaced Japan in the No. 2 ranking.

Given that these countries are our major competitors in the world marketplace, concerns about their productivity growth rates are clearly well-founded. Increased awareness of these trends has galvanized many business leaders in the United States to try to stem the tide. Through a variety of actions, many of which are noted in the following sections, U.S. businesses are working harder than ever to maintain their position in the world economy.[14]

## Improving Productivity

Organizations that wish to enhance productivity at one or more levels can take several different approaches. As shown in Table 22.2, some of these methods are based on operations, and others focus on enhanced motivation and the involvement of employees.

**TABLE 22.2** Methods for Improving Productivity

| IMPROVEMENTS THROUGH OPERATIONS AND MANAGEMENT | IMPROVEMENTS THROUGH MOTIVATION AND INVOLVEMENT |
| --- | --- |
| Improve technology and facilities | Increase training |
| Increase R&D (research and development) spending | Increase employee participation |
| Adopt automated and robotic systems | Improve reward systems |
| Enhance speed | |
| Enhance flexibility | |

© Cengage Learning 2014.

**Operations and Management.** Organizations can often improve productivity through various operations and basic management techniques.[15] For example, improved technological methods and facilities can be a big contributor. In 1992, the average machine tool in a U.S. manufacturing plant was seven years older than the same tool in a comparable Japanese plant. The older the equipment and plants, the less efficient they are likely to be. Thus, by building new plants, installing new technology, and investing in new machinery, productivity may well be increased (Figure 22.4). The construction of new facilities and/or the installation of new equipment have helped improve productivity at Ford, Rubbermaid, and Caterpillar. And schools from kindergarten to college increase maintenance productivity through the use of new equipment.[16]

Likewise, building new distribution systems, information systems, and office buildings may also boost productivity if they result in more efficient ways for employees to do their jobs. Summer Farm's putting records on computers, Walmart's modern

**FIGURE 22.4** A manager observes workers in an attempt to improve productivity by first determining why they are performing below expectations.

and efficient distribution centers, Federal Express' marvelous computerized communication network, and Union Carbide's efficient corporate headquarters building each contribute to the productivity of their respective organizations. Duramet Corporation of Warren, Michigan, doubled sales over three years without an increase in its sales force. Prudential Insurance provided notebook computers to insurance agents and dramatically reduced the time that agents spend sifting through paperwork.[17]

Another approach to enhancing productivity is by increasing spending on research and development (R&D). R&D can create new products, new uses for current products, and new methods for making products. Breakthroughs in any of these areas result in improved productivity.[18] Apple, Bausch & Lomb, 3M Corporation, Merck, and IBM all credit R&D as key ingredients in their overall levels of organizational effectiveness.

Unfortunately, R&D is often a prime target for cutbacks when a business faces a downturn. Since R&D dividends are usually realized "tomorrow," short-sighted managers who worry about today's bottom line may therefore inadvertently hurt a firm's future by trying to maximize short-term profits.[19] Avon, USX, and Texaco all cut R&D spending at least once in the 1990s. On the other hand, a study by Booz Allen Hamilton notes that while spending more does not necessarily help, spending too little will hurt.[20]

Related to both technology and R&D, organizations can also increase productivity by investing in automation and/or robotics. Automated production systems, for example, can make products with much higher levels of precision and at more exacting tolerance standards than can human workers. Similarly, robots do not tire and let their concentration slip. Cummins Engine Corporation has boosted labor productivity significantly by implementing automated production systems.

© Levent Konuk/www.Shutterstock.com

**Motivation and Involvement.** Managers can increase productivity by improving the motivation and involvement of their employees. Table 22.2 notes common methods for doing this.

One approach is to increase training. Sometimes workers fail to perform at their maximum efficiency simply because they don't know how. Training is especially important in conjunction with the operations improvements noted above. For example, installing new production technology and not appropriately training employees in how to use it accomplishes very little. Indeed, some experts speculate that service productivity lags because companies invest billions of dollars in sophisticated computer systems, networks, and workstations but do not adequately train employees and managers how to use them effectively.[21]

Increased employee participation is also often cited as a key to improving productivity. Japanese firms like Hitachi, Sony, and Honda are examples of how important employee participation can be to productivity growth. U.S. firms like Westinghouse, Ford, and General Electric have learned valuable lessons from the Japanese and have each taken steps to dramatically improve the level of participation among their employees. For example, each company now gives operating employees considerably more control over how they do their jobs and makes them directly responsible for monitoring the quality of their work and correcting any defects that they observe. As a result, their productivity has improved.

Finally, organizations can also improve productivity by modifying their reward systems. All too often, workers receive the same rewards regardless of the quality of work they perform. Thus, there are no incentives to work hard or to be responsible for how much one is producing or the quality of that output. On the other hand, firms that tie significant rewards to improvements in productivity often see significant improvements. For example, Du Pont implemented an incentive system whereby employees receive bonuses if unit productivity exceeds previously agreed-upon levels.

## The Importance of Quality

With the preceding introduction to quality and discussion of productivity as a foundation, we are now ready to delve more deeply into a discussion of the quality and its role in businesses. We begin by investigating its fundamental importance to organizations. Indeed, quality is seen as being so important that the U.S. government sponsors an annual award (the Baldridge Award) for firms that improve quality the most.

The basic factors on which the award is based include competition, cost, and long-term effectiveness.[22] Among the previous award winners are Armstrong, AT&T, Boeing, Cargill, Corning, Eastman Chemical, Motorola, Ritz-Carlton, Sunny Fresh Foods, The Bama Companies, and many smaller companies.[23]

## Quality and Competition

In recent years quality has become an important point of competition in virtually every industry, particularly agribusiness.[24] Consumers around the world are increasingly demanding higher quality products and services. Thus a firm, such as Summer Farms, that can argue or demonstrate that its products or services are as good as or better than those offered by competitors will have an upper hand in the marketplace.

Quality has become an especially important point of competition in the food, automobile, airline, and electronics industries. Service quality has become important in competition as well. American Express, for example, advertises that its cardholders get service that is superior to that provided by other credit card companies. Land's End uses customer service to make major inroads in the catalog sales market for outdoor clothing and accessories. L.L. Bean and Federal Express also receive high marks in customer service and satisfaction.[25]

### FOOD FOR THOUGHT 22.3

To avoid a grainy feel in its chocolates, Ghirardelli Chocolate Co. refines most of the chocolate flakes until they are 19 microns rather than the 40 microns of other mass-market chocolates (human hair is 100 microns in diameter).

## Quality and Costs

Our earlier discussion of productivity notes several ways that quality improvement lowers costs (and thus increases profits). Direct improvements in productivity often result in both lower costs as well as higher quality, but quality improvement lowers costs in other ways as well. One major cost factor is replacement parts and products. The higher the quality of original products, the lower the subsequent costs of repairing or replacing them. Following directly from this is also lower overall costs of servicing warranties and product guarantees. And still another cost that is affected includes potential losses from lawsuits from disgruntled customers or customers who are injured by poorly made products.

At one time, Whistler Corporation was using 100 of its 250 employees to repair radar detectors that would not work. After implementing a quality improvement program, the company was able to transfer most of the workers back to the original production department, substantially lowering its labor costs attributable to quality.

## Quality and Long-Term Effectiveness

Finally, quality is also very important because of its role in influencing the long-term effectiveness of an organization. Organizations have found that they can actually boost profits in the short run by neglecting research and development, by using cheap materials, and by cutting corners. But over time these measures often come back to haunt these organizations as their reputations slide and customers shun their products. However, organizations that make continual and sustained commitments to quality improvement usually find just the opposite result. Their images in the marketplace become increasingly positive as customers realize they get better value from the firm's products and services.

## Improving Quality: Strategic Initiatives

As is the case for productivity, organizations and managers have a number of methods they can draw upon to improve quality. Some of these methods fall under the general area of strategic initiatives while others are more specific operational techniques. In this section we discuss the strategic initiatives. Our next section addresses operational techniques.

## The TQM Philosophy

The fundamental strategic initiative that organizations pursue when they become truly committed to quality is the adoption of total quality management. **Total quality management (TQM)** is a comprehensive, organization-wide strategy for improving product and/or service quality on a systematic and continuous basis.[26]

Figure 22.5 illustrates the basic components of a TQM strategy. TQM is based on a strategic commitment to quality and relies heavily on employee involvement, materials, methods, and technology to achieve improved quality. As such, it is a good fit for agribusiness.[27]

**Strategic Commitment.** The starting point for any real TQM effort is a strategic commitment by top management of an organization to make quality a top priority in every aspect of operations. Without such a commitment, quality is most likely to get only superficial consideration. And if a firm tries to promote itself on the basis of quality but that quality is eventually found to be lacking, the result will do more harm than good.[28]

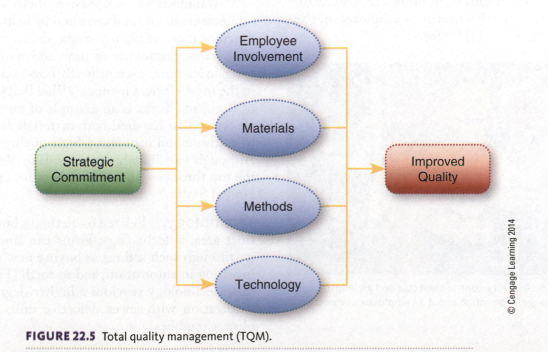

**FIGURE 22.5** Total quality management (TQM).

© Cengage Learning 2014

When James Houghton took over as CEO of Corning Glass Works in 1983, he stated publicly, clearly, and without equivocation that quality was a top priority. He then proceeded to back his assertion by committing the funds necessary to increase quality, to reward other managers on the basis of their quality improvements, and so forth. Thus, he made a firm and clear strategic commitment to quality. As a result, the TQM program at Corning has been a resounding success. Executives at Harley-Davidson also made a strategic commitment to quality following their buyout of the Harley-Davidson firm.

### Employee Involvement.

Again, companies often use employee involvement to increase productivity. In similar fashion, employee involvement and participation is also necessary to improve quality. Without such involvement, and without the commitment and acceptance of the program by employees, any attempt to enhance quality is not likely to succeed.

Organizations use various terms to describe employee involvement, including participation and empowerment. And it is often operationalized through work teams, as we discuss in Chapter 19. Regardless of the terms used, involvement generally focuses on giving employees throughout the organization more information about what the firm is doing and more autonomy over managing their own contributions. Each work team is usually responsible for scheduling its own work, for instance, and for assessing the quality of its output (Figure 22.6).[29] Harley-Davidson relies heavily on employee involvement as part of its TQM efforts.

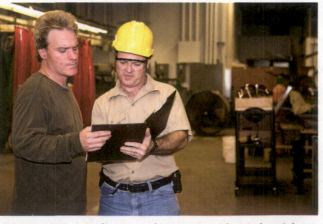

**FIGURE 22.6** A quality-control inspector and an industrial auditor exchange information about an employee's inspection results.

### Materials.

Another part of improving quality through TQM is by using better and higher quality materials. It stands to reason, for example, that if a food processer uses poor quality inputs in its products, it is likely to have quality problems. On the other hand, by demanding better inputs from suppliers, the quality of the food products will improve.

Many firms today are demanding that their suppliers adopt new and more stringent quality standards. Ford, for example, mandates higher quality standards for virtually all of the parts it buys from other suppliers. And some U.S. suppliers have found that by improving the quality of what they make, they have a better opportunity to sell to Japanese manufacturers. We noted in Chapter 21, for instance, that Cooper Tire sells rubber parts to Japanese automakers.

### Methods.

In similar fashion, companies can enhance quality through the use of more efficient and effective methods of operation. For example, L.L. Bean has an enviable reputation of customer loyalty and satisfaction. Part of the reason for this is because L.L. Bean trains its telephone operators to listen carefully to what customers want and then to do everything possible to satisfy those needs. In contrast, some mail-order houses suffer because their operators are rude, ill-informed, inaccurate, or otherwise unable to help customers. Such businesses may make the improvement of telephone assistance methods a first step in enhancing quality.[30]

Manufacturers emphasize methods through refinements in jobs and tasks and by helping workers function more efficiently. Some firms, for example, are making greater use of time-and-motion studies to again learn more scientifically how to perform jobs in the most efficient manner.[31] Blue Bell Creameries, of Brenham, Texas, is an example of one small food processor that has used both materials and methods to cultivate and maintain a high-quality reputation. Although sold in only 19 states, Blue Bell is one of the top three best-selling brands of ice cream in the United States.[32]

### Technology.

Related to methods, but also a distinct area, is technology. Firms can improve quality through such actions as buying new equipment, investing in automation, and so forth (Figure 22.7). Such technology provides a higher degree of standardization with fewer defective units or incomplete assemblies.

© Lisa F. Young/www.Shutterstock.com

**FIGURE 22.7** Installing new irrigation equipment with water-saving technology will help this agribusiness owner produce a higher quality product that sells at a higher price.

**PHASE 1**
- Train heavily
- Promote teamwork
- Emulate competitors
- Choose suppliers based on price and reliability
- Focus on learning about quality

**PHASE 2**
- Encourage efficiency
- Emulate market leaders and world-class companies
- Choose suppliers based on quality, then price
- Strive to make quality a component of culture

**PHASE 3**
- Implement self-managed work teams
- Promote self-training
- Emulate only world-class companies
- Choose suppliers based on technology and quality
- Make continuous quality your way of doing business

© Cengage Learning 2014

**FIGURE 22.8** Implementing total quality management.

Service firms can also use technology to enhance quality. For example, restaurants can learn how to serve food faster and more efficiently. And airlines can learn to how to process and transfer baggage more accurately and quickly.[33] We return to technology in Chapter 23.

## Using TQM

Assuming an organization wants to adopt the TQM philosophy, where does it start? We noted above that a strategic commitment is the first step. But after taking that step, what happens next? A recent study suggests that organizations should actually avoid moving too quickly.[34] Instead, as shown in Figure 22.8, they should proceed through three phases.

**Phase 1.** During Phase 1, the organization moving toward TQM is primarily concerned with learning about quality. Thus, it begins to invest heavily in training and tries to emulate its major competitors. It selects suppliers on the basis of price and reliability and concentrates on the fundamentals of quality enhancement.

**Phase 2.** After the organization learns the basics of TQM, it moves to Phase 2. During this phase it encourages workers to find more efficient ways to do their own jobs and starts to emulate market leaders and selected world-class companies. It selects suppliers primarily on the basis of quality, with price a secondary consideration. And the firm moves to make quality enhancement a fundamental part of its culture.

**Phase 3.** Finally, the firm that is ready to enter Phase 3 of TQM adopts self-managed work teams for most of its operations. Team members train themselves, with other training mainly reserved for new hires. Emulating the world's best companies becomes the standard way to do business. And continuous improvement becomes a routine, ongoing part of how the firm does business.

## A FOCUS ON AGRIBUSINESS
## TQM

Statistical quality control began when R. A. Fisher developed techniques and charts to help British farmers understand optimal approaches to planting and rotating crops. Refined by Walter Shewhart and W. Edwards Deming, statistical quality control applications in industry and the military have been extensive and profound. Its application in agribusiness, however, has been slower to develop despite its obvious importance.

While quality has always been important in agribusiness, the total quality management (TQM) approach has been slow to take root. When employed, however, its impact is pronounced. One important factor underlying the growing use

of TQM in agribusiness is food safety. Consumers want more assurance that the food they eat is safe. To help make certain that is the case, more and more firms are implementing the tools and philosophy of TQM.

Training personnel, involving suppliers, performing internal and external audits, conducting customer surveys, and developing quality circles are all recommended techniques for successful implementation of TQM. The keys to success, however, are to keep the system lean and the motivation high. A complex system with numerous forms and reporting cycles is not likely to contribute much or for long.[35]

# Improving Quality: Operational Techniques

In addition to the strategic initiatives described above, organizations can draw upon a number of operational techniques to boost quality.

## Statistical Quality Control

One important operational approach to quality enhancement is **statistical quality control**, which consists of a set of mathematical and/or statistical methods and procedures for measuring and adjusting quality levels. Control charts and other problem-solving tools are used to monitor processes and reduce variability. In some cases experiments are designed in an effort to discover key factors that influence process performance so that the process can be optimized for best performance.

For example, **acceptance sampling** is a process whereby finished goods are sampled to determine what proportion of them is of acceptable quality. By pre-establishing the desired quality level and then statistically determining confidence levels, the manager can determine what percentage of finished goods must be checked to achieve the target level of quality.

Similarly, **in-process sampling** involves testing products as they are being made rather than after

they are finished. In-process sampling works best for products like beverages, chemicals, commodity processing, paint, and so forth that go through various transformation steps (Figure 22.9). In-process sampling enables an organization to make corrections in production before producing a whole batch, thus reducing rejects and waste.

Beyond the various statistical quality control techniques, other useful operational procedures for enhancing quality can be drawn from any of the different decision-making, control, and operations management methods discussed in Chapters 10, 11,

**FIGURE 22.9** A QC technician for a cereal manufacturer examines a wheat ear during an in-field quality-control inspection.

© Rido/www.Shutterstock.com

and 23, respectively. The key, of course, is to match the technique with the situation to facilitate quality.

## Benchmarking

Benchmarking is a relatively new approach to improving quality. **Benchmarking** is the process of finding out in a legal and ethical manner how other firms do something and then either imitating it or improving upon it.[36] Sometimes benchmarking involves buying competing equipment and taking it apart to see how it works (sometimes referred to as "reverse engineering"). In other situations, it involves simply talking to managers at other firms to determine how they do things.

For example, a manager at Convex Computer Corporation spent a week with Disney executives learning how the firm trains its workers and schedules its maintenance. Many firms have visited L.L. Bean to learn more about its mail-order procedures. And Xerox routinely buys competing copiers and disassembles them to discover how they are manufactured.[37]

A recent survey of executives identified speed and time as the number one competitive issue of the 1990s.[38] It should come as no surprise, then, that many organizations are searching for ways to get things done faster. General Electric provides a good case-in-point. In 1985, the company needed three weeks from the time the order was received to deliver a custom-made circuit breaker. At the time, GE had six plants and several hundred workers involved. Today the company can deliver the same order in only three days —using only one plant and 129 workers.[39]

This speedup was accomplished primarily by totally revamping one of its plants (start from scratch) and eliminating most supervisory jobs (minimize the number of approvals). Given the dramatic successes achieved by GE and similar organizations that have also promoted speed and time as major concerns, it is very likely indeed that these issues will be of major importance in the years ahead.

## Flexibility

Finally, **flexibility** is an organization's ability to adapt to different conditions and circumstances. Organizations are realizing that the greater the flexibility they maintain, the easier it is for them to adopt new methods and approaches and to respond to shifts in technology, consumer tastes, and so forth. For example, if a firm invests millions of dollars in a plant that is capable of making only a single product and making it in only one way, it will be very vulnerable to obsolescence. If a new and less costly method of making the same product is discovered, for example, the firm will need either to continue to produce at a cost disadvantage or else to scrap the plant and start over.

## CHAPTER SUMMARY

Quality refers to the total set of features and characteristics of a product or service that bears on its ability to satisfy stated or implied needs. Quality can be assessed both in absolute and relative terms and is closely linked with productivity. Productivity, in turn, can be assessed at several levels. The growth rate of productivity in the United States is falling behind that of other countries. There are several techniques managers can use to boost productivity.

Quality is important for several reasons. The three most important reasons are due to its role in competition, in reducing costs, and in improving long-term effectiveness.

Total quality management, or TQM, represents the fundamental strategic initiative organizations can use to enhance quality. TQM relies on a strategic commitment and uses employee involvement, methods, materials, and technology. Organizations should implement TQM gradually.

Several operational techniques are also available to organizations for boosting quality. These include statistical quality control, benchmarking, speed and time, and flexibility.

## CHAPTER ACTIVITIES

### REVIEW QUESTIONS

1. What are some examples of different levels of productivity?
2. What has been the trend in productivity growth in the United States compared to that of other countries in recent years? What differences exist within the U.S. economy?
3. How are productivity and quality related to one another?
4. How can managers go about trying to increase quality?
5. What roles do speed, time, and flexibility play in quality?

### ANALYSIS QUESTIONS

1. Think of a few manufacturing jobs and a few service jobs you are familiar with. What factors can you identify that may account for productivity differences between them?
2. Given that productivity and quality are related, efforts to improve them might also be related. Identify compatible ways that managers might try to boost productivity and quality simultaneously.
3. Brainstorm a list of products that U.S. companies seem to make better than foreign competitors. Now do the same for products that foreign manufacturers seem to do a better job with. What trends can you identify?
4. Are there limits to quality improvement? That is, is it possible to attain absolute quality?
5. Cite some instances where productivity and quality are not related.

### FILL IN THE BLANK

1. The total set of features and characteristics of a product or service that bears on its ability to satisfy stated or implied needs is called _____ .
2. An economic index of the value or amount of what is created relative to the resources necessary to create it is known as _____ .
3. A comprehensive, organization-wide strategy for improving product and/or service quality on a systematic and continuous basis is called _____ _____ _____ .
4. _____ quality refers to the quality of an intangible service provided by an organization.
5. The process of finding out, in a legal and ethical manner, how other firms do something and then either imitating or improving on it is called _____ .
6. How easily an organization can react, respond, or change is known as _____ .
7. The generally understood level of quality that a product or system needs to fulfill its intended purpose is _____ quality.
8. A category of operational techniques that consists of a set of mathematical and/or statistical methods for measuring and adjusting quality levels is known as statistical _____ _____ .
9. _____ productivity is determined by dividing outputs by the sum of all inputs used by the organization.
10. _____ productivity is determined by dividing outputs by direct labor.

# ▼ CHAPTER 22 CASE STUDY
## "IT ISN'T EASY BEING GREEN"

When Howard Schultz bought the company in 1987, it was a small Seattle coffee bean retailer. It has since grown to nearly 18,000 stores in 60 countries. Not satisfied to be the largest and most recognized coffee purveyor in the world, Starbucks is also positioning itself as a leader in quality broadly defined to include corporate social responsibility. It wants to demonstrate that it is possible to grow, be profitable, have high quality products, and be responsible. Or at least these are the goals set by Mr. Schultz.

As Kermit the Frog™, put it so well in song, "It's not easy being green." This was especially true between 2008–2010 when an unfavorable economic environment forced store closings, corporate restructuring, and acknowledging stronger competition from McDonald's and Dunkin' Donuts. However, Schultz has retained the corporate quality and responsibility agenda regardless of economic circumstances.

Starbucks' basic product—coffee—is not a given. Its coffee purchases account for 3 percent of total global Arabica coffee bean supplies, but it is not the world's largest coffee buyer. Coffee prices in 2010 hit historical levels because of grower problems. This meant coffee purveyors raised prices or took a margin hit. Initially, Starbucks chose the margin hit; McDonald's raised prices; later Starbucks relented and changed some prices, too.

The coffee world has grown complicated. Brazil added to coffee grower problems by demanding inclusion on the Intercontinental Exchange (ICE). It wants to join the 19 other Arabica bean growers on the exchange. Until 2010, Brazil was content to be an outsider, but its inclusion on ICE's futures exchange means its bean would be included on all "C" futures contracts.

Nestlé, a Starbucks' competitor on the grocery shelf, is initiating a $487 million cultivation project in Mexico. It will train thousands of Mexican farmers to grow new coffee trees over the next ten years. "We shouldn't just be the world's biggest coffee buyer, we should be involved upstream," said Nestlé's CEO Paul Bulcke. Vertical integration will affect the competitive environment on the grocery store coffee shelf—a place where Starbucks (through an agreement with Kraft Foods) and Nestlé compete. These situations would be complicated and involved anyway, but the problems are compounded by Howard Schultz's demand that Starbucks be 100 percent fair-traded and Coffee and Farmer Equity certified by 2015.

Then there are problems with the company-owned stores. Not only are the customer lines longer than in the past, many consumers are feeling that quality is slipping since Starbucks began its move to make the stores operate more efficiently. Coffee making was becoming a mechanized process with all the romance of an assembly line. The perception of customers may not be far from the truth. For the past few years Starbucks has been applying the "lean manufacturing techniques car makers have long used." At least lean rhymes with green.

Finally, there is "the cup." Introduced in 1984, the Starbucks logo cup has become a cultural icon. It is also a monumental problem. Approximately three billion of the nation's 200-plus billion paper cups in municipal trash dumps have the Starbucks logo. "From the customer's standpoint, the cup is our No.1 environmental liability," says Jim Hanna, Starbucks' Director of Environmental Impact. Mr. Hanna is feeling pressure because Howard Schultz pledged that Starbucks would have 100 percent recyclable cups by 2012. Starbucks is working with its competitors Tim Hortons, Dunkin' Donuts, McDonald's, and Green Mountain to solve this monumental problem.

The cup problem in a nutshell is that Starbucks has no control over where its customers put their cups. It is not a problem of waste disposal or recycling; it calls for cultural change. That cultural change is expected to result in the customer being provided an incentive or otherwise induced either (a) to dispose of the cup appropriately or (b) to bring a personal re-usable cup with them to the store. Culture change is not easy, and the goal is 2012.

About Howard Schultz's 2012 goal, Jim Hanna says: "We want to put in place the mechanisms to achieve recyclability by 2012, but 2015 is when we expect to accomplish it. It's always been 2015." Peter Senge, an MIT professor involved in the project, hedged when asked about the 2015 goal, "It's not what your vision is. It's what your vision does." And finally, from Jim Hanna again, "We were talking about it in the '80s and we're still talking about it 30 years later. The complexity is what it is."

It really isn't easy being green.[40]

### ▶ Case Study Questions

1. Is "being green" realistically a part of quality for organizations? Did Starbucks reach its recyclability goal for 2012?

2. How are quality and productivity related at Starbucks?

3. What strategic and operational approaches to quality and responsibility does Starbucks use?

## REFERENCES

1. Rich Nuheisel, Interview, ABC television, UCLA vs University of Texas telecast, September 25, 2010; J. M. Juran, Frank M. Gryna, and R. S. Bingham, *Quality Control Handbook* (New York: McGraw-Hill Book Company, 1974); Merle D. Pierson Donald A. Corlett, *HACCP: Principles and Applications* (New York: AVI Books (Van Nostrand Reinhold), 1992; Mina Kimes, "Why J&J's Headache Won't Go Away," *Fortune,* September 6, 2010, 100–108; Jonathan D. Rockoff and Peter Loftus, "J&J Consumer Chief to Retire," *The Wall Street Journal* (September 17, 2010) at wsjonline.com (accessed September 28, 2010); Jonathan D. Rockoff, "Beetles Spur Recall of Baby Formula," *The Wall Street Journal* September 23, 2010, B1; Valerie Bauerlein, "Firm Fined in Explosion," *The Wall Street Journal,* July 8, 2010, B3: James T. Areddy, "Accidents Plague China's Workplaces," *The Wall Street Journal* (July 29, 2010) at wsjonline.com (accessed September 29, 2010).

2. C. K. Prahalad and M. S. Krishnan, "The New Meaning of Quality in the Information Age," *Harvard Business Review* (1999): 109–120.

3. "Quality Isn't Just for Widgets," *Business Week*, 2002, 72–73.

4. "When Service Means Survival," *Business Week*, 2009, 26–40.

5. Ross Johnson and William O. Winchell, *Management and Quality* (Milwaukee, WI: American Society for Quality Control, 1989).

6. Ludwig Theuvsen, Achim Spiller, Martina Peupert, and Gabriele Jahn eds., *Quality Management in Food Chains* (Wageningen, Netherlands: Wageningen Academic Publishers, 2007); and Gerrit Willem Ziggers and Jacques Trienekens, "Quality Assurance in Food and Agribusiness Supply Chains: Developing Successful Partnerships," *International Journal of Production Economics* (1999): 271–279.

7. John W. Kendrick, *Understanding Productivity: An Introduction to the Dynamics of Productivity* (Baltimore, MD: Johns Hopkins, 1977).

8. Brian O'Reilly, "America's Place in World Competition," *Fortune*, November 6, 1989, 83–88.

9. "International Comparisons of GDP Per Capita and Per Employed Person," U.S. Department of Labor, Bureau of Labor Statistics (July 28, 2009) at www.bls.gov (accessed August 2, 2010).

10. "Productivity's Second Wind," *Business Week*, February 17, 2003, 36–37; and "Productivity Grows in Spite of Recession," *USA Today*, July 29, 2002, 1B, 2B.

11. "Agricultural Productivity in the United States," United States Department of Agriculture, Economic Research Service at www.ers.usda.gov (accessed August 1, 2010).

12. Ronald Henkoff, "Make Your Office More Productive," *Fortune*, February 25, 1991, 72–84.

13. "International Labor Comparisons," U.S. Bureau of Labor Statistics (October 27, 2009) at www.bls.go (accessed August 2, 2010).

14. Michael E. Porter, "Why Nations Triumph," *Fortune*, March 12, 1990, 94–108.

15. Robert S. Kaufman, "Why Operations Improvement Programs Fail: Four Managerial Conditions," *Sloan Management Review* (Fall 1992): 83–94.

16. Alan S. Bigger, "More Power Leads to Greater Productivity," *Sanitary Maintenance* (July 2001) at www.cleanlink.com (accessed July 26, 2010).

17. Joe Vanden Plas, "Study Affirms Information Technology-Productivity Link," *WTN News* (March 22, 2007) at wistechnology.com (accessed July 28, 2010); Marianne Kolbasuk McGee, "It's Official: IT Adds Up," *Information Week* (April 17, 2000) at www.informationweek.com (accessed July 30, 2010); and Tim R. V. David, "Information Technology and White-Collar Productivity," *The Academy* of *Management Executive* (February 1991): 55–63.

18. Gene Bylinsky, "Turning R&D into Real Products," *Fortune*, July 2, 1990, 72–77.

19. Gary Hector, "Yes, You Can Manage Long Term," *Fortune*, November 21, 1988, 64–76.

20. Barry Jaruzelski, Kevin Dehoff, and Rakesh Bordia, "Money Isn't Everything," *strategy+business* (2005) at www.strategy-business.com (accessed August 1, 2010).

21. William Bowen, "The Puny Payoff from Office Computers," *Fortune*, May 26, 1986, 20–24.

22. Genichi Taguchi and Don Clausing, "Robust Quality," *Harvard Business Review* (January–February 1990): 65–75.

23. "Baldrige Award Recipients' Contacts and Profiles," *National Institute of Standards and Technology* at www.baldrige.nist.gov (accessed July 26, 2010).

24. Julie A. Caswell and Siny Joseph, "Consumer Demand for Quality: Major Determinant for Agricultural and Food Trade in the Future?" *Journal of International Agricultural Trade and Development* (2008): 99–116.

25. "Beg, Borrow, and Benchmark," *Business Week*, November 30, 1992, 74–75.

26. James Dean and David Bowen, "Management Theory and Total Quality: Improving Research and Practice Through Theory Development," *Academy of Management Review* (1994): 392–418; and "Quality," *Business Week*, November 30, 1992, 66–72.

27. Carl Glass, "TQM: A Good Fit for the Food Industry … If the Program is Tailored with Conviction," *Food Processing* (July 1, 1993) at www.allbusiness.com (accessed August 1, 2010).

28. Richard J. Schonberger, "Is Strategy Strategic? Impact of Total Quality Management on Strategy," *The Academy of Management Executive* (August 1992): 80–86.

29. "Building a Self-Directed Work Team," *Training & Development*, December 1992, 24–28.

30. "King Customer," *Business Week*, March 12, 1990, 88–94.

31. Paul S. Adler, "Time-and-Motion Regained," *Harvard Business Review* (January–February 1993): 97–108.

32. "Our History," *Blue Bell* website at www.bluebell.com (accessed August 2, 2010); "The Ice Cream Man Cometh," *Forbes,* January 22, 1990, 52–56; and C. Kevin Swisher, "Just Desserts," *Texas Highways,* August 1991, 12–17.

33. "Quality is Becoming Job One in the Office, Too," *Business Week*, April 29, 1991, 52–56.

34. "Quality," *Business Week*, November 30, 1992, 66–72.

35. Clemens Morath and Reiner Doluschitz," Current Situation and Potential of Total Quality Management in the Food Industry," *4th Aspects and Visions of Applied Economics and Informatics* (2009): 864–869; Clemens Morath and Reiner Doluschitz, "Total Quality Management in the Food Industry—Current Situation and Potential in Germany," *APSTRACT: Applied Studies in Agribusiness and Commerce* (2009): 83–87; Ludwig Theuvsen ed., *Quality Management in Food Chains* (Weimar, TX: C.H.I.P.S. Culinary and Hospitality Industry Publications Services, 2007); Steven A. Schulz and Robert J. Masters, "Total Quality Management: The Deming-Food Processing Connection," *Journal of Food Products Marketing* (1997): 61–70.

36. Jeremy Main, "How to Steal the Best Ideas Around," *Fortune,* October 19, 1992, 102–106.

37. "Beg, Borrow, and Benchmark." Gary Hamel and C. K. Prahalad, *Competing for the Future* (Boston: Harvard Business School Press, 1994); George Stalk Jr. and Thomas M. Hout, *Competing Against Time* (New York: The Free Press, 1990).

38. Brian Dumaine, "How Managers Can Succeed Through Speed," *Fortune*, February 13, 1989, 54–59.

39. Ibid.

40. Anya Kamenetz, "The Starbucks Cup Dilemma," *Fast Company* (October 20, 2010) at www.fastcompany.com (accessed November 10, 2010); Nancy F. Koehn, "Howard Schultz and Starbucks Coffee Company," *Harvard Business School Case No.9-801-361* (September 30, 2005) at www.scribd.com (accessed November 10, 2010); Julie Jargon and Paul Ziobro, "Starbucks Will Boost Prices on Some Drinks as Costs Rise," *The Wall Street Journal* (September 23, 2010), B7; Julie Jargon, "For Starbucks, a New Retail Mix," *The Wall Street Journal* (August 19, 2010), B9; Julie Jargon, "At Starbucks, Baristas Told No More than Two Drinks," *The Wall Street Journal* (October 13, 2010) at online.wsj.com (accessed November 8, 2010); Tony Danby and Tom Sellen, "No Relief in Sight for the Coffee Market," *The Wall Street Journal* (August 23, 2010) at online.wsj.com (accessed November 8, 2010); Tom Sellen, "Coffee Approaches a 13-Year High," *The Wall Street Journal* (August 30, 2010), C6; Anna Raff, "Brazil Closer to Joining ICE's Futures Blend," *The Wall Street Journal* (October 15, 2010) at online.wsj.com (accessed November 9, 2010); Christina Passariello and Laurence Iliff, "Nestle Plans Ground Attack Over Coffee Beans," *The Wall Street Journal* (August 26, 2010) at online.wsj.com (accessed November 9, 2010).

# Operations and Technology Management

## LEARNING OBJECTIVES

After studying this chapter, you should be able to:

- Discuss the nature, meaning, and importance of operations management.

- Describe operations decisions and operations planning.

- Indicate what is involved in the management and organization of operations, and describe the relation of change to operations.

- Discuss operations control in its most important forms: inventory control, quality control, scheduling control, and cost control.

- Identify basic operations control techniques.

- Discuss the nature of technology management.

## Designer Algae and Scum

A hypothesis generated by biochemists had "life" originating from a primordial "soup." This soup consisted of carbon compounds generated from the interaction of heat, light, and proximity. Some of this soup would be trapped in little rock indentations, and over time this coacervate (accumulated, concentrated soup) would coalesce into something living. Farfetched, possibly, but at least it provides a story for the origin for pond scum; and pond scum, as it turns out, is important to the operation of agribusiness (Figure 23.1).

**FIGURE 23.1** "Yucky," unsightly green algae pond scum will be more attractive when successfully "farmed" as an input for biofuel.

© Dominique Landau/www.Shutterstock.com

Operations in agribusiness use energy and lots of it. Energy is needed to plant, cultivate, and harvest crops; to process the commodities into final consumer-oriented products; and to transport these products to the end user. While many forms of technology for the production of energy have been suggested, our planet remains dependent upon energy derived from fossil fuels—oil and natural gas. They are referred to as *fossil fuels* because today's oil and natural gas were yesterday's plants. Well, many yesterdays, that is—several hundred million years of yesterdays to be precise. Prehistoric forests, swamps, or other organic matter were trapped by a geologic event, and the weight of the Earth's layers and the heat generated by its core transformed plant material into "oil," or "black gold." This oil, after some processing, is our gasoline and diesel fuel. Together they are the lifeblood of operations in agribusiness.

Agribusiness has always been intrigued by the origin of today's fossil fuel and wondered (often out loud) if there were a technology that would continue this process in some controlled fashion using renewable plants. Under the heading "biofuels," the agribusiness industry has responded to the allure of biofuel potential. Unfortunately, it is also a history of "false starts" and "supernovae." The false starts were dead ends or limited-potential ends, like the use of vegetable oils in diesel engines. These "supernovae" received much more attention but their promise went unrealized or was compromised by chemical reality—ethanol. Biodiesel fuel is compromised by the need to collect and transform the material prior to use; the energy equation is not balanced. Ethanol is not a substitute but an additive, an extender; a "hamburger helper™" for gasoline. Its corrosive nature, low-energy yield compared to gasoline and ability to absorb water provided a limit to its potential as a gasoline engine additive.

Now a new post-alcohol world is emerging in biofuels. Instead of ethanol, the plan is to use emerging technologies to make hydrocarbons—compounds similar to those extracted from the ground. These are referred to as "drop-in fuels," that is, direct substitutes for current fuels. Codexis, a California company, has joined with Shell Oil Company, an Anglo-Dutch company, and Cosan, a Brazilian sugar producer, to produce 656 million gallons of "drop-in" fuel. Enzymes and bacteria produced by Codexis will do the work by acting on sugar cane to produce 12–16 carbon chain molecules—alkanes. We call it *diesel fuel*. According to Dr. Alan Shaw, CEO of Codexis, his industry's problem has been a "failure to think big."

Unilever, an Anglo-Dutch agribusiness food company, has a slightly different technology driven by a different problem. The company uses palm oil to make Dove soap, Vaseline lotion, and Magnum ice cream. Environmentalists are putting pressure on Unilever to stop the use of palm oil to preserve the rain forests. Its solution is a partnership with Solazyme Inc., a California company that harvests algal oil. In fact, its micro-algae in huge bioreactors will ferment switch grass (cellulose source), beets (sugar source), wood waste (cellulose source), and sugar cane (sugar source) into biodiesel, jet fuel, biocrude oil, and edible oils.

Unlike the past, companies today approach these projects more realistically. "This isn't just a niche application. This is something which we believe has tremendous capability," says Phil Giesler, director of innovation at Unilever. Not to be left out in the cold, Chevron Corporation is also an investor in Solazyme. Another company, Sapphire Energy Inc., is working on similar algal oils for the transportation industry, with financial support derived from Bill Gates, Microsoft's founder.

Perhaps the single most illustrious participant in this movement, however, is J. Craig Venter. His company, Celera Genomics, was the private industry participant in the Human Genome project that led to the decoding of the human genome. It was his company that drove the U.S. Government portion of the project to move more quickly. He has formed a new company, Synthetic Genomics, to use bacteria, algae, and even higher plants to carry out industrial tasks such as producing drop-in fuel products. The progress being made in this area has led ExxonMobil Corp. to commit $600 million for further research, half of which would go to Synthetic Genomics.

The difference in the "post alcohol" world includes the participants and the technology. This is no longer an academic exercise; it is serious business. Academic research will continue unencumbered by a need for application, but business product development requires an endpoint and a return on investment (ROI).

Pond scum has come a long way. It may not have arrived yet; however, this time it will not be due to lack of commitment, interest, or real resources.[1]

## INTRODUCTION

By definition, business organizations provide goods and services to customers. Monsanto Company is in the business of manufacturing and selling pharmaceuticals, agricultural chemicals, food additives, and other specialty products. In similar fashion, Chevron produces and sells gasoline, Compaq combines thousands of component parts into computers, Green Things designs and installs aesthetically pleasing landscapes, and Summer Farms grows vegetables for human consumption plus other agricultural products for processing. The various processes, decisions, and systems involved in the acquisition of resources and the transformation of those resources into the firm's products or services is the domain of operations management.

But as you can tell by reading the above vignette, operations management in today's companies is becoming unbelievably complex as companies must step out of their comfort zones and adapt to a changing "product world." These organizations cannot survive if they fail to manger their operations effectively. Thus, in this chapter we explore the nature of operations management and discuss the connection between planning and operations. We describe managing operations, and then go on to discuss operations control and to identify a number of operations control techniques.[2] We conclude the chapter with a discussion of technology management, a related part of effective operations management.

## The Nature of Operations Management

The management of operations is an extremely complex and important function. Indeed, without effective operations management, few organizations could survive for any length of time. Thus, it is important to have a clear understanding of both the meaning and the importance of operations management.

### The Meaning of Operations Management

We define **operations management** as the total set of managerial activities an organization uses to transform resource inputs into products and services.[3] Figure 23.2 illustrates the essential nature of operations management. Think back to our discussion of systems theory in Chapter 2. Systems theory holds that organizations consist of four basic parts: inputs, transformation processes, outputs, and feedback. As shown in the exhibit, the concern of operations management is primarily with the transformation processes themselves. At a secondary level, its concern is with inputs, outputs, and feedback from the environment.

**Manufacturing** is a form of business that combines and transforms resources into tangible outcomes that are then sold to others. Because manufacturing was once the dominate industry in the United States, the entire area of operations management was formerly called production management. Monsanto Company is a manufacturer because it uses chemicals to create products. Similarly, The Goodyear Tire & Rubber Company is a manufacturer because it combines rubber and chemical compounds and uses

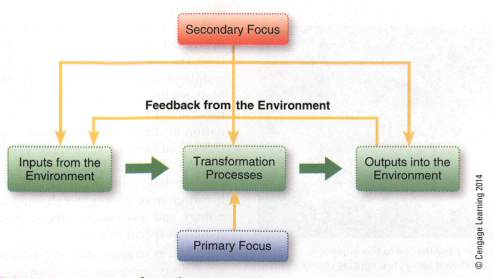

**FIGURE 23.2** The nature of operations management.

blending equipment and molding machines to create tires. And Caterpillar and John Deere are manufacturers because they buy metal components and combine them into equipment.

During the 1970s, manufacturing entered a long period of decline in the United States, primarily because of foreign competition. U.S. firms had grown lax and sluggish, and new foreign competitors came onto the scene with new equipment, lower wages, and much higher levels of efficiency. For example, steel companies in the Far East were able to produce high-quality steel for much lower prices than were U.S. companies, such as Bethlehem Steel and U.S. Steel (now USX Corporation). Faced with a battle for survival, many companies underwent a long and difficult period of change by eliminating waste and transforming themselves into leaner and more efficient and responsive entities. They reduced their workforces dramatically, closed antiquated or unnecessary plants, and modernized their remaining plants. In recent years, their efforts have started to pay dividends as U.S. business regains its competitive position in many different industries. Although manufacturers from other parts of the world are still formidable competitors and U.S. firms may never again be competitive in some markets, the overall picture is much better than it was just a few years ago. Prospects continue to look bright.[4]

During the decline of the manufacturing sector, a tremendous growth in the service sector kept the U.S. economy from declining at the same rate.[5] A **service organization** is one that transforms resources into an intangible output and creates time or place utility for its customers (Figure 23.3). For example, Merrill Lynch Co. Inc. makes stock transactions for its customers, Avis leases cars to its customers, and your local barber or hairdresser cuts your hair. In 1947, the service sector was responsible for less than half of the U.S. gross national product (GNP), but recently it has exceeded 75 percent. Especially since women have entered the workforce, we have been moving away from a do-it-yourself society. The service sector is expected to create 14.5 million new wage and salary jobs from 2008 to 2018.[6] Managers have come to see that many of the tools, techniques, and methods in use in factories are also useful to a service firm.[7] For example, managers of food processing plants and hair salons must each decide how to design their facilities, identify the best locations for those facilities, determine optimal capacity, make decisions about inventory storage, set procedures for purchasing raw materials, and set standards for productivity and quality.

---

### FOOD FOR THOUGHT 23.1

There are seven stages involved in getting chicken to the consumer—breeder flock, pullet farm, breeder house, hatchery, broiler farm, processing, and distribution.

---

## The Importance of Operations Management

It should be clear by this point that operations management is quite important to organizations. Beyond its direct impact on quality and productivity, it also directly influences the organization's need for capital and its overall level of effectiveness. Obviously, then, operations management should be addressed at every level, starting at the top. For example, the deceivingly simple strategic decision of whether to emphasize high quality regardless of cost, lowest possible cost regardless of quality, or some intermediate combination of the two has numerous important implications. A highest-possible-quality strategy will dictate state-of-the-art technology and rigorous control of product design and materials specifications. A combination strategy might call for lower-grade technology and less concern about product design and materials specifications.

Just as strategy affects operations management, operations management likewise affects strategy. Suppose that a firm decides to upgrade the quality

**FIGURE 23.3** The company that owns this airplane is performing a service when it sprays fire-fighting chemicals on a forest.

© Mark III Photonics/www.Shutterstock.com

of its products or services. The organization's ability to implement the decision is dependent in part on current production capabilities and other resources. If existing technology will not permit higher-quality work and if the organization lacks the resources to replace its technology, increasing quality to the desired new standards will be difficult. Coleman, well known for its camping equipment, used operations management as part of its strategy to upgrade product quality. By adopting a just-in-time inventory system, the company reduced its inventory costs by $10 million. By putting workers in charge of inspecting their own work and stopping the assembly line, if necessary, Coleman reduced its scrappage rates by 60 percent and increased its productivity by 35 percent.[8]

Operations management is also a critically important function for specific activities within an organization. Monsanto faces problems if it cannot produce pharmaceuticals as efficiently as can its foreign competitors. Walmart has problems if it has too much inventory on hand (carrying costs, spoilage, warehousing expenses) or too little inventory (customer complaints, lost sales). It also has problems if it has the wrong selection of merchandise or pays too much for its merchandise. The same is true of Summer Farms. Chevron has problems if it cannot refine gasoline efficiently, if it cannot deliver the gas to service stations efficiently, or if it must charge too much for its products. Tibbals' Hartco flooring plant faced similar problems. A Pizza Hut restaurant has problems if it has too many or too few ingredients to make pizza, if it produces poor-quality pizzas, or if its service is bad.

The hallmarks of operations management, then, are efficiency and effectiveness—doing things in a way that gets the maximum value from resources, and doing the right things to begin with. Operations management helps Ford determine what parts to make and what parts to buy, when to have them delivered, how they should be combined, and how the finished product will be delivered to the showroom floor. Without an effective operations management system, few organizations would survive.[9]

## Planning for Operations

The first stage in effective operations management includes both operations decisions and operations planning.[10]

## Operations Decisions

Operations decisions encompass virtually all aspects of the operations management system. Eight areas in particular are especially important.

**Product or Service Line.** The **product or service line** decision is one of the most crucial decisions an organization makes. It is almost always made by top managers from all relevant functional areas because the organization's overall strategy determines the general products and/or services on which it will concentrate. That is, Sony deals in consumer electronics, whereas Pepperidge Farm concentrates on baked goods. Summer Farms must plant intelligently and avoid concentrating entirely on products that become ready for harvesting at the same time or products that have an especially short shelf life.

Marketing and operations managers then work together to define more precisely what the product line should, can, and will be. Marketing managers at Pepperidge Farm may use consumer research to get information about what products customers want and then ask operations managers to figure out how best to produce those products. Similarly, operations managers at Hormel may realize that producing another version of an existing product would be very easy and efficient. They would then ask marketing managers to find out whether such a variation would sell at acceptable levels. Thus, product or service line decisions are almost always made jointly by managers from different areas.

**Capacity.** **Capacity** refers to the ability of the organization to produce goods or services. Capacity decisions then involve determining the amount of product or service the organization needs to meet demand efficiently and effectively. For example, how many birds does Pilgrim's Pride want to process each year? How many tables does a certain McDonald's restaurant need? How many acres should Summer Farms devote to a particular crop? These are capacity decisions.

Capacity decisions must be made with great care. If a firm has too much capacity resulting from a decision to have more of some resource than it actually needs, the result can be a dramatic underutilization of resources. If Walmart builds a new store with 150,000 square feet of floor space when 100,000 would have sufficed, the company will need to find uses for the extra space, maintain it, and heat and cool it. The same is true of Summer Farms, which must not devote too much acreage to crops that will not sell on time. Many recent plant closings and cutbacks in U.S. businesses have been the result of such excess capacity.

Too little capacity can be just as damaging, however. For example, a food processing plant with insufficient capacity cannot provide an adequate quantity of merchandise and will subsequently lose customers to competitors. Likewise, if grocery customers have to wait in line too long, they will get discouraged and shop elsewhere. Food establishments with an insufficient amount of proper storage will lose perishable inventory to spoilage.

In general, the key is to optimize; but that may be quite difficult as demand fluctuates. A restaurant, for instance, might be able to fill 80 tables on Friday night and 100 on Saturday night. The rest of the week, however, it will fill only 30 tables. Having 100 tables may not be efficient; instead, the restaurant should probably have 50 or 60 tables. The excess capacity during the week will not create a problem, and on weekends the tables will be filled, although some people will have to wait to be seated and a few will leave and go elsewhere. On balance, the organization will likely be most effective if it has more than the minimum capacity it needs but less than the maximum it occasionally will be able to use.

**Planning System.** The **planning system decision** involves determining how operations managers will get the information they need and how they will provide information to other managers. For example, assume an organization uses sales forecasts as a basis for deciding how much of a particular product to make. In the middle of January the operations manager receives a sales forecast for the month of February. He or she then needs to check current inventory, including work-in-process, and set production levels for February. In some cases, however, the information can also go in the other direction—the operations manager may need to keep the marketing manager informed about current cost and inventory levels. These figures will help the marketing manager decide what discounts to grant, which products to push, and so forth.

**Organization.** Another operations management decision involves the organization of the operations function; that is, where operations activities should be housed and administered in the overall design of the organization. We explore these issues in a later section.

**Human Resources.** Working in conjunction with human resource managers, operations managers must decide what kinds of employees they need. These decisions involve both the quantity and the quality of the workforce. For example, suppose a manufacturer needs to increase its workforce prior to the Christmas season or the spring and summer seasons. Its managers will make decisions about how many new workers will be needed, what skills they should have, and when they should be hired. Recent trends toward automation and high-tech manufacturing have made these decisions even more critical, in that managers must take greater care to select just the right kinds of employees to work in such settings.[11]

**Technology.** **Technology** involves the actual processes used in transforming raw materials and other inputs into appropriate outputs. Decisions related to what form of technology to use and when to change technology are critical parts of operations management. Some organizations, such as home cleaning services, athletic teams, and small farms tend to be **labor-intensive**, which means that people do most of the work. Other firms, such as General Mills, are **capital-intensive**, which means that machines do almost all of the work. Still other companies use a balance of people and machines. We discuss technology management later in this chapter.

**Facilities.** Another important set of operations decision involves **facilities**—the means of accomplishing production. Should a firm have one or two large plants or several smaller plants? Should it make all its own components and then assemble them, or buy some or all of them from other manufacturers? Where should the plants be located? And how should they be arranged? Summer Farms faced such decisions when deciding on the location of new fields; e.g., how important is it to locate a new cornfield adjacent to previous cornfields?

Each of these decisions requires considerable research and consideration. Suppose that a company is searching for a new plant site. It may be efficient to locate the plant near major suppliers, or it may be more efficient to put it close to major customers. The company also needs information about land costs, the supply of labor, construction costs, tax rates, utility rates, quality of life for employees, and so forth. And considerable pressure is often put on companies to locate plants in certain cities. Small towns and major cities throughout the United States have engaged in high-stakes competition as potential sites for new manufacturing facilities. To get the predicted economic benefits, such communities offer reduced tax rates, free land, and other incentives.[12]

**Controls.** Decisions also must be made about operations control, including inventory control, quality control, and scheduling control. We will discuss each of these in more detail later in this chapter.

## Operations Planning

Operations planning relates to the day-to-day, ongoing activities of operations management. There are many associated questions that must be answered by managers responsible for carrying out operations activities. How much should the organization produce? When should it be produced? How often should it be produced? In general, operations planning involves the following four basic steps.

1. The first step in operations planning is to select a **planning horizon**. This decision will vary, of course, depending on the kind of products and/or services the company provides. Research and development managers at Monsanto have a long-term planning horizon. A manufacturer of heavy construction equipment may plan over a twelve-month cycle, but a restaurant manager may plan only a few days ahead.

2. Once the appropriate time horizon has been chosen, the manager then estimates demand for that period. For example, the manufacturer's plant manager might use marketing data to estimate construction equipment demand for each month of the next year. The restaurant manager, in contrast, will forecast customers for each meal for each of the next seven days, with higher estimates developed for special occasions and weekends, and lower estimates developed for most weekdays.

3. The third step in operations planning is to compare projected demand with current capacity for each meaningful block of time (Figure 23.4). In the plant, this could mean making comparisons on a monthly basis. As a consequence, materials may be ordered and labor hours scheduled in one-month blocks. For the restaurant, on the other hand, comparisons may be made for each meal. As a result, it may be necessary to order fresh ingredients twice a day (once for lunch and once for dinner) and to schedule five table servers for lunch and nine servers for dinner.

**FIGURE 23.4** The capacity of a highly mechanized dairy is limited by the number and speed of the milker machines, but mostly by the number of cows.

© www.Shutterstock.com

4. The last step in operations planning is to adjust capacity to demand. If demand exceeds capacity in the plant, the manager has several options: schedule overtime, add more workers, add another shift, subcontract some of the work out to other manufacturers, or shift work to other plants in the same company. If capacity exceeds demand, the manager can lay off some workers and/or shut down some of the plant. Similarly, the restaurant manager may increase or decrease staff, order more or less food, and so on. Summer Farms may find it advantageous to place vegetable farm workers "on holiday" or "on hiatus" between crops, or reassign them temporarily to, say, the corn or hay divisions. Managers may also decide to adjust hours of operation.

### FOOD FOR THOUGHT 23.2

"If you can find a path with no obstacles, it probably doesn't lead anywhere."

—*Frank Howard Clark, B-Western screenwriter and a former cowhand and trail guide*

## Managing Operations

Managing the various elements of an organization's operations systems is an important consideration for managers. In addition to the elements we have already discussed, two major areas of concern in managing operations are organizing and organizational charge.

## Organizing for Operations

The principal issue in organizing for operations is defining where the operations management function fits into the overall structure and design of the organization.[13] Directly or indirectly, operations management affects and is affected by all of the dimensions of organizational design described in Chapters 11–12.

For example, job design considerations permeate operations management. Specialization is one approach to designing jobs, but operations management is affected when managers turn to alternative approaches, such as rotation, enlargement, or enrichment. And new efforts to use participative work teams are relevant for both organization design and operations management.

Perhaps even more significant is the link between departmentalization and operations management. Figure 23.5 shows how the operations function looks when the organization is departmentalized by function and by product. As we can see, operations are centralized at the top under the control of a vice president or similar top manager when departmentalization is functional. When departmentalization is by product, as in a divisional structure like Summer Farms, operations for each product group or division are decentralized under the manager responsible for the division. Other variations are

**Functional Approach**

**Divisional Approach by Product**

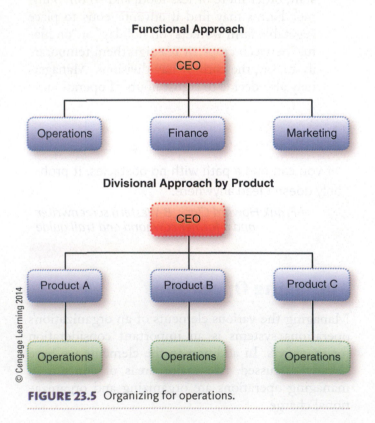

© Cengage Learning 2014

**FIGURE 23.5** Organizing for operations.

also possible. Each plant location may be thought of as a separate department (departmentalization by location), or departmentalization by time may be used for different shifts within a plant location.

Operations management must also address delegation and decentralization issues. Firms, such as General Electric and Westinghouse, practice relatively high levels of decentralization. As a result, their plant managers have considerable discretion and autonomy and can make fairly significant decisions without approval from corporate managers. In settings like the distribution division of Walmart, more centralization is the norm. Warehouse managers there, for example, have relatively little discretion in decision making. In most cases they have to follow established procedures and regulations or they must consult with higher-level managers before making decisions.[14]

New product development is important to all companies with different focuses, depending on the company. In recent years, managers have become increasingly interested in how to structure their organizations to facilitate operations management, especially new-product development. Kraft Foods and Medisyn Technologies, Inc., collaborate to develop new ingredients that could improve product quality, performance, safety and/or health and wellness.[15] General Mills, for instance, thinks about consumers who eat while driving and tries to develop products that can be handled with only one hand.[16] Innovation and the creation of new products and services, and the ways to create those products and services, are becoming ever more critical. When Monsanto refocused its research and development program to concentrate more on major projects, it used its organization design to implement the new approach.

## Change and Operations

Another important dimension of the management of operations relates to organizational change. As we discussed in Chapter 14, two of the major reasons for change are technology and competition. Technology, of course, relates directly to operations management, so being aware of new technology and adopting it when it is appropriate are important to operations managers. Similarly, many (but not all) of the reasons for responding to competitors involve operations. Changes in packaging, product design, product quality, and so forth may be undertaken for competitive reasons.

Technology also provides a major way to change organizations. Changes in work processes and sequences, for instance, are common forms of organizational change that have direct implications for operations management. Several structural (for example, coordination and decentralization) and people-focused (such as selection and training) change techniques are also related to operations management.[17]

# Operations Control

A major concern of operations management is the control function. In fact, many people consider operations management to be almost totally concerned with control. Although this view may be too narrow, operations and control are certainly interrelated.[18] Four areas of operations management that are especially critical are inventory control, quality control, scheduling control, and cost control.

## Inventory Control

**Inventory control** is essential for effective operations management because inventories represent a major investment for all organizations. The goals of inventory control are to make sure the organization has an adequate supply of raw materials to transform into products, that there are enough finished goods to ship to customers, and that in-process inventory is adequate to meet future needs.[19]

As shown in Figure 23.6, there are four basic forms of inventory. **Raw materials inventory** is the supply of materials, parts, and supplies the organization needs to do its work. Ford's raw materials include mechanical parts, electrical parts, paint, belts and hoses, upholstery fabric, and so forth. Pizza Hut's raw materials include flour, sausage, tomato paste, cheese, and other ingredients. The most important thing about raw materials inventory is to make sure that enough materials are on hand to meet production needs but not so much that materials spoil, get broken, or are stolen—or tie up too many funds that the organization may need elsewhere.

**Work-in-process inventory** refers to the inventory of parts and supplies that are currently being used to produce the final products, which are still incomplete at this time. For example, at any given point in time Ford has millions of dollars' worth of materials at various stages of completion on the assembly lines of its factories. At Pizza Hut, in contrast, work-in-process is fairly minimal and consists only of pizzas as they are being cooked.

**Finished goods inventory** is the set of products that have been completely assembled but have not yet been shipped. The key concern here is to have enough goods on hand to meet customer demands but not so much that the products become obsolete, spoiled, or damaged. Boeing does not maintain an inventory of 747 jumbo jets, but instead makes each plane to customer specifications whereas Ford often has thousands of automobiles in finished goods inventory, awaiting dealer orders. Pizza Hut may step up production just prior to its busy serving times so that it has more work-in-process inventory but cannot accumulate a large finished goods inventory that will deteriorate in quality before it is ordered. Obviously, Summer Farms must be concerned about moving its inventories quickly, although the time varies significantly for vegetable products versus corn or bales of hay.

Finally, **in-transit inventory** includes goods that have been shipped from the company but have not yet been delivered to the customer. Again, Ford is likely always to have hundreds of automobiles on trains and trucks headed for dealer showrooms while Pizza Hut has only a few dozen pizzas in transit to local customers.

## Quality Control

Another important part of operations management control is quality control—insuring that the inputs and outputs of the organization meet desired levels of quality. As we discussed in Chapter 22, quality control has become increasingly important in recent years, primarily because of the growing recognition that Japanese success is often due to the high quality of its companies' products.[21]

**Quality control** begins with strategic planning, which, for most firms is determining how they want the marketplace to perceive their products.

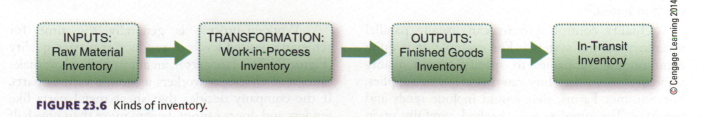

**FIGURE 23.6** Kinds of inventory.

© Cengage Learning 2014

## A FOCUS ON AGRIBUSINESS
### Food-Trak

System Concepts, Inc. (SCI), of Scottsdale, AZ, is one of the first companies to specialize in the field of food and beverage management automation. SCI's primary product is the FOOD-TRAK® System. Founded by Bill Schwartz in 1980, SCI now has more than 10,000 clients in 18 countries.

The FOOD-TRAK® software system and consulting services are designed to assist food service operations of all sizes. SCI tailors that system to its clients through personal, on-site consulting. It uses a project approach to customize a diverse set of documented steps and procedures for each client. Beginning with a total assessment of the client's operation and objectives, SCI focuses on integrating the food and beverage operation into the entire organization structure. If the client so chooses, the company will implement a fully automated system producing maximum benefit and immediate positive impact on profitability.

SCI provides software for all types of food businesses—casinos, hotels, resorts, clubs, restaurants, sports facilities, institutions, and amusement parks. Its clients include Copper Mountain Resort, Whistler Blackcomb Ski Resort, Cherokee Town and Country Club, Boca West Country Club, Keg Restaurants, Outpost Natural Foods, Le Cordon Blue Schools North America, Culinary Institute of America at Greystone, Brookfield Zoo, Canada's Wonderland, Soaring Eagle Casino & Resort, Lucky Eagle Casino, Prairie Meadows Race Track and Casino, and Denver Museum of Nature & Science, to name just a few.

A client in Seattle, WA, Brian Hames, noted that inventorying 1,000 items required six hours prior to using the FOOD-TRAK Mobile Partner. "With the FOOD-TRAK handheld device, our monthly inventory now takes just an hour," said Hanes. In addition to reduced inventory time, some of the other benefits that clients report for FOOD-TRAK® include recipe cost awareness, theft and waste reduction, and more accurate accounting.

As a result of its performance, in 2008 SCI received the *BoardRoom Magazine* Excellence in Achievement Award for Technology/Software Product of the Year.[20]

---

Mercedes-Benz, F.A.O. Schwarz, Rolex, and Neiman-Marcus have all decided that quality is to be their hallmark. Thus price becomes a secondary consideration. Honda, Seiko, Sears, Ford, and Sony have all determined to strike a balance between good quality and reasonable prices. At the low end, Kmart, Timex, and Radio Shack strive for acceptable quality but emphasize low prices. Firms that have developed brands with strong reputations for high quality of service to their customers include Ace Hardware, Apple Computers, Lexus automobiles, and the Ritz-Carlton hotels.[22]

Quality control systems can usually parallel inventory control systems. For example, it is quite common for companies to specify certain quality standards when they buy raw materials or supplies. For Summer Farms, that would include seeds and fertilizer. The supplies are checked carefully upon delivery, and if they do not meet standards or if the shipment is damaged, the company either refuses the materials or accepts them subject to further review.

### FOOD FOR THOUGHT 23.3

White Castle burgers have five holes because the holes help the patties cook faster and more evenly while eliminating the need to turn them over.

Work-in-process is generally the time for achieving desired levels of product or service quality (Figure 23.7). Managers in an automobile plant make numerous checks as workers assemble various parts. If the company decides that sheet-metal parts like fenders and doors cannot deviate more than one-half

**FIGURE 23.7** Factory employee taking a marmalade sample for process monitoring.

inch from target fittings in order to be acceptable, the managers can detect and correct a problem in a fender or door before the problem is compounded. For Summer Farms, irrigation is a crucial factor that affects quality and amount of production amounts.

Finished goods are also checked for quality. For example, as each car at an auto plant is driven from the end of the line to a holding area, the worker can check each automobile for things like the lights, radio, and brakes. In other instances, the quality of finished goods can only be assessed via sampling. Obviously, checking a flashcube or a bottle of wine makes the product unusable, so managers check only a sample, maybe every 100th item, to determine overall quality. Farm products are inspected visually, usually at different intervals before harvesting.

We should also note that quality control is as important for a service company as it is for a manufacturer. A restaurant that sells poor-quality foods, a barber who gives bad haircuts, or a university that does not care about teaching will have problems. Even a landscape architect must be concerned with quality; for example, the amount of care and the survival of the plants he suggests are indicators to clients of the designer's service quality.

## Scheduling Control

Still another important aspect of operations management control is **scheduling control**—having the

right things arrive and depart from the organization at the right time.[23] Consider the case of a contractor who is building a new house. He will need a supply of $2 \times 4$ studs to use in framing the house. If the shipment arrives too soon, the wood may be stolen or scattered; but if it arrives too late, it will delay work. Thus, one of the contractor's scheduling control concerns is to have the lumber delivered on time. The same situation exists for Green Things: If the plants do not arrive from the grower, the planting crew cannot complete the job. In fact, the crew may not even be able to start the planting job, as leaving unfilled holes in the ground is a potential safety hazard.

A recent innovation in scheduling control is the **just-in-time** method, or **JIT**. Used by Robert E. Wood in the construction of the Panama Canal, JIT was pioneered by the Japanese but has become increasingly useful in the United States.[24] As Figure 23.8 illustrates, the traditional approach to scheduling incoming materials is to order a relatively few large shipments of materials and then store them in warehouses until they are needed for production.

Under a JIT system, the company makes more frequent, and therefore smaller, orders of raw materials. The idea is to have resources arrive just as they are needed—just in time. Some materials go straight to the job while others are maintained (at a low level) in small warehouse facilities. Today, United States firms have begun to use this system. For example, automobile manufacturers deliver automobile engines to many automotive plants daily and take them directly to assembly lines. Computers and advanced communication systems facilitate this practice a great deal. Summer Farms may be expected to deliver their corn, soybean, and hay to their large customers who practice JIT.

Another dimension of scheduling applies to manufacturing itself. In traditional systems, machines in a plant are fixed in terms of both where they are and what they do. Parts flow through the plant and are assembled into a final product, so changes in products necessitate major changes in the facilities as well. New types of manufacturing systems promise to change this approach, however. These systems rely on computers to adjust machine placements and settings automatically. This greatly enhances both the complexity and the flexibility of scheduling in a manufacturing system.

## Cost Control

Finally, operations management control also gives considerable attention to **cost control**. Costs are

**Traditional Approach**

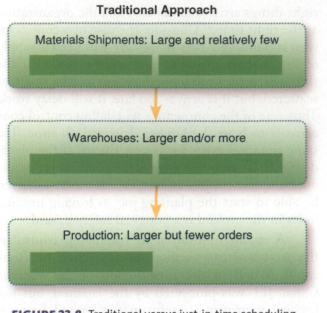

**JIT Approach**

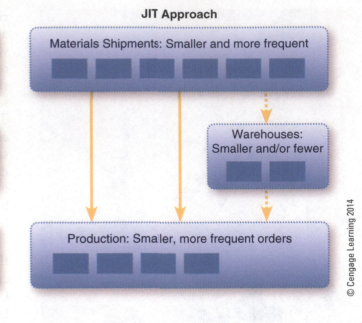

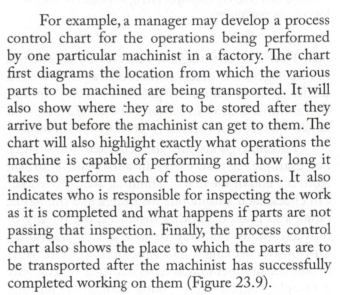

© Cengage Learning 2014

**FIGURE 23.8** Traditional versus just-in-time scheduling.

the expenses incurred by the organization as it conducts its business. Obviously, payments for labor and materials frequently represent major costs, so managers using cost control attempt to identify areas in which costs are excessive and to find ways to reduce them appropriately. Cost control has become a major area of managerial attention. For example, Alcoa, DuPont, and General Electric have all undertaken major cost-cutting programs in recent years. This was what Joshua Summer was trying to accomplish years ago when he reduced the quantity of water Summer Farms was using to irrigate some crops.

## Operations Control Techniques

To carry out operations control most efficiently, managers rely on various control techniques. Several of these techniques were discussed earlier in Chapter 10. Two additional techniques that are particularly relevant to operations management, however, are process control charts and materials requirements planning.

### Process Control Charts

A **process control chart** is a visual representation of an operations function. It illustrates the operation itself, shows how materials are transported to and from workstations, highlights inspection points, and details inventory and storage arrangements. The chart also shows the time and resources associated with each element of the operations function.

For example, a manager may develop a process control chart for the operations being performed by one particular machinist in a factory. The chart first diagrams the location from which the various parts to be machined are being transported. It will also show where they are to be stored after they arrive but before the machinist can get to them. The chart will also highlight exactly what operations the machine is capable of performing and how long it takes to perform each of those operations. It also indicates who is responsible for inspecting the work as it is completed and what happens if parts are not passing that inspection. Finally, the process control chart also shows the place to which the parts are to be transported after the machinist has successfully completed working on them (Figure 23.9).

By carefully studying such charts, operations managers can develop a better understanding of work flows within a particular facility, spot inefficiencies, and make adjustments to improve the overall operations management system. Thus, process control charts are continuously being updated and refined as new and better ways of doing things are identified.

### Materials Requirements Planning (MRP)

Another useful control technique for operations management is **materials requirements planning**, or **MRP**,[25] which companies use to manage complex delivery schedules so that materials arrive as needed

**FIGURE 23.9** Food processing uses technology, but there are still tasks that must be done by hand. After several rounds of filtering, this processed maple syrup is ready to be poured into bottles.

and in the proper quantities. Table 23.1 outlines the steps in the MRP process.

MRP is actually performed with computer software. First, the manager specifies the parts and supplies necessary for a project and determines when they should arrive. Then he determines existing inventories. The MRP system then specifies an ordering and delivery system for materials and parts that are not currently in inventory. Finally, the software generates reports that tell the manager when to place orders and the quantities of each material and part to specify in each order.

Perhaps the greatest value of MRP is its ability to handle different delivery systems and lead times effectively. When a company needs hundreds of parts in vastly different quantities and when delivery times range from a day to several months in the future,

coordination by an individual manager may well be impossible. A properly designed MRP system, however, can cope with such factors fairly easily.

---

**FOOD FOR THOUGHT 23.4**

Bioinformatics is the application of statistics and computer science to the field of molecular biology.

---

## Enterprise Resource Planning

Enterprise Resource Planning (ERP; also known as enterprise requirement planning) is an extension of MRP. ERP is an integrated information system that enables the user to monitor all organizational processes—production, inventories, sales, marketing, accounting, and human resources. Generally ERP involves computer integrated systems due to the volume and complexity of the information involved. Agribusiness users range from individual farmers, growers, and ranchers to government agencies and global enterprises like Cargill.[26]

## Technology Management

Again, the importance of managing technology is an integral part of operations management. In this section we first discuss manufacturing technology and then service technology.

## Manufacturing Technology

Organizations use numerous forms of manufacturing technology. Two especially important new forms are automation and computer-assisted manufacturing.

**TABLE 23.1** The MRP Process

| STEP | EXAMPLE |
|------|---------|
| 1. Manager specifies the needed resources and decides when they must be available. | Plant manager determines the need for 240 steel wire casings as well as component nuts, bolts, wires, and screws to make 240 products during the next six weeks. Production is to be spaced evenly throughout the period. |
| 2. Manager determines the existing inventory. | Current inventory includes 17 wire casings as well as specified numbers of nuts and bolts, wires, and screws. |
| 3. Computer specifies an ordering and delivery system for parts not currently in inventory. | Specifications include details as to amounts needed, suppliers, schedules, and the like. Suppliers are all local. |
| 4. Computer generates reports that tell the manager what and when to order. | Manager gets a report summarizing orders for 23 casings and sets of component items for delivery next Monday, and 200 additional casings and sets of component items for delivery each of the following five Mondays. |

© iStockphoto/Image Innovation

**Automation.** **Automation** is the process of designing work so that it can be completely or almost completely performed by machines. Because automated machines operate quickly and make few errors, they increase the amount of acceptable work that can be done. Thus, automation helps to improve products and services, and it fosters innovation. Automation is the most recent step in the development of machines and machine-controlling devices.

Machine-controlling devices have been around since the 1700s. James Watt, a Scottish engineer, invented a mechanical speed control to regulate the speed of steam engines in 1787. The Jacquard loom, developed by a French inventor, was controlled by paper cards with holes punched in them; early accounting and computing equipment was controlled by similar punched cards. These early machines were primitive, and the use of automation was relatively slow to develop. The value of automation in agribusiness was clearly evident and led to developments such as the cotton gin, hay balers, threshers, and milking machines.

Automation today relies on sensors, information, a control mechanism, and feedback, as Figure 23.10 illustrates. *Sensors* are the parts of the system that gather *information* and compare it to some preset standards. The *control mechanism* is the device that sends instructions to the automatic machine. *Feedback* is the flow of information from the machine back to the sensor. A thermostat, for example, has sensors that monitor air temperature and compare it to a preset low value. If the air temperature falls below the preset value, the thermostat sends an electrical signal to the furnace, turning it on. The furnace heats the air. When the sensors detect that the air temperature has reached a value higher than the low preset value, the thermostat stops the furnace. The last step (shutting off the furnace) is known as feedback, a critical component of any automated operation.

The big move to automate factories began during World War II. The shortage of skilled workers and the development of high-speed computers combined to bring about a tremendous interest in automation. **Programmable automation** (the use of computers to control machines) was introduced during this era, far outstripping **conventional automation** (the use of mechanical or electromechanical devices to control machines).[27] The automobile industry began to use automatic machines for a variety of jobs. In fact, the term *automation* came into use in the 1950s in the automobile industry. The chemical and oil-refining industries also began to use computers to regulate production. It is this computerized, or programmable, automation that presents the greatest opportunities and challenges for management today.

The impact of automation on people in the workplace is complex. In the short term, people whose jobs are automated find themselves without jobs. In the long term, however, more jobs are created than are lost. Nevertheless, not all companies are able to help displaced workers find new jobs, so the human costs are sometimes high. In the coal industry, for instance, automation has been used primarily in mining. The output per miner has risen dramatically from the 1950s on, but the demand for coal has decreased and productivity gains resulting from automation have lessened the need for miners.

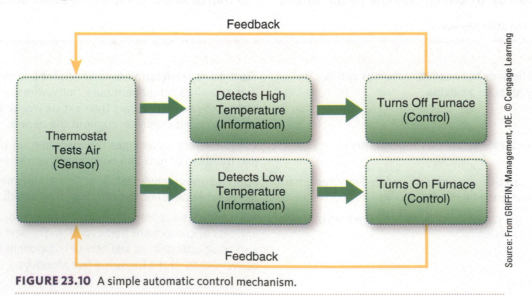

**FIGURE 23.10** A simple automatic control mechanism.

Source: From GRIFFIN, Management, 10E. © Cengage Learning

Consequently, many workers have lost their jobs, and the industry has not been able to absorb them. Automation and offshoring have also cut employment in the electronics and other industries despite rising demand for products.

**Computer-Assisted Manufacturing.** Current extensions of automation generally revolve around computer-assisted manufacturing. **Computer-assisted manufacturing** is technology that relies on computers to design or manufacture products. One type of computer-assisted manufacturing is **computer-aided design (CAD)**—the use of computers to design parts and complete products and to simulate performance without having to construct prototypes. McDonnell Douglas uses CAD to study hydraulic tubing in DC-10s. Japan's automotive industry uses it to speed up car design. GE used CAD to change the design of circuit breakers. And Benetton uses CAD to design new clothing styles and products. Oneida Ltd., the table flatware firm, used CAD to design a new spoon in only two days.[28]

CAD is usually combined with **computer-aided manufacturing (CAM)** to ensure that the design moves smoothly to production. The production computer shares the design computer's information and is able to have machines with the proper settings ready when production is needed. A CAM system is especially useful for handling re-orders because the computer can quickly produce the desired products, prepare labels and copies of orders, and send the products wherever they are wanted.

Closely aligned with this approach is **computer-integrated manufacturing (CIM)**. In CIM, CAD and CAM are linked together and computers adjust machine placements and settings automatically to enhance both the complexity and the flexibility of scheduling. Computers control all manufacturing activities. Because the computer can access the company's other information systems, CIM is a powerful and complex management control tool.[29]

**Flexible manufacturing systems (FMS)** usually have robotic work units or work stations, assembly lines, and robotic carts or some other form of computer-controlled transport system to move material as needed from one part of the system to another. FMS such as the one at IBM's manufacturing facility in Lexington, Kentucky, rely on computers to coordinate and integrate automated production and materials-handling facilities.[30]

These systems are not without disadvantages, however.[31] For example, because they represent fundamental change, they also generate resistance. Additionally, because of their tremendous complexity, CAD systems are not always reliable. CIM systems are so expensive that they raise the breakeven point for firms using them.[32] This means that the firm must operate at high levels of production and sales to afford the systems.

**Robotics.** One of the newest trends in manufacturing technology is robotics. **Robotics** refers to the science and technology of the construction, maintenance, and use of robots. A **robot** is any artificial device that can perform functions ordinarily thought to be appropriate for human beings. The use of industrial robots has been increasing steadily since 1980 and is expected to continue increasing slowly as more companies recognize the benefits that accrue to users of industrial robots.

Welding was one of the first applications for robots, and it continues to be the area for most applications. In second place and close behind is materials handling. Other applications include machine loading and unloading, painting and finishing, assembly, casting, and machining applications such as cutting, grinding, polishing, drilling, sanding, buffing, and deburring. Chrysler, for instance, replaced about 200 welders with 50 robots on an assembly line and increased productivity about 20 percent.[33] The use of robots in inspection work is also increasing. They can check for cracks and holes, and they can be equipped with vision systems to perform visual inspections.

Robots are also beginning to move from the factory floor to all manner of other applications. The Dallas police used a robot to apprehend a suspect who had barricaded himself in an apartment building. The robot smashed a window and reached with its mechanical arm into the building, scaring the suspect into running outside. At the Long Beach Memorial Hospital in California, robot arms assist brain surgeons by drilling into patients' skulls with excellent precision.[34] Newer robot applications involve remote work. For example, using robot submersibles controlled from the ocean surface can help divers in remote locations. BP used robots underwater to monitor and repair the Deepwater Horizon well after it ruptured in 2010. In other applications, surveillance robots fitted with microwave sensors can do things that a human guard cannot do such as "seeing" through nonmetallic walls and in the dark. And automated farming (agrimation) uses robot harvesters to pick fruit from a variety of trees.[35]

Small manufacturers also use robots. One robot slices carpeting to fit the inside of custom vans in an upholstery shop. Another stretches balloons flat so that they can be spray-painted with slogans at a novelties company. At a jewelry company, a robot holds class rings while a laser engraves them. These robots are lighter, faster, stronger, and more intelligent than those used in heavy manufacturing and are the types that more and more organizations will be using in the future.[36] Dairies are increasingly using robotic for milking cows as well as palletizing (loading cases of dairy product onto wooden pallets).[37]

### FOOD FOR THOUGHT 23.5

"Oh, my goodness! Shut me down! Machines building machines. How perverse."

—C-3PO

## Service Technology

Service technology is also changing rapidly; and it, too, is moving more and more toward automated systems and procedures. In banking, for example, new technological breakthroughs have led to automated teller machines and made it much easier to move funds between accounts or between different banks. Many people now have their paychecks deposited directly into checking accounts from which many of their bills are then paid automatically. And credit card transactions are recorded and billed electronically.

Hotels use increasingly sophisticated technology to accept and record room reservations. Universities use new technologies to electronically store and provide access to all manner of books, scientific journals, government reports, and articles. Hospitals and other health care organizations use new forms of service technology to manage patient records, dispatch ambulances, and monitor vital signs. Restaurants use technology to record and fill customer orders, order food and supplies, and prepare food. Given the increased role that service organizations play in today's economy, even more technological innovations are likely to be developed in the years to come.

## CHAPTER SUMMARY

Operations management deals with the transformation of inputs into outputs; the inputs and outputs themselves are secondary components. Operations management is critically important for both strategic and operational reasons.

Planning for operations is the first stage in effective operations management. It involves both operations decisions and operations planning. Operations decisions must be made in eight primary areas, which affect all aspects of operations management but relate most significantly to the transformation process. Operations planning relates to the more day-to-day activities of operations management. There are four basic steps in operations management.

The management and organization of operations involve all of the aspects discussed under planning for operations. Managers must decide how to organize the operations function within the context of organization design. There is also a very important link between change and operations management.

Operations control involves four areas that must be carefully controlled: inventories, which represent a major investment for an organization; quality;

scheduling, which involves having the right things arrive and depart from the organization at the right time; and cost, which deals with the expenses the organization incurs in conducting its business.

Managers can use numerous techniques in operations management. Two major tools are process control charts and material requirements planning (MRP) or enterprise resource planning (ERP).

Technology management is a related area that is also important to organizations. Manufacturing technology has been influenced by CAD, CAM, and CIM, as well as robotics. Service technology is also an important consideration for organizations.

## CHAPTER ACTIVITIES

### REVIEW QUESTIONS

1. Why has the term "operations management" gradually replaced the term "production management?"
2. What are the eight key operations decisions managers must make?
3. Identify the four key areas of operations control.
4. Name and describe the four types of inventory most organizations maintain.
5. What is "just-in-time" scheduling?

### ANALYSIS QUESTIONS

1. How do you use operations management in your day-to-day activities?
2. Which operations decisions are most related to other managerial activities and which are "purely" operations management?
3. How is operations planning similar to and different from more general types of planning activities as described earlier in the planning section?
4. Are some forms of operations control more important for some businesses than for others? Support your answer with examples.
5. What steps might an organization go through in converting from a traditional to a just-in-time scheduling system?

### FILL IN THE BLANKS

1. The total set of activities used to transform resources into products and services is called _____ _____ .
2. The set of products and/or services that an organization sells make up its _____/_____ _____ .
3. The set of processes an organization uses to transform raw materials and other inputs into appropriate outputs is known as _____ .
4. The time span across which operations managers plan is called _____ _____ .
5. Making sure the organization has an adequate supply of raw materials to transform into products, there are enough finished goods to ship to customers, and inventory in process is adequate is known as _____ _____ .
6. The attempt to make sure inputs and outputs meet desired levels of quality is called _____ _____ .
7. Having things arrive at designated spots when they are needed, rather than being held in inventory before they are needed is called _____ scheduling.

8.  The process of designing work so it can be performed by machines is known as _____ .

9.  Firms that rely on machines rather than people to do almost all of their work are said to be _____ _____ .

10. Technology that relies on computers to design or manufacture products is _____ _____ manufacturing.

## ▼ CHAPTER 23 CASE STUDY
### ANTS IN THE SUPERMARKET

Food retailing is an extremely competitive environment. In the United States, along with Kroger and Walmart there are competitors from France (Carrefour), Netherlands (Ahold), United Kingdom (Tesco), Japan (7-11), and Canada (Alimentation Couche-Tard). Each is competing fiercely for every food-retailing dollar, each trying to identify a profitable niche to exploit. CEO Craig Herkert, Supervalue Inc., observed that the supermarket sales environment is "extraordinarily competitive," as he commented on the activity of an international competitor in the United States, Koninklijke Ahold NV.

Perhaps the most "interesting" competitor in the food-retailing industry is Trader Joe's with 367 stores in 2012. In the words of *Fortune* magazine, "It's an offbeat, fun discovery zone that elevates food shopping from a chore to a cultural experience. It stocks its shelves with a winning combination of low-cost, yuppie-friendly staples … and exotic affordable luxuries …. Its niche is to appear to be your friendly neighborhood store with the buying power of a grocery chain.

Trader Joe's, which started in 1967 by Stanford University MBA Joe Coulombe, is an effective, highly competitive marketer that is also very secretive about who manufactures its in-house brands and who owns it. You see, Trader Joe's is actually *Herr* Trader Joe's. In 1979 Joe Coulombe sold the chain to Theo Albrecht. The Albrecht family owns the Aldi-Nord supermarket chain in Germany. So Trader Joe's is another foreign competitor in the U.S. market. It is a "stealth" competitor—a foreign competitor that appears to be your local food retailer.

What makes the food retailing industry so competitive are consumers' shopping habits. Most consumers are weekly shoppers with an occasional "foray" into the store for forgotten items on their "mess" list. This weekly pattern makes the average consumer highly aware of prices and price changes. Even if they did not pay attention to prices, the weekly "barrage" of food coupons arriving with the daily newspaper and daily "onslaught" of television advertisements would generate it. This creates a competitive environment for very slim profit margins, generally around 2–3 percent. The difference between victory and defeat is very small.

Marketing students learn early in their careers to master the "Four Ps of Marketing"—Product, Price, Promotion and Place. Each is important; but in the food industry, "place" becomes the focal point of the other three Ps. All the work on product development, pricing strategies, and promotional campaigns is focused on one place—the food retail establishment. All work is meaningless if the product is not on the shelf.

The food retail industry goes to great lengths to differentiate where these shelves are located. The "mom and pop" store, convenience store, supermarket, warehouse store, "niche market" store, and hypermarket are all places that recognize the importance of shelves. This passage from *The Wall Street Journal* illustrates that point: "In a newly renovated Carrefour hyper-market outside Lyon (Ecully, France), shelves stocked ceiling-high and narrow aisles have given way to eye level yogurt displays, gleaming cosmetics counters, and color-coded shopping areas. It is Carrefour's attempt to entice reluctant shoppers back into its biggest stores."

Business literature is filled with military jargon that exemplifies this competitiveness or aggressiveness. Companies do not initiate marketing activities; they "launch campaigns." CEO's do not present ideas; they initiate "strategies" and "order tactical maneuvers"; and they set-up "war rooms" to monitor developing situations. All of this jargon is designed to convey a sense of urgency, to create a need to work together, to achieve the objective, and to celebrate victory or to retreat and learn from defeat. The military lingo adds color. Marketing students study the U.S. domestic "Cola Wars" or the international "beer wars." Case studies with subjects like this are popular and effective. However, the part that is often overlooked in these studies is the battlefield for these "wars"—the supermarket.

This competitive environment also means that the food retailers or the food manufacturers will employ any technology that might supply a "tactical edge," no matter how slim it may be. Enter the ants.

Scientists have been studying ants for decades, led and encouraged by Harvard's E.O. Wilson. Ants are efficient. Once an ant discovers a food source, it finds its way back to the colony to report it. As it races back to home base, it lays down a trail of pheromones (ant hormone). The other ants follow the pheromone trail to the food; but when an ant finds a quicker way to the food or a swifter return, its trail takes precedent and the pheromones of the original trail fade. The other ants follow the freshest and strongest pheromone scent.

This phenomenon has been captured in an algorithm designed by Dr. Marco Dorigo at the Free University of Brussels. His algorithm, *Ant Colony Optimization*, is used to solve complex supply problems. It has been used successfully in solving the classical "traveling salesman problem," logistics and other complex tasks. Migros, a Swiss supermarket chain, and Barilla, Italy's leading pasta maker, use a similar system to manage their daily deliveries from central warehouses to local retailers. The name of the logistics program is *Ant Route*. Ant Route is part of a software program, AntOptima, developed by a competitor of Dr. Dorigo, the Dalle Molle Institute for Artificial Intelligence.

Next time you are stuck in a traffic jam behind a food retailer's delivery truck, it might be some consolation to know that even busy, efficient little ants sometimes face obstacles that they cannot control and must find their way around.[38]

### ▶ Case Study Questions

1. Explain how inventory control and quality control apply to non-factory organizations such as supermarkets.

2. Robots are not common sights in supermarkets, so explain how technology management may affect supermarkets in other ways.

3. How does the competitive nature of supermarkets affect their organizing for operations?

## REFERENCES

1. "The Post-Alcohol World," *The Economist* (October 30, 2010) at www.economist.com (accessed November 10, 2010); Paul Sonne, "To Wash Hands of Palm Oil Unilever Embraces Algae" *The Wall Street Journal* (September 8, 2010), B1; Andrew Pollack, "His Corporate Strategy: The Scientific Method," *The New York Times* (September 5, 2010), B1.

2. Martin Starr, *Production and Operations Management, 2nd ed.* (Mason, OH: South-Western, Cengage, 2008).

3. Everett E. Adam, Jr. and Ronald J Ebert, *Production and Operations Management, 5th ed.* (New Delhi: Prentice Hall of India, 2004).

4. "The Myth of U.S. Manufacturing's Decline," *Forbes*, January 18, 1993, 40–41.

5. Richard B. Chase and Warren J. Erikson, "The Service Factory," *The Academy of Management Executive* (August 1988): 191–196. See also Richard B. Chase and Robert H. Hayes, "Beefing up Operations in Service Firms," *Sloan Management Review* (Fall 1991): 15–24.

6. *Occupational Outlook Handbook, 2010-11 Edition*, U.S. Department of Labor, Bureau of Labor Statistics (December 17, 2009) at www.bls.gov (accessed July 30, 2010).

7. James Brian Quinn and Christopher E. Gagnon, "Will Service Follow Manufacturing Into Decline?" *Harvard Business Review* (November–December 1986): 95–103.

8. Brian Dumaine, "Earning More by Moving Faster," *Fortune*, October 7, 1991, 89–94; "Coleman is Glowing Overseas," *The New York Times,* December 8, 1991, D1, D2.

9. Everett E. Adam, "Towards a Typology of Production and Operations Management Systems," *Academy of Management Review* (July 1983): 365–375.

10. Adam and Ebert, *Production and Operations Management.*

11. Cynthia Fisher, Lyle Schoenfeldt, and James Shaw, *Human Resource Management, 6th ed.* (Mason, OH: South-Western, Cengage, 2005).

12. Louis Kraar, "Japan's Gung-Ho U.S. Car Plants," *Fortune*, January 30, 1989, 98–108.

13. Ricky W. Griffin, *Management, 10th ed.* (Mason, OH: South-Western, Cengage, 2011).

14. For a detailed discussion of other organization design issues that are applicable for operations management, see Richard Daft, *Organization Theory and Design, 10th ed.* (Mason, OH: South-Western, Cengage, 2010).

15. "Kraft Foods and Medisyn Technologies Expand Collaboration to Discover New Ingredients," *Financial News Release* (May 18, 2010) at phx.corporate-ir.net (accessed July 1, 2010).

16. "General Mills Intends to Reshape Doughboy in Its Own Image," *The Wall Street Journal*, July 18, 2000, A1, A8.

17. Shawn Tully, "The Modular Corporation," *Fortune*, February 8, 1993, 106–113; "The Virtual Corporation," *Business Week*, February 8, 1993, 98–102.

18. For a more detailed discussion of change and its relationship with operations management, see Barbara Senior and Stephen Swailes, *Organizational Change* (Toronto: Pearson Education Canada, 2010); W. Warner Burke, *Organizational Change* (Thousand Oaks, CA: Sage, 2002); Michael Beer, *Organizational Change and Development—A System View* (Santa Monica, CA: Goodyear, 1980).

19. For an example of inventory management, see John Kanet, "Inventory Planning at Black and Decker," *Production and Inventory Management* (3rd quarter 1984): 9–22.

20. Various Press Releases, System Concepts, Inc. (SCI) at www.foodtrak.com (accessed July 7, 2010; Beth Kormanik, "What a Waste!" *Hotel Interactive* (February 18, 2009) at www.hotelinteractive.com (accessed July 7, 2010); "New Products and Services," *Indian Gaming*, November 2006, 105.

21. David Garvin, "Product Quality: An Important Strategic Weapon," *Business Horizons*, March–April 1984, 31–36.

22. "When Service Means Survival," *Business Week*, March 2, 2009, 26–40.

23. Adam and Ebert, *Production and Operations Management*.

24. James C. Worthy, *Shaping an American Institution* (Urbana, IL: University of Illinois Press, 1984), 6–8; and Jeanette A. Davy, Richard E. White, Nancy J. Merritt, and Karen Gritzmacher, "A Derivation of the Underlying Constructs of Just-In-Time Management Systems," *Academy of Management Journal* (September 1992): 653–670.

25. Adam and Ebert, *Production and Operations Management*.

26. Norbert Turek, "Cargill: On Top Of App Integration," *Information Week* (June 21, 1999) at www.informationweek.com/739/cargill.htm (accessed October 29, 2011).

27. Paul D. Collins, Jerald Hage, and Frank M. Hull, "Organizational and Technological Predictors of Change in Automaticity," *Academy of Management Journal* (September 1988): 512–543.

28. "Computers Speed the Design of More Workaday Products," *The Wall Street Journal*, January 18, 1985, 25.

29. Robert Bonsack, "Executive Checklist: Are You Ready for CIM?," *CIM Review* (Summer 1987): 35–38.

30. M. Sepehri, "IBM's Automated Lexington Factory Focuses on Quality and Cost Effectiveness," *Industrial Engineering* (February 1987): 66–74.

31. "Computers Speed the Design of More Workaday Products."

32. "How Automation Could Save the Day," *Business Week*, March 3, 1986, 72–74.

33. Otto Friedrich, "The Robot Revolution," *Time*, December 8, 1980, 72–83.

34. Gene Bylinsky, "Invasion of the Service Robots," *Fortune*, September 14, 1987, 81–88.

35. "Robots Head for the Farm," *Business Week*, September 8, 1986, 66–67.

36. "Boldly Going Where No Robot Has Gone Before," *Business Week*, December 22, 1986, 45.

37. Ron Johnson, "Robotic Milkers Make Wiegels' Dairying Easier," *Agri-View* (January 8, 2009) at www.agriview.com (accessed July 29, 2010); "Farm Installs Robotic Milkers," (April 18, 2006) at minnesota.publicradio.org (accessed July 28, 2010); Darrin Youker, "Pa. Dairyman Goes High Tech with Robotic Milkers and Digesters," *Farm and Dairy* (June 18, 2010) at www.farmanddairy.com (accessed August 1, 2010); Amanda Nolz, "High Volume Dairy Implements Robotic Palletizing," *World Dairy Daily* (October 27, 2009) at www.wdexpo.org (accessed July 28, 2010).

38. Christina Passariello, "Carrefour Tries a Booster for Tiring Hypermarkets," *The Wall Street Journal* (August 25, 2010) at wsjonline.com (accessed September 29, 2010); Anna Marli Van der Meulen, "Dutch Retailer Ahold's U.S. Sales, Euro's Gain Drive Profit Increase," *The Wall Street Journal* (August 25, 2010) at wsjonline.com (accessed September 28, 2010); Beth Kowitt, "Inside Trader Joe's," *Fortune* (September 6, 2010), 86–96; "Riders on a Swarm," *The Economist* (August 14, 2010) at www.economist.com (accessed September 28, 2010).

# Information Systems

## LEARNING OBJECTIVES

After studying this chapter, you should be able to:

- Describe the nature of information and information systems, including their effects on a manager's job and the characteristics of effective information.

- Identify the basic components of information systems.

- Discuss the types of information systems available.

- Discuss how to determine information system needs and how to match needs with systems.

- Describe the impact of information systems on organizations.

## A Valley Too Deep

**W**e humans are peculiar creatures: We are curious and inventive; we pride ourselves on our flexibility and inventiveness, and on our openness. Well, almost—it seems we are "hard-wired" to accept almost anything except duplicating our own images (Figure 24.1). Many ancient religions worried about stolen spirits.

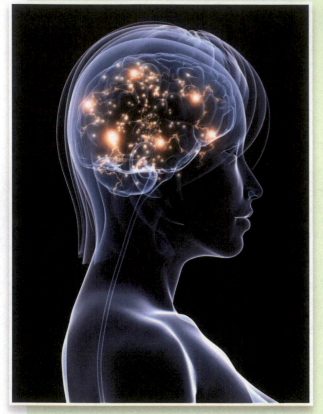

**FIGURE 24.1** The active brain provides rich information for marketing research but also presents ethical dilemmas, perhaps leading to "uncanny valleys" for management to cross.
© Sebastian Kaulitzki/www.Shutterstock.com

As the ability of computer graphics continues to develop and produce amazing results, we revel in simulations. Scenario planning and playing have taken on a new meaning at the corporate level. With the help of clever algorithms, computer graphics, and exceptional programming, it is possible for strategic planners to populate alternate scenarios to test their theories. Information technology or IT has morphed from hardware and software supply and maintenance to a management participant.

It seems that the potential is almost limitless. The operant word in the last sentence is the word "almost." As computer graphics improve, an unspoken or unrecognized limit or barrier is approached. This barrier was coined "the uncanny valley" and was first described by Japanese roboticist Dr. Mashiro Mori in 1970. He noticed that as engineers, scientists, and computer graphic artists began to develop avatars that more closely resemble real humans, our comfort level increased—up to a point. However, as they became "very" realistic, we real humans were "repulsed." In other words, as their creations that are intended to look like real people approached looking like real people, we real people found the "real people" creepy or eerie. So comfort levels increase as human likeness increases, then drops off sharply and rises again only for "real" people. That drop is the "uncanny valley" that has to be crossed by us. Once crossed and the "simulated people" are identical with us real people, we are more apt to accept them.

Two scientists at Indiana University's School of Informatics recently validated Dr. Mori's work. The research of Drs. Chin-Chang Ho and Karl MacDorman asked undergraduates to rate their comfort levels with animations while viewing clips of "The Incredibles" (2004), "The Polar Express" (2004), and an animation of Orville Redenbacher, a businessman who improved microwaveable popcorn and who has been deceased since 1995. The results of their research almost duplicated Dr. Mori's results.

As we use more and more data to analyze consumer-shopping patterns, we are comfortable with algorithms and equations. They are useful in attempting to discern underlying patterns, and manufacturers and retailers need to discover these patterns to meet consumer expectations. The potential is limitless—almost.

A Harvard case study provided a scenario where loyalty-card customers of a supermarket could have their purchases correlated with data from an insurance company to detect lifestyle patterns that might influence insurability. The point of the case study was to decide whether the food retailer and insurance company should enter into a business partnership to share data and determine if there are ethical issues involved. Another valley was being defined. This valley may not be eerie but it does raise ethical concerns.

As information technology expands and we humans increase our potential to explore, the great unexplored places may not be "The Valley of the Dinosaurs" of vintage filmdom as much as crossing the "valley of the uncanny." It might not be scary—just eerie and perhaps ethically uncomfortable.[1]

## INTRODUCTION

Another aspect of organizational control that is useful to managers is information and information systems.[2] A great many organizations already use high technology to manage their information. Walgreens, for example, found that rapid, accurate information about not only prescriptions but also sales of all products carried in its stores enabled it to respond to customers quickly.[3] Information systems much like those at Walgreens could be and are being used by many other organizations. But as information systems become more sophisticated and complicated, they also have begun to raise public concern over the potential negative fallout from such systems—a topic that is beyond the scope of this book.

In this chapter we first describe the nature and basic components of information systems. Then we examine how various information systems are designed to match managers' needs. Next, we look at different types of information systems, and finally, we discuss managing information systems, along with the impact of these systems on organizations.

## The Nature of Information

**Information** consists of data organized in a meaningful way. **Data** are merely facts and figures—unorganized pieces of information.[4] They are useless until they are processed and organized in some way. If Nabisco, for example, has a list of figures that show the monthly sales for a product, those data are made more useful—changed into information—when they are analyzed and organized to show seasonal fluctuations and annual trends.

Remember from Chapter 2 that a **system** is an interrelated set of elements that function as a whole. A system, then, consists of a set of components so arranged as to accomplish some purpose. If Nabisco had built a system to produce marketing reports from its sales data, it would have built an information system. If it further arranged for those reports to go automatically to managers who needed the information, it would have a management information system such as those introduced in Chapter 2. The keys to systems, then, are interrelatedness and purpose.

An information system must accomplish a purpose through the interaction of its component parts.

## Information Needs of Managers

As discussed in Chapters 2–3, a manager's job consists of functions, roles, and skills. Managers have always had to be skilled in using information. The nature and importance of information, however, has changed dramatically in recent years because the amount and the variety of data coming to managers have grown tremendously.[5] Managers have had to become information processors (Figure 24.2): They must decide which information to combine to form new information, which information to discard, which to pass along to others, which to put to immediate use, and which to retain for possible use later.[6]

An accounting manager for a medium-sized beverage firm could serve as an example. During the course of her normal day, she participates in formal and informal meetings on both job- and non-job-related related topics. In addition, she receives letters, memos, notes, and other forms of written communication; she gets and makes telephone calls; and she may send or receive a FAX (electronic facsimile) message or use her company's electronic mail (email) system. All of the data and information she receives from these different sources must be processed in some way.[7]

**FIGURE 24.2**  Obtaining and processing information is vital to a manager's job, although also overwhelming sometimes.

## Effective Information

All managers use information, and the information they get should be effective; that is, it should provide them with what they need to carry out their tasks successfully. To be effective, information must be accurate, timely, complete, and relevant.[8]

These characteristics seem obvious. Yet time after time managers make decisions based on information that is inaccurate. For example, a Japanese company overbid for a piece of land in Great Britain and then found that it could not use the land because a building on it had been declared historic. The information they used to make their decision to buy was inaccurate.[9]

Benetton, on the other hand, has a computerized information system that clearly provides timely as well as complete information that paper systems or centralized systems cannot provide. Knowing only the profit per item, for instance, is not enough for the retailer to decide how many items to stock since a product that is less profitable per unit could sell far more units and hence earn more total profit for the firm.[10]

Finally, information must be relevant. Data or information on sales in one region may not be as relevant to sales managers in other regions, and yet many small-business managers try to function with such irrelevant information on the mistaken assumption that it is better than no information at all.[11] U.S. grocery stores collect tons of demographic data, which they use in building new stores and to some degree in arranging and pricing their products. But the data does not tell them, for example, why the customer is not buying particular products or brands.

## The Information Age

While computers clearly had origins dating back to the 1800s, the first true electronic computer was developed in the 1940s. Even so, it was not until the 1960s, when second-generation computers using transistors came into use, that the computer age was born. With the third generation using integrated circuits and the fourth using microprocessors, developments in the hardware of computing dominated its role. In the 1980s, as the emphasis began to shift from the technology to its applications, the **information age** was born. Now, in the information age the computer is simply a tool which helps managers use information as a corporate resource; it is the focus on information that is important.[12] Information, then, is becoming a business competitive advantage.

© iStockphoto/Xavier Arnau

The use of computers appears to go through the following five stages:

Stage 1. Management is enthusiastic about computers and uses them primarily for accounting applications.

Stage 2. Management becomes aware of some of the limitations of computers and is a bit more cautious about extending applications.

Stage 3. Information systems tend to become overloaded and management becomes cynical about the ability of computers to solve all of its problems.

Stage 4. Managers begin to learn how to use computers and information systems to bring about needed changes in the organization.

Stage 5. Management begins understanding and controlling information systems and computers in all areas of the organization. This, then, is the stage in which the information age comes about.[13]

Many organizations have already entered the information age, and the impact and importance of information and information systems is clear. An understanding of systems is vital to effectiveness in the global marketplace.[14]

## FOOD FOR THOUGHT 24.1

In 2004 the Food and Drug Administration approved computer chips that store personal information and can be implanted in humans. The information can be retrieved with a scan of the person's arm.

# Information and Information Systems

Information and information systems, then, are increasingly important parts of the manager's job. But just what do information systems look like, and how do managers use them?

## Components of Information Systems

Every information system has five basic interrelated components. There must be some way to get the data into the system, to analyze or process the data, to store the data and information, and to make the information available to users. There must also be some overall control of the system itself. In a simple, non-computerized system, the data are recorded on paper forms or memoranda that are stored somewhere (boxes, drawers, file cabinets). People complete the analysis and then prepare reports that are sent to those who need them. Control is basically by exception; that is, once the system is in place it continues as it is unless a problem is detected.

A computerized information system, on the other hand, looks like the one shown in Figure 24.3. Getting the data into the system involves one or more **input devices**. A **central processor** analyzes and processes the data. **Storage** is also accomplished through one or more of several media (in many cases more than one medium is used in case something should happen to one storage device). The information is made available to users through a variety of **output devices**. The central processor and these devices are known as the computer **hardware**. **Software** refers to the instructions (programs) that enable the hardware to function. Finally, the **control system** usually involves some form of computer software as well as human monitoring to ensure that the software and the system are functioning as planned. Management must assure that the information system is integrated with all operating and control systems so individual managers and groups of managers receive relevant and timely information.[15]

## Designing Information Systems

The purpose of an information system is to ensure that proper information is available when needed so that managers do not need to rely on chance or guesswork.[16] Designing a good information system, then, involves knowing the information system needs as well as the kinds of systems that exist or may be developed in the future.

**Information System Needs.** The information system needs of an organization are determined by the kind of organization, its environment, and its size. A high-technology organization, for instance, has greater information system needs than does a low-technology organization. The more uncertain and complex the environment, the greater is the need for a formal information system. And, all other things being equal, larger organizations have greater

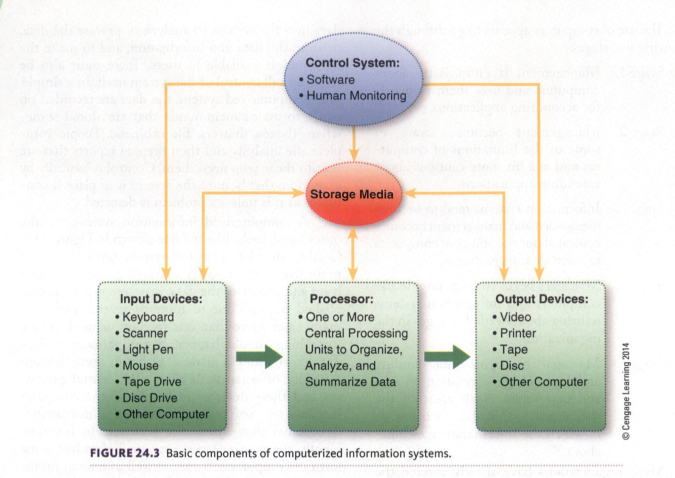

**FIGURE 24.3** Basic components of computerized information systems.

© Cengage Learning 2014

information systems needs than do smaller ones. Thus, General Mills' information needs are greater than those of a local produce company.

Within an organization, information needs are influenced by the area and level of management involved. The information systems needs of a production department are different from those of a human resources department, although each usually needs a formal system. Likewise, executive-level managers have different information needs than do supervisors. Each manager has unique information needs, and a well-designed system will tend to be tailored to each of its users rather than merely provide general information to all users.

The process of determining what information each individual needs to perform his or her job is called **information requirements analysis**.[17] This process should be conducted before the information system is developed and then periodically repeated as part of the updating and maintenance of the system. Normally it involves interviewing individuals about their views on what information they need, in what form, and when. These views must then be verified through the use of some

other technique, such as a paper simulation of the individuals' activities or direct observation of them at work. If many people in similar situations are involved, surveys could also be used. In any event, once the information requirements analysis has been completed, it should be integrated into the system and periodically examined to ensure that it is still accurate.

## FOOD FOR THOUGHT 24.2

Hens laid 77 billion table eggs in 2010 at commercial farms across the country. A hen starts laying eggs at 19 weeks of age and lays an average of 300 eggs annually.

## Implementing Information Systems

Implementation of information systems, then, follows from design. Organizations must be careful to match systems with needs since there are a variety of needs for and uses of information as well as numerous different kinds of information systems.

## Matching Systems with Needs.

Matching involves working through a series of questions like the following: For what goals is information needed, and what information is needed? In what way can that information be readily obtained, stored, analyzed, and reported? (In a computer system this involves determining the hardware and software to use.) What are the costs and benefits of the various ways of meeting those needs? How might the information and the technology for dealing with it be integrated?

After obtaining answers to these questions, the system itself must be actually designed, tested, implemented, and then monitored, maintained, and perhaps improved. As shown in Figure 24.4, matching can be thought of as a process itself.

## Centralized versus Distributed Systems.

Just as decision making in an organization can be centralized or decentralized, so information systems also can be centralized or distributed. In the early days of computer technology, virtually all information systems were centralized. They consisted of data-processing or computer services departments built around large, expensive, mainframe computers. As computer technology changed, it became more and more possible to have the information system components scattered or distributed throughout the organization. Prudential Insurance used a centralized approach, for instance, whereas Travelers Corporation used a distributed system.[18] Centralized information systems are better coordinated but slower; distributed ones are faster but may result in

some duplication of effort. As the power of smaller computers increased and as networking software improved, information systems have made great strides in obtaining the best of both worlds.

Most members of organizations are increasingly computer literate and can have access to computers that can be interfaced to the organization's information system. This opens up information and alters organizations in ways that we are only beginning to understand, including privacy issues for both the organization and its members. Nevertheless, organizations must decide whether the implementation of their information systems will involve relatively centralized or distributed systems.

## Types of Information Systems

Information systems can be formal or informal. A system of informal, unstructured information is a major factor in every manager's life.[19] A manager gathers impressions through interactions with others and through travels in the organization. And there are always emotional reactions to that information. Although this kind of information system is important, it is so situation-specific that few generalizations can be developed about it. More formal, structured information systems involve record keeping of one sort or another. Formal information systems are ancient managerial devices. The Sumerians (2900–1800 BC) had a complicated tax and governance system with records maintained on clay tablets. The Egyptians used papyri for record keeping. Today, virtually all formal systems are computerized. The three most

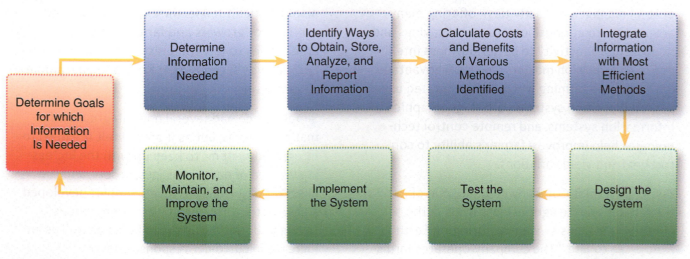

**FIGURE 24.4** Matching information needs with a system.

**TABLE 24.1** Kinds of Information Systems

| SYSTEM | DESCRIPTION |
|---|---|
| 1. Informal | Unstructured information is obtained from day-to-day interactions and impressions, notes, and/or diaries. |
| 2. Formal | Structured information is gathered and communicated either on paper or by computer. Information can be either centralized or distributed. |
| a. Transaction-Processing | Handles routine and recurring transactions. |
| b. Management Information | Gathers, organizes, summarizes, and reports data for use by managers. |
| c. Decision Support | Searches for, analyzes, summarizes, and reports information needed by a manager for a particular decision. |

© Cengage Learning 2014

common ones are transaction-processing systems, basic management information systems, and decision support systems. Table 24.1 summarizes these different kinds of information systems.

## Transaction-Processing Systems

A system designed to handle routine and recurring transactions within an organization is known as a **transaction-processing system**, or **TPS**. These were the first formal, structured information systems and therefore the first to be computerized. A TPS is most useful for tasks that involve a large number of highly similar transactions—filing employee or customer records, handling charge cards, dues notices, subscriptions, and so on.

Recently transaction-processing systems have been combined with optical scanning equipment to strengthen their value to organizations. When your local grocery store uses automated scanners to record each unit sold and its corresponding selling price, the information is part of a TPS. Bankcards such as Discover, MasterCard, and VISA also use TPSs to handle the huge volume of transactions with which they must deal.

## Management Information Systems

A **management information system**, or **MIS**, is a system that gathers, organizes, summarizes, and reports data for use by managers. Sometimes called *information reporting systems*, MISs serve to help

## A FOCUS ON AGRIBUSINESS
### MISs

Management Information Systems (MISs) support all management activities—offering many advantages to agribusiness from cost savings and improved productivity to better business intelligence and decision making. One such advantage is in precision farming where the integrated use of global position systems (GPSs), geographic information systems, and remote control technologies help improve a farmer's ability to control field and yield variations.

The J. R. Simplot Company in Idaho, one of the largest private agribusiness firms in the United States, has a vision of "Bringing Earth's Resources to Life." The company operates farms

and ranches, potato and vegetable processing plants, agricultural fertilizer production, and services for growers and farmers. It supplies French fries to fast-food restaurants like McDonald's, Wendy's, and Burger King, and guacamole to Taco Bell.

Simplot once had several non-integrated information systems, but as it grew it felt the need for integrated systems to keep up with the increasing volume of data, messages, orders, and the like. Working with Microsoft, Simplot developed a new system. The result was a tremendous saving in both time and man hours as well as an increase in customer responsiveness.[20]

link the several parts of an organization together. When adapted to particular use in agribusiness, an MIS may be referred to as an Agribusiness Information System, or AIS. An AIS, then, has all of the advantages and disadvantages of an MIS with the additional advantage of being focused exclusively on the agribusiness organization. For a manufacturing firm like the J.M. Smucker Company, for example, a computerized inventory system might track finished goods, work in progress, and beginning materials to ensure that customer orders can be met. A sales representative working with a customer can access the system to determine fairly precisely when an order will be shipped to that customer. Managers can also access the system regularly to obtain information necessary for effectively operating the organization.

## Decision Support Systems

A newer and very powerful form of information system is known as a **decision support system**, or **DSS**. A DSS automatically searches for, analyzes, summarizes, and reports information a manager needs for making a particular decision. In a sense, a DSS combines a TPS and an MIS with features that make the system even more responsive to the needs of its users. A finance officer of a company like PepsiCo, for example, might need to know the capital recovery periods and tax consequences of several alternative investment opportunities. The DSS would have the relevant information and be able to present it quickly and in a useful way so that the financial officer's decision process could be both fast and accurate. The real basis of the connection between managers and information systems, then, is turning data into decisions, which has come to be known as *decision processing*.[21]

## Other System Technologies

The computer is at the heart of most information systems today. As Figure 24.5 indicates, several new information technologies are currently being developed for organizations. Customer relationship management (CRM) systems have been developed to integrate sales, marketing, and customer and technical service. Specific agribusiness systems have been developed to integrate information from the Agricultural Research Service (ARS), the Economic Research Service (ERS), and the National Agricultural Statistics Service (NASS).

**Computer Software.**  Whether the computer is a notebook, desktop, or mainframe, data can be stored and manipulated by computers through software. Software for use in information systems comes in a variety of types: databases, spreadsheets, word processors, and electronic mail, to name the most common ones. Databases permit users to organize and manipulate primarily numerical

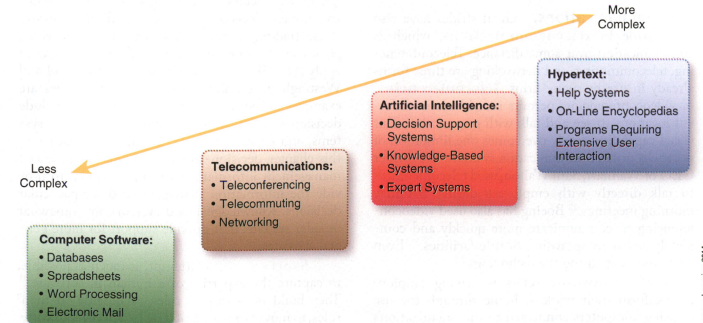

**FIGURE 24.5**  New information technologies.

data in interconnected ways. Spreadsheets arrange numerical data in a matrix of rows and columns. Word processors and electronic mail deal primarily with text data.

Each of these systems can, to some degree, interrelate with others, and the power to do so increases every year. Most word-processing programs (Word and WordPerfect, for instance) can work directly with database programs (dBase, FoxPro, and Rbase, for instance). Likewise, graphic presentation programs like PowerPoint, Overhead Express, and Harvard Graphics can import data from word-processing programs or spreadsheet programs such as Excel or Lotus 1-2-3. As networked systems have become more prevalent and with the increased use of modems to send files over long distances, integrated software and hardware will continue to become even more important to enable information systems to evolve into the highly effective control and communication tools of which they are capable.

As companies place more and more data into databases of one kind or another, certain issues develop. The data must be backed up so that it is not lost or damaged if something unexpected happens. The data must be made secure, and confidentiality must be assured. The use of electronic firewalls and encryption may well be employed to help provide this increased level of safety and security. Summer Farms, for example, has a firewall on its computer as well as antivirus protection that it updates regularly.

## Telecommunications.

Great strides have also been made in **telecommunications**, which is communication over some distance. Teleconferencing, telecommuting, and networking are three forms already in use by many firms. **Teleconferencing**, or videoconferencing, permits individuals in different locations to see and talk with one another. This visual capability clearly overcomes one limitation of other electronic communication systems. The former CEO of Walmart, Sam Walton, used teleconferences to talk directly with employees during Saturday morning meetings.[22] Boeing has also used videoconferencing to communicate more quickly and completely across its sprawling Seattle facilities.[23] Even trade shows are using the technology.[24]

**Telecommuting** refers to having employees perform their work at home through the use of using computers connected to the organization's computer. This is an electronic version of cottage industries, where people produce products in their homes and take them to central locations for distribution. Capital One, IBM, American Express, Johnson & Johnson, J.C. Penney, Wells Fargo, Blue-Cross Blue-Shield, and many other companies find that this approach saves them office space and enables them to obtain the productivity of good people who otherwise might resist commuting to a central workplace.[25]

**Networking** involves connecting independent computers directly together to function in interrelated ways. In this way one gains direct access to common software and databases more easily than through the use of telephone connections between computers. Electronic communications can also be made quicker and more responsive to individuals through the use of a network. Networks at one organizational level can then be linked to others to establish a true information system for the organization. In this way, Local Area Networks (LANs) can be linked. Integrated Services Digital Networks (ISDNs) link computers and other machines through the use of digital cabling with all jacks much like telephones, and Electronic Data Interchanges (EDIs) enable the transmission of exact copies of a company's forms from one unit to another.[26] Agronet is an agricultural and food related network service in Finland, although its services are available only in Finnish.[27]

## Artificial Intelligence.

**Artificial intelligence (AI)** refers to enabling computers to simulate human decision processes. Medical diagnosis, stock trading, robotics, science, and even video games are among the variety of ways companies apply AI.[28] The maintenance system at Bechtel and Westinghouse's online diagnostics for turbines are examples.[29] The most common AI systems include decision-support systems, knowledge-based systems, and expert systems.[30] As discussed earlier in this chapter, decision-support systems are advanced management information systems designed to provide the information managers need for particular decisions. Knowledge-based systems are somewhat broader systems that provide support for more general activities.

**Expert systems** attempt, as much as possible, to capture the expertise of a human in software.[31] They build on series of rules, frequently "if-then" rules, to move from a set of data to a decision recommendation. Boeing has been developing expert systems for various uses for some time. One is known as CASE (connector assembly specification expert). CASE produces assembly procedure instructions for each of the 5,000 electrical connectors on an airplane. What once required more than 40 minutes of searching through 20,000 pages of printed material now takes only a few minutes to get a computer printout for a specific connector.[32] Another is GOS-SYM/COMAX developed for use in the cotton industry.[33] Working with Texas Instruments, Campbell Soup Company developed an expert system to capture the expertise of a manager in one of its soup kettle operations.[34] In another example, Martin Marietta developed an expert system to assist in air traffic control.[35] Expert systems are being developed to aid managers in a variety of tasks, including more interpersonal tasks such as providing performance feedback to subordinates.[36]

Executive information technology (EIT) is being developed to help managers improve product and process quality as well as customer service. Motorola Codex, for instance, is using such a system.[37] This technology is kind of a combination of an expert system and a decision-support system to specifically assist the executive level of an organization with monitoring and managing the activities of the organization.

**Hypertext.**   In most information systems, a user may progress from one level to another in linear fashion. A hypertext system is a database management system that allows the user to move in any direction through the information to acquire what is necessary for the particular task at hand. Although the technology behind hypertext possesses many elements associated with artificial intelligence, it should be considered a separate information technology. Hypertext systems have been used in advanced help systems, online encyclopedias, and some programs that require a lot of user interaction. Hypertext systems tend to be complicated to develop; however, because they are extremely easy to learn and use, they hold tremendous promise for the future.

## Managing Information Systems

Like all other aspects of organizations, once developed, information systems must be managed.[38] In this section of the chapter, we look at that issue.

### Integrating Information Systems

Throughout this chapter, we use the plural *information systems* because most organizations actually use more than one system. Most middle- to large-sized organizations, for instance, have a marketing system, a production system, and a human resources system.[39] As Figure 24.6 indicates, these different

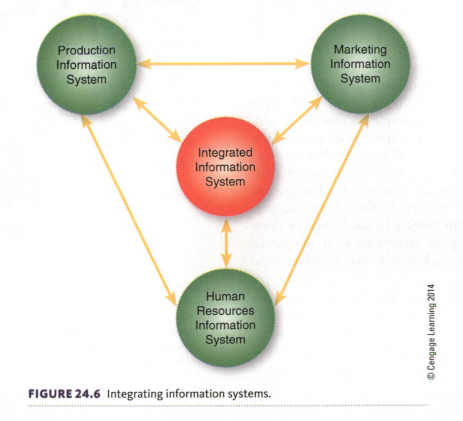

**FIGURE 24.6** Integrating information systems.

information systems must be integrated or linked so that the different kinds of information can merge to form information that is even more useful.

Integrating systems is not easy. The production system may have been installed on an IBM computer while the marketing system may use special-purpose computers developed by Intel, and the human resources department may be using Sony computers. Linking these different systems that were developed and are running on different hardware and with different software may be quite difficult or even impossible.[40]

While companies can avoid this problem by developing all of their systems at the same time, doing so is expensive, and information needs rarely occur at the same time and intensity. Also, both hardware and software change and improve so rapidly that agreeing on a standard for even a few years at a time can be very difficult. Therefore, most systems are developed piecemeal. Luckily, some more recent developments in hardware and software are making it easier to develop the necessary links across information systems. Linked systems, which virtually any company can use, enable managers to use electronic mail, access information needed to make decisions, and handle practically all routine business using only a single, integrated information system.[41] Some of these systems are discussed elsewhere in the chapter.

## Using Systems

Information systems must be in use to be of value. You might think that information should somehow be exactly where you want it when you want it. The reality, of course, is that information must be sought, and learning to use the system to obtain information is critical to the success of the system. For that reason, then, organizations should spend considerable time and effort to ensure that information systems are easy to use, or user friendly. Well-designed, user-friendly systems can quickly be used, even by those who have had no prior experience with computers, such as very senior workers who may be reluctant to use them.[42]

**User-friendly information systems** are typically available for users to examine and easily modify information to maximize its usefulness. The Travelers Corporation uses a team of trained nurses to review health insurance claims. They can access the system and review the medical diagnoses provided with each claim. Based on the information, nurses can determine whether a second medical opinion is warranted before approving a surgical procedure. They can add their decisions to the data on the claim forms for other users of the information.[43]

The use of an information system is one measure of its effectiveness. Many such systems are not, in fact, used because they require considerable computer fluency (Figure 24.7). However, recent developments are changing that situation.

Robert Kidder, CEO of Duracell, can do more in an hour using his current information system than he previously could in several hours. The system lets him begin at one level and progress into more and more detail as he desires. He can examine performance across divisions and then probe one division that is out of line with others to try to determine why. The system makes it possible to do

© www.Shutterstock.com

**FIGURE 24.7** Information can be accessed rapidly from a QR (quick response), a two-dimensional bar code consisting of black modules arranged in a square pattern on a white background. Designed originally for the auto industry, it is now widely used, due to its fast readability and comparatively large storage capacity.

all this quickly and easily on a computer in his office without upsetting the people in the division he is investigating.[44]

The use of a remote control, a touch screen, or a mouse seems particularly appropriate where the user is not a touch typist.[45] Such devices, however, may not be needed in the future since many schools are training virtually all students to operate keyboards. Keyboard mastery, which was once considered a symbol of lower "clerical staff" status, is becoming a symbol of high status for managers. Further, voice command systems which enable users to access most commands orally are becoming more accessible. More importantly, though, software that enables users to browse through information in almost any direction based on their experience or even hunches makes systems more friendly and powerful. Such software, known as hypertext, is just beginning to be used for information systems and promises to be a powerful aid in the future.

## FOOD FOR THOUGHT 24.3

In 2010, more than half of total spending on all organic products was made by just 8 percent of households. Organic food sales are forecast by Farmers Weekly to fall by 7 percent between the years 2010 and 2015 as consumer household budgets tighten further.

# Organizations and Information Systems

Whether informal or formal, information systems are clearly an important part of organizations. They are not, however, panaceas. They are only tools of management; and although they can have very beneficial effects, they are also limited. Table 24.2 outlines some of those effects and limitations.

## Effects of Information Systems on Organizations

Information systems have effects at several different levels, including the organization's performance, structure and design, and the people within the organization. Organizations use a variety of measures to assess the performance of information systems and their impacts on the organizations, and the appropriate measures must match the goals of each individual organization.[46]

**Performance.** There seems to be a growing consensus that information systems enhance company performance. The U.S. Forest Service recently installed a system to assist it in evaluating how to respond to forest fires. In the past every fire was attacked the morning after it was reported. The costs of such efforts would run to $10 million for a major fire. The new information system, however, automatically figures in such natural firebreaks as rivers, enabling containment at far lower costs. General

**TABLE 24.2** Effects and Limitations of Information Systems

| EFFECTS | LIMITATIONS |
|---|---|
| 1. On Performance:<br>  a. Tend to save money.<br>  b. Generally improve performance.<br><br>2. On Organizational Structure:<br>  a. Result in the creation of a separate unit to oversee the information systems.<br>  b. Result in a flatter organization because fewer managers are needed.<br><br>3. On Behavior:<br>  a. Some employees feel isolated.<br>  b. Some employees experience job enrichment.<br>  c. Some employees can work at home more.<br>  d. Some new work groups may be formed. | 1. Costly to develop.<br>2. May be difficult to learn.<br>3. Information can be overly valued.<br>4. Cannot handle all complex problems.<br>5. Operators may misuse them.<br>6. Operators may become discouraged and reject them.<br>7. Dependent on electric power sources. |

© Cengage Learning 2014

Electric and American Airlines also are among companies reporting high levels of satisfaction with the performance of their information systems.[47]

## Organizational Structure.

Since most organizations create separate units to oversee their information systems, the first impact of these systems on organizational structure is the creation of those units. In some cases, the head of the new unit or the person overseeing all information needs of the firm is in a newly created position, the chief information officer, or CIO (discussed more in Chapter 20).[48] Katherine Summer filled this role, not with the fancy title though, at Summer Farms. The more profound effect of information systems on organizational structures is that, with more and better information available throughout the organization, fewer managers, particularly those at middle levels, are needed. IBM, for instance, has eliminated a layer of management in this way, and its span of management may be greatly enlarged through the impact of information systems.[49]

## Behavior.

The behavioral effects of information systems are not yet well understood. Some individuals feel isolated because they spend their time interacting with computers instead of people. Others enjoy the new technology and are excited about having their jobs enriched in this way. Some can work at home rather than at a central office, allowing them to spend more time with their families with no detrimental performance effects. New groupings of personnel may form around electronic bulletin boards and social networking systems. Since there appear to be considerable learning experiences with such systems, however, the long-lasting behavioral effects will not be fully evaluated or understood for several years to come.

## Limitations of Information Systems

The most obvious limitation of information systems is their cost. They are expensive to develop, largely because each one must be tailored to fit particular organizational needs. A close second to this limitation is learning difficulties. Since most information systems involve doing things differently, and hopefully better, everyone involved must learn the new and different way. Although these two limitations

are relatively obvious, they are by no means the only ones.

The information derived from an information system, for example, may be overly valued. Some people assign great credibility to information from computers even when the information is actually very rough. Moreover, information systems simply cannot always handle all tasks or problems. Highly complex tasks or problems may necessitate human intervention; and if people depend upon their systems too much, the information from those systems can become dysfunctional.

Another problem is that some managers use information systems improperly. Many electronic mail systems when first installed become cluttered with graffiti, messages, and notes left from unknown parties saying all manner of things. This initial period of using the system as much for sport as for business may last weeks or even months. Some managers may become annoyed at the constant interruptions (which also can be counterproductive) if they feel they must constantly monitor their e-mail rather than setting aside some times during the day to check and respond to messages. Other managers cease using the system if it is not user friendly. The manager who gives up too soon may fail to eventually realize the power of the new communication system, which may be substantial. And, of course, any organization that does not provide for electrical power failures is asking for trouble.

## Implications of Information Systems

We have considered some of the dramatic changes and effects on management caused by innovations in computers and information systems. Three implications are especially relevant.

1.  The advent of computer networks and electronic mail promises greatly to enhance productivity in the workplace. Managers will have much greater access to information, will be able to sort and process that information rapidly, and will be able to communicate with others quickly.

2.  It will be even more important for managers to remain abreast of breakthroughs and changes. As new developments in information

systems are unveiled and new applications developed, managers will need to assess how and to what degree their organizations can use these effectively.

3. Every organization, whether small or large, will have to contend with issues of control and coordination. For instance, the Hartford Insurance Group tried a program that allowed employees to work at home on computers but soon abandoned it because supervisors had trouble coordinating work and complained that they were losing touch with their work groups. Although widespread telecommuting has been slow to develop, its potential advantages—allowing the disabled to work, using less office space, providing greater work flexibility, reducing energy costs of communicating to the workplace, and so forth—suggest that managers will have to work even harder in the future to make it possible.[50]

FIGURE 24.8 There are various benefits of social networking.

---

### FOOD FOR THOUGHT 24.4

"As I experience certain sensory input patterns, my mental pathways become accustomed to them. The inputs eventually are anticipated and even 'missed' when absent."

—*Star Trek's Troi, speaking to Riker*

---

## Internet and Social Media

An important aspect of information management is the use of the Internet and, in particular, social networking (Figure 24.8). Every organization, no matter how small it is, should have a website connecting the organization with not only its customers but also the community in which it operates, whether that community is local or international. Using photos, links, recipes, and other attention-getting components, the organization can also use its website to promote the business, introduce new products, provide location maps, and announce changes, such as hours, locations, and personnel. A

fully developed website can enable a company to transact business over the web, buying and selling articles or information. Green Things, for example, transmits parts of its in-progress landscape designs to customers by electronic mail instead of waiting until the design is complete to ask for customer feedback. It should soon be expanding its website to post both in-progress and completed designs that customers can view by accessing their online accounts.

In addition to having a web presence, social networking provides unique opportunities for organizations to stay informed, monitor the environment, and interact with stakeholders. An increasing number of sites are available to assist agribusinesses in keeping up with aspects of their environments, such as AgHaven.com, Agrimates.com, Chatterbarn.com, and Farmsphere.com. Courses in using social media in agribusiness are also springing up in extension services and community colleges.

One of the most important uses of social media, however, is for marketing. Twitter, Facebook, LinkedIn, YouTube, and others, allow companies to tell the world, literally, about their organizations and products. Existing and potential customers can also tell the organization what they think, want, or need. Small businesses are finding that using social media helps them to compete with much larger companies as anyone and everyone can find them on the web.[51]

# CHAPTER SUMMARY

Information is data organized in a meaningful way. Data are merely facts and figures—unorganized pieces of information. A system is an interrelated set of elements that function as a whole to accomplish some purpose. Because information is so important to contemporary management, it must be accurate, timely, complete, and relevant.

There are five basic interrelated components of any information system. In a computerized system, getting the data into the system involves input devices. A central processor provides analysis and processing. Storage is accomplished through one or more media, and in many cases more than one medium is used for backup purposes. Making the information available to users is also accomplished through a variety of output devices. Finally, the control system usually involves some form of computer software as well as human monitoring.

Information systems should ensure that proper information is available when needed. Designing information systems involves knowing one's information system needs as well as the kinds of systems that exist or might be developed. Information system needs are determined by the kind of organization, its environment, and its size.

Most managers maintain informal, unstructured systems of information; but formal, structured systems predominate in large organizations. Systems designed to handle routine and recurring transactions within an organization are known as transaction-processing systems. A management information system, or MIS, gathers, organizes, summarizes, and reports data for use by managers. A decision support system, or DSS, automatically searches for, analyzes, summarizes, and reports information a manager needs for making a particular decision. All these systems can further be classified as central or distributed. Organizations must be careful to match their various needs with the many systems available.

Although integrating systems is not easy, it must be done if the total information system is to be effective. Effectiveness also increases when the systems are user friendly. Many new systems are becoming easier to learn and use. Information systems affect an organization's performance, its structure and design, and the people within the organization. There seems to be a growing consensus that information systems enhance performance. Information systems also have some limitations, including being expensive and difficult to learn to use. In addition, the information derived from them may be overvalued, or they may be improperly used or not used at all. And, of course, computerized information systems depend upon electricity to operate and so they are vulnerable to power outages.

## CHAPTER ACTIVITIES

### REVIEW QUESTIONS

1. What is the difference between information and data? Why is that difference important?
2. What are the major components of information systems?
3. What are the different types of information systems and technologies?
4. What are the key considerations in managing information systems?
5. What are the effects and limitations of information systems?

### ANALYSIS QUESTIONS

1. Should computerized information systems be duplicates of paper information systems? Why or why not?
2. Do the information needs differ so much across organizational levels that different information systems have to be developed, or could an organization really get by with just one system?
3. Many people refer to the current period of economic development as the *information age*. Do you think we are really in the information age? Why or why not?
4. Do you think that the chief information officer of an organization can become too powerful? If so, how could that be prevented or corrected? If not, why not?
5. Comment on this quotation:

"The Turing Test of a computer system is designed to evaluate whether or not a user can tell that it is a computer system rather than a human being providing advice. A system that passes is then a computer system that is indistinguishable from a human. But who would want such a system? Humans are cheaper. What I want is a computer that is readily distinguishable from a human because it is better—it has no emotional problems, does not complain about hard work or overtime, wants no bonuses or retirement, makes no errors in judgment, and won't talk back."

### FILL IN THE BLANKS

1. _____ are facts and figures, unorganized pieces of information, useless until processed or organized in some way, whereas _____ consists of data organized in a meaningful way.
2. A system designed to handle routine and recurring transactions within an organization is known as a/an _____ _____ _____.
3. A system built on a series of rules to move from a set of data to a decision recommendation is called a/an _____ system.
4. In the information age, the focus is on _____ rather than the computer.
5. A/An _____ _____ _____ gathers, organizes, summarizes, and reports information for use by managers.
6. A/An _____ _____ _____ is a new form of information system that automatically searches for, analyzes, summarizes, and reports information a manager needs for making a particular decision.
7. Teleconferencing, telecommuting, and networking are forms of _____.
8. Connecting independent computers directly together so they can function in interrelated ways is called _____.
9. An attempt to have a computer simulate human decision processes is called _____ _____.
10. An information system that is designed so users can easily examine and modify information is known as a/an _____ _____ system.

# CHAPTER 24 CASE STUDY
## THE JUST-IN-TIME CONSUMER

Just about everyone involved with supply chain management or manufacturing has copied the Toyota operations model of just-in-time (JIT) manufacturing. Indeed, it has become a business textbook or case study staple. The point of this operational procedure was to have your supplier carry your inventory or, more importantly, get inventory off your books. Of course, this technique, along with other factors, catapulted Toyota to the number one car manufacturer in 2009, finally surpassing General Motors. Unfortunately, this victory was short-lived and uncovered the weakness of JIT—supplier inability to keep pace with growth, to maintain quality, and to retain margins and profitability. Apparently JIT has its limitations.

Or perhaps not, as JIT now manifests itself in a very different venue. It seems that the same economic deflection point—read, Great Recession—that caused Toyota to stumble may have awakened an "ah-ha" moment in consumers. "I had eight boxes of lasagna in there (kitchen pantry) and a year's worth of paper towels," said Rebecca Seabern, a mother of two and an accountant. She added, "I've stopped purchasing things just to have them on hand." These statements sum up a change in American consumerism. For decades American consumers were convinced that they should buy big and stock-up, so large-volume purchases represented the best value for their money. It was this consumer philosophy that fostered the rise of such warehouse chains as Costco, Sam's Club, and BJ's Wholesale Stores. However, the sentiments expressed by Ms. Seabern may augur a change in merchandising and subsequently fortunes.

This change in consumer attitude is changing the way retailers and manufacturers respond. "Consumers are saying, 'I'm going to buy what I need for a specific period of time,' rather than loading up and buying two or three extra units just because they can get a good price on it," said Del Monte CEO, Richard Wolford. The manner in which products are made, packaged, priced, and delivered is changing.

IT, information technology, is and will be the beneficiary of this "sea change." It was not enough to simply make smaller sizes, which manufacturers preferred because it meant higher profit margins; attempting to anticipate and reformulate "correct" product mix, size, and delivery schedules became essential and problematic. Product purchase cycles were disrupted by this JIT approach. Rather than "track" product purchase cycles from retailers, it became essential to "track" the shoppers or consumers. IT was required to generate algorithms to monitor two purchasing patterns: the retailer anticipating the consumer, and the consumer adjusting purchasing patterns based on personal needs and retailer stocking patterns.

Where do IT people for food manufacturers and food retailers get the information to feed into these algorithms? This information already exists for other types of retailers. They have been tracking shoppers for over a decade. Shopper tracking is a business. ShopperTrak Inc. has been using mall cameras to monitor shopper traffic patterns for years and shares this information with its retail clients. For example in 2010, U.S. shoppers made 695 million individual shopper visits to malls over the Thanksgiving weekend (one three-day weekend). Thompson Reuters, another consumer "tracking" firm, goes one step further and uses satellite imagery of mall parking lots to calculate shopper car traffic. These data are provided by Remote Sensing Metrics to retail and marketing clients.

The satellite systems that were launched to monitor crop production and to oversee the use of genetically modified seeds around the world or to monitor missile silos are now ready to count cars in parking lots. What the clothing retailers use routinely is now "new territory" for the food retailers.

Is it time to count cars in the supermarket parking lot?

"There is a palpable change in consumer-buying behavior that is unlike anything we have experienced certainly for a few decades," said Campbell Soup CEO, Doug Conant. "They (consumers) are being more surgical in their shopping." Some food retailers responded in traditional fashion to this trend; they increased advertising rather than shopper tracking. This is the more traditional approach. The increased spending has not yielded any immediate feedback. Sometimes not everyone sees the "sea change" in an industry.

"We're pretty certain that the more intense value mind-set of consumers is here to stay," said CEO Gary Rodkin of ConAgra Foods Inc. Sounds like a headache for management and a challenge for marketing and job security for IT.[52]

### ▶ Case Study Questions

1. What information systems approaches seem applicable to this change in consumer behavior?
2. How might an organization make its information system even better in the future?
3. What are the strengths and weaknesses of tracking consumers as suggested in this case?

## REFERENCES

1. "Crossing the Uncanny Valley," *The Economist* (November 20, 2010) at www.economist.com (accessed December 8, 2010); Kris Hudson, "Shopper Tracking Goes High Tech—Really High," *The Wall Street Journal* (November 24, 2010), C8; Thomas H. Davenport and Jeanne G. Harris, "The Dark Side of Customer Analytics," *Harvard Business Review* (May 2007), 37–41.

2. Charlie Feld and Donna Stoddard, "Getting IT Right," *Harvard Business Review* (February 2005), 72–80.

3. Ronald Henkoff, "Walgreen: A High-Tech Rx for Profits," *Fortune*, March 23, 1992, 106–107.

4. Ralph M. Stair and George Reynolds, *Principles of Information Systems: A Managerial Approach, 10th ed.* (Mason, OH: Delmar, Cengage Learning, 2012); Kenneth C. Laudon and Jane P. Laudon, *Essentials of Management Information Systems, 11th ed.* (Upper Saddle River, NJ: Prentice-Hall, 2009).

5. Donald A. Marchand, William J. Ketinger, and John D. Rollins, "Information Orientation: People, Technology, and the Bottom Line," *Sloan Management Review* (2000): 69–79.

6. William B. Stevenson and Mary C. Gilly, "Information Processing and Problem Solving: The Migration of Problems Through Formal Positions and Networks of Ties," *Academy of Management Journal* (December 1991): 918–928.

7. Lynda M. Applegate, James I. Cash, Jr., and D. Quinn Mills, "Information Technology and Tomorrow's Manager," *Harvard Business Review* (November–December 1988): 128–136.

8. Charles A. O'Reilly, "Variations in Decision Makers' Use of Information Sources: The Impact of Quality and Accessibility of Information," *Academy of Management Journal* (December 1982): 756–771.

9. Carla Rapoport, "Great Japanese Mistakes," *Fortune*, February 13, 1989, 108–111.

10. "Fashionable Tech: How Benetton Keeps Costs Down," *Information Week*, February 12, 1990, 24–25; Janette Martin, "Benetton's IS Instinct," *Datamation*, July 1, 1989, 68:15–16.

11. George Huber, "A Theory of the Effects of Advanced Information Technologies on Organizational Design, Intelligence, and Decision Making," *Academy of Management Review* (January 1990): 47–71; Carol Saunders and Jack William Jones, "Temporal Sequences in Information Acquisition for Decision Making: A Focus on Source and Medium," *Academy of Management Review* (January 1990): 29–46.

12. Paul L. Tom, *Managing Information as a Corporate Resource, 2nd ed.* (New York: Harper Collins, 1991).

13. Ibid; See also Cyrus Gibson and Richard Nolan, "Managing the Four Stages of EDP Growth," *Harvard Business Review* (January–February 1974).

14. Steven Cavaleri and Krzysztof Obloj, *Management Systems: A Global Perspective* (Belmont, CA: Wadsworth, 1993).

15. Robert G. Lord and Karen J. Maher, "Alternative Information-Processing Models and their Implications for Theory, Research, and Practice," *Academy of Management Review* (January 1990): 9–28.

16. Robert G. Murdick and Joel E. Ross, *Introduction to Management Information Systems, 3rd ed.* (Englewood Cliffs, NJ: Prentice-Hall, 1984).

17. Albert L. Lederer, "Information Requirements Analysis," *Journal of Systems Management* (December 1981): 15–19.

18. "Managing Information: Two Insurance Giants Forge Divergent Paths," *Business Week*, October 8, 1984, 121.

19. J. F. Rockart, "Chief Executives Define Their Own Data Needs," *Harvard Business Review* March–April 1979): 81–93.

20. "J.R. Simplot Company," Microsoft Case Study (September 18, 2007) at www.microsoft.com/casestudies (accessed July 5, 2010); Robert P. King, "Management Information Systems for Agribusiness Firms: Managerial Problems and Research Opportunities," *Agribusiness* (2006): 455–466; N. Zhang, M. Wang, and N. Wang, "A Precision Agriculture—A Worldwide Overview," *Computers and Electronics in Agriculture* (2002): 113–132.

21. Gil Press, "Decision Processing," *Information Systems Management* (Winter 1993): 40–46.

22. John Huey, "Walmart—Will It Take Over the World?" *Fortune*, January 30, 1989, 52–61.

23. "Videoconferencing: No Longer Just a Sideshow," *Business Week*, November 12, 1984, 117.

24. Roger Yu, "Companies Turn to Virtual Trade Shows to Save Money," *USA Today* (January 5, 2010) at www.usatoday.com (accessed August 2, 2010).

25. Michelle Conlin, "Telecommuting: Once a Perk, Now a Necessity," *Work-Life Balance* (February 26, 2009) at www.businessweek.com (accessed July 30, 20100; Janice Castro, "Staying Home Is Paying Off," *Time*, October 26, 1987, 112–113.

26. Mike Baker, "Making Digital Advertising More Efficient," *Forbes* (July 16, 2010) at blogs.forbes.com (accessed August 2, 2010); "E-Procurement—Electronic Data Integration Comes of Age," *FDE Finance Director Europe* (October 30, 2007) at www.the-financedirector.com (accessed July 29, 2010); "Where Retailers Shop for Savings," *BusinessWeek Online* (April 15, 2002) at www.businessweek.com (accessed July 30, 2010).

27. "Agronet," (January 8, 2008) at www.agronet.fi/english.html (accessed August 1, 2010).

28. Wayne Morris, "Business Gets Smart, Video Game Style," *Fortune* (January 20, 2010) at tech.fortune.cnn.com (accessed August 1, 2010).

29. "Artificial Intelligence and Expert Systems," *Power Engineering*, January 1, 1989, 26–31.

30. Tim O. Peterson and David D. Van Fleet, "Casting Managerial Skills into a Knowledge Based System," in *Organization, Management, and Expert Systems*, ed. Michael Masuch (New York: Walter de Gruyter, 1990), 171–183.

31. "Computer Applications: Software—A Risky Business," *Fairplay International Shipping Weekly*, January 4, 1990, 23–34.

32. Andrew Kupfer, "Now, Live Experts on a Floppy Disc," *Fortune*, October 12, 1987, 117.

33. "General Model Information," at ecobas.org (accessed August 1, 2010).

34. "Turning an Expert's Skills into Computer Software," *Business Week*, October 7, 1985, 104–108.

35. "Martin Marietta: Computers Lead the Way to Inbound Control," *Traffic Management*, January 1, 1990, 50–68.

36. David D. Van Fleet, T. O. Peterson, and Ella W. Van Fleet, "Closing the Performance Feedback Gap with Expert Systems," *Academy of Management Executive* (2005): 38–53; and Peter H. Lewis, "I'm Sorry; My Machine Doesn't Like Your Work," *The New York Times*, February 4, 1990, F27.

37. Chris Carroll and Chris Larkin, "Executive Information Technology," *Information Systems Management* (Summer 1992): 21–29.

38. Philip Elmer-Dewitt, "A Portable Office That Fits in Your Palm," *Time*, February 15, 1993, 56–57.

39. Cornelius H. Sullivan Jr., and John R. Smart, "Planning for Information Networks," *Sloan Management Review* (Winter 1987): 39–44.

40. "Linking All the Company Data: We're Not There Yet," *Business Week*, May 11, 1987, 151.

41. Malcolm Cole, "Network Your Way to the Automated Office," *Accountancy* (October 1, 1988): 92–94; Malcolm Cole, "Less Paper—The First Step to No Paper," *Accountancy* (October 1, 1988): 88–91.

42. Wendall Hahm and Tora Bikson, "Retirees Using EMail and Networked Computers," *International Journal of Technology and Aging* (Fall 1989): 113-114.

43. "Office Automation: Making It Pay Off," *Business Week*, October 12, 1987, 134–146.

44. Jeremy Main, "At Last, Software CEOs Can Use," *Fortune*, March 13, 1989, 77–83.

45. Ibid.

46. Adolph I. Katz, "Measuring Technology's Business Value," *Information Systems Management* (Winter 1993): 33–38.

47. "Office Automation: Making It Pay Off."

48. John J. Donovan, "Beyond Chief Information Officer to Network Managers," *Harvard Business Review* (September–October 1988): 134–140.

49. Jeremy Main, "The Winning Organization," *Fortune*, September 26, 1988, pp. 50–60.

50. "When Employees Work at Home, Management Problems Often Arise," *The Wall Street Journal*, April 20, 1987, 221.

51. Tricia Simon, "Online social networking and agribusiness," *The Peace Country Sun* at www.peacecountrysun.com (accessed June 26, 2011); Jonathan Knutson, "Ag turns to social media to make its case," *Agweek*, June 26, 2011 at www.agweek.com (accessed June 26, 2011).

52. Ellen Byron, "The Just-in-Time Consumer," *The Wall Street Journal* (November 23, 2010) at online.wsj.com (accessed December 7, 2010); Kelly Evans, "Reasons to be Wary of Warehouse Chains," *The Wall Street Journal* (October 6, 2010) at online.wsj.com (accessed December 7, 2010); Kris Hudson, "Shopper Tracking Goes High Tech—Really High," *The Wall Street Journal* (November 24, 2010) at topics.wsj.com (accessed December 8, 2010); "High-Tech Harvest— Genetically Modified Crops are Growing Up," *The Wall Street Journal* (September 25, 2010), A8; Anjali Cordeiro, "Campbell's Profit Up, but Sales Stew," *The Wall Street Journal* (September 4, 2010) at online.wsj.com (accessed December 8, 2010); Paul Ziobro, "Grocers Crank Up Their Advertising," *The Wall Street Journal* (October 14, 2010), B6.

# APPENDIX A

# A Cursory List of Agribusiness Firms/Businesses

## APPAREL COMPANIES

ANN, Inc.
Brown Shoe Company
Metersbonwe Group
Next plc
Nike, Inc.
PVH Corp.
Warnaco Group, Inc.

## BEVERAGE COMPANIES

AmBev
BevMo!
Brown-Forman Corporation
CadburyCanadaigua Wine Company Inc.
The Coca-Cola Company
Constellation Brands, Inc.
Cott Corporation
Diageo Danone plc
Hansen Beverage Company
Mackinaw Valley Vineyard
Molson Coors Brewing Company
National Beverage Corp.
Ocean Spray Cranberries, Inc.
PepsiCo Inc.
Philip Morris USA Inc.
Quaker Oats Company
SABMiller plc
Seagram Company Ltd
Siemens AG
Southern Wine & Spirits of America, Inc.

## CONFECTIONERY & SUGAR COMPANIES

Albanese Confectionery
American Crystal Sugar Company
American Licorice Company
Asher's Chocolates
Atkinson Candy Company
Blue Diamond Growers
Blueberry Hill Foods
C&H Pure Cane Sugar
Cadbury
Cadbury Adams
Farley's & Sathers Candy Company
Ferrara Pan Candy Company
Ferrero U.S.A., Inc.
Ghirardelli Chocolate Company
Gimbal's Fine Candies
Goetze's Candy Company, Inc.
GumRunners LLC
Haribo of America Inc.
Hawaiian Host Chocolates
The Hershey Company
Mars, Inc.
My Favorite Company
Nestlé
Russell Stover Candies, Inc./ Whitman's
Sconza Candy Co.
Signature Snacks Company
Snyder's of Hanover
Spangler Candy Company
Storck
Tate & Lyle PLC
The Topps Company, Inc.
Tootsie Roll Industries
Tropical Nut & Fruit Co.
Wal-Mart Stores, Inc.
Wm. Wrigley Jr. Company
World's Finest Chocolate Inc.

## DAIRY COMPANIES

Associated Milk Producers Inc.
Berkeley Farms
Clover Farms Dairy
ConAgra Foods, Inc.
Dairy Crest Group
Danone
Darigold
Dean Foods
Glanbia plc
Good Humor-Breyers Ice Cream Company
Horizon Organic
Kraft Foods Inc.
Land O'Lakes Inc.
Prairie Farms Dairy, Inc.
Saputo Inc.

## FERTILIZER COMPANIES

Agrium Inc.
CF Industries
The Mosaic Company
PotashCorp
Terra Industries

## FOOD PROCESSING COMPANIES

Archer Daniels Midland Company
Best Foods, Inc.
Campbell Soup Company
Cargill, Inc.
The Coca-Cola Company
ConAgra Foods Inc.
Dairy Farmers of America
Dole Food Company, Inc.
General Mills Inc.

H.J. Heinz CompanyThe Hershey
    Company
Hain Celestial Group
Hormel Foods Corp.
Ingredion Incorporated (formerly Corn
    Products International, Inc.)
Interstate Bakeries Corporation
Kellogg Company
Land O'Lakes, Inc.
Mars, Inc.
McCormick & Company, Inc.
Nabisco, Inc.
NestléPenford Products Co.
PepsiCo Inc.
Philip Morris USA Inc.
The Procter & Gamble Company
Quaker Oats Company
Ralcorp
Ralston Purina Company
Sanderson Farms
Sara Lee Corporation
Seaboard Corporation
Tyson Foods, Inc.

## FRUIT & VEGETABLE COMPANIES

Chiquita Brands L.L.C.
Dean Foods
Del Monte Foods
Dole Food Company
J.M. Smucker Company
Pillsbury
Sunkist Growers
Welch Foods Inc.

## GRAINS AND SEED BUSINESSES

The Andersons Grain Group
Archer Daniels Midland Company
Bunge Limited
Cargill, Inc.
Conagra Foods Inc.
DuPont
GROWMARK, Inc.
Monsanto Company
Nufarm Americas Inc.
Pioneer Grain Company
Syngenta AG

## GROCERY RETAILERS

A&P
Ahold USA
Albertsons
ALDI
Associated Wholesale Grocers
Bruno's Supermarkets, LLC
Carrefour Group
Dominick's Finer Foods
Food Lion LLC

Fred Meyer, Inc.
Giant Eagle
Giant Food Stores
H.E. Butt Grocery Co.
Hannaford Bros. Co.
Hy-Vee, Inc.
Jewel-Osco
Kmart
The Kroger Co.
Meijer
Nash Finch Company
Pathmark Stores, Inc.
Penn Traffic Company
Publix Super Markets
Raley's Supermarkets
Randalls Food Stores
Roundy's Supermarkets
Safeway Inc.
Schnucks
Shaw's Supermarkets, Inc.
Spartan Stores Inc.
SuperValuWakefern Food Corp.
Wal-Mart Stores, Inc.
Wegmans Food Markets
Winn-Dixie Stores, Inc.

## HORTICULTURE BUSINESSES

The Brickman Group, Ltd
Chicago Botanic Garden
Davey Tree Expert Co.
FS Custom Turf
Green View
Illinois Arborist Association
Kane Brothers, Inc.
Kickapoo Creek Nursery
McAdam Landscaping, Inc.
Midwest Groundcovers
Owen Nursery & Florist
Sunburst Nursery and Landscaping Inc.
TruGreen Lawn Care

## MEAT & POULTRY PROCESSORS

Cargill Meat Solutions Corporation
ConAgra Refrigerated Foods
Farmland Foods
Gold Kist
Hillshire Brands Company
Hormel Foods Corporation
John Morrell & Co.
National Beef Packing Company LLC
Oscar Mayer Foods Corporation
Perdue Farms
Pilgrim's Corporation
Smithfield Foods, Inc.
Tyson Foods, Inc.
Wampler Foods Inc.
Wayne Farms LLC

## SNACK COMPANIES

Beer Nuts
ConAgra Foods
Diamond Foods, Inc.
Frito-Lay, Inc.
General Mills, Inc.
George Weston Foods
Hillshire Brands
J&J Snack Foods Corp.
John B. Sanfilippo & Son, Inc.
Kellogg's Company
Kraft Foods Inc.
Lance, Inc.
Morrison's
Pillsbury
The Procter & Gamble Company
Tasty Baking Company

## TEXTILE AND FIBER COMPANIES

Albany International Group
American & Efirda LLC
BASF Corporation
Berkshire Hathaway, Inc.
Dan River, Inc.
DuPont
International Textile Group
Milliken & Company
Mohawk Industries
Parkdale Mills, Inc.
Polymer Group
R. B. Pamplin Corporation

## WATER-RELATED COMPANIES

American Water Technology, Inc.
Culligan International Company
Global Water Technologies, Inc.
Kinetico, Inc.
Safe Water Technologies, Inc.
Siemens AG
Water Technology, Inc.
Xylem, Inc.

## WOOD- AND TIMBER-RELATED COMPANIES

Anderson Wood Floors
Armstrong World Industries, Inc.
Drexel Heritage Furniture
    Industries Inc.
Ethan Allen Global, Inc.
Georgia-Pacific Corporation
International Paper Company
Plum Creek Timber Co., Inc.
Rayonier, Inc.
Stora Enso AB
Weyerhaeuser Company

# APPENDIX

# B Control Techniques and Methods

## MANAGER'S VOCABULARY

- absenteeism
- assets
- audit
- balance sheet
- balance sheet budget
- budget
- budgeting process
- capital expenditure budget
- cash flow budget
- current ratio
- debt ratio
- expense budget
- external audit
- financial budgets

- fixed costs
- income statement
- internal audit
- labor budget
- liabilities
- liquidity ratio
- market share
- marketing
- nonmonetary budgets
- operating ratio
- operations budgets
- output budget
- owners' equity
- performance appraisal

- profit budget
- profit margin on sales
- profitability ratio
- ratio analysis
- return on assets (ROA)
- return on equity (ROE)
- revenue budget
- semivariable costs
- space budget
- test marketing
- turnover
- variable costs
- Workforce composition ratios

This appendix is designed to give you a better understanding of the techniques and methods that managers use to maintain adequate control. For instance, a company could have a radically different earnings picture depending on such things as the period of time in which it charges goodwill and whether it handles one-time fees more as short streams of payments or as one-time revenue sources.

We first examine the role of control techniques in agribusiness management. Then we explore one of the most important and widespread techniques, the budget, before going on to investigate financial analysis and other control techniques.

## The Nature of Control Techniques

Techniques are tools; control techniques are handy tools that managers can use to enhance organizational effectiveness in general and organizational control in particular.

### The Importance of Control Techniques

As we discuss in the text, adequate control is important for numerous reasons. The proper use of the right techniques facilitates control and also helps managers communicate with others both inside and

513

outside the organization. Many of the techniques are in use by virtually all organizations and so represent standard business practice. The techniques involve different levels of management in the control process and enable managers to know what is happening in the organization.

Financial techniques provide the opportunity to keep up with the assets and debts of the firm, and budgetary techniques can facilitate planning to ensure that the firm will not overspend its resources. Control can be particularly important to small businesses.[1] A sole-proprietorship farmer who hired workers to help him expand his business soon went out of business because he forgot that, although he had to pay his workers weekly, he did not get paid until the harvest was completed. Had he used budgets and financial control techniques, he would have foreseen his cash-flow problem and perhaps been able to borrow enough money to keep from going bankrupt or losing his employees.

## Strengths and Weaknesses

Like tools used in any discipline or subject matter, control techniques have various strengths and weaknesses. When used properly, they are great assets to any manager. When used improperly, however, they can do more harm than good.

On the plus side, control techniques provide objective indicators of an organization's performance. Budgets and various financial ratios, for example, are quantifiable and verifiable indications of an organization's financial situation.[2] Control techniques also provide useful road maps for action, as budgets indicate clearly where and how to allocate resources.

On the other hand, two significant weaknesses characterize these techniques. First, they can be rigid and uncompromising, especially when managers use them with no regard for flexibility or for the appropriate context. A ranch manager should not, for instance, continue to rely on a budget that had been developed on the basis of outdated information about the Endangered Species Act. Control techniques must be flexible to be truly effective.[3]

Second, control techniques can be misused. In particular, individual managers might be able to distort information to mask problems. A manager could report that, historically, employee turnover is greater for personnel recruited from one region of the country. She could use these data to support an argument for not recruiting from that region when,

in fact, the turnover stemmed only from her treatment of those employees.

Thus, when using control techniques, managers must remember that while they are valuable tools, they do not replace individual judgment and insight into organizational dynamics.

## Budgets

Budgets are perhaps the most widely used and universally known control techniques available.[4] A **budget** is simply a plan expressed in quantitative terms. (Budgeting, then, is the process of developing a budget.) Budgets serve four basic purposes: (1) help coordinate resources and projects; (2) define standards used in other control systems and activities; (3) provide clear and unambiguous guidelines about resources; and (4) facilitate appraisals of managerial and departmental performance (Figure B.1).

Because of the importance and pervasiveness of budgets in organizations, we will explore them in detail. First, we will look at the budgeting process. We can then identify types of budgets and fixed and variable costs, which are important parts of most budgets. Finally, we will show how to manage the budgeting process effectively.

### The Budgeting Process

The **budgeting process** is unique for every organization; that is, every organization comes to follow a budgeting process that fits its own culture and style. Thus, we cannot identify a process that is

**FIGURE B.1** Having a well prepared budget can help to avoid last minute tax and financial woes.

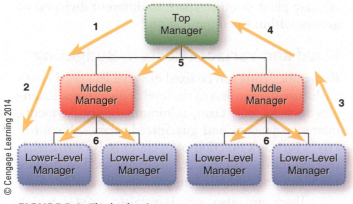

**FIGURE B.2** The budgeting process.

representative of all organizations. However, many organizations follow the general pattern of budgeting shown in Figure B.2.

As a starting point, top management usually issues a call for budget requests, a step indicated in Figure B.2 by Arrow #1. As we can see, the first demand usually goes from top management to middle-level managers. The top managers usually accompany this call with some indication of what resources are available during the coming time period. For example, they might point out that sales and profits are increasing and therefore encourage requests for budget increases. Alternatively, they could ask for cutbacks, such as 10 percent less than last year. As a part of strategic planning, the top managers might also indicate that some units will have higher or lower priorities than in the past.

Step 2 parallels Step 1 in that middle managers request budget proposals from lower-level managers. They provide the same basic information, but that information pertains more to individual subunits than to the overall division.

In Step 3, lower-level managers prepare budget requests and forward them to the appropriate middle managers. In general, these requests summarize the unit's current resources, account for how those resources have been used, and point out what resources the unit will need for the next period (usually one year) and how the unit will use those resources.

The middle manager then coordinates and integrates the various requests. For instance, two department heads in a pickle processing plant might each request $10,000 to buy five computers. The plant manager knows that an order of ten computers carries a 10 percent discount, so the plant will request $18,000 to buy ten computers.

As Arrow #4 indicates, middle managers then forward their division requests to top management, which in turn coordinates and integrates these requests. Some will be returned to middle management for further work. Others will be approved as is. And top management may modify others with little or no input from below.

The last step, as Arrows #5 and #6 indicate, is to pass final budgets back down the organization. In some cases middle managers will still have the option to modify their subunits' budgets, whereas in other cases the final budgets are provided all the way down the hierarchy.[5]

## Types of Budgets

Most organizations produce several different types of budgets.[6] The most common of these are shown in Table B.1. You should recognize, of course, that there are budgets for each part of an organization as well, including marketing, information services, and so on.

**Financial Budgets.**    **Financial budgets** detail where the organization intends to get its cash for the coming period and how it intends to use it. Three types of financial budgets are cash flow, capital expenditure, and balance sheet. Most money comes from sales revenues, the sale of assets, loans, and the sale of stock. This money pays for expenses, repays debt, purchases new assets, and pays dividends to stockholders.

A **cash flow budget** focuses on time. It outlines precisely where the money will come from and when, plus how and when the company will use it, usually for the coming quarter or a year. A cash flow budget is essential for all businesses, regardless of size. It can show, for example, whether Summer Farms should be able to meet its payroll and pay its bills on time each month. If incoming funds will not

**TABLE B.1** Types of Budgets

| MAJOR TYPE | SUBTYPES |
|---|---|
| *Financial Budgets* | Cash flow budget<br>Capital expenditure budget<br>Balance sheet budget |
| *Operations Budgets* | Revenue budget<br>Expense budget<br>Profit budget |
| *Nonmonetary Budgets* | Output budget<br>Labor budget<br>Space budget |

© Cengage Learning 2014

cover needs, the organization must make other plans, such as getting loans or deferring payments. Surplus cash, in contrast, can be invested. Agribusiness firms and even large defense contractors like Martin Marietta use cash flow budgets effectively as tools.[7]

Companies use **capital expenditure budgets** to plan for the acquisition of such major assets as new equipment, entire plant facilities, and land. Such assets are often paid for with borrowed funds, so both small agribusiness firms such as Summer Farms and large firms like Archer-Daniels-Midland and Cargill must control their use carefully.

The **balance sheet budget** is a projection of how the firm's assets (what it owns) and liabilities (what it owes) will look at the end of the coming period. In this section we will discuss the balance sheet in more detail.[8]

### Operations Budgets.

**Operations budgets** relate to all or a portion of the organization's operations for the coming period. In particular, they present the various details of operations in financial terms. The revenue budget and the expense budget are examples of operations budgets. Companies such as Summer Farms rely on revenue that comes in seasonally often much later than bills come due, thus making it absolutely essential for management to develop proper budgets.

The **revenue budget** is an extremely important budget for all organizations. For large businesses, such as Colgate-Palmolive or smaller ones, such as Summer Farms, revenues come from sales, but a university derives its revenues from legislative appropriations, tuition and fees, and government contracts for research. Regardless of the source, the revenue budget is important because it is the starting point for virtually all other budgeting.

The **expense budget** is the counterpart of the revenue budget. It summarizes the projected expenses for the organization over the coming period. The **profit budget** simply presents the difference between projected revenues and projected expenses.

### Nonmonetary Budgets.

**Nonmonetary budgets** express important variables in terms other than dollars. For example, an **output budget** may project the number of units of each product the organization will produce. Obviously, this is an important source of information for contracting sales with customers. A **labor budget** details the number of direct hours of labor that are available, and a **space budget** helps allocate plant or office space to different divisions or groups within the organization.[9]

## Fixed and Variable Costs in Budgeting

Remember, costs can be fixed or variable. **Fixed costs** are incurred regardless of the level of operations. Examples include rent, taxes, minimum utility payments, interest payments, and guaranteed salaries. **Variable costs** vary as a function of operations. Raw materials that go into each product, electricity used to operate machines, and commissions paid are all variable costs.

There are also **semivariable costs**, which vary as a function of output but not necessarily in a direct fashion. Advertising, for instance, varies in response to competition and seasonal sales patterns. Equipment repairs, maintenance, and labor are also semivariable costs.

When developing budgets, managers must try to account for all three kinds of costs. Obviously, fixed costs are the easiest to assess and semivariable costs the most difficult. If the managers don't give each kind of cost adequate consideration, however, the budget may not serve its intended purpose, and the organization could end up with unexpected cash shortfalls at the end of the period. Because these types of costs are so important, managers try to find ways to control them.[10]

## Managing the Budgeting Process

Budgets are an important part of organizational life in general and of control in particular. Table B.2 summarizes the strengths and weaknesses of budgets, which managers must understand to manage the budgeting process effectively.[11] Culture also affects budget practices and so it must get careful consideration, especially by multinational organizations and any organization that sells products that are specifically targeted to a particular market or markets.[12]

Four basic strengths characterize budgets. First, they facilitate control, as the foregoing discussion should make obvious.

**TABLE B.2**  Managing the Budgeting Process

| STRENGTHS | WEAKNESSES |
| --- | --- |
| Facilitates control | Subject to rigidity |
| Facilitates coordination | Time consuming |
| Facilitates documentation | Limits innovation |
| Facilitates planning | Reinforces inefficiency |

© Cengage Learning 2014

Budgets also facilitate coordination. This stems from middle managers integrating the budgets of lower-level managers, and top managers subsequently integrating the budgets that middle managers submit.

Budgets also aid documentation. They are almost always written down, so there is a permanent record of expectations and actual performance.

Finally, budgets help the planning process. As we have seen, planning and control are linked; given the pervasive role of budgeting in control, it is logical that budgets are also an important part of planning. Indeed, a budget has been defined as a plan.

Four basic weaknesses also help to characterize budgets.[13] First, some managers make the mistake of using them too rigidly. As indicated earlier, budgets must be flexible to be really valuable.[14] If the manager fails to consider the situation and looks at the numbers in a mechanical way, he or she may lose important related information. For example, assumptions upon which the budget was based could change, resulting in a subsequent change in cash flow and maybe the need for outside financing; for example, after crops have been damaged or competition has changed for any number of reasons.

Budgeting is also a time-consuming process. To do it right, every manager must invest considerable time, effort, and energy in making the budget as effective as possible.

Third, budgets occasionally limit innovation. In a company like ConAgra, for example, if all funds are allocated to operating groups and divisions, the organization could experience difficulty raising money for unexpected opportunities.

Fourth, the way budgets are used frequently reinforces inefficiency. A manager who is efficient (who spends less than her budget) may find that her budget is reduced in the next cycle. On the other hand, a manager who is inefficient (who spends over his budget) may find his budget increased in the next cycle. This common practice occurs because higher-level managers are not carefully examining the assumptions and conditions under which the budget was developed.

A key method many managers have come to use to enhance the strengths and to minimize the weaknesses inherent in budgeting is participation. The greater the role that lower-level managers have in developing their budgets, the more likely they are to make sure that the numbers are useful. They will work hard to justify the funds they want and to use those funds wisely. Of course, the same holds true for middle managers as they develop budgets for entire divisions or operating groups.[15]

# Financial Analysis

Another important technique for control is financial analysis. Whereas budgets are used to plan and control an organization's future financial expenditures, financial analysis helps managers study an organization's financial status at a later point in time. Using financial analysis, which consists of several techniques, managers are better able to understand and control the monetary aspects of an organization. Two of the most important techniques are ratio analysis and audits.[16]

## Ratio Analysis

**Ratio analysis** involves calculating and evaluating any number of ratios of figures obtained from an organization's balance sheet and income statement.[17] This analysis can be used internally or externally to obtain information about companies' financial positions.[18]

**Balance Sheet and Income Statement.** A **balance sheet** is a cross-sectional picture of an organization's financial position at a given time. Figure B.3 presents a simplified (modified) balance sheet for Summer Farms.

The left side of the sheet summarizes the **assets**, those things the company has that are of value. *Current assets* are cash or are easily convertible into cash, such as accounts receivable and inventory. *Noncurrent assets* are less liquid and play a longer-term role in the organization, such as land, plant, and equipment.

The right side of the balance sheet summarizes the firm's **liabilities**—debts and other financial obligations—and **owners' equity**, or claims against the assets. *Current liabilities* must be paid in the near future and include accounts payable and accrued expenses, such as salaries earned by workers but not yet paid. *Noncurrent liabilities* are bank loans amortized over a several-year period and payments on bonds. Owner's (stockholders' for corporations) equity consists of *contributed capital* (stock for corporations) and *retained earnings*. Retained earnings are profits held by the company for expansion, research and development, or debt servicing. As shown in Figure B.3, the totals on each side of the balance sheet must be equal (balanced).

## SUMMER FARMS
### BALANCE SHEET (MODIFIED*)
### DECEMBER 31, 1985

| Assets | | Liabilities | |
|---|---|---|---|
| **Current Assets:** | | **Current Liabilities:** | |
| Cash | $ 30,000 | Accounts Payable | $ 60,000 |
| Receivables | 10,000 | Short-Term Notes Payable | 2,500 |
| Inventories: | | Current Portion of Long-Term Debt | 2,000 |
| Crop 1 | 50,000 | Accrued expenses | 20,000 |
| Crop 2 | 30,000 | Total Current Liabilities | $ 84,500 |
| Supplies | 10,000 | | |
| Prepaid Expenses | 1,500 | **Noncurrent Liabilities:** | |
| Total Current Assets | $ 131,500 | Notes Payable | $ 39,000 |
| | | Mortgages | 150,000 |
| **Noncurrent Assets:** | | Total Noncurrent Liabilities | $ 189,000 |
| Farm Land | $ 290,000 | | |
| Timber Land | 100,000 | **Total Liabilities** | $ 273,000 |
| Equipment | 100,000 | | |
| Buildings | 42,000 | **Owner's Equity** | |
| Total Noncurrent Assets | $ 532,000 | Contributed Capital | $ 200,000 |
| | | Retained Earnings | 190,000 |
| **Total Assets** | $ 663,500 | Total Equity | $ 390,000 |
| | | | |
| **Total Assets** | $ 663,500 | **Total Liabilities and Equity** | $ 663,500 |

\* Numbers were modified to protect privacy of Summer Farms.

**FIGURE B.3** A sample balance sheet.

© Cengage Learning 2014

**Income Statement.** Whereas the balance sheet reflects a point in time, the **income statement** summarizes several activities over a period of time. In general, a company prepares an income statement on an annual basis and a balance sheet for the time at which the income statement ends. For instance, the balance sheet in Figure B.3 reflects Summer Farm's position (modified) as of December 31, 1985. Figure B.4 presents a simplified income statement for the same organization for the year ending December 31, 1985. Essentially, the income statement shows how the accountant adds up all the revenues of the organization and then subtracts all the expenses and other liabilities. The so-called bottom line is the profit or loss realized by the firm.

**Ratios.** Data from both the balance sheet and the income statement provide useful benchmarks for assessing an organization's overall financial health, especially when expressed as any of several commonly used ratios.

**Liquidity ratios** assess how easy it would be to convert the assets of the organization into cash. When a bank lends money to a small business, for example, it might want to know how quickly it could recover its money if the business were to fold. The **current ratio**, the most commonly used liquidity ratio, is determined by dividing current assets by current liabilities. Thus the current ratio for Summer Farms (see Figure B.3) is $161,500 ÷ $84,500, or 1.55. The ratio is expressed in the form of 1.55:1, which means that Summer's has a little more than $1.50 of liquid assets for every dollar of short-term liability. This is a fairly healthy ratio.

**Debt ratios** reflect a firm's ability to handle its long-term debt. The most common debt ratio is found by dividing total liabilities by total assets. This debt ratio for Summer's is $273,500 ÷ $663,500, or 0.41, which indicates that the organization has less than one-half dollar of debt for each dollar of assets. The higher this ratio is, the poorer the financial health of the organization.

| SUMMER FARMS INCOME STATEMENT (MODIFIED*) FOR THE YEAR ENDING DECEMBER 31, 1985 | | |
|---|---|---|
| **Revenue:** | | |
| Crop #1 Sales | | $ 247,500 |
| Crop #2 Sales | | 175,000 |
| Timber Sales | | 253,000 |
| **Total Gross Sales** | | **$ 675,500** |
| Less Sales Discounts & Spoilage | $ 60,000 | |
| **Net Sales** | | **$ 615,500** |
| | | |
| **Expenses:** | | |
| Seed and Fertilizer | $ 45,000 | |
| Gasoline and Other Cash Expenses | $ 50,000 | |
| Labor | $ 60,000 | |
| Depreciation Allowance | $ 10,000 | |
| Insurance | $ 12,000 | |
| Set-Aside Allowance for new property | $ 300,000 | |
| **Total Expenses and Set-Asides** | | **$ 477,000** |
| | | |
| **Operating Profit** | | **$ 138,500** |
| | | |
| Interest Expense | $ 9,500 | |
| Taxable Income | | $ 129,000 |
| Less Taxes | $ 23,220 | |
| | | |
| Net Income | | **$ 105,780** |

\* Numbers were modified to protect privacy of Summer Farms.

**FIGURE B.4**   A sample income statement.

The **return on assets (ROA)** tells how effectively an organization is using its assets to earn additional profits. This ratio is usually of most interest to potential investors. The normal method for calculating this ratio is to divide net income by total assets (see Figures B.3 and B.4). Thus, the ROA for Summer is $105,780 ÷ $663,500, or 0.16. This figure is the percentage return achieved by Summer over the 12-month period. It means that the company earned $0.16 of profit for each $1 in assets it controlled. Since most savings accounts earn less than 5 percent, an ROA of 0.19 or 19 percent is very good. Therefore, investors would probably think that Summer Farms is a good investment.

Companies occasionally use other ratios, of course. **Return on equity (ROE)** is net income divided by owners' equity. For Summer, the ROE is $105,780 ÷ $390,000, or 0.27. The company can then compare this with its previous figures and to other companies in the industry to get an indication of how well Summer is doing.

**Profitability ratios** indicate the relative effectiveness of an organization. For example, $1 million in profits from sales of $10 million (0.1) is quite good, whereas a profit of $1 million on sales of $100 million (0.01) is not so good. For Summer Farms the profitability ratio of 0.17 ($105,780 ÷ $615,500) shows that it is performing satisfactorily.

**Operating ratios** can also be useful. The index obtained by dividing the total cost of goods sold by the average daily inventory, for instance, provides a good indication of how efficiently the firm is forecasting sales and ordering merchandise.

## Audits

Another important part of financial analysis is the **audit**, an independent appraisal of an organization's accounting, financial, and operational systems. An audit may be either external or internal.[19] Care must be taken to ensure that audits are of high quality.[20] An audit can focus on not only the monetary but also the nonmonetary aspects of the organization.[21]

**External Audits.**   **External audits** are conducted by experts who do not work directly for the organization. Most often these experts are employed by an accounting firm. Their purpose is to evaluate closely the appropriateness of a company's controls and reporting procedures and to report their findings to relevant parties, such as stockholders and the IRS. Publicly held corporations (corporations whose stock is traded on public markets and is available for purchase by anyone) must conduct external audits on a regular basis.

External auditors are extremely thorough. In some cases they even visit warehouses to count inventory in order to verify the accuracy of the firm's balance sheets. Auditors who are found to have made mistakes may lose their reputations and even their licenses, so they are generally very careful. Virtually all smaller firms, and many medium-sized ones, conduct external audits at least on an annual basis.

**Internal Audits.**   **Internal audits** are conducted by people who work directly for the organization. Their purpose is the same as that of external auditors: to verify the accuracy of the organization's reporting system. Internal audits tend to go further than external audits, however, and also deal with matters of efficiency. For example, if an organization's accounting system is technically accurate but somewhat inefficient, an external audit will verify the system's accuracy but ignore the efficiency problem, whereas an internal audit will deal with both.

Most large firms have internal auditing staffs. It is considerably cheaper to maintain these staffs than to rely totally on the services of an external accounting firm.[22] Internal auditors spend much of their time auditing various subunits or divisions of the organization. In many ways, internal auditors are also more valuable than external auditors. Besides uncovering a variety of problems, they tend to be very familiar with all the inner workings of the organization.

Even when an organization has an internal auditing staff, though, it periodically chooses or is required to use an external auditing group to provide an independent assessment of its practices. Accurate, well-kept records are crucial to both internal and external auditors and can influence a company's fortunes in many ways (Figure B.5).

**FIGURE B.5** Idetifying problem areas during an audit can improve performance.

## Other Control Techniques

Although financial and operations control are the primary concerns of most businesses, other important areas of control should not be neglected. Two of these are human resource control and marketing control.

### Human Resource Control

Human resource control focuses on an organization's workforce. In particular, it is concerned with the extent to which members of the workforce are productive and the extent to which the organization is effectively managing them.

**Performance Appraisal.**   The primary way in which an organization controls the performance of its employees is through performance appraisal. **Performance appraisal** is the way the organization determines each individual's level of performance.[23] From a control standpoint, it helps the manager monitor the performance of employees, compare that performance with desired standards, and address any problems or deficiencies that she finds.

Indeed, the basic steps in performance appraisal parallel those in control very closely. First, using job analysis, the organization determines the exact content of each job and then assesses how much each person should produce. For example, the work of a machine operator might be defined as adjusting machine settings, working metal parts through the machine, and inspecting the final product. Management might further decide that the desired level of output per operator is 144 dozen parts, which becomes the standard that is then used in the performance appraisal. For example, the supervisor checks the output of each operator under his control every month.

The actual performance is then measured against the standard, and appropriate action is taken. Jack may be packing more than 160 dozen boxes of produce per day, so he should be complimented and perhaps given a small pay raise. Laverne may be packing around 146 dozen boxes per day; the supervisor can tell her she's doing a good job and leave it at that. If Joe is found to be producing only 98 dozen boxes per day, appropriate action might include inquiring as to whether he has a physical problem, observing to see if he is wasting time, counseling, encouraging him to work harder, and training. If things don't improve, the supervisor can transfer him to another job or perhaps even fire him.

**Human Resource Ratios.** Managers can also use any of several human resource ratios to assess the degree to which the organization is managing its workforce properly. Three of the more important human resource ratios are turnover, absenteeism, and workforce composition.[24]

**Turnover** is the percentage of the organization's workforce that leaves and must be replaced over a period of time, usually one year. Companies like McDonald's and Pizza Hut experience high levels of turnover, occasionally as high as 100 percent, whereas companies like Kmart have moderate turnover and still others (for example, family businesses) have relatively low turnover. The manager needs to know the acceptable turnover for both her firm and the industry. If turnover increases or is higher than the industry average, she will need to take steps, such as increasing wages to meet the industry standard, to get things back in line.

**Absenteeism** is the percentage of an organization's workforce that is absent on a given day. Absenteeism may range from only a few percent to as high as 20 or 30 percent. It is especially troublesome on Mondays and Fridays, when employees use sick time to extend their weekends. Again, the key for the manager is to know what the acceptable level of absenteeism is and to take steps if it gets out of line. He might issue warnings about excessive absences, for instance, or offer incentives to those who do not call in sick. The company could instead plan for a certain percentage of absenteeism on Fridays and encourage employees to schedule the time off rather than just not reporting to work.

**Workforce composition ratios** are indicators of how many of the organization's employees fall into various groups. For example, for Federal reporting purposes the manager may need to know how many of the organization's employees are Black, Hispanic, female, handicapped, or over the age of 45. If the general labor market from which the firm hires its employees contains a higher proportion than the organization hires, the company may be charged with discrimination. That is, if Black workers make up 25 percent of the potential workforce in a particular city or area but only 10 percent of the firm's employees are Black, discrimination may be suspected. The obvious step in this case is to eliminate such discriminatory practices.

## Marketing Control

Another area that is important for control is **marketing**—the set of activities involved in getting consumers to want the goods and services provided by the organization and getting those goods and services to the consumer.[25] Two common approaches to marketing control are test marketing and assessing important marketing ratios.

**Test Marketing.** **Test marketing** involves introducing a new product on a limited basis to assess consumer reaction on a small scale. For example, Wendy's creates a new sandwich and sells and advertises it only in Missouri and Kansas. If it is successful, Wendy's will introduce it in all Wendy's restaurants. If it is unsuccessful, the company might decide not to introduce it anywhere else and to drop it from the menu in Missouri and Kansas as well.

The advantage of test marketing is that it minimizes the risk of losing large sums of money by introducing new products nationally and then watching them fail. Test marketing also allows firms to make adjustments and refinements based on limited consumer responses before committing the products to national introduction.

In recent years there has been a trend away from test marketing because it is slow and gives competitors time to copy and perhaps get ahead of the company with new products. Figure B.6 shows an alternative approach being tried by some companies.

Under the traditional approach, also shown in Figure B.6, new products are developed and then test marketed. The results are assessed quantitatively (that is, actual sales are measured), modifications are made, and the product is then introduced via a full-scale marketing effort. Under

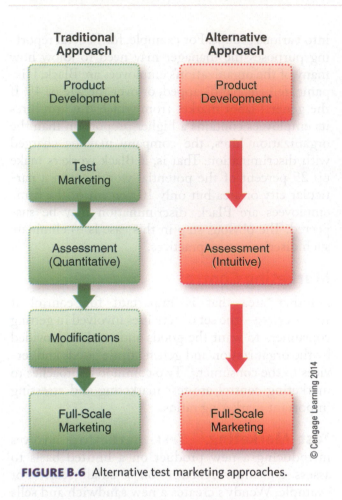

**Traditional Approach**

- Product Development
- Test Marketing
- Assessment (Quantitative)
- Modifications
- Full-Scale Marketing

**Alternative Approach**

- Product Development
- Assessment (Intuitive)
- Full-Scale Marketing

© Cengage Learning 2014

**FIGURE B.6** Alternative test marketing approaches.

the new approach, products are developed and then assessed intuitively by the firm's managers and perhaps by very limited consumer panels. If things look positive, the product is introduced full-scale, bypassing the test market introduction.

Sara Lee, Ralston Purina, and Frito-Lay have all adopted this approach.[26] The risk, of course, is that if the product fails, the company loses a lot of money. On the other hand, the company may be able to get a big jump on the competition and save the expenses of test marketing if it guesses right.

**Marketing Ratios.** There are also several marketing ratios that are useful in marketing control. Market share and profit margin are two of the most common.

**Market share**, the most common marketing ratio, refers to the proportion of the total market controlled by the firm's products. If the total market for cut Christmas trees in the "Southeast Christmas tree market" is one million units per year and Smoky Mountain Wholesale Nursery sells 100,000 units, its share of that market is 10 percent. A common goal of many organizations is to increase market share through increased advertising, promotion, and so forth. Decreases in market share, of course, are a source of considerable concern.

**Profit margin on sales** is another useful marketing ratio. This is determined by dividing net income by sales. The higher the ratio, the more effective the organization is in managing its marketing function. When the profit margin is calculated for each product a firm sells, managers can learn which products are contributing the most in profits.

## SUMMARY

Control techniques are the tools that managers use to enhance organizational effectiveness in general, and organizational control in particular. The major strengths of control techniques are that they provide objective indicators of how the organization is doing and road maps for action. The weaknesses include being rigid, uncompromising, and subject to misuse.

Although the particulars of the budget process are unique for every organization, the process usually starts at the top of the organization with a call for budgets, which are then transmitted downward throughout the whole organization. Working from the lowest levels upward, budget requests are then developed and moved up through the organization. After one or more attempts, the final budgets are sent back down.

There are three basic types of budgets: financial, operations, and nonmonetary. Budgeting involves fixed, variable, and semivariable costs.

Budgeting also has some weaknesses: Some managers make the mistake of using budgets too rigidly; they are a time-consuming process; they occasionally limit innovation; and they can be used to reward inefficiency. As for strengths, budgeting makes planning, coordination and control easier and also provides documentation in case of controversy and for future reference.

Financial analysis refers to techniques and methods managers use to understand and control monetary aspects of an organization. The most important are ratio analysis and audits. Ratio analysis involves calculating and evaluating any number of ratios obtained from an organization's balance sheet and income statement. Audits are independent appraisals of the company's accounting, financial, and operational systems. Experts who do not work directly for the organization perform external audits. Internal audits are conducted by those who do work for the organization.

Control techniques are available for every area of organizational performance, not just financial and operational performance. Two of the more important areas are human resource control, which focuses on the members of the organization, and marketing, which refers to activities that get consumers to purchase the goods and services of the company.

## APPENDIX ACTIVITIES

### REVIEW QUESTIONS

1. What are the strengths and weaknesses of control techniques?
2. Describe the budgeting process. What are the several types of budgets?
3. What is financial analysis, and how is it used by organizations?
4. What are the major types of human resource control? Marketing control?
5. What are the advantages and disadvantages of using computers in control?

### ANALYSIS QUESTIONS

1. If a frequent misuse of budgets reinforces inefficiency, why do they continue to be misused?
2. Do all organizations need to use all of the different types of budgets? Why or why not?
3. Comment on this statement: "If you want low turnover, hire incompetent people and pay them well."
4. Comment on this statement: "The only real test of a product is to offer it for sale."
5. Despite careful market research, many products fail. Why is this true? What does this suggest about marketing controls?

### FILL IN THE BLANKS

1. A statement showing a company's financial position in terms of assets, liabilities, and equity at a given time is known as a/an _____ _____ .
2. An organization's plan expressed in quantitative terms is called a/an _____ .
3. Raw materials that go into each product, electricity used to operate machines, and commissions paid are examples of _____ _____ .

4. _____ _____ assess how easily the assets of the organization can be converted into cash.

5. A/An _____ _____ summarizes income and expenditures over a period of time.

6. The proportion of the total market controlled by the firm's products is known as _____ _____ .

7. _____ _____ detail where and when the organization intends to get its cash for the coming period and how and when it intends to use it.

8. Dividing net income by total assets produces a ratio known as _____ _____ _____ , which indicates how effectively the organization is using its assets to earn additional profits.

9. _____ _____ is a marketing ratio that is determined by dividing net income by sales.

10. Turnover, absenteeism, and workforce composition are examples of _____ _____ ratios.

## REFERENCES

1. Harland E. Hodges and Thomas W. Kent, "Impact of Planning and Control Sophistication in Small Business," *Journal of Small Business Strategy* 17(2): 75–87; "Small Business," *Director* (January 1, 1990), 87–88; Terry J. Engle and David M. Dennis, "Benefits of an Internal Control Structure Evaluation in a Small Business Audit," *The Ohio CPA Journal* (Spring 1989), 5–11; "Cashing Out and Maintaining Control; Have Your Cake and Eat it Too," *Small Business Report* (December 1, 1989), 27–41; "Case History: Growing Pains," *Small Business Report* (July 1, 1989): 34–44; Ralph M. Stair, William F. Crittenden, and Vicky L. Crittenden, "The Use, Operation, and Control of the Small Business Computer," *Information and Management* (March 1, 1989), 125–130; A. Shapero and L. Sokol, "Exits and Entries: A Study in Yellow Pages Journalism," in *Frontiers on Entrepreneurship Research 1982,* ed. Karl H. Vesper (Englewood Cliffs, NJ: Prentice-Hall, Inc., 1982), 72–90.

2. Remember that the organization may be public as well as private. Clearly, budget issues are a major concern in the public sector; see, for example, Aman Kahn and W. Bartley Hildreth, *Budget Theory in the Public Sector* (Westport, CT: Quorum Books, 2002); Sharon Randall, "Top Issues For 1993: Budget, Education, Health Care," *State Legislatures* (January 1, 1992), 20–21; Muhlis Bagdigen, "Budgeting Systems and Their Applicability in Public Sector," *EU IIBF Dergisi* (2001), 17, 17–36.

3. Hugh Coombs, David Hobbs, and Ellis Jenkins, *Management Accounting* (London: Sage, 2005), 151–156; "Flexible Budget System—A Practical Approach to Cost Management," *Healthcare Financial Management* (January 1, 1989), 38–53.

4. Belverd E. Needles, Jr., Henry R. Anderson, and James C. Caldwell, *Principles of Accounting, 2002 ed.* (Boston: Houghton Mifflin, 2002).

5. Jay W. Lorsch, et al., *Understanding Management* (New York: Harper & Row, 1978).

6. Glenn A. Welsch, Ronald W. Hilton, and Paul N. Gordon, *Budgeting: Profit Planning and Control, 6th ed.* (Englewood Cliffs, NJ: Prentice-Hall, 2005); Wesley J. Obst, Rob Graham, and Graham Christie, *Financial Management for Agribusiness* (Collingwood, Victoria: Landlinks Press, 2007).

7. Paul De Ionno, et al. "A Bioeconomic Evaluation of a Commercial Scale Recirculating Finfish Growout System—An Australian Perspective," *Aquaculture* (2006), 259: 315–327; Thomas Moore, "Why Martin Marietta Loves Mary Cunningham," *Fortune,* March 16, 1987, 66–70.

8. Harry Ernst, "New Balance Sheet for Managing Liquidity and Growth," *Harvard Business Review* (March–April 1984), 122–135.

9. Welsch, *Budgeting.*

10. Robin Cooper, "You Need a New Cost System When . . . ." *Harvard Business Review* (January–February 1989), 77–82.

11. J. M. Argiles and E. J. Slof, "New Opportunities for Farm accounting," *The European Accounting Review* (2001), 10:2. 361–383; Bess Ritter May, "How Any Supervisor Can Control Company Costs: Be a Budget Watcher," *Supervision* (April 1, 1990), 3–5.

12. Susumu Ueno and Uma Sekaran, "The Influence of Culture on Budget Control Practices in the USA and Japan: An Empirical Study," *Journal of International Business Studies* (1992), 23: 659–674.

13. Andy Neely, Mike Bourne, and Chris Adams, "Better Budgeting or Beyond Budgeting?" *Measuring Business Excellence* (2003), 7:3, 22–28; Thomas A. Stewart, "Why Budgets Are Bad for Business," *Fortune,* June 4, 1990, 179–190.

14. David Solomons, "Flexible Budgets and the Analysis of Overhead Variances," *Management International Review, Special Issue* (1992), 32:83–91.

15. Henry L. Tosi, Jr., "The Human Effects of Budgeting Systems on Management," *MSU Business Topics,* Autumn 1974, 53–63.

16. Eugene F. Brigham and Michael C. Ehrhardt, *Financial Management: Theory and Practice, 13th ed.* (Mason, OH: Cengage Learning, 2011).

17. For an example of the use of ratios in the public sector, see "Selected Financial and Operating Ratios," *Public Power,* January 1, 1993, 46–55.

18. Kenneth D. Duft, "Financial Ratio Analysis: An Aid to Agribusiness Management." Department of Agricultural Economics, W.S.U. (November 1968, Appendix I); Raphael Amit and Joshua Livnat, "Grouping of Conglomerates by Their Segments' Economic Attributes: Towards a More Meaningful Ratio Analysis," *Journal of Business Finance & Accounting* (Spring 1990), 85–100.

19. Brigham and Ehrhardt, *Financial Management.*

20. Alejandra Diaz and Rosario Uría, *Good Manufacturing Practices: A Guide for Small and Medium-Scale Agribusiness Operators* (San Jose, Costa Rica: Inter-American Institute for Cooperation on Agriculture, 2009); Douglas A. Clarke and William R. Pasewark, "Establishing Quality Control for Audit Services," *National Public Accountant* (December 1, 1989), 40–44.

21. "Auditors of Corporate Legal Bills Thrive," *The Wall Street Journal,* February 13, 1991, B1.

22. Wanda Wallace, "Internal Auditors Can Cut Outside CPA Costs," *Harvard Business Review* (March–April 1984), 16.

23. Wendell French, *Human Resource Management, 6th ed.* (Boston: Houghton Mifflin, 2006).

24. *Ibid.*

25. William M. Pride and O. C. Ferrell, *Marketing 2008, 14th ed.* (Mason, OH: Cengage South-Western, 2007).

26. "Companies Get on Fast Track to Roll Out Hot New Brands," *The Wall Street Journal,* July 10, 1986, p. 25.

# GLOSSARY: MANAGER'S VOCABULARY

## A

**absenteeism** The percentage of an organization's workforce that does not come to work on a given day. (Ch. 16, App. B)

**absolute quality** The generally understood level of quality that a product or system needs to be capable of fulfilling its intended purpose, rather than its quality relative to other alternatives. (Ch. 22)

**acceptance sampling** A quality-control process whereby finished goods are sampled to determine what proportion of them meet the pre-established quality level. (Ch. 22)

**accountability** The answerability for actions, decisions, and performance. (Chs. 12, 21)

**achievement need** The desire to excel or to accomplish some goal more effectively than in the past. (Ch. 18)

**action plan** A blueprint or framework that specifies decisions that call for actions to complete a specific project. (Ch. 10)

**adaptation model** An approach to business strategy that suggests managers should focus on solving three basic managerial problems: entrepreneurial, engineering, and administrative problems. (Ch. 9)

**adjourning** The final stage of a group, when it completes a project or breaks up for some other reason. (Ch. 19)

**administrative management** The subarea of classical management theory that is concerned with how organizations should be put together. (Chs. 2, 3)

**administrative managers** Generalists who oversee a variety of activities in several different areas of the organization. (Ch. 2)

**administrative problem** A managerial situation in strategic planning that involves structuring an organization. (Ch. 9)

**affiliation need** The psychological need to work with others, to interact, and to have friends. (Ch. 18)

**agribusiness** Economic activity in the food and fiber system, related to the production, distribution, and consumption of food, clothing, and even shelter; encompasses the input supply industries, agricultural production industries, and post-harvest, value-added activities such as commodity processing, food manufacturing, and food distribution. (Ch. 1)

**agribusiness value chain** Agribusiness activities that begin with raw materials and commodities of relatively low value, with each step in the chain then modifying the material or commodity to add value to the product. (Ch. 1)

**analyzing** One of the three strategic business alternatives in the adaptation model, it involves keeping a core set of products that provide predictable revenues (**defending**) while systematically looking for new opportunities (**prospecting**). (Ch. 9)

**application form** A standardized form for collecting information from job applicants about their background, education, experience, and so on. (Ch. 15)

**arbitrator** A labor law specialist paid jointly by a union and an organization to listen to both sides of a labor dispute and then decide how to settle the dispute. (Ch. 15)

**artificial intelligence (AI)** Attempts to have computers simulate human-decision processes. (Ch. 24)

**assessment centers** Employee selection techniques that allow human-resource managers to observe and evaluate a prospective employee's performance on simulated tasks, such as decision making and time management. (Ch. 15)

**assets** Things of value that a company owns. (App. B)

**attitudes** Sets of beliefs and feelings that individuals have about specific ideas, situations, or other people. (Chs. 16, 20)

**audit** An independent appraisal, internally or externally, of an organization's accounting, financial, and operational systems. (App. B)

**authoritarianism** The extent to which a person believes that power and status differences are appropriate within organizations. (Ch. 16)

**authority** The power to carry out an assignment. (Ch. 12)

**automation** The use of machinery, especially computers and robots, to replace human labor as part of the technological transformation process. (Chs. 5, 23)

**avoidance or negative reinforcement** A method of reinforcement that allows an employee, because of otherwise good performance, to escape from an unpleasant situation. (Ch. 18)

## B

**balance sheet** A cross-sectional picture of an organization's financial position (assets, liabilities, and equity) at a given time. (App. B)

**balance sheet budget** A projection of what the assets and liabilities of a firm will look like at the end of the coming period. (App. B)

**behavioral model** An approach to decision making recognizing that managers do not always have complete information and are limited in their abilities to make rational decisions. (Ch. 11)

**behavioral school** The school of management thought that focuses on the role of the individual in the workplace. (Ch. 2)

**benchmarking** The process of finding out, in a legal and ethical manner, how other firms do something and then either imitating or improving on it. (Ch. 22)

**benefits** Payments to employees beyond wages or salaries, such as healthcare, life insurance, vacation time, sick leave. (Ch. 15)

**boundary spanner** Someone working for an organization, like a sales representative or a purchasing agent, who spends much of her or his time in contact with others outside the organization. (Ch. 4)

**bounded rationality** The notion that, in decision making, a manager's attempts at rationality are constrained by values and experiences and by unconscious reflexes, skills, and habits. (Ch. 11)

525

**brainstorming** The process of bringing people together and encouraging a free and open discussion of creative solutions to a problem. (Ch. 11)

**breakeven analysis** A planning technique that determines the point at which revenues and costs will be equal. (Ch. 10)

**budget** A financial plan expressed in quantitative terms. (App. B)

**budgeting process** The manner in which an organization goes about obtaining numbers from various departments for developing a budget. (App. B)

**bureaucracy** An organizational form based on rational rules and guidelines, with an emphasis on mechanical competence. (Chs. 2, 13)

**burnout** A feeling of psychological exhaustion that may develop when someone experiences too much stress for an extended period of time. (Ch. 16)

**business strategy** Strategic plans that chart the course for each individual business or division within a company. (Ch. 9)

**"buy national" legislation** Laws that give preference to domestic producers through content or price restrictions. (Ch. 6)

## C

**capacity** The space that an organization has available to create products and/or services. (Ch. 23)

**capital expenditure budget** A plan that details how an organization will pay for the acquisition of major assets, such as new equipment, entire plant facilities, and land. (App. B)

**capital-intensive technology** Using machines rather than labor to do most all of the work.

**cash cow** A product that controls a large share of a low-growth market. (Ch. 9)

**cash flow budget** A budget that outlines precisely where money will come from and when, plus how and when it will be used. (App. B)

**central processor** That hardware within a computer system that carries out the instructions of a computer program. (Ch. 24)

**centralization** Keeping power and control at the top level of the organization. (Ch. 12)

**certainty** Knowing exactly what the alternatives are and that each alternative is guaranteed. (Ch. 11)

**channels** The means of transmitting an idea, such as a face-to-face meeting, letter, telephone call, electronic message, or facial expression. (Ch. 20)

**charisma** An intangible attribute in a leader's personality that inspires loyalty and enthusiasm. (Ch. 17)

**chief information officer (CIO)** The executive who oversees all aspects of information technology, such as computing, office systems, and telecommunications. (Ch. 20)

**classical school** The school of management thought that emerged around the turn on the century and is composed of two subareas: scientific management and administrative management. (Ch. 2)

**codes of conduct** Meaningful symbolic statements about the importance of ethical behavior. (Ch. 7)

**coercive power** The ability to force compliance through psychological, emotional, or physical threat. (Ch. 17)

**cognitive dissonance** A situation that occurs when there is a conflict among the three parts of an attitude. (Ch. 16)

**cohesiveness** The extent to which members of a group are motivated to remain together. (Ch. 19)

**collective bargaining** Negotiating a written, mutually-binding contract covering all relevant aspects of the relationships between an organization and members of a union. (Ch. 15)

**commitment** The favorable response of persons confronted with an attempt by their leader to influence them. (Ch. 17)

**committee** A special type of task group. (Ch. 19)

**commodity** A fungible good with little or no qualitative differences; that is, it is essentially the same no matter who produces it. (Ch. 1)

**communication** The process of transmitting information. (Ch. 20)

**communication overload** A communication difficulty that occurs when the sender is transmitting too much information for the receiver to process adequately. (Ch. 20)

**company productivity** The amount produced or created by all organization members and/or units combined. (Ch. 22)

**compensation** Wages or salaries paid to employees for their services. (Ch. 15)

**competitive challenges** Difficulties in efforts to gain an advantage in acquiring scarce resources. (Ch. 5)

**competitive forces** Influences on an organization that include the threat of new entrants, jockeying among contestants, the threat of substitute products, the power of buyers, and the power of suppliers. (Ch. 4)

**competitor intelligence** The active approach of scanning public information about competitors. (Ch. 10)

**competitors** Other organizations that compete with your organization for resources. (Ch. 4)

**compliance** Going along with the boss's request but without any stake in the results. (Ch. 17)

**computer-aided design (CAD)** The use of computers to design parts and complete products and to simulate performance so that prototypes need not be constructed. (Ch. 23)

**computer-aided manufacturing (CAM)** Manufacturing that ensures that a design moves smoothly to production. (Ch. 23)

**computer-assisted manufacturing** Technology that relies on computers to design or manufacture products. (Ch. 23)

**computer-integrated manufacturing (CIM)** The use of computers to adjust machine placements and settings automatically to enhance both the complexity and the flexibility of scheduling (Ch. 23)

**conceptual modeling** The formation of patterns and models in an efficient way for problem solving, especially in looking for and handling information. (Ch. 10)

**conceptual skills** The skills a manager needs for thinking in the abstract, such as seeing relationships between forces, understanding how a variety of factors interrelate, taking a global perspective of the organization and its environment, and defining and understanding situations. (Ch. 3)

**concurrent control** Monitoring the activities that occur as inputs are being transformed into outputs; also known as yes/no control and screening control. (Ch. 21)

**conflict of interest** A situation where a person's decision may be compromised because of competing loyalties. (Ch. 7)

**conglomerate design** An organization design that uses the product/process form of departmentalization; typically found in an organization that has grown through the development of new and perhaps relatively unrelated product lines. It is also called the **H-Form** because such conglomerates may be holding companies for groups of diverse products. (Ch. 13)

**contingency approach** A management theory stating that appropriate managerial actions depend or are contingent on the elements of each situation; attempts to specify circumstances under which different kinds of leadership behavior are appropriate. It is also called **situational approach**. (Chs. 2, 17)

**contingency planning** Identifying alternative courses of action that may be followed if various conditions arise. (Chs. 8, 9)

**continuous-process technology** The set of processes used when making a product requires that the composition of the raw materials be changed mechanically or chemically. (Ch. 13)

**continuous reinforcement** Rewarding a person's behavior after every occurrence of the desired behavior. (Ch. 18)

**control system** Part of the computerized information system that involves some form of computer software as well as human monitoring to ensure that the software and the system are functioning as planned. (Ch. 21, 24)

**controller** A high-level manager who has specific responsibility for control; helps line managers, coordinates the overall control system, and gathers important information and relays it to all managers. (Ch. 21)

**controlling** The process of monitoring and adjusting organizational activities toward goal attainment. (Ch. 21)

**conventional automation** The use of mechanical or electromechanical devices to control machines. (Ch. 23)

**cooperatives (co-ops)** Organizations owned and controlled by members who pool their resources to maximize their benefits; all members share profits , which are usually distributed as patronage-based refunds at the end of the fiscal year. (Ch. 13)

**core industries** The input supply industries, production agriculture industries, and value-added activities in agribusiness; they use the materials and services of certain ancillary industries. (Ch. 1)

**corporate culture** The shared experiences, stories, beliefs, norms, and actions that characterize an organization. (Ch. 13)

**corporate ownership** Owning a share of the business by purchasing the company's stock. (Ch. 5)

**corporate social audit** A formal and thorough analysis of the effectiveness of a firm's social performance. (Ch. 7)

**corporate strategy** Strategic plans that chart the course for the entire organization and attempt to answer the question of what businesses the organization should be in. (Ch. 9)

**cost control** Identifying areas in which costs are too high and reducing them appropriately. (Ch. 23)

**country productivity** The amount created or produced by an entire country. (Ch. 22)

**creativity** A way of thinking that generates new ideas or concepts. (Ch. 11)

**critical path** The longest path of connecting tasks in a PERT chart as a project moves toward completion. (Ch. 10)

**culture** The set of values that helps members understand what the organization stands for, how it does things, and what it considers important. (Ch. 4)

**current ratio** A liquidity ratio that is determined by dividing current assets by current liabilities to determine whether the firm can meet its short-term liabilities. (App. B)

**customers** Individuals or companies who pay money to acquire an organization's product or service. (Ch. 4)

**cutbacks** Reductions in the scope of an organization's operations. (Ch. 5)

## D

**data** Facts and figures, unorganized pieces of information, useless until processed or organized in some way. (Ch. 24)

**debt ratio** A ratio that reflects a firm's ability to handle its long-term debt. The most common debt ratio is found by dividing total liabilities by total assets. (App. B)

**decentralization** The result of shifting, through delegation, some power and control to lower levels in an organization. (Ch. 12)

**decision framing** The tendency of people to let the way that a decision is stated, either as a potential gain or loss, affect their choice of risky or cautious alternatives. (Ch. 11)

**decision making** The process of choosing one alternative from among a set of potentially feasible alternatives. (Ch. 11)

**decision support system (DSS)** A newer form of information system that automatically searches for, analyzes, summarizes, and reports information needed by a manager for a particular decision. (Ch. 24)

**decision tree** A decision-making technique, basically an extension of payoff matrices that diagrams alternatives including first-, second-, and third-level outcomes that can result. (Ch. 11)

**decisional roles** The roles managers play when making decisions, including entrepreneur, disturbance handler, resource allocator, and negotiator. (Ch. 3)

**declarative knowledge** Comprehension of facts and definitions. (Ch. 3)

**decoding** Translating an encoded message. (Ch. 20)

**defending** A conservative strategic business alternative that involves defining a market niche and then defending it from competitors. (Ch. 9)

**delegation** Assigning a portion of one's tasks to subordinates; involves giving an employee the responsibility for a job, the authority to perform it, and accountability for seeing that it gets done. (Ch. 12)

**Delphi forecasting** The systematic refinement of forecasting that takes advantage of expert opinion to make various predictions. (Chs. 10, 19)

**departmentalization** Grouping jobs according to function, product, or location. (Ch. 12)

**developing economy** An economy that is relatively underdeveloped and immature, characterized by weak industry, weak currency, and relatively poor consumers. (Ch. 6)

**development** Training for managers or other employees. (Ch. 15)

**devil's advocate strategy** Taking issue with the actions of the rest of a group to prevent groupthink and to stimulate creative ideas. (Ch. 19)

**differentiation** The process of setting the firm's products apart from those of other companies on some basis, such as quality, style, or service. (Ch. 9)

**direct investment** Building or purchasing operating facilities or subsidiaries in a different country from where the firm is headquartered. (Ch. 6)

**disruptive technology** A technological innovation that radically transforms markets, creates wholly new markets, or destroys existing markets. (Ch. 1)

**disseminator** The role a manager plays when relaying information from monitoring to the appropriate people in the organization. (Ch. 3)

**distinctive competency** The advantage or advantages the firm holds relative to its competitors. (Ch. 9)

**distribution model** A quantitative decision-making technique that helps managers plan routes for distributing products by minimizing travel time, fuel expenses, and so on. (Ch. 11)

**disturbance handler** The role that a manager plays when resolving conflicts between groups of employees, between a sales representative and an important customer, or between another manager and a union representative. (Ch. 3)

**diversity training** The acquisition of knowledge, skills, and competencies that is specifically designed to better enable members of an organization to function in a diverse workplace. (Ch. 5)

**divisional design** An organization design that establishes fairly autonomous product departments that operate as strategic business units; also called the **M-Form** for its multi-divisional characteristics. (Ch. 13)

**dogmatism** The rigidity of a person's beliefs and his or her openness to other viewpoints. (Ch. 16)

**dogs** Products with a small share of a stable market and hence may not be salvageable since their growth must come at the expense of competing products or businesses. (Ch. 9)

**domestic business** A company that acquires essentially all of its resources and sells all of its products or services within a single country. (Ch. 6)

**downsizing** A planned reduction in organizational size. (Ch. 5)

## E

**economic challenges** Various forces and dynamics associated with the economic system within which the organization competes. (Ch. 5)

**economic environment** The totality of economic factors, such as employment, income, inflation, interest rates, productivity, and wealth that influence the buying behavior of consumers and firms. (Ch. 4)

**effectiveness** Doing the right things in the right way at the right times. (Ch. 2)

**efficiency** Operating in a way that does not waste resources. (Ch. 2)

**efficiency emphasis** Seeking stability and direction by helping managers use resources wisely to make the organization as efficiently as possible. (Ch. 3)

**employee stock ownership plan (ESOP)** The transfer of stock ownership to employees in an effort to increase their commitment, involvement, and motivation. (Ch. 13)

**enacted role** How a person in a group actually behaves, not how others perceive him or her to behave. (Ch. 19)

**encoding** Translating a message into the exact mix of words, phrases, sentences, pictures, or other symbols that best reflect the content of a message. (Ch. 20)

**engineering problem** The managerial problem in strategic planning that involves determining which business opportunities to undertake, which to ignore, and so forth. (Ch. 9)

**entrepreneur** The role a manager plays when taking the lead in looking for opportunities that the organization can pursue. (Ch. 3)

**entrepreneurial problem** The managerial problem in strategic planning that involves determining which business opportunities to undertake and which to ignore. (Ch. 9)

**entrepreneurship** The capacity and willingness to undertake the conception, organization, and management of a new venture with all attendant risks while seeking profit as a reward. (Ch. 5)

**entropy** The negative result, faltering and dying, that occurs when an organization takes a closed-system approach to management. (Ch. 2)

**environmental change** The extent to which forces outside an organization alter and the rate of such alterations. (Ch. 13)

**environmental complexity** The condition that exists when the organization's environment contains many different elements, which in turn leads to a high level of complexity and therefore more environmental uncertainty. (Ch. 13)

**environmental opportunities** Aspects of an organization's environment that, if acted upon properly, would enable the organization to achieve higher than planned levels of performance. (Ch. 9)

**environmental scanning** A proactive approach to monitoring the environment through observation, reading, and so forth, to spot trends and anticipate changes that may affect an organization at the strategic and tactical levels of planning. (Chs. 4, 10)

**environmental threats** Aspects of an organization's environment that, if not countered in some way, would impede the organization's progress in achieving its goals. (Ch. 9)

**environmental uncertainty** The condition that exists when an organizational environment is changing frequently or contains many different elements. (Ch. 13)

**equifinality** The idea that two or more paths may lead to the same goal by pursuing different means. (Ch. 2)

**equity** Fairness in the workplace (Ch. 18) or Capital contributed to an organization in exchange for sharing the firm's profits. (App. B)

**equity positions** Ownership positions obtained through the purchase of significant portions of stocks. (Ch. 13)

**escalation of commitment** The tendency of people to continue to pursue an ineffective course of action even when current information indicates that the project will fail. (Ch. 11)

**ESOP or employee stock ownership plan** The transfer of stock ownership to employees in an effort to increase their commitment, involvement, and motivation. (Ch. 13)

**esteem needs** The need for recognition and respect from others, and the needs for self-respect and a positive self-image. (Ch. 18)

**Ethical compliance** The extent to which the members of an organization follow basic ethical (and legal) standards of behavior. (Ch. 7)

**ethics** Standards or morals that a person sets for himself or herself regarding what is good and bad or right and wrong. (Ch. 7)

**ethnicity** The ethnic (cultural, national, racial, tribal) composition of a group or organization. (Ch. 5)

**European Union (EU)** The mature market system that consists of Denmark, the United Kingdom, Portugal, the Netherlands, Belgium, Spain, Ireland, Luxembourg, France, Germany, Italy, and Greece. (Ch. 6)

**events** Points in time on a PERT chart when a project begins, when a set of tasks are completed, or when the project is completed. (Ch. 10)

**expectancy model** A comprehensive model of employee motivation based on the assumption that motivation is determined by how much we want something and how likely we think we are to get it. (Ch. 18)

**expected role** How others in a group expect a given person to behave. (Ch. 19)

**expected value** The sum of all possible outcomes of an alternative multiplied by their respective probabilities. (Ch. 11)

**expense budget** A summary of the projected expenses for the organization over the coming period. (App. B)

**expert power** Power that is based on knowledge and expertise. (Ch. 17)

**expert systems** Computer-based programs that build on a series of rules (frequently "if-then" rules) to move from a set of data to a decision recommendation. (Chs. 10, 24)

**export restraint agreements** Agreements designed to convince other governments to voluntarily limit the volume or value of goods exported to a particular country. (Ch. 6)

**exporting** Making the product in the firm's domestic marketplace and selling it in another country. (Ch. 6)

**external audit** An audit conducted by experts who do not work directly for the particular organization, but rather for an accounting firm. (App. B)

**external environment** Everything outside an organization that might affect it. (Ch. 4)

**external focus** Attempting to make an organization competitive by focusing on the external needs or forces of its environment, such as organization growth and productivity. (Ch. 3)

**extinction** Weakening someone's behavior by no longer reinforcing (e.g., ignoring) previously reinforced behavior. (Ch. 18)

## F

**facilities** The physical means an organization uses to create products and/or services. (Ch. 23)

**family orientation** Starting a business with an intent to keep ownership of the business in the family. (Ch. 1)

**feedback (performance)** Telling an employee the results of his or her job performance appraisal. (Ch. 15)

**feedback (communication)** Response from the receiver of a message to the sender of that message. (Ch. 20)

**figurehead** The symbolic role a manager plays when appearing as a representative of an organization to perform routine duties of a legal or social nature. (Ch. 3)

**final consumption activities** Agribusiness services or activities such as those performed by restaurants and groceries to make outputs available to buyers; related to the final use of agribusiness products. (Ch. 1)

**finance manager** The individual responsible for the financial assets of an organization, including overseeing the accounting systems, managing investments, controlling disbursements, and providing relevant information to the CEO about the firm's financial condition. (Ch. 2)

**financial budgets** Details of where an organization intends to get its cash for the coming period and how it intends to use it; that is, cash flow budget, capital expenditure budget, and balance sheet. (App. B)

**financial resources control** Monitoring and ensuring that enough but not too much cash is on hand and that other finance-related areas are under control (inventory, salaries, etc.). (Ch. 21)

**financial strategy** The functional strategy related to monetary matters such as dividends, retained earnings, debt financing, and equity financing. (Ch. 9)

**finished goods inventory** Products that have been completely assembled but have not yet been shipped. (Ch. 23)

**first-line managers** People who manage operating employees, including supervisors, department managers, office managers, and foremen. (Ch. 2)

**fixed costs** Costs that are incurred regardless of the level of operations (e.g., rent, taxes, and contracted salaries). (App. B)

**fixed-interval schedule** Providing reinforcement rewards on a periodic basis, regardless of performance (e.g., giving a paycheck every Friday). (Ch. 18)

**fixed-ratio schedule** Providing reinforcement rewards on the basis of number of behaviors rather than the basis of time. (Ch. 18)

**flat organization** An organization that has relatively few levels of management. (Ch. 12)

**flexibility** How easily an organization can react, respond, or change. (Ch. 22)

**flexibility emphasis** Using means such as employee participation and innovation in seeking to keep an organization flexible enough to adapt to environmental forces. (Ch. 3)

**flexible manufacturing system (FMS)** Systems that have robotic work units or work stations, assembly lines, and robotic carts to move material. (Ch. 23)

**force-field analysis** Systematically looking at the pluses and minuses associated with a planned organizational change from the standpoint of the employees, and then attempting to increase the pluses and decrease the minuses. (Ch. 14)

**forecasting** Systematically developing predictions about the future. (Ch. 10)

**forming** The initial stage of group development; involves the coming together of would-be members to form a group. (Ch. 19)

**front-end-back-end organization** Organizations that simultaneously use product departmentalization and customer departmentalization. (Ch. 13)

**functional departmentalization** Grouping together employees who are involved in the same or very similar functions, such as marketing or finance. (Ch. 12)

**functional design** The most common organization design; based on functional specialization. It is also known as the U-Form because it uses a unitary or uniform approach to design. (Ch. 13)

**functional group** A group created by an organization to accomplish a range of goals within an indefinite time horizon. (Ch. 19)

**functional strategies** Strategic plans that correspond to each of the functional areas within an organization: marketing, finance, production, human resources, organization design, and research and development. (Ch. 9)

## G

**Gantt chart** A bar chart with each bar representing a project task, depicting the overlap of scheduled tasks. (Ch. 10)

**GAP, GHP, GMP** Recommended food-safety guidelines intended to reduce contamination of consumable products during growing, manufacturing, and other handling stages. (Ch. 8, 21)

**General Adaptation Syndrome (GAS)** A cycle of alarm, resistance, and exhaustion that stress usually follows. (Ch. 16)

**general environment** The broad dimensions and forces in an organization's surroundings that provide opportunities and impose constraints on the organization. (Ch. 4)

**generic strategy** An overall framework for action developed at the corporate level; the three basic generic strategies are growth, retrenchment, and stability. It is also called grand strategy. (Ch. 9)

**global business** A company that transcends national boundaries and is not committed to a single home country. (Ch. 6)

**global challenges** Difficulties or challenges that come from international competition. (Ch. 5)

**global design** A relatively decentralized organization design that allows the necessary coordination and integration for worldwide business while also enabling the flexibility and autonomy necessary to compete in regional and local markets. (Ch. 13)

**goal** A desired state or condition that an organization wants to achieve. It provides direction for an organization. (Ch. 8)

**goal optimization** The process of trading off among different goals and achieving an effective balance among the different goals of an organization. (Ch. 8)

**goal setting** A six-part process resulting in a set of consistent and logical goals that permeate an entire organization. The process involves scanning the environment for opportunities and threats, assessing organizational strengths and weaknesses, establishing general organizational goals, setting unit and subunit goals, and monitoring programs toward goal attainment at all organizational levels. (Ch. 8)

**goal-setting theory** The use of goal setting to increase individual motivation. (Ch. 18)

**grapevine** Informal communication network within a company. (Ch. 20)

**grievance** A written statement or complaint filed with the union by an employee concerning his or her alleged mistreatment by the company. (Ch. 15)

**group** Two or more people who interact regularly to accomplish a common goal. (Ch. 19)

**groupthink** The phenomenon that happens when the maintenance of cohesion and good feelings of group members becomes more important than the original purpose of the group. (Ch. 19)

**growth strategy** A strategic plan of actively seeking to acquire other related businesses when an organization wants to generate high levels of growth in one or more areas of operation. (Ch. 9)

## H

**H-form** An organization design that uses the product/process form of departmentalization; typically found in an organization that has grown through the development of new and perhaps relatively unrelated product lines. It is also called the conglomerate design because such conglomerates may be holding companies for groups of diverse products. (Ch. 13)

**hardware** The central processor and the output devices of a computer. (Ch. 24)

**Hawthorne studies** A series of early research studies (conducted at the Hawthorne plant of Western Electric between 1927 and 1932) of the human element in the workplace, that provided the catalyst for the behavioral school. (Ch. 2)

**high-involvement management** All forms of participative management, including reliance on self-control and self-management at the lowest levels and such techniques as quality circles, work teams, and new design plants. (Chs. 2, 18)

**high-involvement organization** Organization design based on a process, orientation, open communications, a low level of functional specialization and standardization, and cooperation. (Ch. 13)

**hollow corporation** An emerging organizational design for companies that usually engage in outsourcing (the contracting of many of their usual functions to other firms); also called network organization and value-added partnership. (Ch. 13)

**horizontal communication** Transmission of messages between two or more colleagues or peers at the same level in the organization. (Ch. 20)

**hostile takeover** A corporation or group of investors buys or trades for enough stock to gain control over a company that does not wish to be taken over. (Ch. 5)

**human relations** Recognizes that people have their own unique needs and motives that they bring to the workplace with them. (Ch. 2)

**Human resource (HR) managers** Managers responsible for determining and filling human resource needs. (Ch. 2)

**human resource strategy** The functional strategy that deals with employee-related issues such as hiring and retention options, unions, employee development, and compliance with federal employment regulations. (Ch. 9)

**human resources** The department that recruits applicants, trains and develops new employees, and evaluates performance; formerly called personnel. (Ch. 2, 15)

**human resources control** Monitoring and ensuring proper focus on such activities as employee selection and placement, training, performance appraisal, and compensation (paying neither too much nor too little). (Ch. 21)

**human resources model** Emphasizes flexibility in adapting to internal organizational changes by focusing on the people inside the organization. (Ch. 3)

**human skills** The skills a manager needs to work with other people. (Ch. 3)

**Hybrid** Refers to combining different aspects in the same object or organization. (Ch.13)

**hybrid designs** Organizational designs in which different parts of the same organization have different bases of departmentalization. (Ch. 13)

## I

**importing** Bringing a product, service, or capital into the home country from abroad. (Ch. 6)

**in-process sampling** A quality-control process that involves testing products as they are being made rather than after they are finished. (Ch. 22)

**in-transit inventory** Goods that have been shipped from the company but have not yet been delivered to the customer. (Ch. 23)

**inappropriate focus** Occurs when the control is too narrowly focused or does not provide a reasonable balance between different outcomes that are important.

**income statement** A summary of income and expenses over a period of time. (App. B)

**incremental innovation** A product or technology that modifies an existing product or technology. (Ch. 14)

**individual differences** Personal attributes that vary from one person to another, such as personality, attitudes, and perception. (Ch. 16)

**individual productivity** The amount produced or created by a single individual relative to his or her costs to the organization. (Ch. 22)

**industry productivity** The amount produced or created by all companies in a particular industry. (Ch. 22)

**informal groups** A group created by members of the organization for purposes that may or may not be related to the organization and has an unspecified time horizon; also called an interest group. (Ch. 19)

**informal organization** The overall pattern of influence and interaction defined by the total set of informal groups within the organization. (Ch. 19)

**information** Data organized in a meaningful way. (Ch. 24)

**information age** The period of time when the focus began to shift from computers and technology (1940s and 1960s) to the applications for information (1980s). (Ch. 24)

**information requirement analysis** The process of determining what information each individual needs to perform his or her job. (Ch. 24)

**information resources control** Monitoring and ensuring that various forecasts and projections are prepared accurately and on a timely basis, that managers have access to the information they need to make decisions effectively, and that the proper image of the organization is projected to the environment. (Ch. 21)

**informational role** The role a manager plays when serving as a monitor, disseminator of information, or spokesperson. (Ch. 3)

**infrastructure** The basic physical and organizational structures needed to facilitate the production of goods and services for the operation of a society or enterprise; communications, transportation, roads and rail network. (Chs. 1, 6)

**innovation** The process of creating and developing new products or services and/or identifying new uses for existing products or services. (Chs. 5, 11, 14)

**innovation decline** The stage during which demand for an innovation decreases and substitute innovations are developed and applied. (Ch.14)

**innovation development** The stage in which an organization evaluates, modifies, and improves a potential innovation before turning that idea into a product or service to sell. (Ch.14)

**innovation maturity** The stage in which most organizations in an industry have access to an innovation and are applying it in approximately the same way. (Ch.14)

**input device** A piece of equipment that puts data into a computerized information system where it can be analyzed or processed, stored, and made available to users. (Ch. 24)

**inputs** Raw materials, seeds, fertilizer, capital, equipment, and other things that are used in the production of agribusiness products. (Ch. 1)

**integrative framework** A process model to help managers better understand how they must move along two continua: internal-external focus and the efficiency-effectiveness emphasis. (Ch. 3)

**interest group** Members who organize as a group in an attempt to influence their organization. (Ch. 4)

**intermediate activities** Grading, storage, processing, packaging, distribution, pricing, and marketing. (Ch.1)

**intermediate planning** Formulating goals and plans for a time horizon of one to five years. (Ch. 8)

**internal audit** Audits conducted by people who work directly for the organization. (App. B)

**internal environment** An organization's culture. (Ch. 4)

**internal focus** Attempting to maintain an organization by focusing on the internal needs or forces of its environment with activities such as employee participation in decision-making and the management of information. (Ch. 3)

**internal recruiting** Finding current employees who would like to change jobs rather than searching for qualified applicants outside the company. (Ch. 15)

**international business** A company that is primarily based in a single country but acquires some meaningful share of its resources or revenues from other countries. (Ch. 6)

**international economic community** A set of countries that agrees to significantly reduce or eliminate trade barriers among its member nations. (Ch. 6)

**international environment** Forces that extend beyond national boundaries. (Ch. 4)

**internships** Temporary jobs that provide work experience for students who are potential employees. (Ch. 15)

**interpersonal communication** The exchange and flow of information and ideas between people. (Ch. 20)

**interpersonal roles** The role that a manager plays when serving as a figurehead, leader, or liaison. (Ch. 3)

**interview** Face-to-face talk between a manager and a prospective employee during the employee selection process. (Ch. 15)

**intrapreneur** Similar to entrepreneurs, except that they work in the context of a large organization. (Ch. 14)

**intrapreneurship** The process of starting new ventures within a larger organization. (Ch. 5)

**inventory control** Monitoring inventory to ensure that there is an adequate supply of raw materials to transform into products, an adequate supply of inventory in process, and enough finished goods to ship to customers. (Ch. 23)

**inventory models** A quantitative decision-making technique that helps the manager plan the optimal level of inventory to carry. (Ch. 11)

## J

**jargon** Words that have specific meaning within a profession or group of people. (Ch. 20)

**job analysis** The systematic investigation of the nature of a job, which results in a job description and an identification of the skills and credentials needed to perform the job, known as a job specification. (Ch. 15)

**job description** A summarization of the duties encompassed by a job; the working conditions where the job is performed; and the tools, materials, and equipment used on the job. (Ch. 15)

**job design** Determining what procedures and operations are to be performed by the employee in each position. (Ch. 12)

**job enlargement** Expanding a job by adding more activities for the worker to perform. (Ch. 12)

**job enrichment** Attempting to make a job more satisfying by giving the worker more activities to perform and more discretion in deciding how to perform various activities. (Ch. 12)

**job evaluation** Determining the relative value of a job within an organization, primarily to establish the proper wage structure. (Ch. 15)

**job posting** Listing of job openings inside the company to find current employees who would like to transfer or be promoted rather than searching for qualified applicants outside the company. It is also called **internal recruiting.** (Ch. 15)

**job rotation** Systematically moving employees from one job to another. (Ch. 12)

**job satisfaction or dissatisfaction** An attitude that reflects the extent to which a person is or is not gratified by (fulfilled) in his or her work. (Ch. 16)

**job specialization** Defining the tasks that set one job apart from others. (Ch. 12)

**job specification** A description of the skills, abilities, and other credentials necessary to perform a job. (Ch. 15)

**joint venture** Two firms sharing control and ownership of a new enterprise. (Chs. 6, 13)

**judgmental methods** Subjective, non-quantifiable evaluations of how well an employee is doing in his or her job. (Ch. 15)

**just-in-time (JIT) scheduling** Planning so products arrive at designated spots just as they are needed—to prevent having the purchaser hold them in inventory. (Ch. 23)

## L

**labor** People who work for an organization, especially when they are organized into unions. (Ch. 4)

**labor budget** A detailed listing of the number of direct hours of labor that will be available for the coming period. (App. B)

**labor productivity** A measure of productivity determined by dividing outputs by direct labor. (Ch. 22)

**labor relations** Dealing with employees when they are organized in a labor union. (Ch. 15)

**labor-intensive** Manufacturing where people, not machines, do most of the work. (Ch. 23)

**large-batch technology** The set of processes used when a product is made in assembly-line fashion by combining component parts into a finished product; also known as mass-production technology. (Ch. 13)

**leader** The role a manager plays when hiring employees, motivating them, or dealing with behavioral processes. (Ch. 3)

**leader–member relations** Relationship between the leader and the members of a group. (Ch. 17)

**leadership** An influence process that is directed at shaping the behavior of others. (Ch. 17)

**leadership behaviors** Actions that differentiate effective leaders from ineffective leaders. (Ch. 17)

**leadership style** Combinations of skills and behaviors that are used by a manager in shaping the behavior of others. (Ch. 17)

**leadership traits** Distinguishing features that were once thought to be stable and enduring characteristics that set leaders apart from nonleaders. (Ch. 17)

**leading** Guiding and directing employees toward goal attainment by motivating employees, managing group processes, and dealing with conflict and change. (Ch. 2)

**legal challenges** Forces and dynamics that reflect the judicial context in which an organization operates. (Ch. 5)

**legal compliance** The extent to which an organization conforms to local, state, federal, or international laws. (Ch. 7)

**legitimate power** Power that is created and conveyed by an organization; the same as authority. (Ch. 17)

**liabilities** A listing of an organization's debts and other financial obligations. (App. B)

**liaison** The role a manager plays when dealing with people outside the organization on a regular basis. (Ch. 3)

**licensing agreement** The allowance by one company to another company to use its brand name, trademark, technology, patent, copyrights, or other assets. (Ch. 6)

**life cycle** The stages of an organization's life over a period of years: development, growth, maturation, and decline. (Ch.13)

**line positions** Jobs that are in the direct chain of command with specific responsibility for accomplishing the goals of the organization. (Ch. 12)

**linear programming (LP)** A method for determining the optional combination of resources and activities for certain types of problems. (Ch. 10)

**liquidity ratio** An assessment of how easily the firm's assets may be converted into cash. (App. B)

**listening** The process of receiving encoded symbols from a sender and decoding them into a message for interpretation. (Ch. 20)

**lobbyist** A person who is paid to influence legislators, agencies, groups, or committees. (Ch. 7)

**locational departmentalization** Grouping together jobs that are in the same place or in nearby locations. (Ch. 12)

**locus of control** The degree to which a person believes that behavior has a direct impact on its consequences. (Ch. 16)

**long-range planning** Formulating goals and plans that span several years to several decades. (Ch. 8)

**LPC or Least Preferred Coworker model** A contingency model of leadership suggesting that appropriate leadership behavior varies as a function of the favorableness of the situation, including leader-member relations, task structure, and position power. (Ch. 17)

## M

**M-Form** An organization design known for it is multi-divisional characteristics; also called the **divisional design**; establishes fairly autonomous product departments that operate as strategic business units. (Ch. 13)

**management** A set of activities directed at the efficient and effective use of resources in the pursuit of one or more goals. (Ch. 2)

**management by objectives** A technique specifically developed to facilitate the goal-setting process in organizations. (Ch. 15)

**management information system (MIS)** An integrated and organized data bank created specifically to gather, organize, summarize, and report relevant and timely information for use by managers. (Chs. 2, 24)

**management science** The branch of the quantitative school of management that develops advanced mathematical and statistical tools and techniques for managers to enhance efficiency. (Ch. 2)

**managerial ethics** Standards or morals applied to the management of an organization. (Ch. 7)

**managerial grid** An organizational development technique that is used to assess current leadership styles in an organization and then to train leaders to practice an ideal style of behavior. (Ch. 14)

**managerial innovation** A change in the management process by which products and services are conceived, built, and delivered to customers. These changes do not necessarily affect the appearance or performance of products or services directly. (Ch. 14)

**managerial knowledge** The special information and mental activity a manager uses to decide how to behave. (Ch. 3)

**managerial models** Different schools of thought about management, each of which represents a set of assumptions and explains one part of the management process. (Ch. 3)

**managerial roles** Actions that a manager is expected to perform and ways in which he or she is expected to behave. (Ch. 3)

**managerial skills** The ability to perform the various behaviors necessary for managers to execute their roles. (Ch. 3)

**manufacturing** Combining and transforming resources into tangible outcomes or products that are then sold to others. (Ch. 23)

**maquiladoras** Light assembly plants built in northern Mexico close to the U.S. border. (Ch. 6)

**market economy** An economic system based on the private ownership of business and allowing market factors such as supply and demand to determine business strategy. (Ch. 6)

**market share** The proportion of the total market controlled by the firm's product. (App. B)

**market system** Clusters of countries that engage in high levels of trade with each other. (Ch. 6)

**marketing** The business activities involved in getting buyers to want goods and services provided by organizations, then getting those goods and services to the buyer. (App. B)

**marketing managers** Individuals who are responsible for pricing, promoting, and distributing the products and services of a firm. (Ch. 2)

**marketing strategy** The functional strategy that relates to promotion, pricing, and distribution of products and services. (Ch. 9)

**mass-production technology** The set of processes a company uses when a product is made in assembly-line fashion by combining component parts into a finished product; also known as large-batch technology. (Ch.13)

**materials requirements planning (MRP)** A technique that enables managers to organize complex delivery schedules. (Ch. 23)

**matrix design** An organization design that allows a firm to retain the efficiency of functional departments and gain the advantages of product departmentalization. (Ch. 13).

**mechanistic design** An organization design that is based on limited communication systems, a relatively high level of specialization and standardization, and more independence than cooperation. (Ch. 13)

**middle managers** Executives between top management and first-line management, including plant managers, division managers, and operation managers who implement the policies and strategies set up by top management and coordinate the work of lower-level managers. (Ch. 2)

**MIS or management information system** See **management information system (MIS)**.

**mission** The way in which an organization attempts to fulfill its purpose, which is its reason to exist. (Ch. 8)

**MNE or multinational enterprise** A company (multinational business) that has a worldwide marketplace from which it buys raw materials, borrows money, manufactures its products, and subsequently sells its products. (Ch. 6)

**models** Representations of a more complex reality. (Ch. 3)

**monitor** The role that a manager plays when actively watching the environment for information that may be relevant to the organization. (Ch. 3)

**motivation** The set of processes that determines the choices people make about their behaviors—that which makes people decide what to do. (Ch. 18)

**multinational business** A company that has a worldwide marketplace from which it buys raw materials, borrows money, manufactures its products, and subsequently sells its products. (Ch. 6)

**multiple constituencies** Different people and groups that have different interests and may want different things from the organization. (Ch. 17)

## N

**NAFTA** See North American Free Trade Agreement (NAFTA).

**need** A drive or force that initiates behavior, causes people to do things. (Ch. 18)

**need hierarchy** A variety of human needs classified into five specific groups and then arranged in a hierarchy of importance: physiological, security, social, esteem, and self-actualization. (Ch. 18)

**negotiator** The role a manager plays when attempting to work out agreements and contracts either within or outside an organization. (Ch. 3)

**network organization** An emerging organizational design for companies that usually engage in outsourcing (the contracting to other firms of many of their usual functions); also called value-added partnership or a hollow corporation. (Ch. 13)

**networking** Connecting independent computers directly together so they can function in interrelated ways. (Ch. 24)

**new plant design** A high-involvement approach that includes a long selection process of management, a physical layout designed to suggest an egalitarian and team approach, autonomous job teams, salaried employees where pay is based on skills and performance, high decentralization, and extensive training. (Ch. 13)

**new venture units** Small, semi-autonomous, voluntary work units that develop new products or ventures for companies. It is also known as "skunkworks." (Ch. 13)

**noise** Anything that disrupts the communication process. (Ch. 20)

**nominal group technique** A structured process whereby group members individually suggest alternatives that are then discussed and ranked until everyone agrees. (Ch. 19)

**nonmonetary budget** A budget that expresses important variables in terms other than dollars; for example, number of products, number of labor hours, amount of space. (App. B)

**nonprogrammed decisions** Decisions that have significant or expensive consequences or that have not occurred in the past and for which there is no established decision rule or procedure. (Ch. 11)

**nonverbal communication** Transmitting messages through body movements, facial expressions, and gestures rather than words or sounds. (Ch. 20)

**norm** A standard of behavior that a group develops for its members. (Ch. 19)

**norming** The third stage of group formation, characterized by the resolution of conflict and the development of roles. (Ch. 19)

**North American Free Trade Agreement (NAFTA)** An economic agreement between the United States, Canada, and Mexico that aims to form a unified North American market to allow easier movement of goods and services among the three countries. (Ch. 6)

## O

**objective measures of performance appraisal** Quantifiable indicators of how well an employee is performing his or her job. (Ch. 15)

**online searching** Performing computerized literature searches using public databases. (Ch. 10)

**open systems model** A managerial model that emphasizes flexibility in adapting to external changes. (Ch. 3)

**operating ratio** The total cost of goods sold relative to the average daily inventory to determine how efficiently the firm is forecasting and ordering merchandise. (App. B)

**operational planning** Narrow-focused, short time-frame planning supervised by middle managers and executed by supervisory managers; the two basic kinds of operational plans are standing plans and single-use plans. (Ch. 8)

**operations budgets** Details of the coming year's operations in financial terms; examples are the revenue budget and the expense budget. (App. B)

**operations management** The total set of managerial activities that an organization uses to transform resources into finished products and services. Focuses on the application of mathematical and statistical tools to managing an organization's processes and systems. (Chs. 2, 23)

**operations managers** Individuals who are responsible for actually creating the goods and services of the organization. (Ch. 2)

**oral communication** Transmitting messages by means of spoken word. (Ch. 20)

**organic design** An organization design based on open communication systems, a low level of specialization and standardization, and cooperation (Ch. 13)

**organization change** Alteration of the organization brought about by people, technology, communication, and competition. (Ch. 14)

**organization chart** A vertical chart composed of a series of boxes Pictures or maps of organizations—may be useful to small companies for clarifying relationships. (Ch. 13)

**organization design** The overall configuration or arrangement of positions and their interrelationships within an organization. (Ch. 13)

**organization design strategy** The functional design strategy that is concerned with how the various positions and divisions within an organization will be arranged. (Ch. 9)

**organization development** A planned, organization-wide effort intended to improve the health of the organization by systematically applying behavioral science techniques. Common techniques include the managerial grid, team building, survey feedback, third-party peacemaking, and process consultation. (Ch. 14)

**organization revitalization** A planned effort to bring new energy, vitality, and strength to an organization. (Ch. 14)

**organizational citizenship** The behavior of individuals that makes a positive overall contribution to the organization. (Ch. 16)

**organizational commitment** An attitude that reflects an individual's identification with and attachment to the firm. (Ch. 16)

**organizational constituents** People and organizations that are directly affected by the practices of an organization and that have a stake in its performance. (Ch. 7)

**organizational innovation process** The process of developing, applying, launching, growing, and managing the maturity and decline of a creative idea. (Ch. 14)

**organizational resilience** An organization's ability to recover from unexpected change. (Ch. 14)

**organizational strengths** Aspects of an organization that let it compete effectively. (Ch. 9)

**organizational weaknesses** Aspects of an organization that prevent or deter it from competing effectively. (Ch. 9)

**organizing** Grouping activities and resources in a logical and appropriate fashion. (Chs. 2, 12)

**orientation** A process wherein new employees are introduced to various types of information about their jobs and the organization. (Ch. 15)

**output budget** A projection of an organization's product output for the coming period. (App. B)

**output devices** Means, methods, devices, or equipment to make system information available to users. (Ch. 24)

**outsourcing** The contracting of many of the usual function of one firm to other firms. (Chs. 6, 13)

**overall cost leadership** A strategy of keeping costs as low as possible so the company can increase sales volume and/or market share. (Ch. 9)

**overall productivity** A measure of efficiency determined by dividing all outputs by the sum of all inputs: labor, capital, materials, information, and energy costs. It is also called total factor productivity. (Ch. 22)

**overcontrol** So much control that the independence and autonomy of employees are limited. (Ch. 21)

**owners** People, organizations, and institutions who legally control an organization, most commonly through stock in a corporation. (Ch. 4)

**owners' equity** Claims against the company's assets. (App. B)

## P

**PAC** See political action committee (PAC).

**Pacific Rim** Countries that lie along the Pacific Ocean (the Pacific Rim) and make up the Asia–Pacific Economic Cooperation (APEC), a mature market system. (Ch. 6)

**paradigm shift** A radical change in somebody's basic assumptions about or approach to something. (Ch. 1)

**participation model** A leadership model that addresses the question of how much participation employees should be allowed in making various kinds of decisions. (Ch. 17)

**path-goal model** A leadership model suggesting that leaders should attempt to determine their group members' goals and then clarify the paths to achieving those goals. (Ch. 17)

**payoff matrix** A decision-making technique that involves calculating the expected values for two or more alternatives, each of which is associated with a probability estimate. (Ch. 11)

**peak performers** Individuals who are committed to achieving results and excelling at their performances. They are not synonymous with individuals who are obsessed with or addicted to their jobs. (Ch. 10)

**people-focused change** Strategic changes that focus on the skills and performances of employees or their attitudes, perceptions, behaviors, and expectations. (Ch. 14)

**perceived role** How an individual thinks that he or she should behave in a group. (Ch. 19)

**perception** How an individual recognizes and interprets information about the environment. (Chs. 16, 20)

**performance appraisal** A review or evaluation of how well an individual is carrying out the tasks associated with his or her job. (Ch. 15, App. B)

**performance behaviors** The set of work-related behaviors that a firm expects people to display. (Ch. 16)

**performing** The final stage of group formation, when the group moves toward accomplishing its goals. (Ch. 19)

**perishable** Not capable of remaining fresh and edible. (Ch. 1)

**person-job fit** The extent to which the contributions made by an individual match the inducements offered by an organization. (Ch. 16)

**person-role conflict** The conflict a person feels when the demands of a role are incongruent with the person's preferences or values. (Ch. 19)

**personality** The relatively stable set of psychological and behavioral attributes that distinguish one person from another. (Ch. 16)

**personnel** A department that recruits qualified applicants, trains and develops new employees, and evaluates performance; also called human resources. (Ch. 15)

**PERT chart** See Program Evaluation and Review Technique (PERT chart).

**philanthropic giving** Awarding funds or other gifts to charities or other social programs. (Ch. 7)

**physical resources control** Monitoring and adjusting to ensure the right amount of inventory, production of quality products, and having the proper equipment to do the job. (Ch. 21)

**physiological needs** Things we need to survive: food, air, warmth, and clothing. (Ch. 18)

**plan** Blueprint or framework that an organization uses to describe how it expects to achieve its goals. (Ch. 8)

**planned organizational change** Anticipating possible changes in the environment and planning how the organization should most likely respond to those changes. (Ch. 14)

**planning** Establishing goals and objectives, developing strategic plans, and developing tactical plans. (Ch. 8)

**planning and decision making** Determining an organization's goals and deciding how best to achieve them. (Ch. 2)

**planning horizon** The time span across which operations managers plan. (Ch. 23)

**planning system decision** Determining how operations managers get the information they need and then provide information to others. (Ch. 23)

**planning-control link** The way in which planning and control are integrated within the business cycle. (Ch. 21)

**point system** A job evaluation method that involves assigning or awarding points to each job according to the factors that characterize that particular job. (Ch. 15)

**policies** General guidelines that govern relatively important actions. (Ch. 8)

**political action committee (PAC)** An organization that solicits funds from various individuals and other organizations, then makes contributions to political candidates to gain their favor. (Ch. 7)

**political-legal environment** That part of the general environment pertaining to the relationship between business and government, including government regulation of business, policies toward foreign trade, and investment incentives. (Ch. 4)

**portfolio approach** An approach to business strategy that views an organization as a collection of different businesses, involves identifying strategic business units (SBUs), and then classifying them in some meaningful framework. (Ch. 9)

**portfolio matrix** The portfolio approach that classifies strategic business units (SBUs) along two dimensions: market growth rate and relative market share. SBUs are classified as stars, cash cows, question marks, or dogs. (Ch. 9)

**position power** Power that comes from a leader's position within a company. (Ch. 17)

**positive reinforcement** A reward or desirable outcome that is given after a particular behavior for the purpose of obtaining that behavior again in the future. (Ch. 18)

**postaction control** Monitoring the quality and/or quantity of outputs or products as they leave a company's system. (Ch. 21)

**probability** The likelihood that an event will or will not occur. (Ch. 11)

**problem solving** Determining a course of action when faced with a nonroutine situation for which there are no established procedures that specify how to handle the problem. (Ch. 11)

**procedural knowledge** The "how to" of managerial knowledge; outlines a set of steps to be taken to accomplish a task. (Ch. 3)

**process control chart** A visual representation of an operations function. (Ch. 23)

**process knowledge** The "how something works" part of managerial knowledge; provides a manager with a mental map for a specific topic. (Ch. 3)

**product life cycle** How a product's sales volume changes over its lifetime—from development to growth, competitive shake-out, maturity, saturation, and decline. (Ch. 9)

**product or process departmentalization** A basis of organizing by grouping together financial, marketing, and production activities that are associated with individual products or processes. (Ch. 12)

**product quality** The quality of a real or tangible item. (Ch. 22)

**product or service line** The set of products or services on which an organization will concentrate. (Ch. 23)

**production strategy** The functional strategy that addresses questions concerning product quality, costs, techniques, location, efficiency, and compliance with governmental regulations. (Ch. 9)

**productivity** A measure of efficiency that indicates what is created relative to the resources used to create it. (Chs. 5, 22)

**profit budget** A presentation of the difference between projected revenues and projected expenses. (App. B)

**profit margin on sales** Dividing net income by sales to determine which products are contributing the most in profits. (App. B)

**profitability ratio** An organization's profits relative to its sales. (App. B)

**program** A single-use plan for a large set of activities. (Ch. 8)

**Program Evaluation and Review Technique (PERT chart)** A charting technique that involves identifying the various activities necessary in a project, developing a network that specifies the interrelationships among those activities, determining how much time each activity will take, and refining and controlling the implementation of the project using the network. (Ch. 10)

**programmable automation** The use of computers to control machines. (Ch. 23)

**programmed decisions** Situations which occur frequently when the decision maker can use a decision rule or company procedure to make the decision. (Ch. 11)

**project** Single-use program similar to a program but usually with a narrower focus. (Ch. 8)

**project planning tools** Techniques designed to assist in the development of an acceptable solution to a problem within a reasonable time frame and at minimum cost. (Ch. 10)

**prospecting** A strategic business alternative that involves seeking and exploring new market opportunities. (Ch. 9)

**psychological contract** The set of expectations held by an individual about what he or she will contribute to the organization and what it will provide them in return. (Ch. 16)

**punishment** Reprimands, discipline, fines, and so on that are used to shape behavior by causing a reduction in unwanted behaviors. (Ch. 18)

**purpose** An organization's reason for existence. (Ch. 8)

## Q

**quality** A measure of value; the total set of features and characteristics of a product or service that bear on its ability to satisfy stated or implied needs. (Chs. 5, 22)

**quality circle (QC)** A group of employees that focuses on how to improve the quality of products. (Chs. 13, 19)

**quality control** The attempt to make sure inputs and outputs meet desired levels of quality. (Ch. 23)

**quantitative school** The school of management thought that focuses on quantitative, or measurement, techniques and concepts of interest to managers. (Ch. 2)

**question mark** A product with a small share of a growing market, creating a question as to whether more resources should be invested in the hope of transforming the product into a star. (Ch. 9)

**queuing models** A quantitative decision-making technique that helps managers solve problems involving waiting lines to determine, for instance, the best number of operators or checkout clerks to have on duty at various times of day. (Ch. 11)

**quota** A limit on the number or value of goods that can be traded. (Ch. 6)

## R

**radical innovation** A new product or technology developed by an organization that completely replaces an existing product or technology. (Ch. 14)

**ranking** An approach to performance appraisal wherein the supervisor compares subordinates with one another and then places them in their relative positions in a continuum from high to low performance; measures employees against one another. (Ch. 15)

**rating** An approach to performance appraisal wherein the supervisor compares a subordinate with one or more absolute standards and then places that employee somewhere in relation to that standard; compares employees with a standard of performance rather than against one another. (Ch. 15)

**ratio analysis** Calculating and evaluating any number of ratios of figures obtained from an organization's balance sheet and income statement. (App. B)

**rational model** An approach to decision making that assumes that managers are objective, have perfect information, and consider all alternatives and consequences. (Ch. 11)

**raw materials inventory** The materials, parts, and supplies an organization requires to meet its production needs. (Ch. 23)

**reactive change** Modifications that occurs within an organization as a result of external (environmental) events rather than being planned. (Ch. 14)

**receiver** Person who receives a message from a sender. (Ch. 20)

**recruiting** The process of attracting a pool of qualified applicants who are interested in working for the organization. (Ch. 15)

**referent power** Authority and capacity based on personal identification, imitation, and charisma. (Ch. 17)

**regulators** Units in the task environment that have the potential to control, regulate, or influence an organization's policies and practices. (Ch. 4)

**regulatory agencies** Groups created by the government to protect the public from certain business practices or to protect organizations from one another. (Ch. 4)

**reinforcement** Rewarding people's current behavior to motivate them to continue that behavior. (Ch. 18)

**research and development limited partnerships (RDLPs)** Consortia, usually among high technology firms, designed to do basic research. (Ch. 13)

**research and development strategy** The functional strategy that relates to the invention and development of new products and services as well as the exploration of new and better ways to produce and distribute existing ones. (Ch. 9)

**resistance** The negative, uncooperative response of persons when their bosses attempt to influence them. (Ch. 17)

**resource allocator** The role a manager plays when determining how resources (e.g., dollars, personnel, space) will be divided among different areas within the organization. (Ch. 3)

**resource deployment** The component of strategy that indicates how an organization intends to allocate resources. (Ch. 9)

**responsibility** A duty or obligation to carry out an assignment. (Ch. 12)

**retrenchment strategy** The cutting back of resources, as in worker layoffs and plant closings. (Ch. 9)

**return on assets (ROA)** Dividing net income by total assets to determine how effectively the organization is using its assets. (App. B)

**return on equity (ROE)** Dividing net income by owners' equity to see how profitably the owners' equity is being used. (App. B)

**revenue budget** A summary of an organization's projected revenue for the coming period. (App. B)

**reward** Anything an organization provides to employees in exchange for their services. (Ch. 18)

**reward power** The authority, capability, and position to grant and withhold various kinds of rewards. (Ch. 17)

**risk** A potential danger, peril, or loss; a decision-making condition wherein the manager understands the available options and estimates the probabilities associated with each. (Ch. 11)

**risk management** An approach to control uncertainties and potential dangers by assessing what those are, then developing strategies and tactics to reduce them. (Ch. 1)

**risk propensity** The degree to which an individual is willing to take chances and make risky decisions. (Ch. 16)

**ROA (return on assets)** See **return on assets (ROA)**.

**robot** Any artificial device that can perform functions ordinarily thought to be appropriate for human beings to perform. (Ch. 23)

**robotics** The science and technology of the construction, maintenance, and use of robots. (Ch. 23)

**ROE (return on equity)** See **return on equity (ROE)**.

**role** The part that a person plays in a group. (Ch. 19)

**role ambiguity** Lack of clarity as to how an individual in a group is expected to behave. (Ch. 19)

**role conflict** Inconsistency or contradiction in messages about a role that a person is to play in a group. (Ch. 19)

**role dynamics** The process whereby a person's expected role is transformed to his or her enacted role. (Ch. 19)

**rules and regulations** Statements of how to perform specific activities. (Ch. 8)

## S

**satisficing** Selecting the first minimally acceptable alternative without conducting a more thorough search. (Ch. 11)

**scheduling control** Making sure that the right things arrive and depart an organization at the right time. (Ch. 23)

**scientific management** The subarea of classical management theory that focuses on the work of individuals, primarily defining the steps needed to complete tasks and training employees to perform them efficiently while the manager assumes all planning and organizing responsibilities. (Ch. 2)

**scientific management model** A set of assumptions and way of viewing management that focuses on making the best product in the most efficient way. (Ch. 3)

**scope** The component of strategy that specifies the position the firm wants to have in relation to its environment; details the markets or industries in which the firm wants to compete. (Ch. 9)

**security needs** The needs for a safe physical and emotional environment. (Ch. 18)

**selection process** A systematic attempt to determine how well the skills, abilities, and aspirations of a job applicant match the needs, requirements, and opportunities within the organization. (Ch. 15)

**selective perception** The process of screening out information that we are uncomfortable with or which contradicts our beliefs. (Ch. 16)

**self-actualization needs** The needs to grow, develop, and expand our capabilities. (Ch. 18)

**self-esteem** The extent to which an individual believes that he or she is a worthwhile and deserving person. (Ch. 16)

**semantics problems** Communication difficulties due to problems with word meanings. (Ch. 20)

**semi-autonomous work groups** Workers who operate with no direct supervision to perform some specific task. (Ch. 13)

**semivariable costs** Costs that vary as a function of output, but not necessarily in a direct fashion. (App. B)

**sender** The person who transmits a message to a receiver. (Ch. 20)

**sensitivity analysis** A technique that enables the decision maker to vary the values of critical assumptions and re-do an analysis. (Ch. 10)

**sent role** The role that others in a group communicate that they expect a given person to play. (Ch. 19)

**serial entrepreneurs** Lone individuals who develop application-oriented companies not with a family orientation but to demonstrate the feasibility of the business concept, seek out investors, sell the company, and move on to another entrepreneurial venture. (Ch. 1)

**service organization** A company that transforms resources into intangible outputs that create utility for customers. (Ch. 23)

**service quality** The quality of an intangible "product" or utility, known as a service. (Ch. 22)

**services** Actions that provide some sort of utility rather than a tangible product for consumers. (Ch. 5)

**shelf life** The usable or safe life of a product, after which the product should be removed from the shelf to prevent its sale. (Ch. 1)

**short-range planning** Formulating goals and plans covering a period of less than one year. (Ch. 8)

**single-use plan** A program, project, policy, or method developed to handle events that happen only once. (Ch. 8)

**situational approach** An approach to leadership that recognizes that the same form of leadership is not appropriate in all circumstances. It is also called contingency approach. (Ch. 17)

**skunkworks** Small, autonomous, often voluntary work units that operate outside the normal process and/or premises to expedite it or to keep it a secret; also known as new venture units. (Ch. 13)

**small-batch technology** The set of processes used when a product is made in small quantities, usually in response to customer orders; also known as unit technology. (Ch. 13)

**social challenges** Forces and changes that are related to prevailing social customs and mores. (Ch. 5)

**social involvement** Approach to social responsibility that involves not just fulfilling obligations and requests but also actively seeking ways to benefit society. (Ch. 7)

**social need** The need for belongingness. (Ch. 18)

**social obligation** An approach to social responsibility in which the company meets its economic and legal responsibilities but does not go beyond them. (Ch. 7)

**social obstruction** Doing as little as possible to solve social or environmental problems. (Ch. 7)

**social reaction** A social responsibility approach that goes beyond social obligation by also being willing to react to appropriate societal requests and demands. (Ch. 7)

**social responsibility** Obligations of an organization to protect and/or enhance the society in which it functions; basic areas include organizational constituents, the natural environment, and general social welfare. (Ch. 7)

**sociocultural environment** The customs, mores, values, and demographic characteristics of the society in which an organization functions. (Ch. 4)

**software** The computer instructions (programs) that enable hardware to function. (Ch. 24)

**SOP** See standard operating procedure.

**space budget** The amount of plant or office space that is available to be allocated to different divisions or groups within an organization. (App. B)

**span of management** The number of subordinates who report directly to a given manager. A wide span of management means that there are relatively few managerial levels; therefore, the organization is flat. A narrow span leads to tall organizations. (Ch. 12)

**specificity** The extent to which goals are precise or general. (Ch. 8)

**spokesperson** The role that a manager plays when acting as a company representative while presenting information of meaningful content and/or answering questions on the firm's behalf. (Ch. 3)

**stability strategy** A plan to maintain the status quo of an organization. (Ch. 9)

**staff positions** Jobs that are outside the direct chain of command; primarily advisory or supportive in nature. (Ch. 12)

**staffing** Procuring and managing the human resources of an organization. (Ch. 15)

**standard** A measure or target against which performance will be compared. (Ch. 21)

**standard operating procedure (SOP)** A standing plan that serves as a specific guideline for handling a series of recurring activities. (Ch. 8)

**standing plans** Operational plans that are designed to handle recurring and relatively routine situations. Three forms are policies, standard operating procedures, and rules and regulations. (Ch. 8)

**stars** Products that have a high share of a fast-growing market, thus generating large amounts of revenue. (Ch. 9)

**statistical quality control** A category of operational techniques that consists of a set of mathematical and/or statistical methods for improving quality by measuring and adjusting quality levels. (Ch. 22)

**steering control** Monitoring the quality and/or quantity of resources before they enter a company's system; also called preliminary control and feed-forward control. (Ch. 21)

**stereotyping** Categorizing people on the basis of a single attribute. (Chs. 16, 20)

**storage** Any one of several types of media (tapes, disks, floppies, CDs, DVDs, and so on (used to hold information during and after it has been processed in a computer. (Ch. 24)

**storming** The second stage of group formation, when members begin to pull apart as they disagree over what needs to be done and how best to do it. (Ch. 19)

**strategic alliance** A cooperative agreement that does not necessarily involve ownership. (Ch. 6)

**strategic allies** Two or more companies that work together in joint ventures. (Ch. 4)

**strategic business unit (SBU)** An autonomous division or set of divisions (within another firm) that has its own competitors, a distinct mission, and a unique strategy. (Ch. 9)

**strategic change** A modification of an existing strategy or the adoption of a new one; occurs when an organization modifies or adopts a new strategy. (Ch. 14)

**strategic control** The process whereby management assures that the strategic planning process itself is effective. (Ch. 9)

**strategic planning** Formulating the broad goals and plans developed by top managers to guide the general directions of an organization. (Ch. 8)

**strategic plans** Broad goals and plans developed by top managers to guide the general directions of an organization. (Ch. 9)

**strategy formulation** The set of processes involved in creating or developing strategic plans. (Ch. 9)

**strategy implementation** The set of processes involved in executing strategic plans; involves tactical planning, contingency planning, and integration with organization design. (Ch. 9)

**stress** The condition that occurs when a person is subjected to a strong stimulus, such as unusual situations, difficult demands, or extreme pressures. (Ch. 16)

**stressor** A stimulus that is strong enough to cause stress. (Ch. 16)

**structural change** Any organizational change directed at a part of the formal organization system, including its structural components, its overall organization design, or related systems such as the reward system. (Ch. 14)

**subsystem interdependencies** The dependence of subsystems within a parent system on one another, such that a change in one part of an organization affects other parts. (Ch. 2)

**suppliers** Organizations that provide resources to the firm. (Ch. 4)

**support industries** Ancillary industries used by agribusiness core industries to produce and deliver their products and services; for example, banking, transportation, government, insurance, equipment, supply/service, industry associations, and education. (Ch. 1)

**survey feedback** Asking subordinates about their perceptions of their leader and then feeding back that information to the entire group. (Ch. 14)

**symbolic leadership** A leadership style associated with creating and maintaining a strong organizational culture; recognizing that the leader's behavior serves as a symbol of behavior for others to follow. (Ch. 17)

**synergy** The extra results that occur when two or more people or business units working together can draw from one another to produce more than they could when working alone. (Chs. 2, 9)

**system** Interrelated parts or elements that function as a whole. (Ch. 24)

**System 1** A hierarchical organization design, largely bureaucratic in nature. (Ch. 13)

**System 4** An organization design approach based on the premise that most organizations start out as bureaucracies and can be transformed to more appropriate models through a series of prescribed steps called Systems 1, 2, 3, and 4. (Ch. 13)

**systems theory** An approach to understanding how the different elements of an organization function and operate by considering the process by which an organization receives inputs, transforms them into outputs, produces outcomes, and receives feedback. (Ch. 2)

## T

**tactical planning** Establishing how to implement the strategic plans developed by top management; tends to focus on people and action and to deal with specific resources and time constraints; has a moderate scope and an intermediate time frame. (Ch. 8)

**takeover** When one corporation or group of investors buys or trades for enough stock in a company to gain control over it. (Ch. 5)

**tall organization** An organization that has several levels of management. (Ch. 12)

**targeting** Identifying and focusing on a clearly defined and highly specialized market, such as regional markets or special categories of consumers. (Ch. 9)

**tariff** A tax collected on goods shipped across national boundaries. (Ch. 6)

**task environment** Other specific organizations or groups that are likely to influence an organization. (Ch. 4)

**task force** A temporary group within an organization created to accomplish a specific purpose (task) by integrating existing functional areas. (Ch. 19)

**task group** A group created by an organization to accomplish a limited number of goals within a stated or implied time. (Ch. 19)

**task structure** The degree to which a group's task is well defined and understood by everyone. (Ch. 17)

**team building** A series of activities and exercises designed to enhance the motivation and satisfaction of people in groups by fostering mutual understanding, acceptance, and group cohesion. (Ch. 14)

**technical innovation** A change in the physical appearance or performance of a product or service, or in the physical process through which a product or service is manufactured. (Ch. 14)

**technical skills** The skills a manager needs to perform specialized tasks within the organization. (Ch. 3)

**technological change** Alterations related to technology, such as new equipment, new work processes or sequences, automation, and revised information-processing equipment. (Ch. 14)

**technological environment** The methods available for converting resources into products or services. (Ch. 4)

**technology** Processes and steps used to transform various inputs such as raw materials and component parts into appropriate outputs. (Chs. 5, 13, 23)

**telecommunications** Communication over some distance, usually by electronic means; includes teleconferencing, telecommuting, and networking. (Ch. 24)

**telecommuting** Employees performing their work at home through the use of computers connected to the organization's computer. (Ch. 24)

**teleconferencing** Videoconferencing, which permits individuals in different locations to see and talk with one another. (Ch. 24)

**test marketing** Introducing a new product on a limited basis to assess consumer reaction on a small scale. (App. B)

**tests** Examinations that are used to help managers select employees. (Ch. 15)

**Theory X** A pessimistic view of managerial thinking that assumes that workers dislike work and responsibility, thus requiring managers to control, direct, coerce, and threaten employees. (Ch. 2)

**Theory Y** An optimistic view of managerial thinking that assumes that workers enjoy work, seek responsibility, are bright and innovative, and are internally motivated. (Ch. 2)

**time management** The act of setting priorities for how our time will be used in achieving our needs and desires. (Ch. 10)

**top management** Managers at the upper level of an organization, including the chief executive officer and the vice presidents. (Ch. 2)

**total factor productivity** A measure of efficiency determined by dividing all outputs by all inputs: labor, capital, materials, information, and energy costs. It is also called overall productivity. (Ch. 22)

**total quality management (TQM)** A comprehensive, organization-wide strategy for improving product and/or service quality on a systematic and continuous basis; achieved through a combination of strategic commitment, employee involvement, materials, methods, and technology. (Ch. 22)

**trade agreement** A legal arrangement between two or more nations indicating they will cooperate in trading. (Ch. 6)

**training** Preparing employees for specific job skills rather than developing general abilities. (Ch. 15)

**transaction-processing system (TPS)** Computerized system designed to handle routine and recurring transactions within the organization. (Ch. 24)

**transformational leadership** A leadership perspective that attempts to foster innovation and vision; also called entrepreneurial leadership. (Ch. 17)

**turnover** Percentage of an organization's workforce that leaves and must be replaced over a period of time. (Ch. 16, App. B)

**two-factor view** A way of describing employee needs viewing satisfaction and dissatisfaction as two distinct dimensions affected by two different and independent sets of factors. (Ch. 18)

**Type A** Personality type of individuals who are extremely competitive, very devoted to work, and have a strong sense of time urgency. (Ch. 16)

**Type B** Personality type of individuals who are less competitive, less devoted to work, and have a weaker sense of time urgency. (Ch. 16)

## U

**U-Form** The most common organization design; uses a unitary or uniform approach to design. It is also known as the functional design because it is based on functional specialization. (Ch. 13)

**uncertainty** Not being sure of the alternatives or their probabilities; created by the degree of change and the degree of homogeneity that characterizes an organization's environments. (Ch. 11)

**unit productivity** The amount produced or created by a single unit, department, or group within an organization. (Ch. 22)

**unit technology** The set of processes used when a product is made in small quantities, usually in response to customer orders; also known as small-batch technology. (Ch. 13)

**user-friendly information** A system that is easy to understand and use; designed so that users can examine and modify information easily. (Ch. 24)

## V

**value-added partnership** Another name for **network organization**, which is an emerging organizational design for companies that usually engage in outsourcing, which is the contracting to other firms of many of their usual functions; also called hollow corporations. (Ch. 13)

**variable costs** Costs, such as materials and electricity to operate production equipment, that vary as a function of operations. (App. B)

**variable interval schedule** Providing reinforcement on a time-interval basis, but the time intervals between reinforcement vary in length. (Ch. 18)

**variable ratio schedule** Providing reinforcement on the basis of behaviors but varying the number of behaviors an employee must display to get the reinforcement. (Ch. 18)

**vertical communication** Transmission of messages between bosses and their subordinates. (Ch. 20)

**VRIO framework** In strategic analysis, asking for each strength whether it is **V**aluable, **R**are, difficult to **I**mitate, and expl**O**itable. (Ch. 9)

## W

**wage level** A company's wages relative to the prevailing local or industrial wages in the local economy. (Ch. 15)

**wage structure** The comparison of wages for different jobs within the company. (Ch. 15)

**whistle blowing** The disclosure by an employee of illegal or unethical conduct on the part of another or others within an organization. (Ch. 7)

**work-in-process inventory** The parts and supplies that are currently being used to produce the final product or service, which is not yet complete. (Ch. 23)

**work team** A small, self-managed group of organization members who are responsible for a set of tasks. (Ch. 19)

**workforce composition ratio** The percentages of an organization's employees that fall into various groups. (App. B)

**workforce diversity** A situation that exists in a group or organization when its members differ from one another along one or more important dimensions, such as race, age, or ethnicity. (Ch. 5)

**workplace challenges** The challenges that involve the relationships among organizations, their managers, and their operating employees; for example, workforce diversity, employee expectations and rights, and workplace democracy. (Ch. 5)

**written communication** Messages transmitted by means of memos, e-mail, text-messaging, letters, reports, and notes. (Ch. 20)

# INDEX